STATIST

Concepts and Methods

Fourth Edition

Ditlev Monrad
University of Illinois

E. James Harner
West Virginia University

William F. Stout
University of Illinois

ISBN: 978-1-61549-114-8

Produced in the United States of America.
10 9 8 7 6 5 4

CONTENTS

PREFACE

The authors strongly believe that the most effective way for beginning statistics students to learn statistics well is for them to directly experience its central concepts and standard procedures by closely observing and actively interacting with simulated and real data sets. The particular advantage of having a student simulate data sets is that the student then actually experiences the underlying mechanism—that is, the probability model—producing his or her data. Then students experience firsthand the process of statistically analyzing data, and in this way their statistical expertise develops. Actively learning statistics by direct contact and interaction with data is facilitated in this textbook by the demonstration of statistical concepts using the five-step simulation method throughout. Carrying out these five steps, students design simulation studies, and often also solve their statistical problems, facilitated by the textbook's powerful supporting software. This software is fast, Web-based, point-and-click, and instructional. It produces data-driven demonstrations of fundamental statistical concepts and illustrates highly accurate simulation-based statistical analyses that are heavily used by professional statisticians in the modern era of statistics.

Instructors will use either the printed textbook option or the Great River Technologies supported e-learning option, where the textbook will be delivered in online chunks, with hypertext links to appropriate applets and other features sprinkled throughout. Moreover, for both versions, an electronic grade book is provided for the instructor. Also, many of the homework questions can, at the option of the instructor, be computer-graded. Finally, a supplementary workbook is provided for use with both the traditional printed textbook option and the e-learning option. These various ancillaries should be especially helpful for teaching large sections. The e-learning package has taken many years to develop and will, of course, be periodically upgraded and renewed in years to come.

A vital complement to the experiential, simulation-based, data-driven presentation of statistical concepts and methods is a clear and understandable combined verbal and formula-based presentation of the formal, logical, and deductive aspects of the subject, using only the moderate amount of mathematical formalism required to make the concepts and procedures understandable. In particular, normal population sampling, the central limit theorem, and other elements of large-sample approaches to statistical inference are given prominence just as they are in a traditional statistics textbook. The modern simulation-based resampling approach to inference is also clearly and completely presented. Inference thus has three distinct approaches: normal population, central limit theorem/large sample, and simulation-based.

In the textbook, the theoretical formal aspects of each major topic are intermingled with rich real-data-based and, when appropriate, simulation-based examples of concepts and methods. The basic aspects of probability that are so important to a statistics course (and often very difficult for beginning students to master) are introduced as empirical phenomena observable via simulation and then abstracted and expanded into a body of useful probability concepts and accompanying formulas in Chapters 5, 6, and 7. The capstone of this empirical emphasis is the five-step computer-based simulation that is used throughout. Simulation allows students to transcend (and often avoid) abstract mathematical formalism and yet gain a deep understanding of the empirical role that probability plays in statistical thinking. In Chapter 7 the core probability models (normal, binomial, geometric, and Poisson) are thoroughly discussed with an emphasis on their roles in providing descriptions or models of real-world random phenomena.

Pedagogically, the textbook stands on three legs: (1) carefully crafted and thorough verbal explanations of concepts and procedures; (2) clear, formula-based descriptions and explanations of statistical concepts and procedures, including both traditional procedures and the more recently emphasized simulation-based resampling procedures; and (3) immersion in the world of probability model–produced data sets and their statistical analyses. This immersion is achieved through the use of (i) textbook-supplied and student-created simulation studies, made possible by the textbook's fast, sophisticated, and easy-to-use software; and (ii) interesting and compelling examples rich in real-world data.

To be most effective, an introductory statistics textbook must strike a balance: On the one hand, it must help the student to learn basic statistical concepts in depth and, ultimately, to think statistically; on the other hand, it must expose the student to the evolving body of statistical procedures that are widely used in practice and are considered to constitute statistical literacy. This balance, a central goal of the book, equips students with both the faculty of statistical discernment and the needed familiarity with the body of statistical procedures so widely used in science, business, and government, and whose applications are so often reported in the media. Without this balance, the course becomes only a "liberal arts" introduction to statistics, failing to equip students with the capability to analyze data sets. The textbook is divided into three parts: (1) using numerical and graphical techniques to explore and discover possibly meaningful and important patterns in data, and obtaining quality data for a planned statistical study via experimental designs that use randomization and via surveys that use random sampling; (2) actively experiencing the fundamental role of probability models in statistics, with the mathematical formalism of probability nicely interwoven with the rich and active experiencing of randomness via simulations; and (3) using statistical inference as a tool to draw valid and useful conclusions about important real-world settings from quality data via estimation and hypothesis testing.

In today's technology-rich learning environments, an introductory statistics textbook must be designed to make profound and meaningful use of available computer-based technology, both to be pedagogically sound and to help students carry out statistical inferences using real and sometimes sizable data sets. Instructional software must be platform independent, extremely easy for all to use (point-and-click access is vital, because there is no time available in a typical jam-packed statistics course syllabus for a week or more of software training!), and thoroughly integrated with the textbook. The software should make the concepts and methods of the course come alive, be a rich source of data, and make complex statistical computations and demonstrations and inferences involving large data sets fast and easy to carry out. The software must be an "active learning" tool, to use an overworked but apt phrase. The accompanying simulation-based, Web-delivered, customized instructional software does all of these things, using the five-step method to allow students to construct simulations that facilitate learning.

The textbook perhaps breaks new ground with its in-depth treatment of bootstrap-based resampling and other inferential procedures that require simulation. The bootstrap approach is one of the most important advances of the past 25 years in statistical methodology and has become an essential component of the practicing statistician's toolbox. Hence, it needs to be taught at the elementary level. Consistent with the instructional approach of the book, the bootstrap method is simulation-based and, as such, meshes seamlessly with the five-step method used throughout. The bootstrap approach of this textbook is natural and intuitive, which makes it easy for the beginning statistics student to understand. Bootstrapping equips the student with a powerful, cutting-edge, and widely used method of statistical inference. The bootstrap procedure, together with the traditional small-sample, normal population inferential approaches (such as the t-test) and the large-sample inferential approaches (based on the central limit theorem and on the associated large-sample result for the chi-square statistic), provide the student with enormous inferential power.

The book is written to be accessible to all students having had at least a modest exposure to algebra; intermediate algebra will suffice. Although mathematical concepts and formulas appear frequently, and students are encouraged to seriously think about what they are learning from such formulas, the amount of formal mathematical background needed is minimal. The book should work very well in any non-calculus-based introductory statistics course. The instructor has the luxury, if he or she so desires, of downplaying large-sample and normal population–based inference and the formal structure of probability theory. In fact, the instructor can even design a course that puts most of its focus on the simulation-based approach to statistical inference. The textbook offers a choice between (1) a course design where the major stress is on simulation-based inference and (2) a balanced approach to inference that is sometimes simulation-based, somties normal-population-based, and someties large-sample-theory-based.

Several influential, nationally circulated reports have stressed the need for new emphases in the teaching of statistics. Most notably, the widely respected American Statistical Association/Mathematical Association of America Cobb report recommends that the teaching of statistics be heavily based on data and proposes that more emphasis be placed on statistical concepts than on abstract theory. Further, it stresses active learning and simulation-based learning. This textbook is tailored to address these valid recommendations, not only by means of its simulation-based approach but also through the use of numerous data-driven examples and exercises (provided at the end of each section), many encouraging students to use the instructional software. Thus, students have ample opportunity to practice using the concepts and methods presented in the text.

Part I of the book begins with three chapters describing how one explores and summarizes important patterns occurring in data—data being the focus of statistics—with tables and graphs and by means of numerical (statistical) indices. This coverage includes a descriptive introduction to linear regression in Chapter 3. Chapter 4 teaches the student how one obtains quality data for a planned statistical study through probability sampling of real populations and the use of well-designed randomized experiments.

Part II presents probability modeling—both empirically, using simulations, and formally, using a set of axioms and definitions. Chapters 5 and 6 provide a heavily empirical and simulation-based introduction to probability with an emphasis on probability distributions, and on the mean and standard deviation of a random variable, that is, its probability distribution. This is appropriate because probability is the logical underpinning of inferential statistics. In particular, probability, expected value, and standard deviation are introduced as empirical concepts, with large-scale simulated samples (routinely involving 10,000 simulations, and even 100,000 occasionally) often employed to accurately estimate unknown theoretical probabilities, expected values, and standard deviations. This is facilitated by student usage of the supporting instructional software.

Chapter 7 supplements the basic introduction to probability in Chapters 5 and 6 with four core probability distributions—normal, binomial, geometric, and Poisson—and the standard large-sample relationships involving them, such as the normal approximation to the binomial. Some instructors may wish to deemphasize or omit portions of Chapter 7.

Part III covers statistical inference, focusing on confidence intervals and hypothesis testing. Chapter 8 presents an introduction to estimation, beginning with a discussion of the two sources of estimation error, namely bias and chance variation. Then the large-sample confidence interval estimation of a population mean μ and of a population proportion p are developed using the central limit theorem (CLT) for $\overline{X}$.

Chapter 9 introduces hypothesis testing. After an explanation of fundamental concepts, testing of H_0: $p = p_0$ is demonstrated via both the simulation approach and the large-sample CLT for $\hat{p}$ approach. The chapter closes with a novel simulation approach to randomized controlled experiment hypothesis testing when population members each display one of two possible characteristics (e.g., recovery from illness or death).

Chapters 10 and 11 cover the important confidence interval estimation and hypothesis-testing procedures in the standard one- and two-population settings involving means and proportions, stressing normal population inference, large-sample inference, and simulation-based inference.

Chapter 12 presents chi-square testing, beginning with the more widely applicable (not requiring a large sample) simulation-based approach. Then the traditional large-sample, chi-square distribution approach is presented. Coverage includes both the usual specified multinominal distribution chi-square test (null hypothesis of a specified many-sided, possibly unfair die) and contingency table–based chi-square tests of independence and multiple-population homogeneity.

Chapter 13 complements the descriptive treatment of linear regression in Chapter 3 with an inference-based treatment of linear regression.

The entire book can easily be covered in a two-semester course. However, a rich one-semester course can be nicely carried out, with careful planning and judicious choice of topics covered. If desired, the instructor can omit Chapters 12 and 13 with no loss of continuity and can deemphasize Chapter 7. In this regard, starred (*) sections can also be omitted without loss of continuity.

This textbook has been heavily influenced by instructor and student input based on classroom use of earlier editions and other textbooks written by some of the authors. In particular, Chapter 4 on data collection was developed in response to such input.

One unique aspect of the textbook is the integration of the printed material with easy-to-use simulation software. This integration enriches the learning experience and also enables easy access to a large set of simulation-based inferential procedures often used in modern statistical practice. When use of the software is especially appropriate, a computer symbol sometimes appears in the margin. A hand-held calculator will often prove useful for ordinary data compilation, and the image of a calculator may appear in the margin in such contexts. A key graphic indicates material related to the chapter's Key Problem.

ABOUT THE AUTHORS

Ditlev Monrad is an Associate Professor in the Department of Statistics and the Department of Mathematics at the University of Illinois at Urbana-Champaign (UIUC). His research is in the area of theoretical probability. A dedicated instructor, Prof. Monrad has served for many years as an undergraduate and graduate advisor for statistics majors at UIUC. He has extensive experience in teaching introductory statistics and probability at the undergraduate level.

E. James Harner is a Professor and Chair of the Department of Statistics at West Virginia University. His principal research efforts are in environmental statistics, statistical computing environments, and dynamic graphics, although he is increasingly focusing on learning-based probability models and bioinformatics. He has over 40 published research papers and has co-developed several statistical software and Web-based learning environments. Currently, he is directing the development of IDEAL (Intelligent Distributed Environment for Adaptive Learning), a Web-based learning environment that will incorporate advanced tutoring and assessment subsystems into college instruction.

William F. Stout is a Professor of Statistics at UIUC, where he has been on the faculty since 1967. He is an internationally acclaimed researcher in the application of statistics to the fields of educational and psychological measurement. He has been the thesis research advisor of 17 Ph.D. students working on the interface of statistica and educational measurement. As the founder of the University of Illinois Statistical Laboratory for Educational and Psychological Measurement, Prof. Stout has led the development of several new and widely applicable probability model–based methods for improving standardized testing. He is a past president of the International Psychometric Society and was the director of an Education Testing Service research group that developed statistical tools to carry out skills-level formative assessments of students using standardized tests. Prof. Stout is the author of over 60 books and published research papers. He has received National Science Foundation research grants continuously for almost 40 years. Recently he received the Purdue University 2009 College of Science Department of Statistic Distiguished Science Alumnus Award.

ACKNOWLEDGMENTS

The authors wish to acknowledge the superb, in-depth intellectual and technical support provided by publisher Möbius Communications and textbook manufacturer Publication Services in the development and manufacture of previous editions of this textbook. The authors especially wish to acknowledge Publication Services (PS) in general and PS Project Manager Kelly Applegate in particular; and our publisher, Great River Technologies (GRT), in general and Tim Shade, GRT Instructional Technology Consultant, and Jade Sprecher, GRT Web Project Editor, in particular. Their superb oversight and judgment concerning the many details and decisions required was essential in enabling us to turn the third edition manuscript into this fourth edition of our textbook. Foremost, we thank GRT for the constant support and expertise required for the development of this learning project with its many interacting components, especially the print and e-learning versions of the textbook, the simulation-based instructional software, the computer-graded homework component, and the GRT website that will serve as the instructional "headquarters" for future students of *Statistics: Concepts and Methods*.

We thank Louise Toft, Möbius Communications Product Manager, for developing the nicely written historical profiles that open the chapters. We also thank University of Illinois Department of Statistics Ph.D. student Nathan Hirtz for his informed assistance in developing the computer-graded homework component.

Finally, the authors wish to express heartfelt thanks to their wives for all their support and encouragement in the writing this textbook—specifically, thanks to Siva from Ditlev Monrad, thanks to Molly from James Harner, and thanks to Ann from William Stout.

I

Describing and Obtaining Data

PROFESSIONAL PROFILE

Florence Nightingale
1820–1910
Nurse,
Health Care Pioneer,
Statistician

A polar area chart created by Florence Nightingale as an illustration to her report to Queen Victoria and members of Parliament.

Florence Nightingale about 1860.

Florence Nightingale showed a gift for math early in life. Guided by her father, William Edward Nightingale (an expert in the then-new field of epidemiology), the young Florence had mastered statistics by her teens.

In the 1840s nursing was a career with very low status, one pursued by poor women and military camp followers. Her parents were shocked when, in 1845, she announced her decision to become a nurse, following a period in which she became a leading proponent of better medical care for poor and homeless patients in infirmaries.

In 1854 she led a group of 38 women volunteers who were sent to care for wounded soldiers in the Crimea. When the group arrived at Selimiye Barracks, they found conditions deplorable. Their first steps were a thorough cleaning and a complete reorganization of patient care. Initially, this had little impact. Some 4,077 soldiers had died the first winter, 90% from disease rather than battle wounds. With the help of a British sanitary commission sent the next spring, sewers were cleaned and ventilation improved. The death rate began going down. Nightingale's reputation soared. She became a heroine.

Originally a believer in the miasma (noxious "bad air") theory of disease causation, Nightingale came to realize that most soldiers died because of unsanitary, crowded living conditions. This realization came during her statistical review of death rates prepared for the Royal Commission on the Health of the Army, Parliament members, and civil servants. In the report, she pioneered the graphic presentation of statistical information, drawing on the pie charts developed by William Playfair in 1801. Nightingale called her graphics "coxcombs." They improved understanding among officials at all levels. The report prompted improvements in military living conditions and health care.

In 1859 Nightingale became the first female member of the Royal Statistical Society and later was made an honorary member of the American Statistical Society.

1

Exploring Data by Graphical Methods

Statistical thinking will one day be as necessary for efficient citizenship as the ability to read and write.

H.G. Wells

Objectives

After studying this chapter, you will understand the following:

- The role of statistics
- Types of statistical data
- Graphical methods of summarizing data, including dotplots, stem-and-leaf plots, pie charts, bar charts, frequency histograms, and density histograms
- How to use a density histogram to summarize the shape of a data set
- How to recognize misleading graphical descriptions of data
- Simpson's paradox

KEY PROBLEM

Will the Real Author Please Stand Up?

It can be very important to find out who actually wrote a book or document. After billionaire Howard Hughes died, several different wills came to light, and the courts had to decide which one was actually written by Hughes. Other questions of authorship may not have as much money riding on them but are very interesting. For example, various scholars have suggested that some or all of the plays and poems attributed to William Shakespeare were actually written by Sir Walter Raleigh, Sir Francis Bacon, or the 4th Earl of Oxford.

Davy Crockett (1786–1836) is now a legend of the American frontier, but between his days as a pioneer and hunter in Tennessee and his death in battle at the Alamo during the Texas War of Independence, he also served in the U.S. Congress. There he opposed President Andrew Jackson's plans to seize the lands of the Native Americans east of the Mississippi that had been granted by treaties and deport the people along the Trail of Tears to what is now Oklahoma. Two books were published in Crockett's name in the 1830s in support of his political activities, and another was published in his name after his death about his experiences in the Texas Revolution. Scholars have doubted whether the frontiersman actually wrote the books, because he did not learn to read and write until his early 30s. They especially doubted that he wrote the Texas war memoir, which lacks the plain-speaking style that Crockett was admired for.

How can we identify the author of a text that was published more than a century ago? We cannot cross-examine possible authors in court. We may not have any evidence except the words of the text in question and other texts known to have been written by the various possible authors. By using statistical techniques, we may be able to determine whether the text in question was written by the same author as one or another of the known texts.

Statisticians Frederick Mosteller and David Wallace invented a method that uses a set of words called "contentless" because they can be used anywhere in a piece of writing (see Table 1.1). Researchers count the rates at which each of these words occurs in the text in question and in the writing samples for which authorship is known and compare the patterns of these word counts. If the distribution of the rates of these words in the text in question is very different from the distribution in a sample known to be written by the author, the text in question was probably written by someone else. Mosteller and Wallace first used their method to conclude that 12 of the *Federalist Papers* (a set of 85 essays in support of the U.S. Constitution, published during 1787 and 1788 under the pen name "Publius") had been written by James Madison rather than Alexander Hamilton.

David Salsburg and Dena Salsburg applied the method of Mosteller and Wallace to the three books attributed to Davy Crockett: *A Narrative of the Life of David Crockett, Written by Himself* (called *Narrative* for short); *An Account of Col. Crockett's Tour of the North and Down East* (called *Tour*); and *Col. Crockett's Exploits and Adventures in Texas* (called *Texas*). As a comparison sample, Salsburg and Salsburg used Crockett's speeches in Congress. Only 9 of the 30 contentless words were used consistently in each of the four texts. Their distributions are shown in Table 1.2.

At first glance the differences among texts do not look very great. But one notes that the frequencies of *to* and *this* are lower in *Texas* than in the *Narrative*, the *Tour*, and the speeches; the frequency of *an* is much higher in *Texas* than in others. These differences suggest that Crockett could have written *Narrative* and *Tour* and may not have written *Texas*. But this is only a vague impression!

Table 1.1 Contentless Words

according	considerable	probability
also	direction	there
although	enough	this
always	innovation	though
an	kind	to
apt	language	upon
both	matter	vigor
by	of	while
commonly	on	whilst
consquently	particularly	works

Source: D. Salsburg and D. Salsburg. 1999. Searching for the "real" Davy Crockett. *Chance.* 12 (2, Spring): 31.

Table 1.2 Rates of Contentless Words (per 1000 Total Words) in Texts Attributed to Davy Crockett

Word	Narrative	Tour	Texas	Speeches
also	0.75	0.32	0.41	0.51
always	0.13	0.79	0.55	0.00
an	2.63	2.21	4.93	2.54
both	1.00	0.47	0.14	0.00
there	3.75	4.90	3.43	1.52
this	4.75	5.22	2.60	7.10
though	0.25	0.00	1.51	0.00
to	36.79	33.21	28.78	30.93
while	1.13	0.63	0.68	2.54

Source: Ibid, 32.

Applying a statistical hypothesis-testing (hypothesis testing is covered in Chapter 9) technique to these word distributions, Salsburg and Salsburg found that the *Narrative,* the *Tour,* and the speeches were very possibly written by the same person, but that *Texas* was most likely written by someone else.

Salsburg and Salsburg were careful *not* to say, "Davy Crockett wrote the *Narrative,* the *Tour,* and the speeches." Such a statistical inference would have gone beyond what their statistical methods could reliably tell us. At most we can say that if Crockett composed the speeches, he could well have written the *Narrative* and the *Tour* but probably did not write *Texas.*

The underlying concepts and methods of such statistical inference are the central focus of this book. The science of statistics is essential: Statistical reasoning enables us to draw conclusions based on data patterns that are often hidden from the statistically uninformed observer of the data. It also forces us to not exceed the limits of what we can determine from the data. Author identification problems usually involve both of these aspects.

1.1 THE SCIENCE OF STATISTICS

statistics
data

As a science, **statistics** is the discipline of gathering data, describing data, and drawing conclusions (statistical inferences) from data. **Data** are pieces of information obtained by counting, measuring, or categorizing, using some kind of observational process. In this book we will study the basic ideas in statistics and learn how to obtain, summarize, and interpret data.

statistics

Statistics also refers to numbers that somehow inform us about the world around us. For example, a baseball fan might consider Barry Bonds's 73 home runs in 2001 an interesting statistic. The word *statistics* originally meant the kind of information that a state (i.e., a government) can use in managing public affairs. Accordingly, all around us there are statistics about taxes collected, prices of goods, weather patterns, and so forth.

statistical inference

The daily newspaper is filled with statistics, presented in many clever ways, about topics ranging from sports and entertainment to politics and health. Here are some examples of statistics in both senses of the word: as numerical information and as the science of reasoning statistically from data to reach conclusions about the unknown character of the real world (this is known as **statistical inference**).

Statistics in Real-World Studies

The Rise in Undergraduate Tuition On April 2, 2006, a front-page article in *The Ann Arbor News* was titled "U-M budget gap drives tuition." The article discussed how tuition at the University of Michigan had increased at a much faster pace than the inflation rate from 1990 to 2005. The rise in annual tuition was illustrated by the graph in Figure 1.1.

The reason given in the article for the increase in tuition was that the university budget grew steadily over the past 15 years but state funding did not keep pace and, in fact, declined over the last 5 years. This was illustrated by the graph in Figure 1.2. The article noted that whereas the university budget had increased 115% since 1990, state funding had only increased 28% and had not even kept pace with inflation. As a result the university relied more and more on tuition and student fees. In the 2005–2006 school year tuition and fees paid for nearly 60% of the university budget, compared to 46% in 1990–1991. This was illustrated by the graph in Figure 1.3.

Statisticians look for patterns in data and attempt to link different trends, such as the increase in tuition at state schools and the decrease in state funding for education.

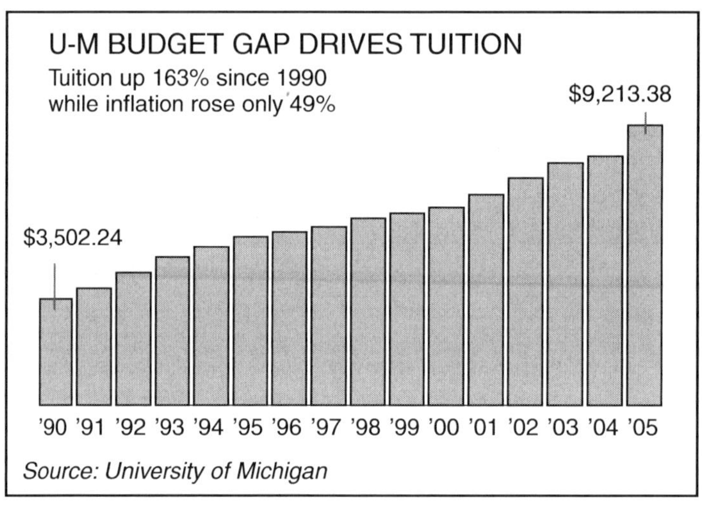

Figure 1.1 Annual in-state tuition at the University of Michigan from 1990 to 2005 (*Source: The Ann Arbor News,* April 2, 2006. All rights reserved. Reprinted with permission).

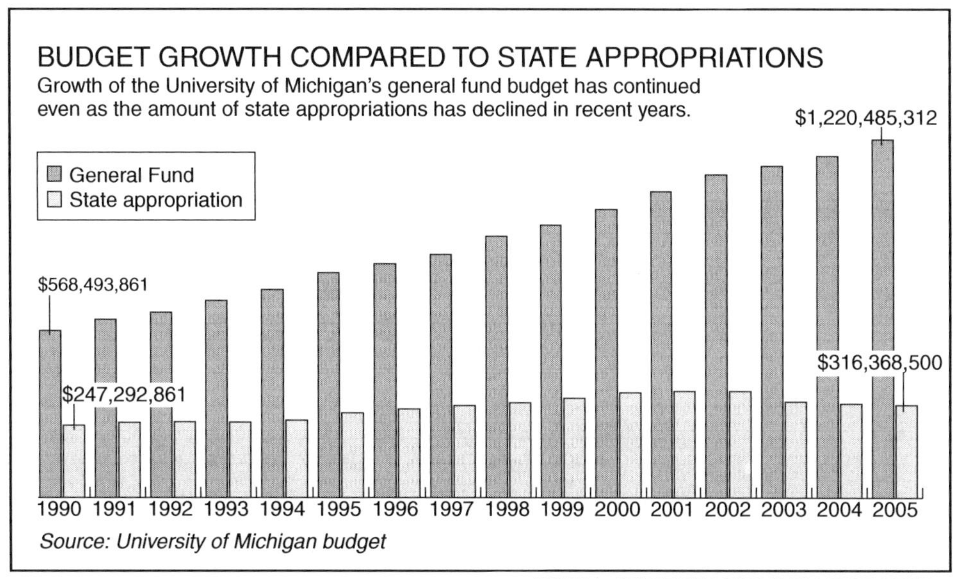

Figure 1.2 University of Michigan's budget and state funding from 1990 to 2005 (*Source: The Ann Arbor News,* April 2, 2006. All rights reserved. Reprinted with permission).

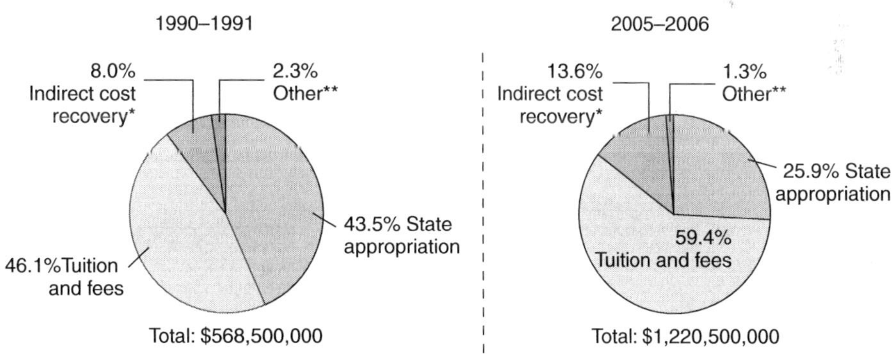

Figure 1.3 Revenue sources for the University of Michigan in 1990–1991 and in 2005–2006 (*Source: The Ann Arbor News,* April 2, 2006. All rights reserved. Reprinted with permission).

"American Idol" Hits Record Heights On March 21, 2006, a record 33.36 million TV viewers watched Fox's 2-hour installment of *American Idol*. That week's second-most-watched program was a half-hour installment of *American Idol*, which drew 27.68 million viewers, followed by *Desperate Housewives*, which had 21.41 million viewers, according to figures released by Nielsen Media Research a week later. These numbers are only **estimates**. Nielsen did not obtain information from the entire **population**, only from a **sample** of approximately 25,000 metered households. The Nielsen TV households are a cross section of households from all over the United States, carefully selected so that the viewing habits of the 25,000 sample households mirror the viewing habits of the more than 100 million households in the United States. In Chapters 7 and 8 we shall discuss how a surprisingly small but carefully selected sample can give quite accurate information about the much larger population the sample was drawn from.

estimates
population, sample

Is Coffee Bad for Your Heart? Some medical studies in the 1970s found a connection between coffee drinking and heart disease: The rate of heart disease was higher for coffee drinkers than for nondrinkers. Did this mean that coffee drinking *causes* heart disease? Further studies revealed that the coffee drinkers smoked more than the nondrinkers. It was smoking that contributed to the coffee drinkers' risk of heart disease, not the coffee. In Chapter 3 and 7 we shall discuss how statisticians investigate whether there is a **cause-and-effect** relationship between a **treatment** (such as coffee drinking) and a **response** (such as heart disease).

cause-and-effect
treatment, response

Does More Math Mean More Money? One of the most extensive studies ever made of the careers of high school graduates confirms the conventional wisdom that those who take more mathematics courses earn considerably more in the marketplace (*Education Week*, Jan. 11, 1989, 9). On the surface, it seems plausible that taking more math courses should increase one's earning potential. On the other hand, other characteristics of those who take lots of math (such as having the motivation to take on a difficult challenge, or being of high intelligence) may be the major causes of their higher incomes. Therefore, taking more math courses might not be the *cause* of higher earning potential. The science of statistics can tell us when two quantities are statistically **positively associated**—that is, they are large or small together. It is dangerous and incorrect, however, to conclude automatically that an increase in one quantity *causes* an increase in the other (see also Section 3.4 and the lung cancer/smoking debate in Chapter 7).

positively associated

Want to Avoid a Cold? Don't Get Stressed Out! Survey research carried out in England indicates that persons under stress have a greater tendency to catch a cold than those who are not stressed (Psychological Stress and Susceptibility to the Common Cold. *New England Journal of Medicine*, Aug. 29, 1991, 606–612). Provided the experimental work that produced these data was done properly, it may be valid to conclude that stress increases one's susceptibility to colds. For example, did the study balance out or adjust for other factors that make catching a cold more likely and that go together with being stressed? If not, these other factors (such as lack of exercise or poor diet) could be the *real* cause of stressed people catching more colds.

Do Headaches Run in the Family? If you suffer from migraine headaches, perhaps blame your genes: Migraine headaches may be a hereditary condition. Data show that if both the father and mother have had them, then a son or daughter has a 75% chance of migraines. If only one parent has had migraine headaches, data show that the chance of the condition occurring in offspring is 50%. Even having distant relatives with migraine headaches increases the risk of having them by about 20% over the general population rate. Of course, concluding from these data that heredity causes migraine headaches is risky. For example, stress in the family environment could really be causing migraine headaches to occur more frequently in all family members. For that matter, the parents' frequent migraines could put stress on the family that then contributes to the children having migraine headaches (*Headway*, vol. 3, Dec. 1993).

All of these news items deal with topics of general interest. Each contains numerical information (i.e., statistics) or presents statistical inferences drawn (rightly or wrongly) from numerical information. Here are some questions these statistically based discussions may help us with:

- *Entertainment:* The popularity of one TV program as contrasted to another, such as "American Idol" versus "Desperate Housewives"
- *Health:* The relationship between stress and susceptibility to colds
- *Economics:* Whether the study of mathematics may increase one's earning power

We've said that statistics is the discipline of gathering numerical data, describing these data, and drawing conclusions from the data. It can help us answer questions such as:

- Does the future economic growth of the United States depend on continued immigration?
- Why do casinos make a profit at roulette?
- How accurately can a person's future income be predicted from his or her education?
- How do medical researchers determine whether a new drug works or not?
- How accurately can a poll of 1500 randomly selected eligible voters predict the outcome of an election?

This book contains many examples of the statistical analysis of real data sets as well as the basic ideas and techniques of gathering, describing, and analyzing data. After studying this book, you will be able to draw from data sound conclusions that are not necessarily obvious to the casual observer. Further, you will often be able to tell when the reported conclusions of a study cannot be trusted to be valid, in spite of a glitzy graph or compelling words to the contrary.

The techniques of graphically representing data and numerically summarizing them with appropriate statistics in ways that make it easy to recognize the essence of the information in the data are called **descriptive statistics**. On the other hand, the techniques used to draw conclusions from data using the theory of probability and the methods of statistics are called **inferential statistics**. Good inferential statistics often needs to be preceded by a good descriptive statistical analysis of a data set. This textbook presents both of these aspects of the discipline of statistics.

descriptive statistics

inferential statistics

This chapter presents some commonly used graphical techniques of descriptive statistics, beginning with the dotplot. The last section will cover some of the common ways in which the techniques of descriptive statistics have been abused to make *misleading* presentations.

We will first introduce some terminology and consider the different types of statistical data we are going to study.

population
units
subjects
variables
numerical
continuous variable
quantitative
discrete variable
qualitative
categorical
ordered categorical
ranked
levels

Individuals and Variables The **population** in a statistical study is the collection of objects or individuals of interest. The population members are called **units** or **subjects** (if the population consists of people). We are interested in certain attributes or characteristics that vary from one individual to the next. For example, eye color, height, and number of siblings vary from person to person. A corporation's profit varies from quarter to quarter. The bond rating (AAA through C) varies from corporation to corporation. Such characteristics of people or other units of interest that vary from one individual to another are called **variables**. Some variables such as height and number of siblings take numerical values and are called **numerical** or **quantitative**. Numerical variables may be discrete or continuous. A **continuous variable**, such as height, can be any value in a given range. A **discrete variable**, such as number of siblings, can only be certain distinct values, in this case 0, 1, 2, 3, and so on. No value in between the distinct values is possible. A variable such as eye color cannot be expressed as a number. Such variables are called **qualitative** or **categorical**. We call a categorical variable **ordered categorical** or **ranked** if the categories imply some order or relative position. Military rank, for example, is an ordered categorical variable. Corporate bond rating is an ordered (or ranked) categorical variable taking on the ordered **levels** AAA, AA, A, BBB, BB, B, CCC, CC, and C (using S&P rating from highest to lowest).

We shall give a couple of examples in which we identify the individual units of interest, the type of variable of interest, and the possible values for the variable.

- If we are looking at the highest elevation (in feet) for each of the 50 states in the United States, then the individual units of interest are the states. Elevation is a numerical variable. It is a continuous variable, but when we round it to the nearest value in feet, elevation becomes a discrete variable with values between −282 feet (in Death Valley) and 20,320 feet (Mt. McKinley).
- If we are looking at the size of households in the United States, then the individual units of interest are the households. Household size is a discrete numerical variable with possible values 1, 2, 3, 4, . . . , and so forth.
- If we are looking at the letter grades of the students in a statistics class, then the individuals are the students. Letter grade is a ranked categorical variable with ordered levels: A+, A, A–, B+, B, B–, C+, C, C–, D+, D, D–, and F.

As well as describing variables, the terms *numerical, categorical, ordered categorical, discrete,* and *continuous* are also used to describe data.

distribution

The pattern of variation of data for a variable in a study is called its **distribution**. It is usually described by a table, graph, or formula that tells us what values the variable takes and how often they occur. Distributions provide the shape of a data set, such as being bell-shaped.

Section 1.1 Exercises

1. In each of the following examples, identify the individuals/units of interest, the variable of interest, the type of variable, and indicate the possible values of the variable.
   a. A public opinion poll asks a representative cross section of 1541 adults whether they favor a constitutional amendment banning same-sex marriages.
   b. A car buyer compares the highway mileage (miles per gallon) of several car models.
   c. A population survey finds that 60% of American adults are married, 23% have never been married, 7% are widowed, and 10% are divorced.
   d. A weigh station along the freeway weighs all passing trucks.

e. A study is conducted to find out whether weekend binge drinkers cause more fatal automobile accidents than everyday drunks. The study looks at 185 fatal-accident reports involving drunk drivers and notes the day of the week each accident occurred.
f. A blood bank records the ABO blood type and Rh factor (O+, for example) of each donated unit of blood.
g. In 1882 Simon Newcomb conducted an experiment to determine the speed of light. Sixty-six times he measured how long it took a light beam to travel a known distance from his laboratory on the Potomac River across the river to a reflecting mirror at the base of the Washington Monument and back.

2. Classify each of the following variables as either nonranked categorical, ranked categorical, discrete numerical, or continuous numerical.
   a. race
   b. weight
   c. sex
   d. religious preference
   e. age
   f. employment status (employed, unemployed, not in the labor force)
   g. family size
   h. marital status (married, never married, widowed, divorced)
   i. college degree (bachelor, master, doctor)
   j. Olympic medal (bronze, silver, gold)

3. Classify each of the following variables as either discrete or continuous numerical, nonranked categorical, or ranked categorical.
   a. blood type
   b. grade-point average
   c. IQ
   d. number of children
   e. occupation
   f. state in which a person's legal residence is
   g. systolic blood pressure
   h. letter grade in a college course
   i. college major

4. Charles Darwin conducted an experiment to determine whether cross-pollinated plants (i.e., another plant is involved in the breeding of a seedling) or self-pollinated plants (seedling produced without genetic influence of another plant) have greater vigor. Darwin took 15 pairs of seedlings, each pair matched by age. One seedling in each pair was produced by cross-pollination and the other by self-pollination. The members of each pair were grown in as nearly identical conditions as possible. The following data show the final height (in inches) of each plant after a fixed period of time.
   a. What is the fraction of the 15 pairs of seedlings for which the cross-pollinated seedling exhibited greater growth than the self-pollinated seedling?
   b. What seems reasonable to infer statistically from the data?
   c. Identify the three major components of the science of statistics in the statement of the problem and in your answers to parts (a) and (b).
   d. If you had the option of purchasing peas guaranteed to have been produced by cross-pollination, what would your decision be as a result of your (informal) statistical analysis conducted in part (b)?
   e. What general name does this chapter use for the kind of number computed in part (a)?
   f. Suppose the fraction had been 9/15 in part (a). Is your decision in part (b) then obvious? The lesson: We need to understand how to do statistical inference when the data do not give us an obvious answer. That is one reason for studying this textbook.

Pair number	Cross-pollinated	Self-pollinated
1	23.5	17.4
2	12	20.4
3	21	20
4	22	20
5	19.1	18.4
6	21.5	18.6
7	22.1	18.6
8	20.4	15.3
9	18.3	16.5
10	21.6	18
11	23.3	16.3
12	21	18
13	22.1	12.8
14	23	15.5
15	12	18

Source: Charles Darwin. 1976. *The Effect of Cross- and Self-Fertilization in the Vegetable Kingdom*, 2nd ed. London: John Murray.

5. A survey was conducted by the *British Medical Journal* to study the relationship between snoring and heart disease.

Heart disease	Nonsnorers	Occasional snorers	Snore nearly every night	Snore every night	Total
Yes	24	35	21	30	110
No	1355	603	192	224	2374
Total	1379	638	213	254	2484

Source: P. G. Norton and E. V. Dunn. 1985. Snoring as a risk factor for the disease: An epidemiological survey. *British Medical Journal*, 291: 630-632.

a. Which of the four categories of snorers is most prone to heart disease? *Hint:* Would it be right to say that occasional snorers are most prone to heart disease because out of the 110 persons who have heart disease, the greatest number (35) are occasional snorers and no other category has such a high number of snorers with heart disease? Explain.
b. Is there any kind of relationship between the degree of snoring and being prone to heart disease?
c. Suppose a behavioral psychologist tells a person who snores often that he can condition him not to snore. Is this likely to reduce that person's risk of heart disease? In other words, is it plausible that snoring is the cause of heart disease? If you do not think so, give another possible explanation of why frequent snorers have more heart disease than infrequent snorers.

6. The following data show the brain weights (in grams) and body weights (in kilograms) of selected animals (extant and extinct).
   a. Rank the animals from largest to smallest according to the body weight and then do the same for the brain weight. (For example, the African elephant ranks fourth (rank = 4) in body weight but first (rank = 1) in brain weight.)
   b. Are the two rankings positively associated?
   c. Do you notice that a certain group of animals seems very different in the closeness of matching of their ranks? (Statisticians call data points that are very different from the main body of the data **outliers.**)

Species	Body weight (kg)	Brain weight (g)
Mountain beaver	1.35	8.1
Grey wolf	36.33	119.5
Guinea pig	1.04	5.5
Diplodocus	11,700	50
Asian elephant	2547	4603
Potar monkey	10	115
Giraffe	529	680
Gorilla	207	406
Human	62	1320
African elephant	6654	5712
Triceratops	9400	70
Rhesus monkey	6.8	179
Mouse	0.023	0.4
Rabbit	2.5	12.1
Jaguar	100	157
Chimpanzee	52.16	440
Brachiosaurus	87,000	154.5
Rat	0.28	1.9
Mole	0.122	3

Source: H. J. Jerison. 1973. *Evolution of the Brain and Intelligence.* New York: Academic Press.

7. The table below gives the road distance and the shortest distance between 15 different pairs of points in a city. Plot the data on graph paper and discuss the nature of the relationship between the road and the shortest distance in this city. Why do the points not all lie on a straight line?

Road distance	Shortest distance	Road distance	Shortest distance
5.1	4.3	10.4	7.9
5.8	5.2	9.5	6.4
11.1	9.5	10.3	7.7
6.6	4.1	4.2	3.5
5.9	4.3	3.8	3.4
2.3	1.7	10.5	8.2
9.2	7.8	7.0	4.2
5.9	4.3		

8. A survey interviewed a representative cross section of several hundred Canadians, asking about their level of education and their level of participation in physical activity. "Active" persons were defined as those taking part in 3 or more hours of physical activity per week for 9 months or more during the year.

Amount of education	Percentage who are active
Elementary school only	41
Some secondary school	53
Secondary certificate or diploma	58
University degree	63

a. What relationship do you see between educational level (a ranked categorical variable) and physical activity?
b. Between which two *consecutive* educational levels was the increase in percentage of physical activity the largest?

9. *Infant mortality* in the following table refers to the number of infants who die during the first year of life per 1000 live births. *Life expectancy* is the number of years, on average, that an infant can expect to live. Here are the infant mortality and the life expectancy rates in the United States for people born in the years 1920 through 2000:

Year of birth	Infant mortality	Life expectancy
1920	86	54
1930	65	60
1940	47	63
1950	29	68
1960	26	70
1970	20	71
1980	13	74
1990	9	76
2000	7	77

a. Look at infant mortality throughout the years. What trend do you see?
b. What trend do you see for life expectancy?
c. Which improved more dramatically from 1920 to 2000, infant mortality or life expectancy?
d. What may be a cause of both of the trends you see?
e. Can we conclude that lowering infant mortality is the sole cause of the increase in life expectancy?

1.2 DISPLAYING SMALL SETS OF NUMBERS: DOTPLOTS AND STEM-AND-LEAF DISPLAYS

dotplot

The simplest way to visualize a small set of numbers is to construct a **dotplot**—a plot that displays a set of data on a number line by using a dot to represent each member of the data set. If multiple values in the data set are the same, they are stacked vertically in the dotplot.

Example 1.1 Heights of 11-Year-Old Girls

Eleven-year-old Julie is short for her age. She is only 138 cm tall (about 4 1/2 feet tall). Should her parents worry that she is not growing properly? Table 1.3 gives the heights of a representative sample of forty 11-year-old girls.

Looking at Table 1.3 it is difficult for us to tell whether Julie is unusually short or just short. We have to display these numbers graphically. The simplest way to represent 40 heights graphically is to construct a dotplot.

The data in Table 1.3 are displayed as a dotplot in Figure 1.4. Each observation is represented by a dot. The dotplot reveals that 2 of the 40 girls in the sample were shorter than Julie. So although Julie is short, her height is within the normal range.

The dotplot reveals a lot about the data in Table 1.3. For example, we see that half of the girls in the sample were shorter than 147.5 cm and half were taller than 147.5 cm. We express this fact by saying that the median height of the girls in the sample is 147.5 cm. (see Section 2.1). In Figure 1.5 we have superimposed a smooth curve on the dotplot indicating the overall shape of the height distribution. The smooth curve helps us focus on the overall pattern of the data by ignoring minor accidental deviations from the overall pattern.

Table 1.3 Heights (cm) of Forty 11-Year-Old Girls

144	152	147	146	151	145	142	148
146	141	160	148	153	150	149	139
157	138	146	144	155	148	149	135
137	158	142	152	145	150	156	147
143	151	150	147	144	153	146	149

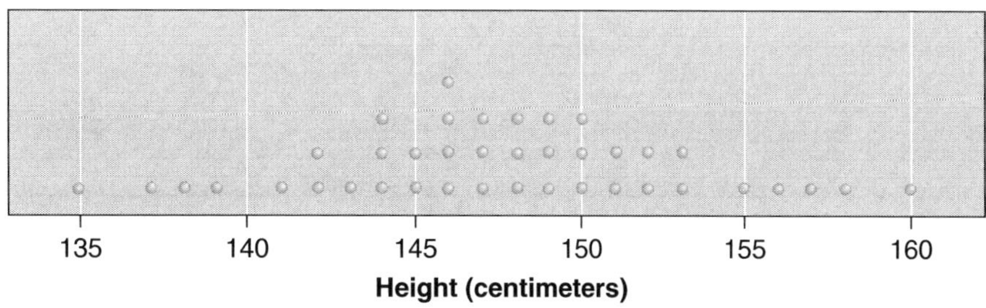

Figure 1.4 Dotplot for the data in Table 1.3.

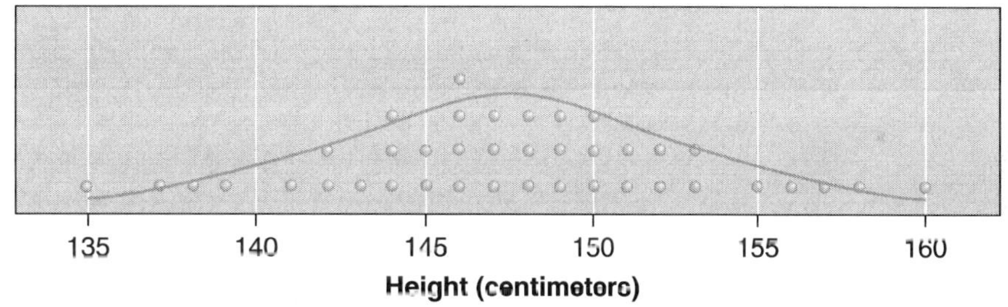

Figure 1.5 Heights of 11-year-old girls with curve.

Dotplots work well for many small sets of numbers that, like the data in Table 1.3, (a) have repeated values and (b) are not too spread out. The dotplot helps us discover and display the pattern of variation or distribution of the data.

stem-and-leaf plot
stemplot

Another graphical tool for describing the distribution of a small set of numbers is the **stem-and-leaf plot** (also called a **stemplot**). Such a plot is a mixture of a table and a graph.

Example 1.2 Luminosities (Brightnesses) of Stars

Table 1.4 lists 40 stars in a certain section of the sky in the constellation Taurus. (They were selected from the stars that can be seen without a telescope in that part of the sky; they are the ones whose distances from Earth are known most accurately.) Quite possibly this sample of stars can be viewed as representative of the population of all visible stars. If so, properties of the sample can be inferred to be approximately true for the population of all visible stars.

Table 1.4 Star Luminosities

Star ID	Luminosity (Sun = 1)	Star ID	Luminosity (Sun = 1)
18735	7	20661	4
19038	43	20711	34
19076	1	20713	29
19261	7	20842	9
19877	5	20877	29
19990	7	20885	53
20087	13	20889	67
20205	60	20894	71
20219	9	20901	18
20255	6	20995	9
20261	14	21029	19
20284	6	21036	11
20400	8	21039	10
20455	53	21137	6
20484	10	21273	23
20542	19	21402	33
20614	8	21421	138
20635	35	21459	6
20641	12	21588	8
20648	30	21589	32

The numbers in the left (and third) column in the table were used by software on the satellite Hipparcos to identify the stars and have no meaning for us. The right (and second) column shows how much light each star puts out in comparison to our own star, the Sun. For example, star 18735 puts out seven times as much light as the Sun does. That is, if it were the same distance away from Earth as the Sun is, it would appear seven times as bright as the Sun does. Astronomers call this number the luminosity. These numbers are rounded to the nearest integer; to illustrate, a star with a measured luminosity of 0.79 would be listed with luminosity 1 in this table.

The stars are listed in order of their identification numbers. Thus, such a table is not designed to enable us to see important patterns in the data. We cannot easily see how the luminosities are grouped or any special characteristics of their distribution: For example, are they tightly clustered around one value, or are they spread out over a wide range of values? In other words, what is the pattern of variation of the data? Are there any highly unusual values (i.e., much different from the majority of the values)? We cannot easily tell from the table. We would have to reorder the data at least.

❐ ❐

A stem-and-leaf plot is a way of listing a data set that displays the basic distribution or pattern of variation of the data. It is easy to construct and makes the data easier to interpret. This way of organizing data was developed by John Tukey, the founder of a modern approach to descriptive statistics called **exploratory data analysis (EDA)**.

exploratory data analysis (EDA)

Here we present only the simplest method for constructing a stem-and-leaf diagram. Variations of this method enable us to decrease or increase the number of stems, as is sometimes desirable (see, for example, the textbook entitled *The ABCs of EDA* by Velleman and Hoaglin). These techniques are beyond the scope of the textbook.

Constructing a Basic Stem-and-Leaf Plot

The starting point for constructing a stem-and-leaf plot is a table of the numbers whose distribution (shape) we wish to display. Suppose we wish to display all of the individual values as well, because those values are interesting in addition to the general shape. This dual goal of wanting to display the general shape of the data while retaining all of the individual values sometimes arises, for example, when one examines sports data. Some athletes have superb lifelong careers, performing at a high level for many years, whereas other athletes have brief moments in the sun when they really shine, and then rapidly fade away. A stem-and-leaf plot can clearly contrast the two types of careers. We will consider Roger Maris's career as a Major League Baseball home run hitter and, in particular, construct a stem-and-leaf plot to see whether he falls clearly into either of these two categories. Consider the raw data of the number of home runs per year, ordered from year to year, for his 12 years as a Major League Baseball player:

14	28	16	39	61	33	23	26	8	13	9	5

stem
leaf

In a stem-and-leaf plot every value to be plotted is represented by its **stem** and its **leaf**. The basic idea is that the last digit (farthest to the right) of a number provides its leaf and the rest of the number (all numbers to the left of this last digit) supplies the stem. To illustrate:

138 yields a stem of 13 and a leaf of 8

33 yields a stem of 3 and a leaf of 3

9 yields a stem of 0 and a leaf of 9

To construct the simplest kind of stemplot, this is all you have to know about finding the stem and leaf of a number. Let us apply this knowledge and build the stem-and-leaf plot for the Maris career home run data. Clearly the stems run from 0, 1, . . . , 6. We create a stem column that has as entries 0, 1, . . . , 6. To the right of it we create a leaf column, where we will list the leaves of all numbers having the same stem, separated by commas.

Let us deal with the numbers with stem 1, namely the data points 13, 14, 16. We obtain

Stem	Leaf
0	
1	3, 4, 6
2	
3	
4	
5	
6	

This completes the tens row. Now we simply do the same for all of the other stems to complete the stemplot.

Stem	Leaf
0	5, 8, 9
1	3, 4, 6
2	3, 6, 8
3	3, 9
4	
5	
6	1

This graphical approach (in the sense that it displays the shape of the data) is often very useful. It shows at a glance the general shape of the data while retaining the values of each individual data point. For example, any visual representation of the Maris home run data that fails to identify the value 61, which at that time broke Babe Ruth's record of 60 home runs in one season, loses much of its interest. From the view of doing descriptive statistics, we see from the stemplot that over his career Maris fairly uniformly hit home runs in the 5–39 range and that 61 is an extreme value, called an **outlier** by statisticians, that made Roger Maris famous. It would be interesting to compare the stem-and-leaf plot with that for Babe Ruth's home run career. We shall do so in Exercise 3.

outlier

Because each data value is represented by a one-leaf digit, the length of the string of leaves is proportional to the total number of data items that fall within that stem. This visual property, displayed in the Maris stem-and-leaf plot, is why stem-and-leaf plots show us the shape (distribution) of the data. Useful stem-and-leaf plots usually have between 5 and 15 stems, with larger data sets allowing the use of more stems.

Let's return to the star luminosity data of Table 1.4. Here is how the stem-and-leaf plot can be constructed for this data set. A brief inspection of Table 1.4 shows that the values range from 1 to the 70s, with one highly unusual value of 138. It is standard practice in a stemplot to list unusual values (outliers) separately, putting unusually small values in a group labeled LO at the top and unusually high values in a group labeled HI at the bottom of the stemplot. So, grouping the data by tens (0–9, 10–19, etc.), we will have eight stems, which is a reasonable number. Now we write down the list of the stems. We start with a 0, which stands for the group of stars whose luminosities are less than 10 times the Sun's, namely in the range 0–9. In the next row we put a 1 for the group having luminosities from 10 to 19 times the Sun's, and so on.

Then we start making the leaves. Let's start with the first two stars in the table:

Star 18735	luminosity 7 times the Sun's
Star 19038	luminosity 43 times the Sun's

We list those two data points as shown in the stemplot below. That is, for star 19038 with its luminosity of 43, we separate the 4, the stem part (which in this table stands for 40), from 3, the leaf.

Stem	Leaf
0	7
1	
2	
3	
4	3
5	
6	
7	

To read stem-and-leaf plots correctly, in general, we need a key. We write a key under the table:

Key: "4 3" stands for a luminosity 43 times that of the Sun.

The key tells us that in this table the stems are put in the tens place and the leaves are put in the units place. When you make a stem-and-leaf plot, you should always provide a key. For example, in another application "4 3" might stand for 0.043 or 4300. The key is needed to tell us which.

After the rest of the data are added, the completed stem-and-leaf plot is as presented in Table 1.5. Notice that we have put the leaves of the table in order. That is a desirable step, and it is easy to do, especially by computer.

The unusual value, or outlier, is listed below the main portion of the table. If there were another outlying large value (say, 3100) we would have written "HI 138, 3100." If there were an outlying small value (which would not happen in this problem) we would have written it in a group labeled "LO" above the main portion of the table.

range

A stem-and-leaf plot has two very attractive properties. First, it shows the shape of the distribution of the data *at a glance*. We can see the **range** of the data (between 1 and 138, with all the values falling between 1 and 71 except outlier value 138). Half of the stars have luminosities under 12.5 and half have luminosities greater than 12.5. We express this fact by saying that the median luminosity of the stars in the sample is 12.5 times the Sun's (see Section 2.1.) Moreover, we can distinguish several important aspects concerning the shape of the distribution of this data set:

- Is the distribution flat, or does it have a definite peak? As shown by its stemplot in Table 1.5, star luminosity distribution is rather peaked, being high near 0 and steadily decreasing toward 70. By contrast, the Maris distribution is rather flat (or uniform) from 0 to 40.

Table 1.5 Stem-and-Leaf Plot of Star Luminosities

Stem	Leaf
0	1,4,5,6,6,6,6,7,7,7,8,8,8,9,9,9
1	0,0,1,2,3,4,8,9,9
2	3,9,9
3	0,2,3,4,5
4	3
5	3,3
6	0,7
7	1
HI	138

Key: "1 0" in this stem-and-leaf plot indicates a luminosity 10 times the Sun's.

symmetric distribution

- Is the distribution **symmetric** around some central value, that is, are the distributions on the two sides of the central value mirror images of each other? Or is the distribution skewed to one side or the other? The distribution of the luminosities is strongly skewed, which means much more stretched out in one direction than the other, here toward the high end of the range at the bottom of the plot. Imagine that the stemplot in Table 1.5 is turned counterclockwise 90 degrees. It will then have a long tail to the right, and we say that the distribution is **skewed** *to the right* (following the usual convention based on other kinds of displays, such as the histograms introduced in Section 1.4, where data values increase from left to right).

skewed

gaps
clusters

- Does the distribution contain one or more **gaps**? That is, sometimes a data set splits into two or more distinct chunks, often called **clusters** (see Example 1.7). In Table 1.5 there do not appear to be any large gaps in the 1–71 range.
- Does the distribution have any outliers (observations that are widely separated from the rest of the data)? The star with luminosity 138 and, for the Maris data, the year with 61 home runs are outliers.

Second, the stem-and-leaf plot represents every point of the numerical data that was in the original table. That is, we can recover (know the exact values) all of the data points from the stem-and-leaf diagram. In some other kinds of plots, such as the box-and-whisker plot introduced in the next chapter, the numbers that were used to make the plot are not recoverable from the plot—to see the numbers, you must go back to the original list of data, such as a scientist's lab notebook or a computer file.

Comparing Populations: Back-to-Back Stem-and-Leaf Plots

One of the most common tasks in statistics is to compare two data sets that differ in some important characteristic, such as the annual incomes of male and female college graduates, the sizes of U.S. families and Brazilian families, or the cost of two-bedroom homes in San Francisco and in St. Louis, to name a few examples. The two data sets are usually regarded as representative samples from two populations we want to compare. Statisticians look for differences between the samples to determine whether they can conclude that the populations differ. Trustworthy methods for comparing populations based on their representative samples will be presented later in this book, when statistical inference is studied.

One way to compare two small data sets visually is to write their stem-and-leaf plots back to back. The stems are written in the middle of the table, the leaves for one data set are written to the left of the stem and extend leftward, and the leaves for the other data set are written to the right and extend rightward. Be sure to arrange the leaves for each data set in increasing order out from the stem.

Example 1.3 Infant Mortality Rates in Europe

Tables 1.6 and 1.7 show the infant mortality rates per thousand births, measured over the years 1995 through 1999, for the countries in Europe that had not been under communist rule during the Cold War and those that had been, respectively. (Germany, listed among the noncommunist countries here, is a special case, because about a quarter of the country had been communist. But both parts had almost the same infant mortality rate.)

The completed back-to-back stem-and-leaf plots are shown in Table 1.8. The differences between the two data sets are clear. The noncommunist countries' infant mortality data from Table 1.6 are tightly concentrated, between 5 and 10, with a single outlier. The rates from the former communist countries in Table 1.6, on the other hand, are mainly distributed between 10 and 26. The back-to-back stem-and-leaf plots reveal dramatically how much higher on average and how much more spread out the infant mortality rates in the former communist countries are.

Table 1.6 Infant Mortality Rates in Europe 1995–1999: Noncommunist Countries

Country	Infant mortality (deaths per thousand live births)
Austria	6
Belgium	7
Cyprus	8
Denmark	7
Finland	5
France	7
Germany	6
Greece	8
Ireland	6
Italy	7
Luxembourg	9
Malta	10
Monaco	58
Netherlands	6
Norway	5
Portugal	8
Spain	7
Sweden	5
Switzerland	5
United Kingdom	6

Source: New York Times Almanac 2000, 490–492.

Table 1.7 Infant Mortality Rates in Europe 1995–1999: Former Communist Countries

Country	Infant mortality (deaths per thousand live births)
Albania	32
Belarus	15
Bosnia	13
Bulgaria	16
Croatia	10
Czech Republic	9
Estonia	12
Hungary	14
Latvia	16
Lithuania	13
Macedonia	23
Moldova	26
Poland	13
Romania	24
Russia	19
Slovakia	12
Slovenia	7
Yugoslavia	19

Source: New York Times Almanac 2000, 490–492.

Table 1.8 Back-to-Back Stem-and-Leaf Plot for Infant Mortality in Europe, 1995–1999

Leaf (Table 1.6 Data)	Stem	Leaf (Table 1.7 Data)
9,8,8,8,7,7,7,7,7,6,6,6,6,6,5,5,5,5	0	7,9
0	1	0,2,2,3,3,3,4,5,6,6,9,9
	2	3,4,6
	3	2
58	HI	

Key: "1 2" indicates 12 infant deaths per 1000 live births.

Also, the Monaco outlier deserves careful investigation. What is so unusual about Monaco (a very small country)? In this case, the outlier "58" is probably a misprint in the book from which the values were taken. The correct value is most likely 5.8, which we would round to 6. Sometimes outliers result from mistakes like this; in other cases they are real (not caused by error) and, therefore, important values. They must be checked carefully in each case to determine whether they are mistakes or not.

Most standard statistical computer software packages for practitioners have a stem-and-leaf program that enables the operations we have described: constructing a standard stem-and-leaf plot, back-to-back plotting, representation of high- and low-valued outliers, and so on. Further refinements of the standard approach allow additional choices of stem intervals. It is important for you as a student in an introductory statistics course to be familiar with the basics of this powerful and often-used tool.

Section 1.2 Exercises

1. The origins of the Etruscan empire (centered in Italy around 800 B.C.) are a bit of a mystery to anthropologists. In a study, observations on the maximum head width (measured in millimeters) were taken on 84 skulls of Etruscan males and 70 modern Italian males. (Source: N. A. Barnicot and D. R. Brothwell. 1959. The evaluation of metrical data in the comparison of ancient and modern bones. In *Medical Biology and Etruscan Origins*, edited by G. E. W. Wolstenholme and C. M. O'Connor. Little, Brown, 136.)

Etruscan Skulls

141	148	132	138	154	142	150	146	155	158	150	140
147	148	144	150	149	145	149	158	143	141	144	144
126	140	144	142	141	140	145	135	147	146	141	136
140	146	142	137	148	154	137	139	143	140	131	143
141	149	148	135	148	152	143	144	141	143	147	146
150	132	142	142	143	153	149	146	149	138	142	149
142	137	134	144	146	147	140	142	140	137	152	145

Italian Skulls

133	138	130	138	134	127	128	138	136	131	126	120
124	132	132	125	139	127	133	136	121	131	125	130
129	125	136	131	132	127	129	132	116	134	125	128
139	132	130	132	128	139	135	133	128	130	130	143
144	137	140	136	135	126	139	131	133	138	133	137
140	130	137	134	130	148	135	138	135	138		

a. Compare a dotplot for the Etruscan skulls with a dotplot for the Italian skulls.
b. Construct back-to-back stem-and-leaf plots for the Etruscan skulls and the Italian skulls.
c. Is the hypothesis that Etruscans and modern Italians are of the same origins supported by the data (as it would be if the two plots seem quite similar)?

2. The ages of the 30 Oscar-winning Best Actors and the 30 Oscar-winning Best Actresses from 1976 to 2005 are as follows:

Actors

38	60	30	40	42	37	76	39	52	45	35	61	43	51	32
42	54	52	37	38	31	45	60	46	40	36	47	29	43	37

Actresses

41	35	31	41	33	31	74	31	49	38	61	21	41	26	82
42	29	33	35	45	49	38	34	26	25	33	33	35	28	30

Source: Dr. Anne E. Lincoln of Rice University

a. Construct back-to-back stem-and-leaf plots for actors and actresses.
b. What can you infer?

3. The numbers of home runs hit by Babe Ruth during his 15 years with the Yankees, from 1920 to 1934, are as follows:

54	59	35	41	46	25	47	60	54	46	49	46	41	34	22

The home run record of Roger Maris during his 10 years in the American League is as follows:

14	28	16	39	61	33	23	26	8	13

a. Construct a back-to-back stem-and-leaf plot.
b. Who was the better home run hitter from the viewpoint of his entire career? Explain.
c. In any discussion of famous home run hitters, Hank Aaron certainly must be mentioned as well. Over 21 years with the Braves, Aaron's record was

13	27	26	44	30	39	40
34	45	44	24	32	44	39
29	44	38	47	34	40	20

Construct a back-to-back stem-and-leaf plot comparing Ruth and Aaron. Discuss it.

4. The following are the test scores of students in a class:

77	70	65	77	60	79	71	86	82	78	76	65	81	59	100
62	81	67	74	71	62	84	75	70	76	72	66	73	91	49
66	70	81	79	72	64	95	71	66	80	78	73	69	58	89

Construct a stem-and-leaf plot.

5. The following stem-and-leaf plot gives final scores on a statistics test for 20 students. From the table, write down each student's score.

Stem	Leaf
7	2,3,4,5,9
8	0,2,3,4,4,5,7,8
9	0,1,2,3,5,6,6

Key: "8 2" stands for 82.

6. Using the stem-and-leaf plot of Exercise 4, answer the following questions:
   a. What was the lowest score obtained?
   b. What was the highest score obtained?
   c. How far apart were the lowest and highest scores? (Recall that this statistic is called the *range* of the data.)
   d. How would you describe the general shape of the distributions of the data?

7. The following table gives the high temperatures in degrees Fahrenheit for 20 cities on April 3, 1997.
   a. Construct a stem-and-leaf plot of the high temperatures.
   b. How far apart were the lowest and highest temperatures for these cities?
   c. Describe the distribution of temperatures.

Asheville, North Carolina	72	Miami, Florida	77
Champaign, Illinois	68	Birmingham, Alabama	74
Indianapolis, Indiana	69	Iowa City, Iowa	63
Abilene, Texas	67	Detroit, Michigan	58
Las Cruces, New Mexico	72	Molokai, Hawaii	77
Los Angeles, California	71	St. Louis, Missouri	76
Billings, Montana	56	Lincoln, Nebraska	63
Chicago, Illinois	65	Boston, Massachusetts	53
Albany, New York	53	Baltimore, Maryland	66
Charleston, South Carolina	77	Boise, Idaho	60

8. The following table gives the unemployment rates for the so-called G7 nations (Group of Seven Leading Industrial Nations) for the years 1975 and 2004 (from *The World Almanac* 2006).

	1975	2004
United States	8.5	5.5
Canada	6.9	6.4
Japan	1.9	4.8
France	4.2	9.8
Germany	3.4	9.8
Italy	3.4	8.1
Great Britain	4.6	4.8

a. Construct a stem-and-leaf plot for the unemployment rates in 1975.
b. What was the highest rate in 1975? Which country had it? What was the lowest? Which country had it?
c. Construct the stem-and-leaf plot for the unemployment rates in 2004.
d. What was the highest rate in 2004? Which country had it? What was the lowest? Which country had it?
e. Line up the stem-and-leaf plots for 1975 and 2004 back-to-back. Make a general comparison of the rates in 1975 to those in 2004.

9. Assume that the stars in Table 1.4 of Example 1.2 are representative (typical) of the roughly 2000 stars visible to the naked eye on a dark night away from city lights. Do you think that the stars in Table 1.4 also are representative of the 200 billion stars in our galaxy that are not visible to the naked eye? Explain.

1.3 GRAPHING CATEGORICAL DATA

We cannot draw a stem-and-leaf plot for categorical data. We cannot say anything about the "shape" of a table of categorical data, because it would depend on the order of the categories, and we could put them in any order. For example, there would be no preferred order for the six continents from which a person's ancestors may have come. (They *could* be put in alphabetical order, but that order indicates nothing about the continents themselves and would not be the same when written in a different language, such as Russian.)

Two kinds of graphical displays are particularly useful for making comparisons involving categorical data: the bar chart and the pie chart.

The Bar Charts

bar chart
frequency

A **bar chart** is constructed by drawing a vertical bar for each of the possible category values. All bars are of equal width. The height of each bar equals the number (or **frequency**) of persons (or objects) in that category. A vertical scale, usually on the left, helps the reader see the frequency. Sometimes the bars touch each other, and sometimes an equal space is left between each pair of bars (see Figure 1.20). When the categories are ordered, then this order determines the order of the bars from left to right. One example of such a bar chart would be the number of traffic accident fatalities in Boston on each day of the week. A bar chart is often called a **bar graph**.

bar graph

Pareto chart

Often the categories have no special order. In this case, if the categories are arranged so that the tallest bar is to the left, then the second tallest, and so forth, the resulting graph is called a **Pareto chart**. By arranging the bars in order of decreasing frequency, we focus attention on the more important categories. Thus, a Pareto chart enables us to easily compare the numbers of objects in the various categories. Pareto charts are often used in engineering studies of causes of failure.

Example 1.4 Causes of Death

frequency table

For Americans under 45 years of age, accidents are the leading cause of death. But for the entire U.S. population, accidents are only the fifth most common cause of death. Table 1.9 lists the number of deaths due to the eight leading causes of death in the United States in 1997. The cause of death is a categorical variable characterizing each death. Table 1.9 is a **frequency table**. The data in Table 1.9 are displayed graphically in Figure 1.6. Clearly, the vast majority of deaths are due to heart disease or cancer. Often in newspapers and magazines, the bars are separated for artistic reasons, as in Figure 1.20.

Table 1.9 Leading Causes of Death in the United States, 1997

Cause of Death	Number of Deaths
Heart disease	726,000
Cancer	537,000
Stroke	160,000
Pulmonary diseases	111,000
Accidents	92,000
Pneumonia and influenza	88,000
Diabetes	62,000
AIDS	30,000

Provisional data, estimated from a 10 percent sample of deaths.
Source: The New York Times Almanac 2000, 392.

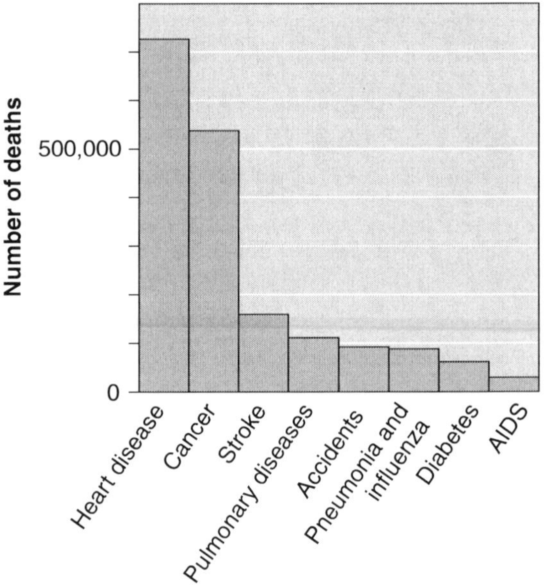

Figure 1.6 Pareto chart for leading causes of death in the United States.

The Pie Chart

If we want to draw attention to the percentage that each category contributes to the whole, we might present a so-called *pie chart*. (See Figure 1.3 and Figure 1.7.) In a pie chart, each category value is represented by a wedge-shaped section of a circle between two radii (lines drawn from the center of the circle to its circumference). The size of each sector is determined by the angle between the radii:

$$\text{Angle between radii} = \text{Relative frequency of this category} \times 360^\circ$$

Just as the frequencies of all the categories add up to the total number of individuals being counted, the angles of the sections of the pie chart add up to 360 degrees, the number of degrees in a complete circle. Note that we cannot obtain the total frequencies in the various categories from a pie chart the way we can from a bar graph.

Example 1.5 The U.S. Population by Marital Status

Marital status is a categorical variable that classifies persons. Each adult is either *married, never married, divorced,* or *widowed.* The first two columns in Table 1.10 give the total number of occurrences (in millions) of each of these category values—in other words, the total number of persons having each marital status—in the United States in 1998.

The data in Table 1.10 could be displayed in a Pareto chart, but we shall instead use a pie chart. The third and fourth columns in Table 1.10 give the proportions and the resulting sizes (in degrees) of the sections needed to draw a circle graph from these data. For example, the number of persons never married is 44.6 million. The proportion of the total number of adults is 44.6 million divided by the total, 195.5 million, or 0.228. The angle between the radii needed to draw the corresponding wedge is 0.228×360 degrees, rounded to 82 degrees.

The complete pie chart is shown in Figure 1.7. The number of adults who have never been married is nearly a quarter of the total; the number of adults who are currently married is somewhat over half the total.

Table 1.10 U.S. Population (Age 18 or Over) by Marital Status, 1998

Status	Millions of persons	Proportion of total	Angle (degrees)
Married	117.9	0.603	217
Never married	44.6	0.228	82
Widowed	13.6	0.070	25
Divorced	19.4	0.099	36
Total	195.5	1.000	360

Source: The New York Times Almanac 2000, 294.

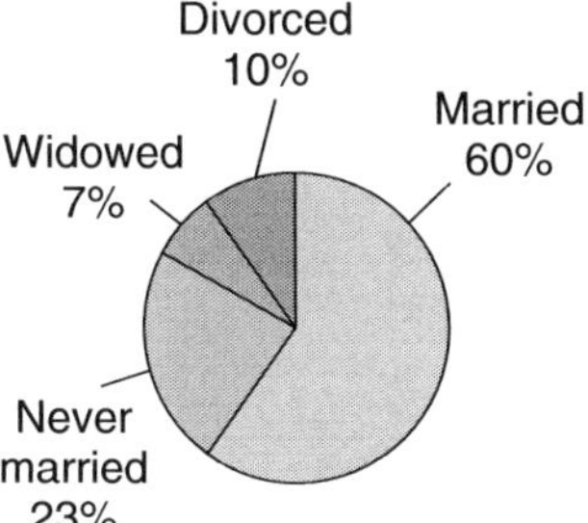

Figure 1.7 Pie chart of the marital statuses of U.S. adults in 1998 from Table 1.10.

Example 1.6 Authorship of Three Books Attributed to Davy Crockett

In the Key Problem we learned that the rate at which authors use "contentless" words can help establish the authorship of a disputed book. In particular, statistical evidence suggests that the *Narrative* and *Tour* writing is consistent with that of Davy Crockett's congressional speeches (presumed written by him). In contrast, the writing of the *Texas* book is inconsistent with both the *Narrative* writing and the writing in the speeches.

We cannot present the sophisticated hypothesis-testing analysis that produced these conclusions, but we can present back-to-back bar charts that suggest the conclusions drawn. Using the rate data of Table 1.2, in Figure 1.8 the *Narrative-Tour* and the *Narrative-Texas* bar chart comparisons are made. Statistically the bar charts for *Narrative* versus *Tour* were judged entirely similar (shown shaded dark): There were no pairs of bars too different to occur just by chance if the same author wrote both. In sharp contrast, the differences in heights for the "an," "both," and "this" pairs of bars (shown shaded light in the bottom chart) are each too large to have easily arisen by chance if the same author wrote *Texas* as wrote *Narrative*. A similar bar chart comparison of *Narrative* with the speeches shows no rate disparities, whereas the *Texas*-speeches comparison shows rate disparities for "this" and "through" inconsistent with a common authorship (see Exercise 5 below).

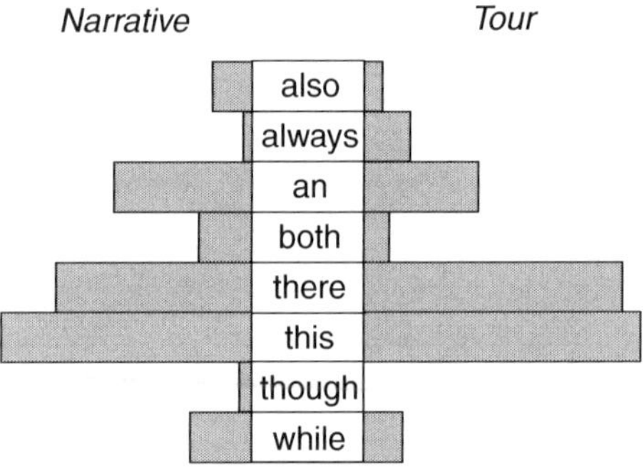

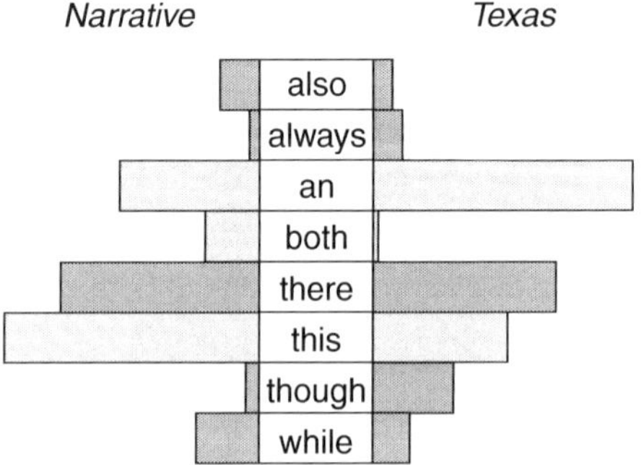

Figure 1.8 Back-to-back bar charts of *Narrative-Tour* and *Narrative-Texas* comparisons.

Section 1.3 Exercises

1. Bar graphs are sometimes used to compare categories even when we are looking at data other than frequencies. The table below shows the percentage of foreigners (from the perspective of the listed country) in the total labor force in different countries in 1991. (For more details on this table, see *Statistical Abstract of the United States, 1997*.)

Country	Percentage	Country	Percentage
United States	9.3	Germany	8.9
Australia	24.8	Japan	0.9
Austria	8.9	Luxembourg	33.3
Belgium	7.4	Netherlands	3.9
Canada	18.5	Norway	4.4
Denmark	2.5	Spain	0.4
France	6.2	Sweden	5.5

   a. Depict these data using a bar graph. Then make comments on the distribution of the data.
   b. Explain why the data cannot be displayed in a pie chart.

2. The following table gives the breakdown of total U.S. oil consumption (by percentage) by the purposes for which oil is used:

Use	Percentage
Gasoline	43.0
Heating oil	17.0
Industrial fuel oil	12.0
Jet fuel	7.0
Diesel fuel	5.5
Petrochemical	3.5
Other	12.0
Total	100.0

   a. Construct a pie chart from the data.
   b. Construct a Pareto chart.
   c. Which is more useful, or are they both useful for different purposes?
   d. What conclusions do you draw?

3. Forty randomly chosen customers were asked to name the brand of toothpaste (labeled A–E) they liked the most. The preferences are as follows:

D	A	B	E	C	B	D	A	A	B	A	D	A	E	A	D	A	A	D	A
D	E	D	A	D	B	D	C	A	E	D	B	C	E	D	B	A	B	C	E

a. Use an appropriate graphical tool that facilitates comparison among brands from most to least popular.
b. Use an appropriate graphical tool that shows clearly the relative preference for each brand of toothpaste compared to the total for all brands.

4. A sample of employees at an organization start work at the following times in the morning:

9:00	9:00	8:30	9:00	8:00	9:30	9:00	10:00	8:30	9:00	9:30	10:00	8:00
8:00	8:30	9:00	9:00	8:00	8:00	9:00	8:30	8:30	8:30	9:00	9:30	10:00
9:30	8:30	9:00	8:30	9:30	8:30	9:30	9:00	9:00	9:00	9:00	9:30	9:00

Use an appropriate graphical tool that facilitates an accurate comparison of the proportion of employees starting their work at two different times, such as 9:00 A.M. and 9:30 A.M.

5. a. Construct two back-to-back bar charts: one for the *Narrative*-speeches comparison and the second for the *Texas*-speeches comparison in the Key Problem and Example 1.6.
b. In part (a), for the *Texas*-speeches comparison, do the differences in height for the "this" bars seem unusually large? For the "though" bars?

1.4 FREQUENCY HISTOGRAMS

In Section 1.3 we constructed bar graphs for categorical (qualitative) data. We can do the same thing for quantitative (numerical) data in order to study the distribution (shape) of the data. In this case, each bar represents the number of persons (or things, observations, etc.) for which the measured value falls within a certain interval. Such a bar graph for measured data is called a **frequency histogram**, or just a **histogram**. (The word *histogram* comes from the Greek word *histos,* meaning pole, because the bars in the histogram look somewhat like a row of poles. Histograms appeared as early as 1786.)

frequency histogram

The horizontal axis of such a graph shows the intervals of the observed data values, and the vertical axis indicates the number of observations, or the **frequency**, within each interval in the range.

frequency

Example 1.7 The Highest Elevation in Each of the 50 States

Consider the highest elevation for each state, shown in Table 1.11. We shall construct a *frequency histogram* of these elevations in three steps.

Step 1. *Divide the range of the data into class intervals of equal width.* The data in Table 1.11 range from 345 (Florida) to 20,320 (Alaska). We choose as our class intervals 0 to 1500−, 1500 to 3000−, 3000 to 4500−, and so on, each of width 1500. The minus signs indicate that the value 1500 is not included in the first interval (note it is included in the second); the value 3000 is not included in the second interval but is included in the third; and so forth.

Table 1.11 Highest Elevation in Feet

State	Elevation (feet)	State	Elevation (feet)	State	Elevation (feet)	State	Elevation (feet)
AL	2,405	IN	1,257	NE	5,425	RI	812
AK	20,320	IA	1,670	NV	13,140	SC	3,560
AZ	12,633	KS	4,039	NH	6,288	SD	7,242
AR	2,753	KY	4,139	NJ	1,803	TE	6,643
CA	14,494	LA	535	NM	1,161	TX	8,749
CO	14,433	ME	5,267	NY	5,344	UT	13,528
CT	2,380	MD	3,360	NC	6,684	VT	4,393
DE	442	MA	3,487	ND	3,506	VA	5,729
FL	345	MI	1,979	OH	1,549	WA	14,410
GA	4,784	MN	2,301	OK	4,973	WV	4,861
HA	13,796	MS	806	OR	11,239	WI	1,951
ID	12,662	MO	1,772	PA	3,213	WY	13,804
IL	1,235	MT	12,799				

Source: The Universal Almanac.

Step 2. *Count the number of observations in each class interval.* The resulting table is a *frequency table*. Here are the counts (frequencies):

Class interval	Tallies	Frequency
0 to 1,500–	卌 II	7
1,500 to 3,000–	卌 卌	10
3,000 to 4,500–	卌 III	8
4,500 to 6,000–	卌 II	7
6,000 to 7,500–	IIII	4
7,500 to 9,000–	I	1
9,000 to 10,500–		0
10,500 to 12,000–	I	1
12,000 to 13,500–	卌	5
13,500 to 15,000–	卌 I	6
15,000 to 16,500–		0
16,500 to 18,000–		0
18,000 to 19,500–		0
19,500 to 21,000–	I	1

The highest elevation in Alabama is 2405, so we put a tally mark next to the class interval 1500 to 3000–. The highest elevation in Alaska is 20,320, so we put a tally mark next to the class interval 19,500 to 21,000–, and so forth.

Step 3. *Draw the histogram.* In Figure 1.9 we mark elevations on the horizontal axis. The horizontal scale runs from 0 to 21,000. We mark frequency (number of states) on the vertical axis. For each class interval we draw a bar whose base is the class interval and whose height is the corresponding frequency.

We notice that the 50 states fall in three separate groups: Alaska with Mt. McKinley (elevation 20,320 feet) is in an outlier group by itself. The middle group (with maximum elevations between 11,000 feet and 14,500 feet) consists of the Rocky Mountain states, the Pacific states, and Hawaii. Finally, the largest group (with maximum elevations under 9000 feet) consists of the 37 states east of the Rocky Mountains. Within this group, only Texas ("TX" in Table 1.11), which borders on the Rocky Mountains, has an elevation above 7500 feet. This is an example of a data set that has explainable clusters with gaps between them.

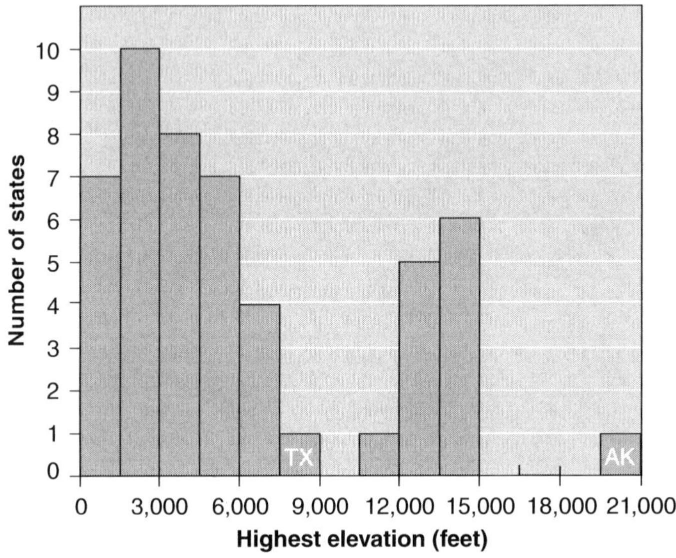

Figure 1.9 Frequency histogram for highest elevation in the 50 states.

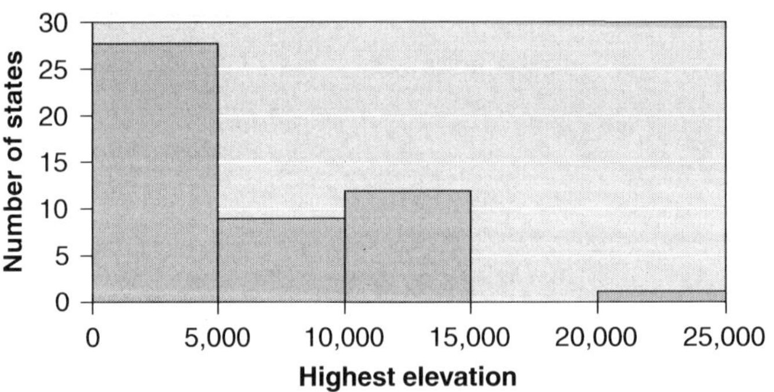

Figure 1.10 Frequency histogram for highest elevation in the 50 states.

You may ask how we chose our class intervals in Step 1. It is largely a matter of common sense. The number of class intervals should be large enough to display the important characteristics of the shape of data but small enough to provide an effective summary of the basic shape of the data. Not counting the 4 empty class intervals, we grouped the data in 10 class intervals. With fewer observations we would have used fewer class intervals. As a general rule of thumb, 5 to 20 class intervals are usually appropriate. In this example we found the range of the data, 345 to 20,320, then we picked a slightly larger interval, 0 to 21,000, and divided it into 14 class intervals of equal width of 1500. We could just as well have used class intervals of slightly greater width, say 2000 or 2500. But if we use too few class intervals, then we throw away too much information about the shape of the distribution. Unlike a stem-and-leaf plot, the frequency table of a histogram records only the number of observations in each class interval, not the exact value of each data point.

Figure 1.10 illustrates what happens if we use too few class intervals. Figure 1.10 is not very informative. Important and interesting information about how the data are distributed is lost. We can still see that a majority of the states have maximum elevation under 5000 feet. But we have blurred the distinction between the mountain states and the states east of the Rocky Mountains. The gap between the two groups has disappeared. Figure 1.10 simply has too few bars to tell the real story of the data.

The next example illustrates the fact that every stem-and-leaf plot can be converted into a frequency histogram.

Example 1.8 **Annual Salaries**

Look at the stem-and-leaf plot of annual salaries at a certain company to the nearest $1000 in Table 1.12. By simply enclosing each row (set of leaves) within a bar, we have a respectable histogram (see Figure 1.11(*a*)).

We can also build a frequency table by counting how many leaves are in each stem (see Table 1.12) and then draw the histogram from the frequency table, showing the frequencies (the number of leaves) along one axis of the graph (see Figure 1.11(*b*)).

Table 1.12 Stem-and-Leaf Plot of Annual Earnings (Thousands of Dollars)

Stem	**Leaf**
1	6,7
2	9,9,9,2,7,7,8,3,5
3	4,5,1,2,6,6,3,9
4	6,9,7,7,7,5,9,8,6,3,0,3,4,8,3
5	3,5,3,7,7,9
6	3,2,2,4,4,0,0
7	7,4

Histograms are usually shown with vertical bars rather than horizontal bars. In that case, our histogram of the annual earnings becomes Figure 1.12. In this graph the horizontal axis scale (salaries in $1000s) has been included to make it clear which interval in the range of incomes each bar represents, with the first interval from $10,000 to $20,000–, and so on.

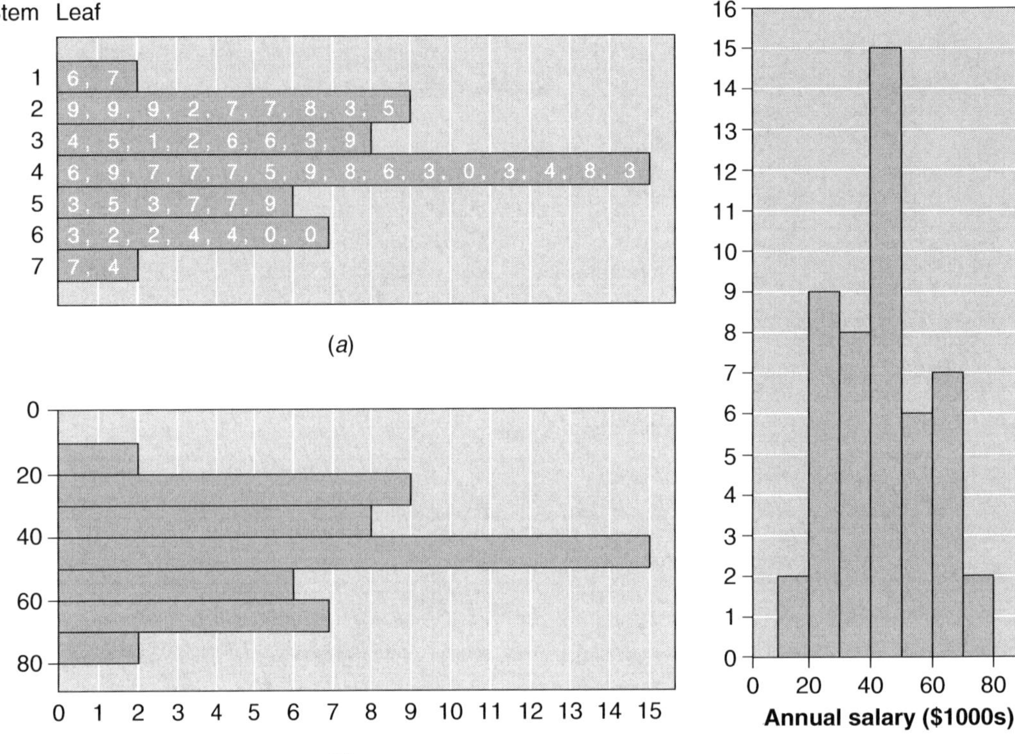

Figure 1.11 Frequency histogram derived from stem-and-leaf plot of Table 1.12.

Figure 1.12 Frequency histogram of Figure 1.11 put in usual position.

Building a histogram from a frequency table whose class intervals we chose rather than from a stem-and-leaf plot has a major advantage. We can freely choose the width of the class intervals in a frequency table. In this way we can choose a number of class intervals that does a nice job of showing interesting and important aspects of the shape of the data. Stem-and-leaf plots do not offer the same freedom of choice. For example, the frequency histogram in Figure 1.9 cannot be derived from a stem-and-leaf plot. Another limitation to the use of stem-and-leaf plots is that they are effective only for small data sets.

Distribution Shapes

We have already seen that stem-and-leaf plots help reveal the underlying shape of a data set. Now we will see how useful histograms are for this purpose.

In 1903 Pearson and Lee published a paper on the laws of inheritance in humans. They studied over 1000 English families, relating the height, span of arms, and length of left forearm of the various family members. Figure 1.13 displays a frequency histogram for the heights of the 1052 mothers in the study. The heights were recorded to the nearest quarter of an inch. We have superimposed a smooth curve approximating the overall shape of the histogram. The smooth curve makes the shape easier to see.

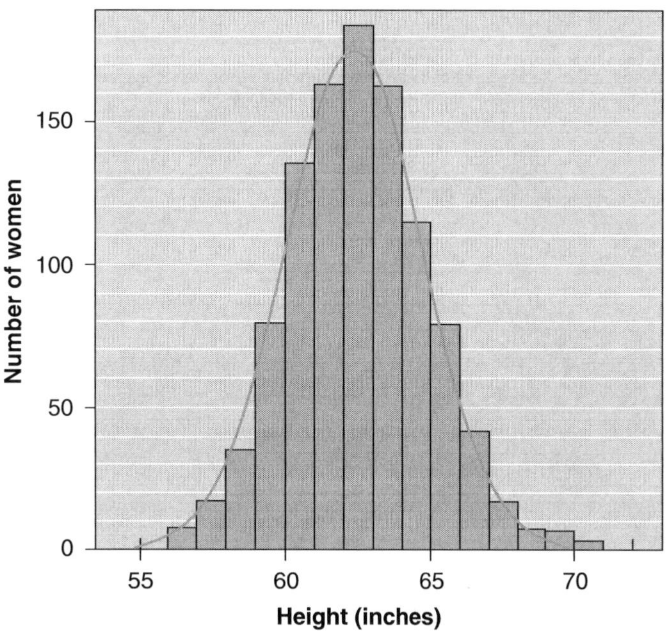

Figure 1.13 Heights of mothers.

bell-shaped
uniform
U-shaped
right-skewed
left-skewed

The histogram in Figure 1.13 is "bell-shaped," a very commonly occurring distribution shape. Figure 1.14 displays some common distribution shapes: **bell-shaped**, **uniform**, **U-shaped**, **right-skewed** (or **skewed to the right**), and **left-skewed** (or **skewed to the left**). We say that figures (*d*) and (*e*) are "skewed" in the direction, right or left, of the long tail. Figures (*a*), (*b*), and (*c*) are symmetric.

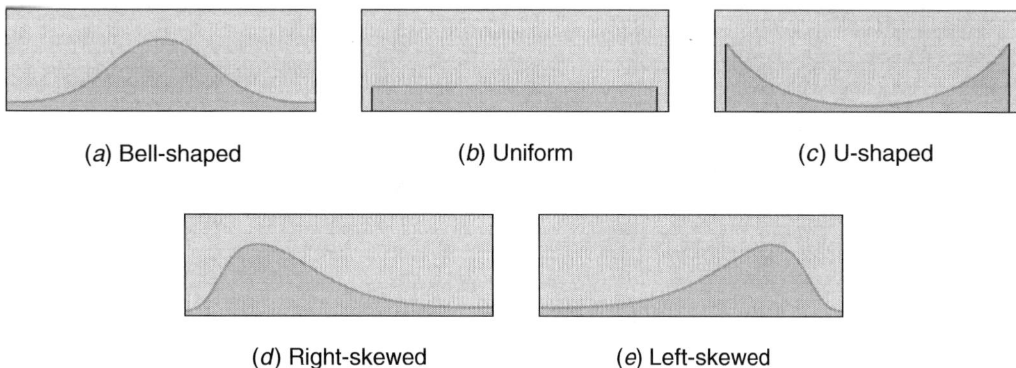

Figure 1.14 Common distribution shapes.

When we superimpose a smooth curve on a histogram we focus on the overall shape of the distribution and ignore minor irregularities.

Section 1.4 Exercises

1. One of the histograms sketched below shows the distribution of age at death from natural causes (heart disease, cancer, stroke, and so forth). The other shows the distribution of age at death from trauma (accident, suicide, murder). Which is which? Explain.

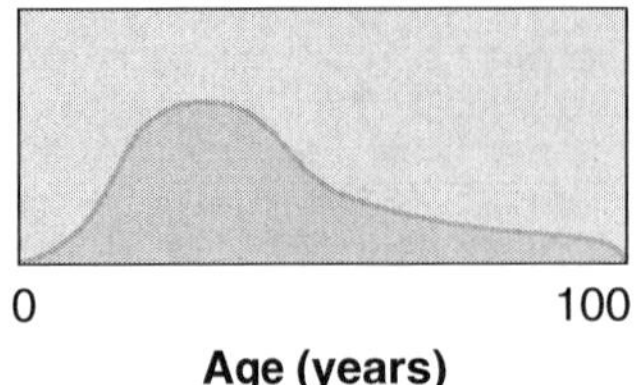

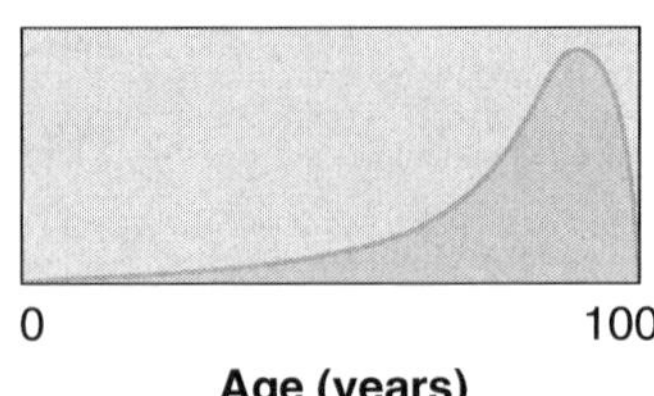

2. The following data give the per capita (per person) spending distribution in federal dollars for the 50 states. (For example, 7 states receive an outlay in the $2,200–$2,399 interval).
   a. Draw a frequency histogram.
   b. What are your observations?

Dollars per capita	Number of states	Dollars per capita	Number of states
2000–2199	1	3400–3599	2
2200–2399	7	3600–3799	1
2400–2599	9	3800–3999	0
2600–2799	6	4000–4199	3
2800–2999	9	4200–4399	3
3000–3199	6	4400–4599	0
3200–3399	2	4600–4799	1
		Total	50

Source: U.S. Bureau of Census, *Statistical Abstract of the United States*, 1997.

3. The following data show the time in days between 63 successive major earthquakes. An earthquake is included in this data set only if its magnitude was at least 7.5 on the Richter scale, or if over 1000 people were killed.

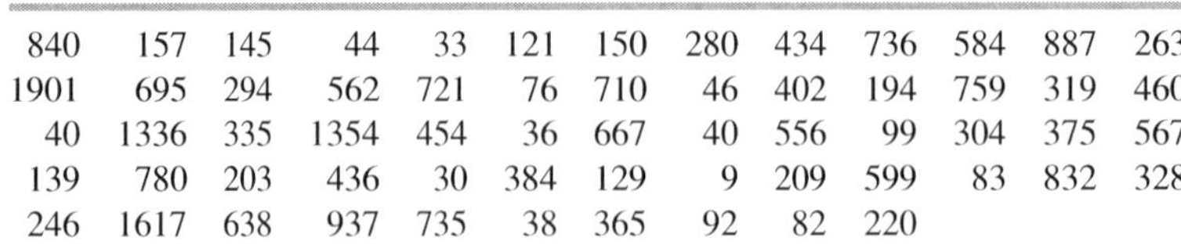

840	157	145	44	33	121	150	280	434	736	584	887	263
1901	695	294	562	721	76	710	46	402	194	759	319	460
40	1336	335	1354	454	36	667	40	556	99	304	375	567
139	780	203	436	30	384	129	9	209	599	83	832	328
246	1617	638	937	735	38	365	92	82	220			

   a. Draw a frequency histogram, using the class intervals 0 to 100–, 100 to 200–, and so on. List the values over 1000 as HI values.
   b. What are your observations about the distribution?

4. Given below are measurements resulting from repeatedly carrying out an experiment.

75	95	100	10	16	50	26	84	29	80	35	57	27	50	12	51	12	80	29	68
27	84	28	52	53	61	40	94	88	52	54	14	36	64	40	87	28	52	65	31
101	40	93	33	65	33	64	69	44	102	46	95	34	66	99	39	84	44	79	47
37	70	55	72	59	66	44	49	70	56	62	32	62	41	78	49	61	37	80	59
63	42	92	76	67	67	70	73	74	85	77	86	90	76	103	103	90	90	100	

a. Construct a histogram using an interval size of 5.
b. Construct a histogram using an interval size of 10.
c. Which of the two interval sizes is more appropriate and why?

5. Given below are the observations from random sampling of a population.

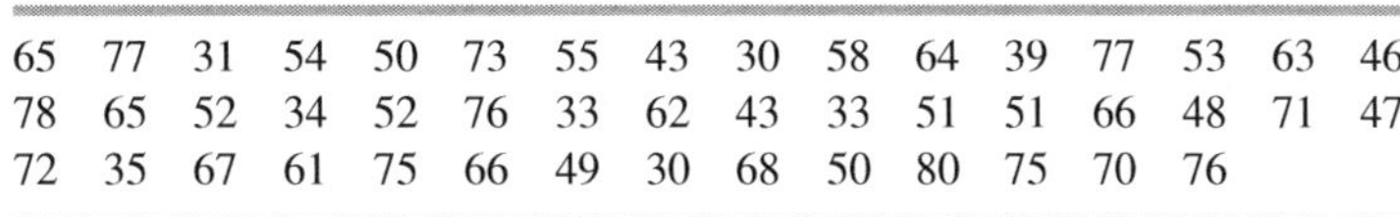

65	77	31	54	50	73	55	43	30	58	64	39	77	53	63	46
78	65	52	34	52	76	33	62	43	33	51	51	66	48	71	47
72	35	67	61	75	66	49	30	68	50	80	75	70	76		

a. Construct a histogram using an interval size of 5.
b. Construct a histogram using an interval size of 10.
c. Which of the two interval sizes is more appropriate and why?

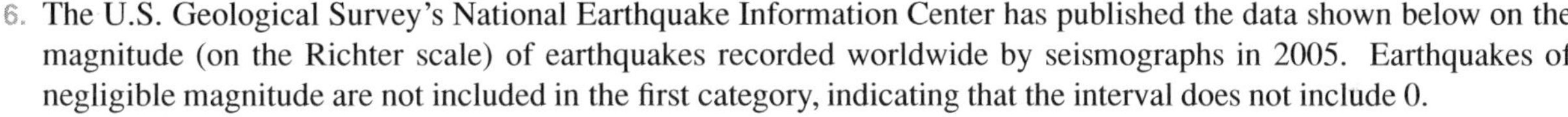

6. The U.S. Geological Survey's National Earthquake Information Center has published the data shown below on the magnitude (on the Richter scale) of earthquakes recorded worldwide by seismographs in 2005. Earthquakes of negligible magnitude are not included in the first category, indicating that the interval does not include 0.

Magnitude	Number of earthquakes
0^+–1^-	0
1–2^-	26
2–3^-	4,636
3–4^-	9,154
4–5^-	13,898
5–6^-	1,721
6–7^-	148
7–8^-	10
8–9^-	1

a. Construct a histogram.
b. Make your observations.
c. The USGS estimates that several million earthquakes occur worldwide each year. Explain why the 30,000 recorded earthquakes are not a representative cross section of the millions of earthquakes that occur each year.

7. The percentage of the residents of each of the 50 states who were at least 65 years old in 2004 are as follows:

Percentage of Population at Least 65 Years Old, by State, 2004

State	Percent	State	Percent
Alabama	13.2	Montana	13.7
Alaska	6.4	Nebraska	13.3
Arizona	12.7	Nevada	11.2
Arkansas	13.8	New Hampshire	12.1
California	10.7	New Jersey	12.9
Colorado	9.8	New Mexico	12.1
Connecticut	13.5	New York	13.0
Delaware	13.1	North Carolina	12.1
Florida	16.8	North Dakota	14.7
Georgia	9.6	Ohio	13.3
Hawaii	13.6	Oklahoma	13.2
Idaho	11.4	Oregon	12.8
Illinois	12.0	Pennsylvania	15.3
Indiana	12.4	Rhode Island	13.9
Iowa	14.7	South Carolina	12.4
Kansas	13.0	South Dakota	14.2
Kentucky	12.5	Tennessee	12.5
Louisiana	11.7	Texas	9.9
Maine	14.4	Utah	8.7
Maryland	11.4	Vermont	13.0
Massachusetts	13.3	Virginia	11.4
Michigan	12.3	Washington	11.3
Minnesota	12.1	West Virginia	15.3
Mississippi	12.2	Wisconsin	13.0
Missouri	13.3	Wyoming	12.1

Source: Statistical Abstract of the United States, 2006.

a. Draw a frequency histogram, using the class intervals 5.6 to 6.5 (inclusive), 6.6 to 7.5, 7.6 to 8.5, and so on.
b. Are there any gaps or outliers?
c. Comment on the smallest and the largest values.

1.5 DENSITY HISTOGRAMS

counting variables

Consider a variable such as the size of a family, the number of children in the family under 16 years of age, or the number of rooms in the family's home. The possible values for such **counting variables** are the numbers 0, 1, 2, 3, 4, ... (called integers). When we plot histograms for such variables, we center the histogram bars at the possible integer values. To do so, we use the following class intervals: –0.5 to 0.5, 0.5 to 1.5, 1.5 to 2.5, and so forth, even though the variables cannot have a value like 0.5.

Example 1.9 Household Sizes

According to the U.S. Census Bureau, a household is any group of persons living together. In 1999 the Current Population Survey studied a representative cross section of a little more than 50,000 households, obtaining the following distribution.

Size	Class interval	Frequency	Relative frequency (proportion)	Percentage
1	0.5 to 1.5	12,709	0.2503	25.03
2	1.5 to 2.5	16,517	0.3252	32.52
3	2.5 to 3.5	8,435	0.1661	16.61
4	3.5 to 4.5	7,521	0.1481	14.81
5	4.5 to 5.5	3,640	0.0717	7.17
6	5.5 to 6.5	1,251	0.0246	2.46
7	6.5 to 7.5	425	0.0084	0.84
8	7.5 to 8.5	169	0.0033	0.33
Over 8	8.5 and above	118	0.0023	0.23
Total		50,785	1	100.00

Source: U.S. Census Bureau.

Because the sample of 50,000 is so large, the percentage of households of size 1 in the Current Population Survey (25.03%) is likely to be very close to the percentage of the population of all 100 million U.S. households that are of size 1. The percentage of households of size 2 in the survey (32.52%) is likely to be very close to the percentage of all U.S. households that are of size 2, and so forth. We can therefore use the survey data to approximate very closely the distribution by size of all U.S. households. Figure 1.15 is the resulting **density histogram**. We note that household size data are right-skewed. The graph is called a density histogram because the areas of its rectangles add to 100%. We could just as well have displayed relative frequencies (proportions) for the heights, in which case the areas of the rectangles add to 1. If the areas of the rectangles add to 1, it is still called a density histogram.

density histogram

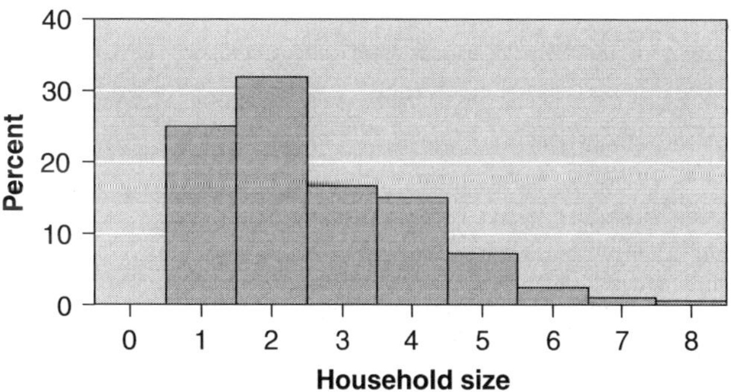

Figure 1.15 Density histogram of U.S. households by size.

In Section 1.4 we learned how to construct frequency histograms displaying the general shape, or distribution, of data sets. Often we want to compare two or more distributions. To illustrate, we compare the age distribution of the population of Germany with the age distribution of the population of Kenya. We want to know whether the German population is older than the population of Kenya. We shall use the data in Table 1.13, taken from the U.S. Bureau of the Census, International Data Base.

Table 1.13 Distribution of German and Kenyan Populations by Age, 2000

Age group	Millions of Germans	Millions of Kenyans
0 to 15–	13.01	12.99
15 to 30–	14.41	9.59
30 to 50–	26.29	5.21
50 to 65–	15.64	1.72
65 to 80–	10.54	0.72
80 to 95–	2.91	0.11
Total	82.80	30.34

We first note that, because the population of Germany is almost three times as large as the population of Kenya, there are more Germans than Kenyans in each of the age groups in Table 1.13. But it is not the *number* of Germans of a given age we want to compare with the *number* of Kenyans of the same age. Rather, we want to compare the *percentage* of the total number of Germans in each age group with the *percentage* of the total number of Kenyans in the same age group. To make this comparison, we shall construct a density histogram for each population. In this density histograph the class intervals are not of length 1 as in Example 1.9. In such density histograms, the areas of the blocks, rather than the heights, represent percentages. We proceed in three steps.

Step 1. Compute the percentage of Germans in each age group by dividing the number of Germans in that age group by the total number of Germans and then multiplying by 100%. For example, the percentage of Germans under 15 years of age is just the relative frequency times 100%:

$$\frac{13.01}{82.80} \times 100\% = 15.7\%.$$

Step 2. Find the height of the histogram bar over each class interval. Using the principle that Area = Percentage, we use the formula

$$\text{Percentage} = \text{Area of rectangle} = \text{Width} \times \text{Height},$$

where Width denotes the width of the class interval. In other words,

$$\text{Height} = \text{Percent} / \text{Width}.$$

For example, for the age group from 0 to 15 years (class interval), we get for the German population

$$\text{Height} = 15.7\% / 15 \text{ years} = 1.05\% \text{ per year}$$

Combining the two steps for the German population, we get Table 1.14.

Table 1.14 Density Histogram Heights for Age Distribution of German Population, 2000

Age group	Percentage	Height
0 to 15–	15.7	1.05
15 to 30–	17.4	1.16
30 to 50–	31.8	1.59
50 to 65–	18.9	1.26
65 to 80–	12.7	0.85
80 to 95–	3.5	0.23

We note that for a density histogram the class intervals do *not* have to be of the same size. Here the class intervals vary in size.

Step 3. Draw the density histogram. We mark age on the horizontal axis. The horizontal scale runs from 0 to 100 years. The unit on the vertical axis is "percentage per year."

We note that the height of the appropriate rectangle of the histogram tells us the approximate percentage of Germans of any specified age in the rectangle: For example, the percentage of 10-year-olds is roughly 1.05%, and the percentage of 40-year-olds is roughly 1.59%. The percentage of 60-year-olds is approximately 1.26%. And the percentage of all Germans who are between their 5th and their 10th birthdays is roughly $5 \times 1.05\% = 5.25\%$, namely, the area of the shaded block:

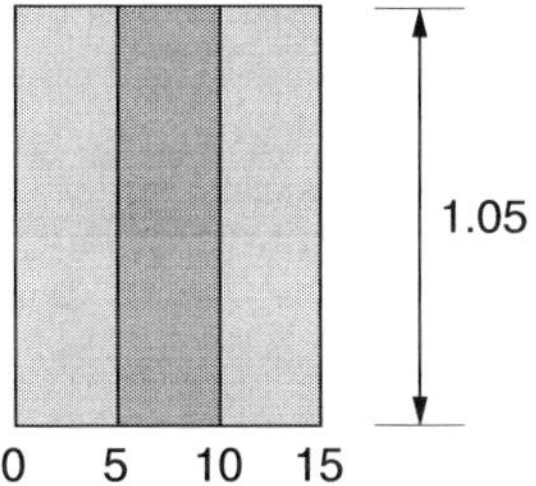

Looking at Figure 1.16, we notice that the two blocks covering the ages from 0 to 30 years are not as tall as the two blocks covering the ages from 30 to 65 years. Thus, assuming a uniform age distribution within each rectangle, the percentage of Germans of any particular age between 30 years and 60 years is greater than the percentage of Germans of any particular age under 30. For example, the percentage of 40-year-olds is greater than the percentage of 10-year-olds. Essentially, this is because Germans are having fewer children today than they had in the past. At the same time, people are living longer. As a result, in 20 years there will be fewer workers and more retired people than there are today. This becomes a serious social and political problem. It will be difficult for the working people to pay for the health care and the pensions of the retired people. To maintain their economic growth and standard of living, the Germans will have to recruit foreign workers—a prospect they do not yet seem to have come to terms with. Many European countries face the same problem. Unlike the United States, many of the European countries do not have a tradition of recruiting skilled foreign labor.

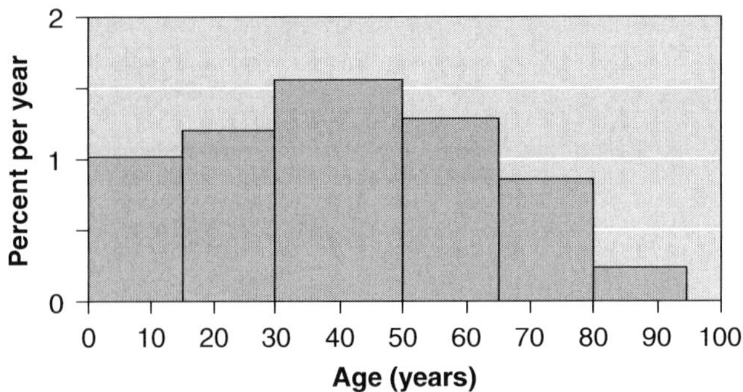

Figure 1.16 Distribution of German population by age, 2000, as shown by a density histogram.

The age distribution of Kenya tells a very different story. The numbers in Table 1.15 are derived from Table 1.13. Again, the height of the density histogram tells us the approximate percentage of Kenyans of any age. The percentage of 5-year-olds is roughly 2.85%; the percentage of 20-year-olds is roughly 2.11%; and so on. Figure 1.17 shows that the population of Kenya is very young. As can be computed from the graph, more than half of all Kenyans are under the age of 20 years. The population of Germany is much older: Around half of all Germans are 40 years old or older. While the population of Germany is shrinking, the population of Kenya is growing rapidly, mostly because the birth rate is high. If the current rate of growth persists, the population of Kenya will double in size in 30 years.

Table 1.15 Density Histogram Heights for Age Distribution of Kenyan Population, 2000

Age group	Percentage	Height
0 to 15–	42.8	2.85
15 to 30–	31.6	2.11
30 to 50–	17.2	0.86
50 to 65–	5.7	0.38
65 to 80–	2.4	0.16
80 to 95–	0.4	0.03

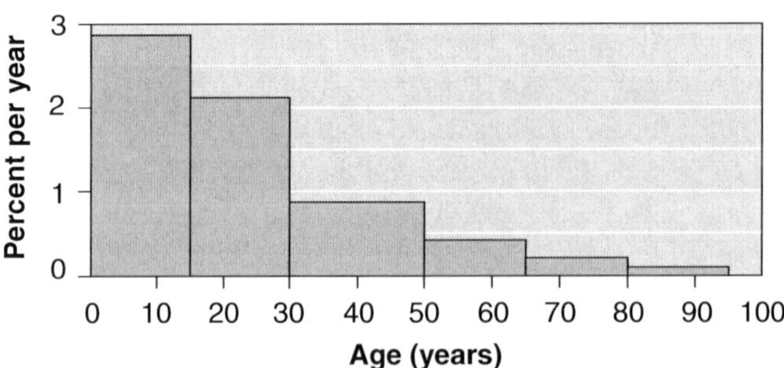

Figure 1.17 Distribution of Kenyan population by age, 2000, as shown by a density histogram.

The statistical lesson here is that the density histogram is an ideal tool for comparing the distributions of a variable in two settings, such as the age distributions of Germany and Kenya.

Section 1.5 Exercises

1. Use Table 1.14 and Figure 1.16 to answer the following questions about the population of Germany.
   a. What percentage of Germans are under 30 years of age?
   b. Estimate the percentage of Germans who are at least 30 years old, but not yet 40 years old.
   c. Estimate the percentage of Germans under 40.
   d. Are there more 10-year-olds than 40-year-olds?

2. Use Table 1.15 and Figure 1.17 to answer the following questions about the population of Kenya.
   a. Is the percentage of 10-year-olds in Kenya much smaller than, about the same as, or much greater than the percentage of 40-year-olds?
   b. What percentage of Kenyans are under 30 years of age (assuming a uniform age distribution within a rectangle)?
   c. Estimate the percentage of 16-year-olds.
   d. Estimate the percentage who are at least 15 years old, but not yet 18 years old.
   e. Estimate the percentage who are under 18.

3. Below is a density histogram for the age distribution of the U.S. population in 2000:

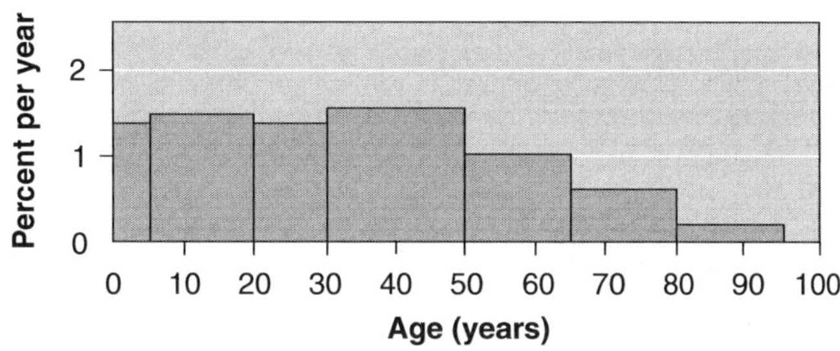

Source: U.S. Census Bureau, Year 2000.

   a. The percentage of 40-year-olds is around
      1% 1.5% 15% 30% 45%
      (Choose one option and explain.)
   b. The percentage who are at least 30, but not yet 50, is around
      1% 1.5% 15% 30% 45%
   c. The percentage of 55-year-olds is around
      1% 1.5% 15% 30% 45%
   d. The percentage who are at least 50, but not yet 65, is around
      1% 1.5% 15% 30% 45%
   e. The percentage who are at least 30, but not yet 65, is around
      1% 1.5% 15% 30% 45%
   f. Does the United States have rapid population growth like Kenya, slow growth, negative growth like Germany, or something else?

4. In a medical study the systolic blood pressure (i.e., the pressure while the heart is contracting and pumping blood) for a couple of thousand young women was measured. The unit for such measurements is "millimeter of mercury" (mmHg). The figure below is the density histogram for the data.

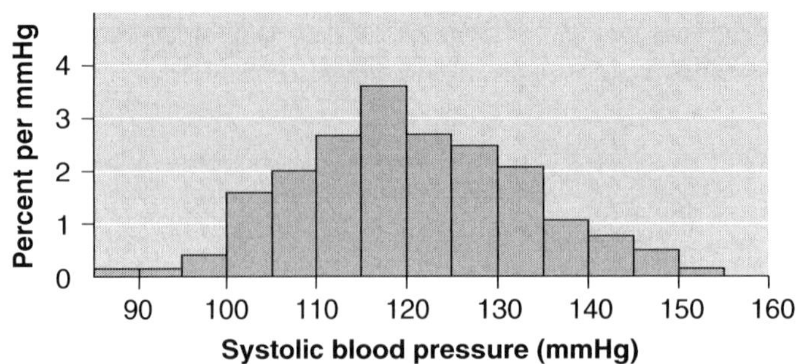

a. How would you describe the shape of the histogram?
b. The percentage of women with blood pressure between 90 mmHg and 150 mmHg was around
   1% 20% 50% 80% 99%
   (Choose one option.)
c. The percentage of women with blood pressure under 120 mmHg was around
   1% 20% 50% 80% 99%
d. The percentage of women with blood pressure over 130 mmHg was around
   1% 20% 50% 80% 99%
e. The percentage of women with blood pressure between 105 mmHg and 110 mmHg was around
   1% 2% 5% 10% 20%

5. The projected age distribution of U.S. residents in 2050, projected by the Census Bureau, is shown below:

Age group (in years)	Number of people (in millions)
Under 10 years	55.6
10–19	53.5
20–29	52.6
30–39	52.8
40–49	50.0
50–59	46.2
60–69	42.8
70–79	32.6
80–89	23.1
90–99	9.5
100–109	1.2
Total	419.9

a. Construct a density histogram.
b. Compare the density histogram with the density histogram in Exercise 3. Is the population getting older or younger?

6. A major attraction at Yellowstone National Park is the eruption of the geyser Old Faithful. The following table summarizes a sample of 400 times (in minutes) between eruptions.

a. Construct a density histogram.
b. What percentage of the time was the waiting time less than an hour?

Time (minutes)	Frequency
30–39	4
40–49	15
50–59	85
60–69	49
70–79	15
80–89	211
90–99	18
100–109	3

7. Major active volcanoes in Alaska have the following heights above sea level:

4,265	3,490	4,450	3,995	150	5,315	2,560	7,985	5,710	3,545	5,370	5,775	10
140	8,185	2,945	4,450	7,540	885	4,025	4,885	3,945	6,050	7,295	6,720	2,759
3,540	8,960	7,050	5,030	7,545	10,265	2,015	3,465	9,430	11,070	7,015	11,413	6,830
8,450	6,965	5,055	1,980									

Construct a density histogram using the class intervals 0 to 1000−, 1000 to 2000−, and so forth.

8. The following are the IQ scores of 50 randomly sampled undergraduates:

124	106	108	87	114	102	132	117	99	102	128	119	105	84	93	135	110
97	103	130	92	103	102	116	93	111	88	109	108	142	90	130	107	109
114	98	103	136	110	107	104	109	115	147	113	119	115	125	128	98	

Construct a density histogram using the class intervals 80 to 90−, 90 to 100−, and so forth.

9. A company had 105 sales last year, grouped by price as follows:

Price range (in thousands of dollars)	Number of products sold
0–50−	21
50–100−	47
100–150−	15
150–200−	12
200–250−	10

a. Construct a density histogram. Describe the shape.
b. Estimate the total annual sales. *Hint:* Make the rough approximation that the cost of every item is the midpoint of its interval.

10. The distribution of a company's stock to shareholders according to the number of shares held is as shown in the following table:
a. Construct a density histogram.
b. How would you describe the shape of the distribution?

Number of shares held	Percentage of stockholders
0–25–	39
25–50–	27
50–100–	20
100–500–	12
500–700–	2

1.6 MISUSING STATISTICS AND SIMPSON'S PARADOX

Graphical displays can easily be used to give misleading or wrong impressions about the world. This happens all too frequently in books, reports, politics, advertising, and the media. In this section we present examples of how statistical techniques have been misused, so you will not be misled as easily when you encounter statistics in newspapers or magazines or on radio or television. Hopefully, you will also be better able to interpret them correctly.

Many of the misuses of statistics are achieved by graphical distortions, but there are other ways as well.

The Area Fallacy

In most of the bar charts and histograms constructed in this chapter, the widths of the bars were equal. Only the heights differed from bar to bar. Thus, the area of each bar represented the relative contribution of that bar, just as the height did. One of the most common forms of distorted graphical presentation is a figure in which the *length* (or height) of each symbol accurately indicates the quantity the figure is supposed to represent, but the *area* does not. This type of distortion often occurs in media stories in which the statistics of interest are numbers of people. For example, suppose that the story is about the different numbers of police officers per thousand residents in three cities:

City X	City Y	City Z
1.1	2.2	3.3

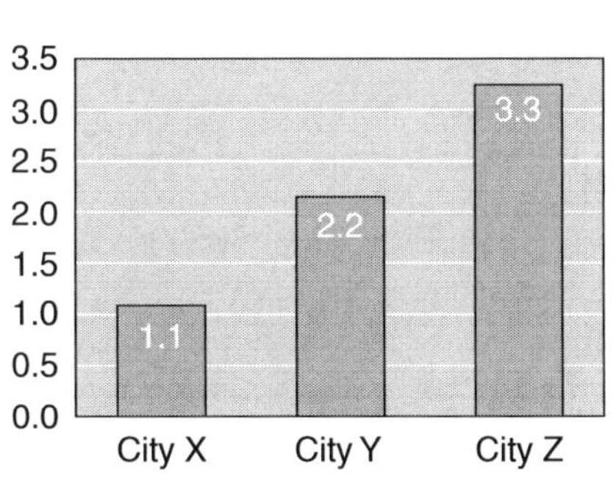

(*a*) **Police officers per thousand residents**

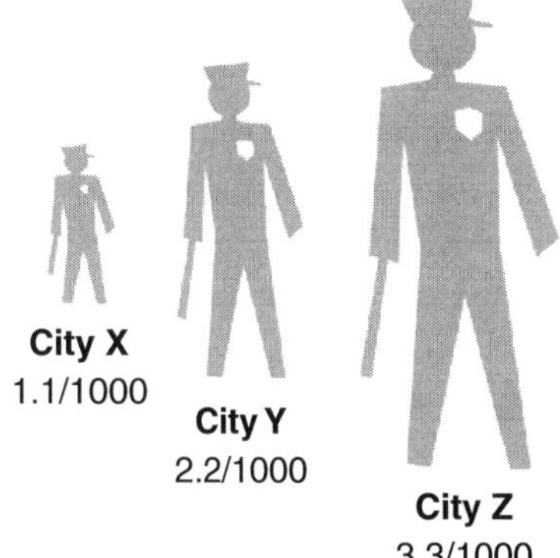

(*b*) **Police staffing: A tale of three cities**

Figure 1.18 Two ways to display statistics about numbers of people: (*a*) truthful but boring bar chart; (*b*) lively but misleading pictorial.

A fair bar chart of these data is shown in Figure 1.18(*a*), but many newspapers would instead use a "livelier" display of drawings of people, such as a chart using stylized figures of police officers with heights proportional to the numbers, as in Figure 1.18(*b*). The officer for City Z is correctly three times as tall as the officer for City X but (because the figures are similarly shaped) three times as wide as well, and therefore the officer for City Z looks *nine* times as large. Worse yet, the figures are not placed on the same baseline, obscuring the proper (height) relationship among them even further.

This kind of distortion is often not a conscious attempt to deceive the reader, but rather a thoughtless attempt to make the graphic display eye-catching by substituting pictures for text and abstract bar shapes.

The Missing Baseline

A reader glancing at a bar graph is likely to base a first impression on the assumption that the vertical scale in the graph extends all the way to zero, because that is the way we think of graphs. Often, however, much of the bottom of the vertical scale is cut off. Such a graph makes a difference look far more dramatic than it really is. We illustrate with an example.

Example 1.10 The Most Expensive Big Cities for Drivers

Figure 1.19 appeared in *USA Today,* May 3, 1995. Once you have *found* the bar graph among all the cartoon cars, buildings, and seagulls (called "chart clutter" by Edward R. Tufte, one of the world's experts on the efficient and appropriate display of information), you might think that you could cut your driving expenses by almost two-thirds by living and working in Hartford rather than Los Angeles. In fact, you would save only about 14% (see Figure 1.20 for an undistorted impression).

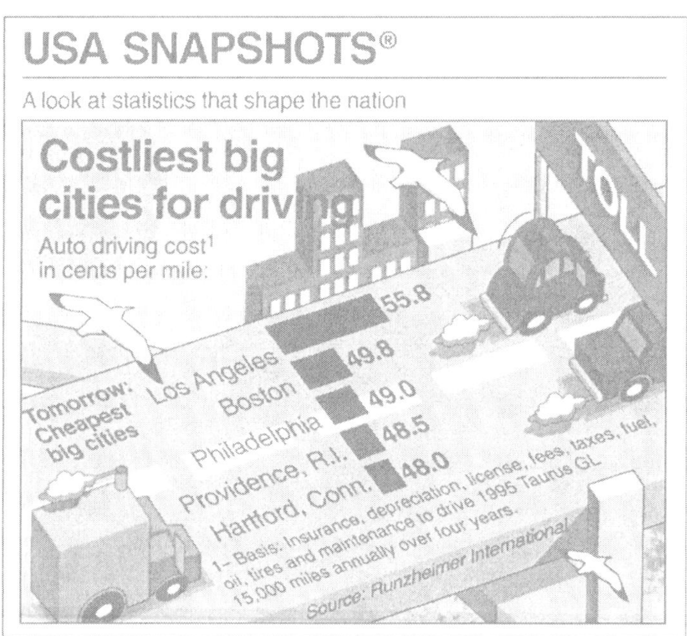

Figure 1.19 Bar graph of driving expenses (cents per mile) in various U.S. cities (*Source: USA Today,* May 3, 1995).

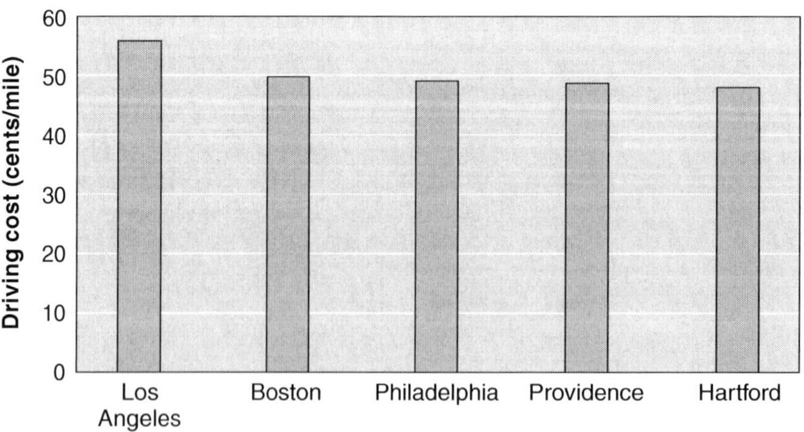

Figure 1.20 Correctly drawn bar graph of the driving expense data in Figure 1.19.

❐ ❐

The intention behind these graphs, as with the graphs that use pictures instead of bars, is often not malicious. Graphic designers are trying first of all to make their displays attract the reader's attention. Parts of a graph that show a change catch the eye. Parts that do not show a change are often regarded as boring and a waste of display space, even though they are needed if they are to make an impression that agrees with reality. In fact, the missing baseline approach is so universally used in the media that one simply must train oneself not to be fooled by it.

The Combination Graph

combination graph

Sometimes the changes in two different measurements over time are shown on a single graph, called a **combination graph**. Such a graph can be very useful for giving a view of how one quantity may be affecting another (later in this book we will discuss statistical techniques for determining the amount of influence of one variable on another). Each of the two graphs, however, is susceptible to the zero point fallacy just described. By carefully choosing the scales and vertical placement for the two plots, a presenter can make an impression, or its opposite, that has nothing to do with the data.

Example 1.11 SAT Scores versus Public School Funding

Suppose one wishes to assess the influence of public education spending on student preparedness for college. Table 1.16 shows the amount of money spent per pupil on public education in the United States for school years ending between 1990 and 1995. Table 1.17 shows the national average composite SAT (Scholastic Aptitude Test then, now called the Scholastic Assessment Test) scores for college-bound students taking the tests over the same range of years.

Table 1.16 Per-Pupil Expenditures on Public Education

School year ending	Per-pupil expenditures on elementary and secondary education,1997–1998 dollars
1990	6591
1991	6626
1992	6587
1993	6587
1994	6633
1995	6676

Source: New York Times Almanac 2000, 368.

Table 1.17 National Average Composite SAT Scores

Year	Average composite SAT score, all students
1990	900
1991	896
1992	899
1993	902
1994	902
1995	910

Source: U.S. Department of Education, *The Condition of Education 1996*, Indicator 22.

Figure 1.21 shows three combination graphs constructed from the data in Tables 1.16 and 1.17. Figure 1.21(*a*) might have been prepared to support an article opposing increases in public school funding by suggesting that SAT scores have changed little. Figure 1.21(*b*), on the other hand, might have been prepared to support an article in praise of increased per-pupil spending by suggesting that SAT scores have greatly increased as a consequence. In both graphs, a wide range is used for the measurement the presenter wants to look unchanged, whereas a narrow range is used for the measurement that the presenter wants to make dramatic. Figure 1.21(*c*) is an impartial plot of the same data, achieved by starting both scales at 0. Neither measurement changed dramatically in this graph, which is the point to be made.

One might wonder whether these two indices are the best indices of whether public school funding directly influences how well students are prepared for college. Rather than the per-pupil expenditure for only the year in which the test is taken, one might construct an index based on averaging the secondary school expenditures (in constant dollars) for the 4 years before the student took the test, for example. Further, as an index of college preparedness in a given year, one might use an index based on college performance rather than on a test used in college admissions. See also Example 1.15.

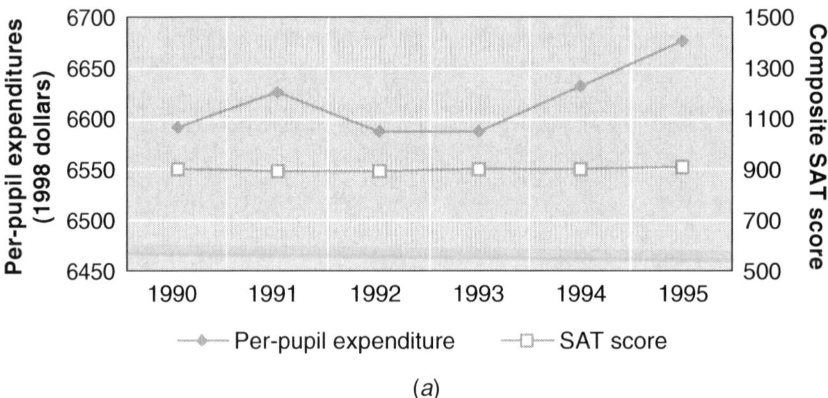

(*a*)

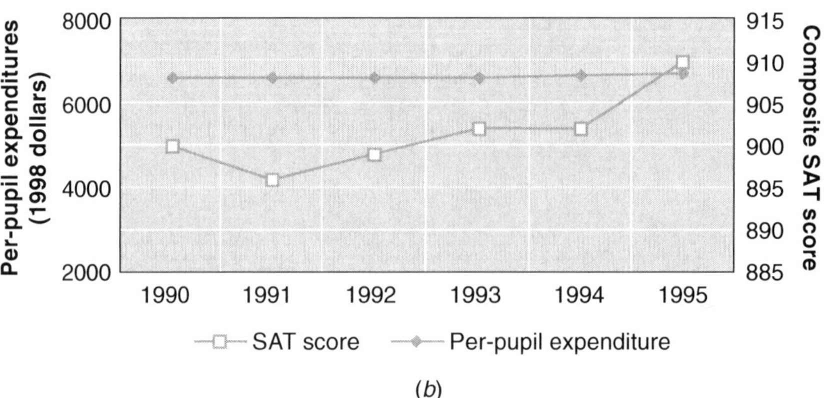

(*b*)

Figure 1.21 Combination graphs of per-pupil public education expenditures in constant dollars versus composite SAT scores.

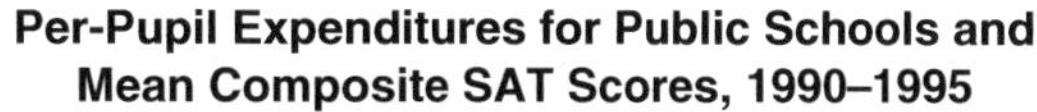

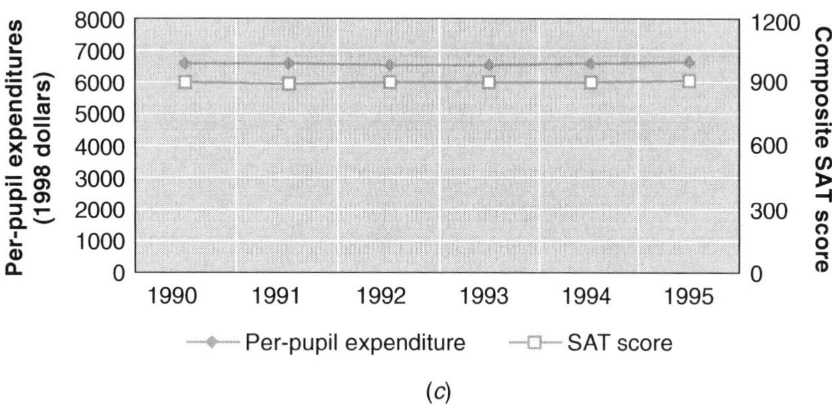

(*c*)

Figure 1.21 (*continued*)

Other Tricks

There are a few other ways a graph can be misleading in addition to the ways we have discussed. For example, a pie chart is often shown at an angle to look more like a pie. The wedges on the front facing the reader then look wider than the wedges on the sides.

Sometimes the distortion does not depend on the graph. Numerical indices can be computed in a variety of ways. It is not difficult to construct an index that makes the point one wants to make "statistically" no matter what the facts of the situation are.

Example 1.12 **How Fast Can You Read?**

A speed reading training school once used in its advertisements a reading index that measured reading level by the score on a test of "comprehension" multiplied by the reading speed.

Measured by that index, "readers" who learned to skim over the text extremely rapidly could show much higher reading level scores after completing the course than before they took the course, even if they comprehended far less of what was on the page than they did when reading at their ordinary pace before taking the course.

We shall finally illustrate how a comparison of overall averages for two (or more) data sets can give misleading impressions.

SIMPSON'S PARADOX

In 1987, the U.S. Department of Transportation made it a requirement that airlines report each month what percentage of all their flights into the nation's 30 busiest airports arrived on time. Major newspapers published these statistics, and airlines with a high percentage of on-time arrivals began advertising the fact. But a comparison of two airlines' overall on-time percentages can be misleading.

Each airline's overall on-time rating depends on its performance at 30 airports, but the rating is determined mainly by the airline's performance at the airports it serves frequently. The ratings therefore favor airlines that mostly fly in and out of fair-weather airports, and the ratings are disadvantageous to airlines that serve cities that often have bad weather.

We shall illustrate this by comparing the on-time performance in June 1991 of America West Airlines with that of Alaska Airlines at 5 of the 30 busiest airports that Alaska Airlines served in Table 1.18.

Table 1.18 Number of Flights on Time and Delayed for Alaska Airlines and America West Airlines in June 1991

Airline	Alaska Airlines		America West	
Destination	**On Time**	**Delayed**	**On Time**	**Delayed**
Los Angeles	497	62	694	117
Phoenix	221	12	4840	415
San Diego	212	20	383	65
San Francisco	503	102	320	129
Seattle	1841	305	201	61

Source: Arnold Barnett. 1994. How numbers can trick you. *Technology Review.* 97 (7): 38–45

We shall compute each airline's on-time percentage at each of the five airports in Table 1.19.

Table 1.19 On-Time Percentages for Alaska Airlines and America West Airlines in June 1991

Airline	Alaska Airlines	America West
Destination	**Relative Frequency On Time**	**Relative Frequency On Time**
Los Angeles	$\frac{497}{559} = 88.9\%$	$\frac{694}{811} = 85.6\%$
Phoenix	$\frac{221}{233} = 94.8\%$	$\frac{4840}{5255} = 92.1\%$
San Diego	$\frac{212}{232} = 91.4\%$	$\frac{383}{448} = 85.5\%$
San Francisco	$\frac{503}{605} = 83.1\%$	$\frac{320}{449} = 71.3\%$
Seattle	$\frac{1841}{2146} = 85.8\%$	$\frac{201}{262} = 76.7\%$
Five-City Total	$\frac{3274}{3775} = 86.7\%$	$\frac{6438}{7225} = 89.1\%$

We notice the surprising fact that, whereas Alaska Airlines outperformed America West Airlines in on-time arrivals at each of the five airports, America West had a better overall on-time percentage when we combined the five cities. This counterintuitive result is due to the fact that 73% of America West Airlines' flights landed in sun-drenched Phoenix, whereas 73% of Alaska Airlines' flights landed either in foggy San Francisco or in rainy Seattle.

As a result, America West Airlines' overall on-time percentage (89.1%) is close to America West Airlines' on-time percentage in Phoenix (92.1%), and Alaska Airlines' overall on-time percentage (86.7%) is close to Alaska Airlines' on-time percentage in Seattle (85.8%). The fact that Alaska Airlines' flights even outperformed America West Airlines in Phoenix has little effect on Alaska Airlines' five-city average because only 6% of Alaska Airlines' flights landed in Phoenix. What we have here is an example of *Simpson's paradox.* Simpson's paradox occurs when the combined data show one pattern, but subgroups show a different pattern.

Simpson's Paradox

Simpson's paradox refers to the reversal of the direction of a comparison when data from several groups are combined to form a single group.

lurking variable

When we compare the on-time performance of Alaska Airlines and America West Airlines, the airport being served is a so-called **lurking variable**—a factor lurking in the background, influencing the comparison, but not taken into account in overall ratings. If we ignore the lurking variable and look at overall on-time performance, America West Airlines seems to do better than Alaska Airlines. But when we take the lurking variable into account and look at the individual airports, we discover that Alaska Airlines is the one with the better on-time performance.

We shall give another example of Simpson's paradox.

Example 1.13 **Is Smoking Good for You?**

A study on thyroid and heart disease among women in Whickham, an area in England, was conducted from 1972 to 1974. Twenty years later, in 1994, a follow-up study of the same people highlighted an interesting relationship between smoking and death. Of the 1314 women in the original study, 28% (369) were dead 20 years later. However, of those women who were smokers originally (582), only 24% (139) had died, while 31% (230) of the nonsmokers (732) had died.

The study seems to imply that smoking helps people stay alive, but that is obviously not true. The lurking variable turns out to be age. There were many more old people among the nonsmokers than among the smokers. In fact, 26% of the nonsmokers were 65 or older, whereas only 8% of the smokers were 65 or older. So, many more nonsmokers than smokers died of causes related to old age. This accounts for the higher overall death rate for the nonsmokers.

controlling for age

In Table 1.20 we categorize the women as either young (18 to 34 years old), middle-aged (35 to 64 years old), or old (65 years or older). We can then compare the young smokers with the young nonsmokers, the middle-aged smokers with the middle-aged nonsmokers, and the old smokers with the old nonsmokers. Statisticians call this **controlling for age**.

Look first at the women who were young (18–34) in 1974. Almost all were still alive 20 years later, whether they smoked or not. Look next at the women over 65. Most of them (about 85–86%) were dead 20 years later, again, whether they smoked or not. However, there is a sizeable difference in the death rate between the smoking and nonsmoking women who were middle-aged (35–64) in 1974; that is, 26% of the smokers, compared to 18.4% of the nonsmokers, were dead.

The original article broke the middle-aged group down into three groups: 35 to 44 years old, 45 to 54 years old, and 55 to 64 years old. In each age group the smokers had a higher death rate than the nonsmokers. But when we combine all the smokers in the study and combine all the nonsmokers in the study, then the nonsmokers have a higher death rate than the smokers. This reversal is an example of Simpson's paradox.

Table 1.20 20-Year Death Rates of Smoking and Nonsmoking Women

		Smokers		
Age Category in 1974	**Number in Category**	**Proportion in Category**	**Number Dead in 1994**	**Proportion Dead in 1994**
18–34	179	0.31	5	0.028
35–64	354	0.61	92	0.260
65+	49	0.08	42	0.857
Totals	582	1.00	139	0.239
		Nonsmokers		
Age Category in 1974	**Number in Category**	**Proportion in Category**	**Number Dead in 1994**	**Proportion Dead in 1994**
18–34	219	0.30	6	0.027
35–64	320	0.44	59	0.184
65+	193	0.26	165	0.855
Totals	732	1.00	230	0.314

Source: D. R. Appleton, J. M. French, and P. J. Vanderpump. 1996.
Ignoring a Covariate: An Example of Simpson's Paradox, *The American Statistician*, 50: 340–341.

Our next example looks at a comparison of two medical treatments for kidney stones. The lurking variable turns out to be the size of the kidney stone.

Example 1.14

A medical study compared the success rates of two methods of removing kidney stones. The investigators looked at 350 patients who had undergone open surgery (Treatment A). They compared them with 350 patients who had had a hollow tube inserted into their kidneys through small incisions in their backs and had had the kidney stones removed through the tube (Treatment B).

Open surgery had a success rate of 78% (273/350), whereas Treatment B had a success rate of 83% (289/350), an improvement over open surgery. So, Treatment B seems to be more effective than Treatment A. But, if we take the size of the kidney stone into account, we get a different result, as shown in Table 1.21.

Table 1.21 Success Rates, Controlling for Size of Kidney Stone

	Treatment A	Treatment B
Small stones	$\frac{81}{87} = 93\%$	$\frac{234}{270} = 87\%$
Large stones	$\frac{192}{263} = 73\%$	$\frac{55}{80} = 69\%$

Source: S. A. Julious and M. A. Mullee. 1994. Confounding and Simpson's Paradox. *British Medical Journal*, 309: 1480–1481.

According to the data presented in this study, open surgery (Treatment A) is more effective than Treatment B, both for patients with small kidney stones and for patients with large kidney stones. The reason that the overall success rates gave the opposite impression is that most of the large kidney stones were removed using open surgery, whereas most of the small kidney stones were given Treatment B. So open surgery was used to handle the more difficult cases (the large stones), whereas Treatment B was used to treat most of the easy cases (the small stones). That obviously biases the comparison against open surgery and in favor of Treatment B.

❒ ❒

Our final example has important implications for educational policy.

Example 1.15 Those Misleading SAT and NAEP Trends

In 2002 the average SAT Verbal score had not improved over what it had been in 1981. The NAEP (National Assessment of Educational Progress) reading scores for 9-year-olds, for 13-year-olds, and for 17-year-olds were up only marginally over the same period. Critics of the American educational system took this as evidence that most of the money invested in public education between 1981 and 2002 had been wasted. In his March 2, 2003, column, George F. Will stated in his usual erudite manner, "This refutes the durable delusion that schools' cognitive outputs vary directly with financial inputs."

But when we break down the SAT and NAEP data into ethnic subgroups, a different picture emerges: Minorities have improved their averages considerably. In fact, each of the major ethnic categories used by the College Board shows an increase in scores from 1981 to 2002. For example, although the overall average Verbal SAT score was unchanged from 1981 to 2002, Whites were up 8 points on average, African Americans were up 19 points on average, Asian Americans were up 27 points on average, Mexican Americans were up 8 points on average, Puerto Ricans were up 18 points on average, and Native Americans were up 8 points on average. This is exactly the kind of improvement that the money invested in public education was supposed to achieve!

But how can it be that all the ethnic groups that make up the national average have improved, but the overall national average score is unchanged? This is yet another example of Simpson's paradox. In this case, the explanation is the changing composition of the SAT test-takers. Minority students now make up a much larger percentage of college-bound seniors than they did in 1981. A lot of minority students who would not have dreamed of going to college 25 years ago, had they been of college age then, are today planning to go to college.

The reason that the overall SAT averages have not improved is that the group of college-bound seniors has become more inclusive than it was 25 years ago. And although minority scores have improved significantly over the years, they are still relatively low (except for the Math scores of Asian Americans).

❐ ❐

statistical literacy

Of course, the specific ways presented here of distorting the underlying truth of the data are just a sampling of the many ways that casually or dishonestly presented and casually examined statistical information can mislead us. The goal of this textbook is to impart **statistical literacy**. One aspect of statistical literacy is not to be easily misled by improper conclusions drawn from data, often supported by false ways of interpreting and displaying data.

Section 1.6 Exercises

1. A commercial released by an automobile company says, "Ninety percent of all our cars sold in this country in the last 10 years are still on the road." This gives the impression that the company's cars are built to last. Explain how this commercial can mislead, especially if the company's sales have rapidly increased from year to year over the last 10 years.

2. The average weekly salary for men in a certain company is $800 and that for women is $710. Draw a bar chart that visually exaggerates the difference in the salaries.

3. A manager who has achieved a small growth in sales over the past 2 years wants to show his performance in a better light.
   a. What trick might he adopt when he is showing his performance graphically?
   b. If he has had a steep decrease in his sales, what trick might he use when showing his performance graphically?

4. The following U.S. map (distributed recently by First National Bank of Boston) purports to show what portion of income earned by U.S. citizens was being taken and spent by the federal government. The shading is to indicate that federal spending has become equal to the total incomes of all people residing in the states west of the Mississippi River, except Louisiana and Arkansas. This map certainly gives a very distorted picture. Can you figure out why? Is the federal government in fact spending an amount that is over half the entire income of its citizens? *Hint:* What do you know about the population densities of various states?

5. The market share of a toothpaste company for the past few years is as shown in the accompanying figure. This is an honest bar graph with no intention to mislead. But how can the performance be represented so that it seems better than it actually was? *Hint:* Would you begin the vertical axis at zero?

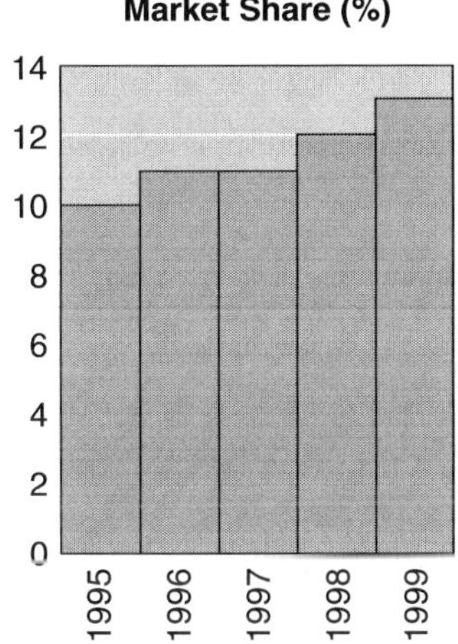

6. The profit or loss made by a company in the last 5 years is as follows:

Year	Profit or loss ($)
1995	−97,563
1996	28,576
1997	30,379
1998	41,274
1999	49,273

Note that the company had a loss in the year 1995. The chart the company publishes to show this data is shown in the accompanying figure.

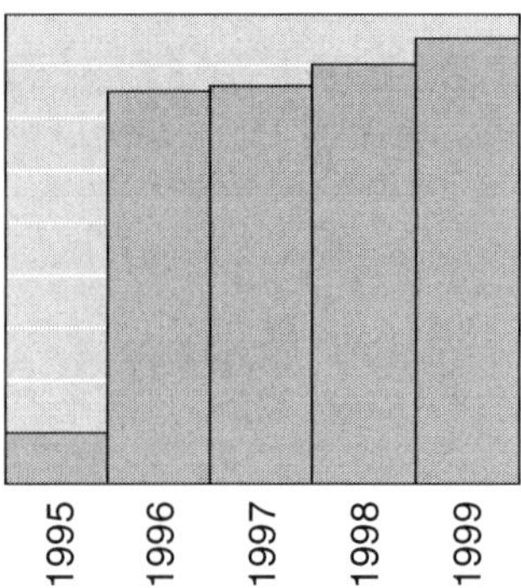

a. What is misleading about this bar graph?
b. Draw an appropriate graph. (*Hint:* For a negative amount, let the bar lie below the axis.)

7. A government agency reports population trends and value of agency services on the same graph, similar to those shown for SAT scores and school funding in Figure 1.21. The data are as shown in the following table.
a. Graph these data in a way that makes the government agency look better than it should.
b. Graph it honestly.

	1960	1970	1980	1990	2000
Government spending (in billions of $)	6.4	6.45	6.51	6.54	6.57
Population (in millions)	92.1	96.4	100.1	112.5	120.1

c. Propose an appropriate index (a statistic) to measure value of agency services adjusted for population growth.

8. Find examples from newspapers, magazines, and similar materials of presentations of statistics that are subject to the kinds of misinterpretation presented in this section.

9. Imagine that the marketing department of a corporation wants to mislead people as to how the increasing market share of the firm has changed over the decades. The graph they produced is shown here.
a. Why is this graph misleading?
b. Redraw the graph so that a fair visual image is presented.
c. Think of another example where such a trick can be used.

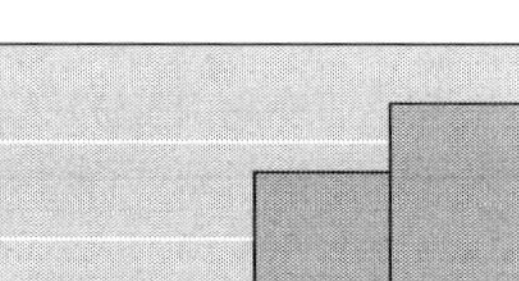

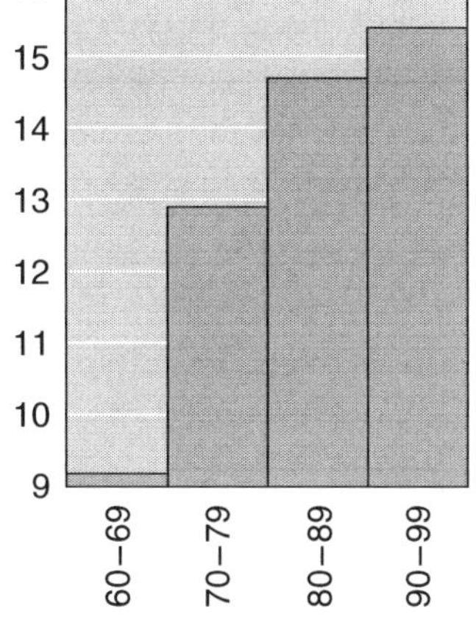

10. Discuss whether the accompanying graph (a similar graph appeared in a major national magazine) fairly represents the price increase of gasoline at the pump over several years.

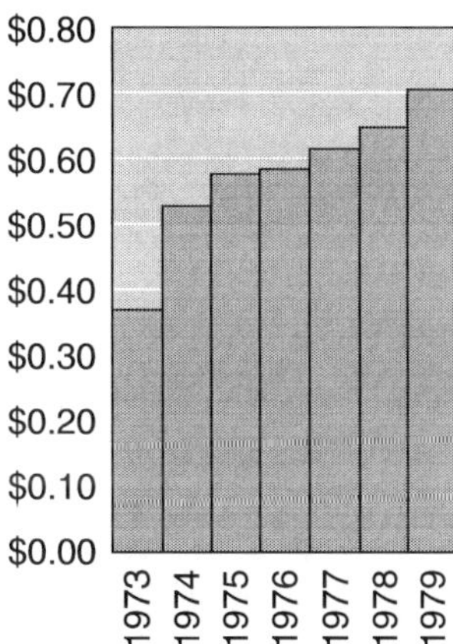

11. Here are the batting performances of Barry Bonds and Mark McGwire for their rookie season, 1986, and for 1998, when both players were at the top of their game. A player's batting average is his number of hits, divided by his number of times at bat.

	Barry Bonds		Mark McGwire	
Year	**At-bats**	**Hits**	**At-bats**	**Hits**
1986	413	92	53	10
1998	552	167	509	152
Combined	965	259	562	162

a. Compute Barry Bonds's batting average for 1986.
b. Compute Mark McGwire's batting average for 1986.
c. Who had the higher batting average in 1986?
d. Who had the higher batting average in 1998?
e. If we combine the scores for the two seasons, who had the higher overall batting average?
f. Explain the reversal in terms of how much each player played during his rookie season.

12. We shall compare the survival rates for surgery patients at a smaller rural hospital and those at a large teaching hospital in a big city. We control for the difficulty of the surgery.

	Simple surgery		Complicated surgery	
Outcome	**Rural**	**City**	**Rural**	**City**
Died	36	60	15	160
Survived	864	1940	85	1840
Total	900	2000	100	2000

a. What is the overall survival rate for surgery patients at the rural hospital?
b. What is the overall survival rate for surgery patients at the city hospital?
c. Which hospital has the higher overall survival rate?
d. Which hospital has the higher overall survival rate for simple surgery patients?
e. Which hospital has the higher overall survival rate for complicated surgery patients?
f. Explain the reversal in terms of the percentage of the patients at each hospital who require complicated surgery.

CHAPTER 1 SUMMARY

A **variable** is a characteristic that changes from individual to individual in a study.

Numerical variables take numerical values. (Examples: height, income, IQ.)

Categorical variables place each object (or person) in one of several groups or categories. (Examples: gender, religion, state of residence.) If the categories are **ranked**, then the variable is said to be **ordered** or **ranked categorical.**

The pattern of variation of a variable in a study is called its **distribution.** It is usually described by a table, graph, or formula that tells us what values the variable takes and how often they occur.

Categorical data are best displayed in **bar charts** (sometimes **Pareto charts** when the bars are ordered from tallest to shortest) or **pie charts.**

Quantitative data are displayed in **dotplots, stem-and-leaf plots,** or **frequency histograms.** We look for **symmetry** or **skewness.** We also look for **gaps** and resulting **clusters** in the distribution and exceptional values (**outliers**).

Histograms for **counting variables** center their rectangles of width 1 at the integers 0, 1, 2, 3, ...

A histogram is **symmetric** if its left and right sides are mirror images of each other. (Examples: bell-shaped, uniform, U-shaped.)

The histogram is **right-skewed** (skewed to the right, positively skewed) if it has a long tail to the right.

The histogram is **left-skewed** (skewed to the left, negatively skewed) if it has a long tail to the left.

Density histograms are very useful for comparing how a variable is distributed in different settings.

Displaying a data set is not an end in itself. The purpose of such displays is to help us recognize the overall pattern of the data so that we can understand the information contained in the data.

Beware of picture graphs and distorted scales!

A **lurking variable** in a comparison is a factor lurking in the background, influencing the comparison, but not initially taken into account.

Simpson's paradox refers to the reversal of the direction of a comparison when data from several groups are combined to form a single group. More generally, we have an example of Simpson's paradox if the combined data show one pattern, but the subgroups show a different pattern.

CHAPTER REVIEW EXERCISES

1. Here are the speeds of cars traveling down a street with a restricted speed zone when school children are present:

28	26	29	27	22	24	23	29	34	24	23	22	27	26	22
36	54	29	28	16	41	34	27	29	20	21	34	29	21	28

a. Construct a stem-and-leaf plot.
b. Is the shape what you would expect?

2. The following data were collected to determine whether the chance that a horse will win a race is influenced by its starting position. Data represent 144 races, each with exactly eight horses in the starting lineup. Starting position 1 is closest to the rail on the inside of the track.

	Starting Position							
	1	2	3	4	5	6	7	8
Number of wins	29	19	18	25	17	10	15	11

Source: New York Post, August 30, 1955, 42.

a. Draw a pie chart for the data.
b. Draw a bar chart.
c. Can you see any apparent relationship between the chances of winning and the starting position?

3. Identify whether the following statements are true or false, and give reasons why they are true or false.
   a. In a frequency histogram, the height of a bar is always equal to the frequency of the class the bar represents.
   b. In a density histogram, the area of each bar equals the percentage of observations in the corresponding class interval.

4. A sequence of 300 pseudo-random digits (the 10 digits are to appear random and hence each should have the same likelihood of occurring) was generated on a Casio calculator.

Digit	0	1	2	3	4	5	6	7	8	9
Frequency	25	28	29	35	35	31	27	33	32	25

Source: I. R. Dunsmore, F. Daly, and the M345 Course Team. 1987. M345 Statistical Methods, Unit 9: Categorical data: In Milton Keynes, *The Open University,* Table 1.5.

   a. Construct a pie chart and a bar chart.
   b. What can you infer, if anything?

5. People have noticed that sports teams tend to do better at home in their own stadiums than when they play in their opponents' stadiums. Part of the reason may be the fans' effect on the referees. To study this possibility, videotape of a particular soccer match was shown to 11 experts, 5 of whom were shown the videotape with the crowd noise (the "noise" group), and 6 of whom watched without any sound (the "no noise" group). Each person judged 52 incidents in which there may have been a foul, and all were asked to decide whether a foul had occurred. (A. Neville, N. Balmer, and M. Williams. 1999. Crowd influence on decisions in Association football. *The Lancet,* Vol. 353.)

   In half of the incidents, all the experts agreed. For the other half of the incidents, the following were the results:
   - Potential fouls committed by the home team: 61% were deemed fouls by the "no noise" group, and 47% were deemed fouls by the "noise" group.
   - Potential fouls committed by the visiting team: 36% were deemed fouls by the "no noise" group, and 58% were deemed fouls by the "noise" group.

   a. Look at the data on the potential fouls committed by the home team. What does the fact that the "no noise" group decided there were more fouls than the "noise" group suggest about the effect of crowd noise on the experts' judgment?
   b. Look at the data on the potential fouls committed by the visiting team. What does the fact that the "no noise" group decided there were fewer fouls than the "noise" group suggest about the effect of crowd noise on the experts' judgment?
   c. Does it appear that crowd noise biases the experts' judgment in favor of the home team, in favor of the visiting team, or neither?
   d. For the potential fouls committed by the home team, the referees in the actual game decided that 53% were fouls. Were the referees' decisions closer to those of the "noise" group or "no noise" group?
   e. For the potential fouls committed by the visiting team, the referees in the actual game decided that 60% were fouls. Were the referees' decisions closer to those of the "noise" group or "no noise" group?

f. Overall, does it appear that the crowd noise tends to lead the referees to favor the home team?

6. In a histogram, explain how the selected width of the class interval affects
   a. the preservation of important details contained in the original data.
   b. accidental irregularities in the heights of the bars that distort the true overall pattern or shape of the data.

7. The frequency table that follows shows the number of children per household for 50 households interviewed in a survey.

Number of children	f	Percentage
0	3	
1	8	
2	26	
3	10	
4	2	
5	0	
6	1	
Total	50	

   a. Complete the third column to show the percentage of families with each number of children.
   b. Construct a density histogram of the data. (Center each rectangle at an integer: 0, 1, ..., 6.)
   c. From your graph in part (b), find the proportion of families having two or fewer children.

8. Here are data on average fuel economy and average vehicle purchase price for all cars made by a certain automobile manufacturer in selected years.

	1970	1975	1980	1985	1990	1995
Fuel economy (MPG)	16.8	19.2	23.1	25.1	25.9	25.7
Price (thousands of 1995 dollars)	15	16.5	18.1	19.2	20.5	15

   a. Using combination graphing (see Figure 1.21), draw a graph that stresses gains in fuel economy while deemphasizing rising prices.
   b. Draw a combination graph in a way that a political group wanting to stress poor advances in fuel economy and the high price of cars might draw the graph.
   c. Draw it honestly.

9. Identify the true and false statements among the following statements, and give reasons why they are true or false.
   a. In a bar chart for categorical data, the width of the bars is unimportant, but the choice of width is important in the case of a histogram for quantitative data.
   b. A histogram is used to display the distribution of quantitative data such as measurements or counts.
   c. In a histogram, all the bars must touch each other, but in a bar graph for categorical data they need not do so (see Figure 1.20).

10. The following data concern 40 heavy men, each weighing at least 225 pounds. Their cholesterol levels (measured in milligrams per 100 milliliters) were recorded and were categorized as belonging to either the Type A behavior category or the Type B behavior category. (Type A behavior is characterized by urgency, aggression, and ambition, whereas Type B behavior is characterized as relaxed, noncompetitive, and less hurried.)

Type A behavior	233, 291, 312, 250, 246, 197, 268, 224, 239, 239, 254, 276, 234, 181, 248, 252, 202, 218, 212, 325
Type B behavior	344, 185, 263, 246, 224, 212, 188, 250, 148, 169, 226, 175, 242, 252, 153, 183, 137, 202, 194, 213

Source: S. Selvin. 1991. *Statistical Analysis of Epidemiological Data*. New York: Oxford University Press, Table 2.1.

a. Construct a back-to-back stem-and-leaf plot for Type A behavior and Type B behavior. *Hint:* Drop the ones digits. For example, 188 becomes 180; then 180 is assigned a hundreds stem of 1 and a tens leaf of 8.
b. What do you infer?

11. In a national survey, men were asked why they exercise. Here are the reasons they gave:

Reason	**Percentage responding**
Health	51
Stress relief	25
Weight loss	20
Other	4

Draw a pie chart of these data.

12. You have learned that usually the heights in a picture-based "histogram" are accurate and the distortion arises because our perception keys in on area.
a. Is this the case with the following picture?

Comparative Annual Cost per Capita for Care of Insane in Pittsburgh City Homes and Pennsylvania State Hospitals

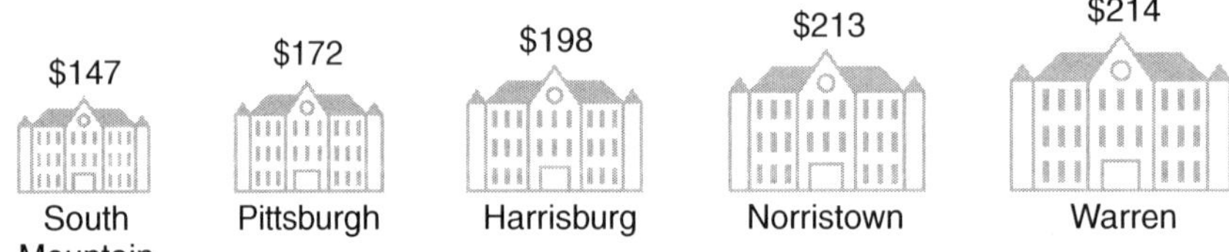

Source: Pittsburgh Civic Commission, *Report on Expenditures of the Department of Charities* (Pittsburgh, 1911), 7.

b. Draw an appropriate bar graph for the given institutions. (Note that this was drawn in 1911. Would such a graph contain the word "insane" today?)

13. A six-sided die is rolled 120 times, and the following outcomes are obtained:

Outcome	*f*	Proportion
1	15	
2	21	
3	23	
4	19	
5	17	
6	25	
Total	120	

a. Draw a density histogram of the data.
b. From the graph, find (i) the percentage of rolls that give 3 or less, (ii) the percentage of rolls that give 5 or more, and (iii) the percentage of rolls that give an even number.
c. How would you roughly describe the shape of the density histogram?

14. The following table gives the percentage breakdown of the cigarette smoking habits of young adults (ages 18–24). They are broken into three groups: (i) those who currently smoke, (ii) those who are former smokers, and (iii) those who have never smoked. Two years (1965 and 1991) and gender are considered (from the 1998 paper "Trends in Cigarette Smoking" by the American Lung Association).

Smoking status	1965, Males	1965, Females	1991, Males	1991, Females
Current	54.1	38.1	23.5	22.4
Former	7.6	6.2	8.0	7.5
Never	38.3	55.7	68.5	70.1

a. Create pie charts for the 1965 males and the 1965 females. On which categories do males and females differ the most?
b. Create a pie chart for the 1991 males. What changes do you see from 1965 to 1991 for the males?
c. Create a pie chart for the 1991 females. What changes do you see from 1965 to 1991 for the females?
d. Now compare the males and females in 1991. What differences do you see? Are the differences larger than in 1965 (see part (a)) or smaller?
e. Briefly summarize what these data indicate.

PROFESSIONAL PROFILE

AGES.	A CHARENTON AVANT 1829. ADMISSIONS.	A CHARENTON AVANT 1829. GUÉRISONS.	RAPPORT.	ALIÉNÉS EN TENANT COMPTE DE LA POPULATION.	A CHARENTON 1829 A 1833. HOMMES.	A CHARENTON 1829 A 1833. FEMMES.	RAPPORT.
15 à 20	22	11	2,0	24	24	11	2,2
20 à 25	67	30	2,2	79	65	23	2,9
25 à 30	86	40	2,2	109	78	31	2,5
30 à 35	98	36	2.7	134	79	47	1,7
35 à 40	81	25	3,3	125	65	64	1,0
40 à 45	79	21	3,8	129	64	39	1,1
45 à 50	72	14	5.1	131	52	44	1,2
50 à 55	52	12	4,3	108	54	37	1,2
55 à 60	21	6	3,5	51	32	20	1,6
60 à 65	21	9	2,3	63	33	18	1,8
65 à 70	6	1	6,0	24	14	9	1,6
70 et plus	14	4	3.5	45	6	7	0,9

(1) D'après un ouvrage de M. Klotz : *de Vesaniæ prognosi*, le rapport

Aliénés, en tenant compte de la population.

20 40 60

Adolphe Quetelet (1796–1874), Belgian pioneer in statistics and sociology and originator of the concept of the "average man."

Bell curve and table on mental illness patients in Charenton asylum, from volume 2 of *Physique sociale, ou Essai sur le développement des facultés de l'homme*, by Adolphe Quetelet.

Adolphe Quetelet. Steel engraving by Joseph-Arnold Demannez from the 1875 *Annuaire de l'Académie royale de Belgique*. Scan from the Library of Congress, Prints and Photographs Division.

When your doctor tells you you're overweight and you need to lose so many pounds, that assessment is based on your body mass index, a number first developed by the Belgian mathematician Adolphe Quetelet. The body mass index is sometimes called the Quetelet index and is only one of Quetelet's innovations that influence our lives today.

Quetelet, who began his professional life as a mathematician and astronomer, became convinced that probability influenced human lives more than his contemporaries believed. He began to apply the law of error (originally used to study observational errors in astronomy) to human beings and societies. Based on his belief that it was possible to determine from gathered facts the average physical and intellectual characteristics of a given population, he began data collection and graphically arranged this data in bell-shaped curves to identify what he called the "average man." He thus became a pioneer in the new field he called "social mechanics" (*physique sociale*) and a pioneer in the field of sociology.

In the field of statistics, Quetelet was the first to use a normal bell curve in the study of humans and their characteristics. He was interested in the analysis of crime, mental health, mortality, and physical characteristics. He collected data in these areas for the Belgian government and made recommendations for improvements in census-taking. Among his correspondents was United States President James A. Garfield who, during his six months in office, sought Quetelet's input on improving the American census.

In his studies, Quetelet developed two central principles. First, causes are proportional to the effects they produce. And, second, exact conclusions are possible only with large numbers. Quetelet believed that the greater the number of individuals studied, the more the influence of any single individual is reduced. When large numbers of individuals are involved, a series of general facts emerges that is based on the general causes within a society.

Today, statisticians make heavy use of the principles that Quetelet developed in the mid-19th century and rely on what can be learned about populations from the collection of data and its graphic display.

2

Summarizing Data by Numerical Measures: Center and Spread

Society suffers from the curse of the arithmetic mean. Such compression necessarily ignores the fact that most measurements involve a spread of values around the mean—arising from real differences and random fluctuations.

R. E. Beard

Objectives

After studying this chapter, you will understand the following:

- ❒ The importance of assessing the center and the spread of a data set
- ❒ The two most important measures of center: the mean and the median
- ❒ The most widely used measure of spread: the standard deviation
- ❒ The 68-95-99.7% Rule for interpreting and using the standard deviation
- ❒ Using the normal distribution to find the proportion of a bell-shaped data set in any interval
- ❒ z-scores and the standard normal curve
- ❒ A resistant measure of spread: the interquartile range
- ❒ How to use and construct boxplots
- ❒ How to apply the descriptive tools of Chapters 1 and 2 to describe and compare data sets
- ❒ The cumulative distribution and percentiles

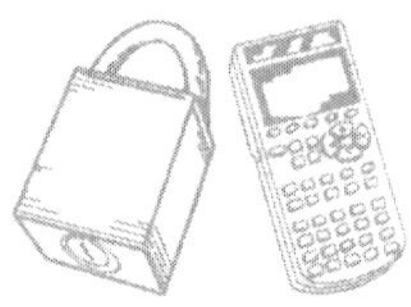

KEY PROBLEM

The ages of the actors and the actresses who have won Best Actor/Best Actress Academy Awards from 1976 to 2006 are as follows:

Year	Best Actor	Age	Year	Best Actor	Age	Year	Best Actress	Age	Year	Best Actress	Age
1976	Jack Nicholson	38	1992	Anthony Hopkins	54	1976	Louise Fletcher	41	1992	Jodie Foster	29
1977	Peter Finch	60	1993	Al Pacino	52	1977	Faye Dunaway	35	1993	Emma Thompson	33
1978	Richard Dreyfuss	30	1994	Tom Hanks	37	1978	Diane Keaton	31	1994	Holly Hunter	35
1979	Jon Voight	40	1995	Tom Hanks	38	1979	Jane Fonda	41	1995	Jessica Lange	45
1980	Dustin Hoffman	42	1996	Nicolas Cage	31	1980	Sally Field	33	1996	Susan Sarandon	49
1981	Robert De Niro	37	1997	Geoffrey Rush	45	1981	Sissy Spacek	31	1997	Frances McDormand	38
1982	Henry Fonda	76	1998	Jack Nicholson	60	1982	Katherine Hepburn	74	1998	Helen Hunt	34
1983	Ben Kingsley	39	1999	Roberto Benigni	46	1983	Meryl Streep	31	1999	Gwyneth Paltrow	26
1984	Robert Duvall	52	2000	Kevin Spacey	40	1984	Shirley Maclaine	49	2000	Hilary Swank	25
1985	F. Murray Abraham	45	2001	Russell Crowe	36	1985	Sally Field	38	2001	Julia Roberts	33
1986	William Hurt	35	2002	Denzel Washington	47	1986	Geraldine Page	61	2002	Halle Berry	33
1987	Paul Newman	61	2003	Adrien Brody	29	1987	Marlee Matlin	21	2003	Nicole Kidman	35
1988	Michael Douglas	43	2004	Sean Penn	43	1988	Cher	41	2004	Charlize Theron	28
1989	Dustin Hoffman	51	2005	Jamie Foxx	37	1989	Jodie Foster	26	2005	Hilary Swank	30
1990	Daniel Day-Lewis	32	2006	Philip Seymour Hoffman	38	1990	Jessica Tandy	82	2006	Reese Witherspoon	29
1991	Jeremy Irons	42				1991	Kathy Bates	42			

Source: Dr. Anne E. Lincoln of Rice University.

Do winning actresses tend to be younger than winning actors? Is there less spread in the age distribution for actresses than for actors? How do we summarize the data so we can answer such comparative questions?

The data are presented graphically in Figure 2.1 in graphs called **boxplots** or **box-and-whiskers plots**. In the boxplot, the vertical line inside the box locates the **median** (a measure of center that will be explained in Section 2.1), and the ends of the box locate the first and third **quartiles** (two descriptive statistics that give the range covered by the middle half of the data, as will be explained in Section 2.5). Outside the boxes, four data points are marked with special symbols (○ or ∗), indicating values so far from the bulk of the data that they are viewed as not really belonging with the rest of the data and hence are called extreme values or outliers (see Section 1.2). We extend lines (the whiskers) from the ends of the box to the minimum and maximum values of the data, excluding the outliers.

Do you think these two boxplots help answer the two questions posed above about differences between male and female Academy Award winners? Do they help us compare the distribution of the two data sets?

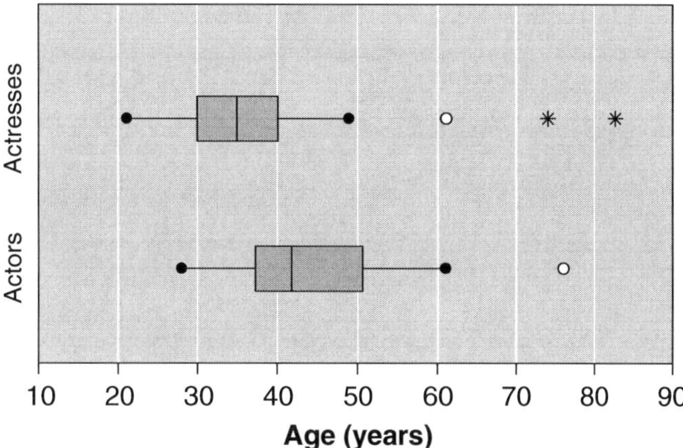

Figure 2.1 Boxplot comparison of ages of male and female Academy Award winners.

descriptive statistics

The goal of **descriptive statistics** is to summarize and describe data so that we can recognize the information the data contain. In Chapter 1 we learned how to organize raw data and display them graphically. Descriptive statistics always begins with visual displays. For quantitative data we try to characterize the shape and location of the distribution, and we look for gaps and exceptional observations (outliers).

center
spread
skewness

In this chapter we will add summarizing numbers to our description of data. These numbers indicate where the **center** of the data set lies, indicate the **spread** of the data, or measure the **skewness** of the data, to list a few possibilities.

2.1 THE CENTER OF A DATA SET

In this section we will discuss three commonly used measures of center: the median, the mean, and the mode.

Example 2.1

In 1850 the population of the United States was much younger than it is today. Typically, a family had many children, and people only lived half as long as they do today. The age distribution of the United States looked in 1850 the way the age distribution of Kenya looks today (see Figure 1.17). The median age was 19 years: Half of all Americans were under the age of 19, and half were 19 years old or older. The life expectancy at birth was 39 years. (The life expectancy for a particular year is the average number of years that everyone born in that year would live if death rates remained constant.)

The twentieth century brought dramatic improvements in public health, and life expectancy doubled between 1850 and 2000. This, combined with a decline in the number of children born, caused a doubling of the median age in the United States (see Table 2.1). The line graphs in Figure 2.2 display the change over time in median age and life expectancy.

Table 2.1 Median Age and Life Expectancy for the United States

Year	Median age (years)	Life expectancy (years)
1850	19	39
1900	23	47
1950	30	68
2000	35	77

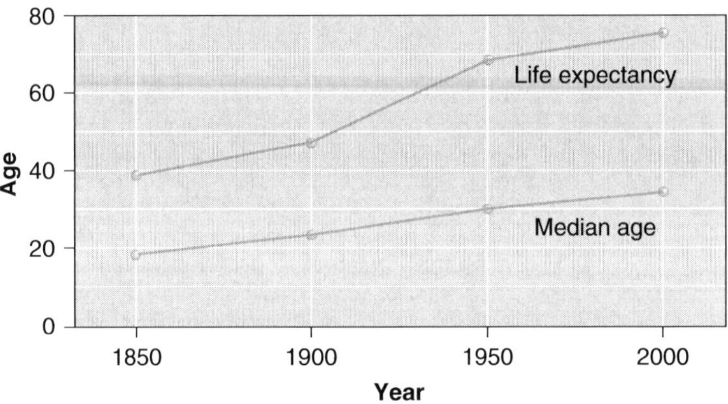

Figure 2.2 Median age and life expectancy at birth from 1850 to 2000.

Median

median

The **median** M of a data set is the middle value when the data are ordered from smallest to largest. Half of the observed values are smaller than the median, and half are larger. The median of a list of numbers can be found by the following rule:

Step 1. Arrange the observations according to size, from the smallest to the largest value.

Step 2. Record the median M as the observed value located at the exact middle of the ordered list when the number of observations is odd. When the number of observations is even, record the median M as the number halfway between the two middle numbers on the list.

The median is often interpreted as the "typical" value of a data set.

Example 2.2

The number of home runs Babe Ruth hit during his 15 years with the New York Yankees, from 1920 to 1934, are as follows:

1920	1921	1922	1923	1924	1925	1926	1927	1928	1929	1930	1931	1932	1933	1934
54	59	35	41	46	25	47	60	54	46	49	46	41	34	22

Arranged in increasing order, the data are

22 25 34 35 41 41 46 46 46 47 49 54 54 59 60
↑
M

Here we have 15 observations, an odd number. The median is the exact middle number: $M = 46$. The median is a good indicator of Ruth's "typical" annual home run production because he hit between 40 and 50 home runs almost half of the seasons he played for the Yankees.

Ruth's 1927 record of 60 home runs in a single season remained unbroken until 1961, when Roger Maris, another Yankee, hit 61 home runs. Maris's home run totals for his 10 years in the American League are, in increasing order:

8 13 14 16 23 26 28 33 39 61
↑
M

Here we have 10 observations, an even number. The median is halfway between the fifth smallest and the fifth largest observations; that is, we take the average of the middle two numbers:

$$M = \frac{23 + 26}{2} = \frac{49}{2} = 24.5$$

The median is a fairly good indicator of Maris's typical annual home run production. His record of 61 home runs in 1961 is an exceptional achievement—an outlier. Overall, Ruth was a much better home run hitter than Maris, as we can see by comparing their medians.

Mean

arithmetic mean

mean

average

The **arithmetic mean** is the most commonly used measure of the "center" of a data set. We derive it by adding up all the values and dividing the sum by the number of observations. The arithmetic mean is usually observed as simply the **mean** or the **average**. Because the word *average* can sometimes be misinterpreted, we prefer the term *mean*.

Example 2.3

During the summer of 1998, the attention of sports fans was caught by a race between the Chicago Cubs' Sammy Sosa and the St. Louis Cardinals' Mark McGwire to beat Roger Maris's single season home run record. Sosa got 66 home runs and McGwire hit 70, a record that stood until 2001 when Barry Bonds hit 73 home runs. To compare the overall home run productions of Sosa and McGwire, we shall compute the mean annual number of home runs for each.

The number of home runs hit by Sosa during his years with the Chicago Cubs, from 1992 to 2004, are as follows:

1992	1993	1994	1995	1996	1997	1998	1999	2000	2001	2002	2003	2004
8	33	25	36	40	36	66	63	50	64	49	40	35

The mean number of home runs hit by Sosa is

$$\frac{8+33+25+36+40+36+66+63+50+64+49+40+35}{13} = \frac{545}{13} = 41.9$$

The number of home runs hit by McGwire from 1987 to 2001 are as follows:

1987	1988	1989	1990	1991	1992	1993	1994	1995	1996	1997	1998	1999	2000	2001
49	32	33	39	22	42	9	9	39	52	58	70	65	32	29

McGwire's mean number of home runs is

$$\frac{49+32+33+39+22+42+9+9+39+52+58+70+65+32+29}{15} = \frac{580}{15}$$
$$= 38.7$$

We note that the two means are close. The two players' median numbers of home runs are even closer: 40 for Sosa and 39 for McGwire.

Notation and Terminology

We shall introduce some shorthand notation to remind us how we compute the mean of a set of observations. The mean of n observations $x_1, x_2, \ldots, x_n$ is

$$\overline{x} = \frac{x_1 + x_2 + \cdots + x_n}{n}.$$

The symbol $\overline{x}$ is read "x-bar." Using even more compact notation, we write

$$\overline{x} = \frac{\sum x_i}{n}.$$

Here the symbol Σ stands for "sum" and indicates that all the x_i-values should be added up. (The symbol Σ is the capital Greek letter sigma, corresponding to the Roman letter S.)

Mode

mode

The **mode** is another measure of the typical value. It is the value (or values) that occurs most frequently. There may be more than one mode, and the mode(s) may be far from the center of the distribution of the values. For that reason, the mode is not often used as a measure of center for quantitative data, even though the mode sometimes happens to be near the center. Note, however, that the mode is the only measure of center that we can compute for categorical data. (See the end of Section 1.1 and Section 1.3.)

Example 2.4

Figure 2.3 displays a density histogram that shows the distribution of educational levels (number of years of schooling completed) for people 25 years old or older in 1998. The histogram's spike at 12 years of schooling represents all those who quit school after high school. The class intervals were intentionally chosen to be 0–8, 9–11, 12, 13–15, 16, 17–20. (Why?) The mode is 12: More people have 12 years of schooling than any other number, as one might expect. (Why?)

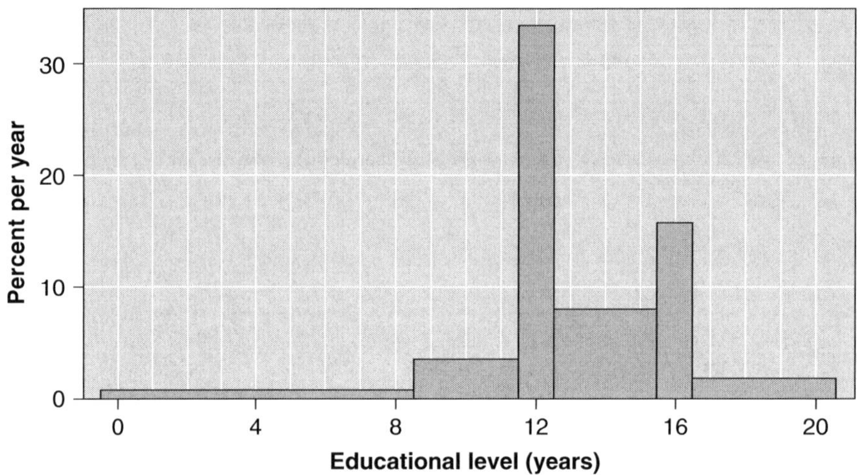

Figure 2.3 Distribution of people age 25 and older in the United States by educational level in 1998.

Section 2.1 Exercises

1. Determine the mean, the median, and the mode of each of the following four lists:

List A:	1	1	4	9	20	
List B:	1	1	4	9	420	
List C:	1	4	9	16	30	30
List D:	10	40	90	160	300	300

2. A study reported the following litter sizes for 23 lions:

3	4	2	2	2	1	1	3
3	2	2	2	2	2	3	2
1	1	4	2	3	1	5	

a. Construct a frequency table for these data.
b. Determine the mean and the median litter sizes.
c. Is there a typical litter size?

3. Some ornithologists were interested in the clutch size of the Common Moorhen, nesting in Louisiana rice fields. They counted the number of eggs in each of 105 nests, obtaining the following data:

7	5	13	7	7	8	9	9	9	8	8	9	9	7	7
5	9	7	7	4	9	8	8	10	9	7	8	8	8	7
9	7	7	10	8	7	9	7	10	8	9	7	11	10	9
9	4	8	6	8	9	9	9	8	8	5	8	8	9	9
14	10	8	9	9	9	8	7	9	7	9	10	10	7	6
11	7	7	6	9	7	7	6	8	9	4	6	9	8	9
7	9	9	9	9	8	8	8	9	9	9	8	10	9	9

a. Construct a frequency table for these data.
b. Determine the median clutch size.
c. What is the mode (the typical clutch size)?
d. What is the mean clutch size?

4. Below is a stem-and-leaf plot of the annual earnings in thousands of dollars of a sample of 50 recent college graduates (some working only part-time).
   a. Find the mode, the mean, and the median earnings. Compare the three values.
   b. Which measure (mode, mean, or median) is the best indicator of typical earnings?

Stem	Leaf
1	2,6,7,8,9,9
2	2,3,5,7,7,7,7,8,9,9
3	1,2,3,4,5,6,6,7,7,7,9
4	0,3,3,3,4,5,6,6,7
5	3,3,7,7,8
6	0,2,6
7	4,7
8	7
9	
10	4
HI	156,248

Key: "16" stands for $16,000.

5. The histogram in Figure 2.3 has a second spike at 16 years of schooling.
   a. Which group of people does this spike represent?
   b. Is the percentage of people age 25 and over who are not high school graduates closest to (choose one) 17%, 24%, or 34%?
   c. Is the percentage of people age 25 and over with a bachelor's degree (earned by completing 4 years of college) or higher closest to (choose one) 17%, 24%, or 34%?

6. The salaries (in millions of dollars) of the San Diego Padres for the 2006 season were as follows:

Player	Salary (millions $)	Player	Salary (millions $)
Chan Ho Park	15.505	Eric Young	0.700
Ryan Klesko	9.000	Jim Brower	0.700
Brian Giles	7.667	Geoff Blum	0.650
Mike Cameron	7.333	Chris Young	0.500
Woody Williams	5.000	Dewon Brazelton	0.500
Trevor Hoffman	4.500	Khalil Greene	0.405
Vinny Castilla	3.200	Josh Bard	0.353
Jake Peavy	2.500	Termel Sledge	0.347
Dave Roberts	2.250	Brian Sweeney	0.335
Scott Linebrink	1.365	Clay Hensley	0.329
Mike Piazza	1.250	Scott Cassidy	0.328
Shawn Estes	1.100	Adrian Gonzalez	0.327
Doug Brocail	1.000	Josh Barfield	0.327
Alan Embree	0.850	Ben Johnson	0.327
Mark Bellhorn	0.800		

a. Construct a frequency histogram using the intervals 0–1−, 1–2−, 2–3−, . . .
b. Find the mean and the median salaries.
c. Do you think the mean or the median is the better statistic to use to report the typical salary for the Padres? What about the mode?

7. The number of campsites in each of five campgrounds in Yoho National Park is

Campground	Number of campsites
Kicking Horse	92
Hoodoo Creek	106
Chancellor Park	64
Takakkaw Falls	35
Lake O'Hara	30

a. What is the mean number of campsites per campground?
b. What is the median number of campsites per campground?

8. The number of calories in a serving of various kinds of cheese is given by the following table:

Kind of cheese	Number of calories
American	106
Cream	99
Feta	75
Monterey	106
Ricotta (whole milk)	216
Swiss	107

a. What is the mean number of calories per serving?
b. What is the median number of calories per serving?
c. Compare the mean and the median.
d. Which type of cheese corresponds to the mode?

9. a. For the Key Problem find the mean and median ages of actors and actresses who won Academy Awards.
   b. What do you learn from these measures of central tendency?

10. Several years ago the numbers of millionaires per 1000 persons in the 10 states with the highest density of millionaires were reported as shown in the accompanying table.
   a. Without calculating, state which is larger: the mean or the median of the 10 numbers.
   b. Find the mean and the median.
   c. How many of these numbers are less than the mean?
   d. Is the mean or the median a better indicator of the center of these data?

State	Number of millionaires per 1000 persons	State	Number of millionaires per 1000 persons
Idaho	27	Indiana	5
Maine	8	Wisconsin	4
North Dakota	7	Iowa	4
Nebraska	7	New Jersey	4
Minnesota	6	Connecticut	3

11. The numbers of earthquakes that were recorded as 7 or greater on the Richter scale during the years 1900–1909 and 1980–1989 are as follows:

Year	Number of earthquakes	Year	Number of earthquakes
1900	13	1980	18
1901	14	1981	14
1902	8	1982	10
1903	10	1983	15
1904	16	1984	8
1905	26	1985	15
1906	32	1986	6
1907	27	1987	11
1908	18	1988	8
1909	32	1989	7

a. Make a back-to-back stem-and-leaf plot for the data.
b. Compare the years 1900–1909 to the years 1980–1989 using the means or medians as appropriate. Which decade tended to have more earthquakes?

12. The frequency histogram for the durations, in minutes, of 230 eruptions of the geyser Old Faithful is given in the following graph. The intervals are 0–0.5–, 0.5–1–, and so forth. The mean of these durations is 3.26 minutes.
a. Is that a good measure of the typical duration? Explain your answer.
b. Discuss the tendency of the data to form distinct **clusters.** Is there essentially a **gap**?

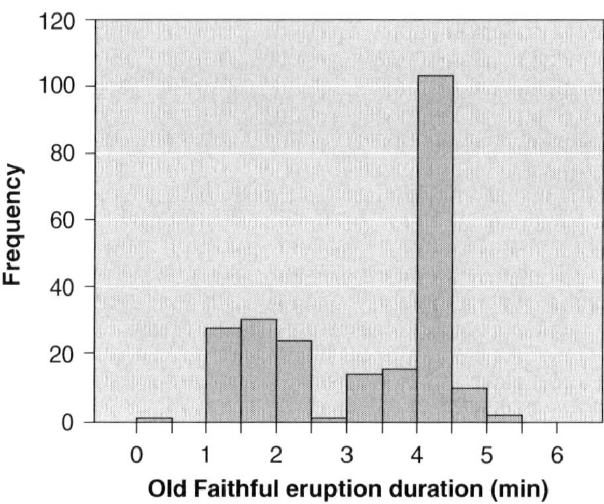

13. a. Use the data in Example 1.9 in Section 1.5 to estimate the mean U.S. household size in 1999. (Make the simplifying assumption that no household had over nine people.) *Hint:* Our best estimate of the mean size of all 100 million U.S. households in 1999 is the mean size of the households in the survey, which equals

$$\frac{\text{number of persons in the sample}}{\text{number of households in the sample}}$$

The number of households is 50,785. To find the total number of people in the sample, first count the number of people in all the sample households of size 1 $(12,709)$. Then count the number of people in all the sample households of size 2 $(2 \times 16,517 = 33,034)$ and add that number to the first. Likewise, add the number of people in all the sample households of size 3 $(3 \times 8,435 = 25,305)$ to the total, and so forth.
b. What was the median household size?
c. What was the mode?

2.2 MEAN VS. MEDIAN VS. MODE AS A MEASURE OF CENTER

When we look at categorical data, the mode is a key statistic. For example, in Example 1.4 we learned that heart disease is the most common cause of death in the United States. The mode is also of interest when we look at "counting variables" such as household size (see Example 1.9) or the number of years of schooling completed (see Example 2.4). But the mode is not really a measure of the center of a data set—even though the mode sometimes happens to be near the center.

The most common measures of the center of a distribution (of data) are the median and the mean. If the distribution is roughly symmetric and without outliers, then the mean and the median will be close together and either one can be used to indicate the center.

Example 2.5

The ages of American presidents when they died are as follows:

Washington	67	Pierce	64	Wilson	67
J. Adams	90	Buchanan	77	Harding	57
Jefferson	83	Lincoln	56	Coolidge	60
Madison	85	A. Johnson	66	Hoover	90
Monroe	73	Grant	63	F. Roosevelt	63
J. Q. Adams	80	Hayes	70	Truman	88
Jackson	78	Garfield	49	Eisenhower	78
Van Buren	79	Arthur	57	Kennedy	46
W. Harrison	68	Cleveland	71	L. Johnson	64
Tyler	71	B. Harrison	67	Nixon	81
Polk	53	McKinley	58	Gerald Ford	93
Taylor	65	T. Roosevelt	60	Ronald Reagan	93
Fillmore	74	Taft	72		

We can summarize the data in the following stem-and-leaf plot (here we use such a plot, rather than an ordinary histogram, so we can still see the actual ages):

Stem	**Leaf**
4	6,9
5	3,6,7,7,8
6	0,0,3,3,4,4,5,6,7,7,7,8
7	0,1,1,2,3,4,7,8,8,9
8	0,1,3,5,8
9	0,0,3,3

The mean age at death of the 38 presidents is $2676/38 = 70.4$ years. The median age at death is the average of the nineteenth and twentieth youngest ages, that is, $(68+70)/2 = 69$ years. For this rather symmetric and outlier-free data set, the mean and the median are close.

Since three presidents died at age 67 and no more than two died at any other age, the mode is 67. In this example, the mode happens to be a "central value." Note, however, that if another president dies at age 78, then there will be two modes because 67 and 78 will each occur three times. And if another president dies at age 90, then 90 will also be a mode.

❐ ❐

Notice that it is easy to find the median from a stem-and-leaf plot because the data are arranged in increasing order. If a distribution is skewed or if there are outliers, then the mean and the median need not be close together. The reason is that *the mean is very sensitive to the "pull" of observations far from the center*. To illustrate this pull, consider the home run data for Roger Maris in Example 2.2. His total of 61 home runs in 1961 is an exceptional performance, an outlier. If we exclude this score as atypical, then the remaining nine scores have a roughly symmetric distribution with median 23 and mean 22.2. If we include his record-setting score of 61 home runs, then the median increases a little bit from 23 to 24.5, but the mean jumps from 22.2 to 26.1. A single outlier increases Roger Maris's mean annual home run production by almost four home runs!

influential
extreme values
resistant

An observation whose presence greatly changes the value of a statistic (in this case, the mean) is said to be **influential**. Roger Maris's total of 61 homers in 1961 is an influential observation. Because the mean cannot resist the influence of **extreme values** (very large or very small), we say that the mean is not a **resistant** measure of center. In a skewed distribution, the extreme values pull the mean away from the median in the direction of the skew, toward the long tail.

The following examples illustrate that income distributions are usually skewed to the right and therefore have higher mean incomes than median incomes.

Example 2.6

The salaries of the Miami Heat, the NBA champions for the 2005–2006 season, were as follows:

Player	Salary	Player	Salary
Shaquille O'Neal	$20,000,000	Gary Payton	$ 1,138,500
Jason Williams	$ 7,562,500	Shandon Anderson	$ 1,035,000
Antoine Walker	$ 7,000,000	Dorell Wright	$ 1,032,200
James Posey	$ 5,900,400	Jason Kapono	$ 1,000,000
Udonis Haslem	$ 5,000,000	Wayne Simien	$ 867,720
Dwayne Wade	$ 3,031,920	Andre Emmett	$ 641,748
Michael Doleac	$ 2,640,000	Earl Barron	$ 641,748
Wesley Person	$ 1,800,000	Matt Walsh	$ 200,000
Derek Anderson	$ 1,670,000	Gerald Fitch	$ 200,000
Alonzo Mourning	$ 1,138,500		

Shaq's salary is clearly an outlier, and the Heat did not pay Dwayne Wade what he was worth. We can display the salaries (in millions of dollars) in a frequency histogram (Figure 2.4) with class intervals 0–1–, 1–2–, and so on. The median salary is $1,138,500, the mean salary is $3.29 million. The reason that the mean is so much higher than the median is clear: Most players have salaries in the low end, but a few players have very high salaries. Thus, the distribution is skewed to the right. Two-thirds of the Heat each earn $3 million or less. Shaq alone earns more than these 14 players combined! Both the skewness and the outlying Shaq salary pull the mean to the right.

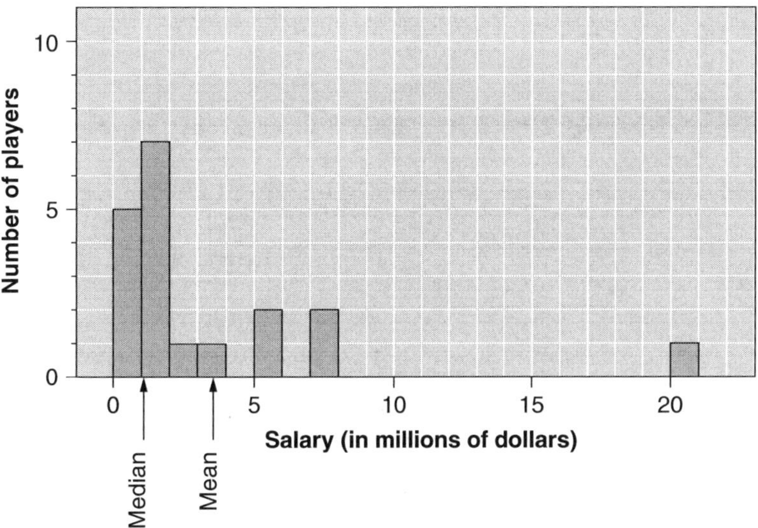

Figure 2.4 Distribution of salaries of the Miami Heat for the 2005–2006 season.

The salary distribution of Major League Baseball players is also skewed to the right. (See, for example, Exercise 6 in Section 2.1.) During the 1994–1995 Major League Baseball players' strike, players and owners used different statistics to represent the typical player's salary. In the public relations war, the players wanted to demonstrate that they were underpaid, so they used the median salary. The owners wanted to show that the players were well paid, so the owners used the higher number, the mean salary. Because even the "low" median salary was over 10 times higher than the typical family income in the United States, the players lost the public relations battle.

Besides public relations, the players and owners also had legitimate reasons for using different measures of the typical salary. The median salary is a good indicator of roughly what the vast majority of players will earn during their first few years as professional athletes. For the sports team owners who pay the salaries, however, it is the mean salary that counts, not the median. For example, the total payroll for the Miami Heat in 2005–2006 was:

$$\begin{aligned}\text{total payroll} &= (\text{number of players}) \times (\text{mean salary}) \\ &= 19 \times \$3.29 \text{ million} \\ &= \$62.5 \text{ million.}\end{aligned}$$

Concerning the total payroll, the computation (number of players) × (median salary) does *not* give us the total payroll.

Example 2.7

Each year, the Current Population Survey estimates the previous year's distribution of household incomes (see Example 1.9). The income distribution for 1998 is displayed in a density histogram in Figure 2.5 (note that the interval 0–10 has been split in two to better display the shape of the data, and that the range 100–160 is split into only two class intervals, namely, 100–125– and 125–160, because of limited data in the high-income range).

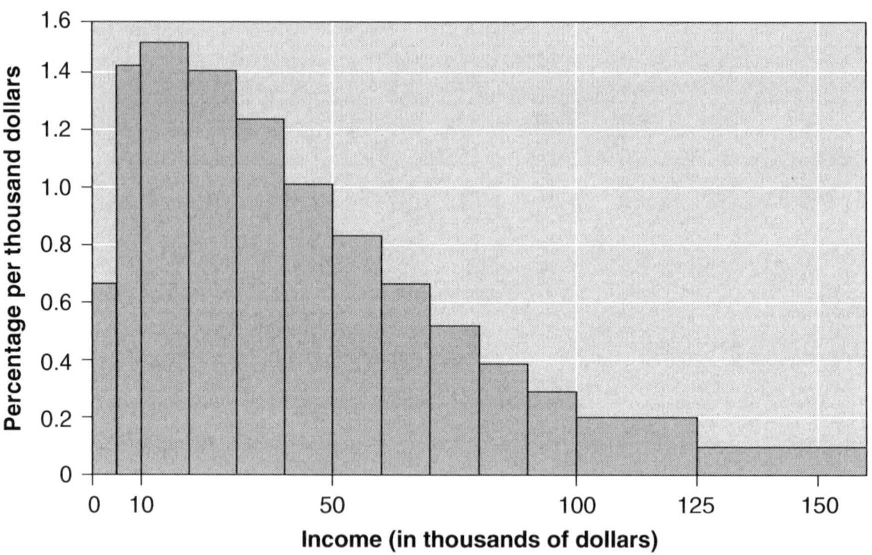

Figure 2.5 Distribution of households in the United States by income, 1998.

The income distribution is clearly skewed to the right. In 1998 the median household income was around \$39,000. The mean household income was around \$52,000, which, as expected, is much larger than the median.

The distribution of house prices in a community is also usually skewed to the right. There will be many moderately priced houses and a smaller number of expensive houses. The few expensive houses pull the mean up but have little effect on the median. So the mean house price will be higher than the median house price.

Which number is more relevant, the median price or the mean price? That depends on your point of view. A young couple moving into a community will probably be more interested in the median home price. The median tells them how much it will typically cost to buy a home. But if you are a member of the school board, then you will be more interested in the mean home price. The amount of property taxes that can be raised to support local schools is determined by the total market value of all the homes, which can be computed from the mean:

$$\text{Total market value of all homes} = (\text{Number of homes}) \cdot (\text{Mean home price}).$$

In symbols,

$$\sum x = n \cdot \overline{x}.$$

This property of recovering the total from the mean is often useful.

Figure 2.6 shows the relative positions of the median and the mean for right-skewed, symmetric, and left-skewed distributions. Notice that in a skewed distribution, the extreme values

in the tail pull the mean away from the median toward the long tail. Note that the mean and the median are the same for a symmetric distribution regardless of whether it is high, flat, or low in the middle.

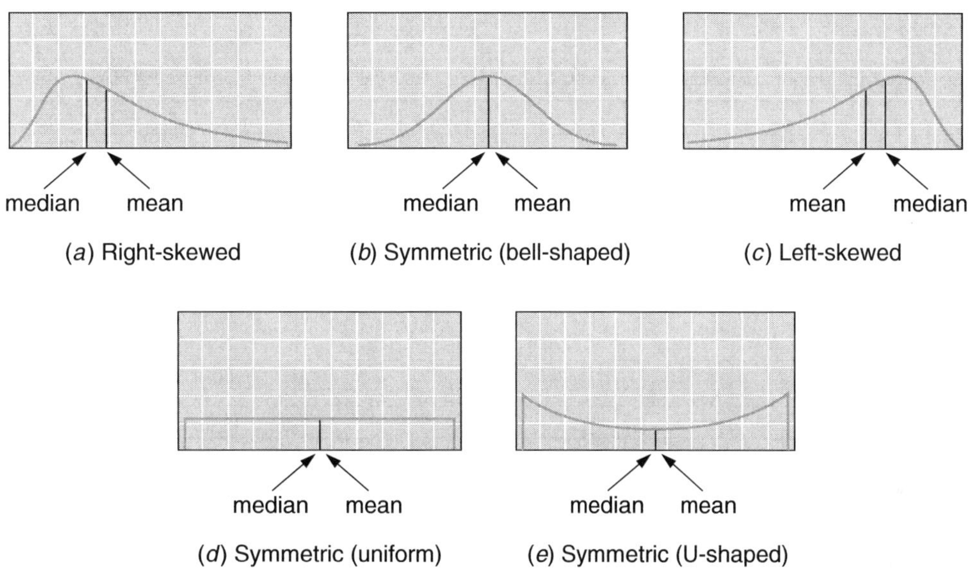

Figure 2.6 Relative positions of the median and the mean for right-skewed, symmetric, and left-skewed distributions.

Outliers

Shaq's salary in Example 2.6 and Roger Maris's record-setting 61 home runs in 1961 (see Example 2.2) raise the question: What do we do with outliers? Do we include them, or do we leave them out because they are atypical and distort the mean as an indicator of the typical value? As a rule, we do not remove an outlier unless the outlier is the result of a mistake. Shaq's salary is not a mistake. It is part of the story on how the Heat put together a winning team. Shaq's salary should not be removed. But because of this outlier the median salary is a better indicator of the typical salary than the mean salary. In Roger Maris's case also we have no excuse for leaving out his exceptional achievement in 1961. It did not come completely out of the blue: In 1960 he had hit 39 home runs, his second highest total overall; in 1962 he hit 33 home runs, his third highest total overall. Besides, the 61 home runs is certainly the most interesting number in the data set! The line graph in Figure 2.7 illustrates that over three seasons, Maris peaked and then faded as a home run hitter (sidelined by injuries).

The best way to characterize Roger Maris as a home run hitter is to note his record-setting 61 home runs in 1961, but to use the median as the indicator of his typical annual home run production over his career. What is unusual about him is that he peaked and then faded so rapidly, like a supernova in the sky. An outlier is removed from a data set only when its inclusion distorts the story needing to be told by the data set.

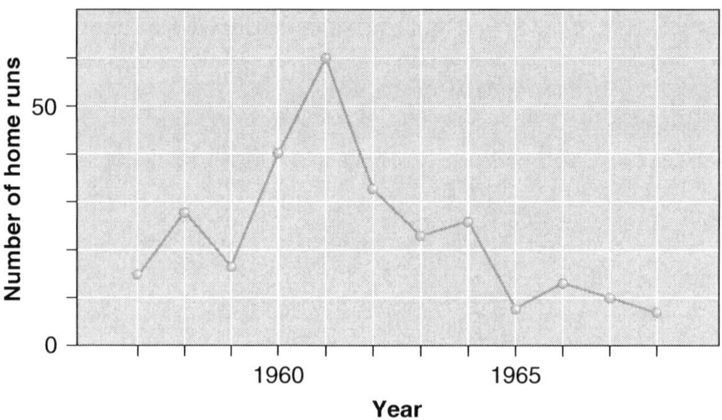

Figure 2.7 Roger Maris's annual home run production from 1957 to 1968.

The next example looks at measurement errors. As anyone who has stepped on a bathroom scale knows, repeated measurements of the same thing differ slightly from each other. One of the first scientists to deal with this problem was the Danish astronomer Tycho Brahe. His careful observations were used by Johannes Kepler to derive his three fundamental planetary laws, which later enabled Isaac Newton to formulate his theory of gravitational force.

Example 2.8

It has been known since the late seventeenth century (the 1600s) that light travels at a finite speed and is not transmitted instantaneously. It takes about 8 minutes for light from the sun to reach Earth and about 1 second for light to travel from the moon to Earth. Around 1880 Simon Newcomb and A. A. Michelson conducted experiments that gave the first accurate determination of the speed of light.

Newcomb's experiment consisted of bouncing a light beam off a rapidly counter-clockwise rotating mirror at a right angle (the rotating mirror being at 45°), to a distant fixed mirror, and back to the rotating mirror (see Figure 2.8). Because the rotating mirror would have turned a little bit by the time the light beam is bounced back to the rotating mirror, the returning light would be reflected in a slightly different direction from its original source. Newcomb measured the angle between the returning light beam and the source. This angle is proportional to the passage time (the time the light beam took to travel from the rotating mirror to the fixed mirror and back again). That is, if the light takes three times as long, the angle is three times as large. This relation, together with knowing the rotational speed of the mirror and the measured angle, allows measurement of the passage time. The speed of light can then be computed:

$$\text{Speed of light} = (\text{Distance traveled})/(\text{Passage time})$$

Newcomb made 66 determinations of the time a light beam takes to travel to the fixed mirror 3721 meters (over 2 miles) away and back. His passage times in nanoseconds (billionths of a second) are 24,800 plus the values in Table 2.2. Thus, the number 28 in Table 2.2 stands for 24,828 nanoseconds, and the number –44 stands for 24,756 nanoseconds.

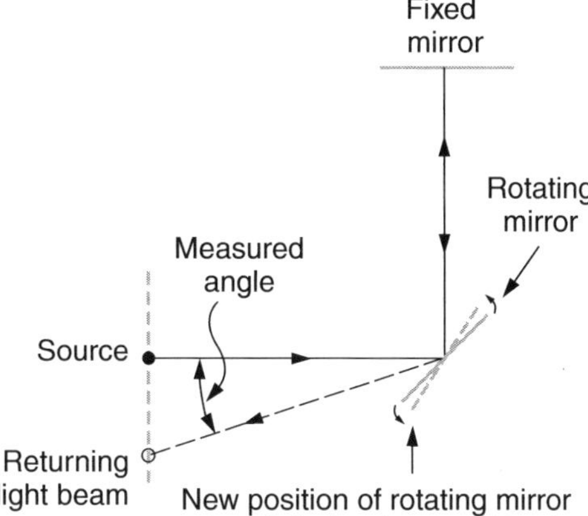

Figure 2.8 Schematic of Newcomb's experiment.

Table 2.2 Newcomb's Measurements of the Passage Time of Light

28	22	36	26	28	28
26	24	32	30	27	24
33	21	36	32	31	25
24	25	28	36	27	32
34	30	25	26	26	25
−44	23	21	30	33	29
27	29	28	22	26	27
16	31	29	36	32	28
40	19	37	23	32	29
−2	24	25	27	24	16
29	20	28	27	39	23

Source: Stigler, S. M. (1977). Do robust estimators work with real data? Ann. Stat. 5, No. 6, 1055–1098.

Why didn't Newcomb get the same number each time? Part of the reason is that it is impossible to measure the rotational speed of the mirror or the angle made by the returning light beam with complete accuracy. Furthermore, the speed of light traveling through air varies a tiny bit with the atmospheric conditions. Newcomb's data are displayed in the histogram of Figure 2.9, where the data have been grouped into the class intervals: −45+ to − 42, −42+ to − 39, and so forth. Newcomb's data seem to follow a bell-shaped distribution except for two outliers: −2 and −44.

He was unable to identify any equipment failure that might have caused these two outliers.

So how do we estimate the true passage time? The mean of the 66 observations is 26.21. Newcomb discarded the most extreme outlier (−44) and computed the mean of the remaining 65 observations. He got 27.29. Another possibility is to use the median of all 66 observations, which equals 27.

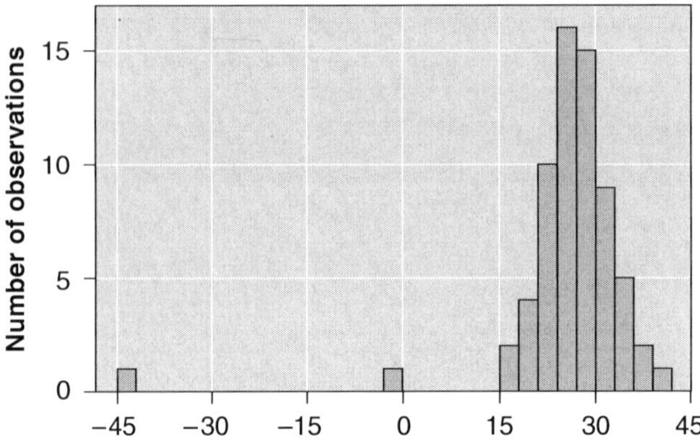

Figure 2.9 Newcomb's passage times (in nanoseconds) minus 24,800.

❐ ❐

Influential extreme values pose a problem in statistics. The problem is deciding what to do with them. Some outliers are simply mistakes. For example, the infant mortality rate of Monaco in Table 1.6 (copied from the *New York Times Almanac 2000*) is almost surely an error. Such outliers should be discarded. Other outliers, such as Roger Maris's 61 home runs (see Example 2.2 and Figure 2.7) represent significant events. Such outliers should not be discarded but should be studied separately. Finally, we may need to make an informed decision about if and when to include unexplained outliers in our analyses, such as Simon Newcomb's two lowest passage times in Table 2.2. Are they the result of error or carelessness; are they factors external to the observation, or are they valid in some way? For example, even at the National Institute of Standards and Technology (NIST), where things are measured as carefully as possible, the occasional outlier is unavoidable. NIST discards outliers only "for cause, such as door-slam or equipment malfunction." That leaves a small number of outliers that cannot be explained away and therefore cannot be discarded.

Section 2.2 Exercises

Technically, *average* is another word for the mean of a set of data, but in Exercises 1 and 2 the word *average* is used in its everyday, nontechnical interpretation as "central or typical value," and your answer can be either mean, mode, or median.

1. Which measure of center (mean, median, or mode) would be the most appropriate measure of center to use in each of the following cases? Explain your answer. In some cases, both the mean and the median can be appropriate, depending on the purpose, or because they are about the same value.

a. The average number of students in statistics classes at the College of Lake County, Illinois, is 27.4.
b. The average home value in Lake County, Illinois, is $285,000.
c. The most common shoe size sold in Macy's department store is 8.
d. The average yearly wage in a company is $53,000.
e. The average age of new faculty members at the University of Illinois is 28.3.
f. The average weight of adult men is 169 pounds.

2. Describe which measure (mean, median, or mode) was used in each situation (or state that it could have been either of two measures).
   a. Half of the workers in the company make more than $30,000, and half make less.
   b. The average number of children per family in Champaign, Illinois, is 1.9. (*Hint:* The only possible values are integers. If so, how could 1.9 occur as the average?)
   c. More people prefer red cars than any other color.
   d. The average age of college students is 20.3 years.
   e. More people vote Republican in Lake County, Illinois, than Democratic or Independent.

3. Three students, Bruce, Ross, and Carmen, are taking a statistics course. Their scores on five exams are given in the following table:

	Exam I	**Exam II**	**Exam III**	**Exam IV**	**Exam V**
Bruce	84	88	68	68	95
Ross	81	79	81	79	81
Carmen	79	80	78	90	80

Each student wants to claim the best performance in the class. Compute the mean, median, and mode for each student. Which measure would each use to compare the three students to prove his or her claim of being the best? Which student's claim is clearly without validity?

4. Three instructors are comparing scores on their finals. Each instructor had 21 students.
   - In Class A, one student received a score of 40 points, one received a score of 60 points, and the rest received a score of 50 points.
   - In Class B, one student received a score of 40 points, one student received a score of 41 points, one student received a score of 42 points, and so forth, all the way through 60 points.
   - In Class C, 10 students received a score of 40 points, one student received a score of 50 points, and 10 students received a score of 60 points.

   The following dotplots illustrate the three distributions.
   a. True or false: Each of the three distributions of scores is symmetric around 50.
   b. Determine the median score for each of the three classes.
   c. Explain why the mean score for each class must equal the median score.

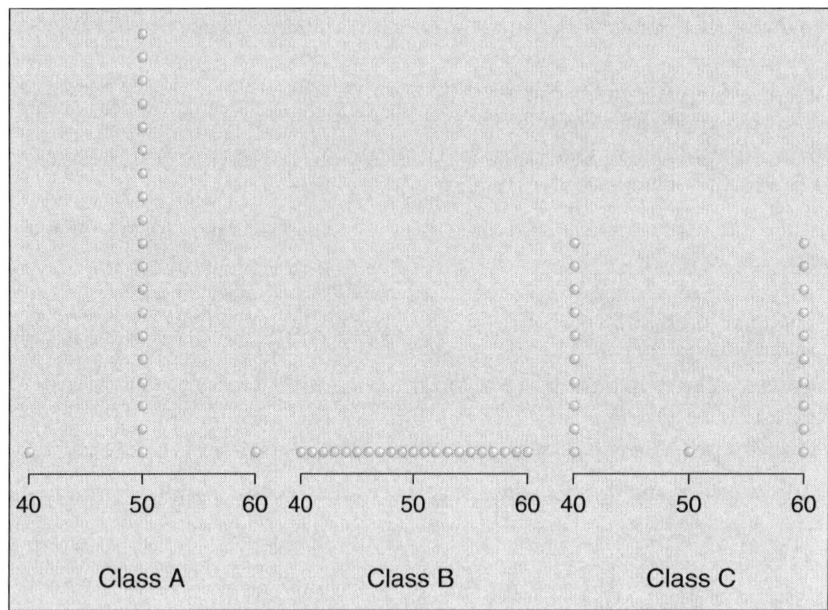

5. The mean salary for 920 Major League Baseball players in 1999 was $1,567,873. What was the total salary for all the 920 Major League Baseball players?

6. A physician's office reported that the mean cost of a physical exam was $150. If 90 patients were examined last month, how much money did the physician collect on these exams?

7. According to *Monitoring the Future,* a study published by the University of Michigan Institute for Social Research and National Institute on Drug Abuse, the following data represent the percentages of high school seniors in the United States who had ever used marijuana or hashish:

1990	1991	1992	1993	1994	1995	1996	1997	1998	1999	2000	2001	2002	2003	2004
40.7	36.7	32.6	35.3	38.2	41.7	44.9	49.6	49.1	49.7	48.8	49.0	47.8	46.1	45.7

a. Compute the mean and the median percentage of high school seniors using marijuana or hashish.
b. Which is larger and why?

8. The following represent actual values from a sample of white blood cell counts (thousands of cells/microliter) for 25 different patients in a northern Illinois hospital:

4.1	19.5	7.0	7.0	1.5	13.9	4.5	4.2	10.2	6.0	12.9	25.3	6.2
5.6	10.0	6.7	54.1	12.9	3.2	5.5	11.0	9.8	9.9	72.6	50.3	

a. Construct a histogram using the intervals 0 to 5.0–, 5.0 to 10.0–, 10.0 to 15.0–, ...
b. Compute the mean and median for these counts. Is the distribution symmetric, skewed to the left, or skewed to

the right? Is the location of the mean relative to the median represented by one of the graphs in Figure 2.6? If so, which one?

c. Compute the mean and median if the values 19.5, 25.3, 54.1, 72.6, and 50.3 are removed. How do the values of the mean and median compare with those of part (*b*)?

9. Suppose a doctor measures a patient's diastolic blood pressure on five different occasions and gets the following results:

85	79	93	82	86

a. Compute the mean and median.
b. Suppose the doctor inadvertently recorded 97 instead of 79. What would the mean and median be now?
c. If the doctor was having a really bad day, the blood pressure might have been recorded as 790 instead of 79. In that case, what would the values be for the mean and median, assuming that the calculation is automatically done by computer and that nobody has noticed this absurd error?
d. What can you say about the influence of extreme values on the mean and median?

10. According to *Consumer Reports,* April 1995, the following represents the manufacturer's ratings of electrical power used (in amps) for each of 22 upright vacuum cleaners.

7.8	7	12	10	12	7	7.2	7.3	11	7	12
11	9.5	9	9	10	7.2	9.5	11	10	7.5	10

a. Construct a stem-and-leaf plot for the data.
b. Compute the mean and median. Are the two values close? Explain your answer in terms of the amount of skewness that seems to exist in the data.
c. Suppose the four values of 10 had been inadvertently recorded as 19 instead. Now what would the mean and median be? Which measure would be most affected by the change?

11. A study in Switzerland examined the number of cesarean sections in deliveries of babies performed in a year by 15 doctors.

27	50	33	25	86	25	85	31
37	44	20	36	59	34	28	

a. Draw a stem-and-leaf plot of the data. Do you see any potential outliers?
b. Compute the mean and median.
c. Remove the two largest values from the data set and compute the mean and median again. What can you say about the resistance of each measure to extreme values?

12. In 1997 the mean and median values of the top 1% of all incomes in the United States were \$330,000 and \$675,000, in some order. Assume this top 1% is skewed to the right, as in fact it is. Which of the values represents the mean and which represents the median?

13. Repeat Exercise 10 of Section 2.1, noting that your understanding of the exercise is now much greater.

14. Below is sketched the age distribution for the people who died of heart disease in 1996.
    a. Is the distribution skewed to the left or to the right?
    b. Which is greater, the mean or the median age?

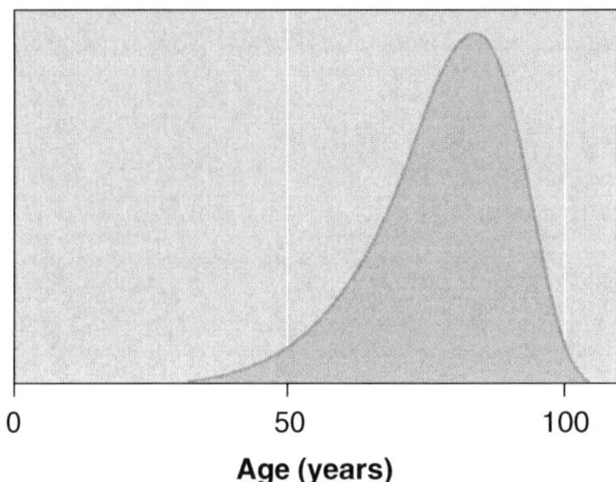

15. Use Newcomb's data in Example 2.8 to estimate the speed of light. *Hint:* There is more than one reasonable way to proceed (mean versus median, one versus two outliers removed). Note also that the distance the light travels is 2×3721 meters.

2.3 MEASURING THE SPREAD OF A DATA SET: THE STANDARD DEVIATION

One mystery that has fascinated baseball fans for half a century is the disappearance of the so-called .400 hitter. A baseball player's batting average for a season is his total number of hits divided by his number of times at bat. Between 1901 and 1930, seven players had season batting averages exceeding .400. In 1941 Ted Williams hit .406. Since 1941, however, no player has managed to hit above .400. Why is that? Are there no great hitters anymore? Williams himself concluded that today's players have plenty of power but lack finesse. In other words, today's players are not smart enough to hit above .400.

Harvard's Stephen Jay Gould had a different explanation for the disappearance of the .400 hitter. In the book *Full House*, he looks at the distribution of batting averages for all Major League Baseball players in a given year.

bell-shaped histogram

mean

spread

standard deviation

Figure 2.10 displays the distribution of the batting averages of 263 Major League Baseball players in 1985 via a histogram. The histogram is without outliers and is roughly bell-shaped, as shown by the superimposed idealized histogram curve. A roughly **bell-shaped histogram** can be effectively described using just two numbers (two statistics): its center, indicated by the **mean** $\overline{x}$, and the **spread** of the data around the mean, measured by the **standard deviation**, denoted s. We will shortly learn how to compute s, but first we learn how to interpret s. The two idealized histograms in Figure 2.11 show that *the more spread out a roughly bell-shaped data set is, the larger its standard deviation is.* Beyond this fact, we need to understand in detail what information about the spread of a data set we get from the specific value of s. For example, what does $s = 0.029$ tell us about the spread of the batting averages around $\overline{x} = 0.263$ in Figure 2.10?

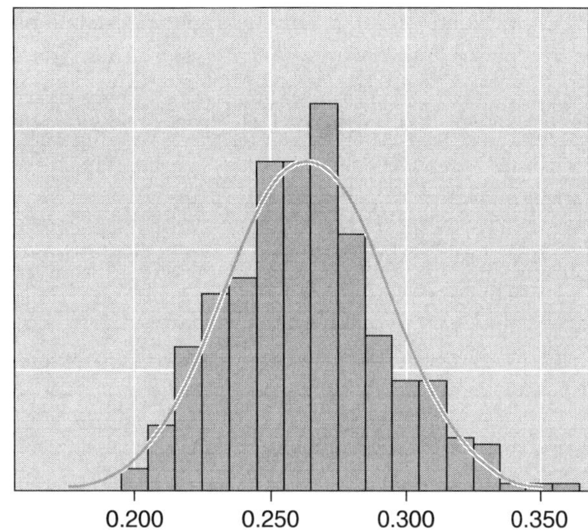

Figure 2.10 Batting averages of 263 Major League Baseball players, 1985.

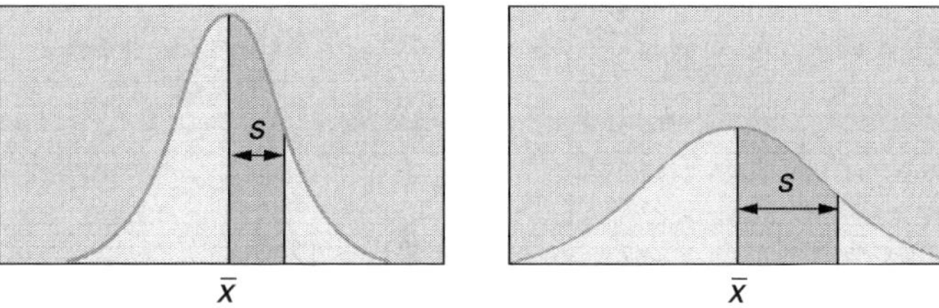

Figure 2.11 Two bell-shaped idealized histograms showing the mean $\bar{x}$ and the standard deviation s.

Empirical Rule for bell-shaped distributions

If the histogram of a data set roughly follows a bell curve and the data set is without outliers, then a majority of the observations will be at most one standard deviation away from the mean. Very few observations will be more than two standard deviations away from the mean. The **Empirical Rule for bell-shaped distributions** (or 68-95-99.7% Rule) makes this quantitatively precise.

Empirical Rule for Roughly Bell-Shaped Histograms Without Outliers (or 68-95-99.7% Rule)

1. About 68% of the data will lie within 1 standard deviation (in either direction) of the mean.
2. About 95% of the data will lie within 2 standard deviations of the mean.
3. About 99.7% of the data (for all practical purposes, all of the data!) will lie within 3 standard deviations of the mean.

Note: Because many data sets are roughly bell-shaped, this rule is widely useful.

Figure 2.12 illustrates the Empirical Rule. For example, for the batting averages in Figure 2.10, $\overline{x} = 0.263$ and $s = 0.029$, so

$$\overline{x} - s = 0.263 - 0.029 = 0.234$$

$$\overline{x} + s = 0.263 + 0.029 = 0.292$$

It turns out that 178 of the 263 players had batting averages between .234 and .292, which is 67.7%—very close to the 68% predicted by the Empirical Rule (Figure 2.12).

$$\overline{x} - 2s = 0.263 - 0.058 = 0.205$$

$$\overline{x} + 2s = 0.263 + 0.058 = 0.321$$

Of the 263 players, 251 had batting averages between .205 and .321, which is 95.4%—very close to the 95% predicted by the Empirical Rule.

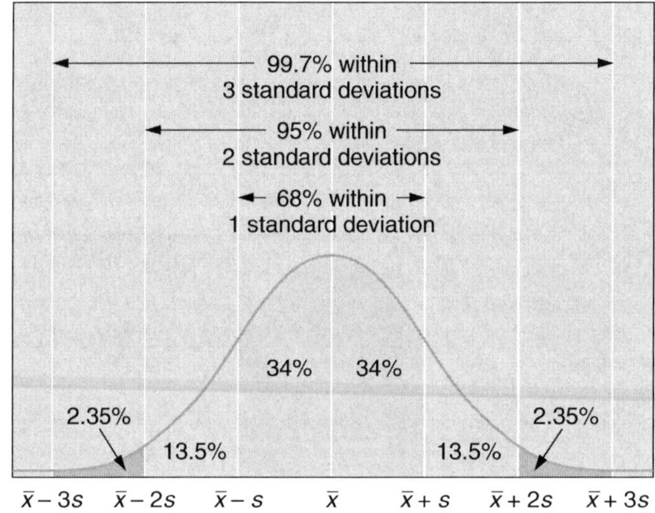

Figure 2.12 The Empirical Rule (or 68-95-99.7% Rule) for bell-shaped histograms without outliers.

$$\overline{x} - 3s = 0.263 - 0.087 = 0.176$$

$$\overline{x} + 3s = 0.263 + 0.087 = 0.350$$

Of the 263 players, no player hit below .176, two players hit above .350, and 261 players had batting averages between .176 and .350, which is 99.2%—very close to the 99.7% predicted by the Empirical Rule.

Many sportswriters have speculated that the reason for the disappearance of the .400 hitter in baseball is that tougher competition and more distractions have tilted the balance between hitting and pitching against the hitters. It is true that pitching has improved dramatically over the last 50 years, but Gould argues that the balance has not tilted. If it had, then the mean batting average $\overline{x}$ should have declined, and that has not happened. Gould notes that the mean batting average has been fairly steady at .260 for over 100 years. So what is going on? Gould argues that as baseball has come of age, *both* pitching and hitting have improved to near the optimal level, maintaining a balance between hitting and pitching.

As overall play has improved, the difference between the highest and the lowest season batting average has shrunk: The best hitters today consistently face much better-trained pitchers than their counterparts did 75 years ago. At the same time, the worst hitters today are much better than their counterparts 75 years ago. The result is that the bell curve we are using to describe the distribution of the batting averages in a given year has shrunk around its mean: As an examination of batting average data over the years shows, the standard deviation has decreased from around 0.05 in the 1870s to around 0.03 today, a 40% drop. Figure 2.13 compares the distribution of batting averages today (solid curve with its histogram included) with the distribution of batting averages at the beginning of the twentieth century (dashed curve with the histogram it approximates not included).

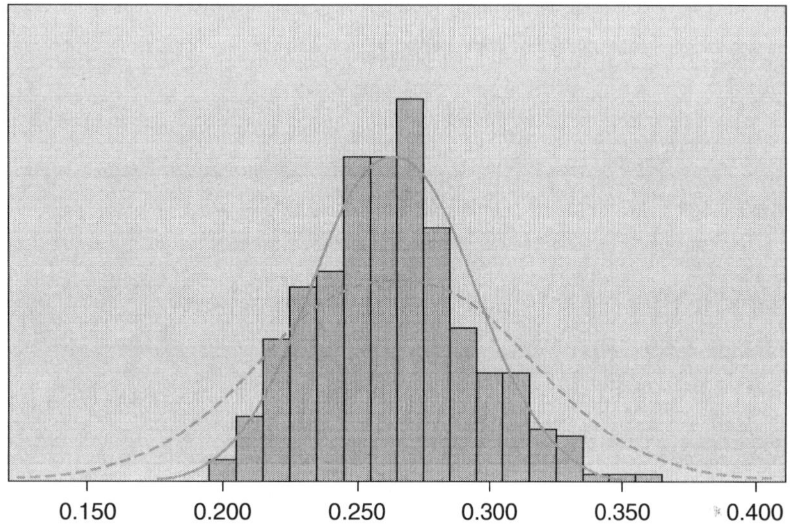

Figure 2.13 Distribution of batting averages in 1985 (solid curve) and batting averages at the beginning of the twentieth century (dashed curve). For the past 50 years, there has been less variation (spread) than there was 100 years ago.

We can now explain the disappearance of the .400 hitter in terms of the 68-95-99.7% Rule: A hundred years ago a hitter had to score 3 standard deviations above the mean to hit above .400. According to the 68-95-99.7% Rule, each year roughly $100\% - 99.7\% = 0.3\%$, or 3 out of 1000 batting averages, will be more than 3 standard deviations away from the mean. Half of these, or roughly 1 out of 700 batting averages, will be more than 3 standard deviations above the mean. So at the beginning of the twentieth century, it was not surprising that a couple of times per decade, some player hit over .400. Today, however, a hitter has to score more than 4 standard deviations above the mean to hit above .400, and that happens so very rarely that we might have to wait 100 years for just one to occur. This example illustrates that in order to understand a bell-shaped distribution, we need to know both the center and the variation (spread) of the data.

Variation is also the issue in the next example. Many banks used to have their customers wait in separate lines at each teller's window. It frustrated the customers to have to guess which line would move faster. (And the fastest lines by chance often tend to move much faster than the slow lines!) Therefore the banks switched to a single waiting line for all the tellers. The change neither sped up nor slowed down the tellers, so the mean waiting time for the customers did not change. Because there was no longer a fast line or a slow line, however, everyone waited in line for about the same amount of time. In other words, the variation in waiting time from person to person was reduced, and the customers were spared a lot of frustration. The new standard deviation for the customers' waiting time will be a lot less, but the new mean waiting time will be about the same.

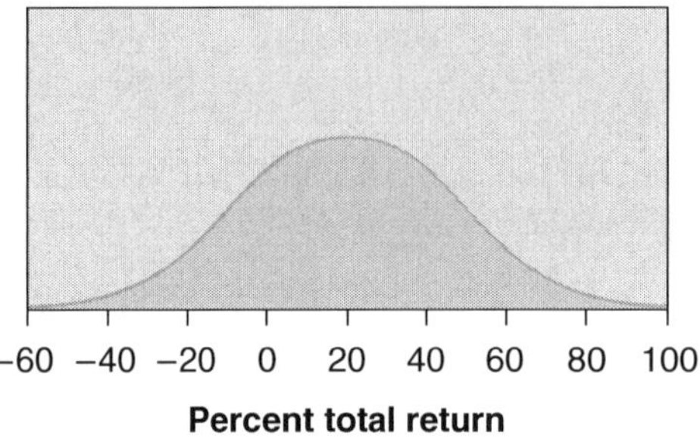

Figure 2.14 Distribution of percent total NYSE return.

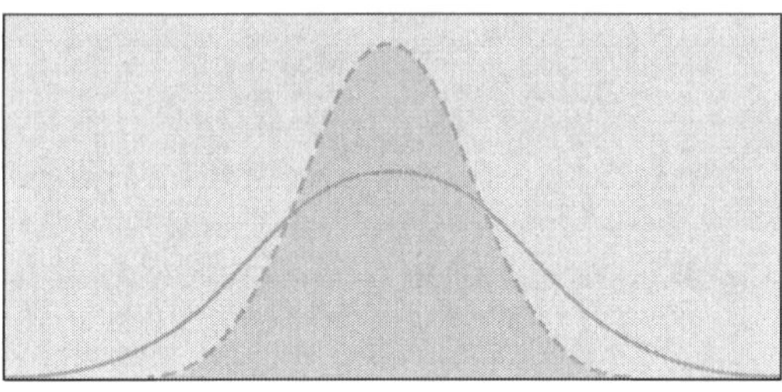

Figure 2.15 Distribution of percent returns on one-stock portfolios (solid curve) and diversified portfolios (dashed curve).

Consider next the (idealized) distribution of percent total return for all common stocks on the New York Stock Exchange (NYSE) over a recent year (see Figure 2.14). The total return on a stock over a year is the change in its market price (up or down) plus any dividend payments made. It is expressed as a percentage of the beginning price. A return less than zero means a loss. The variation in the returns of all stocks over a year reflects the "firm-specific risk" or "unique risk" of holding one individual firm's stock. One could easily lose 30% or gain 70%, going by Figure 2.14. The risk can be reduced by diversification, that is, by splitting one's investment into a portfolio of 20 to 30 stocks. The risk that remains even after extensive diversification is called "market risk." It basically represents the potential variation of the stock market as a whole from one year to the next and is also called "systematic risk." It comes from major economic factors such as the business cycle (expansion, depression, recession, etc.), the inflation rate, and interest rates.

Figure 2.15 illustrates that fully diversified portfolios (dashed curve) have about half the risk (variation) of one-stock portfolios (solid curve). Can you imagine why some people would prefer a one-stock investment strategy while most would prefer a portfolio strategy?

For data sets with roughly bell-shaped histograms and no outliers, we can use the 68-95-99.7% Rule to roughly estimate $\bar{x}$ and s from the histogram. A casual inspection of the histogram won't tell us the exact values of $\bar{x}$ and s, but we can get approximate values and thus be able to apply the Empirical Rule even if we only have the histogram to work with.

To illustrate how we do this, consider the histogram in Figure 2.16, which represents the heights of a couple of thousand college women. Since $\bar{x}$ is near the center of a symmetric histogram, we estimate that $\bar{x} \cong 64.5$ inches. We will use the 95% rule to obtain an approximate value of s. According to the rule, as Figure 2.12 shows, approximately 95% of the data lie in the interval from $\overline{x} - 2s$ to $\overline{x} + 2s$. It follows that if we can determine an interval centered at $\overline{x}$ that contains roughly 95% of the data, then the length of this interval should be roughly $4s$. We shall apply this method to Figure 2.16.

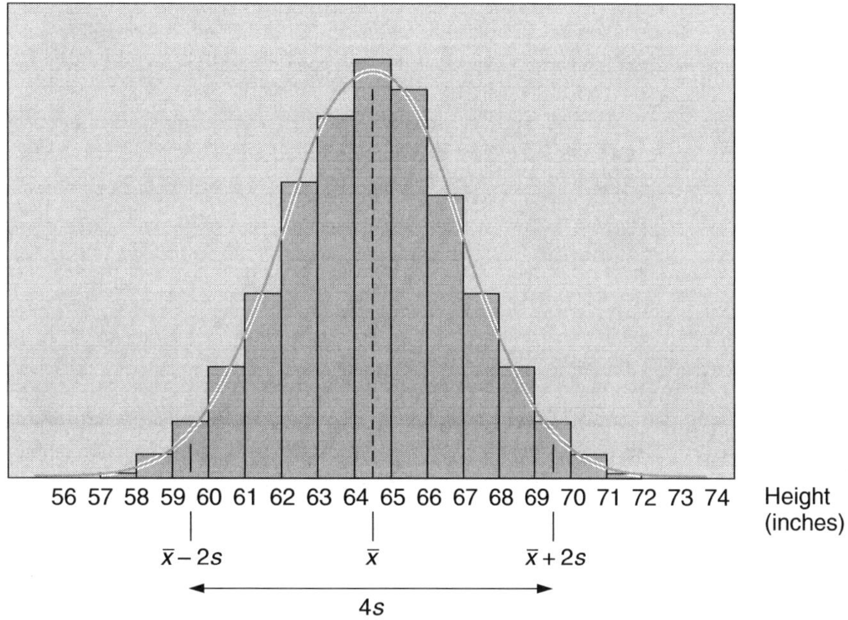

Figure 2.16 Estimating $\overline{x}$ and s from a histogram of women's heights.

Clearly, the combined area of the histogram rectangles starting at 58 inches and ending at 71 inches is more than 95% of the total area of the histogram. So more than 95% of the data is between 58 inches and 71 inches. Therefore, the computation

$$4s = 71 - 58 = 13,$$

which yields $s = 3.25$, gives us too large a value for s.

On the other hand, the combined area of the histogram rectangles starting at 61 inches and ending at 68 inches is less than 95% of the total area of the histogram: The combined area of the rectangles below 61 inches and those above 68 inches is greater than 5%. (It looks closer to 10%.) So less than 95% of the data is between 61 inches and 68 inches. Therefore the computation

$$4s = 68 - 61 = 7,$$

which yields $s = 1.75$, gives us too small a value for s.

We can get a rough estimate of s by "splitting the difference" between a reasonable underestimate, 1.75 inches, and a reasonable overestimate, 3.25 inches. We get

$$s \approx 2.5 \text{ inches.}$$

This is only a rough approximation. However, we can say with confidence that s is between 2 and 3 inches. Later in this section we will see how to compute s exactly from the data.

We shall now consider some numerical examples.

Example 2.9

Two imaginary patients, Dylan and Laverne, are hospitalized in Mercy Community Hospital. In order to monitor their conditions, the doctor has the nurse check their pulse rates three times a day. The results of one day's readings are shown in Table 2.3.

If we order the data, we see that the median pulse rate of each patient is 72. In fact, if we compute the mean for each, we see that each mean is also 72. Obviously there is more that needs to be described than the centers.

The basic difference between the two patients is the variation of the two data sets. Laverne's readings are clustered together much more closely than are Dylan's.

Table 2.3 Pulse Readings

	Morning	Afternoon	Evening
Dylan	72	58	86
Laverne	70	74	72

Why should we be concerned about variation? We sometimes think of variation as a measure of consistency. Here, the large variation or inconsistency in readings for Dylan may indicate a medical problem that needs attention.

❒ ❒

As another example where variation is important, consider an aerospace company that manufactures parts for the space shuttle. The dimensions of the parts must fall within specific ranges, or tolerances, to ensure that everything fits and works properly. If a part is too large, it will not fit; if it is too small, it will not work. Many of the supplied parts could fail to meet these requirements, even if they average out to their target value. Their actual measurements must also be very close to this mean value (thus making the mean value truly typical of almost all of the values). That is, variation must be acceptably small.

range

How do we obtain a measure that tells us the spread or variation of a data set? The simplest measure is the **range**, which is the difference between the highest value and the lowest value.

$$\text{Range} = (\text{highest value}) - (\text{lowest value})$$

The range is often used in weather reports, which give us the low and high temperatures of the day. For the two data sets in Example 2.9, the ranges are easily computed:

$$\text{Dylan: Range} = 86 - 58 = 28$$
$$\text{Laverne: Range} = 74 - 70 = 4.$$

Although the range is easy to obtain, it is generally not used outside weather reporting. The reason is that since the range is computed from the two most extreme values, the range often exaggerates the spread of large data sets. For example, the vast majority of the college women in Figure 2.16 were between 60 inches and 69 inches tall, a spread of 9 inches. But the tallest woman was 73 inches and the shortest 56 inches. So the range was 17 inches. The range gives us the misleading impression that college women vary more in height than they typically do. We are (usually) interested in measuring the *typical variation*, rather than the *most extreme variation*. The range does not tell us the typical variation. The range is also very sensitive to extreme values, because it is computed using *only* the most extreme values. Therefore, any extreme, and hence influential, value that is in error could totally change our measure of the spread when using the range. The range is not a resistant measure of spread. We will not use the range as a measure of spread.

deviation from the mean.

Another approach that we could take, which leads us to the formula for s, is to see how far away each data value is from the mean. Let us try this approach with the data from Example 2.9. The mean, $\overline{x}$, was computed to be 72 in each case. Then we find $x - \overline{x}$, called the **deviation from the mean,** for each data point x, shown in the mean Table 2.4.

We could attempt to get a measure of spread by adding up the deviations. But in each case the sum of the deviations is 0. In fact, it can be shown that this will always be the case. Therefore, we do not obtain a good measure of spread by adding or averaging the deviations—they always add to 0 since the positive deviations offset the negative deviations.

Table 2.4 Deviations in Pulse Readings

Dylan		Laverne	
x	$x-\overline{x}$	x	$x-\overline{x}$
72	0	70	−2
58	−14	74	2
86	14	72	0
Sum	0	Sum	0

variance
standard deviation

To overcome this problem, we square each deviation. The **variance** of the data is essentially the average of the squared deviations, and the **standard deviation** is the square root of the variance.

We compute the variance s^2 of n observations $x_1, x_2, \ldots, x_n$, with mean $\overline{x}$ as follows:

$$s^2 = \frac{(x_1-\overline{x})^2 + (x_2-\overline{x})^2 + \cdots + (x_n-\overline{x})^2}{n-1}.$$

Using even more compact notation, we write

$$s^2 = \frac{\Sigma(x_i-\overline{x})^2}{n-1}.$$

The standard deviation s is the square root of the variance s^2:

$$s = \sqrt{\frac{\Sigma(x_i-\overline{x})^2}{n-1}}.$$

Why do we divide by $n-1$ and not n? There is a technical reason, but mainly we are just following tradition. We note that if the number of observations n is at least 30, then it makes little difference whether we divide by n or $n-1$.

You can use a computer or a statistical calculator to compute the standard deviation from keyed-in data. (If your calculator gives you a choice between dividing by n and dividing by $n - 1$, make sure that you choose $n - 1$.) But if you have to compute s^2 yourself, then you should use the following *computational formula for* s^2, which is much quicker than using the definition of s^2 (if you have more than five observations) and is less prone to accumulating round-off errors:

$$s^2 = \frac{\sum x_i^2 - (\sum x_i)^2/n}{n-1}.$$

To illustrate how we can compute the standard deviation s, let us return to the Dylan and Laverne data in Example 2.9. The squared deviations $(x - \overline{x})^2$ and their sums $\sum(x - \overline{x})^2$ are shown in Table 2.5. The standard deviations are

$$\text{Dylan:}\quad s = \sqrt{\frac{392}{2}} = \sqrt{196} = 14; \qquad \text{Laverne:}\quad s = \sqrt{\frac{8}{2}} = \sqrt{4} = 2.$$

Table 2.5 Squared Deviations in Pulse Readings

Dylan			Laverne		
x	$x-\overline{x}$	$(x-\overline{x})^2$	x	$x-\overline{x}$	$(x-\overline{x})^2$
72	0	0	70	−2	4
58	−14	196	74	2	4
86	14	196	72	0	0
	Sum	392		Sum	8

Example 2.10

To illustrate how we use the computational formula for s^2 we shall compute the standard deviation s of the following list of numbers:

1	3	3	4	9

We compute $\sum x$, the sum of the data, and $\sum x^2$, the sum of the squares:

	x	x^2
	1	1
	3	9
	3	9
	4	16
	9	81
Sum	20	116

We note that the mean is

$$\bar{x} = 20/5 = 4.$$

The variance is

$$s^2 = \frac{116 - (20)^2/5}{4} = \frac{116 - 80}{4} = 9.$$

It follows that the standard deviation is

$$s = \sqrt{9} = 3.$$

❐ ❐

The standard deviation is not a resistant measure of spread. It is easily distorted by outliers, as the next example illustrates.

Example 2.11

Consider Newcomb's 66 repeated measurements in Example 2.8. If we exclude the two outliers −44 and −2, then the remaining 64 observations have a roughly bell-shaped distribution (see Figure 2.9). We will use the computational formula for s^2 to save time. For these 64 observations we get

$$\sum x = 1776 \quad \text{and} \quad \sum x^2 = 50{,}912,$$

giving

$$\bar{x} = \frac{1776}{64} = 27.75 \quad \text{and} \quad s = \sqrt{\frac{50{,}912 - (1776)^2/64}{63}} = 5.08.$$

The 64 observations obey the 68-95-99.7% Rule reasonably well: In the interval $\overline{x} \pm s$, we find 46 out of 64 observations, or 72%—which is not too far from 68%. In the interval $\overline{x} \pm 2s$, we find 60 out of 64 observations, or 94%—which is close to 95%.

If we include the two outliers, then we get

$$\sum x = 1730 \quad \text{and} \quad \sum x^2 = 52{,}852,$$

giving

$$\overline{x} = 26.21 \quad \text{and} \quad s = \sqrt{\frac{52{,}852 - 1730^2/66}{65}} = 10.7.$$

Note that the inclusion of the two outliers doubled the standard deviation. Indeed, the standard deviation is even more vulnerable to the influence of outliers than the mean. Also note that if we include the two outliers, then the data set *does not* obey the 68-95-99.7% Rule: In the interval 26.21 ± 10.7, we find 61 out of 66 observations, or 92%—which is far from 68%.

Linear Transformations and Change of Scale; The Effect of Changing Units

Strictly speaking, in Examples 2.8 and 2.11 we found the mean, the median, and the standard deviation for the values in Table 2.2—not for Newcomb's passage times. We can easily determine those now. Because

$$\text{passage time in nanoseconds} = 24{,}800 + (\text{value in Table 2.2})$$

we get

$$\begin{aligned} \text{mean passage time} &= 24{,}800 + \ (\text{mean value for Table 2.2}), \\ \text{median passage time} &= 24{,}800 + \ (\text{median value for Table 2.2}), \text{ and} \\ \text{standard deviation of passage times} &= \ \text{standard deviation of values in Table 2.2.} \end{aligned}$$

Adding 24,800 to each value in Table 2.2 adds 24,800 to the mean and the median but leaves the standard deviation unchanged. Adding 24,800 to each value shifts the center of the distribution (mean or median) by that amount but has no effect on the spread.

What would happen to the mean, the median, and the standard deviation if the passage times were recorded in seconds instead of nanoseconds? Because

$$\text{passage time in seconds} = (\text{passage time in nanoseconds}) \times 10^{-9},$$

we get the new mean, median, and standard deviation simply by multiplying the old ones by 10^{-9}.

We summarize these important rules:

linear transformation

The Effect of a Change of Units (Linear Transformation) on the Mean, Median, and Standard Deviation

1. Adding the same number to each value in a data set or subtracting the same number from each value in the data set shifts the center (mean or median) by the same amount while the spread (standard deviation) remains unchanged.
2. Multiplying each value by a positive constant also multiplies the mean, median, and standard deviation by the same constant.
3. These two rules can be applied simultaneously when the change in units is a linear transformation that involves both a shift (adding or subtracting) and a scale change (multiplying).

In the sciences when one changes the unit of measurement, as frequently happens, this change of scale often requires application of the foregoing rules. Examples of unit changes include feet to inches, centimeters to inches, pounds to kilograms, miles per gallon to kilometers per liter, and so on.

Example 2.12

In 1987 the maximum temperature in Mayville, North Dakota, was recorded each day. For the year, the mean of the daily maximum temperatures was 58.21 °F. The median was 60 °F, and the standard deviation was 24.39 °F. The linear transformation for converting temperatures from Fahrenheit to Celsius is

$$°\mathrm{C} = \frac{5}{9}(°\mathrm{F} - 32).$$

Thus, first 32 is subtracted from each Fahrenheit temperature reading, and then each difference is multiplied by the positive constant $5/9$. If each of the 365 daily maximum temperatures is converted from Fahrenheit to Celsius, we get

$$\begin{aligned}
\text{new mean} &= \frac{5}{9}(\text{old mean} - 32)\ °\mathrm{C} \\
&= \frac{5}{9}(58.21 - 32)\ °\mathrm{C} = 14.56\ °\mathrm{C} \\
\text{new median} &= \frac{5}{9}(\text{old median} - 32)\ °\mathrm{C} \\
&= \frac{5}{9}(60 - 32)\ °\mathrm{C} = 15.56\ °\mathrm{C} \\
\text{new standard deviation} &= \frac{5}{9}(\text{old standard deviation}) \\
&= \frac{5}{9} \times 24.39\ °\mathrm{C} = 13.55\ °\mathrm{C}
\end{aligned}$$

Section 2.3 Exercises

Note: A calculator is needed for many of the exercises.

1. Compute the mean $\overline{x}$ and use the computational formula (see Example 2.10) to compute the standard deviation s of each of the following two lists of numbers:

List A:	1	1	2	2	2	2	4	
List B:	1	2	3	4	4	5	6	7

2. The following are the dotplots of four data sets labeled A, B, C, and D.

A

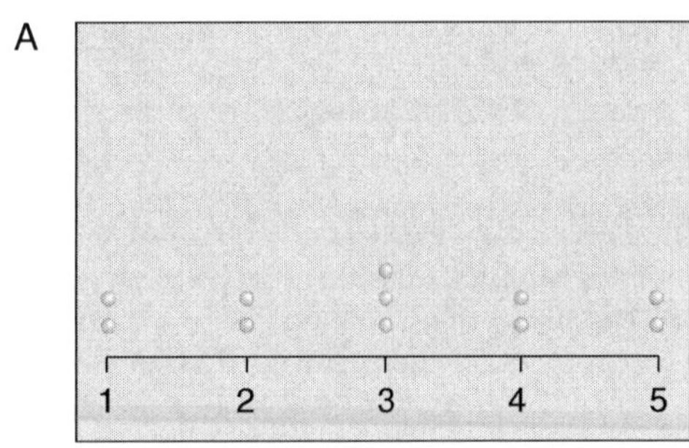

B

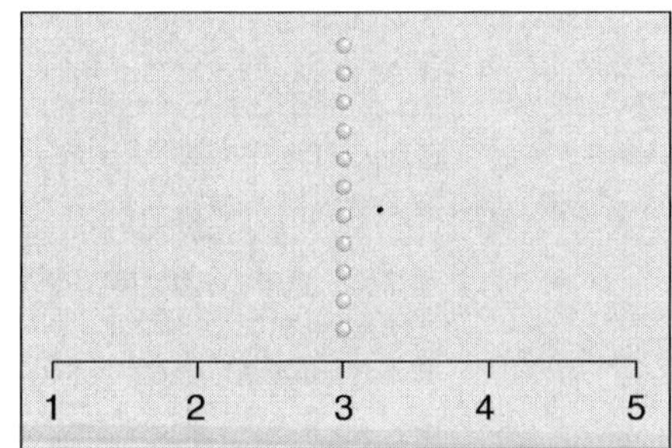

C

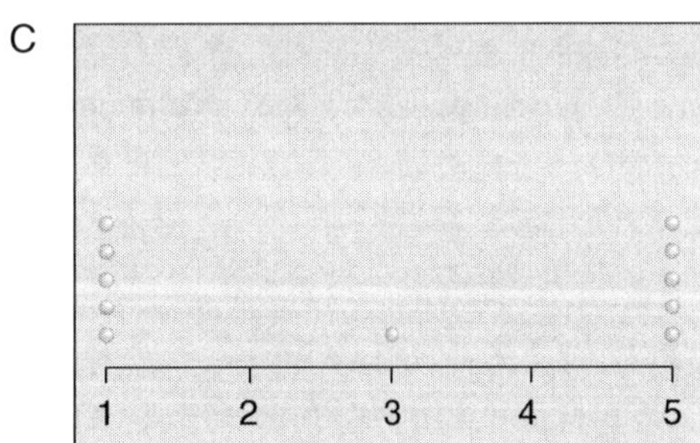

D

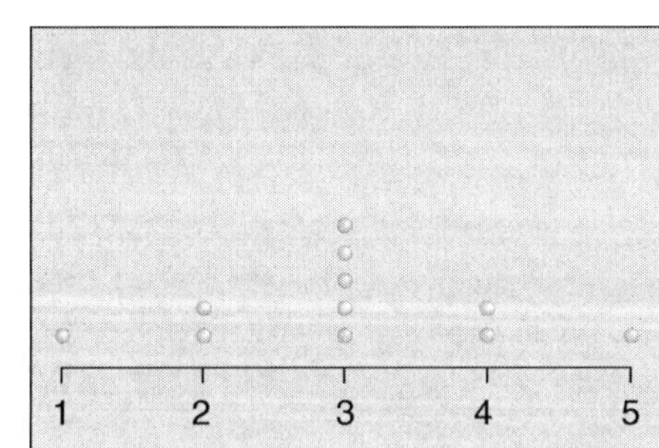

a. True or false: Each data set has mean 3.
b. Which of the four data sets has the smallest standard deviation? (No computation is necessary.)
c. Which of the four data sets has the largest standard deviation? (No computation is necessary.)

3. Consider again the three data sets in Exercise 4 of Section 2.2. The following are the dotplots for the three distributions.

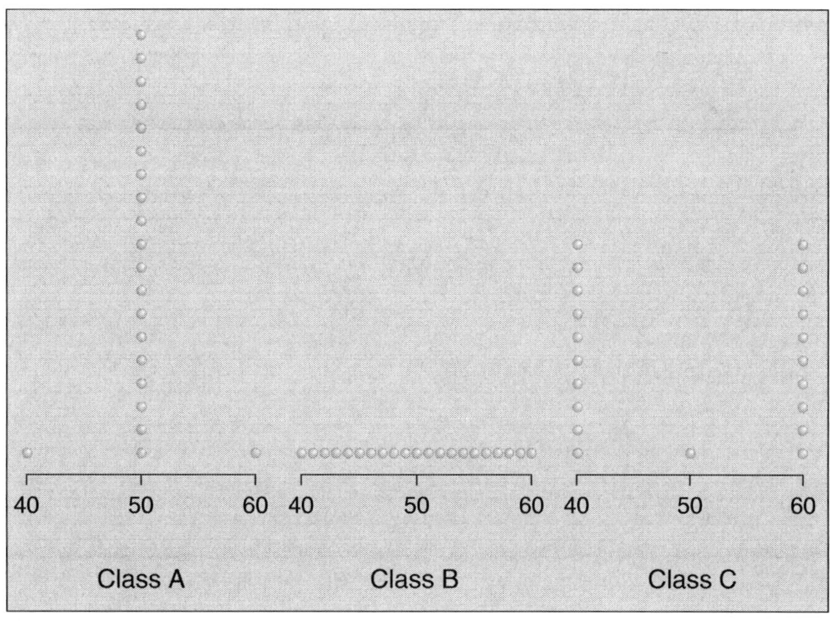

Each of the three classes has mean score 50.

a. Measuring spread by the range, which class has the largest spread of scores? (Choose one option.)

Class A Class B Class C All three are equal

b. Measuring spread by the standard deviation, which class has the largest spread of scores? (Choose one option. No computation is necessary.)

Class A Class B Class C All three are equal

c. Measuring spread by the standard deviation, which class has the smallest spread of scores? (Choose one option. No computation is necessary.)

Class A Class B Class C All three are equal

4. A math instructor gave an exam. A sample of 10 exam grades is given here:

62	54	76	69	43	77	61	73	81	56

a. Compute the mean, median, and standard deviation.

b. Because the test was hard, the instructor decided to adjust the grades by adding 10 points to each score. Without recomputing, what will be the values of the mean, median, and standard deviation for the new scores?

5. An instructor gave an exam with a possible perfect score of 33, with the following results:

25	30	22	14	17	29	33	11	20	22

a. Compute the mean, median, and standard deviation.
b. The instructor rescales the scores by tripling each of the scores. Without recomputing, what are the values of the mean, median, and standard deviation now?

6. The United States Environmental Protection Agency has reported the number of days metropolitan areas failed to meet adequate air quality standards in 1988 and 1997.

Area	1988	1997	Area	1988	1997
Atlanta	44	26	New Haven	26	19
Bakersfield	126	55	New York	57	23
Baltimore	60	30	Orange County, CA	56	3
Boston	28	30	Philadelphia	53	32
Chicago	40	9	Phoenix	29	15
Dallas	37	15	Pittsburgh	43	20
Denver	35	0	Riverside, CA	185	106
Detroit	35	12	Sacramento	88	2
El Paso	15	3	St. Louis	44	15
Fresno	110	50	Salt Lake City	16	1
Hartford	39	16	San Diego	123	14
Houston	72	47	San Francisco	1	0
Las Vegas	30	0	Seattle	20	1
Los Angeles	239	63	Ventura	108	44
Miami	8	3	Washington, DC	56	28
Minneapolis	11	0			

a. Compute the mean, median, and standard deviation for each year. What do the results suggest?

7. Produce five data points with a mean of 12 and a standard deviation of 0. Is there any other set of five points that achieves this?

8. *MacWorld* in September 1996 reported these average costs per megabyte of storage on hard drives available for home computers:

1992	1993	1994	1995	1996
\$5.07	\$2.40	\$1.14	\$0.53	\$0.36

a. Compute the mean, median, and standard deviation of costs given above.
b. Do you think that these are appropriate measures to reflect the real story in these data?

9. A baseball pitcher's earned run average (ERA) is the number of runs that opposing teams score per nine innings. The ERAs of various Chicago Cubs pitchers early in the 2000 season, through April 29, were as follows:

3.86	6.45	8.71	0.00	2.77	2.84	8.53	3.48	5.33	5.75	14.46

a. Compute the mean, median, and standard deviation.
b. Suppose you eliminate 14.46 as an outlier. Recompute the mean, median, and standard deviation. Discuss how each has changed and why.

10. An article in *Farming* (September/October 1994) listed the following in-state yearly tuitions plus fees per school for 14 land-grant universities:

1554	2291	2084	4443	2884	2478	3087
3708	2510	2055	3000	2052	2013	2550

a. Compute the mean, median, and standard deviation.
b. Suppose each university decided to increase its tuition and fees by \$500 in 1996. What would the new values of the mean, median, and standard deviation be?
c. We know that tuitions and fees have steadily risen. Suppose each university had increased its tuition and fees for the year 2000 by 10% (multiply the 1994 tuition by 1.1). Now what would the values be for the mean, median, and standard deviation?

11. Is it possible for the standard deviation to be negative? Explain.

12. Consider a sample of five exam grades in each of two classes:

Class I:	90	92	87	88	95
Class II:	50	70	82	55	80

Without doing any calculations, which set of scores has the higher mean? Which set has the larger standard deviation? Why?

13. Briefly describe the difference between what the mean measures and what the standard deviation measures. Explain why reporting both is important.

14. a. Compute the standard deviation for the presidents data in Example 2.5.
b. What percentage of the data lies within one standard deviation of the mean?
c. What percentage of the data lies within two standard deviations of the mean?
d. Is the 68-95-99.7% Rule accurate? Referring to the stemplot in Example 2.5, is this what you expected?

15. a. Compute the standard deviation for Roger Maris's home runs in Example 2.2.
    b. Leave out Maris's exceptional 61 home runs and compute the standard deviation for the remaining nine scores.
    c. Compare the two answers. Is the standard deviation a resistant measure of spread?

16. Scores on IQ tests have a symmetric distribution that is roughly bell-shaped. Reported IQs are scaled so that the distribution has a mean of 100 and a standard deviation of 15. Suppose that a standard IQ test is given to 2000 students. Use the 68-95-99.7% Rule to answer the following questions:
    a. About how many students in the group will have an IQ between 70 and 130?
    b. About how many students in the group will have an IQ between 100 and 130?
    c. About how many students in the group will have an IQ greater than 145? *Hint:* Find the approximate number between 55 and 145 first.

17. The following observations were made of the weights (in kilograms) of 12-year-old boys.

40.96	33.54	37.64	26.82	42.99	44.39	32.04	31.02	38.53	34.20
31.54	34.66	44.81	39.69	40.73	38.73	37.06	38.15	34.85	33.01
44.57	42.96	49.89	50.05	42.11	49.99	54.18	35.13	41.06	37.14
39.58	41.42	48.64	32.82	41.62	38.07	44.61	40.61	38.93	44.10
36.13	52.28	29.71	35.03	50.87	45.82	34.87	49.15	31.00	41.86

    a. Construct a histogram using the intervals 25.0–27.5–, 27.5–30.0–, 30.0–32.5–, ...
    b. Find the mean and the standard deviation.
    c. What percentage of the data lies within one standard deviation of the mean? What percentage of the data lies within two standard deviations of the mean? How does this compare with what we would expect of a normal (bell-shaped) distribution?

18. The questions in this exercise refer to the histograms that follow, which show the distribution of the daily high temperatures in 1987 for Olga, Washington (top); Minneapolis, Minnesota (middle); and Belle Glade, Florida (bottom). All three histograms use the same temperature scale shown at the bottom of the Belle Glade histogram. *Hint*: Figure 2.16 indicates how $\overline{x}$ and s can be estimated from the histogram.
    a. The average high temperature for Olga is closest to which (choose one)?

    $$45\,^{\circ}\text{F} \quad 60\,^{\circ}\text{F} \quad 75\,^{\circ}\text{F} \quad 83\,^{\circ}\text{F} \quad 95\,^{\circ}\text{F}$$

    b. The standard deviation of the high temperatures for Olga is closest to which (choose one)?

    $$1\,^{\circ}\text{F} \quad 3\,^{\circ}\text{F} \quad 5\,^{\circ}\text{F} \quad 10\,^{\circ}\text{F} \quad 20\,^{\circ}\text{F}$$

    c. Is the average high temperature for Minneapolis at least 20 degrees less than, about the same as, or at least 20 degrees higher than the average high temperature for Olga? (Choose one.)
    d. Is the standard deviation of the high temperatures for Minneapolis about half of, about the same as, or about twice the standard deviation for Olga?
    e. Is the average high temperature for Belle Glade at least 20 degrees less than, about the same as, or at least 20 degrees higher than the average high temperature for Olga?
    f. Is the standard deviation for Belle Glade about half of, about the same as, or about twice the standard deviation for Olga?

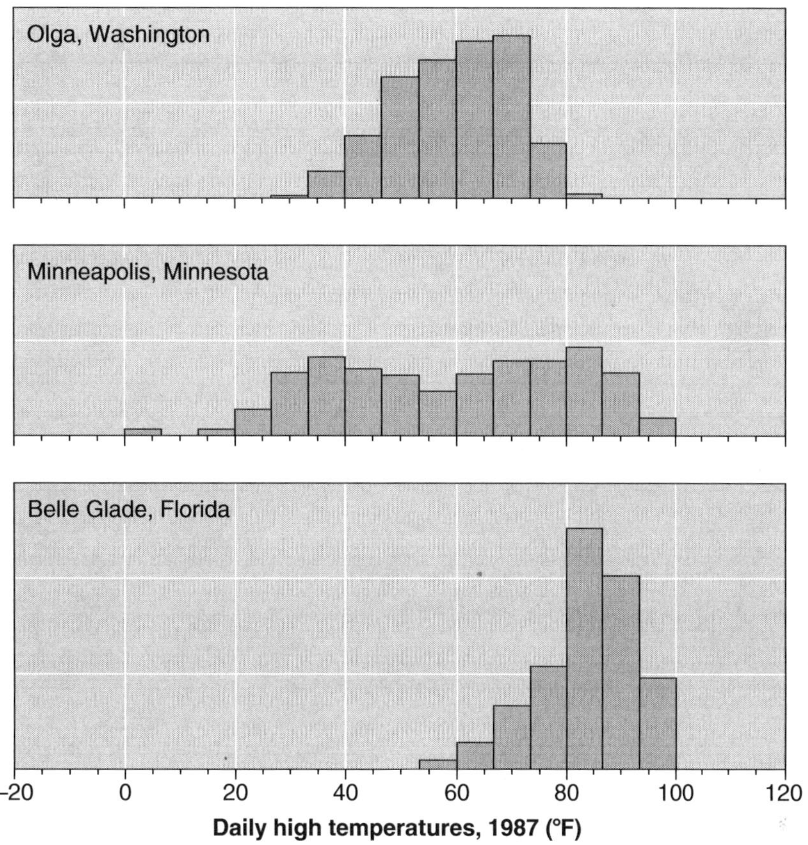

19. a. The average height of the women in Figure 1.13 (at the end of Section 1.4) is closest to which (choose one)?

60″ 62.5″ 65″ 67.5″

(*Hint:* Figure 2.16 indicates how $\overline{x}$ and s can be estimated from the histogram.)

b. The standard deviation of the heights is closest to

0.5″ 1″ 2.5″ 4″ 10″

2.4 THE NORMAL APPROXIMATION FOR DATA

If the histogram of a data set roughly follows a bell curve and the data set is without outliers, then the mean $\overline{x}$ and the standard deviation s summarize the data. All we need to be able to apply the 68-95-99.7% Rule from Section 2.3 is the two numbers $\overline{x}$ and s. We illustrated the 68-95-99.7% Rule using the 1985 baseball batting averages. We shall give two more examples.

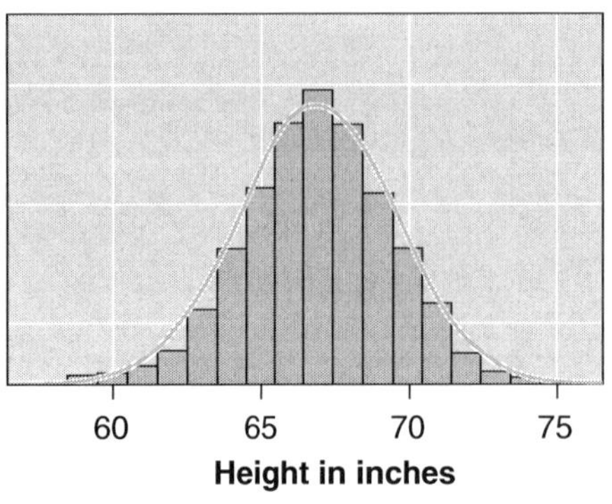

Figure 2.17 Density histogram of the heights of 8585 nineteenth-century British men.

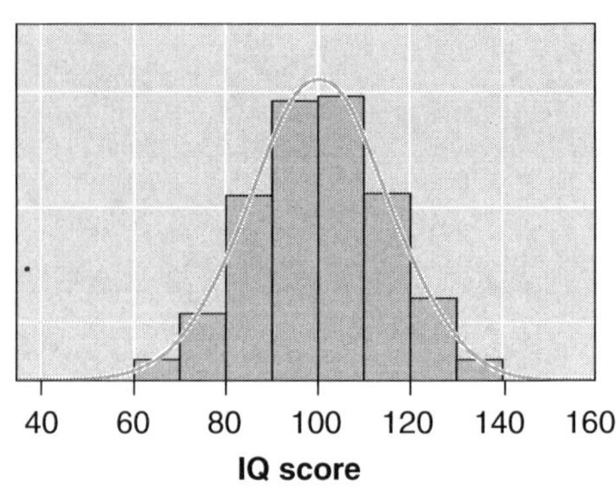

Figure 2.18 Density histogram of 2200 children's Wechsler IQ scores.

Figure 2.17 shows the density histogram of the heights of 8585 nineteenth-century British men (collected by Sir Francis Galton, 1822–1911, a pioneer of statistics) together with the superimposed idealized bell-shaped histogram. Figure 2.18 is the density histogram of 2200 children's IQ scores on a test called the Wechsler intelligence test, together with its idealized histogram. The superimposed bell-shaped curves follow both histograms extremely well.

The Belgian scientist A. Quetelet (1796–1874) first noted that if the heights of a large group of people are measured, the density histogram of these heights will resemble a bell-shaped curve. He also studied the distributions of other physiological measurements, such as weight (in recent years, usually skewed just a bit to the right; can you guess why?), chest girth, and arm length, and found that each approximately follows the same type of bell-shaped distribution.

Because so many physiological data sets are well approximated by bell-shaped distributions, Galton coined the term "normal distributions" for bell-shaped distributions. They are also called Gaussian distributions, after the German mathematician and astronomer Carl Friedrich Gauss (1777–1855), who encountered such distributions in his study of measurement errors in science.

normally distributed

The fact that most physiological characteristics are **normally distributed** (this phrasing is commonly used in describing a bell-shaped data set) has many practical applications. For example, the designer of an airplane cockpit must arrange it so that most pilots are comfortable and can reach all of the controls. Clearly, this requires knowledge of average heights, average arm lengths, and so on, as well as knowledge of the variability around these averages so that most pilots will be accommodated. Since the physiological variables in question are normally distributed, all that the cockpit designers need to know is the mean and the standard deviation of each variable.

Table 2.6 uses the heights displayed in Figure 2.17 and the IQ scores displayed in Figure 2.18 to illustrate the 68% rule. We are also including the baseball data we already looked at in Section 2.3. For each of the three data sets, we have counted the number of observations in the interval from $\overline{x} - s$ to $\overline{x} + s$.

As Table 2.6 shows, 5778 out of the 8585 men in Galton's study, or 67.3%, had heights between $\overline{x} - s$ and $\overline{x} + s$. This is very close to the 68% predicted by the Empirical Rule. The two other data sets also conform to the 68% rule.

Table 2.6 The $(\bar{x} \pm s)$: 68% Rule Empirically Demonstrated

					Between $(\bar{x} - s)$ and $(\bar{x} + s)$		
Data set	$\bar{x}$	s	$\bar{x} - s$	$\bar{x} + s$	**Count**	**Total**	**Proportion**
Batting averages	0.263	0.029	0.234	0.292	178	263	0.677
Heights	67.02	2.564	64.46	69.58	5778	8585	0.673
IQs	100.47	15.189	85.28	115.66	1481	2200	0.673

Table 2.7 illustrates the 95% rule using the same three data sets.

Table 2.7 The $(\bar{x} \pm 2s)$: 95% Rule Empirically Demonstrated

					Between $(\bar{x} - 2s)$ and $(\bar{x} + 2s)$		
Data set	$\bar{x}$	s	$\bar{x} - 2s$	$\bar{x} + 2s$	**Count**	**Total**	**Proportion**
Batting averages	0.263	0.029	0.205	0.321	251	263	0.954
Heights	67.02	2.564	61.89	72.15	8087	8585	0.942
IQs	100.47	15.189	70.09	130.85	2105	2200	0.957

In all three cases, very close to 95% of the observations are between $\bar{x} - 2s$ and $\bar{x} + 2s$. But we can do much more. Using only $\bar{x}$ and s, we can for each of these three data sets (batting averages, male heights, and IQ scores) approximately tell the percentage of observations in *any interval,* not just in the intervals $(\bar{x} - s, \bar{x} + s)$ and $(\bar{x} - 2s, \bar{x} + 2s)$. To do this we first have to introduce **standard scores** or **z-scores**.

standard scores

z-Scores

In high school Yoko scored 29 on the mathematics section of the ACT college entrance exam. Her friend Soledad took the SAT and scored 650 on the mathematics section. Nationwide the ACT scores followed a normal curve with mean 21 and standard deviation 6. The SAT scores were normally distributed with mean 500 and standard deviation 100. If we assume that the ACT and the SAT measured the same kind of ability, who did better?

To compare the two scores we have to convert them into z-scores. The z-score tells us how many standard deviations the original (raw) score is above or below the mean:

$$\textbf{\textit{z}-score} = \frac{\textbf{raw score} - \textbf{mean}}{\textbf{standard deviation}}$$

For Yoko we get

$$z\text{-score} = \frac{29 - 21}{6} = \frac{8}{6} = 1.33.$$

For Soledad we get

$$z\text{-score} = \frac{650 - 500}{100} = \frac{150}{100} = 1.50.$$

Since Soledad's score is 1.50 standard deviations above the mean and Yoko's score is 1.33 standard deviations above the mean, Soledad did better.

Using the Standard Normal Table

Let us return to the Major League Baseball batting averages in 1985 that we discussed in the beginning of Section 2.3 and again in this section. Suppose we want to know the proportion of players whose batting averages were between .225 and .265, for example. Rather than going back to the original data set, counting the number of batting averations between .225 and .265, and then computing the proportion, we will now see that we can estimate this proportion using only the overall mean $\overline{x} = .263$ and the standard deviation $s = 0.029$, plus Table E (in Appendix E) for the **standard normal curve**.

standard normal curve

We have chosen a vertical scale in Figure 2.19 that gives the histogram a total area equal to 1. (This is just the density histogram, with bars of width 0.01 scaled to have total area 1 rather than 100%.) The area under the superimposed normal curve is also equal to 1. The area of each histogram bar equals the proportion of players with batting averages in the corresponding interval on the horizontal axis. The proportion of players with batting averages between .225 and .265 equals the combined area of the four shaded histogram bars in Figure 2.19. Since the superimposed normal curve follows the histogram closely, the sought-after area of the four shaded bars is close to the area under the normal curve from 0.225 to 0.265. This illustrates how we can estimate the proportion of players with batting averages in any interval by finding the area over the interval under the approximating normal curve.

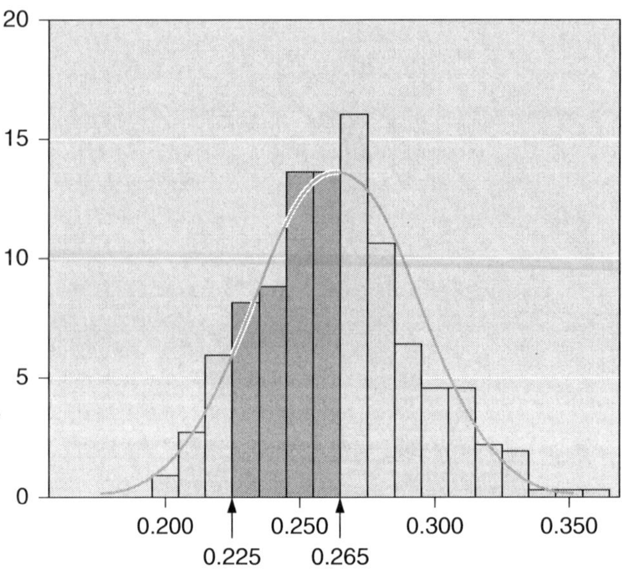

Figure 2.19 Batting averages of 263 Major League Baseball players in 1985, with averages between .225 and .265 emphasized.

Important fact: We can determine the area over an interval under any normal curve in two steps: First we convert the interval endpoints into z-scores. Then we find the area over the z-score interval under the normal curve that has mean 0 and standard deviation 1. The latter is called the **standard normal curve**, often called the **standard normal density**. Table E is used to compute the area.

standard normal curve
standard normal density

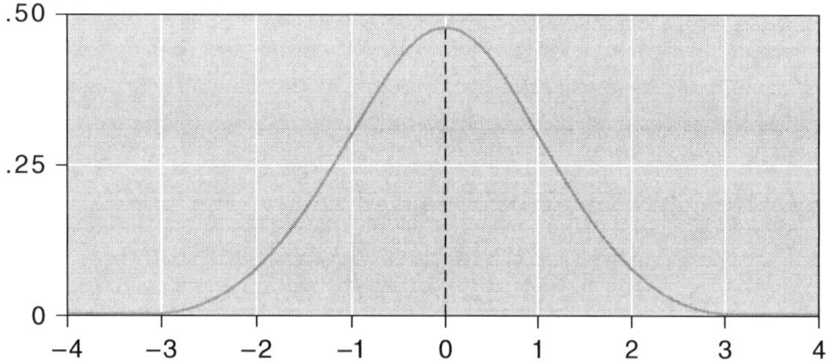

Figure 2.20 The standard normal curve.

The standard normal curve, shown in Figure 2.20, has the following basic properties:

- The curve is bell-shaped and symmetric around 0; the part of the curve to the right of the dashed line in Figure 2.20 is the mirror image of the part to the left of the dashed line.
- The curve extends indefinitely in both directions, always above the horizontal axis, approaching but never touching the axis.
- The total area under the curve is 1.
- Almost all of the area under the curve is between −3 and 3 (the 99.7% rule).

The z-scores corresponding to 0.225 and 0.265 are

$$\frac{0.225 - 0.263}{0.029} = -1.31 \qquad \text{and} \qquad \frac{0.265 - 0.263}{0.029} = 0.07$$

The batting averages between .225 and .265 correspond to the z-scores between −1.31 and 0.07. Therefore, the proportion of batting averages between .225 and .265 equals the proportion of z-scores between −1.31 and 0.07, which in turn approximately equals the *area under the standard normal curve* from −1.31 to 0.07, indicated in Figure 2.21.

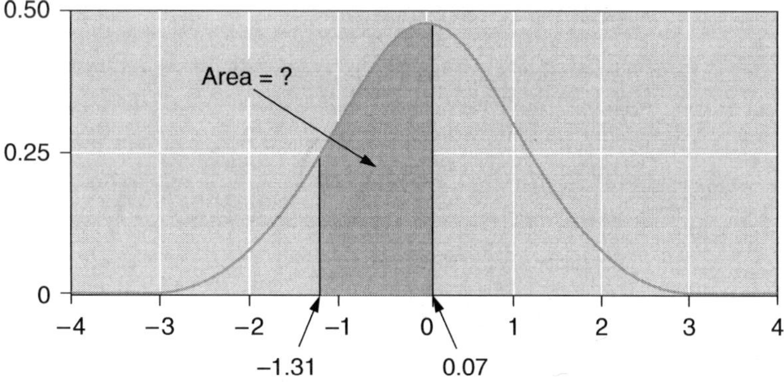

Figure 2.21 The area from −1.31 to 0.07.

Table E in the back of the book gives us the area under the standard normal curve that lies to the left of a specified z-score. A portion of Table E is reproduced as Table 2.8. To find the area to the left of $z = 0.07$, as shown in Figure 2.22, locate 0.0 in the table column under z, enter the table horizontally, and locate the number under 7. We get 0.5279. The area to the left of 0.07 is 0.5279.

The area to the left of 0.07 is the sum of the area to the left of −1.31 and the area between −1.31 and 0.07, as shown in Figure 2.23. Therefore,

$$(\text{Area between } -1.31 \text{ and } 0.07) = (\text{area to the left of } 0.07) - (\text{area to the left of } -1.31).$$

According to the normal Table E in Appendix E, the area under the standard normal curve to the left of −1.31 is 0.0951. Therefore, the area between −1.31 and 0.07 is

$$0.5279 - 0.0951 = 0.4328.$$

Table 2.8 The Relevant Portion of the Standard Normal Table

z	0	1	2	3	4	5	6	7	8	9
0.0	.5000	.5040	.5080	.5120	.5160	.5199	.5239	.5279	.5319	.5359
0.1	.5398	.5438	.5478	.5517	.5557	.5596	.5636	.5675	.5714	.5733
0.2	.5793	.5382	.5871	.5910	.5948	.5987	.6026	.6064	.6103	.6141
0.3	.6179	.6217	.6255	.6293	.6331	.6368	.6406	.6443	.6480	.6517

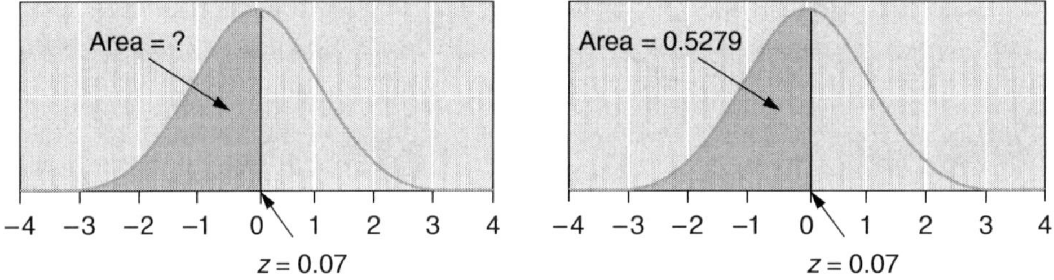

Figure 2.22 Finding the area under the standard normal curve to the left of 0.07.

Based on the z-score of −1.31 and the normal table value of 0.0951 corresponding to −1.31, we estimate that approximately 9.51% of the batting averages were below .225. This happens to be exactly right: 25 out of 263 were, which is 0.0951 to three significant figures! Using the normal curve we further estimate that 43.28% of the batting averages were between .225 and .265. In fact, 115 out of 263 players, or 43.73%, had batting averages in this range. The normal approximation is quite accurate!

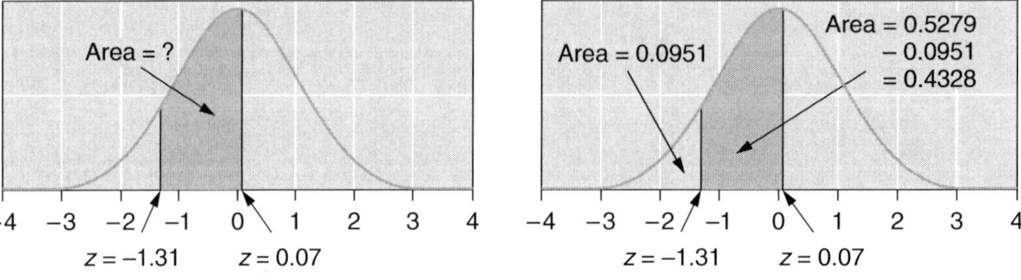

Figure 2.23 Finding the area under the standard normal curve that lies between −1.31 and 0.07.

We can summarize the normal approximation for bell-shaped histograms: We estimate the proportion of the data that are in an interval of interest in three steps.

The Normal Approximation for Data

1. Draw a number line and shade the interval of interest.
2. Convert the interval endpoint(s) into z-scores using the formula

$$z\text{-score} = \frac{\text{raw score} - \text{mean}}{\text{standard deviation}}$$

3. Find the area under the standard normal curve over the z-score interval using Table E.

For example, let us estimate the proportion of players whose batting average was over .295. We shall proceed in three steps.

1. Draw the raw score number line and shade the interval of interest.

.295 Batting average

2. Convert the endpoint .295 into a z-score.

$$z\text{-score} = \frac{\text{raw score} - \text{mean}}{\text{standard deviation}} = \frac{0.295 - 0.263}{0.029} = 1.10$$

3. Find the area under the standard normal curve over the z-score interval.

1.10 z-score

Since the total area under the standard normal curve equals 1, the area under the standard normal curve to the right of 1.10 is 1 minus the area to the left of 1.10, as indicated in Figure 2.24. We estimate that 13.57% of the players had batting averages over .295. In fact, 38 out of 263 (or 14.45%) had.

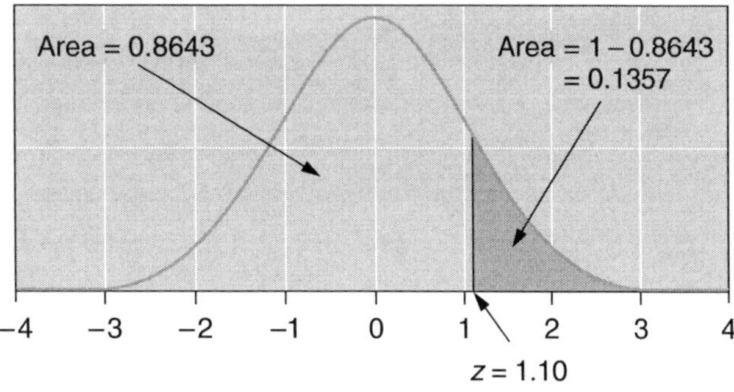

Figure 2.24 Finding the area under the standard normal curve to the right of 1.10.

percentile rank

Let us take another look at Yoko and Soledad's math scores. What was Soledad's **percentile rank** on the math SAT? In other words, what percentage of all the students who took the SAT scored lower than Soledad on the mathematics part? We can find the answer to this question in the standard normal Table E in the back of the book, the relevant portion of which is reproduced in Table 2.9. Since Soledad's z-score is 1.50, locate 1.5 in the table column under z, enter the table horizontally and locate the number in the column under 0. We get 0.9332, that is, 93.32% or about 93%. So around 93% of all students who took the SAT scored lower than Soledad on the math part.

Yoko's z-score was 1.33. We locate 1.3 in the column under z, enter the table horizontally and locate 0.9082 in the column under 3. The tabulated value converted to a percentage becomes 90.82%, or about 91%. So around 91% of all the students who took the ACT scored lower than Yoko on the math section. Therefore, Soledad's relative performance (expressed as a percentile rank) is slightly above Yoko's.

How high did a student have to score on the math SAT to be just in the top 5%? We need a score at the 95th percentile. In the table, .95 corresponds to a z-score of about 1.65. To get the SAT score corresponding to $z = 1.65$, we use the formula

$$\text{raw score} = \text{mean} + [(z\text{-score}) \times (\text{standard deviation})].$$

Therefore,

$$\begin{aligned} \text{SAT score} &= 500 + [1.65 \times 100] \\ &= 500 + 165 \\ &= 665. \end{aligned}$$

Table 2.9 Relevant Portion of the Standard Normal Table

z	1	0	2	3	4	5	6	7	8	9
1.2	.8849	.8869	.8888	.8907	.8925	.8944	.8962	.8980	.8997	.9015
1.3	.9032	.9049	.9066	.9082	.9099	.9115	.9131	.9147	.9162	.9177
1.4	.9192	.9207	.9222	.9236	.9251	.9265	.9279	.9292	.9306	.9319
1.5	.9332	.9345	.9357	.9370	.9382	.9394	.9406	.9418	.9429	.9441
1.6	.9452	.9463	.9474	.9484	.9495	.9505	.9515	.9525	.9535	.9545
1.7	.9554	.9564	.9573	.9582	.9591	.9599	.9608	.9616	.9625	.9633

This section introduced the normal curve as a descriptive tool to summarize the distribution of roughly bell-shaped data sets. Following the introduction of probability modeling in later chapters, the normal curve will be used further as the central theoretical tool in probability modeling and statistical inference.

Section 2.4 Exercises

1. On a midterm exam the mean score was 50 and the standard deviation was 15.
   a. Convert each of the following exam scores to z-scores: 50, 35, 80.
   b. Find the exam scores corresponding to the z-scores 0, 0.6, 1, −2.
2. Convert each entry in the following list into z-scores: 5, 2, 6, 8, 5, 4. (Use the mean and the standard deviation of the list.)
3. IQ scores are normally distributed with mean 100 and standard deviation 15. Find the percentage of IQ scores that are
   a. below 120

b. below 90
c. between 90 and 120

4. How high must a person's IQ score be to fall in the top 5% of all IQ scores? How high to fall in the top 2.5%? (See the previous exercise.)

5. The heights of young women are approximately normally distributed with mean 65 inches and standard deviation 2.5 inches. Find approximately the percentage of women who are
   a. under 69 inches tall
   b. under 62 inches tall
   c. between 62 and 69 inches tall

6. The Public Health Service examined a representative cross section of several thousand American men ages 18 to 74. The systolic blood pressure of these men approximately followed a normal curve with mean 130 and standard deviation 18. Any blood pressure of 160 or above is considered "high." Any blood pressure from 140 to 160 is considered "borderline."
   a. Estimate the percentage of men with high systolic blood pressure.
   b. Estimate the percentage with borderline systolic blood pressure.

7. The length of human pregnancy from conception to birth approximately follows a normal curve with mean 266 days and standard deviation 16 days.
   a. Estimate the percentage of pregnancies that last less than 240 days (which is roughly 8 months).
   b. Estimate the percentage of pregnancies that last between 240 days and 270 days (which is about 8 to 9 months).

8. The systolic blood pressure of men ages 18 to 24 approximately follows a normal curve with mean 124 and standard deviation 14, according to the Public Health Service and the National Center for Health Statistics (see Exercise 6).
   a. Estimate the percentage of men ages 18 to 24 with high systolic blood pressure (160 or above).
   b. Estimate the percentage of men ages 18 to 24 with borderline systolic blood pressure (140 to 160).

2.5 BOXPLOT: THE FIVE-NUMBER SUMMARY

We saw in Sections 2.3 and 2.4 that if a data set has an approximately bell-shaped distribution without outliers, the mean and the standard deviation summarize the main features of the distribution and indeed allow us to estimate the proportion or percentage of the data falling in any interval. For skewed distributions and distributions with outliers, however, the mean and standard deviation will not give us a meaningful summary. The problem is that it is difficult to interpret the mean and the standard deviation when there are outliers or the distribution is skewed. Instead we shall use the **quartiles** to summarize the data. The quartiles Q_1, Q_2, and Q_3 divide the data into four equal parts (Figure 2.25). Q_1 separates the bottom 25% of the data from the top 75%. Q_2 is the median, which separates the bottom 50% from the top 50% of the data. Finally, Q_3 separates the bottom 75% from the top 25% of the data.

quartiles

We will give a very simple method for determining the quartiles. Some computer programs

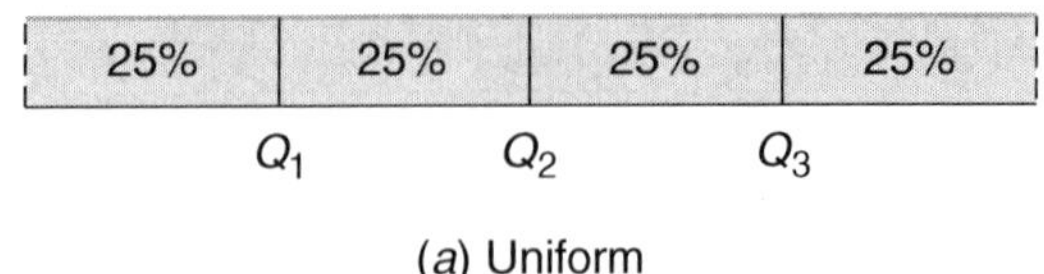

(*a*) Uniform

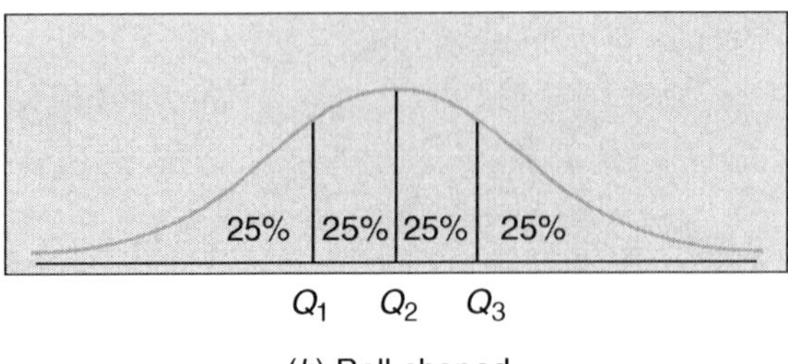

(*b*) Bell-shaped

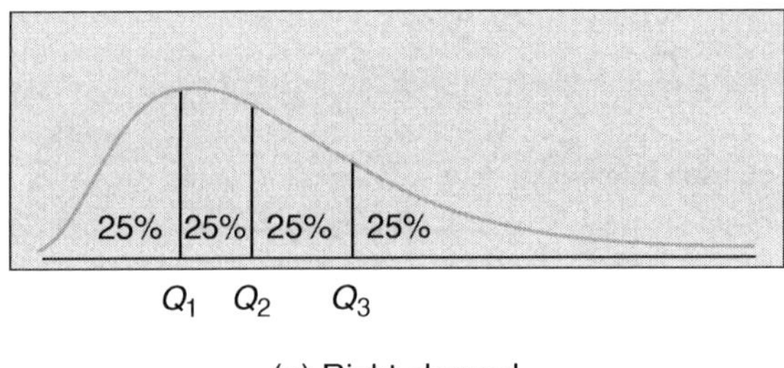

(*c*) Right-skewed

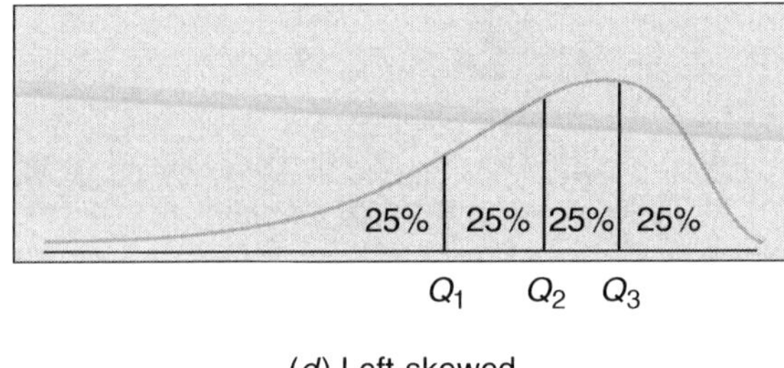

(*d*) Left-skewed

Figure 2.25 Quartiles for uniform, bell-shaped, right-skewed, and left-skewed distributions.

and calculators might compute the values in a slightly different manner, but the results will be close.

second quartile

first quartile

third quartile

1. Sort the data from smallest to largest and compute the median. This is the **second quartile,** Q_2.
2. The **first quartile**, Q_1, is defined to be the median of all the data to the left of Q_2.
3. The **third quartile**, Q_3, is defined to be the median of all the data to the right of Q_2.

Example 2.13 illustrates how the quartiles are determined.

Example 2.13

The number of home runs hit by Babe Ruth during his 15 years with the New York Yankees were, in increasing order, as follows:

22	25	34	35	41	41	46	46	46	47	49	54	54	59	60
			↑				↑				↑			
			Q_1				Q_2				Q_3			

With 15 observations, the median (Q_2) is the eighth smallest observation. The first quartile (Q_1) is the median of the first seven observations, to the left of Q_2, and the third quartile (Q_3) is the median of the last seven observations, to the right of Q_2. Note that even though the seventh, eighth, and ninth smallest observations have the same value (one 46), the seventh observation (one 46) is considered to be to the left of Q_2, and the ninth observation (another 46) is considered to be to the right of Q_2.

From 1987 to 2000 the number of home runs hit by Mark McGwire were, in increasing order, as follows:

9	9	22	32	32	33	39	39	42	49	52	58	65	70
			↑			↑				↑			
			Q_1			Q_2				Q_3			

With 14 observations, the median is the average of the seventh and the eighth smallest observations, the middle two observations. The first quartile (Q_1) is the median of the first seven observations, and the third quartile (Q_3) is the median of the last seven observations. We get $Q_1 = 32$, median $= (39 + 39)/2 = 39$, and $Q_3 = 52$. Here the seventh observation is considered to be to the left of Q_2, and the eighth observation is considered to be to the right of Q_2.

boxplot

To compare the performances of Babe Ruth and Mark McGwire, we could draw back-to-back stem-and-leaf plots, or we could draw side-by-side **boxplots**. We need five summary values to construct a boxplot:

1. The minimum value
2. The first quartile
3. The median
4. The third quartile
5. The maximum value

five-number summary
five-number boxplot

These five values are referred to as the **five-number summary**. The boxplot based on the five-number summary is called a **five-number boxplot** or simply a **boxplot**.

To draw the five-number boxplot, draw a box around the first and third quartiles. Draw a vertical line through the box at the median value, and draw horizontal lines (the whiskers) to connect the edges of the box (at the first and third quartiles) to the minimum and maximum values. Boxplots give a rough idea of the shape of the data set. In Figure 2.26, boxplots are sketched for five common idealized data distributions.

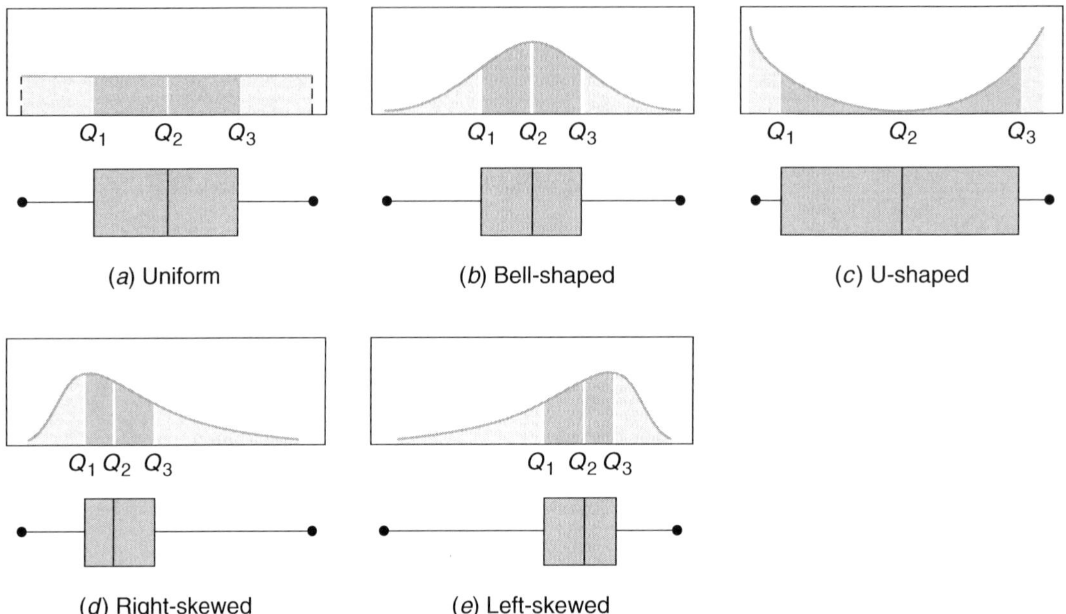

Figure 2.26 Distribution shapes and typical boxplots for uniform, bell-shaped, U-shaped, right-skewed, and left-skewed distributions.

outlier boxplots

In addition to using the five numbers, we sometimes want to specially identify outliers as was done in Figure 2.1, which displays **outlier boxplots**. Outliers are indicated by asterisks (*) and potential outliers by circles (∘). The whiskers extend from the box to the largest and the smallest non-outlier values. (Outliers and potential outliers will be defined below.)

Example 2.14

Table 2.10 gives the number of farms (in thousands) for individual states in 1998. For example, Alabama has 49,000 farms. The five-number summary is

$$\text{Min} = 0.6, Q_1 = 9.6, M = Q_2 = 38.75, Q_3 = 65.0, \text{Max} = 226.0$$

The outlier boxplot is drawn in Figure 2.27. We see that there is one extreme value. It is a potential outlier, with value 226.0 (Texas). (The precise rule for deciding when a point is labeled a potential outlier or an outlier is given in a while.) If we separately label the potential outlier (as was done in the Key Problem at the beginning of this chapter), the maximum of the remaining data is 110.0 (Missouri). Note that the whisker is drawn to this remaining states' maximum and *not* to the potential outlying value (see also Figure 2.1). However, Q_1, Q_2, and Q_3 were computed using all 50 states, including the potential outlying Texas value.

Table 2.10 Number of Farms per State, 1998 (in thousands)

Alabama	49.0	Nebraska	55.0
Alaska	0.6	Nevada	3.0
Arizona	7.9	New Hampshire	3.1
Arkansas	49.5	New Jersey	9.6
California	89.0	New Mexico	16.0
Colorado	29.5	New York	38.0
Connecticut	4.1	N. Carolina	58.0
Delaware	2.7	N. Dakota	31.0
Florida	45.0	Ohio	80.0
Georgia	50.0	Oklahoma	83.0
Hawaii	5.5	Oregon	39.5
Idaho	24.5	Pennsylvania	60.0
Illinois	79.0	Rhode Island	0.8
Indiana	66.0	S. Carolina	25.0
Iowa	97.0	S. Dakota	32.5
Kansas	65.0	Tennessee	91.0
Kentucky	90.0	Texas	226.0
Louisiana	30.0	Utah	15.0
Maine	6.9	Vermont	6.7
Maryland	12.5	Virginia	49.0
Massachusetts	6.0	Washington	40.0
Michigan	52.0	W. Virginia	21.0
Minnesota	80.0	Wisconsin	78.0
Mississippi	42.0	Wyoming	9.2
Missouri	110.0		
Montana	27.5	United States	2192

Source: National Agricultural Statistics Service, U.S. Department of Agriculture.

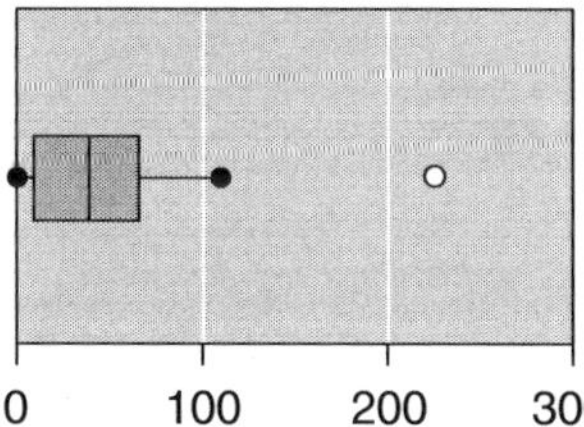

Figure 2.27 Boxplot of number of farms per state.

When we look at the data, we notice that this potential outlying value corresponds to the state of Texas. It is useful to call attention to outliers, like Texas in this example, because we may want to understand why they are so different from the rest of the data.

❐ ❐ ❐ ❐ ❐ ❐ ❐ ❐ ❐ ❐ ❐ ❐ ❐ ❐ ❐ ❐ ❐ ❐ ❐ ❐

interquartile range (IQR)

The boxplot gives us a graphical method for looking in broad strokes at variation within the data (see Figure 2.26). The box itself represents the middle 50% of the data. We define the **interquartile range (IQR)** as the difference between the third and first quartiles:

$$\text{IQR} = Q_3 - Q_1.$$

In Example 2.14, IQR $= 65.0 - 9.6 = 55.4$.

The interquartile range is a much more resistant measure of variation than the standard deviation is, because (unlike the standard deviation) it is not influenced by the largest and smallest values in the data set. We will see later, when we study estimation and hypothesis testing, however, that the standard deviation is a more important measure of variation for statistical inference.

Example 2.15

The following data consist of the percentages of eighth-grade students in various states (all states that reported data) who scored at or above proficiency levels in national tests in 1998 (*National Assessment of Educational Progress*, National Center for Education Statistics, U.S. Department of Education).

Reading	21	28	23	22	30	42	25	12	23	25	19	35	29	
	18	42	31	36	37	19	29	24	24	34	31	29		
	33	30	22	26	28	31	33	32	27	33	29	31		

Science	18	31	23	22	20	32	36	21	5	21	21	15	30	36
	13	41	25	37	32	37	12	28	41	35	19	27	24	41
	32	26	17	22	23	32	34	27	27	21	39	34	27	23

We can compare the two sets of data by using boxplots, as drawn in Figure 2.28.

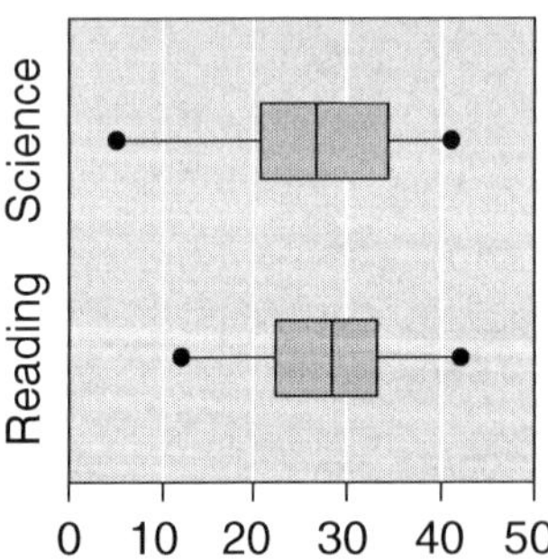

Figure 2.28 Percentages of eighth-grade students scoring at or above proficiency levels on national tests.

There is not a great deal of difference between the medians of the two groups of percentages, but because the science box is wider, there appears to be more variation in the science scores. When we compare the medians, we are comparing the centers of the two groups. However, when we look at the size of each box (the IQR), we are considering the variation (or spread) within each group of percentages.

In this case the difference in variation might be of considerable importance to educators, suggesting the need for more remedial programs in science, perhaps.

❐ ❐

We need a rule for deciding whether a value is so extreme that it needs to be specially labeled as an outlier to call the user's attention to it. The interquartile range gives us a starting point to do this. Here is a commonly used method:

When a data value is an *outlier* or *potential outlier.*

outlier

- Label a value an **outlier** if it is more than 3 times the IQR in distance from the nearer of the first and third quartiles. Denote it by an asterisk (∗).

potential outlier

- If a value is more than 1.5 and less than 3 times the IQR in distance from the nearer of the first and third quartiles, label a **potential outlier** and denote it by a circle (∘).

Figure 2.29 clearly shows the procedure for a typical boxplot. This somewhat arbitrary choice of 1.5 and 3 has become standard practice in defining outliers when doing a boxplot. The TI-83 Plus and TI-89 calculators and many statistical computer programs will also show outliers on a modified boxplot, as was done in the Key Problem. In this case, the whiskers are drawn to the smallest and largest values that are not outliers. Q_1, Q_2, and Q_3 are computed using all the data, including potential and actual outlier values. Keep firmly in mind that the labeling of points as outliers is tricky and subjective; the 1.5-3 rule above is just one approach that statisticians use.

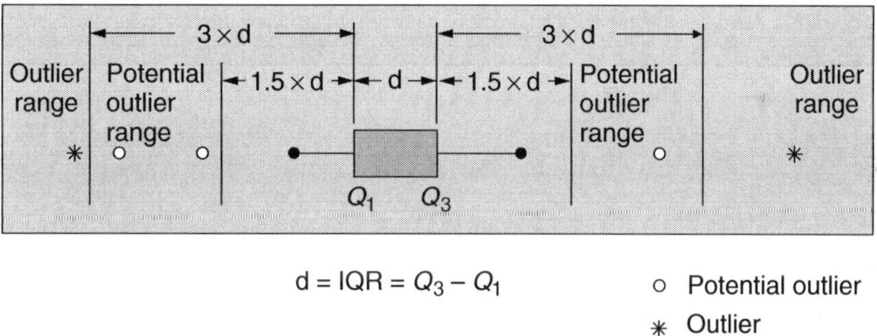

Figure 2.29 How outliers and potential outliers are identified in an outlier boxplot.

Caution: A boxplot displays the center and the spread of a set of numbers. An outlier boxplot also indicates outliers. But a boxplot does not tell us whether the data set has clusters or gaps, which are important characteristics of a data set when they occur. As a tool for determining the shape (or distribution) of a data set, boxplots can be deceptive. For example, consider the back-to-back stem-and-leaf plots in Table 2.11. The first data set has clusters and gaps, whereas the second data set is bell-shaped with no gaps. But the two data sets have the same five-number summary and therefore the same boxplot! Obviously, the boxplot does not tell the whole story.

Table 2.11 Back-to-Back Stem-and-Leaf Plots

Leaf (First Data Set)	Stem	Leaf (Second Data Set)
9,6,5	0	5
	1	6
	2	2,7
9,8,7,7	3	2,5,7
0	4	1,3,5,9
	5	0,2,4,6,8
9,8	6	1,4,6,8
3,3,1	7	3,6,9
3,1	8	4,8
	9	5
1	10	1

Example 2.16

Consider the ages of the male Academy Award winners in the Key Problem at the start of the chapter. The data has been summarized in Table 2.12.

Actors: $\text{IQR} = 51 - 37 = 14$

$$1.5 \times \text{IQR} = 1.5 \times 14 = 21$$
$$Q_1 - (1.5 \times \text{IQR}) = 37 - 21 = 16$$
$$Q_3 + (1.5 \times \text{IQR}) = 51 + 21 = 72$$
$$3 \times \text{IQR} = 3 \times 14 = 42$$
$$Q_1 - (3 \times \text{IQR}) = 37 - 42 = -5$$
$$Q_3 + (3 \times \text{IQR}) = 51 + 42 = 93$$

Table 2.12 Key Problem Statistics

	Minimum	First Quartile	Median	Third Quartile	Maximum
Actors	29	37	42	51	76
Actresses	21	30	34	41	82

Thus, any award-winning actor older than 93 or younger than −5 is an outlier, an impossibility for −5 and extremely unlikely for 93. Any award-winning actor of age between −5 and 16 or between 72 and 93 is a potential outlier. Looking at the data, we see that Henry Fonda at age 76 is a potential outlier for the actors. Maybe Fonda's Best Actor Academy Award was partly a lifetime achievement award, given to a popular actor shortly before his death. Since there have been famous child stars, we could have had a potential low outlier, of course, but in this case we did not. A similar analysis is possible for the actress data.

Cumulative Distributions

cumulative distribution

Table 2.13 gives the **cumulative distribution** of American households by income in 1997. For selected income levels the table gives both the number of households and the percent of all 102.5 million households with income below that level. Note that the tabulated percents, by their definition, are increasing from top to bottom of the last column.

For example, 34.9 million households, or 34% of all 102.5 million households, earned less than $25,000. Figure 2.30 shows the cumulative percents plotted against income, as obtained from Table 2.13.

Table 2.13 Cumulative Distribution of American Households by Income, 1997

Income	Number (in millions)	Percent
Under $10,000	11.3	11.0
Under $15,000	19.6	19.1
Under $25,000	34.9	34.0
Under $35,000	48.5	47.3
Under $50,000	65.2	63.5
Under $75,000	83.7	81.6

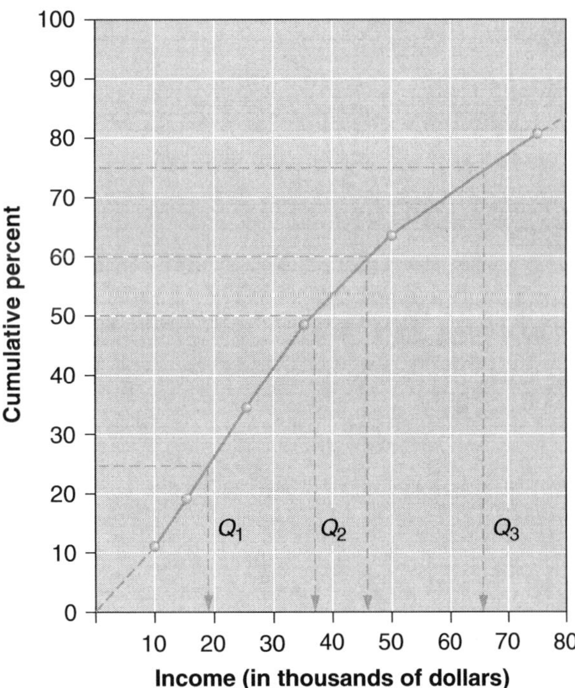

Figure 2.30 Cumulative percents plotted against income.

cumulative distribution function

The **cumulative distribution function** in Figure 2.30 was drawn using Table 2.13. It can be used to estimate the quartiles Q_1, Q_2, and Q_3 of the income distribution. We get

$$Q_1 \approx \$19{,}000, Q_2 \approx \$37{,}000, \text{ and } Q_3 \approx \$66{,}000.$$

qth percentile

It is just as easy to estimate other percentiles in addition to the three quartiles (Q_1 = 25th percentile, Q_2 = 50th percentile, Q_3 = 75th percentile) of the income disribution. Recall that the ***q*th percentile** separates the bottom $q\%$ of the data from the top $(100 - q)\%$ of the data. For example, the 60th percentile was around \$47,000. In other words, we estimate that around 60% of all households earned less than \$47,000 in 1997 and around 40% of all households earned more than \$47,000. A cumulative distribution function is often drawn with the vertical axis being a proportion from 0 to 1 rather than a percentile from 0 to 100%.

Section 2.5 Exercises

1. The following are dotplots of four data sets labeled A, B, C, and D.

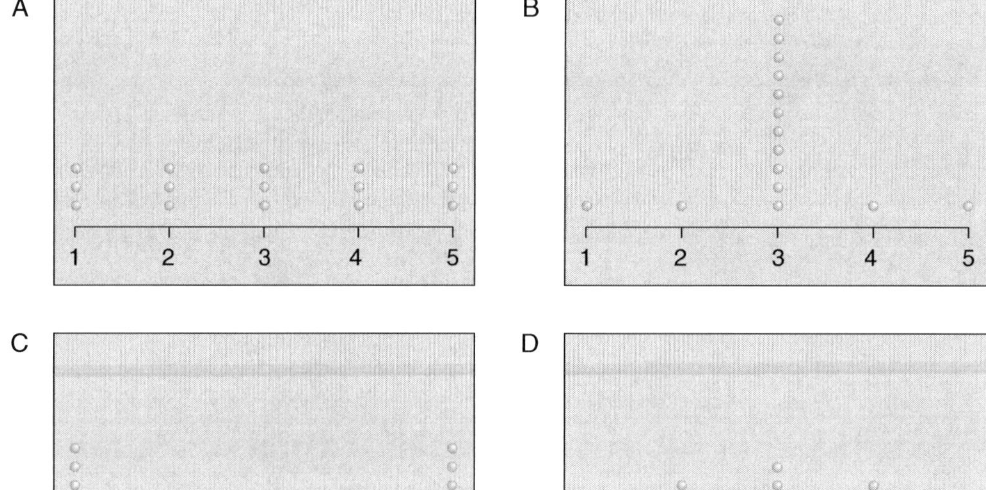

a. Which of the four data sets has the five-number summary (minimum, Q_1, median, Q_3, maximum) = (1, 2, 3, 4, 5)? (Choose one option)

only A only B only C only D both A and D

b. Which of the four data sets has interquartile range (IQR) equal to 0?

only A only B only C only D both A and D

c. Which of the four data sets has interquartile range (IQR) equal to 4?

only A only B only C only D both A and D

2. From 1986 to 2004 the number of home runs hit by Barry Bonds were:

1986	1987	1988	1989	1990	1991	1992	1993	1994	1995
16	25	24	19	33	25	34	46	37	33
1996	**1997**	**1998**	**1999**	**2000**	**2001**	**2002**	**2003**	**2004**	
42	40	37	34	49	73	46	45	45	

Consider the stem-and-leaf plots below:

Plot A

1	6, 9
2	4, 5, 5, 6
3	3, 3, 4, 4, 7, 7
4	0, 2, 5, 5, 6, 9
5	
7	3

Plot B

1	6, 9
2	4, 5, 5
3	3, 3, 4, 4, 7, 7
4	0, 2, 5, 5, 6, 9
5	
7	3

Plot C

1	6, 9
2	4, 5, 5
3	3, 3, 4, 4, 7, 7
4	0, 2, 5, 5, 6, 6, 9
5	
7	3

a. The stemplot for Barry Bonds's home runs is (choose one option)

Plot A Plot B Plot C

b. The median of Barry Bonds's home run scores is

33.5 37 35.5 38.5 34

c. The first quartile Q_1 is

29 24.5 25.5 25 24

d. The third quartile Q_3 is

46 43.5 42 45.5 45

e. The range is

73 34 19.5 20 57

f. The interquartile range (IQR) is

73 34 19.5 20 57

g. According to our criterion, is Barry Bonds's 73 home runs in 2001 a potential outlier?

Yes No

3. The salaries of the Kansas City Royals for the 2006 season were as follows:

Player	Salary	Player	Salary
Mike Sweeney	$11,000,000	Mike MacDougal	$430,000
Reggie Sanders	$ 5,000,000	Zack Greinke	$365,000
Mark Redman	$ 4,500,000	Mike Wood	$350,000
Mark Grudzielanek	$ 4,000,000	Chris Booker	$350,000
Scott Elarton	$ 4,000,000	John Buck	$349,500
Tony Graffanino	$ 2,050,000	Andrew Sisco	$346,500
Angel Berroa	$ 2,000,000	Jimmy Gobble	$345,500
Doug Mientkiewicz	$ 1,850,000	Mark Teahen	$344,500
Emil Brown	$ 1,775,000	Ambiorix Burgos	$339,500
Elmer Dessens	$ 1,700,000	Denny Bautista	$335,500
Matt Stairs	$ 1,350,000	Bobby Madritsch	$335,000
Jeremy Affeldt	$ 1,000,000	Esteban German	$333,000
Paul Bako	$ 700,000	Steve Stemle	$330,000
David DeJesus	$ 500,000	Shane Costa	$327,500

a. Determine the median salary.
b. Find the first quartile Q_1 and the third quartile Q_3.
c. Determine the interquartile range.
d. Is Mike Sweeney's salary an outlier? Explain.
e. Are there any potential outliers? Explain.
f. Draw an outlier boxplot for the salaries.

4. Consider the salaries of the San Diego Padres for the 2006 season. (See Exercise 6 in Section 2.1.)
a. Determine the median salary.
b. Find the first quartile Q_1 and the third quartile Q_3.
c. Determine the interquartile range.
d. Is Chan Ho Park's salary an outlier? Explain.
e. Are there any potential outliers? Explain.
f. Draw an outlier boxplot for the salaries.

5. Construct an outlier boxplot for the white blood cell count data of Exercise 8 of Section 2.2. Be sure to indicate outliers and potential outliers. What do the outliers indicate in this example?

6. Repeat Exercise 5 but using the data on drug use in Exercise 7 of Section 2.2 (high school marijuana use).

7. The U.S. Bureau of Labor Statistics gave unemployment rates for the years 1975 and 1980–2004 as listed in the following table.
a. Draw a line graph. (See Figures 2.2 and 2.7.)
b. Determine the interquartile range.
c. Determine whether there are any outliers or potential outliers.
d. Draw an outlier boxplot and label the corresponding values.

Year	Unemployment rate (percent)	Year	Unemployment rate (percent)	Year	Unemployment rate (percent)
1975	8.5	1988	5.5	1997	4.9
1980	7.1	1989	5.3	1998	4.5
1981	7.6	1990	5.6	1999	4.2
1982	9.7	1991	6.8	2000	4.0
1983	9.6	1992	7.5	2001	4.7
1984	7.5	1993	6.9	2002	5.8
1985	7.2	1994	6.1	2003	6.0
1986	7.0	1995	5.6	2004	5.5
1987	6.2	1996	5.4		

8. The percentages of the popular vote won by the winning candidates in presidential elections from 1948 to 2008 are given below.

1948	1952	1956	1960	1964	1968	1972	1976	1980	1984	1988	1992	1996	2000	2004	2008
49.6	55.1	57.4	49.7	61.1	43.4	60.7	50.1	50.7	58.8	53.4	43.0	49.2	47.9	50.6	52.9

a. What are the mean, median, and standard deviation of the winning percentages?

b. Compute the first and third quartiles. Which elections fall below the first quartile (close elections)? Which fall above the third quartile (landslides)?

9. The U.S. Department of Energy released fuel consumption figures (in miles per gallon) for midsize cars and for four-wheel-drive sport utility vehicles (SUVs) in 1998.

Midsize											
25	26	29	24	26	29	30	28	28	29	27	28
23	23	25	26	33	29	26	24	28	26	16	25
30	25										

SUV											
19	20	19	17	18	19	20	19	19	18	19	16
19	21	20	19	26	26	22					

Discuss any significant differences you can find between fuel consumption results for of midsize cars and SUVs, using indices and techniques you have learned in this chapter.

10. *USA Today* (February 17, 1995) reported the sizes of the police forces in the 10 largest cities in the United States in 1993 (in hundreds) as follows:

29.3	7.6	12.1	4.7	6.2	1.9	3.9	2.8	2.0	1.7

a. Construct an outlier boxplot for the data.
b. Find any outliers and potential outliers.
c. Find the mean and median of the data. What can you tell about the shape of the distribution? Is it skewed to the right, skewed to the left, or symmetric?

11. When should we use the interquartile range as a measure of variability instead of the standard deviation?

12. Draw an outlier boxplot for the salaries of the Miami Heat players for the 2005–2006 season (see Example 2.6). Is Shaq's salary an outlier?

13. Draw first a dotplot and then an outlier boxplot for Newcomb's measurements in Table 2.2 (Example 2.8). Do not include the two negative values in the dotplot; they are too far off to be included. But include the negative values in the boxplot. Determine the interquartile range and identify any outliers.

14. Draw a boxplot for Roger Maris's home run data (see Example 2.2). Determine the interquartile range. Is Maris's record-setting total of 61 homers an outlier, according to our technical definition?

15. Sketch a boxplot for the age distribution of deaths due to heart disease in 1996 (see Exercise 14 of Section 2.2). Make sure that the shape of the boxplot reflects the shape of the histogram (see Figure 2.26).

16. The following quartile boxplots represent four different data sets. One distribution is uniform, one is bell-shaped, one is right-skewed, and one is left-skewed. Which is which? *Hint:* See Figure 2.26.

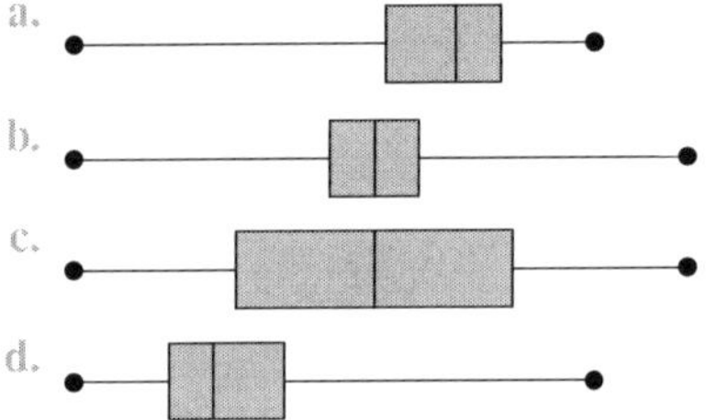

17. Do the analysis of Example 2.16 for the Academy Award actress age data. Determine the IQR. Identify outliers and potential outliers. (See the list of actresses in the Key Problem at the very beginning of the chapter. Use Table 2.12.) Also compute the mean and the standard deviation. Remove the outliers and the potential outliers and compute the median, mean, interquartile range, and the standard deviation for the new data set. Which measures are most resistant to change? (*Hint:* Use the "computational" formula for s^2.)

18. The following table gives the percent of the populations of Alaska and Florida, respectively, below selected ages in 1998.

Age	Alaska	Florida	Age	Alaska	Florida
Under 5 years	8%	6%	Under 55 years	87%	72%
Under 18 years	31%	24%	Under 65 years	94%	82%
Under 25 years	42%	32%	Under 75 years	98%	91%
Under 35 years	55%	45%	Under 85 years	100%	98%
Under 45 years	73%	60%			

For example, 8% of the population of Alaska and 6% of the population of Florida were under 5 years old. The table also shows that 55% of the population of Alaska and 45% of the population of Florida were under 35 years old. The graphed cumulative distributions show the cumulative percents plotted against age for Alaska and Florida. The table and graph are to be used to compare the age distributions of Alaska and Florida in 1998.

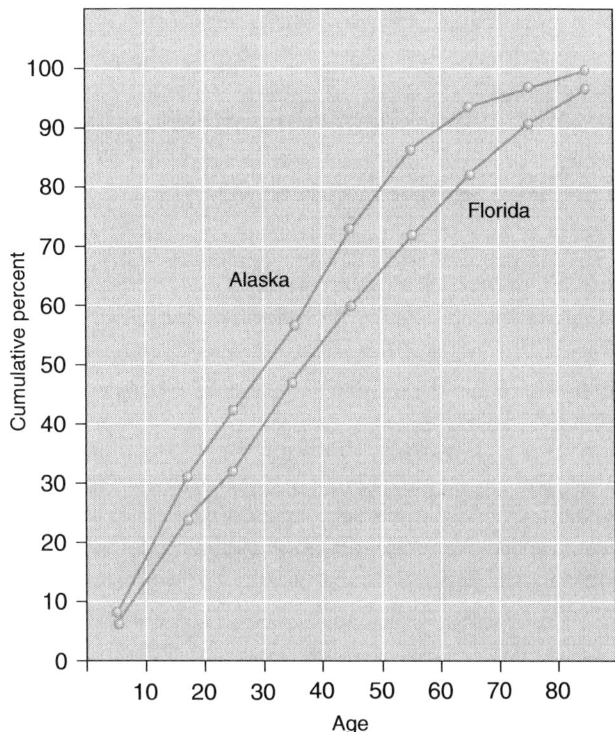

In each of parts (a) through (f), choose one option and explain.

a. The median age in Alaska was around

15 19 31 38 47 58

b. The median age in Florida was around

15 19 31 38 47 58

c. The first quartile of the Alaska age distribution was around

15 19 31 38 47 58

d. The third quartile of the Alaska age distribution was around

15 19 31 38 47 58

e. The first quartile of the Florida age distribution was around

15 19 31 38 47 58

f. The third quartile of the Florida age distribution was around

15 19 31 38 47 58

g. Approximate the interquartile range for each distribution.

h. Write a sentence or two comparing the age distributions of Alaska and Florida.

19. Confirm that Texas is a potential outlier in Example 2.14.

CHAPTER 2 SUMMARY

Descriptive data analysis begins with graphical displays. We then add summarizing numbers to the description. The most common descriptions of center and spread are the **mean** and **standard deviation** and the **five-number summary**. The five-number summary consists of the median, the first and third quartiles, and the smallest and largest observations.

Both the mean and the median are measures of the center. The **median** separates the bottom 50% of the data from the top 50% of the data. The **mean** of the data is the usual arithmetic average. For approximately symmetric distributions without outliers, the mean and the median will be close together and both indicate the center of the distribution. In skewed distributions and distributions with outliers, the mean is affected by the "pull" of extreme (very large or very small) observations. In skewed distributions, the mean is pulled away from the median toward the long tail. For such distributions, the median is a better measure of the center than the mean.

Unlike most distributions, approximately bell-shaped distributions can be summarized by just two numbers, the mean and the standard deviation. For approximately bell-shaped distributions without outliers, the **standard deviation** describes the spread of the data around the mean. The 68-95-99.7% Empirical Rule applies to such distributions.

***z*-scores** tell how many standard deviations each observation is above or below the mean:

$$z\text{-score} = \frac{\text{raw score} - \text{mean}}{\text{standard deviation}}$$

The **percentile rank** of an observation is the percentage of all the data that are less than the observation.

For approximately bell-shaped distributions without outliers, we can estimate the percentage of the data that are in a given interval, using only the mean and the standard deviation (and standard normal tables). For such distributions, the mean and the standard deviation thus hold almost all the information about the data set.

The **first quartile** Q_1 separates the bottom 25% of a data set from the top 75%. The **third quartile** Q_3 separates the bottom 75% of a data set from the top 25%. For skewed distributions and distributions with outliers, the quartiles and the **interquartile range** IQR $= Q_3 - Q_1$ give a better description of the spread of the data than the standard deviation. Graphing the **cumulative distribution** of a data set allows one to find quartiles as well as other percentiles. The qth percentile separates the bottom $q\%$ of the data from the top $(100 - q)\%$ of the data.

Boxplots, based on the five-number summary, are useful for comparing skewed distributions.

The mean and standard deviation can be distorted as measures of center and spread by one or more outliers. We express this by saying that the mean and standard deviation are not **resistant measures**. (They cannot resist the influence of one or more outliers.) In contrast, the median and quartiles are resistant.

Statistic	Resistance to extreme values
Measure of center	
Mean	Sensitive
Median	Resistant
Measure of spread	
Range	Very sensitive
Standard Deviation	Sensitive
Interquartile Range	Resistant

Chapter Review Exercises

1. On April 16, 2000, the winning teams in the National Basketball Association scored the following numbers of points:

95	102	101	114	100	105	85	104	112

   a. Compute the mean, median, standard deviation, and interquartile range.
   b. Suppose the team with 114 points had shot better and scored 133 points instead. Then what would the measures be? What does this say about the resistance of these measures to extreme values?

2. A mathematics teacher is considering retirement. He has been told that San Diego and Las Vegas are both good retirement locations, so he decides to do some research on climate. He dislikes big temperature changes from summer to winter. He obtains the following average high temperatures for each month:

	Jan	Feb	Mar	Apr	May	June	July	Aug	Sept	Oct	Nov	Dec
L.V.	55	60	69	79	88	98	105	102	95	81	66	57
S.D.	65	65	67	68	70	72	76	77	77	73	71	67

   Draw boxplots for each set of data and compare the results. Which city would he prefer?

3. Combine the data for both the actors and actresses in the Key Problem into one data set.
   a. Compute the mean, median, interquartile range, and standard deviation for the combined data set.
   b. Draw a boxplot for the data.

4. Draw a density histogram for the combined data in Exercise 3. What can you determine about the distribution?

5. Which measure of the center of a distribution (mean, median, or mode) is most likely the best to use in each of the following cases?
   a. When we have very large or very small values in relation to the rest of the data
   b. When we are concerned with which value occurs most frequently
   c. When the data distribution is sharply skewed to the right
   d. When the data distribution is sharply skewed to the left
   e. When the data distribution is approximately symmetric

6. Is the mode a good measure for the center of a distribution? Explain your answer.

7. *Monitoring the Future,* from the University of Michigan Institute for Social Research and the National Institute on Drug Abuse, presents the following data on the prevalence of heavy drinking among twelfth-graders. Below are the percentages of white, black, and Hispanic twelfth-graders who had five or more drinks in a row on one or more occasions during the past 2 weeks.

Percentage reporting 5+ drinks in a row on one or more occasions

	Class of														
	1990	1991	1992	1993	1994	1995	1996	1997	1998	1999	2000	2001	2002	2003	2004
White	36.6	34.6	32.1	31.3	31.5	32.3	33.4	35.1	36.4	35.7	34.6	34.5	33.7	32.4	32.5
Black	14.4	11.7	11.3	12.6	14.4	14.9	15.3	13.4	12.3	12.3	11.5	11.8	11.5	10.8	11.4
Hispanic	25.6	27.9	31.1	27.2	24.3	26.6	27.1	27.6	28.1	29.3	31.0	28.4	26.4	25.9	26.0

a. Find the minimum, maximum, median, first and third quartiles, and interquartile range for each of the three groups.
b. Construct a boxplot for each of the three sets of data on the same set of axes.
c. Discuss any differences you see between the three groups. (*Note:* One could ask about possible trends over time as well. That is addressed by Chapter 3.)

8. Discuss why we are interested in resistant measures when doing statistics.

9. The U.S. Federal Trade Commission has released the following data on the amounts of tar and nicotine in 26 brands of cigarettes. All measurements are in milligrams.
a. Compute the mean, median, and standard deviation for both tar and nicotine content.
b. Are there any outliers or potential outliers in either group?

Tar	Nicotine	Tar	Nicotine
16	1.2	11	0.9
16	1.2	2	0.2
9	0.8	18	1.4
1	0.1	15	1.2
8	0.8	13	1.1
10	0.8	15	1.0
16	1.0	17	1.3
14	1.0	9	0.8
13	1.1	12	1.0
15	1.2	14	1.0
16	1.2	5	0.5
9	0.7	6	0.6
8	0.7	18	1.4

10. The *Universal Almanac,* 1993, gave the winning times for the Boston Marathon. The following figures represent the number of minutes over 2 hours for the winning male runners:

Years 1953–1972										
	18	20	18	14	20	25	22	20	23	23
	18	19	16	17	15	22	13	10	18	15

Years 1973–1992										
	16	13	9	20	14	10	9	12	9	8
	9	10	14	7	11	8	9	8	11	8

a. Draw boxplots for both sets of data.
b. Can you determine any appreciable differences between the two different groups?

11. The following data consist of the ages of a random sample of 50 individuals arrested while driving under the influence of alcohol (DUI). These data were given in the *Statistical Abstract of the United States,* 112th edition.

46	16	41	26	22	33	30	22	36	34	63	21	26	18	27	24	31
38	26	55	31	47	27	43	35	22	64	40	58	20	49	37	53	25
29	32	23	49	39	40	24	56	30	51	21	45	27	34	47	35	

a. Construct a boxplot of the data. Are there any outliers or potential outliers?
b. Construct a stem-and-leaf plot of the same data. What can you say about the shape of the distribution?

12. The body weights of mule deer fawns between 1 and 5 months old in Mesa Verde National Park have an approximately bell-shaped distribution with a mean of 27.2 kilograms and a standard deviation of 4.3 kilograms (based on information from *The Mule Deer of Mesa Verde National Park,* 1981). Based on that information, compute the following:
   a. The approximate percentage of mule deer fawns in Mesa Verde that weigh more than 35.8 kilograms or less than 18.6 kilograms.
   b. The approximate percentage of mule deer fawns that weigh between 22.9 kilograms and 31.5 kilograms.

13. Consider the data on the percentage of residents of each state age 65 and older given in Exercise 7 in Section 1.4. Draw a stem-and-leaf plot for the data. Determine the median, the first quartile, the third quartile, and the interquartile range. Identify outliers and potential outliers. Create an outlier boxplot. What can you say about the general shape of the distribution?

14. The peak wind gusts (in miles per hour) at Chicago's O'Hare International Airport and Denver's Stapleton Airport for February 1995 are given in the following table:

Date	Chicago	Denver	Date	Chicago	Denver
1	15	32	15	25	14
2	15	39	16	12	17
3	30	15	17	17	22
4	30	27	18	26	36
5	26	17	19	22	17
6	18	16	20	33	20
7	23	14	21	33	13
8	21	14	22	18	18
9	27	32	23	34	12
10	34	26	24	24	14
11	32	20	25	24	17
12	21	12	26	25	25
13	26	11	27	15	26
14	24	38	28	17	13

a. Find the mean and median of the peak wind gusts for Chicago and Denver for February 1995.
b. Find the interquartile range of the peak wind gusts for each city.
c. Draw two boxplots, one for each city. What can you conclude about the differences between the peak wind gusts for Chicago and Denver for this month?

15. To investigate the work done by income tax preparation specialists, *Money* magazine sent one individual's financial data to 50 tax preparers. Below are the values of the calculated income tax due (in thousands of dollars) by the 50 specialists.

12.5	14.7	16.0	16.6	16.7	16.8	16.9	17.3	17.3	17.5	18.4	18.9	19.1
19.1	19.2	19.7	21.1	21.8	21.9	22.6	22.7	22.7	22.7	22.7	22.8	22.9
22.9	22.9	22.9	23.1	23.3	23.4	23.5	23.5	23.5	23.5	23.6	23.7	24.0
24.0	24.0	24.0	24.2	24.4	24.5	25.0	25.3	25.7	25.9	35.8		

a. Find the mean and median amount of tax due.
b. Find the interquartile range of the tax due.
c. Draw a boxplot of the given data.
d. Suppose the person with this income tax report was audited by the IRS. How could he or she use this study to his or her advantage?

16. The following table gives the ages of presidents of the United States at their inaugurations.
   a. Construct a stem-and-leaf plot of the ages. What can you conclude about the distribution? From the shape of the distribution, do you expect the median and mean to be about the same, the median to be much greater than the mean, or the mean to be much greater than the median?
   b. Find the mean, median, standard deviation, and interquartile range.
   c. Determine any outliers and potential outliers. Was President Clinton especially young when he was inaugurated compared to the others? Was president Obama? Was President Reagan especially old?

Washington	57	Lincoln	52	Hoover	54
J. Adams	61	A. Johnson	56	F. Roosevelt	51
Jefferson	57	Grant	46	Truman	60
Madison	57	Hayes	54	Eisenhower	61
Monroe	58	Garfield	49	Kennedy	43
J.Q. Adams	57	Arthur	51	L. Johnson	55
Jackson	61	Cleveland	47	Nixon	56
Van Buren	54	B. Harrison	55	Ford	61
W. Harrison	68	Cleveland	55	Carter	52
Tyler	51	McKinley	54	Reagan	69
Polk	49	T. Roosevelt	42	G.H.W. Bush	64
Taylor	64	Taft	51	Clinton	46
Fillmore	50	Wilson	56	G.W. Bush	54
Pierce	48	Harding	55	Obama	47
Buchanan	65	Coolidge	51		

17. A large statistics class took an exam. The scores followed a normal curve with mean 78 and standard deviation 10. Students who scored 90 or above got an A on the exam. Those who scored between 80 and 90 got a B, between 70 and 80 a C, between 60 and 70 a D, and below 60 an F. Estimate the percentage of students who got each of the grades A, B, C, D, and F.

18. In 1994 the Math SAT scores of male high school students nationwide followed a normal distribution with mean 500 and standard deviation 120. The Math SAT scores of female high school students followed a normal distribution with mean 460 and standard deviation 110.
   a. Estimate the percentage of the young men who scored 800 or above. (Any score above 800 is reported as 800.)
   b. Estimate the percentage of the young women who scored 800 or above.

19. In 2001 the Math ACT scores followed a normal curve with mean 21 and standard deviation 5. What score would place you in the top 1% (a 99th percentile rank)?

PROFESSIONAL PROFILE

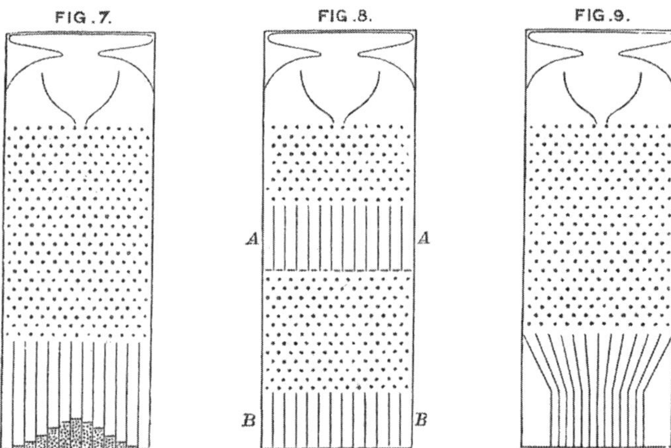

Sir Francis Galton's drawing of his bean machine, also known as a Galton box. The dots in the diagram show the placement of pins on a vertical board. When balls are dropped from the top, they bounce through the pins and fall into one-ball columns at the bottom. The height of the columns of collected balls comes close to a bell curve. The box demonstrates the central limit theorem and the normal distribution.

Sir Francis Galton (18221911), British polymath, inventor, meteorologist, psychometrician, and statistician.

Sir Francis Galton portrait from the 1850s or 1860s.

For the use of fingerprints in crime investigations and the appearance of regular weather reports in your newspaper, you can thank Sir Francis Galton. A wealthy British aristocrat, Galton made many practical contributions to society.

Galton observed fingerprints being used as a means of individual identification by Sir William James Herschel in India. Others had also seen their potential for identification. The problem was the lack of both scientific information to support their uniqueness and a classification system that could match recovered fingerprints to a specific individual. These hurdles were overcome by Galton over a ten-year period. His book *Finger Prints,* published in 1892, set forth a scientific basis for their use in forensics. With this scientific backing, fingerprint evidence became more readily accepted by the courts.

A dedicated man of measurement, Galton had an acute interest in the weather. He was the first to propose the theory of anticyclones, and he developed a complete record of climatic events for the European continent. He used such information to develop the first newspaper weather map (published in *The Times* [of London] on April 1, 1875). Today's newspaper weather maps can be traced to Galton's pioneering achievements in meteorological data collection and analysis.

Because of his development of eugenics, Galton is often overlooked by modern historians of statistics. The astute statistics historian, however, will not make that mistake. Galton introduced the concept of correlation and the use of the regression line. He described and explained the common phenomenon of "regression toward the mean." His interests included both mental abilities and physical traits. Inspired by the work of his cousin, Charles Darwin, he wondered whether human traits were hereditary or due to environmental factors. He coined the phrase "nature versus nurture." A study of the probability of the extinction of aristocratic surnames (conducted in cooperation with Rev. Henry William Watson) resulted in the Galton-Watson stochastic processes.

Although Galton's theories of eugenics are now widely discredited, his contributions to the use and understanding of data are lasting legacies that have influenced the lives of us all.

3

Linear Relationships: Regression and Correlation

One has to draw the line somewhere.

Objectives

After studying this chapter, you will understand the following:

- The scatterplot as a graph of two-variable (bivariate) data
- Correlation as a measure of how well a straight line fits two-variable data
- Finding the best-fitting regression line using the least squares method
- Uses and importance of regression
- Coefficient of determination
- Residual SD
- Correlation does not imply causation
- Residual plots
- Outliers, leverage, and influential points
- Nonlinear relationships in bivariate data

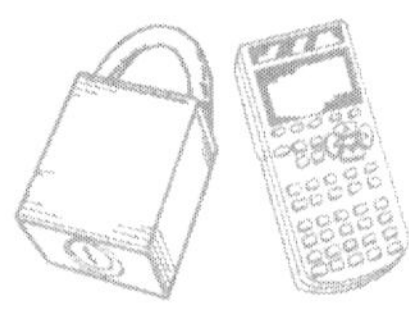

KEY PROBLEM

Building a Linear Model of Data

Americans have a love affair with big, powerful cars. But when the price of gas topped $4 per gallon for a couple of months in 2008, we had to ask: Do heavier cars get poorer gas mileage? We might think so, but can we quantify this relationship between weight and gas mileage? Table 3.1 gives the weight and city gas mileage data for 38 car models. Going through this table of data as it stands, it is hard to tell how the two variables, weight and mileage, are related. In particular, can we estimate the average fuel economy of all vehicles of a particular weight? How much can we trust our estimate? Would the Environmental Protection Agency (EPA) be willing to use our results? As a first step, we graph the data to see if there is an underlying pattern in the data.

Figure 3.1 is a **scatterplot** of the data. Each point in the plot represents one of the models from Table 3.1. For example, the point farthest to the right in the graph is the Cadillac Escalade. From the table we see that the Cadillac Escalade weighs 5814 pounds and gets 13 miles per gallon in city driving. As a review of graphing points, note that one can approximately read these coordinates of the Cadillac Escalade from the graph: about (5800, 13).

We can look at the scatterplot of the data and ask whether there appears to be an approximately linear relationship (straight line) between car weight and gas mileage. We see in the plot that, generally, the heavier the car, the lower the gas mileage, and we see that this trend seems rather linear in nature. Our Key Problem is to find out how to summarize such a linear relationship between two variables by graphing a line in the scatterplot and writing an equation for this line. This so-called **regression** problem has two interrelated components in its solution: (1) the relationship itself, as described by the equation and graph of the regression line, and (2) a measure of how well the regression line fits the data and, hence, how useful it is. For example, the EPA might want to estimate the average gas mileage for all cars weighing 3500 pounds and then assess the accuracy of this estimate.

The scatterplot also shows that there are two outliers in the data set: Two vehicles do not follow the overall pattern of the rest of the data. The Ferrari F430 Spider is a $200,000 sports car that gets only half the miles per gallon that we would expect from its weight. (If you care about fuel economy, then you cannot afford the Ferrari anyway!) And the Ford Escape Hybrid gets almost twice the miles per gallon that we would expect from its weight. The reason for this is that in city driving the Hybrid operates mostly on electricity—which puts the Ford Escape Hybrid in a class by itself among the vehicles considered in Table 3.1.

Table 3.1 Auto Weights and Mileages

Car Model (2006)	Curb Weight (pounds)	City MPG
Acura RSX	2734	27
Audi A4	3428	23
BMW 550i	3803	16
Cadillac Escalade EXT	5814	13
Chevrolet Aveo	2365	26
Chrysler 300	3712	21
Chrysler Town & Country	3899	19
Dodge Caravan	3763	20
Ferrari F430 Spider	3351	11
Ford Crown Victoria	4057	17
Ford Escape XLS	3256	22

Table 3.1 Auto Weights and Mileages

Ford Escape Hybrid 4WD	3792	33
Ford Expedition	5607	14
Ford Focus	2654	26
Ford Fusion	3101	23
Ford Mustang	3300	19
Ford Taurus	3307	20
Honda Accord LX	3197	24
Honda Civic DX	2690	30
Hummer H3	4700	16
Hyundai Accent GLS	2366	32
Infiniti M35	3880	17
Infiniti QX56	5368	13
Jeep Grand Cherokee SRT8	4731	12
Kia Rio SX	2438	32
Lexus LS 430	3990	18
Lincoln Navigator	5555	13
Lincoln Towncar	4502	17
Mercedes-Benz S600	4610	12
Nissan Sentra	2513	28
Pontiac Grand Prix	3477	20
Pontiac Vibe AWD	2980	26
Rolls-Royce Phantom	5577	12
Saab 9-3	3175	18
Saturn ION2	2752	26
Toyota Camry	3108	24
Toyota Corolla	2550	32
Volvo V70R	3646	18

Source: http://www.theautochannel.com

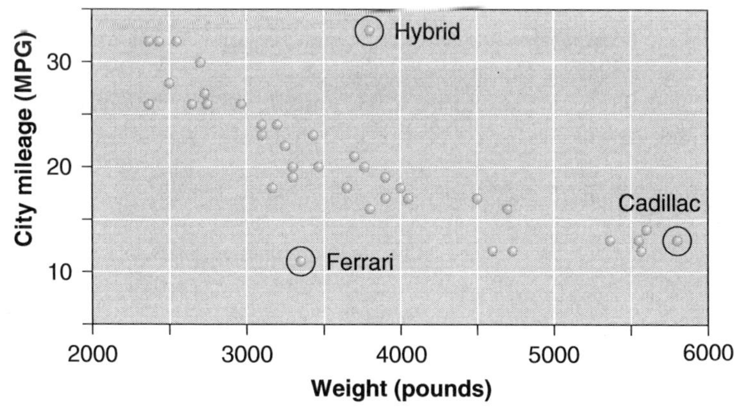

Figure 3.1 Miles per gallon in city driving versus weight for cars in Table 3.1.

3.1 SCATTERPLOTS

So far in this book, we have been dealing with one variable at a time. However, in many situations we want to know how two (or more) variables are related. For example, a pediatrician needs to know how the height and weight of little girls typically increase with age so that she can make sure her young patients are developing normally.

scatterplot
explanatory variable
response variable

The focus of this chapter is to study sample data that come in pairs like the data in Table 3.1. A graph of such data is called a **scatterplot**, such as Figure 3.1. In the Key Problem we think of weight as the **explanatory variable** and city mileage as the **response variable** since the weight of a car helps explain its gas mileage. In Figure 3.1 we plot the explanatory variable on the horizontal axis (the x-axis) and the response variable on the vertical axis (the y-axis).

negatively correlated

Figure 3.1 shows a negative correlation between weight and mileage. Two variables are said to be **negatively correlated** if above-average values of one variable tend to accompany below-average values of the other, and vice versa. The scatterplot slopes downward from left to right.

positively correlated

Two variables are said to be **positively correlated** if above-average values of one variable tend to accompany above-average values of the other and below-average values of one variable tend to accompany below-average values of the other. The scatterplot slopes upward from left to right.

Example 3.1

Cancer is one of the leading causes of death in the United States (see Example 1.4) and worldwide. Although overall cancer rates do not vary much around the world, the types of cancer do. In most affluent countries, cancers of the lung, colon, breast, and prostate are more common. In poorer countries and in the Far East, cancers of the stomach, liver, mouth, esophagus, and uterine cervix are more common. Heredity no doubt plays a role, but migrant studies show that a lot of the variation is attributable to environmental factors. For example, women of Puerto Rican descent living in New York City have a much higher breast cancer rate than women living in Puerto Rico. As people around the world adopt Western eating habits and lifestyles, they develop the same types of cancer as people in the West. Researchers have tried to identify environmental risk factors for cancer. They suspect that a diet rich in animal fat increases a person's risk for colon, prostate, breast, or ovarian cancer. The scatterplot in Figure 3.2 relates the age-adjusted annual breast cancer mortality rates (MR) for 30 countries and the average daily consumption (in grams) of animal fat per capita.

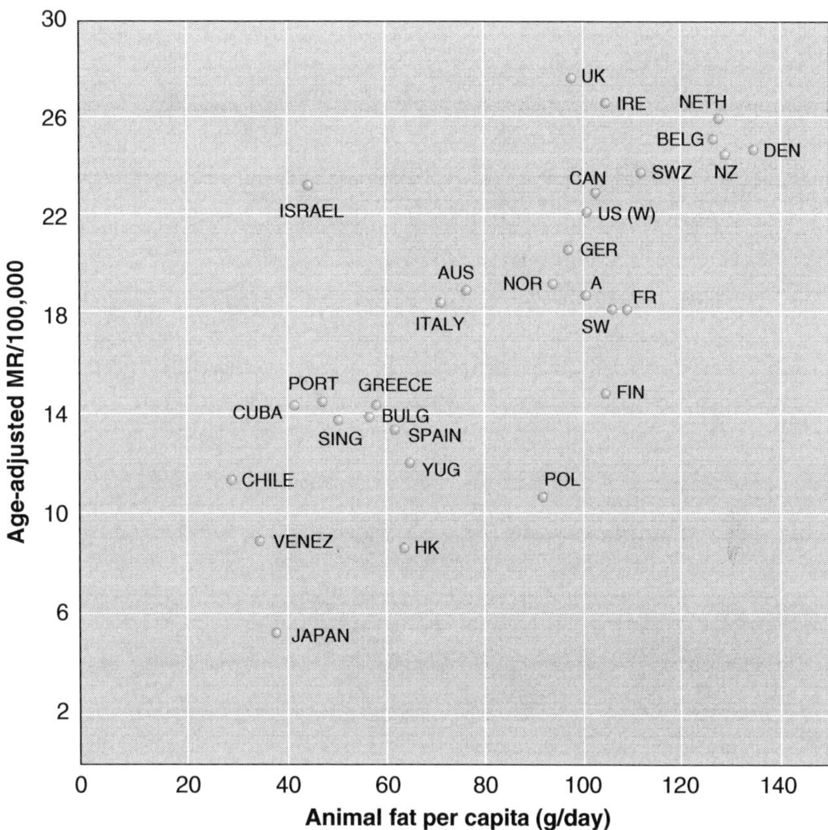

Figure 3.2 Scatterplot of age-adjusted breast cancer mortality rates (per 100,000 population) for 30 countries against average daily consumption of animal fat per capita. (*Source:* Rose et al. Comparisons of Mortality Rates. *Cancer* 1986; 58:2363–2371.)

Figure 3.2 shows a positive correlation between animal fat consumption and breast cancer rates: Countries with low per capita intake of animal fat have lower breast cancer rates than countries with high per capita intake of animal fat. (Israel is an outlier, with a high rate of breast cancer despite low average animal fat consumption.)

Figure 3.3 shows that there appears to be no relationship between breast cancer risk and consumption of vegetable fat. Greek women in particular seem to suffer no ill effects from the large amounts of olive oil they consume.

Epidemiologic studies such as the one summarized in Figure 3.2 make animal fat a suspect in breast cancer deaths. But there are other suspects, such as early onset of puberty, lack of regular physical activity, and excess weight, all of which are common in the West.

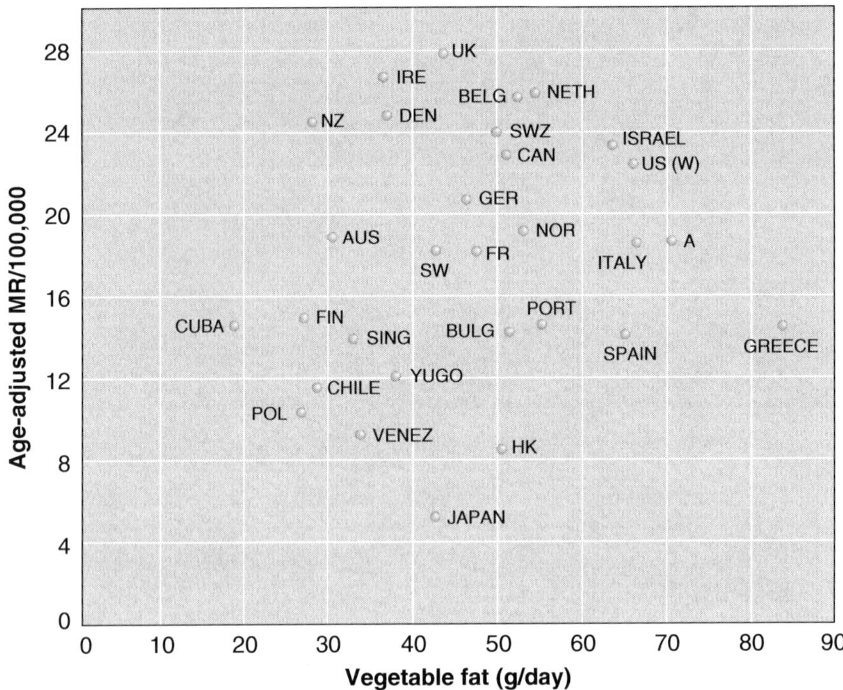

Figure 3.3 Scatterplot of age-adjusted breast cancer mortality rates (per 100,000 population) for 30 countries against average daily consumption of vegetable fat per capita.

Section 3.1 Exercises

1. What two variables are of interest when each of these questions is asked? Identify the explanatory variable and the response variable in each case.
   a. Do children who view a lot of television do poorly in school?
   b. Does a copper bar expand when heated?
   c. Do people who drink more milk become sick less often?
   d. Do tires with higher inflation (more air in them) give better mileage than those with lower inflation?

2. Professor Eron of the University of Illinois did a study of the television viewing habits of 875 third-grade children. The following excerpt reports his conclusions:

   There is a strong positive relationship between the violence rating of favorite programs watched, whether reported by mothers or fathers, and aggression of boys as rated by their peers in the classroom. There was no significant relationship when TV habits of girls were reported either by mothers or fathers. (*Journal of Abnormal and Social Psychology,* vol. 67, 1963, p. 195.)

   What paired variables are of interest in this study?

3. The following extract reports on a relationship between variables. What are the variables, and what is the relationship? (*Hint:* Pick whatever pair of variables you like.)

   As people get older there is a marked decline in sports involvement—except for the viewing of sports on television. The average time devoted to active sports per week is 5.1 hours for those aged 18 to 24, compared with 1.7 hours for those 25 and over. But the time spent watching TV sports increases somewhat for all those over 25.

4. Identify paired variables whose relationships are discussed in the following extract:

   During the months immediately following the 1974 gasoline shortage, it was noted that deaths from vehicular accidents were on a significant decline for the first time in decades.

   The phenomenon was hailed at the time as one of the few benefits from the fuel crunch.

   More recently, notice has been paid to other statistics from that period which indicate a drop in deaths from causes other than highway accidents, but which some researchers believe could have a tie-in to reduced driving. San Francisco County, for one, noted a 13% decline in deaths from all causes during a three-month period at the height of the gasoline shortage.

   A decrease of almost 33% in deaths from chronic lung disease was noted, along with a 16% decline in deaths from cardiovascular diseases. It is possible, of course, that a combination of factors brought about these unexpected results. But a number of driving-related causes also have been suggested: less stress from driving less, fewer pollutants in the air, more walking, and simply more opportunity to relax at home are among suggested possibilities. No definite conclusions have been reached, but the correlations are intriguing. (Champaign-Urbana *News-Gazette,* October 21, 1975.)

5. Imagine that we have a representative sample of American families. For each of the following pairs of variables, state whether the correlation is positive or negative.
   a. Age of husband and age of wife.
   b. Resale value and age of the family car.
   c. Husband's age and his year of birth.
   d. Height of father and (adult) height of son.

6. The following table gives the mean distance R from each of the nine planets to the Sun and the mean length of time T it takes the planet to complete a revolution around the Sun. The distance R is measured in astronomical units. One astronomical unit equals the mean distance between the Earth and Sun, about 149.6 million kilometers, or 93 million miles. The time T is measured in years. In addition, the table gives the natural logarithms of R and T, $\ln(R)$ and $\ln(T)$. On graph paper, draw a scatterplot for the paired values of $\ln(R)$ and $\ln(T)$—nine points in all, one for each planet. Does there appear to be a linear relationship between $\ln(R)$ and $\ln(T)$?

Planet	Distance from Sun *R* (astronomical units)	Revolution time *T* (Years)	ln(*R*)	ln(*T*)
Mercury	0.387	0.241	−0.95	−1.42
Venus	0.723	0.616	−0.32	−0.48
Earth	1.000	1.000	0.00	0.00
Mars	1.523	1.88	0.42	0.63
Jupiter	5.20	11.86	1.65	2.47
Saturn	9.55	29.46	2.26	3.38
Uranus	19.22	84	2.96	4.43
Neptune	30.11	165	3.40	5.11
Pluto	39.4	248	3.67	5.51

Use the scatterplot to estimate the slope m of the equation

$$\ln(T) = m \cdot \ln(R).$$

You are deriving Kepler's third law of planetary motion, a famous milestone in astronomy. Kepler deduced his three planetary laws 400 years ago, using Tycho Brahe's careful astronomical observations—made without the aid of a telescope! Kepler's laws in turn enabled Newton to formulate his theory of gravitational force.

7. Ask at least five of your friends or relatives their weights and heights. Draw a scatterplot for weight versus height.

Is there a linear relationship between the two variables?

8. Find an article in a newspaper or a magazine that deals with a relationship between two variables. What are those variables? What relationship between those two variables is presented or discussed?

9. The following figure shows the amounts of nicotine (in milligrams) and tar (in milligrams) for 25 brands of cigarettes:

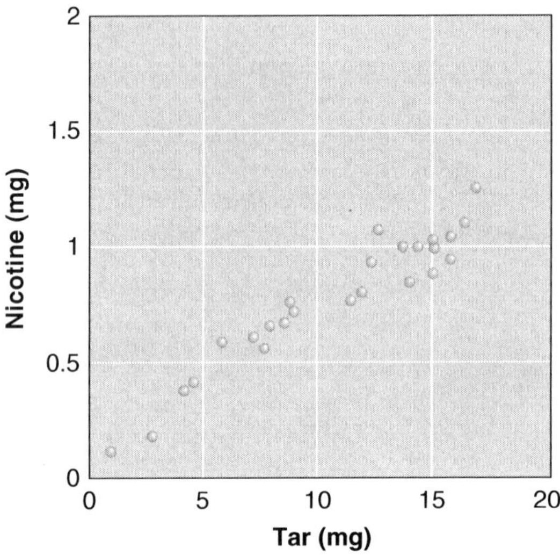

Is the correlation between amounts of tar and nicotine positive or negative?

10. The following figure shows the city and highway mileages (miles per gallon) for 92 cars (1993 models):

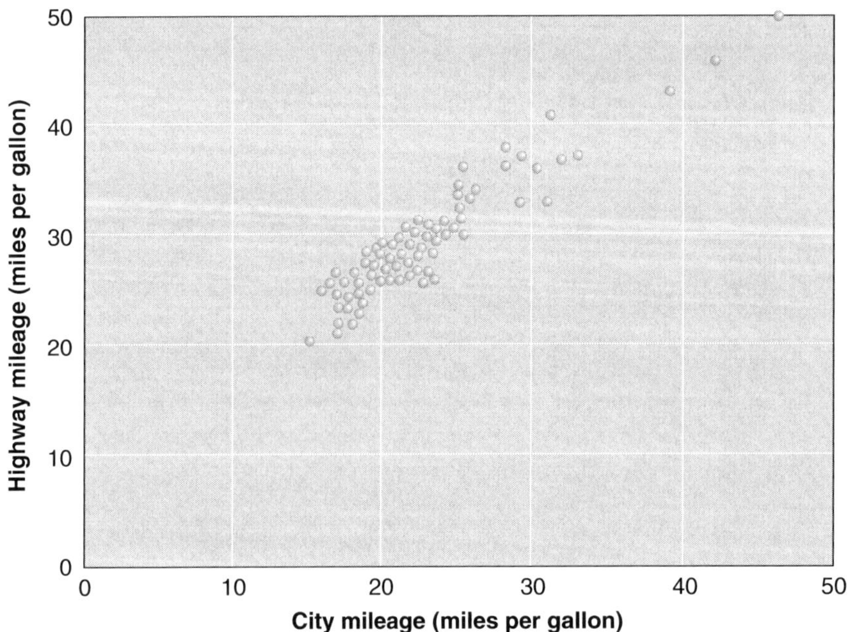

a. Is the correlation between city mileage and highway mileage positive or negative?
b. Is there a cause-and-effect relationship between the two variables? If not, how do you explain the correlation?

11. For each of three response variables—blood pressure, height, and weight—pick at least three possible explanatory variables from the following list: age, height, father's height, mother's height, level of stress, number of calories consumed per day, number of hours of exercise per week, and weight. For each pair of variables, indicate whether the correlation is positive or negative.

12. The following figure shows the infant mortality rate (deaths per 1000 of population for babies less than one year old) and the female life expectancy in years for 97 countries.

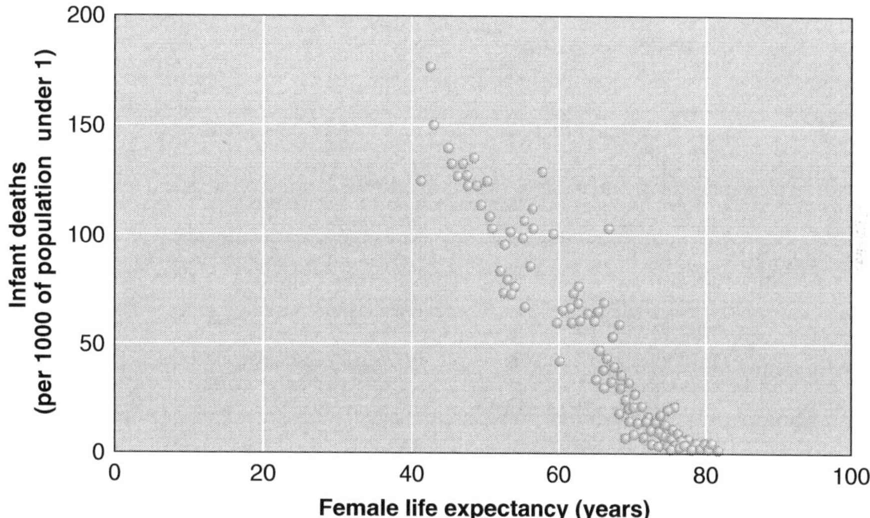

a. What is the range in infant mortality for these countries?
b. Estimate the median infant mortality rate for these countries.
c. What is the range in female life expectancy for these countries?
d. Estimate the median female life expectancy.
e. What is the general relationship between infant mortality and female life expectancy? Is the correlation positive or negative?
f. Does this relationship make sense? Explain.

3.2 THE CORRELATION COEFFICIENT

strong
perfect linear correlation
strong nonlinear relationship
linear correlation

Suppose that the points of a scatterplot can be described as clustering around a sloped straight line as in Figures 3.1 and 3.2. In addition to characterizing the linear correlation as either positive or negative, depending on whether the scatterplot slopes up or down, we would like to characterize the strength of the linear correlation. We say that the linear correlation is **strong** if the points cluster tightly around the line. We have a **perfect linear correlation** if all the points lie on the line. Figure 3.4 illustrates the different types of correlation. Figure 3.4(h) reminds us that the relationship between two variables need not be linear. (Kepler's third law: $T = R^{3/2}$, is an example of a nonlinear relationship. See Exercise 6 of Section 3.1.) In this text, however, we shall only study linear relationships. Although there is a **strong nonlinear relationship** between x and y in Figure 3.4(h), there is no **linear correlation** between the two variables.

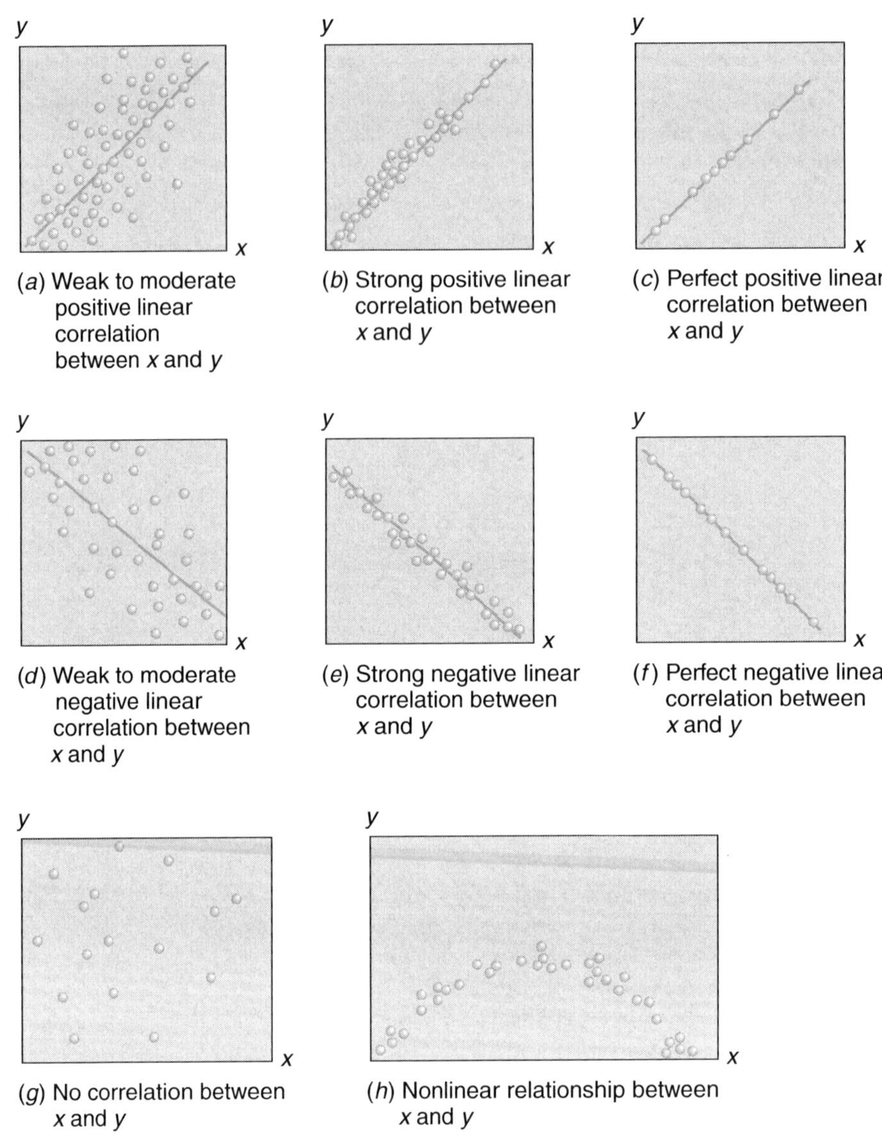

Figure 3.4 Scatterplots of types of linear relationships.

linear correlation coefficient r

To obtain an objective measure of the linear correlation between paired x- and y-values, the statistician Karl Pearson introduced the **linear correlation coefficient r**. Later in this section we shall give the formula for computing r. The value of r is always between -1 and $+1$. If there is a positive correlation between x and y (the scatterplot slopes upward), then r is positive. If there is a negative correlation between x and y (the scatterplot slopes downward), then r is negative. A value of $+1$ corresponds to a perfect positive linear correlation. The tighter the points of a scatterplot cluster around a straight line with a positive slope, the closer r will be to $+1$. A value of -1 corresponds to a perfect negative linear correlation. The tighter the points of a scatterplot cluster around a straight line with negative slope, the closer r will be to -1. A correlation coefficient close to 0 indicates that the points are not clustered around any sloped straight line. In short, r tells us how tightly the points of a scatterplot cluster around a sloped line. Therefore, r measures the strength of the **linear relationship** between the paired x- and y-values. We can classify the strength of the linear correlation as strong, weak to moderate, or negligible, according to the magnitude of r, as indicated in Figure 3.5.

linear relationship

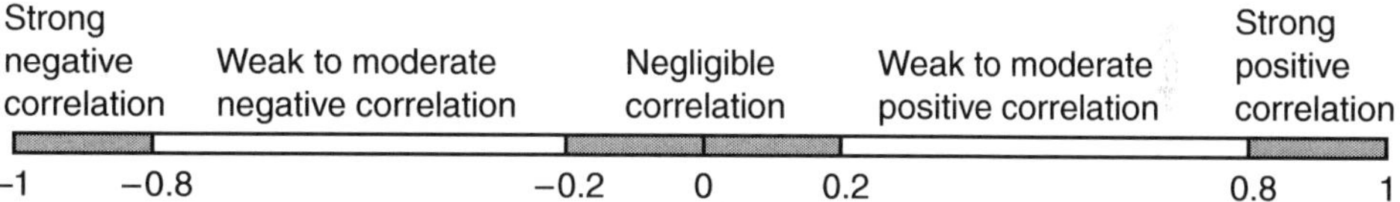

Figure 3.5 Range of values for the correlation coefficient.

As shown in Figure 3.5, there are five natural categories: strong positive, weak to moderate positive, negligible, weak to moderate negative, and strong negative correlation. (Well-informed statisticians might differ among themselves on where to put the cutoff point from one category to the next.) As an illustration, Figure 3.6 contains eight scatterplots. In each scatterplot both variables have mean 0 and standard deviation equal to 1. In the first four plots the variables are positively correlated. In the last three they are negatively correlated.

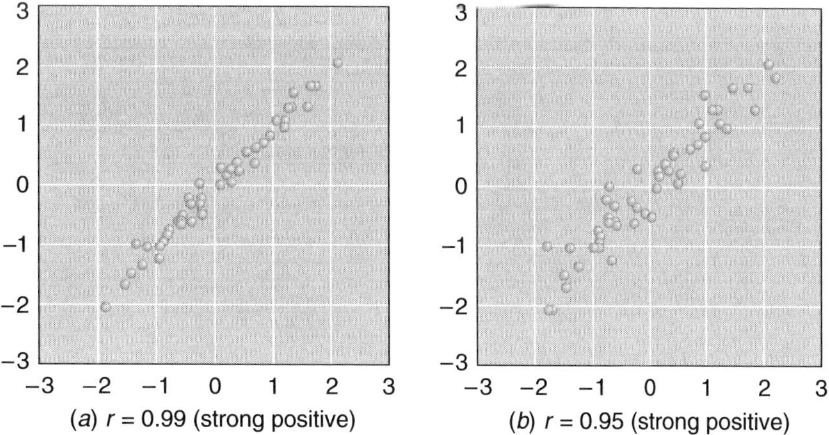

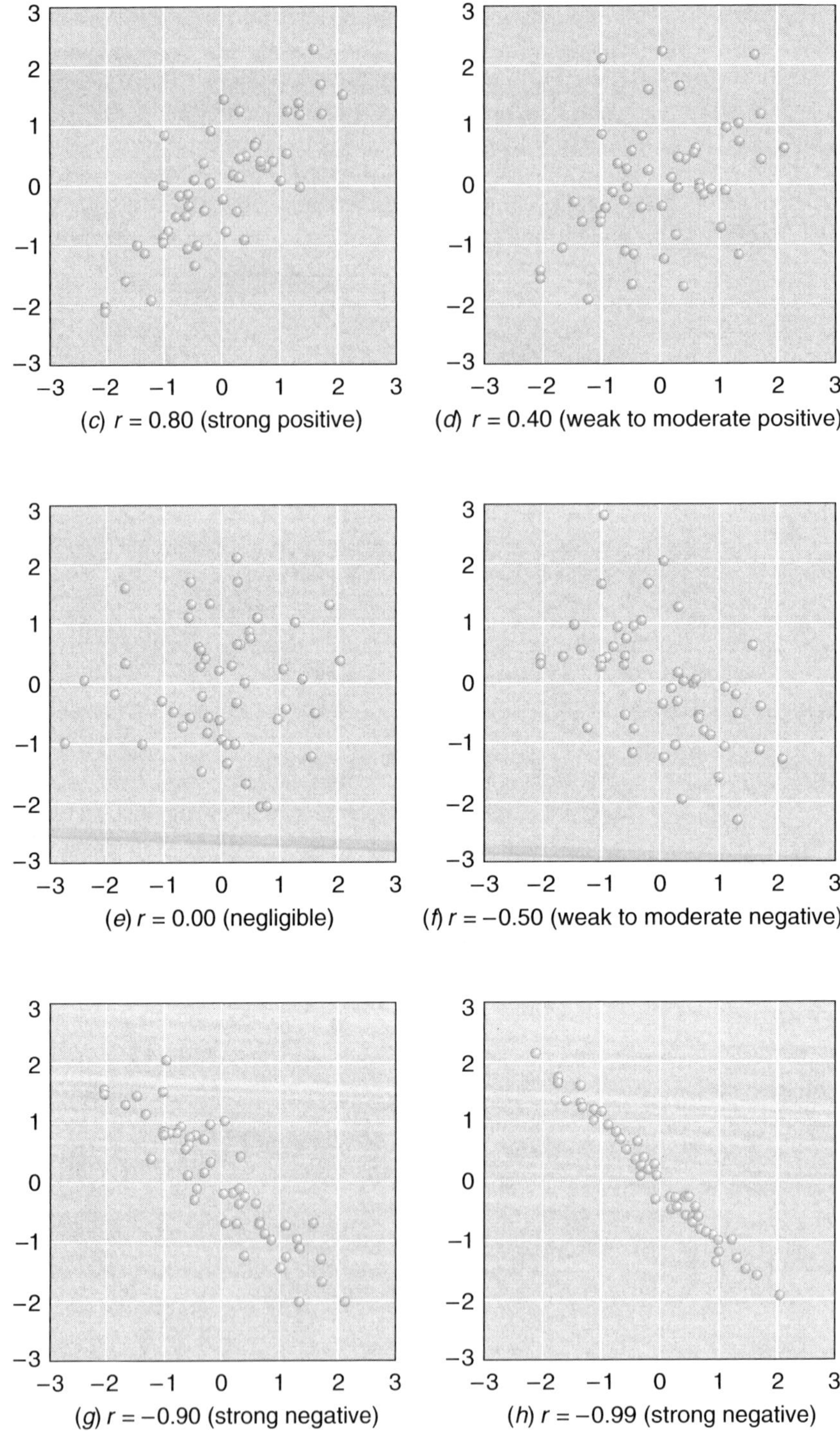

Figure 3.6 Scatterplots illustrating different levels of correlation. In each plot both variables have mean 0 and standard deviation 1.

For the Key Problem, the data in Figure 3.1 seem to lie rather close to a negatively sloped straight line; the correlation $r = -0.802$ is close to -1. This is an example of a strong negative correlation.

After we consider more real data examples, you will learn how to compute the correlation.

Example 3.2

Figure 3.7 shows measurements of the length (x) and width (y) of the palate (the roof of the mouth) of each of 44 male kangaroos. (We would have to ask a zoologist exactly how one defines these two quantities.) As expected, they are positively related. Both reflect the overall size of the kangaroo. So a kangaroo with a larger palate length is likely a larger kangaroo and hence tends to have a larger palate width, as well. The correlation coefficient between x and y, when calculated, is 0.81, indicating a strong positive correlation.

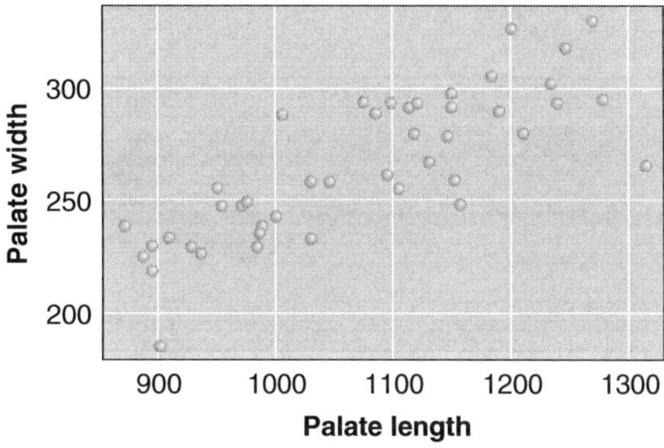

Figure 3.7 Palate width versus palate length of male kangaroos.

Example 3.3

For male high school students, there is a strong positive correlation between performance in high jump and performance in long jump, since both reflect jumping ability, leg strength, and training in events that require jumping: Students with athletic ability and training do well in both, and students without do poorly in both. In the men's decathlon at the Olympics, however, all competitors are both highly trained and very talented. Unusual skill relative to other decathlon contestants in one of the 10 events of the decathlon does not necessarily translate into similar superior skill in the other events at this high level of performance.

Figure 3.8 shows the scatterplot of high jump and long jump results of 34 participants in the men's decathlon in the 1988 Olympics. The variables are measured in meters. There is clearly a positive linear relationship between long jump and high jump performance, but it is not as strong as in Example 3.2. The correlation coefficient between the two variables is only 0.47. This is a weak to moderate positive correlation.

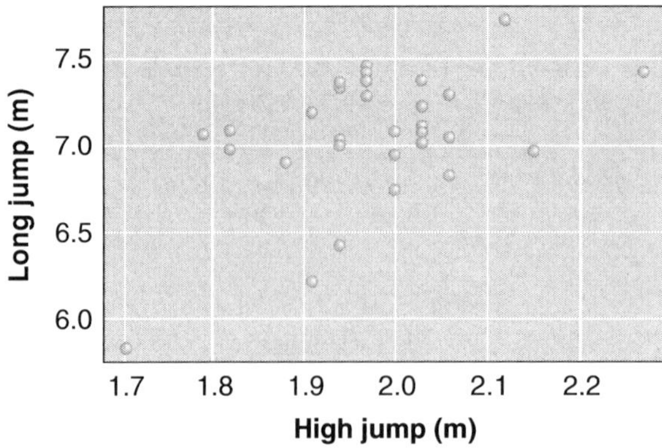

Figure 3.8 Long jump versus high jump results in 1988 men's Olympic decathlon.

Example 3.4

People are always searching for systems to make money by investing in the stock market. The Dow Jones Industrial Average (DJIA), reported continuously, is a widely quoted stock market index based on the combined worth of 30 of America's giant, blue chip companies. One might think that if the DJIA increases by an unusually large percentage one year, the companies it comprises are doing extremely well that year and are likely to continue doing so. Consequently, the DJIA ought to increase by an above-average percentage again the following year. Similarly, a decrease one year should predict a poorer than typical performance the following year. Figure 3.9 shows that we cannot make such predictions.

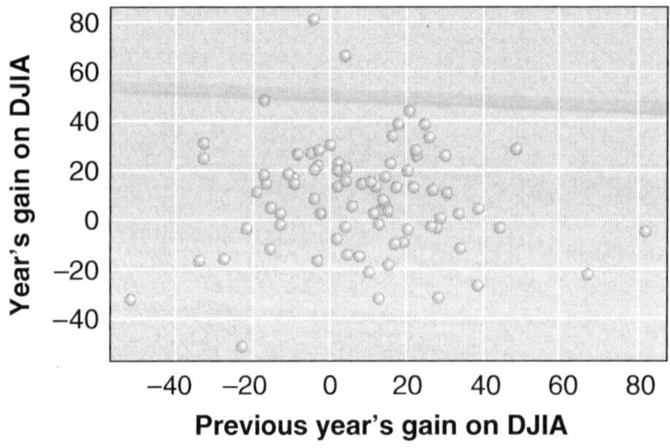

Figure 3.9 Current versus previous year's annual returns on Dow Jones Industrial Average.

Figure 3.9 shows 84 points, one for each year from 1916 through 1999. Each point has the previous year's percentage increase of the DJIA as its x-coordinate and the current year's percentage increase of the DJIA as its y-coordinate. So, for example, the uppermost point of approximately $(-4, 80)$ represents a pair of consecutive years, in the first of which the DJIA lost 4% but in the second of which the DJIA gained 80%. Somewhat surprisingly, perhaps, there appears to be no positively sloped linear relationship in this scatterplot. The correlation coefficient, when calculated, is close to zero $(r = -0.03)$, so it represents a negligible correlation. A big gain in the stock market this year does not tell us anything about the likely overall gain next year! We cannot predict next year's gain or loss from this year's gain or loss. Note, however, that Figure 3.9 shows that the average percentage increase per year $(\overline{y})$ is positive, reflecting the fact that over the long run the stock market has been a good investment.

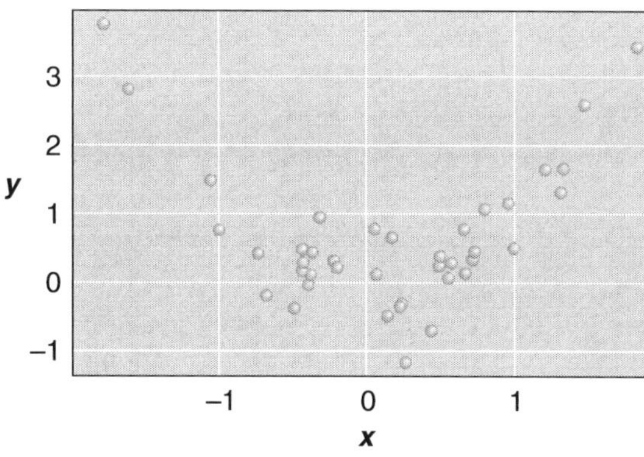

Figure 3.10 A scatterplot with a strong nonlinear relationship but near zero correlation.

Example 3.5

Figure 3.10 shows a scatterplot for the paired variables x and y (no physical interpretation given). There is a strong bowl-shaped (parabolic) relationship between x and y. But the correlation coefficient, when calculated, is close to 0. Why does r fail to detect this strong relationship? Is there a flaw in the correlation coefficient? No. This is because the correlation coefficient measures only the strength of a linear relationship, not other types of relationships. Thus, we should always be careful to refer to the correlation coefficient r as describing the strength of the linear relationship between two variables. We should not jump to the conclusion that there is no relationship between the variables just because r is close to 0.

Note, however, that if we have the scatterplot, we are unlikely to conclude mistakenly that there is no relationship in the case of a small r if the scatterplot shows a significant nonlinear relationship.

Often a strong nonlinear relationship will produce a large correlation coefficient, thus suggesting a linear relationship. But when a correlation coefficient is large, we do not want to report incorrectly that there is a linear relationship if the true relationship is strongly nonlinear. Therefore, the scatterplot is essential in analyzing a relationship.

How to Calculate the Correlation Coefficient r

For a set of paired x- and y-values, we can calculate the means $\overline{x}$ and $\overline{y}$ and the standard deviations s_x and s_y by using the formulas for mean and standard deviation discussed in Chapter 2. Note that we subscript s with either x or y here because there are two standard deviations, one for the x-values and one for the y-values. The subscript helps us distinguish them from each other.

The **correlation coefficient** for a set of paired x- and y-values is defined as

formula for correlation coefficient

$$r = \frac{1}{n-1}\sum\left(\frac{x-\overline{x}}{s_x}\right)\left(\frac{y-\overline{y}}{s_y}\right) = \frac{1}{(n-1)}\frac{1}{s_x s_y}\sum(x-\overline{x})(y-\overline{y}),$$

where n is the number of (x, y) pairs in the data and $\sum$ indicates summation over all the pairs. The following example shows how the computation of r is done. Note that in complicated formulas the multiplication sign is often omitted, so the product of x and y is indicated simply by xy.

Example 3.6

We shall compute r for the following hypothetical set of eight data points (x, y):

x	2	5	8	5	4	6	3	7
y	4	6	16	10	12	8	10	14

It is easily seen that $\overline{x} = 5$, $s_x = 2$, $\overline{y} = 10$, and $s_y = 4$. We standardize each of the x-values by first subtracting $\overline{x}$ and then dividing by s_x. We standardize each of the y-values by first subtracting $\overline{y}$ and then dividing by s_y. Next we compute the products of the corresponding standardized scores, add up the products, and divide by $n - 1$, yielding $r = 5.00/7 = 0.714$.

x	y	$\frac{x-5}{2}$	$\frac{y-10}{4}$	$\left(\frac{x-5}{2}\right)\left(\frac{y-10}{4}\right)$
2	4	−1.5	−1.5	2.25
5	6	0.0	−1.0	0.00
8	16	1.5	1.5	2.25
5	10	0.0	0.0	0.00
4	12	−0.5	0.5	−0.25
6	8	0.5	−0.5	−0.25
3	10	−1.0	0.0	0.00
7	14	1.0	1.0	1.00
				5.00

❐ ❐

Look at the two expressions that are multiplied together to make each term in the summation that defines r:

$$\frac{x - \overline{x}}{s_x} \quad \text{and} \quad \frac{y - \overline{y}}{s_y}$$

standardized score
z-score

Each of these expressions is a **standardized score** or **z-score**.

steps in computing r

Steps in computing the correlation r

Step 1. Standardize the x-values by first subtracting $\overline{x}$ and then dividing by s_x.

Step 2. Standardize the y-values by first subtracting $\overline{y}$ and then dividing by s_y.

Step 3. Multiply each standardized x by its associated standardized y.

Step 4. Divide the sum of the products from Step 3 by $n - 1$ (rather than by n, just as we did when we defined the variance and the standard deviation in Section 2.3).

Note that r is almost the average of the products of the standardized scores; if we divided by n rather than by $n - 1$, it would be exactly the average.

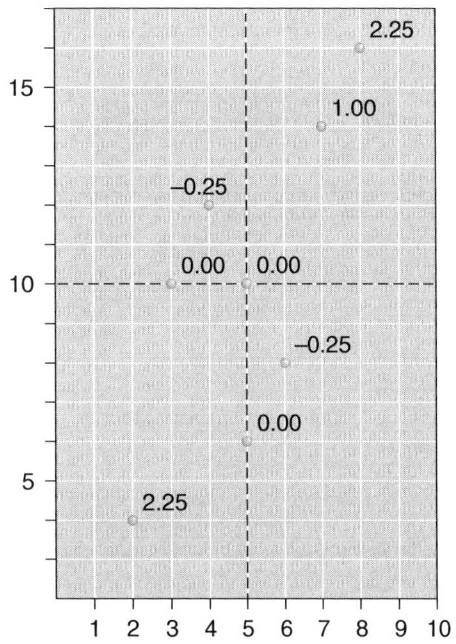

Figure 3.11 Scatterplot for the data in Example 3.6.

Figure 3.11 is the scatterplot for the eight points in Example 3.6. We have drawn a vertical and a horizontal line through the "point of averages" with coordinates $(5, 10)$. The lines divide the scatterplot into four quadrants. Next to each of the eight points we have written the product of the two z-scores, $(x - 5)/2$ and $(y - 10)/4$, that we computed to determine r in Example 3.6. For points in the top right quadrant, both z-scores are positive and hence the product is positive. For points in the bottom left quadrant, both z-scores are negative, and again the product is positive. For points in the top left quadrant and in the bottom right quadrant, one z-score is positive, and the other is negative, so the product is negative. Because the scatterplot slopes upward, the points in the top right quadrant and in the bottom left quadrant predominate. The sum of the products is positive, and therefore r is positive. Figure 3.12 illustrates this general fact.

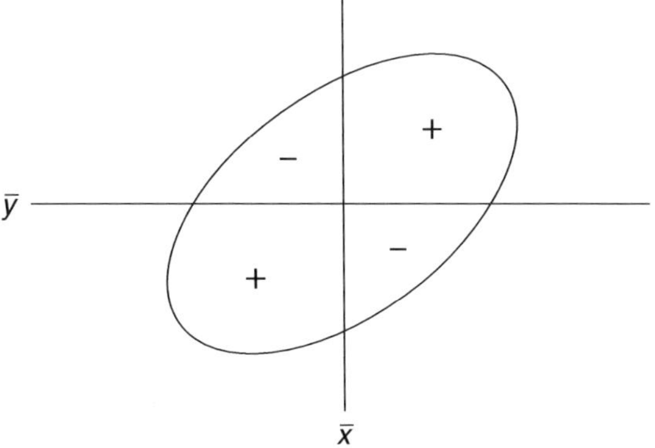

Figure 3.12 The positive products predominate. The correlation coefficient r is positive.

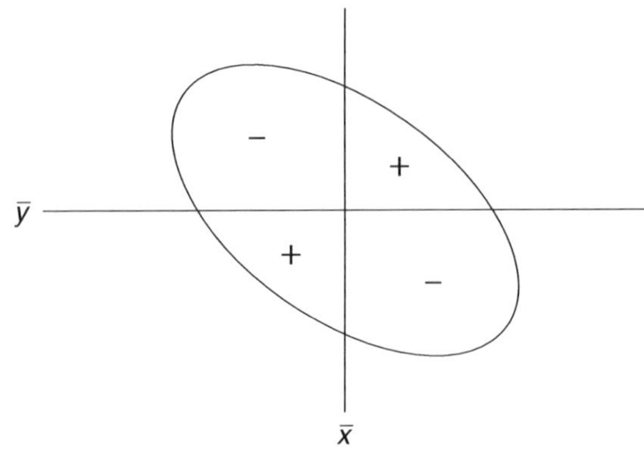

Figure 3.13 The negative products predominate. The correlation coefficient r is negative.

Figure 3.12 is a rough sketch of a scatterplot that slopes upward. We have drawn a vertical and a horizontal line through the "point of averages" with coordinates $(\bar{x}, \bar{y})$. The scatterplot is "centered" at this point. For points in the lower left quadrant and in the upper right quadrant, the product of the z-scores $(x - \bar{x}) / s_x$ and $(y - \bar{y}) / s_y$ is positive. Because these points predominate, the sum of all the products is positive, and therefore r is positive.

Figure 3.13 is a rough sketch of a scatterplot that slopes downward. For points in the upper left quadrant and in the lower right quadrant, the product of the z-scores $(x - \bar{x}) / s_x$ and $(y - \bar{y}) / s_y$ is negative. Because these points predominate, the sum of all the products is negative and therefore r is negative.

Generally you will use computer programs, such as the easy-to-use instructional software available with this textbook, to compute statistics such as r. However, it can be dangerous just to hit a computer key and get an instant "answer" unless you truly understand what the computer is computing from the data. When you learn a new statistical procedure, it *always* helps in the learning process to do a couple of problems "by hand" using a calculator or, perhaps, a spreadsheet.

To compute r using the four steps above takes a great deal of computational work if we have more than 10 paired observations (x, y). There is an easier way, however. Take the definition of the standard deviation of x from Section 2.3:

$$s_x = \sqrt{\frac{\sum(x - \bar{x})^2}{n - 1}}$$

computational formula for s

Applying some algebra converts this formula to the **computational formula for s**:

$$s_x = \sqrt{\frac{\sum x^2 - (\sum x)^2 / n}{n - 1}}$$

This version may look more complicated, but it really requires less computing and saves time. You no longer have to subtract the mean from each entry in the list of numbers. You can set up a spreadsheet-like table with columns for x and x^2, step through the data once to fill out those columns, and use the sums to find the mean and the standard deviation all at once at the end. (See Example 2.10.)

computational formula for the correlation coefficient r

Further algebra work along the same lines leads to the **computational formula for the correlation coefficient *r***:

$$r = \frac{\sum xy - (\sum x)(\sum y)/n}{\sqrt{[\sum x^2 - (\sum x)^2/n][\sum y^2 - (\sum y)^2/n]}}$$

where, of course, $\sum xy$ simply means we multiply each point's coordinates together first and then add these products over all the points. Although this looks complicated (and certainly need not be memorized!), it is determined by five easy-to-compute sums that can be viewed as summary statistics for the paired variables x and y. This expression is faster to compute if only a hand calculator is available. The following example shows the use of this formula to compute r.

Example 3.7

For nine participants in a weight-loss program, find the correlation between the number of weeks spent in the program (x) and the weight lost (y, in pounds) as given in Table 3.2 and shown in Figure 3.14.

Table 3.2 Weight Loss versus Number of Weeks in Program

	Number of weeks in program (*x*)	Total weight loss (*y*)
	12	19
	4	18
	8	8
	4	16
	12	26
	7	12
	5	12
	9	29
	14	31
Total	75	171
Mean	8.33	19.0

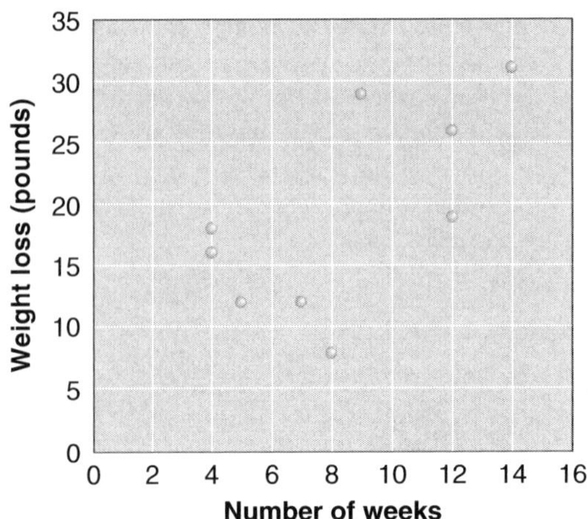

Figure 3.14 Weight loss versus number of weeks for nine people in a weight-loss program.

Solution: We first use the following spreadsheet-like table to compute the five sums with $n = 9$:

Computation of *r* for Weight Loss Data

Person number	x	y	x^2	y^2	xy
1	12	19	144	361	228
2	4	18	16	324	72
3	8	8	64	64	64
4	4	16	16	256	64
5	12	26	144	676	312
6	7	12	49	144	84
7	5	12	25	144	60
8	9	29	81	841	261
9	14	31	196	961	434
Sum	75	171	735	3771	1579
	$= \sum x$	$= \sum y$	$= \sum x^2$	$= \sum y^2$	$= \sum xy$

Now, using the computational formula, we find the correlation:

$$r = \frac{1579 - (75 \times 171/9)}{\sqrt{[735 - (75)^2)/9][3771 - (171)^2/9]}}$$

$$= \frac{154}{\sqrt{110 \times 522}} \approx 0.64.$$

Correlation Is Scale-Free

The correlation coefficient is a pure number, without units, like inches or pounds. It is not affected by

- adding the same number to all the values of one variable
- multiplying all the values of one variable by the same positive number
- switching the two variables

Example 3.8

We can calculate the correlation between car weight and mileage given in Table 3.1 to get $r = -0.802$. (Try it for practice, using our computational formula.) Here, the x variable is weight in pounds and y is miles per gallon. Suppose that a Japanese scientist is studying the same data set but with the weight reported in kilograms and gas mileage reported in kilometers per liter of gas. The numbers will all be different, and in fact there is a scale change (a change in units) in both variables x and y. For example, a weight of 2500 pounds becomes approximately 1136 kilograms, because 1 kilogram $\approx$ 2.2 pounds, and each mile per gallon equals 0.425 kilometers per liter. Would the Japanese scientist obtain a different correlation for the same variables measured in different scales? We certainly hope not! Imagine the chaos that would result if the computed strength of the linear relationship changed based on the units of measurement used. Fortunately, the correlation stays the same regardless of the units used for x and for y. The reason is that, in the definition of r, we multiply x and y converted into z-scores, and z-scores are not affected by scale changes. Therefore, the resulting correlation coefficient is not affected by any scale change in either variable.

scale variance for r

Scale invariance rule for *r*

***r* is the same for all choices of units (linear transformations) for *x* and for *y*.**

Section 3.2 Exercises

1. The following plots show the relationship between two variables. For each graph (*a*) through (*d*), tell by inspection whether the correlation between the two variables is strong positive, weak to moderate positive, negligible, weak to moderate negative, or strong negative.

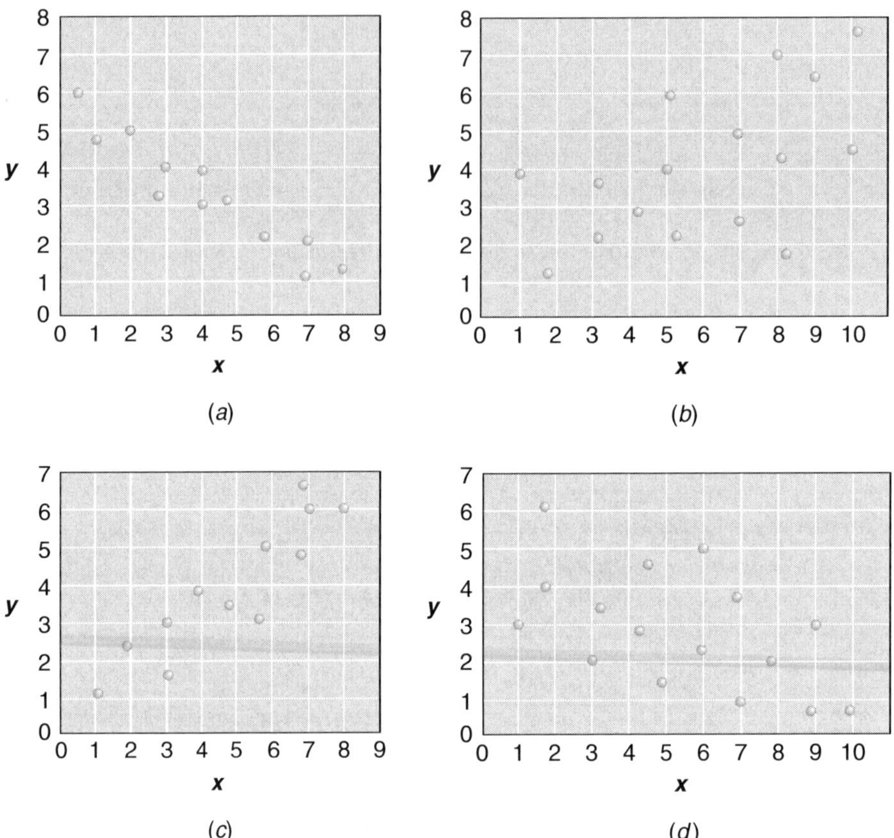

2. Tell whether you would expect the correlation between the two variables in each of the following sets of data, (*a*) through (*g*), to be strong positive, weak to moderate positive, negligible, weak to moderate negative, or strong negative. *Hint:* There may be no clear answer, in that either of two categories of correlation may be reasonable. The point is to get you to think hard about the degree of association.
   a. Height and weight of children from ages 4 to 12 years
   b. Time spent studying for an exam and score on that exam

c. Attitude toward mathematics (whether or not you like mathematics) and final grade in mathematics
d. Popularity in school or college and academic achievement
e. Price of a watch and the number of minutes it gains or loses per week
f. Age of a car and cost per month of repairs
g. Air pressure kept in automobile tires and gasoline mileage associated with the tires

3. The following table gives the daily high and low temperatures in degrees Fahrenheit (°F) in Champaign, Illinois, for 5 consecutive days in April 2000:

Low	35	38	38	40	42
High	63	66	68	69	72

a. Calculate the correlation coefficient between the daily high and low temperatures in two ways. Verify that the answers are the same. *Comment:* Be careful not to round off too much when you compute r using the method in Example 3.6. The computational method used in Example 3.7 is less sensitive to round-off errors.
b. Convert the temperatures to degrees Celsius, $°C = (5/9)(°F - 32)$: this is a change of units. Recompute the correlation between the two variables now measured in Celsius degrees. Is this what you expected? Why? *Comment:* Use at least four significant digits when you convert temperatures in Fahrenheit to Celsius. For example, 35 °F becomes 1.667 °C.

4. The following table gives the midterm and final exam scores of eight students in a mathematics class:

Midterm score x	55	60	69	74	79	88	89	95
Final exam score y	62	74	66	77	82	91	93	60

a. Compute the correlation between the midterm and final exam scores.
b. Draw a scatterplot. Does the correlation summarize the overall relationship between the two exam scores well? Explain.
c. How would you explain the overall relationship between the students' performance in the two exams?
d. Suppose the teacher decides to assign the following simple scoring system to the exam scores: Any score up to 59 is converted into 1, any score between 60 and 69 is converted into 2, any score between 70 and 79 is converted into 3, any score between 80 and 89 is converted into 4, and any score above 90 is converted into 5. What is the correlation between the midterm and the final, using the new exam scores? *Hint:* The new correlation should be close but not identical to the old, because this is only approximately a rescaling, so you actually have to compute it!

5. For each of the four plots that follow, (a) through (d), choose the best description: (i) strong positive correlation, (ii) weak to moderate positive correlation, (iii) very little positive or negative correlation, (iv) weak to moderate negative correlation, (v) strong negative correlation.

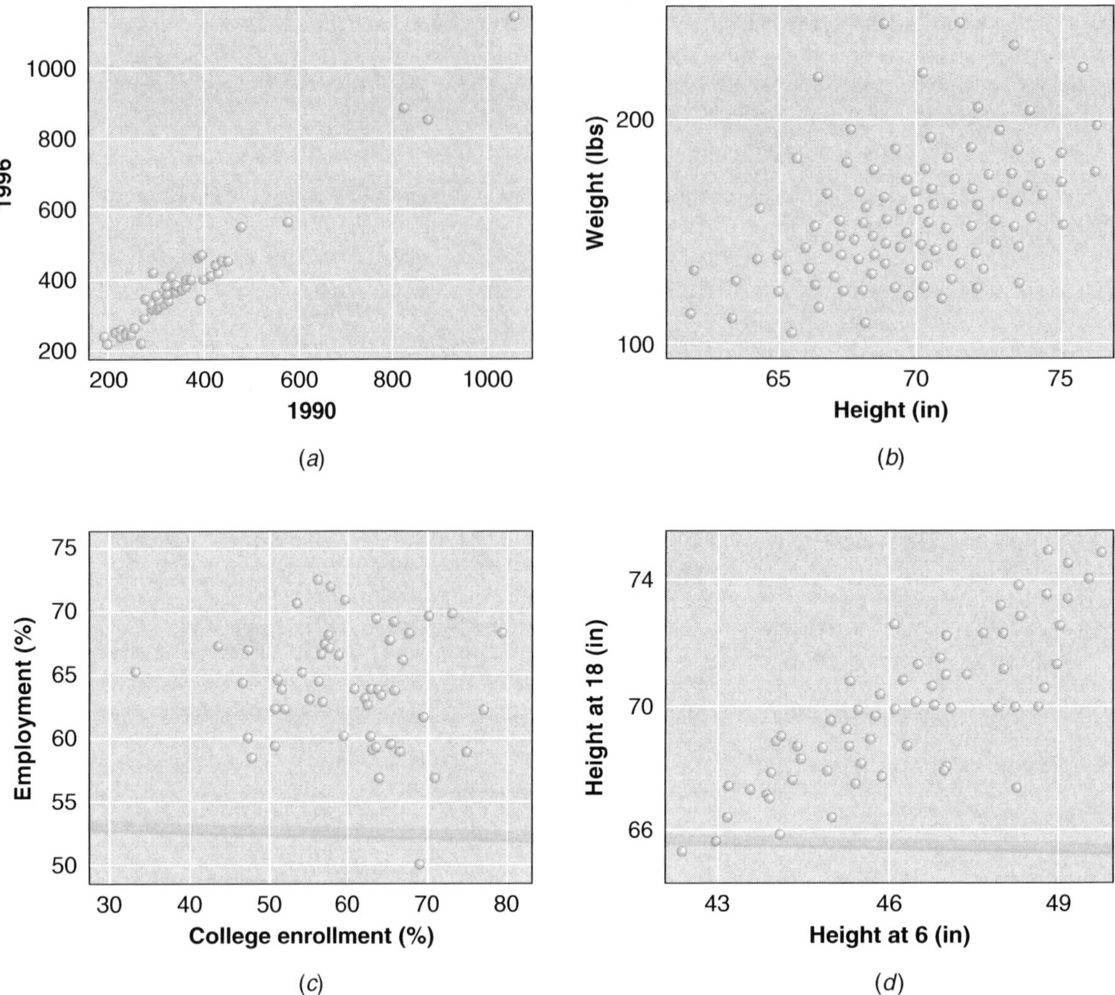

a. Energy usage in 1996 versus energy usage in 1990 for the 50 states
b. Weight versus height for a representative sample of male college students
c. Percentage of adults employed versus percentage of college-age people enrolled in college, for the 50 states (from U.S. Bureau of the Census)
d. Height at age 18 versus height at age 6 for a group of boys participating in a study on human growth

6. Consider the following scatterplots. The correlation coefficients for the graphs are 0.92, –0.05, –0.83, and –0.98, in some order. Match the scatterplots and the correlation coefficients.

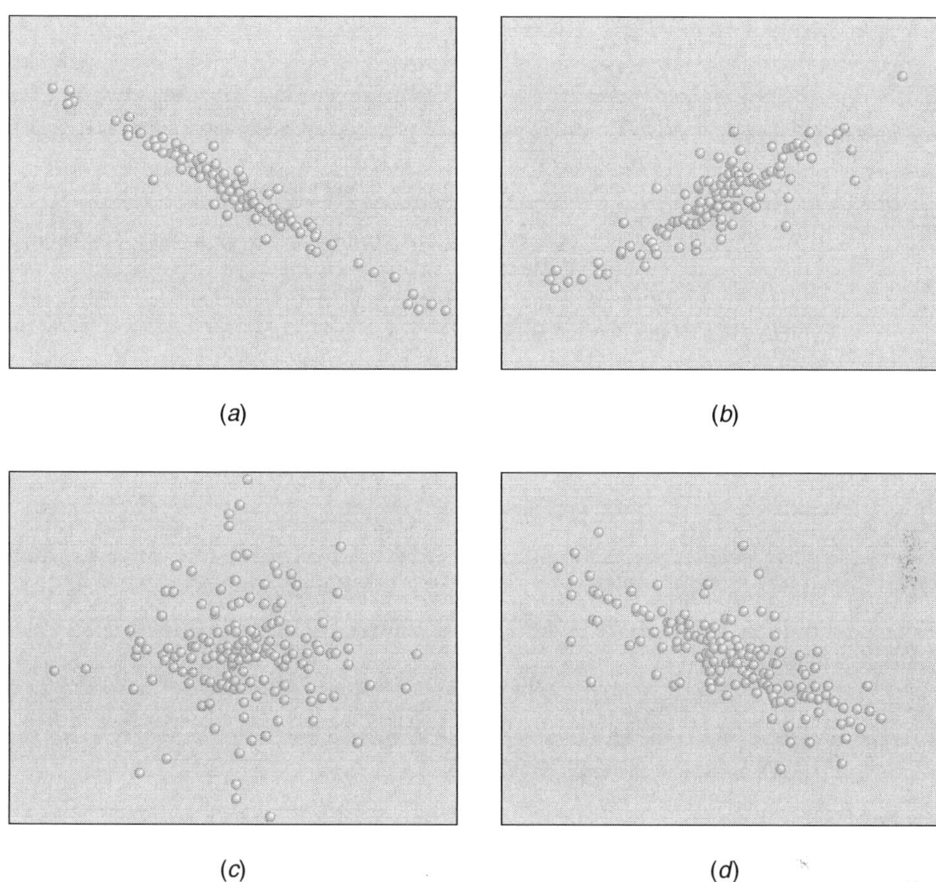

(*a*) (*b*)

(*c*) (*d*)

7. Imagine that we have a large representative sample of American families and that we have computed the following correlation coefficients.
   a. Between age of husband and age of wife
   b. Between height of husband and height of wife
   c. Between height of husband today and his height at age 6
   d. Between height and weight of husband

   The correlation coefficients are 0.25, 0.4, 0.8, and 0.95, in some order. Which is which? The issue is: How tightly do you think the points in the four scatterplots are clustered around a straight line? *Hint:* Look at the scatterplots in Figure 3.6, Figure 3.20, and Exercise 5.

8. An article in the August 1986 issue of the *Journal of Marriage and the Family* studied whether there is a link between the loneliness of college females and that of their parents. The participants in the study were 130 female undergraduates and their parents. Students and parents completed questionnaires measuring loneliness, depression, self-esteem, and social skills.
   a. The correlation between mother's loneliness score and daughter's loneliness score was found to be 0.26. Interpret this value in layman's terms. (How strong is the correlation?)
   b. The correlation between daughter's loneliness score and mother's self-esteem score was found to be −0.14. Interpret this value in layman's terms.
   c. The correlation between daughter's loneliness score and her depression score was found to be 0.69. Interpret this value.

9. Use the instructional software to practice guessing the correlation coefficients of randomly generated scatterplots.

3.3 REGRESSION

Before we discuss linear regression, let us review the *slope-intercept* equation for a straight line,

$$y = mx + b.$$

slope

y-intercept

Every line that is not parallel to the y-axis has such an equation. The **slope** m measures the steepness of the line; it tells us how much the y-value on the line increases (or decreases, if m is negative) when the x-value increases by one unit. The number b, called the **y-intercept**, is the y-value of the point of intersection between the line and the y-axis. Figure 3.15 illustrates this.

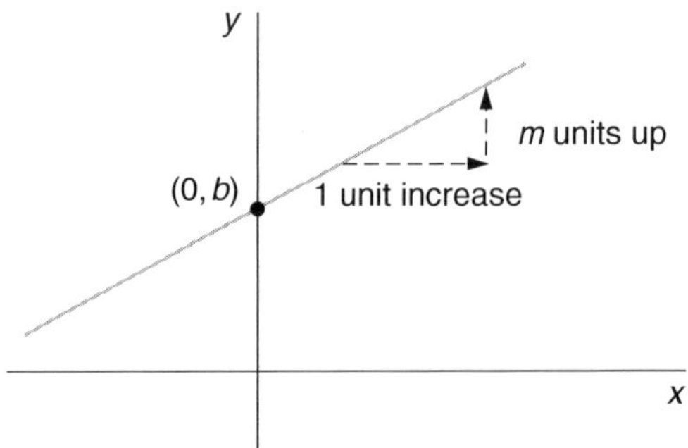

Figure 3.15 The line $y = mx + b$.

Example 3.9

Find the equation for the line with slope $m = 2$ that goes through the point (3,5).

Solution

First we plug in $m = 2$ in the equation $y = mx + b$ and get $y = 2x + b$. Then we plug the point (3,5) into the equation $y = 2x + b$ and get $5 = 2 \times 3 + b$ or $b = 5 - 6 = -1$. So the equation is

$$y = 2x - 1.$$

Some people prefer to memorize the equation of the line with slope m going through the point (x_0, y_0) in the form

$$y - y_0 = m(x - x_0).$$

This equation can of course be rewritten

$$y = mx + (y_0 - mx_0).$$

In other words, the y-intercept is

$$b = y_0 - mx_0.$$

Each student must decide which formula is easier to remember.

In Section 3.2 we learned to use the linear correlation coefficient r to measure the strength of the linear relationship between paired x- and y-values. If there is a strong linear correlation between two variables, then we can use the observed value of the explanatory variable to estimate (or explain) the value of the response variable.

Example 3.10

One of the main attractions in Yellowstone National Park is the geyser Old Faithful, which erupts roughly once every 75 minutes. In fact, the intervals between eruptions last between 45 minutes and 2 hours, and the eruptions themselves last between 1.5 and 5.5 minutes. Since a long eruption depletes more water from the geyser's underground reservoir than a short eruption, it takes longer to refill the reservoir after a long eruption than after a short eruption. So after a long eruption we should expect a longer wait before the next eruption. As a service to park visitors, the Park Service continually posts the estimated time of the next eruption so that visitors know when to gather around the geyser.

Until the Borah Peak earthquake in 1983, the Park Service predicted eruptions using the equation

$$\hat{y} = 10x + 30,$$

where $\hat{y}$ denotes the estimated time interval (in minutes) from the end of the last eruption to the beginning of the next eruption, and x denotes the duration (in minutes) of the previous eruption. About two-thirds of the predictions were correct within 5 minutes.

The 1983 Borah Peak earthquake shifted the underground circulation of hot water away from the geyser. As a result, the intervals between eruptions have become slightly longer and less predictable. The Park Service now makes its predictions with a leeway of 10 minutes. The scatterplot in Figure 3.16 relates the duration of an eruption of Old Faithful to the length of the interval until the next eruption for 150 eruptions in July 1995. As the scatterplot shows, eruptions lasting 2 minutes tended to be followed by an interval lasting around 60 minutes, whereas eruptions lasting 4.5 minutes tended to be followed by an interval lasting around 90 minutes. A line for predicting the length of the wait until the next eruption from the duration of the most recent eruption has been superimposed on the scatterplot. The dashed line indicates how we would predict the length of the wait after an eruption lasting 3.5 minutes. We predict that the wait would be around 80 minutes. In an exercise, we shall return to Old Faithful and look at data from recent eruptions to derive a new formula for predicting the length of the waiting time between eruptions.

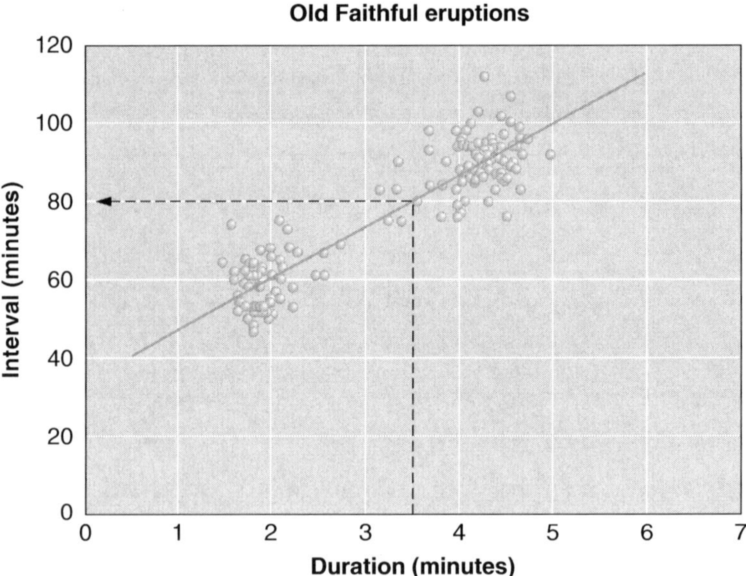

Figure 3.16 Scatterplot for the length of the interval between eruptions of Old Faithful against the duration of the preceding eruption for 150 eruptions in July 1995. A line for predicting the length of the wait until the next eruption from the duration of the last eruption has been superimposed. The dashed line illustrates how we use the line to predict the wait after an eruption lasting 3.5 minutes.

Fitting a Line to Data

In Example 3.10 the Park Service had a linear equation that summarized the relationship between the waiting time until the next eruption and the duration of last eruption. It allowed the park rangers to predict the waiting time to the next eruption with considerable accuracy. Finding such a linear equation that summarizes the relationship between two variables can be accomplished by drawing a line through a scatterplot and determining the equation of that line. Unless we have a perfect linear correlation, no line will pass through all the data points. Fitting a line to the data means drawing a line that comes as close as possible to the points. We now illustrate this basic idea.

The following data were obtained from an experiment that you could easily do yourself. Students in a high school statistics class collected cylindrical objects, such as coffee cans, juice cans, and a rolled oats container. They then measured the circumference (distance around) and the diameter (distance across) of each object and prepared a table like Table 3.3. They used string to measure the circumferences. One purpose of this experiment was to approximately verify the basic relationship between diameter and circumference, namely, circumference $= \pi \times$ diameter. That is, they were using statistical methods with real data to confirm a basic geometric fact. Another purpose was for the students to discover how unreliable actual measurements can be, even for such a clear and simple geometric concept as the relationship between circumference and diameter.

The class prepared a graph to show how the data are related, as shown in Figure 3.17. As you can see in the graph, the data roughly lie along a straight line.

Table 3.3 Diameters and Circumferences of Selected Cylindrical Objects

Object	Diameter (centimeters)	Circumference (centimeters)
Pill bottle	3.0	10.0
Coffee can (small)	5.0	16.0
Tomato juice can	10.8	32.5
Coffee can (large)	13.0	41.0
Rolled oats container	10.0	32.3
Soup can	6.8	21.0
Candle	4.5	18.0

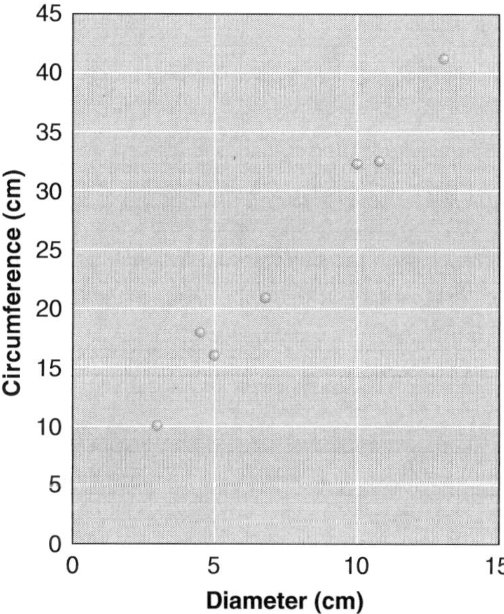

Figure 3.17 Diameters and circumferences of selected cylindrical objects.

Next they looked for a rule (a straight-line equation) that would fit the data. They tried drawing a line through the scatterplot in such a way that it would pass as close to the points as possible. One student took a piece of thread and held it taut over the data points in such a way that it was possible to see where the "best" line should be drawn. When the thread appeared to pass closest to the points, she held the thread as a guide to draw that "best" straight line. Note that she made this judgment informally and subjectively, not guided by any rule or criterion. Imagine, for example, that the student picked the line that goes through the points (3,10) and (13,41). We want to determine the *slope-intercept* equation for this line using the general equation,

$$y = mx + b.$$

We plug the two points into the equation. This gives us

$$10 = 3m + b \qquad \text{and} \qquad 41 = 13m + b.$$

These two equations in the two unknowns m and b have the solution $m = 3.1$ and $b = 0.7$. So the equation for the line through the points (3,10) and (13,41) is

$$y = 3.1x + 0.7.$$

We note that this is obviously not the correct formula for computing the circumference y from the diameter x of a circle. But as anyone who has worked in a lab knows, we cannot expect a formula derived from real data to be 100% accurate.

regression line

Given a scatterplot, a **regression line** "fits the data" by being as close as possible to the points. A regression line summarizes the relationship between the two variables. The standard statistical convention is to denote the points of the scatterplot by (x, y) and to write the equation of the regression line

$$\hat{y} = mx + b.$$

predicted value
estimated value
fitted value
error of estimation

Thus (x, y) is an actual observed data point, whereas $(x, \hat{y})$ is not an observed data point, but instead gives the estimated y-value at x. It is important to remember that, for each observed x, the corresponding value of $\hat{y}$ (called the **predicted value**, **estimated value**, or **fitted value**) may not be the same as the observed value of y. For example, when $x = 4.5$ centimeters, we obtain from the student's constructed line $\hat{y} = 3.1 \cdot 4.5 + 0.7 = 14.65$ centimeters. However, the candle with a measured diameter of 4.5 centimeters had a measured circumference of 18.0 centimeters. The **error of estimation** between the actual observed value and the value estimated by the line in this case is

$$y - \hat{y} = 18.0 - 14.65 = 3.35 \text{ centimeters.}$$

prediction error
residual

This estimation error (sometimes referred to as the **prediction error**), the deviation from the fitted line, is called the **residual**. The residual is the amount that remains unexplained by the estimated rule used to "fit" the data. Note that if we use the correct equation, $\hat{y} = \pi \cdot x$, then $\hat{y} = 14.1$ when $x = 4.5$, and there is an even larger estimation error (18.0 cm − 14.1 cm = 3.9 cm). This demonstrates that the students' measuring process was quite inaccurate. The next section provides a statistical method to find the "best-fitting" line for paired data points (x, y) that do not lie on a straight line.

In Figure 3.18, we see that for each data point the residual $y - \hat{y}$ is the vertical distance from the data point (x, y) to the fitted line. For points above the line, the residual is positive because $y > \hat{y}$, whereas for points below the line, the residual is negative because $y < \hat{y}$.

The Least Squares Regression Line

Our method of fitting a line to the data in the previous example was subjective. Different students might draw different lines by eye through a scatterplot. This is especially true for scatterplots whose points are more widely scattered than those in Figure 3.17. The widely used least squares method gives us an objective way to fit a regression line to the data.

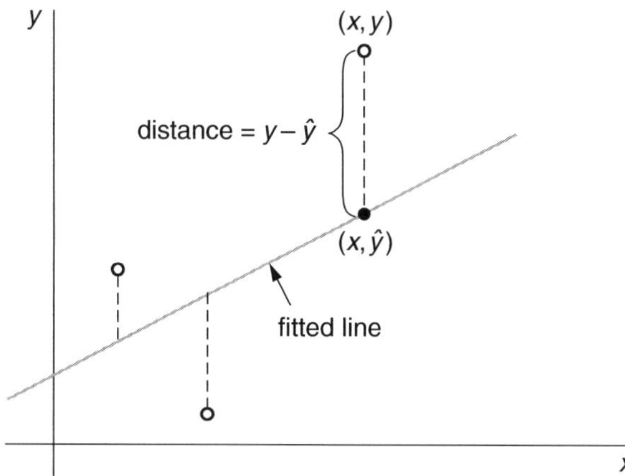

Figure 3.18 The residuals of three points.

How can we tell whether a straight line fits the data well? Clearly, the "overall magnitude of the residuals," measured in some way, should be small. Finding a measure for the "overall" residual size is a problem similar to the one of finding an "overall" deviation from a mean value for a single variable, as discussed in Section 2.3. The most common way of describing the overall size of the residuals is to *square* the estimation errors $y - \overline{y}$, as we did with $x - \overline{x}$ in finding the variance of the x-values, and find the sum of all the squared estimation errors. The resulting measure of overall deviation of the observed points from the line is called the **residual sum of squares**, also called the **error sum of squares**, or **SSE**:

residual sum of squares
error sum of squares
SSE

$$\text{SSE} = \sum (y - \hat{y})^2.$$

Geometrically, this is the sum of the squared vertical distances between the data points (x, y) and the line. (See Figure 3.18.)

For each candidate straight line $\hat{y} = mx + b$, we can calculate the corresponding residual sum of squares. Our best-fitting line is then the line with the smallest possible residual sum of squares. This line is called the **least squares regression line**. The name is appropriate because we are choosing the line that makes the error sum of squares as small as possible. If we consider several lines, we can see that each line will have its own error sum of squares based on its own set of vertical distances to the data points. Consider the following example for data (0,0), (1,0), (2,3):

least squares regression line

For the line $\hat{y} = x$,

$$\text{SSE} = \sum (y - \hat{y})^2 = (0-0)^2 + (0-1)^2 + (3-2)^2 = 2.$$

For the line $\hat{y} = 1.5x - 0.5$,

$$\text{SSE} = \sum (y - \hat{y})^2 = (0+0.5)^2 + (0-1)^2 + (3-2.5)^2 = 1.5.$$

The second line (which happens to be the least squares regression line) has the smaller error sum of squares and, hence, is a better-fitting line than the first line. The residuals for both lines are indicated in Figure 3.19.

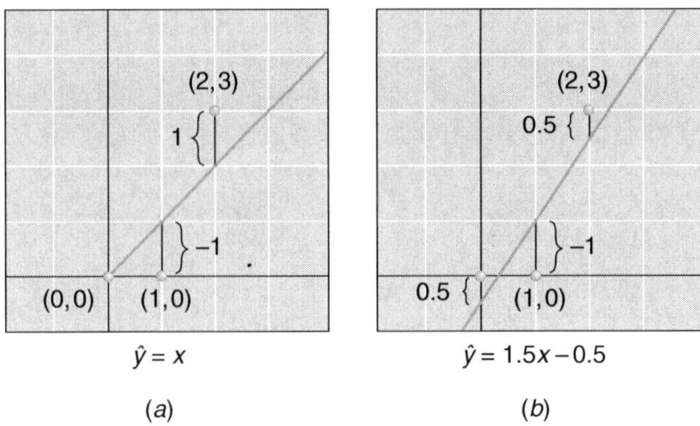

Figure 3.19 Comparison of residuals for two lines.

The following method of finding the least squares regression equation is not the most convenient if you have to use a calculator to determine it from paired data. However, it is easier to understand than the computational version of the equation given later. It assumes r has been found.

When we use the least squares regression line to estimate or predict one variable from another, our first step is to denote the variable whose value we know by x and the variable whose unknown value we are trying to estimate by y:

$$x = \text{known variable}, \quad y = \text{unknown variable}$$

equation for least squares regression line

Equation of the least squares regression line using r, s_x, s_y, $\overline{x}$, and $\overline{y}$

Consider a data set of paired x- and y-values. Let the mean and standard deviation be $\overline{x}$ and s_x for x, and $\overline{y}$ and s_y for y. Let r denote the correlation coefficient. The least squares regression line for estimating y from x goes through the point $(\overline{x}, \overline{y})$ and has slope

$$m = r(s_y / s_x).$$

Its equation is

$$\hat{y} = mx + b,$$

where

$$b = \overline{y} - m\overline{x},$$

as can be see by plugging $(\overline{x}, \overline{y})$ into the regression equation, and then solving for b.

In the next example, the known variable x is the age of the husband, and the unknown variable y is the age of the wife.

Example 3.11

For a representative cross section of several hundred married couples from Illinois, the average age of the husbands was $\overline{x} = 46$ years with a standard deviation of $s_x = 13$ years. The average age of the wives was $\overline{y} = 43$ years with a standard deviation of $s_y = 13$ years as well. The correlation between age of husband and age of wife was $r = 0.95$. The least squares regression line for estimating the age y of the wife from the age x of the husband has the equation

$$\hat{y} = 0.95x - 0.70.$$

Here $m = 0.95$ and $b = -0.70$.

Figure 3.20 displays the scatterplot and the regression line. Note how strong the linear relationship is. If a randomly chosen husband from Illinois happens to be 26 years old, then we would predict the age of his wife to be around $(0.95)(26) - 70 = 24$ years. If he happens to be 66 years old, then we would predict the age of his wife to be around $(0.95)(66) - 0.70$ years $=$ 62 years.

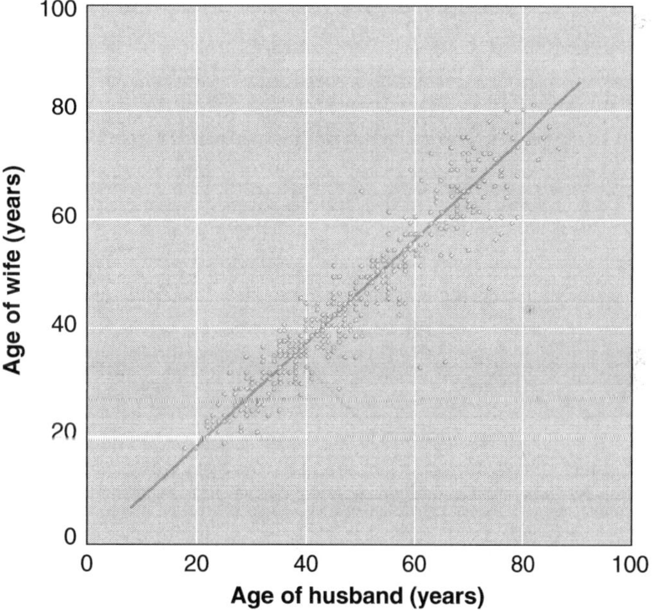

Figure 3.20 The regression line for estimating the age of the wife from the age of her husband.

Finally, imagine that you have to guess the age of a randomly chosen wife from Illinois without knowing the age of her husband. What should you guess? The age distribution of married women in Illinois looks something like Figure 3.21. If we have no information about the married woman whose age we are trying to guess, then we might as well use the mean age $\overline{y}$ of all married women. So we guess that the woman is around 43 years old. This guess could very well be off by 10 to 20 years. In contrast, if we knew the age of her husband and used the regression equation, then our guess of the age of the wife would probably only be off by around 3 to 6 years, a dramatic improvement in accuracy.

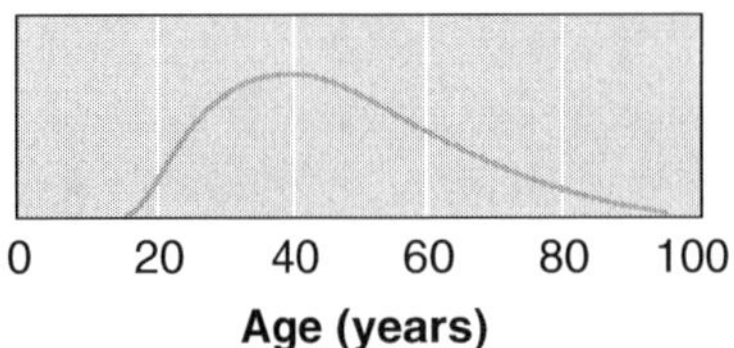

Figure 3.21 Age distribution of married women in Illinois.

❐ ❐

Example 3.11 illustrated the fact that if there is a *strong* linear correlation between two variables, then for a pair (x, y) with x known and y unknown, we can use the known value x to compute a fairly accurate prediction $\hat{y}$ of the unknown value y.

For large data sets of paired x- and y-values, the regression line can also be used to estimate the *average* y-value corresponding to a fixed x-value, as illustrated by the following example where $x =$ height of father and $y =$ height of son.

Example 3.12

In their 1903 paper, "On the Laws of Inheritance in Man," Pearson and Lee studied the correlation between the heights of 1078 fathers and sons. The average height of the fathers was 67.68 inches with a standard deviation of 2.70 inches. The average height of the adult sons was 68.65 inches with a standard deviation of 2.71 inches. The correlation between height of father and son was 0.514. The regression line for estimating the height of a son from the height of his father has slope

$$m = r\frac{s_y}{s_x} = 0.514\frac{2.71}{2.70} = 0.516,$$

and y-intercept

$$b = \overline{y} - m\overline{x} = 68.65 - (0.516)(67.68) = 33.73.$$

The regression equation is

$$\hat{y} = 0.516x + 33.73.$$

We note that r indicates only a weak to moderate correlation. Therefore the prediction of the son's height y for a *particular father* of height x will not be very accurate. (This is not surprising since the height of the son genetically depends just as much on the height of his mother as on the height of his father.) But the regression equation will estimate with great accuracy the *average height* of all the sons of the more than 100 fathers of height $x = 69$ inches (say) in Pearson and Lee's study. Similarly, the regression equation will estimate with great accuracy the *average* height of all the sons whose fathers are 65 inches tall. Figure 3.22 illustrates this. The points in Figure 3.22 show the observed *average* height of the sons, for each value of the father's height (rounded to the nearest inch). The points closely follow the regression line $\hat{y} = 0.516x + 33.73$, which is estimating the average heights of the sons for $x = 59, 60$, and so forth. The regression line is also shown in Figure 3.22.

Figure 3.23 is a rough sketch of the scatterplot for Pearson and Lee's data on the heights of fathers and sons with the ellipse showing where most of the points lie. The regression line goes through the midpoint of each vertical strip within the ellipse, as highlighted for one strip with its midpoint shown.

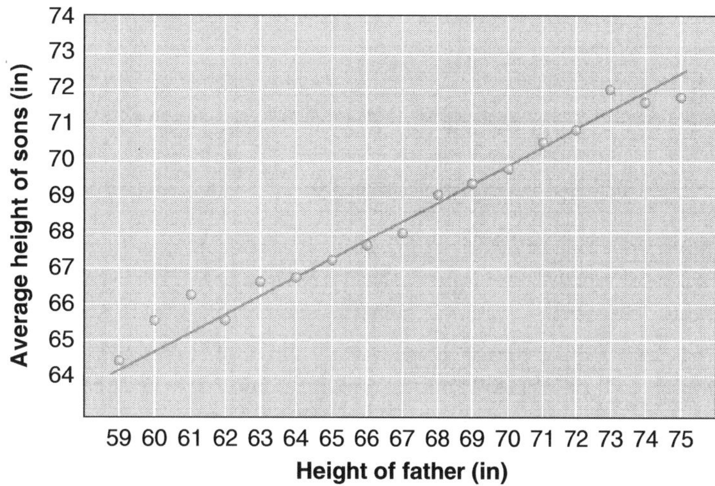

Figure 3.22 Average height of the sons for each value of the father's height.

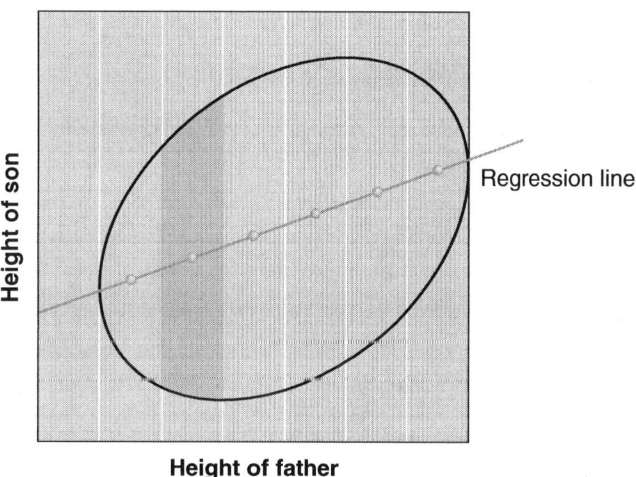

Figure 3.23 The regression line estimates the average value for y corresponding to each value of x.

❐ ❐

We have two closely related and important uses for the regression line:

regression line used for prediction and estimation

The two uses for a regression line

1. predicting a new or future y when its x-value is known (as in Examples 3.10 and 3.11)
2. estimating the average of all the y-values paired with the same specified x-value (as in Example 3.12)

For both uses of regression, the solution is $\hat{y}$, computed from the regression formula

$$\hat{y} = mx + b$$

If we know the correlation coefficient r and the two standard deviations s_x and s_y, then we can compute the slope m of the regression line using the formula $m = r(s_y/s_x)$. A more direct and computationally faster formula for computing the slope m from paired x- and y-values is

computational formula for regression line slope m

$$m = r\frac{s_y}{s_x} = \frac{\sum(x-\bar{x})(y-\bar{y})}{\sum(x-\bar{x})^2} = \frac{\sum xy - (\sum x)(\sum y)/n}{\sum x^2 - (\sum x)^2/n}$$

where n, of course, is the number of observations.

We now return to our Key Problem as an example. We let x denote curb weight and y city mileage. We use Table 3.4 to sketch how we calculate the five sums ($\sum x$, $\sum y$, $\sum x^2$, $\sum y^2$, $\sum xy$) for the two variables x and y. The line is now easy to find via the following steps, based on the computationally friendly formula for m.

Step 1. Find the means of x and y. From the table, we have

$$\bar{x} = 139{,}748/38 = 3677.6$$
$$\bar{y} = 790/38 = 20.789.$$

Note: We may not need so many decimals, but we wait until we have completed our computations before we round off.

Table 3.4 Finding the Regression Line for the Key Problem

x	y	x^2	y^2	xy
2,734	27	7,474,756	729	73,818
3,428	23	11,751,184	529	78,844
⋮	⋮	⋮	⋮	⋮
3,646	18	13,293,316	324	65,628
139,748	790	549,956,508	17,942	2,717,729

Step 2. The slope is

$$m = \frac{2{,}717{,}729 - (139{,}748 \cdot 790)/38}{549{,}956{,}508 - (139{,}748)^2/38} = -0.0052067.$$

Step 3. The y-intercept is

$$b = \bar{y} = m\bar{x} = 20.789 - (-0.0052067) \cdot (3677.6) = 20.789 + 19.148 = 39.937.$$

Step 4. The equation for the least squares regression line is

$$\begin{aligned}\hat{y} &= mx + b \\ &= -0.0052x + 39.9.\end{aligned}$$

Suppose the Environmental Protection Agency would like to know the average city mileage for *all* cars weighing $x = 3500$ pounds. Substituting into the regression equation, we estimate that the average fuel economy will be 21.7 miles per gallon (compute this yourself). This value, 21.7, is also the prediction of fuel economy for a specific car weighing 3500 pounds. The correlation, $r = -0.802$, is strong, and (x, y) points tend to be close to the line. Thus, this prediction will be fairly good. That is, the 3500-pound cars will not vary much in fuel economy, and they will all be rather close to the predicted 21.7 miles per gallon. (See Figure 3.1.)

In this example, we can also use Table 3.4 to compute the standard deviations, $s_x = 986.7$ and $s_y = 6.406$, and the correlation, $r = -0.802$. The slope can then also be computed from its conceptual formula, namely,

$$m = r\frac{s_y}{s_x} = -0.802\,\frac{6.406}{986.7} = -0.0052.$$

*Explained and Unexplained Variation

Tall men tend to weigh more than short men. But a man's weight also depends on factors besides height, such as whether he leads a physically active life. Let us try to find out to what extent a man's weight can be explained by his height.

A study looked at the relationship between height x and weight y for a large, representative cross section of college-age males. The regression formula for computing the estimated weight $\hat{y}$ (in pounds) for a young man of height x (in inches) was

$$\hat{y} = 4.5x - 150.$$

The average height was $\overline{x} = 70$ inches and the average weight was $\overline{y} = 165$ pounds. The correlation was only $r = 0.45$.

The slope of the regression line is $m = 4.5$, which tells us that corresponding to each increase of 1 inch in height there is, on average, an increase of 4.5 pounds in weight. We expect a young man of average height (70 inches) to be of average weight (165 pounds). We expect a 71-inch-tall man to weigh 165 pounds + 4.5 pounds, a 72-inch-tall man to weigh 165 pounds + 9 pounds, a 73-inch-tall man to weigh 165 pounds + 13.5 pounds, and so forth. (To check this, just plug into the regression formula.)

total deviation

If a 72-inch-tall man happens to weigh 165 pounds + 9 pounds = 174 pounds, then we say that his weight is completely explained by his height. If, on the other hand, the 72-inch-tall man happens to weigh 185 pounds, which is 20 pounds above the average $\overline{y}$, then we say that 9 pounds out of the 20 pounds above average are explained by his height, and the remaining 11 pounds are unexplained. Thus we split the **total deviation** of his weight y from the mean weight $\overline{y}$, namely $y - \overline{y}$ (= 20 pounds), into the explained deviation from $\overline{y}$, namely $y - \hat{y}$ (= 9 pounds), and the unexplained deviation from the line (the residual), namely $\hat{y} - \overline{y}$ (= 11 pounds), as illustrated by Figure 3.24. Thus,

$$y - \overline{y} = (y - \hat{y}) + (\hat{y} - \overline{y}),$$

namely,

total deviation = unexplained deviation + explained deviation.

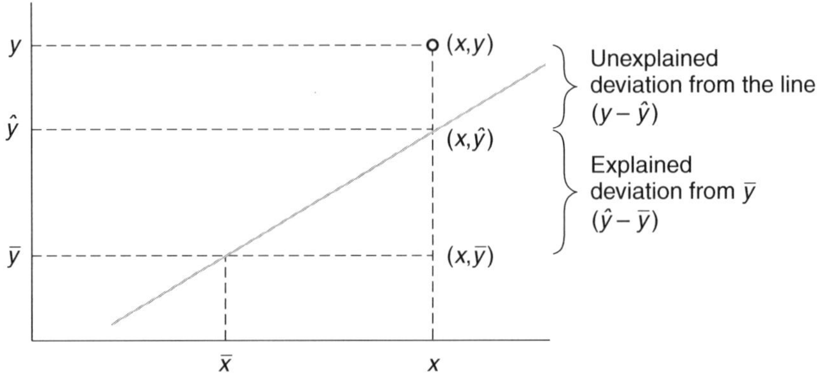

Figure 3.24 Deviations for the regression line.

total variation

The **total variation** (about the mean $\bar{y}$) of all the y-values, namely $\sum(y-\bar{y})^2$, can also be split into two parts:

$$\sum(y-\bar{y})^2 = \sum(y-\hat{y})^2 + \sum(\hat{y}-\bar{y})^2.$$

explained variation

unexplained variation

The last sum, $\sum(\hat{y}-\bar{y})^2$, is called the **explained variation**. It is the part of the variation of the y-values that can be explained by the variation of the corresponding x-values using the regression formula. The sum $\sum(y-\hat{y})^2$ is called the **unexplained variation**. It is the part of the variation of y-values that cannot be explained by the variation of the corresponding x-values. The unexplained deviation $y-\hat{y}$ shown in Figure 3.24 is of course just another name for what was previously referred to as the residual. Thus the unexplained variation is the sum of squares of the residuals.

proportion of total variation explained

It can be shown that **the proportion of the total variation that can be explained by the regression line** equals r^2, the square of Pearson's correlation coefficient:

$$r^2 = \frac{\text{explained variation}}{\text{total variation}} = \frac{\sum(\hat{y}-\bar{y})^2}{\sum(y-\bar{y})^2}.$$

coefficient of determination

percentage of total variation explained

This quantity is called the **coefficient of determination**. The closer r^2 is to 1, the tighter the points of the scatterplot cluster around the line, and the better the fit of the line—in other words, the better the regression line determines y. The quantity $100r^2$ is referred to as **the percentage of the total variation in y that can be explained by the regression on x**. Therefore, when $100r^2\%$ is close to 100%, the regression line fits the point really well, as in Figure 3.20, for example.

In the Key Problem, $r = -0.802$ and $r^2 = 0.64$. So 64% of the total variation in gas mileage can be explained by the regression on weight. In Example 3.12, $r = 0.51$ and $r^2 = 0.26$. So only 26% of the total variation in height of the sons can be explained by the regression on height of the fathers, while 74% of the total variation remains unexplained.

In our example on the heights and weights of college-age men, $r = 0.45$ and $r^2 = 0.20$. So only 20% of the total variation in weight can be explained by the regression on height. Fully 80% of the variation in weight is caused by factors other than height.

The Residuals

Consider again a sample of n paired x- and y-values with least squares regression equation

$$\hat{y} = mx + b$$

for estimating the response variable y from the explanatory variable x. Recall that each $y - \hat{y}$ is called a residual.

properties of residuals

Two basic properties of residuals

Fact 1. The residuals $y - \hat{y}$ have mean 0.

Fact 2. The standard deviation of the residuals, $\sqrt{\Sigma((y - \hat{y})^2/(n-1))}$, can be computed by

residual SD

$$\textbf{residual SD} = s_y\sqrt{1 - r^2}.$$

For *large data sets with elliptical scatterplots,* we can apply the 68-95-99.7% Rule to the residuals: Around 68% of the residuals are within 1 residual SD of 0 and around 95% of the residuals are within 2 residual SDs of 0. If the scatterplot is elliptical, as shown in Figure 3.25, and we use the regression equation to estimate future values of the response variable y from the explanatory variable x, then the residual SD tells us the likely size of future prediction errors, as the next example shows.

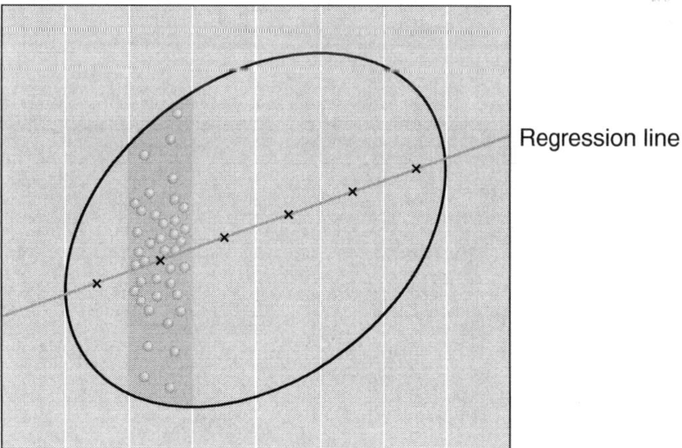

Figure 3.25 For a large data set with an elliptical scatterplot, the residuals in each vertical strip have a mean close to 0 and a standard deviation close to $s_y\sqrt{1 - r^2}$.

Example 3.13

Consider again the data on the ages of husbands and wives in Example 3.11. If we have to guess the age of a randomly chosen wife without knowing the age of her husband, then our best guess is $\bar{y} = 43$ years. The likely size of our estimation error is $s_y = 13$ years. If, on the other hand, we know the age x of her husband and use regression ($\hat{y} = 0.95x - 0.70$) to predict her age y, then the likely size of our prediction error $y - \hat{y}$ is

$$\begin{aligned}\text{residual SD} &= s_y\sqrt{1-r^2} \\ &= 13 \text{ years}\sqrt{1-(0.95)^2} = 4.06 \text{ years.}\end{aligned}$$

By the regression equation, if the husband is 26 years old, then the regression prediction of his wife's age is 24 years. If the husband in 66 years old, then the regression prediction of his wife's age is 62 years. If we predict the age of each Illinois wife from the age of her husband using the regression formula, then approximately 68% of all our predictions will be off by at most 4.06 years. And approximately 95% of all our predictions will be off by at most 8.12 years. Thus, the residual SD is important: It tells us how accurately the regression line predicts y from x.

*Outliers, Leverage Points, and Influential Points

(x, y) a linear regression outlier

influential in regression setting

leverage point

In statistics we call an observation an outlier if it does not follow the pattern *of most of the data*. In a graph of the data, an outlier is separated from the bulk of the data. In the context of linear regression, we call a **paired observation** (x, y) **an outlier** if the pair does not follow the linear pattern of most of the data. Such observations correspond to points in the scatterplot that are either far above or far below the regression line estimated from the bulk of the data.

A paired observation (x, y) is said to be **influential in the linear regression setting** if the observation's removal would change the position of the regression line substantially. Not every outlier is influential. (See Point A in Figure 3.26.) And not every influential point is an outlier. (See Point A in Exercise 18.) There exist diagnostic tools for measuring the degree of influence of each paired observation. Such measures go beyond the scope of this textbook, but it is sufficient to say that points with extreme (unusually large or unusually small) x-coordinates often have a strong influence on the position of the regression line. Such a point with an extreme x-value is called a **leverage point**. A leverage point has the potential to be very influential.

The following example illustrates that an outlier need not be influential if its x-coordinate is close to $\bar{x}$. That is, for an outlier to be influential it usually has to also be a leverage point. However, whether influential or not, an outlier will usually weaken the correlation and increase the residual SD, as shown in the example.

Example 3.14

Consider the 10 paired x- and y-values in the table below.

Point number	x	y	x^2	y^2	xy
1	−4	−4	16	16	16
2	−3	−1	9	1	3
3	−2	−2	4	4	4
4	−1	−1	1	1	1
5	0	−1	0	1	0
6	1	1	1	1	1
7	3	2	9	4	6
8	3	4	9	16	12
9	4	5	16	25	20
10 (Point A)	−1	7	1	49	−7
Sum	0	10	66	118	56

Let us call the tenth point, with coordinates $(-1, 7)$, Point A. As Figure 3.26 shows, Point A is an outlier because it is far from the modified regression line obtained from the other nine points, which form a roughly linear pattern. Since $x = -1$ for Point A and $\overline{x} = 0$, x is close to $\overline{x}$, and thus by definition, Point A is not a leverage point. Because Point A is not a leverage point, it probably is not an influential point. Indeed, the position of the regression line does not change much when we remove Point A: If we include Point A, then the regression line has the equation $\hat{y} = 1 + 0.85x$. If we remove Point A, then the regression equation becomes $\hat{y} = 0.226 + 0.966x$. Notice in Figure 3.26 that the two regression lines are fairly close to each other. Thus Point A is not very influential. Even though Point A is an outlier and hence is very unusual and deserves special attention (is it an error, was something unusual going on in producing Point A, etc.?), it is not seriously distorting the location of the regression line: Not all outliers are influential.

Most outliers decrease the magnitude of the correlation r, sometimes by a lot. Hence, the useful residual SD

$$\text{residual SD} = s_y\sqrt{1 - r^2}$$

will usually be inflated by the presence of one or more outliers, sometimes by a lot, as we now see when the outlier Point A is *removed*:

$$r = 0.943 \quad \text{and} \quad \text{residual SD} = 0.967.$$

However, with the outlier Point A *not removed,*

$$r = 0.663 \quad \text{and} \quad \text{residual SD} = 2.59.$$

Thus, the presence of the outlier (Point A) reduces r and increases the residual SD by a large amount, distorting our measurement of the prediction accuracy attainable by a regression line.

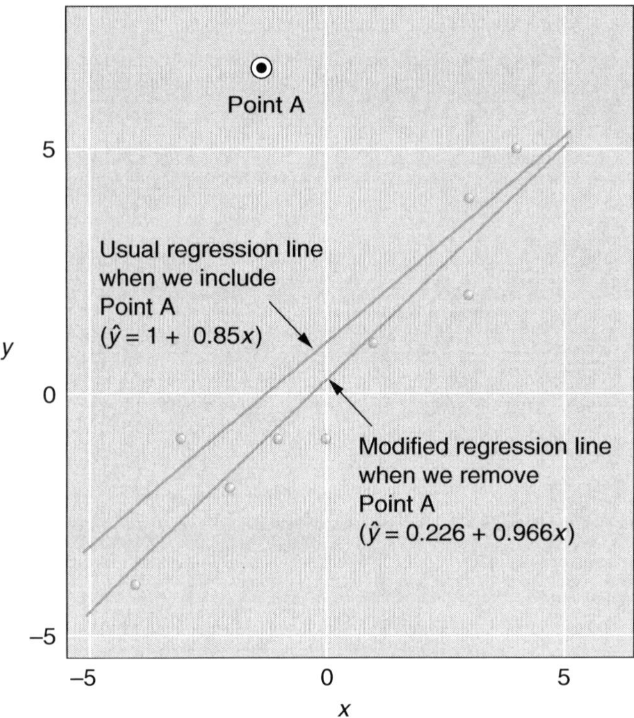

Figure 3.26 An outlier that is not a leverage point and is not influential.

❒ ❒

The next example illustrates how influential a leverage point—that is, a point with an extreme x-coordinate—can be.

Example 3.15

Consider the 10 paired x- and y-values in the table below.

Point number	x	y	x^2	y^2	xy
1 (Point B)	−4	−3	16	9	12
2	−2	−1	4	1	2
3	−2	0	4	0	0
4	−1	0	1	0	0
5	−1	1	1	1	−1
6	0	0	0	0	0
7	0	2	0	4	0
8	1	2	1	4	2
9	1	3	1	9	3
10 (Point C)	8	−8	64	64	−64
Sum	0	−4	92	92	−46

Let us call the first point, with coordinates $(-4, -3)$, Point B, and let us call the tenth point, with coordinates $(8, -8)$, Point C. As Figure 3.27 shows, Point C has a very large x-value compared to the rest of the data. Hence, it is a leverage point. Moreover, it does not follow the linear pattern of the rest of the data. Hence, it is an outlier. We can see this visually from the scatterplot. In particular, we can see that Point C has a large residual relative to the modified regression line computed from the rest of the data. Point C is a very influential point: The removal of Point C completely changes the direction of the regression line. If we include Point C, then the regression line has the equation $\hat{y} = -0.4 - 0.5x$. If we remove Point C, then the regression equation becomes $\hat{y} = 1.36 + 1.03x$.

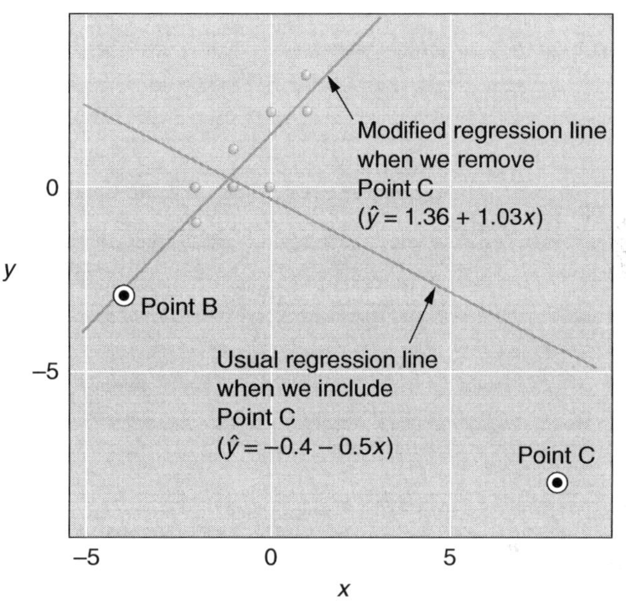

Figure 3.27 A very influential point.

This example again illustrates how sensitive regression analysis and the correlation coefficient r are to outliers. If we remove Point C, then the remaining 9 points are tightly clustered around a straight line. The correlation coefficient for the remaining 9 points is $r = 0.92$. If we include the outlier, Point C, however, then the 10 points do not follow the usual regression line, and the correlation coefficient changes completely to $r = -0.50$. Moreover, the residual SD is greatly inflated by including Point C:

$$\text{residual SD (without C)} = 0.705, \text{ compared to residual SD (with C)} = 2.74.$$

We note finally that Point C is not the point in Figure 3.27 with the largest residual (relative to the regression line for all 10 points). Point B has a larger residual. Point C has simply so much leverage that the regression line for all 10 points is "pulled" so close to it that Point C fails to produce the largest residual. In fact, an outlier that is a leverage point can sometimes be so influential that it has a very small residual. For example, if we move Point C from $(8, -8)$ to $(80, -80)$ we will increase the point's leverage and reduce its residual to roughly -0.5. Finally, note that for the correct regression line obtained by removing the Point C outlier, Point B has a very small residual: It is a leverage point that follows the linear pattern of the bulk of the data.

❐ ❐

For the practitioner of statistics the message is clear: Outliers, especially outliers that are influential, can distort our regression model, and such outliers should be carefully examined and perhaps removed before we do a regression analysis. A single outlier with enough leverage (extreme enough x-value) can completely change the direction of the least squares regression line and the value of the correlation coefficient r. If there are such outliers in our data set of paired observations (x, y), then we face the same dilemma we faced in Chapters 1 and 2: What do we do with influential outliers? Do we discard them as errors, do we study them separately and use linear regression to model the rest of the data, or do we conclude that the outliers invalidate our linear regression model?

*Residual Plots

When we draw a regression line through a scatterplot, it is not always easy to see whether the regression line fits the data points well. In particular, if the regression line is steep, then it may be difficult to detect outliers or nonlinear patterns in the data, which might require us to modify the standard regression analysis.

residual plot

A **residual plot** makes it easier to detect if the regression line and its correlations fail to summarize the relationship between the two variables—either because one or more influential outlying points produce the wrong regression line and distort the correlation coefficient, or because the relationship between the two variables is nonlinear. A residual plot graphs the residuals $y - \hat{y}$ on the vertical axis against the explanatory variable x on the horizontal axis: Corresponding to each point (x, y) in the original scatterplot, we graph the point $(x, y - \hat{y})$ in the residual plot, as indicated in Figure 3.28.

If the regression line in the original scatterplot summarizes the relationship between the two variables well, then there will be no trend or pattern in the residual plot. This is the basic principle in analyzing a residual plot: The absence of any pattern in the residual plot indicates that the regression line summarizes the relationship between x and y well. In this case, the points in the residual plot should be roughly equally scattered above and below the axis, with no outliers far above or below the axis, and there should be no pronounced trend or pattern from left to right (as x gets larger). Figure 3.29 shows one such example. Figure 3.29 provides a strong indication that a linear regression is appropriate for the data from which the residual plot was derived.

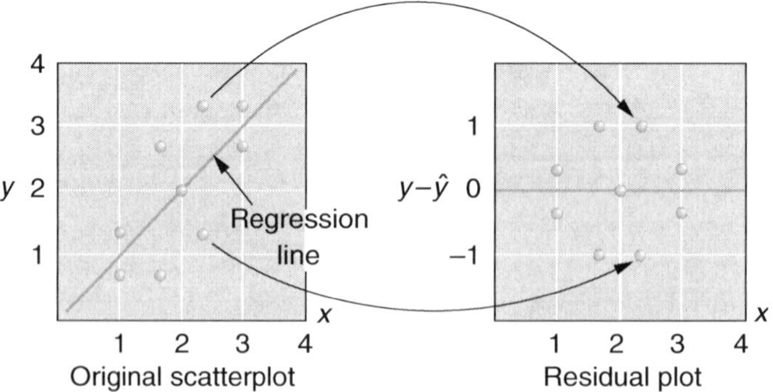

Figure 3.28 Plotting the residuals.

A definite pattern in the residual plot is evidence that the regression lines does not summarize the scatterplot data well. For example, in contrast to Figure 3.29, which suggests a linear regression is appropriate, the nonlinear trend in the residual plot in Figure 3.30 suggests that a linear regression is inappropriate for the data from which Figure 3.30 was derived. (The residual plot for the Key Problem looks a little bit like Figure 3.30. The relationship between curb weight and city mileage may, in fact, best be described as nonlinear.) When a residual plot suggests a nonlinear trend, rather than do a linear regression, one either tries to transform the data into a linear relationship as explained at the very end of this section, or one consults a statistical consultant for help in doing a nonlinear least squares regression.

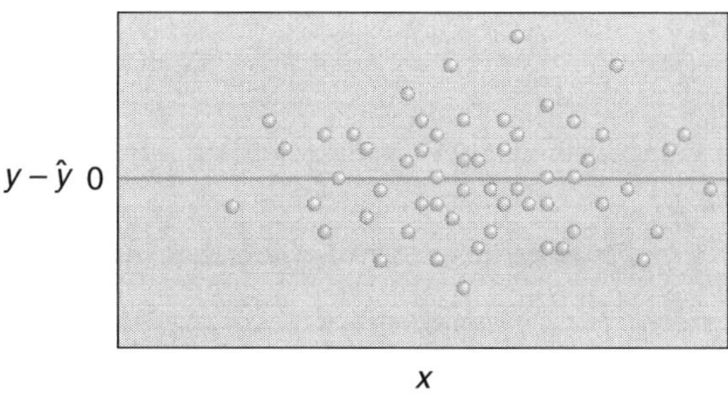

Figure 3.29 A residual plot without trend or outliers.

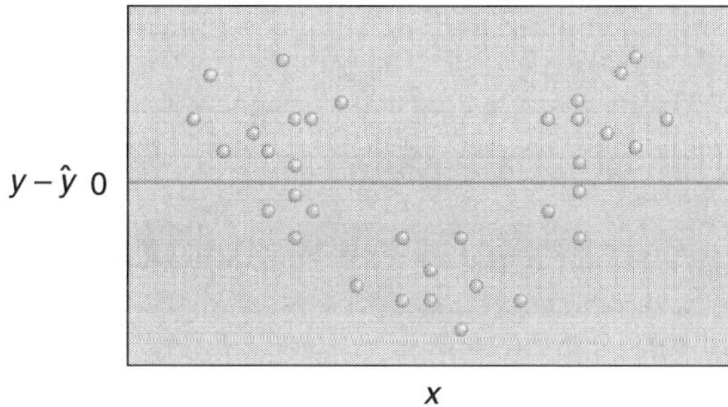

Figure 3.30 A residual plot with nonlinear trend.

Finally, Figure 3.31 illustrates the effect of an influential point on a residual plot. Figure 3.31 is the residual plot corresponding to Figure 3.27 for the regression line based on all 10 points, including Point C in Example 3.15. The nonrandom, linear trend in the residual plot together with the isolated exceptional point shows that the influential outlier corresponding to the exceptional point produced the "wrong" regression line, a line that does not fit the data well.

Even if Point C happened to be almost right on the x-axis of the residual plot of Figure 3.31, the conclusion would be the same—namely, the strong pattern suggests that Point C is a highly influential outlier that should be studied separately.

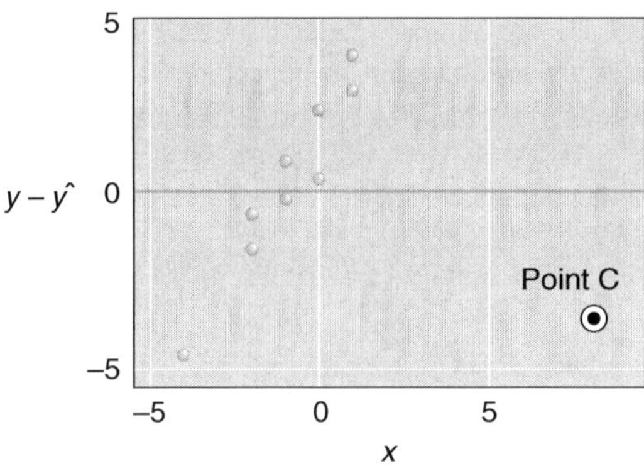

Figure 3.31 A residual plot with a very influential outlier.

The basic steps of residual analysis are as follows:

residual analysis steps in regression

The steps in doing a residual analysis to detect influential outliers or nonlinearity in a regression analysis

1. Construct the residual plot from the regression analysis using all of the data points.
2. Look for patterns or trends in the residual plot as x varies. Look for outliers, isolated points not following the pattern of the rest of the points.
3. If there is no visual pattern or trend, the linear regression that was done is acceptable.
4. If there is a pattern or trend, an influential outlier or a nonlinear trend is likely present. The linear regression that was done is not acceptable.
5. If an outlier(s) seems present, remove it and redo the regression analysis, including the computation of r.
6. If a nonlinear trend seems present, attempt to transform the data to produce a linear scatterplot as explained in the next subsection and using methods given in Chapter 11, or consult a professional statistician.

*Nonlinear Relationships

As we have seen, least squares regression is a powerful tool for analyzing paired variables that are linearly correlated. What do we do if the scatterplot or a residual analysis indicates a nonlinear relationship between the variables? Often a transformation will help—for example, taking logarithms of one or both of the variables. For example, Kepler's third planetary law states that

$$T = R^{3/2}$$

where T denotes the time (measured in years) it takes a planet to complete a revolution around the Sun, and R denotes the average distance (measured in astronomical units) of the planet to the Sun. If we take logarithms, then Kepler's third law becomes a linear relationship between $y = \ln(T)$ and $x = \ln(R)$:

$$\ln(T) = \frac{3}{2}\ln(R),$$

that is,

$$y = \frac{3}{2}x.$$

(See Exercise 6 of Section 3.1.)

In the next example we are also able to transform nonlinear data into linear data.

Example 3.16 Boyle's Gas Law

A high school physics class studied Boyle's gas law by varying the pressure on a confined air mass and measuring the resulting volume. Figure 3.32 is the resulting scatterplot.

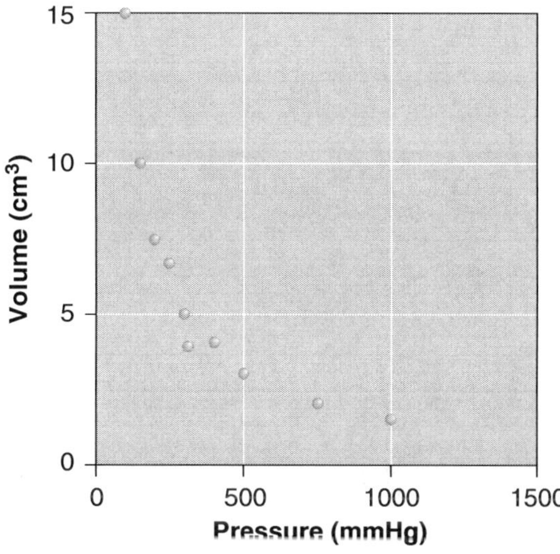

Figure 3.32 Boyle's gas law.

Pressure and volume are clearly negatively correlated, in that increasing pressure decreases the volume. The least squares regression equation for the data is

$$\text{Volume} = 10.35 - (0.01134 \cdot \text{Pressure}).$$

But the corresponding residual plot in Figure 3.33 clearly has a U-shaped pattern, and this clearly shows that the relationship between pressure and volume is not linear. It is worth noting that the nonlinear pattern is only somewhat noticeable in Figure 3.32, while it is clearly visible in Figure 3.33. In fact, residual plots often make patterns of nonlinearity more visually noticeable.

After we take logarithms of both variables, however, we get a nice linear relationship between ln(Volume) and ln(Pressure), as Figure 3.34 shows. Its residual plot (not shown) in fact shows no patterns or trends, as expected.

From Figure 3.34 we derive Boyle's law: A regression analysis can be done. It will show (the line is a 45° line) that

$$\ln(\text{Volume}) + \ln(\text{Pressure}) = \text{constant},$$

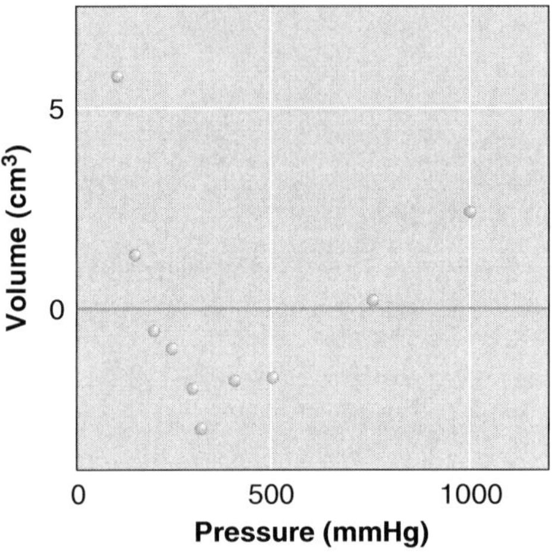

Figure 3.33 Residual plot for measurements.

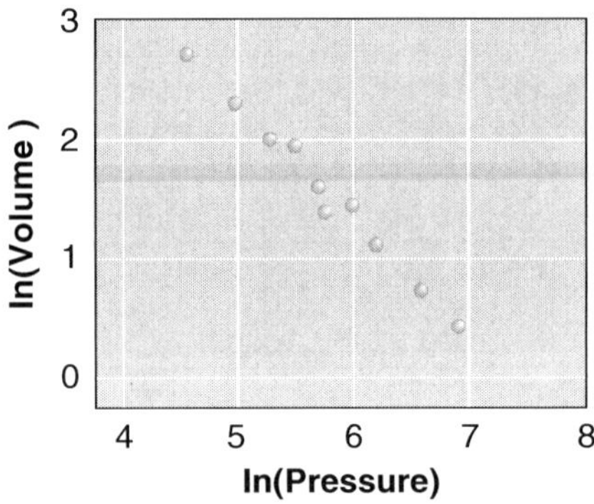

Figure 3.34 Gas measurement from Figure 3.32.

which we can restate (using $\ln(V) + \ln(P) = \ln(V \cdot P)$ on the left-hand side of the above equation)

$$(\text{Volume}) \cdot (\text{Pressure}) = \text{constant},$$

this second constant being different from the first constant.

*Transformation to Achieve Linearity

There are many ways to transform nonlinear bivariate data (x, y) into linear bivariate data. We list below the most common transformations and indicate in which situations transformations might work. In Chapter 11, we will learn more details of how to carry out such transformations and thus be able to use linear regression to estimate nonlinear relationships.

Shape of scatterplot curve	**Example scatterplot**	**Suggested transformation**
Curve goes through $(0, 0)$ and appears to be a power curve ($y = B_0 x^{b_1}$ for $B_0 > 0$, $b_1 > 0$; in fact, $b_1 > 1$ in sketched graph)		$(x, y) \to (\ln x, \ln y)$ for $x > 0$, $y > 0$, producing $\ln y = \ln B_0 + b_1 \ln x$ or $y' = b_0 + b_1 x'$
Curve is asymptotic to both horizontal and vertical axes ($y = B_0 x^{b_1}$ for $B_0 > 0$, $b_1 < 0$)	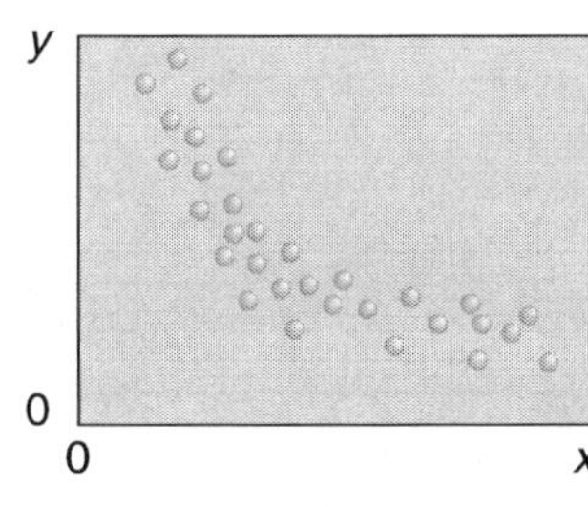	$(x, y) \to (\ln x, \ln y)$ for $x > 0$, $y > 0$, producing $\ln y = \ln B_0 + b_1 \ln x$ or $y' = b_0 + b_1 x'$
Curve has a nonzero y-intercept and appears exponential; either with rapid growth or decay ($y = B_0 B_1^x$ for $B_0 > 0$, $B_1 > 0$; exponential growth shown in graph, namely $B_1 > 1$)	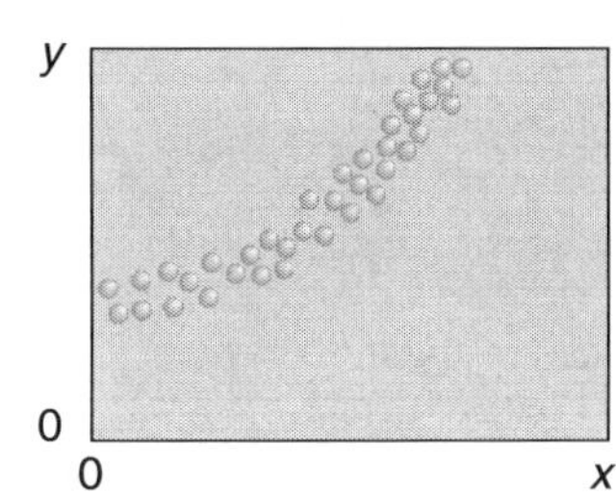	$(x, y) \to (x, \ln y)$ for $y > 0$, producing $\ln y = \ln B_0 + x \ln B_1$ or $y' = b_0 + b_1 x$
Curve has a positive x-intercept and appears logarithmic: grows slowly for large x and decreases rapidly as x goes toward 0 ($y = b_1 \ln(B_0 x)$ for $b_1 > 0$, $B_0 > 0$)	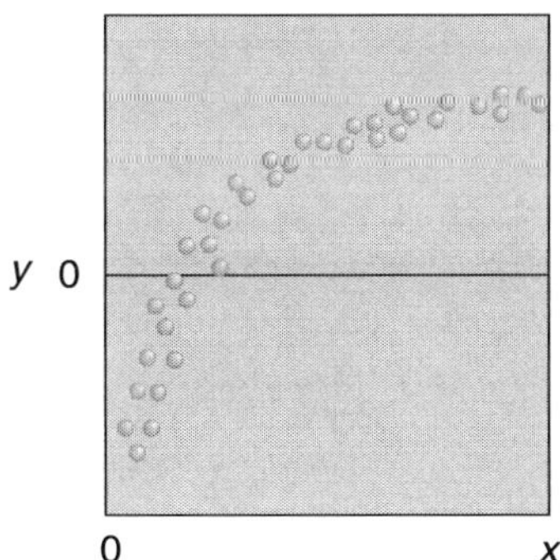	$(x, y) \to (\ln x, y)$ for $x > 0$, producing $y = b_1 \ln(B_0) + b_1 \ln x$ or $y = b_0 + b_1 x'$

Section 3.3 Exercises

Having a calculator or computer software can be very helpful for many of the following problems.

1. a. Find the equation of the line that passes through $(1, 3)$ with slope $4/3$.
   b. Find the equation of the line that passes through the points $(5, 3)$ and $(6, 2)$. *Hint:* Find its slope using the two

points.

c. Which of the following points are on the line $y = 2 - 3x$? $(1, 1), (1, -1), (0, 2)$

2. A researcher measures the lengths (from nose to tail) and weights of eight laboratory mice.

Length (centimeters), x	**Weight (grams), y**
16	32
15	26
20	40
13	27
15	30
17	38
16	34
21	43

x	y	$\hat{y}$	$y - \hat{y}$	$(y - \hat{y})^2$
16	32	32.81	−0.81	0.6561
15	26	31.31	−5.31	28.1961
20	40	38.81	1.19	1.4161
13	27	28.31	−1.31	1.7161
15	30	31.31	−1.31	1.7161
17	38	34.31	3.69	13.6161
16	34	32.81	1.19	1.4161
21	43	40.31	2.69	7.2361

a. A regression equation $\hat{y} = 1.5x + 8.81$ was proposed to predict the weight of a mouse, given its length. The above table was constructed using this regression line. What is the error sum of squares for estimating weight using this equation?

b. The regression equation

$$\hat{y} = 1.9x + 2.16$$

is used in the following table to estimate the weight of a mouse given its length. What is the error sum of squares for this equation?

x	y	$\hat{y}$	$y - \hat{y}$	$(y - \hat{y})^2$
16	32	32.56	−0.56	0.3136
15	26	30.66	−4.66	21.7156
20	40	40.16	−0.16	0.0256
13	27	26.86	0.14	0.0196
15	30	30.66	−0.66	0.4356
17	38	34.46	3.54	12.5316
16	34	32.56	1.44	2.0736
21	43	42.06	0.94	0.8836

c. Which of the two equations

$$\hat{y} = 1.5x + 8.81$$

$$\hat{y} = 1.9x + 2.16$$

fits the mice data better, in terms of error sum of squares (SSE)?

3. Refer back to Exercise 2.

a. Find the least squares regression line for predicting the weights of the laboratory mice from their lengths.

b. Refer to part (a). What percentage of variability in the weight of those mice can be explained by regression on their length? *Hint:* Find r^2.

c. Find the least squares regression line for predicting the lengths of the laboratory mice from their weights.
d. Refer to part (c). What percentage of variability in the lengths of those mice can be explained by regression on their weights?

4. The data in the table below are the numbers of push-ups y that could be done by a sample of 12 male instructors at Newton High School. (The 38-year-old teacher was the track coach.)
   a. Draw the scatterplot. Identify the outlier.
   b. Find the correlation of these data.
   c. Find the slope of the least squares best-fitting line for predicting number of push-ups from age.
   d. Find the equation of the regression line with this slope. Then use the regression line to estimate the number of push-ups that can be done on average by teachers of each of the following ages:
      i. 26 years
      ii. 39 years
      iii. 50 years

Age	Number of push-ups	Age	Number of push-ups
21	10	22	9
25	8	27	6
22	11	44	4
28	6	48	3
30	7	35	8
38	15	48	5

5. For a representative cross section of several thousand American men, the average height was found to be 68.8 inches, with a standard deviation of 2.8 inches. The average weight was 171 pounds, with a standard deviation of 30 pounds. The correlation between height and weight was 0.4.
   a. Find the equation of the regression line for estimating weight (y) from height (x).
   b. Estimate the average weight of all the men who are 66 inches tall.
   c. Guess the weight of a randomly chosen man who happens to be 66 inches tall.
   d. What percentage of the variation in weight can be explained by the variation in height?
   e. Guess the weight of a man who is 72 inches tall.

6. In a study of the stability of IQ scores, a large group of people were tested at age 18 and again at age 35. At both age 18 and age 35, the average IQ score was around 100, with a standard deviation around 15. The scatterplot was elliptical, with a correlation around 0.8.
   a. Find the regression equation for estimating the IQ score at age 35 (y) from the IQ score at age 18 (x).
   b. Estimate the average score at age 35 for all the people who scored 120 at age 18.
   c. Predict the IQ score at age 35 for someone who scored 120 at age 18.

7. The following six statements refer to scatterplots and their regression lines in general. For each statement, determine whether it is true or false.
   a. If $r = 0$, then all of the points are on the regression line.
   b. If all of the residuals are 0, then all the points are on the regression line.
   c. If all of the residuals are 0, then r is either -1 or $+1$.
   d. If the residual standard deviation is 0, then $r = 0$.
   e. If the slope of the regression line is 0, then the residual SD is 0.
   f. If $r = 0$, then the slope of the regression line is 0.

8. The scatterplot below shows the relationship between body temperature (degrees Fahrenheit) and heart rate (beats per minute) for a group of people. Body temperature is the x variable, and heart rate is the y variable. The average body temperature is 98.25 °F, and the SD of body temperature is 0.73 °F. The average heart rate is 73.8 beats per minute, and the SD of the heart rates is 7.03 beats per minute. The correlation coefficient is 0.25. The regression line for predicting heart rate from temperature is shown on the scatterplot.
   a. What percentage of the variation in heart rate can be explained by the variation in temperature?
   b. Find the equation of the regression line for estimating heart rate from temperature.
   c. Estimate the average heart rate (in beats per minute) for all people with a body temperature of 98.75 °F.
   d. Estimate the average heart rate (in beats per minute) for all people with a body temperature of 97.5 °F.
   e. Guess the heart rate (in beats per minute) for a person with a body temperature of 98.75 °F.
   f. About 68% of the heart rate predictions for a given temperature will be off by at most how much?
   g. Answer part (f) for 95%.

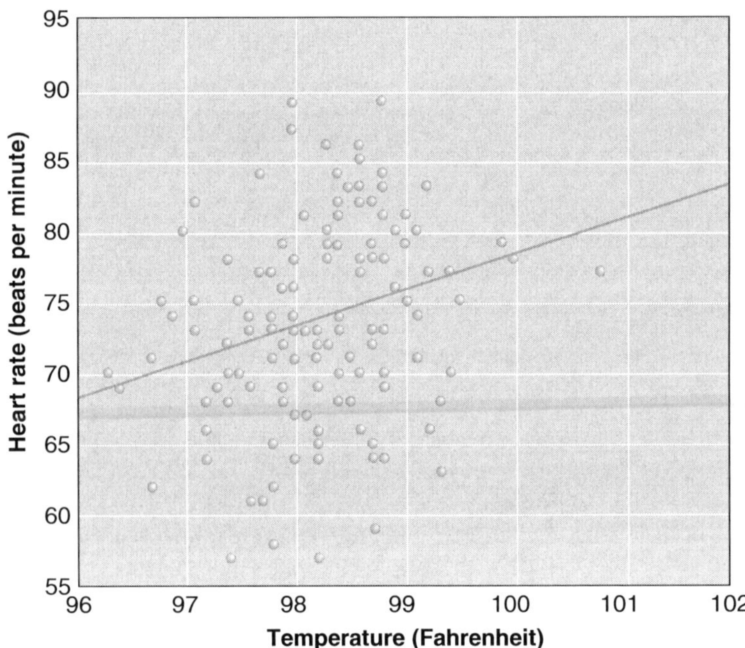

9. In one large study of identical male twins, the average height was found to be about 175 cm, with an SD of about 7.5 cm. The correlation between heights of twins was about 0.95. The scatterplot was elliptical. (In twin studies, the convention is to plot each twin pair twice: once as (x, y) and once as (y, x). That is, the John-Jack twin pair, with John being 175 cm tall and Jack 172 cm tall, produces two data points in the plot: $(175, 172)$ and $(172, 175)$. Thus, $\overline{x} = \overline{y} = 175$, and $s_x = s_y = 7.5$.)
   a. Using the regression line, estimate the average height of those men whose identical twin was 175 cm tall. (Choose one option.)

165 cm   170 cm   160 cm   180 cm   175 cm

b. Using the regression line, estimate the approximate average height of those men whose identical twin was 165 cm tall.

164 cm   164.5 cm   165 cm   165.5 cm   166 cm

c. Using the regression line, estimate the approximate SD of the heights of those men whose identical twin was 175 cm tall.

2.50 cm   5 cm   7.5 cm   10 cm   12.5 cm

Hint: See Figure 3.25.

d. If you had to guess the height of a randomly chosen subject from this study with no information about this person or his twin brother, what would your best guess be?

165 cm   170 cm   160 cm   180 cm   175 cm

e. What is the likely size (i.e., use 68% rule) of the prediction error in part (d)?

2.5 cm   5 cm   7.5 cm   10 cm   12.5 cm

f. A randomly chosen subject stands before you. He is 165 cm tall. You have to guess the height of his twin brother. What should you guess?

164 cm   164.5 cm   165 cm   165.5 cm   166 cm

g. What is the likely size of the prediction error (i.e., use 68% rule) in part (f)?

2.5 cm   5 cm   7.5 cm   10 cm   12.5 cm

Hint: It is the residual SD that tells us the typical size of the prediction error $y - \hat{y}$.

10. The following scatterplot shows the relationship between the duration of an eruption of Old Faithful (in minutes) and the length of the time interval (in minutes) until the next eruption for 150 eruptions in July 1995. The average duration was 3.35 minutes with a standard deviation of 1.157 minutes. The average time interval until the next eruption was 77.94 minutes with a standard deviation of 16.58 minutes. The correlation was 0.9186. The regression line for estimating the time interval until the next eruption from the duration of the last eruption is shown.
    a. Look at the scatterplot. Is the mean duration of 3.35 minutes the "typical" length of an eruption? (Do a lot of eruptions last between 2.5 minutes and 4 minutes?) Use the term "gap" and "clusters" in your answer.
    b. What percentage of the variation of the length of time interval until next eruption can be explained by the duration of the previous eruption?
    c. Find the equation of the regression line for estimating the time interval until the next eruption from the duration of the last eruption.
    d. Predict the time interval until the next eruption if the duration of the last eruption was 2 minutes.
    e. Predict the time interval until the next eruption if the duration of the last eruption was 4.5 minutes.
    f. Compute the residual SD.
    g. If we use the regression equation to predict the time interval until the next eruption, what is the likely size of our prediction error? Apply the 68-95-99.7% Rule to make this precise.

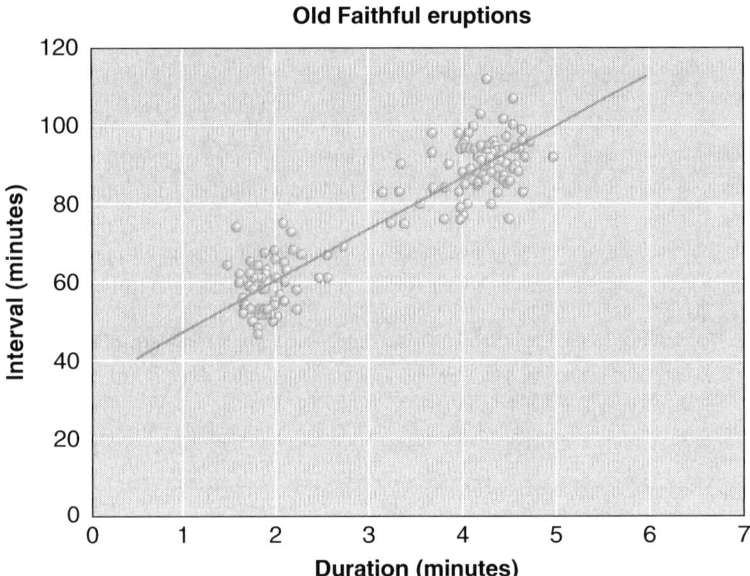

11. The question this exercise is helping us understand is: How does an influential point affect the regression line and how does an outlier affect the correlation coefficient? The four scatterplots below are identical, except for the circled outlier in plots B and D and the leverage point in plot C.

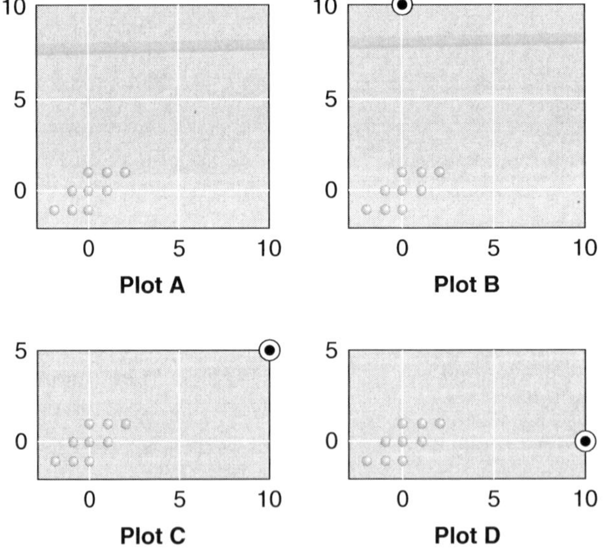

a. Knowing that each regression equation below matches one of the plots above, match each plot with its regression equation and correlation (no computation is necessary).
$\hat{y} = -0.06 + 0.06x, r = 0.24$; $\hat{y} = 0.5x, r = 0.71$; $\hat{y} = 1 + 0.5x, r = 0.18$, $\hat{y} = 0.5x, r = 0.95$
b. Discuss the results, using the concepts of outlier, influential point, and leverage point.

12. As shown below, a least squares regression line was fitted to the winning percentages for a local sports team for the years 1988 through 2000. The percentage for the 2001 season was then plotted (as circled below).

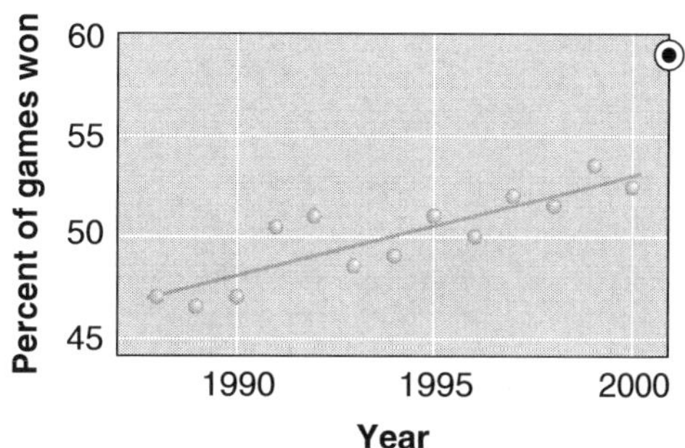

Which of the following statements correctly describes how the winning percentage for the 2001 season will change the least squares regression line and the correlation coefficient if a new regression line is fitted to the 1988 through 2001 data? (No computation is necessary.) Is the 2001 point influential or not?
a. The 2001 point will make the least squares regression line steeper and the correlation stronger.
b. The 2001 point will make the least squares regression line steeper and the correlation weaker.
c. The 2001 point will make the least squares regression line closer to horizontal and the correlation stronger.
d. The 2001 point will make the least squares regression line closer to horizontal and the correlation weaker.
e. The 2001 point will have very little effect on the least squares regression line or the correlation coefficient because the point follows the overall upward trend.

13. Draw the residual plot for the 10 points in Example 3.14. Identify the outlier from the plot. Do you see any pattern?

14. Draw the residual plot for the Key Problem. (See Table 3.1.) Identify the outliers. Which are they? Is there a trend in the residual plot? Describe the trend. Do you detect a nonlinear relationship between curb weight and city mileage?

15. Which of the following conclusions *always* holds?
a. An outlier is a leverage point.
b. An outlier is influential.
c. A leverage point is an outlier.
d. A leverage point is influential.
e. An influential point is an outlier.
f. An influential point is a leverage point.

16. Three points are circled in the scatterplot below.

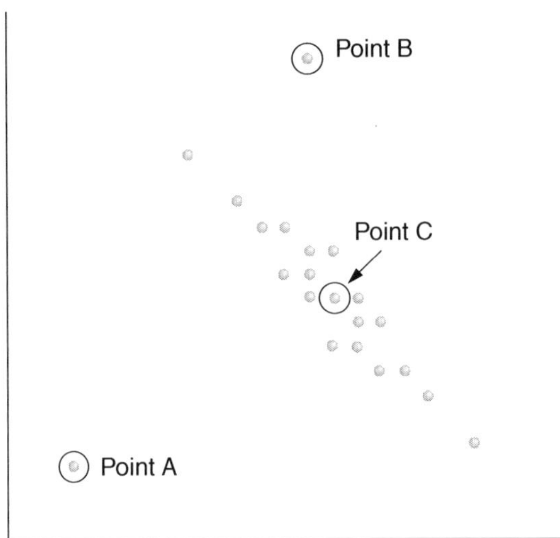

For each of the three points, state whether the point is an outlier, a leverage point, and/or an influential point. Is every outlier influential?

17. Three points are circled in the scatterplot below.

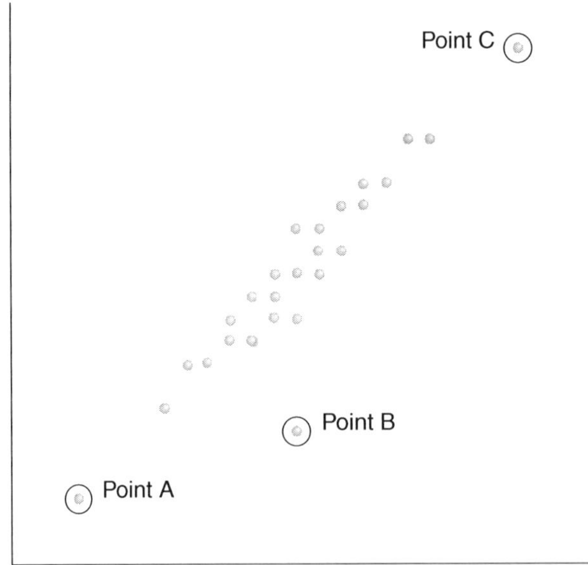

For each of the three points, state whether the point is an outlier, a leverage point, and/or an influential point. Is every leverage point influential?

18. In the scatterplot below, three points are circled and the least squares regression line is drawn.

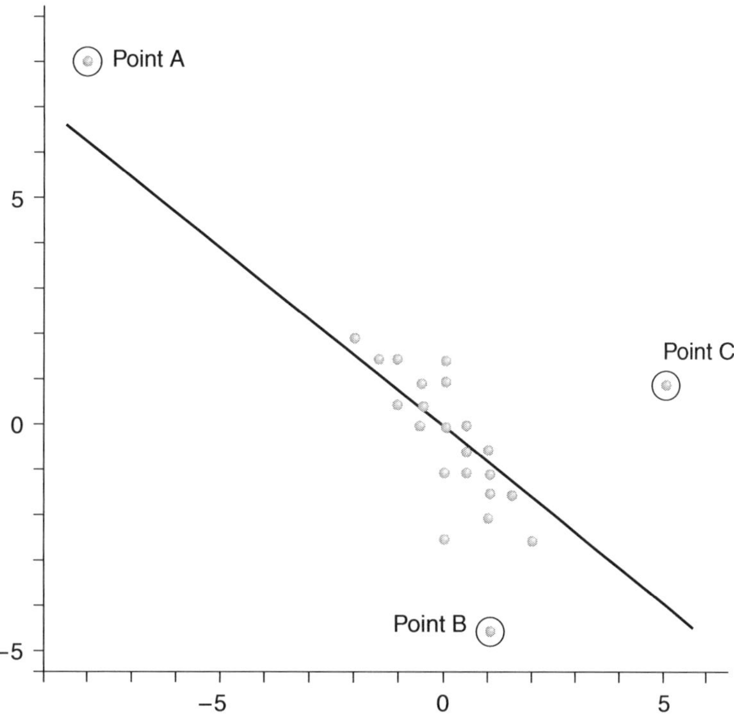

For each of the three points, state whether the point is an outlier, a leverage point, and/or an influential point. Is every influential leverage point an outlier?

3.4 THE QUESTION OF CAUSATION

cause-and-effect relationship

In the Key Problem we found a strong correlation between the weight of a car and its gas mileage, reflecting a **cause-and-effect relationship** between weight and mileage: The heavier the car, the more energy it takes to drive it and, hence, the fewer miles per gallon the car will get. There is also a cause-and-effect relationship between the duration of an eruption of Old Faithful and the length of the time interval until the next eruption (Example 3.10). In Example 3.1 we wondered whether the strong correlation between animal fat consumption and breast cancer rates meant that the consumption of animal fat *causes* breast cancer.

Epidemiologists use statistical methods to search for causes of diseases. Several years before the discovery of the role of bacteria in disease transmission, the nineteenth-century English physician John Snow was able to locate the source of the 1854 outbreak of cholera in London. A week into the outbreak, Snow obtained the addresses of all the people who had died from cholera. He found that nearly all the deaths had taken place in the neighborhood of Broad Street, Golden Square, and the adjoining streets. By interviewing survivors, Snow found that the few cholera deaths outside this area were in households that usually sent for water from the much-frequented street-pump in Broad Street, because they preferred the water from the Broad Street pump to that of nearer pumps. (In those days most people got their water from public wells.) Because the only circumstance all of the cholera victims seemed to share was that they drank water from the Broad Street pump, Snow concluded that the Broad Street pump was somehow the source of the cholera outbreak. He persuaded authorities to remove the handle of the pump to prevent people from using the pump. This brought the cholera outbreak to an end.

John Snow's detective work was classical epidemiology. Even though no one at the time had identified the water-borne bacteria that cause cholera, Snow was able to link people's risk of becoming cholera victims during the 1854 outbreak to whether they drank water from the Broad Street pump. Unfortunately, such positive correlational statistical links between risk factors (drinking water from the Broad Street pump) and diseases (cholera) can be misleading and may not represent cause-and-effect relationships.

Some medical studies in the 1970s found a link between coffee drinking and heart disease: The rate of heart disease was higher for coffee drinkers than for nondrinkers. And among coffee drinkers, the rate was highest for those who drank a lot of coffee. In other words, the researchers found a positive correlation between the amount of coffee people drank and their risk of heart disease. Did this mean that coffee drinking causes heart disease? Further studies revealed that the coffee drinkers tended to smoke more than the nondrinkers. It turned out that it was their smoking that gave the coffee drinkers heart disease, not the coffee. The number of cigarettes smoked per day is the **lurking variable** (or **hidden variable**) that explains the correlation between the number of cups of coffee per day and the risk of heart disease.

lurking variable
hidden variable

A **lurking variable** in a study is one that is not included in the study initially (the number of cigarettes smoked per day) but that affects the response variable (the risk of heart disease) and is strongly correlated with the suspected explanatory variable (the number of cups of coffee per day). Statisticians sometimes find that a hidden variable is influential by graphing the residuals $y - \hat{y}$ as a function of the lurking variable.

confounded

If we cannot separate the effects of two or more variables, then we say that they are **confounded**. For example, studies have found that women who eat a high-fiber diet have a reduced risk of heart disease. Does that mean that dietary fiber protects a woman against heart disease? Or could the explanation be that women who choose their diet carefully also exercise more and avoid smoking, factors that are known to reduce the rate of heart disease? In this example, the effects of a high-fiber diet are confounded with the effects of exercise and an otherwise healthy lifestyle.

Strong correlation suggests but does not prove causation. To establish causation statistically, one must perform randomized, controlled experiments (discussed in Chapter 7), in which the suggested causal variable is varied and the effect is observed. If it is not possible to conduct such experiments, then we must have other scientific proof. One of the best-known examples of the confusion about correlation and cause regards smoking and cancer. For the past 50 years or more, research projects have produced data showing a strong positive correlation between the amount a person smokes and that person's risk of getting cancer. On one side of the debate, medical associations and consumer groups have insisted that smoking causes cancer. On the other side—notably the tobacco industry—it has, until recently, been argued that although smoking and cancer are statistically linked, smoking had not been established scientifically as a cause of cancer because other explanations for the observed correlations between smoking and lung cancer were possible. The problem was that the best way to prove that smoking causes cancer would be to conduct a highly unethical experiment in which scientists tell the participants how much to smoke and then record who gets cancer and who does not. Obviously, nobody would participate in such a highly unethical experiment. Instead, scientists experiment on animals. Carefully designed studies on other mammals have shown strong evidence of causality in these species, which likely applies to humans, too. The study of smoking and lung cancer from the statistical perspective is discussed at length in Chapter 7.

Section 3.4 Exercises

1. Several medical studies have found that snoring is a risk factor for heart disease. The more a person snores, the higher the risk of heart disease. (See Exercise 5 in Section 1.1.) Does this mean that snoring causes heart disease? If not, identify a possible lurking variable.
2. For American cities there is a strong positive correlation between the number of churches in the city and the number of violent crimes reported to the police. Does this mean that if we build more churches, then there will be an increase in violent crime? Identify the lurking variable that explains the correlation.
3. For schoolchildren there is a positive correlation between shoe size and vocabulary size (the number of different words the child uses). Identify the lurking variable that explains the correlation.
4. There is a positive correlation between the weight of a Thanksgiving turkey and the amount of time it takes to cook it. Is this a cause-and-effect relationship, or can you think of a lurking variable?
5. There is a negative correlation between the number of influenza cases reported in Chicago each month of the year and the amount of ice cream sold in Chicago that month. Does this mean that ice cream protects you against the flu? If not, identify the lurking variable.
6. For the world's nations there is a negative correlation between the infant mortality rate (number of infant deaths per 1000 live births) and the number of TV sets per capita. Does this mean that we can lower the infant mortality rate in Ethiopia by increasing the number of TV sets in Ethiopia? If not, identify the lurking variable.
7. People who use artificial sweeteners tend to weigh more than people who use sugar. Does this mean that if you switch from sugar to artificial sweeteners, then you will put on weight? Explain.
8. For the last 30 years, the number of infant deaths in the United States has declined from year to year, whereas the number of deaths due to diabetes has increased from year to year. Is this negative correlation between number of infant deaths and number of deaths due to diabetes a cause-and-effect relationship?

CHAPTER 3 SUMMARY

Paired observations of two numerical variables can be displayed in a **scatterplot.** We use the scatterplot to study the relationship between the two variables. If we suspect a **cause-and-effect** relationship between the two variables, then we plot the suspected **explanatory variable** on the horizontal axis (the x-axis) and the **response variable** on the vertical axis (the y-axis). If there is no explanatory-response distinction, then either variable can go on the horizontal axis.

If the scatterplot slopes upward from left to right, then we say that the variables are **positively correlated**. If the scatterplot slopes downward from left to right, then we say that the variables are **negatively correlated**. If the points of the scatterplot are tightly clustered around a sloped line, then we say that there is a **strong linear correlation** between the variables.

The **linear correlation coefficient** r measures the strength and direction (up or down) of a straight-line relationship. The value of r ranges from -1 (when all of the points lie on a line that slopes down) to $+1$ (when all of the points lie on a line that slopes up). The closer r is to -1, the tighter the points cluster around a line that slopes down. The closer r is to $+1$, the tighter the points cluster around a line that slopes up. If r is close to 0, then the points of the scatterplot are not clustered around any sloped line.

The **least squares regression line** for estimating y from x goes through the point $(\bar{x}, \bar{y})$ and has slope $m = r(s_y / s_x)$. Its equation is

$$\hat{y} = mx + b,$$

where $b = \bar{y} - m\bar{x}$. We call $\hat{y}$ the **fitted** y-value, the **estimated** y-value, or the **predicted** y-value corresponding to x. The regression line (and its equation) can be used to predict *individual* y-values from the corresponding x-values.

For *large* data sets of paired x- and y-values, the regression line estimates the average value of y corresponding to each fixed value of x.

We call $r^2 =$ (explained variation)/(total variation) $= \sum(\hat{y} - \bar{y})^2 / \sum(y - \bar{y})^2$ the **coefficient of determination**. The quantity $r^2 \cdot 100\%$ is **the percentage of the total variation of** y **that can be explained by** x. For each data point (x, y) the **error of prediction**, $y - \hat{y}$, between the actual y-value and the estimated value is also called the **residual**. The residuals have mean 0 and standard deviation

$$\text{residual SD} = s_y\sqrt{1 - r^2}.$$

For a large data set with an elliptical scatterplot, the residual SD tells us the likely size of a future prediction error for a specified x. The 68-95-99.7% Rule applies to the residual SD.

In linear regression, points not following the linear pattern of the rest of the data are considered **outliers**. Outliers often reduce the strength of the correlation, sometimes seriously. Outliers should be removed before computing the regression line and its r. An **influential point** is an observation that has a strong influence on the position of the regression line. The removal of an influential point markedly changes the position of the regression line. Points with extreme x-coordinates are called **leverage points**. They are often very influential. A point that is not a leverage point usually is not influential.

A **residual analysis** of a linear regression can help show the existence of outliers to be removed or a nonlinear trend to be removed by appropriate transformations of x and/or y. We can often transform nonlinear bivariable data (paired observations (x, y)) into linear data by taking logarithms of x or y or both. Then one can use linear regression for the transformed (x', y') data to estimate the nonlinear relationship between x and y.

CHAPTER REVIEW EXERCISES

1. For each of the following questions, identify the two variables of interest, and state whether you would expect the correlation coefficient to be (i) close to 1 (strong positive), (ii) close to 0 (negligible), (iii) close to -1 (strong negative), (iv) positive but close to neither 1 nor 0 (weak to moderate positive), (v) negative but close to neither -1 nor 0 (weak to moderate negative).
   a. Do taller basketball players block more shots?
   b. Will the number of times a student skips a class affect his or her score on the final exam?
   c. Do students from wealthier families perform better on standardized tests?
   d. Do people who have more years of education earn more money?
   e. Is there a relationship between the population of a county and the number of farms in that county?
2. Respond *true* or *false* to each of the following statements. If the statement is false, explain why.
   a. The correlation coefficient can be useful in determining how well a regression line fits the data.
   b. It can happen that the correlation coefficient is negative, and the slope of the least squares best-fitting line is positive.
   c. The method of least squares is usually effective, and it is the most widely used method for finding regression lines.

3. The following table contains verbal (x) and quantitative (y) scores on the SAT from five students. A researcher wants to answer this question: "Is there a relationship between verbal and quantitative scores on the SAT?" For statistical evidence, the researcher wants to calculate the correlation coefficient (r) for these data. This exercise will take you through the four steps of calculating r.

Student	x	y	$x-\bar{x}$	$y-\bar{y}$	$(x-\bar{x})(y-\bar{y})$	$(x-\bar{x})^2$	$(y-\bar{y})^2$
1	670	710					
2	550	500					
3	720	620					
4	410	490					
5	520	560					
Sum	2870	2880					

a. Complete the table.
b. Using the information from the table you just formed, find the variance s_y^2 of y, the variance s_x^2 of x, the standard deviation s_x of x, and the standard deviation s_y of y.
c. Find the correlation between x and y.
d. What kind of relationship between mathematical and verbal scores on the SAT does this value of r imply?
e. Find the least squares regression line.
f. Bob gets a verbal score of 700. Predict his quantitative score.

4. In their study of over 1000 pairs of husbands and wives, Pearson and Lee found the average height of the husbands to be 67.7 inches with a standard deviation of 2.7 inches. The average height of the wives was 62.5 inches with a standard deviation of 2.4 inches. The correlation between height of husband and height of wife was 0.28.
a. Is this a strong or a weak correlation?
b. Determine the regression equation for estimating the height of a wife from the height of her husband.
c. What percentage of the total variation in height of the wives can be explained by the height of their husbands?
d. Estimate the average height of the wives whose husbands were 73 inches tall (to the nearest inch).
e. Predict the average height of a wife whose husband is 65 inches tall.
f. Apply the 68-95-99.7% Rule using the residual SD.

5. The following table is a list of 10 baseball teams and their total runs scored and home runs hit from the 1996 season.
a. The regression line obtained by the method of least squares is $\hat{y} = 2x + 455$. Which of the 10 observations has the largest squared error? Which has the smallest?
b. A team not on the list, the Atlanta Braves, hit 197 home runs. Predict how many runs they scored.

Team	Number of home runs (x)	Total number of runs (y)	$\hat{y}$	$(y-\hat{y})^2$
Colorado Rockies	221	961	897	4096
Pittsburgh Pirates	138	776	731	2025
Los Angeles Dodgers	150	703	755	2704
St. Louis Cardinals	142	759	739	400
Cincinnati Reds	191	778	837	3481
Chicago Cubs	175	772	805	1089
Florida Marlins	150	688	755	4489
San Diego Padres	147	771	749	484
New York Mets	147	746	749	9
Houston Astros	129	753	713	1600

6. Find a newspaper article that cites a study in which two variables are related to each other.
   a. What are the two variables that are being related?
   b. Even if there is a relationship between these two variables, does that mean that one is causing the other? Do you believe that the authors of this article answered this question of causation adequately?

7. Consider the following section from a published health study:

   Among the current smokers who had high fish intake, the number of cigarettes smoked per day had no effect on the incidence of heart disease and the death rates from heart problems. As seen with lung diseases, the consumption of fish seems to protect smokers from the harmful effects of smoking on the heart and blood vessels. It has been suggested that fish oils alter some metabolic processes that result in beneficial effects, including more dilatation (or relaxation) of blood vessels, less platelet adhesiveness, reduced inflammatory response to the injury caused by smoking, lower triglyceride and fibrinogen levels, and lower blood pressure (especially in patients who have mildly elevated blood pressure). (Source: www.healer-inc.com/a039705a.htm.)

   a. What are the two variables being compared in this study?
   b. Could you conclude from this study that if you were a smoker, and you ate a lot of fish, you would not be at the usual increased risk of developing heart disease due to smoking? Why or why not?

8. The following table contains the sizes (in carats) of nine diamonds and their prices.

Size (x)	Price (y)
0.17	$353
0.16	$328
0.17	$350
0.18	$325
0.25	$642
0.16	$342
0.15	$322
0.19	$485
0.21	$483

Source: www2.ncsu.edu/ncsu/pams/stat/info/jse/datasets.index.html.

   a. Draw a scatterplot showing the relationship between these two variables.
   b. Find the least squares regression line. What is the equation for the line?
   c. You find a diamond of size 0.23 carat in a store, and the salesperson is going to sell it to you for $500. Is this likely a good deal, according to the regression line found above?
   d. Which of the nine diamonds listed above was the worst deal for the person who bought it, according to your regression line?

9. What is the correlation between the size of a diamond and its price, using the data from Exercise 8?

10. The chapter discusses two methods for finding the slope of the least squares best-fitting line. Explain both of the methods briefly, stressing advantages when they exist.

11. The following table gives the weights (x) of eight adult men and the amounts of weight they are able to lift (y).

Man	x	y	$x-\bar{x}$	$y-\bar{y}$	$(x-\bar{x})(y-\bar{y})$	$(y-\bar{y})^2$
1	150	172	−21	−12	252	144
2	174	210	3	26	78	676
3	163	159	−8	−25	200	625
4	167	175	−4	−9	36	81
5	210	200	39	16	624	256
6	189	220	18	36	648	1296
7	140	150	−31	−34	1054	1156
8	175	186	4	2	8	4
Sum	1368	1472	0	0	2900	4238

a. What is the correlation between x and y?
b. What is the slope of the least squares best-fitting regression line?
c. Using the fact that the line must pass through $(\bar{x}, \bar{y}) = (171, 184)$, determine the equation for the least squares best-fitting regression line.
d. If another male weighs 156 pounds, what would be the best prediction of how much he will be able to lift?
e. Apply the 95% rule to part (d) using the residual SD.

12. The following table shows the return and risk for selected investments during the period 1960–1990. (The risk variable is a specially defined index intended to summarize in a single number the amount of risk for that type of investment. Higher values indicate higher risk.)

	Return (x)	Risk (y)
Common stock	9.8	15.8
Long-term corporate bonds	6.9	11.1
Long-term Treasury bonds	6.4	10.9
Short-term Treasury bonds	6.4	2.9
Residential real estate	10.6	10.7
Farm real estate	9.9	8.5
Business real estate	8.7	4.9

a. What sort of relationship would you expect between return and risk for investments? Why?
b. What is the correlation between return and risk?
c. Does your answer agree with what you guessed in part (a)?

13. The following table shows the expenditures (in billions of dollars) for advertising and promoting smoking in the United States for the years 1975 to 1996, and the percentages of high school seniors who smoked daily in the subsequent year.
   a. The means for x and y are 2.80 and 21.3, respectively; the standard deviations for x and y are 1.697 and 3.428, respectively; and the correlation between x and y is -0.478. What does this negative correlation suggest?
   b. Find the intercept and slope of the least squares line for estimating the percentage of seniors who smoke from expenditures for advertising.
   c. Plot the points, and draw the least squares line on a plot.
   d. Find the predicted percentage of seniors who smoke for the years 1975, 1983, and 1996 using the least squares line. How far off are the estimates from the actual values?
   e. Is the least squares line appropriate for these data? Why or why not?

	x Expenditures	y % smoking
1975	0.49	28.8
1976	0.64	28.8
1977	0.78	27.5
1978	0.88	25.4
1979	1.08	21.3
1980	1.24	20.3
1981	1.55	21.1
1982	1.80	21.2
1983	1.90	18.7
1984	2.10	19.5
1985	2.48	18.7
1986	2.38	18.7
1987	2.58	18.1
1988	3.27	18.9
1989	3.62	19.1
1990	3.99	18.5
1991	4.65	17.2
1992	5.23	19.0
1993	6.04	19.4
1994	4.83	21.6
1995	4.90	22.2
1996	5.11	24.6

Do these results suggest that higher spending on cigarette advertising causes fewer high school seniors to smoke? Explain.

PROFESSIONAL PROFILE

Charles Sanders Peirce (1839–1914), American philosopher, logician, mathematician, scientist, and pioneer in statistics.

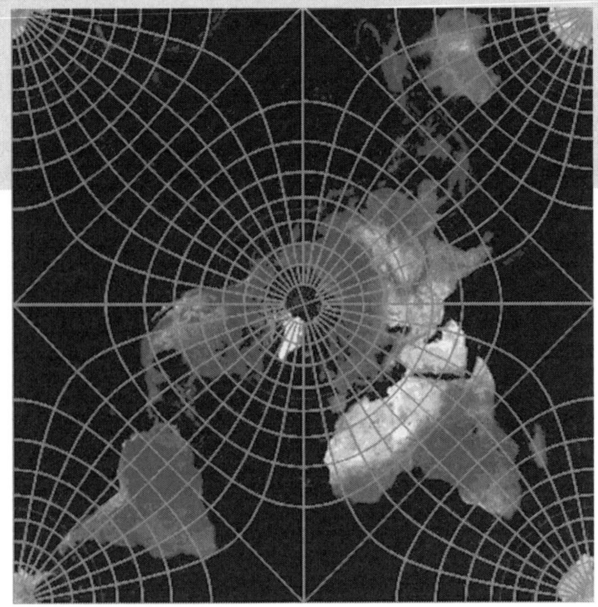

NASA view of the earth presented in the quincuncial projection developed by Charles S. Peirce. This project reduced the distortion compared to other types of projects then in use to present a curved earth on a flat map.

Charles Sanders Peirce

Hailed by many as one of the greatest American minds, Charles Sanders Peirce made many important contributions to science, philosophy, mathematics, and statistics. Among his breakthroughs are the first use of a wavelength of light as a unit of measure, the first known design and theory for an electric switching circuit for computing devices, and the founding of the branch of philosophy known as pragmatism. During his thirty years with the U.S. Coast Survey Office, he earned international renown as a geodesist and for his research on the shape of the Milky Way. Of his books, Illustrations of the Logic of Science (1877–1878) and A Theory of Probable Inference (1883) are among the most important for statisticians.

Today, statisticians regard him as one of the major contributors to the field. His theory of probability is both insightful and important. He saw understanding probability as central to scientific discovery. Peirce defined probability as the limiting relative frequency of an event occurring when repeated trials are conducted under carefully controlled unchanging experimental conditions. To illustrate, the proportion of heads in an increasingly large number of fair coin tosses approaches 0.5, so the probability of heads is 0.5. This repeated trials approach is how this textbook defines probabilities. Peirce was aware that all probabilities are controlled by which conditions hold true. For example, a coin might be unusually hollowed out so that heads tends to occur twice as often as tails for repeated tosses, producing a different limiting relative frequency of heads, namely 0.666.

His contribution to the understanding of probability is enough to make Peirce a major figure in developing the field of statistics. However, he made many other contributions to statistics. Peirce introduced the idea of blinded, controlled, randomized statistical experiments (anticipating the work of Ronald A. Fisher). And he introduced the essential concepts of "confidence" and "likelihood." Modern statistics might have been very different without the contributions of Charles Sanders Peirce.

4

Obtaining Data: Random Sampling and Randomized Experiments

The hypothesis is that 60 is divisible by all positive integers below it. I tried 1, 2, 3, 4, 5, and 6. Encouraged, I tried 10, 20, and 30. Based on this nonrandom sample from the population 1, 2, ..., 60, I infer that the hypothesis is true.

Apocryphal

Objectives

After studying this chapter, you will understand the following:

- ❒ Randomization: the key to carrying out an effective statistical study
- ❒ Probability sampling: sampling from a population using randomization
- ❒ Simple and stratified random sampling; cluster sampling
- ❒ Various nonrandom (and inappropriate) methods for sampling from a population
- ❒ How obtaining data by nonrandom methods can lead to confounding and bias
- ❒ Randomized controlled experiments and their advantages
- ❒ Completely randomized experimental designs
- ❒ Randomized block designs
- ❒ Matched-pairs designs

KEY PROBLEM

Postmenopausal Hormone Replacement Therapy: How Could We Have Been So Wrong?

Heart disease is the leading cause of death among women as well as men in the United States. Since women usually develop heart disease later in life than men do, there is reason to believe that estrogen somehow protects women's hearts. After menopause women lose that protection. So it sounds like a good idea to give postmenopausal women estrogen replacement therapy: Not only would the therapy take care of unpleasant symptoms of menopause, such as hot flashes and fatigue, but the therapy should protect women against heart disease as well.

Between 1970 and 2000, more than 40 medical studies found that postmenopausal women who took estrogen replacement therapy under their doctor's supervision had a 35% to 50% lower risk of coronary heart disease than postmenopausal women who did not take hormone replacement therapy. Hormone replacement became one of the most widely prescribed therapies in the United States, taken by 14 million women annually.

It therefore came as a shock to many women when the monitoring board of the Women's Health Initiative on May 31, 2002, after 5 years of follow-up, stopped a clinical trial of estrogen plus progestin involving 16,000 women because health risks outweighed health benefits of hormone replacement therapy. What was going on? How could this latest trial invalidate three decades of medical studies? What are we to believe?

4.1 INTRODUCTION

descriptive statistics

In this chapter we shall discuss how to obtain useful data from "the real world." In Chapters 1 through 3 we focused on **descriptive statistics,** that is, methods for organizing, displaying, and summarizing data. For instance, in Example 1.7 we looked at the highest elevation in each of the 50 states. The histogram in Figure 1.9 showed that the states fell in 3 clusters separated by gaps, with the largest cluster consisting of the 37 states east of the Rocky Mountains. In Example 2.6 we looked at the salaries of the Miami Heat for the 2005–2006 season. The histogram in Figure 2.4 showed that the salary distribution was skewed to the right with a single outlier at the high end. The skewness of the distribution was reflected in the fact that the mean salary was more than twice the median salary.

population

In a statistical study we use the term **population** to denote the collection of objects or individuals of interest. In Example 1.7 the population of interest was the 50 states. The variable of interest was the maximum elevation of each state. In Example 2.6 the population of interest was the 19 individuals playing for the Heat in 2005–2006 and the variable of interest was their salary. In both of these examples we had complete data from the entire population of interest. We had a **census.** A census is a list of all members of a population along with certain characteristics of each population member.

census

Example 4.1

The Bible tells us that Caesar Augustus wanted to take a census. He wanted to count the number of people living in the various regions of the Roman Empire. Since it is impossible to get an accurate count of millions of people if everybody keeps moving around, the residents of the Roman Empire were told to go to their ancestral home to be counted. That way, it was hoped, few people would be missed and few people would be counted more than once.

Example 4.2 The U.S. Census

The U.S. Constitution requires a census of all residents of the United States every 10 years so that the country can be divided into congressional districts of approximately equal population size. Here the population of interest is the approximately 300 million residents of the United States and the variable of interest is their address. The cost of obtaining the census in 2000 was approximately $6 billion and it took a year to collect and process the data. Even so, it is estimated that several million people were missed, mostly minorities in inner cities.

Sample Surveys

If the population of interest is large, then it is usually not possible to collect information from the entire population. Even the U.S. Census, backed by resources of the U.S. government, misses people. Besides, a census takes too long for most purposes. For example, if the government decided to determine this month's unemployment rate by contacting each of the more than 200 million American adults and asking them about their employment status, then this month's unemployment rate would not be available until next year. That would not be soon enough for policy makers setting interest rates or fiscal policy for the country. Instead they rely on **sample surveys.**

sample surveys

Example 4.3 The Current Population Survey

There are over 100 million households in the United States. Each month the **Current Population Survey (CPS)** interviews a representative cross section of a little more than 55,000 households. With a sample of that size the CPS can provide policy makers fairly accurate information about the economic health of the nation in a timely fashion. Unemployment rates, income distributions, and poverty rates for various sections of American society are all estimated from CPS data. The CPS also records many other economic and social variables. (See Examples 1.9 and 2.7.)

Current Population Survey (CPS)

❐ ❐

In a sample survey we obtain information about a population by examining part of the population. The part of the population of interest from which we actually collect information is called the **sample.** The members of the sample are not selected because they are of special

sample

representative

interest, but because they are thought to be a **representative** cross section of the larger population. We shall give some additional examples of sample surveys.

Example 4.4 A Medical Outcome Study

Each year almost 200,000 men are diagnosed with prostate cancer. Treatment options include radiation and what is known as radical prostatectomy, the surgical removal of the prostate. Lately robotic surgery has become very popular. Many patients insist on it. But no one knows for sure whether robotic surgery has better outcomes than traditional surgery. What we do know is that robotic surgery is more expensive than traditional surgery. For elderly men, who are more likely to die first from other ailments, doctors often recommend "watchful waiting," that is, leaving the cancer untreated.

Anyone considering the various treatment options for prostate cancer would like to know the long-term survival rates and possible side effects. In January 2000 the *Journal of the American Medical Association* published the findings of a survey on the side effects of radical prostatectomy. Researchers had interviewed 1291 men who had undergone the procedure within six months of being diagnosed with prostate cancer. The survey found that nearly 60 percent of the men surveyed were impotent and more than 8 percent lacked bladder control more than 18 months after surgery.

In this example the population of interest consists of the hundreds of thousands of men who have had their cancerous prostate surgically removed. The sample is the 1291 men surveyed. The variables of interest are the various forms of side effects.

Example 4.5 Public Opinion Polls

Should Social Security be privatized? Would people prefer to invest their retirement savings themselves, accepting the risk of losing it all, in the hope of turning their nest egg into a fortune? Public opinion polls attempt to answer such questions by interviewing between 500 and 2000 persons by phone. Most polls are conducted by professional pollsters such as Gallup and Yankelovich and are commissioned by either news organizations or political campaign organizations. News organizations usually publish their polls; politicians don't. Politicians use polls to fine-tune their political message so that they can tell the voters what the voters want to hear. The population of interest consists of all residents of voting age. The sample is the people who are actually interviewed. The variables measured for each individual in the sample are responses to questions about public issues.

Example 4.6 TV Ratings

According to Nielsen Media Research, 10.3 million viewers watched the Miss America Pageant on TV on September 20, 2003. That is less than half the viewers Miss America had as recently as 1995. The Nielsen ratings influence how much advertisers are willing to pay to sponsor a program. The ratings are estimated from a sample survey of about 25,000 households having their TV viewing monitored by so-called people-meters.

Example 4.7 Quality Control

Imagine that a retail chain agrees to buy 50,000 one-time-use cameras from a new supplier. When the shipment arrives, the retailer wants to have some assurance that the disposable cameras meet the agreed-to specifications. The only way to find out for sure would be to test all the cameras. But that would be both costly and time-consuming. Besides, once a disposable camera has been tested it cannot be sold. So the retailer has to decide whether to accept or reject the whole shipment based on the inspection of a small representative sample from the shipment.

Population Parameters and Sample Statistics

sample surveys
population parameters
sample statistics

Samples are used to obtain information about populations of interest. Such studies are called **sample surveys.** Numbers that describe a population are called **population parameters.** For example, the number of children living in poverty in America is a parameter. The proportion of American teenagers who have diabetes is also a parameter, and so is the average size of all American households living in public housing. The exact values of population parameters are usually unknown and have to be estimated from samples. Numbers computed from samples are called **sample statistics.** For example, the median income of the more than 50,000 households surveyed by the Current Population Survey is a sample statistic. Its value can be computed from the sample. We can use this sample median to estimate the median income of the more than 100 million households in the United States, an unknown population parameter.

We also call physical constants of interest parameters. For instance, the speed of light, the age of the universe, the mass of the sun, and the average surface temperature of the planet Venus are all parameters of interest to physicists. Such parameters have to be estimated from experimental data. (See Example 2.8.)

Observational Studies and Designed Experiments

observational studies
treatment group
control group
confounded

Sample surveys are **observational studies.** Such studies are used to describe and compare populations of interest. In observational studies, investigators simply observe the subjects or units being studied and record various variables of interest. For example, in an observational study of the potential links between hormone replacement therapy (HRT) and heart disease or breast cancer, medical investigators might simply examine a representative sample of postmenopausal women taking HRT under their doctor's supervision and compare it with a representative sample of postmenopausal women not receiving HRT. The drawback of such an observational study is that the women in the **treatment group** (the women receiving HRT) might differ from the women in the **control group** (the women not receiving HRT) in many ways besides whether they received HRT or not. The women in the treatment group would probably, on average, be wealthier and more educated than the women in the control group. As a result, the women in the treatment group would probably, on average, be receiving better health care and be spending more time exercising and watching their diet than the women in the control group. (Few poor women have the time and energy to go jogging or the money to be doing aerobic exercises in a health club.) It would be difficult to separate the effects of the hormone treatment on heart disease from the effects of lifestyle and socioeconomic background. The effects are said to be **confounded.** *Confounding* occurs when it is not possible to decide by statistical analysis which of two or more possible causes of an observed effect is the actual cause.

bias

Confounding factors are often overlooked in observational studies. The result is that it is then assumed that the treatment (HRT, for example) is the cause of the observed effect (such as lower heart disease rates). However, the reality may be that other factors (more exercise, better diet) may be the true causes of the effect. If a study is designed in such a way that confounding factors are likely to lead us to draw incorrect conclusions, then we say that the study design is **biased.**

randomized designed experiments

Randomized designed experiments give much better evidence of cause-and-effect than observational studies do because they reduce confounding and resulting bias. In a randomized designed experiment to determine the long-term health effects of hormone replacement therapy, the medical investigators would take a large group of postmenopausal women willing to participate in the study and randomly (by the flip of a coin, for example) assign each woman to either the treatment group (receiving HRT) or the control group (not receiving HRT). The random assignment of the large number of experimental subjects would ensure that at the outset of the experiment, the treatment and control groups would be very similar. The only systematic difference between the two groups would be the treatment. That would make it much easier to separate the effects of the treatment from the effects of lifestyle and socioeconomic background, which would be approximately the same for the treatment and control groups. In this way confounding would be almost eliminated. If the two groups then turned out to have different rates of heart disease or breast cancer we could attribute this difference to the treatment.

Randomization is the Key to Avoiding Bias

In this chapter we shall first discuss how to conduct sample surveys and in later sections how to design statistical experiments. It turns out that if we want to do a sample survey, then the best way to get a representative sample from a population is to use an impartial, objective, probability method, such as drawing tickets out of a hat, (i.e., use randomization) to select the sample, rather than depending on subjective human judgment, even expert judgment. That way we avoid inadvertent (or deliberate) human bias in the selection process.

In a statistical experiment to determine whether different treatments result in different responses, the best way to avoid bias caused by confounding is to assign the available experimental subjects or units at random to the various treatment groups (i.e., use randomization).

Picking the sample at random or assigning the available test subjects at random has the additional benefit that it allows investigators to assess the accuracy and reliability of the inferences (conclusions) they draw from sample surveys or statistical experiments.

In later chapters we shall explain how to make statistical inferences (draw conclusions) from data obtained from random samples from populations or from statistical experiments using randomization.

randomization

Randomization means that the selection of the sample in a sample survey or the assignment of treatments to units in a statistical experiment is carried out by an impartial, objective chance mechanism.

Section 4.1 Summary

observation
experimentation
observational studies
passively observe
designed experiments
actively impose various treatments

In a statistical study we obtain information either by **observation** or by **experimentation.** In **observational studies,** investigators **passively observe** the subjects or units being studied, recording various variables of interest. The investigators do not influence what is being observed. They just describe the results. In **designed experiments,** investigators **actively impose various treatments** on the experimental subjects or units. The goal is to find out whether different treatments result in different responses from the experimental subjects or units. Note the very important difference between observational studies and designed experiments: In a designed experiment, the investigators decide which experimental subjects get which treatment. For example, in a designed experiment to determine the effectiveness of a new vaccine, the medical investigators randomly assign each experimental subject to either the **treatment group** (receiving the vaccine) or the **control group** (not receiving the vaccine). Contrast this with an observational study of the long-term health effects of smoking. In such a study it would be the subjects being studied (the smokers and the nonsmokers) who decide who receives the treatment (smoking) and who does not. An **explanatory variable** in

treatment group
control group
explanatory variable

response variable

an observational study or a designed experiment is a variable that we think influences or explains the outcomes of the study. A variable that measures the outcomes is called a **response variable.**

Survey sampling

Survey sampling is an important kind of observational study. We use the following terminology when we discuss surveys:

- *population* — The **population** in a statistical survey is the collection of objects or individuals of interest.
- *units, subjects* — The population members are called **units** or **subjects** (if the population consists of people).
- *census* — A **census** is a survey of the entire population (or an attempt to survey the entire population).
- *sample survey* — In a **sample survey** we obtain information about a population by examining part of the population).
- *sample* — The **sample** is the part of the population of interest from which we actually get information. The hope is that the sample is a representative cross section of the population so that we can learn something about the population by studying the sample.
- *sampling design* — The **sampling design** describes how the sample is selected from the population.
- *parameter* — A **parameter** is a number that describes some characteristic of a population of interest. The exact values of parameters are usually unknown.
- *sample statistic* — A **sample statistic** or statistic is a number that can be computed from the sample. We use sample statistics to estimate unknown population parameters.

Section 4.1 Exercises

State whether each boldface number in Exercises 1 through 4 is a *parameter* or a *statistic*.

1. The average salinity (saltiness) of open-ocean seawater is known to be **35** parts per thousand. Scientists measured the salinity of Deep Springs Lake in California before and immediately after a major influx of freshwater from storm runoff. The readings were **310** parts per thousand and **25** parts per thousand, respectively.

2. In a random sample of **200** African Americans, **52.5%** were found to have blood type O. Nationwide **49%** of all African Americans have blood type O.

3. Simon Newcomb estimated the speed of light to be **299,750.8** km/sec (see Example 2.8). Today we know that the speed of light (in a vacuum) is **299,792.5** km/sec.

4. The Current Population Survey studied a representative cross section of a little more than **50,000** households out of the slightly more than **100** million U.S. households. The survey found the average size of the sample households to be **2.6** persons.

5. In order to determine customer satisfaction, an airline randomly selects 25 flights during a certain week and hands out mail-back questionnaires to all the passengers on those flights. Eighteen percent of the questionnaires are returned to the airline. Identify the population of interest and the sample, that is, the part of the population that the airline actually got feedback from.

6. A study looked at the potential link between egg consumption and high cholesterol levels in the blood. The subjects in the study were asked how many eggs they ate per week, and their cholesterol levels were measured. The study found that, on average, the more eggs the subjects ate, the higher their cholesterol levels were.
   a. Identify the response variable and the explanatory variable.
   b. Suggest possible confounding factors. (See Section 3.4.)
   c. Was this study a designed experiment or an observational study? Explain.

7. a. After a football game 200 fans leaving the stadium are asked, "Can you taste the difference between regular beer and 'lite' beer?" Is this an observational study or a designed experiment? Explain.
   b. At half-time 200 football fans are asked to perform a taste test in which they drink from two unmarked cups. One cup contains regular beer, the other "lite" beer. After tasting both, each fan is asked to identify the regular beer. Is this an observational study or a designed experiment? Explain.

8. Researchers at the University of Illinois investigated whether a certain gene controls the biological clock in rats. They compared the brain activity of a group of normal rats to the brain activity of a group of rats in which the researchers had blocked the gene in question. The brain activity of the normal rats peaked during the day and waned during the night, whereas the rats in which the gene had been blocked had a constant level of brain activity. Was this an observational study or a designed experiment? Explain.

9. In the 1950s researchers at Johns Hopkins University did a study to determine whether providing public housing to poor people improved their lives. The research was related to important policy questions: Which is better, building low-cost public housing for the poor or providing them with rent subsidies for existing housing? The study was conducted in Baltimore. The residents in a large public housing project had been selected by the Baltimore Housing Authority from several thousand applicants. For this study the researchers selected 300 families that had been accepted by the Housing Authority and 300 families that had been turned down. The two groups were interviewed and compared for 3 years.
   a. What treatment was studied?
   b. Identify the treatment group and the control group.
   c. Was this a designed experiment or an observational study? Explain.

10. Give an example of a designed experiment (look for examples in newspapers, magazines, the Internet, television, etc.). Clearly state the goal of the experiment and the treatment to be given. Note whether randomization seems to have been used or not.

11. Give an example of a survey. Clearly state the question that the survey tries to answer. Note whether randomization seems to have been used or not.

12. Find an example where confounding is a definite possibility (look for examples in newspapers, magazines, the Internet, television, etc.). Identify the variable that is suggested to be the cause of what is observed. Specify a second variable that could also be a partial confounding cause.

4.2 SURVEY SAMPLING FROM A REAL POPULATION: PROBABILITY SAMPLING VERSUS NON-PROBABILITY SAMPLING

survey sampling

Sampling, often called **survey sampling,** is used to obtain information about a large population by examining a small part of the population. Often we are interested in numbers that describe the population, so-called population parameters. When a randomization mechanism is properly used in selecting the sample, the procedure is called probability sampling. If such a randomization mechanism is not used, the selection process will be biased, favoring some members of the population over other members, and the sample will fail to be a representative cross section of the population. The result is likely to be that we draw incorrect conclusions about the population from looking at the sample.

In this section we take a closer look at survey sampling, describing in detail through real-world examples the problems that we may run into, how proper probability sampling techniques can minimize these problems, and the dangers and difficulties to look out for when you choose, or are forced by circumstances to use, non-probability sampling techniques. The following example is a classical illustration of the fact that a sample survey can be very misleading if the sample that we collect information from is not a representative cross section of the larger population.

Preelection Polls

Let us go back to the 1936 presidential election between Franklin Roosevelt (the Democrat) and Alf Landon (the Republican). *Literary Digest* was a popular magazine that for years had accurately predicted the outcome of presidential elections, based on its polls. In 1936 it sent out approximately 10 million sample ballots on postcards asking people for their presidential choices, and received fully 2,376,523 responses. Meanwhile, George Gallup, who had been experimenting with new sampling techniques, sent out only 3000 such cards before the *Literary Digest* poll began. Based on this much smaller sample, Gallup boldly declared that the *Literary Digest* poll would be biased in its prediction about the election. Moreover, Gallup declared that the *Literary Digest* poll would show about 44% for Roosevelt and 56% for Landon. The editor of the *Literary Digest* was understandably peeved at this young upstart, saying, "Never before has anyone foretold what our poll was going to show even before it started.... Our fine statistical friend should be advised that the *Digest* would carry on with those old fashioned methods that have produced correct forecasts exactly one hundred percent of the time." Gallup had the last laugh.[1] The *Digest*'s poll showed an overwhelming win for Landon, with numbers remarkably close to what Gallup predicted:

Roosevelt	42.9%
Landon	57.1%

Gallup was also correct that the *Digest*'s predictions would be wrong. Based on another small but carefully selected sample, Gallup's prediction of the actual result was 55.7% for Roosevelt to 44.3% for Landon. The actual election had Roosevelt winning in a landslide:

Roosevelt	62.5%
Landon	37.5%

The *Literary Digest* was very far off, missing Roosevelt's percentage by 19.6 percentage points. Gallup was much closer. He predicted the correct winner and was only 6.8 percentage points off (a large discrepancy by today's polling standards but tolerable in this election).

A preelection poll can fail to predict the outcome of an election accurately for many reasons: People may change their minds between the polling and the election, the people polled may not be representative of the people who actually vote, people may not answer truthfully when they are polled, and finally, too few people may have been polled to achieve much accuracy. The problem with the *Literary Digest* poll was not that the sample size was too small. Two and a half million is huge compared to typical preelection polling sample sizes of between 500 and 2000 used today for polls that predict with reasonable accuracy. Gallup accurately estimated the results of the *Literary Digest* poll using only 3000 people. Since the *Literary Digest*'s sample size was much larger, could the other reasons stated add up to an error of over 19 percentage points?

[1] George Gallup, *The Sophisticated Poll Watcher's Guide* (Ephrata, Pa.: Princeton Opinion Press, 1972), p. 66.

Selection Bias

A poll will not give an accurate prediction of the outcome of an election if those polled (the sample) are not representative of the voting population. The *Literary Digest* mailed questionnaires to people using telephone directories, lists of magazine subscribers, and lists of automobile owners. That tended to exclude the poor, since in the depths of the Depression, affluent people were much more likely to have telephones and cars. The 10 million people contacted were thus more affluent on average than the electorate at large. This poll is said to suffer from **selection bias.** Selection bias means that in the selection process some people are more likely to be included in the sample than other people. If there is selection bias, the sample may not be representative of the population in important ways.

selection bias

The economy was an important factor in the 1936 election. Roosevelt was much more popular than Landon among the less affluent, and Landon was much more popular among the affluent. A high percentage of wealthy people hated the New Deal, which was Roosevelt's attempt to help mitigate the effects of the Depression. Thus, the selection bias in this poll worked to exclude Roosevelt supporters and to include Landon supporters.

Nonresponse Bias

Notice that 10 million questionnaire cards were sent out, but only about 2.5 million were returned. Could those responding to the questionnaire be different in important ways from those who did not respond? Studies show that less affluent people tend to ignore questionnaires and are less likely to respond than more affluent people. Thus, the *Literary Digest* poll was likely to be still further biased against Roosevelt. This source of bias is termed **nonresponse bias**: The people who did not respond to the poll were more likely to vote for Roosevelt than those who responded.

nonresponse bias

In his effort to predict the results of the *Literary Digest* poll, Gallup sent cards to people on the same list as used by the *Digest* but to only 3000 of them. One of his purposes was to show the power of a small, appropriately selected sample to provide accurate information about a large population. Gallup was being especially clever, for the population he wanted to predict was not the 10 million recipients but rather those who would actually respond to the *Digest* poll. His sample was also subject to nonresponse and selection bias, so it yielded percentages for Landon and Roosevelt similar to those of the huge *Digest* poll. The important statistical lesson we learn from the mistakes of the *Literary Digest* is that if the sampling plan is flawed, it does not help to take a large sample.

sample of convenience
selection bias

The *Literary Digest* chose a **sample of convenience,** consisting of individuals who were easy to contact. That resulted in **selection bias**: Those not selected tended to differ from those selected in important ways. Consequently, those selected were not a representative cross section of the population of interest.

voluntary responses

The *Digest* relied on **voluntary responses.** That resulted in nonresponse bias: Those who did not respond tended to differ from those who responded in important ways. So those who responded were not representative of the people selected by the *Digest* and represented the population of interest even less.

It turns out that the best way to avoid selection bias is to use an impartial, objective probability method to select the sample. The best way to reduce nonresponse bias is to contact the sample members in person.

Probability Sampling

probability sampling method

Any sampling method that gives each population member a known probability of being selected is called a **probability sampling method.** Probability theory will be discussed in detail in the next chapter. Here we shall discuss the following probability sampling methods:

- Simple random sampling
- Stratified random sampling
- Cluster sampling
- Multistage probability sampling

Note that any probability sampling method uses randomization.

margin of error

As the name indicates, simple random sampling is the simplest of the probability methods. Simple random sampling allows easy assessment of the expected **margin of error** of the estimates we derive from the sample. The other probability methods have their advantages in certain situations. But to simplify our mathematical analysis of margins of error we shall in subsequent chapters pretend that every probability sample is a simple random sample.

Simple Random Sampling

To obtain a simple random sample from a population, we can select the individuals to be included in the sample at random, one at a time, in such a way that each time we select an individual, all remaining population members not already selected have an equal chance of being chosen. Simple random sampling is impartial in the following sense: Each individual in the population has the same probability of being included in the sample. Furthermore, any combination of individuals is as likely to be included in the sample as any other combination of the same size.

sampling frame

In practice, to obtain a simple random sample one first constructs a list of everyone in the population, called the **sampling frame.** For example, if the population consists of the students at a particular university, then a roster of the enrolled students could be used. If the population consists of all the households in an area of town, then a list of all addresses in that area could be used. One easy way to proceed is to assign each member of the population a distinct numerical label between 1 and N, where N is the population size. To select a simple random sample (SRS) of size n we then use a computer or a random number table to randomly select n of the numbers 1 to N.

Example 4.8

A class has 30 students, and an outside evaluator wishes to interview a representative group of 10 students to obtain in-depth opinions about the class. Here is a list of the 30 students, with labels attached:

01	Ahn	11	Ito	21	Novikov
02	Beckman	12	Iyer	22	Patel
03	Brooks	13	Jones	23	Pirelli
04	Chang	14	Kim	24	Rabe
05	Cruz	15	Lopez	25	Ramirez
06	Doyle	16	Lorenz	26	Schaeffer
07	Fitzpatrick	17	McCormick	27	Smith
08	Gillespie	18	Mueller	28	Taylor
09	Harris	19	Mustapha	29	Wagner
10	Hossain	20	Nagy	30	Williams

We shall show how we can use random number Table B.3 of Appendix B in the back of the book to randomly select 10 of the 30 students. For example, we can use the first two digits of each row in Table B.3, going from top to bottom. The first 20 rows give us the 20 two-digit numbers.

32	40	88	12	65	50	73	87	48	89
72	02	20	18	11	12	26	74	58	78

We discard those two-digit numbers that are outside the range from 01 to 30, and select the *distinct* numbers in that range. We get

12	02	20	18	11	26

We continue until we have 10 distinct numbers in the range from 01 to 30, getting

12	02	20	18	11	26	17	09	04	25

Our random sample consists of the 10 students with these labels. They are: Beckman, Chang, Harris, Ito, Iyer, McCormick, Mueller, Nagy, Ramirez, and Schaeffer.

Is this a representative cross section of the class? You be the judge! There is no guarantee that a small sample will be representative. But, as we shall see in Chapter 8, a simple random sample of at least 1000 individuals drawn from a large population has a very good chance of being representative of the population it was drawn from, even if the population is many times larger than the sample.

We can select a simple random sample of size n in 3 steps:

Step 1. Obtain a list of all population members, a so-called sampling frame.

Step 2. Assign each population member a distinct numerical label. Make sure that all labels have the same number of digits.

Step 3. Use a computer or a random number table to select n distinct labels.

Accuracy in Simple Random Sampling

Suppose we conduct a telephone poll to estimate the proportion p of all American adults who think that invading Iraq was the right thing to do. Suppose that we interview a simple random sample (SRS) of n adults and determine the proportion $\hat{p}$ of adults in the sample who have that opinion. We then use the sample proportion $\hat{p}$ to estimate the population proportion p. That raises the question: How accurate is our estimate $\hat{p}$ likely to be? We shall discuss estimation accuracy in detail in Chapter 8. Here we shall just give the following preview:

Margin of error for $\hat{p}$ as an Estimator.

Suppose we use the sample proportion $\hat{p}$ derived from an SRS of size n, drawn from a much larger population, to estimate an unknown population proportion p. Then we have about a 95% chance that our estimate will be off by less than $1/\sqrt{n}$ (in either direction). We call $1/\sqrt{n}$ the margin of error.

Example 4.9

Suppose that in an SRS of 1102 adults interviewed in October 2003, 628 thought that invading Iraq was the right thing to do. The proportion $\hat{p}$ of people in the sample who had that opinion was

$$\hat{p} = \frac{628}{1102} = 0.57 \quad \text{(that is, 57\%)}$$

We therefore estimate that the proportion p of all American adults who at the time of the poll had that opinion was around 0.57. We are 95% confident that this estimate is off by at most

$$1/\sqrt{n} = 1/\sqrt{1102} = 0.03 \quad \text{(that is, 3\%)}$$

margin of error

In the news media this would be reported: "A new poll finds that 57% of American adults think that invading Iraq was the right thing to do. The **margin of error** for the poll was 3 percentage points."

❒ ❒

Larger samples have smaller margins of error. But the formula shows that *if we want to cut the margin of error in half, then we have to quadruple the sample size.* Somewhat surprisingly, *the margin of error does not depend on the population size.*

Population Size Does Not Matter

The accuracy of the sample proportion, $\hat{p}$ as an estimator of the population proportion, p, does not depend on the size of the population—as long as the population is at least 20 times greater than the sample.

Example 4.10

In California there are 27 million persons age 16 and over. In Wyoming there are only 400,000 persons age 16 and over. Suppose that the Bureau of Labor Statistics takes a simple random sample of 1500 persons age 16 and over in each state to estimate the unemployment rate in the state. For which state, California or Wyoming, is the estimated unemployment rate likely to be more accurate?

Solution: In California only one out of every 18,000 persons age 16 or over is sampled, while in Wyoming about one out of every 250 persons age 16 or over is sampled. Intuitively, since we are surveying a greater proportion of the population of Wyoming than of the population of California, the survey in Wyoming ought to be more accurate. As it happens, our intuition is not completely wrong. But the added accuracy in Wyoming turns out to be extremely small and can be ignored; the different population sizes of California and Wyoming will have no effect on the margin of error of the estimated unemployment rates for the two states. If the Bureau of Labor Statistics uses the same sample size in the two states, then the estimates for the two states will have the same margin of error.

Stratified Random Sampling

strata

When we sample from a large population we are often interested in parameters that describe smaller sections or **strata** of the population. For example, when a pollster tries to predict the winner of a presidential election, it is not enough to predict who will get the most votes nationwide. The pollster has to predict who will win each state. So the pollster has to estimate the proportion of likely voters in each state who support the various presidential candidates. For another example, when the Bureau of Labor Statistics estimates the monthly unemployment rate for the United States, the rate is broken down by state of residence, marital status, race, age, and sex.

To determine unemployment rates, the Bureau each month interviews a representative cross section of slightly more than 100,000 individuals age 16 and over. If the Bureau took a nationwide simple random sample, then each state would be represented in the sample approximately in proportion to its population size. Since there are 55 times more people in California than in Alaska, a nationwide simple random sample would contain approximately 55 times more Californians than Alaskans. As we discussed, the accuracy of sample estimates is determined by the sample size. So a nationwide simple random sample would estimate the unemployment rate in California with much greater accuracy than the unemployment rate in Alaska. To get roughly the same accuracy in all 50 states, the Bureau uses probability samples of roughly equal sizes drawn from each state. This is called **stratified random sampling.**

stratified random sampling

> **Stratified Random Sampling**
>
> **Step 1.** Divide (or stratify) the population into mutually exclusive subgroups (strata) of similar individuals.
>
> **Step 2.** Select a probability sample from each stratum.

Example 4.11 The Gallup Poll

random digit dialing

Before the elections of 1992, Gallup conducted opinion polls using the following stratified sampling scheme: Each of the four time zones was divided into 3 strata, according to population density, creating a total of 12 strata. Within each stratum, the Gallup Poll selected a simple random sample of phone numbers using **random digit dialing.**

Comment: Unfortunately, nonresponse is becoming an ever-increasing problem in telephone polling. People are tired of telemarketers, and with answering machines and Caller-ID, many people do not pick up the phone when pollsters call. The nonresponse rate in a phone survey usually exceeds 50% of the poll's original sample.

Cluster Sampling

cluster sampling

If a population of interest is large or spread out over a large geographic area, investigators often use **cluster sampling**: They divide the population into clusters and select all the members of some of the clusters. For example, after the 1989 earthquake that shook the San Francisco Bay Area, medical researchers from Stanford University wanted to find out whether there was an increase in heart attacks following the quake. They examined the medical records at five randomly chosen Bay Area hospitals and compared the number of heart attacks recorded the week after the quake to the number the week before. In this example the population of interest was all the heart attacks recorded in the Bay Area during the week before and the week after the quake. The heart attacks recorded at each Bay Area hospital constituted a cluster, and the researchers selected five clusters. The sample consisted of all the heart attacks in the five clusters.

cluster sampling

This is an example of **cluster sampling** in its simplest form.

> **Cluster Sampling**
>
> **Step 1.** Divide the population into mutually exclusive subgroups (clusters).
>
> **Step 2.** Use a probability method to select some of the clusters.
>
> **Step 3.** Use all the members of the chosen clusters as the sample.

Note the difference between stratified sampling and cluster sampling. In both cases we divide the population into mutually exclusive subgroups. In stratified sampling, each subgroup (stratum) is considered important enough to require representation in the sample. By contrast, in cluster sampling, the population is divided into subgroups (clusters) for sampling convenience, and we do not want each cluster to be represented in the sample.

1-in-k Systematic Sampling

systematic sampling

If we do not have a list of all the individuals in the population we are studying, a so called sampling frame, then we cannot draw a simple random sample. One sampling technique that does not require a sampling frame is **systematic sampling.** If the population members are ordered in some way, then we could select every kth individual. For example, a large supermarket might interview every 20th customer leaving the store on a given day. The first customer interviewed could be randomly selected from the first 20. If we select customer number 13, say, we would then interview customers 13, 33, 53, 73, and so on. Systematic sampling is a form of cluster sampling. If the response we are studying is in any way associated with the order of the population, then we probably will not get a representative sample by using systematic sampling.

1-in-k Systematic Sampling from an Ordered Population

Step 1. Choose an integer k.

Step 2. Randomly select an integer x between 1 and k.

Step 3. Use individuals with numbers, x, $x + k$, $x + 2k$ and so on.

Multistage Cluster Sampling

Many surveys use samples selected in several stages using a combination of probability sampling methods. For example, suppose that we want to interview a representative cross section of 1000 graduating college seniors during the month of April to find out how the job market is. The problem is that we do not have a list of the names and phone numbers of all the 1.2 million graduating seniors. We can easily get a list of the 2500 institutions that award bachelor's degrees. If we contacted each of these 2500 institutions, we could obtain a list of the graduating seniors nationwide, giving us a sampling frame that we could draw a simple random sample from. Obviously, this would be a very time-consuming process and would require the right computer software.

The following sampling scheme would be easier to implement: First, randomly select 20 educational institutions in such a way that each institution's chance of being included in the sample is proportional to the size of its senior class. Second, obtain lists of the graduating seniors of the 20 chosen institutions. Finally, select a simple random sample of 50 seniors from each at the 20 chosen institutions. Use the 20×50 chosen graduating seniors as our sample. This is an example of multistage cluster sampling: In Stage 1 we select 20 clusters of graduating seniors. In Stage 2 we select a probability sample from each of the 20 clusters.

Cluster sampling can save time and money. But the members of clusters are often more homogeneous than the population as a whole. As a result, cluster samples often do not reflect the diversity of the population as well as simple random samples of the same size do. In our example, the student bodies at small Christian colleges tend to differ from the student bodies at large state schools and from the student bodies at Ivy League schools. Our cluster sample of college seniors would either contain too many or too few students from Christian colleges, depending on how many Christian colleges were selected in Stage 1.

Most large-scale surveys use a combination of probability sampling methods. We shall use the Current Population Survey as an illustration.

Example 4.12 The Design of the CPS

Each month the Current Population Survey (CPS) interviews a representative cross section of about 55,000 households. To reduce nonresponse, the CPS initially contacts its sample in person before doing later interviews by phone. That rules out simple random sampling. A simple random sample would be so widely scattered over the entire country that it would take forever to contact each sample household in person. The CPS uses a combination of stratified, cluster, and systematic sampling. Stratified sampling ensures that important subsections of the population get adequate representation in the sample. Cluster sampling ensures that the sample is not too widely scattered to be reached in a reasonable amount of time.

The CPS uses independent samples in each of the 50 states and the District of Columbia, tailored to the demographic and labor market conditions of each state. All households in a state have the same chance of being selected. Periodically the CPS is redesigned by the Census Bureau. From 1994 to 2004 the CPS design was based on the 1990 census. Minor design changes were made in 1996 (because of budget cuts) and in 2001 (because of an added congressional mandate to accurately estimate the number of low-income children lacking health insurance in each state).

Primary Sampling Units

At the beginning of each decade the Census Bureau uses the latest census data to divide each state into **Primary Sampling Units** (PSUs)—most of which consist of either a large metropolitan area (like Los Angeles or New York City), a large county, or a group of smaller neighboring counties. Nationwide, the Census Bureau creates roughly 2000 PSUs. The Bureau then selects a representative cross section of the PSUs, and for the next 10 years conducts all CPS interviews in the chosen PSUs. This is the first stage of the selection of the CPS sample.

From 1996 to 2004 all CPS interviews were conducted in the 754 sample areas (PSUs) shaded dark in Figure 4.1. They were selected in the following way: The 428 most populous PSUs (such as Los Angeles and New York City) were automatically included. Within each state the remaining PSUs were grouped together with similar PSUs into strata having approximately equal populations. Nationwide, this created 326 strata. One PSU was selected from each of the 326 strata, using a probability method that gave each PSU within a stratum a selection probability proportional to the PSU's population as of the 1990 census.

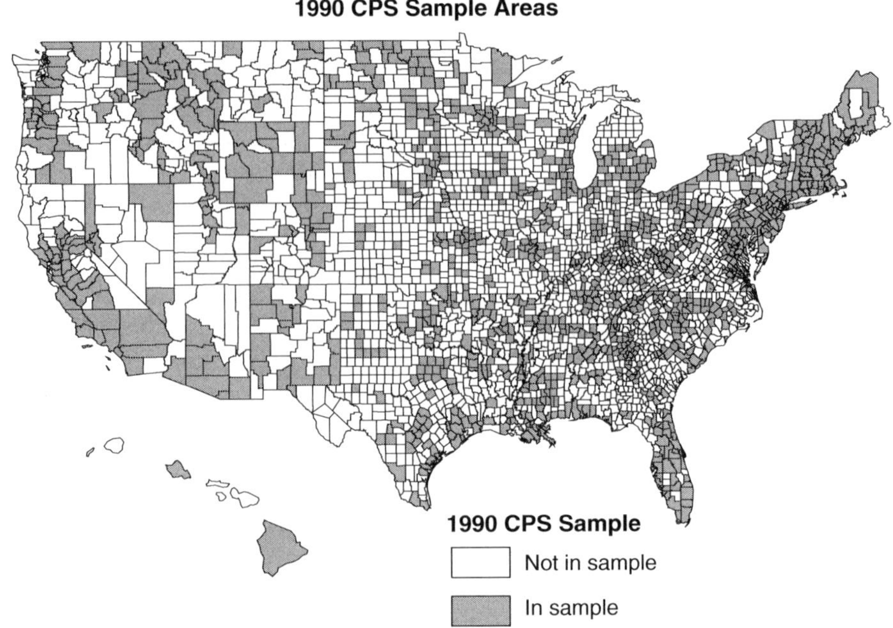

Figure 4.1 Primary Sampling Units for the Current Population Survey: The 1990 sample design.

The second stage of the CPS sample design is the selection of a representative cross section of households within each chosen PSU. To accomplish this, extensive use is made of data from the latest census. Each PSU is divided into census tracts, which are small homogeneous areas with an average population of about 4000. Each tract consists of blocks with a Census Bureau identification number. Each block is divided into **Ultimate Sampling Units** (USUs), consisting of a cluster of about four households each. Using a combination of probability sampling methods, the Census Bureau selects a representative cross section of USUs (clusters). In the end, every person age 16 and over living in a household in a chosen USU cluster gets into the CPS sample. Thus the last step in the selection of the CPS sample is cluster sampling. This example illustrates how complicated actual probability sampling schemes often are.

Ultimate Sampling Units

The key to obtaining an objective probabilistic sample design is that the sampling units (PSUs, strata, USUs) must be defined before any sampling takes place, and there must be a pre-specified random mechanism that will select the units at each stage. Blind chance selects the sample units. No subjective discretion must be allowed on the part of the interviewers regarding whom to interview or where to go to find the people to be interviewed.

In this book we will concentrate on simple random sampling. The principles learned about simple random sampling also apply to more complicated probability sampling approaches, such as the stratified and multistage sampling approach used by the CPS.

Non-Probability Sampling

Probability sampling (that is, a sample design that uses randomization to select units from a well-defined population) has the virtue of yielding results that can be analyzed to determine their accuracy. (Randomized controlled experiments have the same virtue.) However, nonprobability samples abound because they are easy to obtain. They should be looked at with the same critical and skeptical eye that one uses for observational studies. Beware of the results of the magazine reader survey, for example. We next describe some basic types of non-probability samples.

Judgement and Quota Sampling

One of the problems with the *Literary Digest* poll was that no attempt was made to ensure that the sample resembled the population of likely voters. A natural remedy would be to carefully select a sample that resembled the population of interest. Thus, if the population is half male, one would choose a sample with half males; if the population is 13% African American, then the sample should be approximately 13% African American. Similarly, the sample should reflect the population in terms of income, education, residence (urban, suburban, rural), and a number of other factors.

judgment sampling

quota sampling

In **judgment sampling** the samplers depend on their expert wisdom, using their own discretion to select a suitable sample. A particular form of judgment sampling is **quota sampling,** in which the field representatives collecting data have to interview demographically specific types of individuals (for example, three white men over 40 with high school educations who live in the city, two African American women under 40 with college educations who live in the suburbs, and so on) but are otherwise free to choose anyone who fits the descriptions. This "freedom of choice" for the interviewers is a serious flaw of judgment sampling. Within each prescribed category, such samples are samples of convenience, consisting of individuals who are easy to contact. For that reason, judgment samples usually fail to be representative cross sections of the populations of interest. Judgment sampling appears at first glance to be a good method and sometimes produces reasonable results. However, it always results in selection bias, so randomization is better.

The preelection polls conducted by Gallup from 1936 to 1948 used quota sampling and sample sizes varying from 3600 to 50,000. The errors in Gallup's prediction for the Republican candidate—Gallup's prediction minus the actual results—were 6.8%, 3%, 2.3%, and 5.3% for 1936, 1940, 1944, and 1948, respectively. These error sizes are reasonably small (and much smaller than the 19.6% for the *Literary Digest* poll), but they are all positive; that is, each poll was slightly biased in favor of the Republican candidate. For the years 1936, 1940, and 1944, the errors were small enough compared to the Democratic lead that Gallup predicted the correct winner (Roosevelt each time). In 1948, however, the election was much closer. The Democratic nominee was Harry S Truman, who had taken over as president when Roosevelt died in 1945, and the Republican candidate was Thomas E. Dewey. The Gallup Poll predicted Dewey would get 49.5% of the vote and Truman would get 44.5%. Other major polls also predicted a Republican victory, which led to the *Chicago Daily Tribune*'s infamously premature headline on November 3, 1948, announcing before the votes were all counted, "DEWEY DEFEATS TRUMAN."

But on Election Day, Truman scored an upset victory, getting almost 50% of the popular vote. (There were two minor candidates who got 5% of the popular vote.) The pollsters were embarrassed, as was the *Tribune,* and the polls came under quite a bit of criticism. They were off by only about 5 percentage points, a smaller error than Gallup made in 1936, when Gallup was so lauded, but they did get the wrong winner. According to *New York Times* columnist William Safire, "Sampling is no science; ask President Dewey and Prime Minister Peres,"[2] the latter a reference to the 1996 Israeli election. (The authors, however, strongly disagree with Mr. Safire on his main point.) As Mr. Safire points out, there never was a President Dewey. In fact, the actual vote reversed the prediction: Truman had 49.8% and Dewey 44.5%.

---

[2] William Safire, "Sampling Is Not Enumerating," *New York Times,* Dec. 7, 1997.

Let's critique Safire's statement. What went wrong? Were the pollsters simply unlucky? Every poll will be off by a certain amount, and in a close election that amount may be enough to result in the wrong prediction. Another possibility is the polling technique. From Gallup's success in 1936 to the fiasco in 1948, all major polls used quota sampling, following Gallup's lead. Quota sampling is *not* probability sampling; it allows individual discretion in selecting interviewees, which can lead to bias. Even though demographically the sample looks representative, there is still a slight bias in favor of Republicans. It could be that Republicans are more likely to be at home, tend to live in "nicer" neighborhoods, or are "better" dressed and hence are preferred by the interviewer, perhaps unconsciously.

After 1948 Gallup and the other major polling organizations moved to probability sampling. In the years 1952 to 1996, the average size of the error of Gallup's prediction of the winning candidate was 2.23%, compared to 4.35% from 1936 to 1948. In five of the elections, the error was in favor of the Republican, and in seven of them, the error was in favor of the Democrat. Thus, the bias in favor of the Republicans found in the quota sampling has disappeared. Another interesting fact is that the Gallup Poll of 1948 was based on a sample of size 50,000, whereas the probability polls have been based on far smaller samples, around 3000–4000 for the past 20 years. Thus, the probability-based samples are less biased, more accurate, and more efficient than the quota samples. Probability samples of 3000–4000 come within about 2 percentage points of the correct percentage of a population of close to 200 million eligible voters, which seems miraculous.

Safire's criticism of the polls in the 1948 election is on target in that there will always be error if one does anything other than count the voting preferences of everyone in the population who voted. However, as the infamous "hanging chads" of the 2000 presidential election illustrate, sometimes people do not even agree on what it means to "count all the votes." But sampling *is* a science that provides accurate estimates, and it has improved a great deal since 1948. The Israeli elections for prime minister in 1996 pitted Shimon Peres against Benjamin Netanyahu. Peres was favored in the polls, with the last polls before the election predicting Peres with 51.5% and Netanyahu with 48.5%. These polls clearly stated that their results were accurate only to within a margin of error of 3%. The final results showed Peres with 49.5% of the vote and Netanyahu with 50.5%. Thus, the polls predicted the wrong winner but were off by only 2%, which was within the stated margin of error. The science of sampling did not fail; the election was merely so close that the polls had too large a margin of error to make a reliable prediction of who would win.

Haphazard Samples and Samples of Convenience

Haphazard samples and samples of convenience are easy to obtain. For example, one might stand on a street corner and interview anyone who will stop, or just sample people in one's class or workplace or neighborhood. A magazine may ask its readers to fill out a survey and send it in to the magazine: very convenient for the magazine, but hardly representative of any meaningful population! These samples are non-probability samples because there is no objective random mechanism choosing the people for the sample. The standing-on-the-corner poll may seem random, but it is not random in the probabilistic sense; it is just haphazard. Such polls are subject to bias, which is often very large, and there is no statistical way to analyze the accuracy of the poll. In particular, we cannot report a margin of error, as Gallup polling does.

Self-Selected Samples and Voluntary Response Samples

It is fairly common to be watching television and be invited to make a telephone call to register your opinion on some issue. In these polls it is not the pollsters who decide who will be in the sample but the people themselves. Such self-selected sampling polls are subject to all kinds of biases: What type of person watches the show? What type of person is most likely to call? What prevents a group of people from orchestrating a campaign to make hundreds of calls on a particular side of the issue being raised? The polling of magazine readers also suffers from serious self-selection bias.

A similar phenomenon is the Internet-based poll, in which a Web page is put up, inviting people to respond. Only people who happen on the page and decide to vote are in the sample. An example is the Scotty J. survey. One question asks, "Do you believe that marijuana should be legalized?" The following results are based on 5152 responses:

Yes, I strongly support this	70%
Yes, I think this is a good idea	19%
Don't know/not sure	4%
No, I think this is a bad idea	3%
No, I strongly oppose this	4%

Source: http://www.legalize.com/

Thus, 89% believe marijuana should be legalized, and 7% think it should not. Is this representative of the entire population of the United States? Other questions on the poll lend insight into the demographics of those responding: 70% are male and 30% are female; 89% are white, 3% Hispanic, 2% black, and 2% Asian; 51% are under 20 years old, 32% are between 20 and 29, and 10% are between 30 and 39. In this survey, there is an overrepresentation of young white males, the people most likely to be surfing the Web. Looking around further at the Scotty J. Web site makes it clear that it is set up to advocate legalizing marijuana. Scotty J. is upfront about the fact that this is not a scientific poll but one designed to persuade people. It is likely to attract people who are already leaning toward a particular viewpoint. Here lies its most serious bias of all: A disproportionate number of surfers at this site favor the legalization of marijuana.

Section 4.2 Summary

population
units, subjects
sample
design
probability sampling methods

The **population** in a sample survey is the collection of objects or individuals of interest. The population members are called **units** or **subjects** (if the population consists of people). The **sample** is the part of the population of interest from which we actually obtain information. The **design** of a sample survey refers to the method used to select the sample from the population of interest. **Probability sampling methods** use an impartial chance mechanism to select the sample. In probability sampling, each population member has a known probability of being selected. We discuss the following probability methods:

simple random sampling

- **Simple random sampling** is the same as random sampling without replacement. In simple random sampling, any combination of population members of a given size is as likely to be included in the sample as any other combination of the same size.

stratified random sampling

- In **stratified random sampling** the population is divided (or stratified) into mutually exclusive subpopulations (strata). A probability sample is then selected from *each* stratum.

cluster sampling

- In **cluster sampling** the population is divided into mutually exclusive subgroups (clusters). A probability sample of clusters is selected (some clusters thus having *no* members in the sample). Then a probability sample is selected from each of the chosen clusters.

1-in-k systematic sampling

- If the population members are ordered in some way, then we can use **1-in-k systematic sampling.** We randomly select an integer x between 1 and k (the starting point) and use individuals numbered x, $x + k$, $x + 2k$, and so on.
- Multistage probability sampling uses a combination of probability sampling methods to select successively smaller groups within the population in stages. The final stage results in clusters of individuals.

sampling frame

Probability sampling requires a **sampling frame,** a list of all the units or subjects from which the sample is to be drawn. For example, the Current Population Survey uses lists of Primary Sampling Units at the first stage of the selection process and lists of Ultimate Sampling Units (clusters of households) at the final stage of the selection process. Probability sampling guards against **selection bias,** because blind chance is impartial. When we sample from a human population, the best way to reduce **nonresponse bias** is to contact the sample members in person.

selection bias
nonresponse bias

quota sampling

Before Gallup realized that probability sampling is the best method to obtain a representative cross section of a population, he used **quota sampling.** In quota sampling, the interviewers hand-pick the sample so that the sample reflects the population in certain key ways, such as race, gender, age, residence, and income distribution. The approach seems logical but results in inadvertent selection bias.

sample of convenience
self-selected samples
voluntary response

Many surveys use really bad sample designs: A **sample of convenience** simply consists of population members who were conveniently at hand. **Self-selected samples** and **voluntary response** samples consist of individuals who themselves chose to be in the sample. There is no way of judging the accuracy of sample surveys that fail to use probability methods to select their sample. Such samples are rarely representative cross sections of the population of interest.

Section 4.2 Exercises

1. Suppose you are conducting a survey on the spending habits of college freshmen. After taking a simple random sample of 200 students, you have the choice of conducting the interviews in person or sending questionnaires to your sample. Give advantages of each method. Based on your answer, choose a method and explain your choice.
2. Design a survey using convenience sampling to determine whether people approve of the use of Native American names and likenesses as symbols or mascots for high school, college, or professional sports teams. In the design of your survey, consider the following questions: Where will you conduct the survey? How many respondents do you need for the survey? How will you phrase the survey question? Will you ask the question verbally, or will you have the survey respondents fill out a questionnaire? Can your survey be used to make valid and believable inferences?
3. It has been shown by different studies that the characteristics or behavior of the survey taker can influence the outcome of the survey. In Exercise 2, what characteristics of the survey taker could influence the outcome? For each characteristic, will the effect be greater if the questions are asked verbally or if the respondent fills out a questionnaire? How could you minimize the effect of the survey taker on the outcome of the survey?
4. Design a simple random sample survey to determine whether students at your school feel that marijuana should be legalized. How do you choose the sample? What questions do you want to ask? How can you reduce the nonresponse bias of the sample? Complete the survey and report your results. Make sure to include the nonresponse rate for your sample.
5. Design a simple random sample survey to determine whether faculty at your school approve or disapprove of making final exams optional for students. How do you choose the sample? What questions do you want to ask? How can you reduce the nonresponse bias of the sample? Complete the survey and report your results. Make sure to include the nonresponse rate for your sample.

6. Compare stratified random sampling and quota sampling. Explain the similarities and the difference.

In Exercises 7 through 13, identify the sampling method used.

7. A news organization conducts an exit poll in which 50 polling stations are randomly selected and all voters at these stations are interviewed as they leave.
8. Using random digit dialing, a poll interviews 1632 adults by phone.
9. A researcher plans to interview a random sample of 200 rural, 200 suburban, and 200 urban families in Illinois and ask them about the availability of health care.
10. A college student asks 10 of her classmates what they think of a sports team's Indian mascot.
11. A major airline randomly selects 20 flights during a given week and hands out questionnaires to all passengers on those flights.
12. An economist plans to interview a random sample of 50 female MBAs and 50 male MBAs from last year's graduating class.
13. A medical researcher surveys all cardiac patients in five randomly selected hospitals in the New York area.
14. Suppose that there are 150 registered students in an introductory statistics class. Are the students who come regularly to the instructor's office hours a probability sample, a sample of convenience, or a self-selected sample from the class? Are they a representative cross section of the class? Explain.
15. Consider the 30 students in Example 4.8.
    a. Select an SRS of 7 students from the class, using the first two digits of each five-digit number in the third five-digit column in Table B.3. Go from top to bottom, starting with 41949. List the 7 students selected by name.
    b. Select an SRS of 7 students from the class, using the last two digits of each five-digit number in the third five-digit column in Table B.3. Go from top to bottom, starting with 41949. List the 7 students selected by name.
    c. List the students who were in both samples.

Exercises 16 through 20 are multiple choice.

16. Suppose a random sample of 100 University of West Virginia students are sent a survey, to which 53 respond. Which of the following statements is more accurate?
    i. We can trust the sample of 53 to be representative of the population of all UWV students because the students were randomly chosen.
    ii. We cannot trust the sample of 53 to be representative of the population of all UWV students even though the students were randomly chosen. (Choose one option and explain.)
17. A researcher sent a survey to 100 people using expert opinion to choose the sample. Everyone in the sample responded. This sample is subject to (choose one option):
    i. Nonresponse bias
    ii. Selection bias
    iii. Both types of bias
    iv. Neither type of bias
18. A researcher contacted a sample of 100 people drawn from a town population of 10,000 using simple random sampling. Everyone in the sample responded. This sample is subject to (choose one option):
    i. Nonresponse bias
    ii. Selection bias
    iii. Both types of bias
    iv. Neither type of bias
19. A researcher contacted a sample of 100 people using simple random sampling. Forty people in the sample responded. This sample is subject to (choose one option):
    i. Nonresponse bias
    ii. Selection bias
    iii. Both types of bias
    iv. Neither type of bias
20. The author Shere Hite undertook a study of women's attitudes toward sex and love by distributing 100,000 questionnaires through women's groups. Only 4.5% of the questionnaires were returned. Based on this sample of women, Hite wrote *Women and Love,* a best-selling book claiming, among other things, that women today are fed up with

men. Hite's sampling method and conclusions are subject to (choose one option):
i. Nonresponse bias
ii. Selection bias
iii. Both types of bias
iv. Neither type of bias

21. A poll, done for the University of Maryland's Program on International Policy Attitudes in early September 2006, estimated that around 61% of all Iraqis approved of attacks on U.S.-led forces. The poll interviewed a random sample of 1150 Iraqi adults. How many percentage points was the margin of error?

22. In a poll of 501 randomly chosen teenagers, 261 said that they approved of legalized gambling.
a. Estimate the percentage of all teenagers who approve of legalized gambling.
b. How many percentage points is the margin of error?

23. In a survey of 204 Latina women, 143 had never had a mammogram. Pretend that the women surveyed were a simple random sample drawn from the population of all Latina women.
a. Estimate the percent of all Latina women who have never had a mammogram.
b. How many percentage points is the margin of error? (See Example 4.9.)

24. Suppose that in each of the 50 states, the Bureau of Labor Statistics takes a simple random sample amounting to one-hundredth of 1% of the population of the state in order to estimate the unemployment rate. Other things being equal, which of the following three statements is correct?
i. The accuracy to be expected in California (population 36 million) is about the same as the accuracy to be expected in Wyoming (population half a million).
ii. The accuracy to be expected in California is quite a bit lower than the accuracy to be expected in Wyoming.
iii. The accuracy to be expected in California is quite a bit higher than that in Wyoming.
Choose one option and explain. (*Hint:* How many percentage points is the margin of error for California? How many percentage points is the margin of error for Wyoming? See Example 4.9.)

25. There are about 25 times as many voters in California as in Hawaii. Suppose that a polling organization takes a simple random sample of 1000 voters in each state to estimate the proportion of voters in each state who favor same-sex marriages. Other things being equal, the poll in Hawaii is likely to be about
i. half as accurate
ii. as accurate
iii. 5 times as accurate
iv. 25 times as accurate
as the poll in California. Choose one option, and explain. Hint: How many percentage points is the margin of error for California? How many percentage points is the margin of error for Hawaii? (See Example 4.9.)

26. New York City has around four times as many registered voters as Houston, Texas. A simple random sample of registered voters is taken in each city, to estimate the proportion of registered voters in each city who oppose an amnesty program for undocumented aliens. Other things being equal, a sample of 2000 registered voters taken in New York City is likely to produce an estimate that is about
i. 16 times as accurate
ii. 4 times as accurate
iii. twice as accurate
iv. as accurate
v. half as accurate
as a sample of 500 registered voters taken in Houston, Texas. Choose one option, and explain. (*Hint:* How many percentage points is the margin of error for New York City? How many percentage points is the margin of error for Houston? See Example 4.9.)

4.3 EXPERIMENTAL DESIGN: OBSERVATIONAL STUDIES VERSUS RANDOMIZED EXPERIMENTS

observation
experimentation
observational studies
passively observe
designed experiments
actively impose various treatments

When trying to answer a question regarding cause-and-effect (such as "Does smoking cause lung cancer?"), we either obtain information by **observation** or by **experimentation.** In **observational studies,** investigators **passively observe** the subjects or units being studied, recording various variables of interest. The investigators do not influence what is being observed. They just describe it. In **designed experiments,** investigators **actively impose various treatments** on the experimental subjects or units. The goal is to find out whether different treatments result in different responses from the subjects or units being studied.

Both observational studies and designed experiments can be used to investigate cause-and-effect questions, such as: Do breast-fed babies grow up to be smarter than children who were given formula? Can middle-aged people lower their blood pressure and risk of heart disease by doing yoga? Do Head Start programs increase children's chances of graduating from high school?

confounding factors

Observational studies of the effects of a treatment are often difficult to interpret because of **confounding factors** whose effects are not easily separated from the effects of the treatment. For example, observational studies have found that children who were breast-fed in infancy tend to score higher on standardized tests than children who were given formula. But it is not clear whether the higher test scores are due to the breast-feeding (the treatment) or due to the fact that in the West women who breast-feed tend to be slightly older, better educated, and wealthier (confounding factors) than women who give formula to their infants. Confounding is always a problem in observational studies of cause-and-effect. The best way to avoid bias caused by confounding is to conduct a **randomized designed experiment** in which the available experimental subjects or units are randomly assigned to the various treatment groups. If we have enough experimental subjects, then the random assignment will ensure that at the outset of the experiment, the various treatment groups will be very similar. The only large systematic difference between the groups will be the treatments imposed by the investigators.

randomized designed experiment

In this section and the next we shall take a closer look at observational studies and designed experiments. Example 4.13 illustrates how careful we have to be not to jump to conclusions from observational studies. Observational studies can be used to detect association between two or more variables. But observational studies do not prove cause-and-effect.

Example 4.13 Cervical Cancer

Each year over 60,000 women in the U.S. are diagnosed with either advanced or early stages of cervical cancer. Before Pap smears were introduced in the middle of the 20th century, cervical cancer was responsible for more deaths among women than any other type of cancer.

Smoking is a risk factor: Women who smoke have twice the rate of cervical cancer of women who don't smoke. That raises the question: Does smoking *cause* cervical cancer? After all, smoking does cause heart disease and lung cancer. The use of oral contraception, "the pill," is also a risk factor: Women who take the pill have a 50% higher rate of cervical cancer than women who don't take the pill. Does that mean that the hormones in the pill can cause cervical cancer? An even greater risk factor is the lack of a high school diploma! Women who never completed high school have five to six times the rate of cervical cancer of women who completed high school..All these findings are based on observational studies: The investigators did not tell the women whether to smoke or not, whether to use the pill, or whether to drop out of school. The women themselves made those decisions.

Unfortunately, as we discussed in Section 3.4, observational studies can be misleading, due to confounding. These medical studies seem to imply that if a young woman decides to drop out of high school, then she will increase her risk of cervical cancer. But obviously, no woman ever developed cervical cancer as a direct result of dropping out of high school. It must be something in the lifestyle of high school drop-outs that puts them at risk. It turns out that neither smoking, nor the pill, nor dropping out of high school causes cervical cancer. It is a sexually transmitted disease, triggered by certain strains of the human papilloma virus, spread by sexual contact. The more sexually active a woman is and the more partners her partner has, the greater her risk of being exposed to the virus that causes cervical cancer. That appears to be the reason that certain groups of women have higher rates of cervical cancer than other groups.

Example 4.13 illustrates how careful we have to be not to jump to conclusions from observational studies. When observational studies identify an association (between a virus and a disease, say, or between breast-feeding and IQ, or between smoking and lung cancer) additional scientific evidence will be required to prove that the association is cause-and-effect and not due to confounding factors.

Example 4.14 Smoking and Lung Cancer

A major public health controversy of the 20th century was whether smoking causes lung cancer. In 1957 the *British Medical Journal* editorialized that smoking does indeed cause lung cancer, citing "the painstaking investigations of statisticians that seem to have closed every loophole of escape for tobacco."

One relevant study was conducted by British statisticians R. Doll and A. B. Hill in 1948 and 1949. In those days, 5 out of 6 British men age 35 to 59 years smoked. So did 40% of British women in that age group. Doll and Hill examined a selected set of 709 people with lung cancer and a selected matching set of 709 people without lung cancer. In each set there were 649 men and 60 women because the number of men and women were matched. The treatment variable was whether an individual smoked. Among the 1230 men from both groups who smoked (most of the men in both groups smoked), 647, almost 53%, had lung cancer (a very high rate, but note that half of the people in the study were selected because they had lung cancer). However, among the 68 nonsmoking men, only 2 had lung cancer, a percentage of just under 3%. Thus, for the men in the study there was a very strong association between smoking and lung cancer.

For the women the rate of lung cancer among the smokers was about 59%, compared to a 37% lung cancer rate among the nonsmokers. Again the association is clear, but with the intriguing observation that a much higher percentage of the nonsmoking women had lung cancer than their nonsmoking male counterparts. (A possible reason is that the women may have had more exposure to secondhand smoke.) These data, as well as data from many other observational studies, establish a strong association between smoking and lung cancer.

Sir Ronald Fisher (1890–1962) was perhaps the most famous and influential statistician of all time (and a well-known geneticist, as well). In his role as careful scientist and eminent statistician, Fisher was appalled by the *British Medical Journal's* jump to conclusions on the basis of only observational studies linking lung cancer and smoking. He forcefully and continually pointed out that just because smoking and lung cancer are associated, one cannot necessarily conclude that smoking causes lung cancer.[3]

In the absence of further scientific evidence, it could just as well be that lung cancer causes smoking or that a third factor causes both to tend to occur together. For example, people in the early stages of a long-developing disease such as lung cancer could experience severe physical discomfort, and smoking could help alleviate the discomfort. If this explanation should prove true, lung cancer would indeed cause cigarette smoking. Moreover, this rather unusual causal mechanism would produce a positive association between cigarette smoking and lung cancer. To quote Fisher, "And to take the poor chap's cigarettes away from him would be rather like taking away a stick from a blind man. It would make an already unhappy person a little more unhappy than he need be."

lurking variable
hidden variable

As for a third factor (a **lurking** or **hidden variable**) causing both smoking and lung cancer, Fisher went to great lengths to argue that one's genetic makeup could predispose one toward or away from smoking, and this same genetic makeup could also affect susceptibility to lung cancer. To understand this idea, imagine a gene whose possession causes a high rate of lung cancer and facilitates a high rate of cigarette smoking. Interestingly, some geneticists currently claim that a tendency toward addictive behaviors is genetically influenced, and it is well known that certain cancers do have a genetic predisposition. Any positive linkage between genetic characteristics contributing to addictive behaviors and those contributing to cancer would make Fisher's claim at least somewhat true.

Other possible lurking variables include type of employment (people working in coal mines, foundries, and other settings with polluted and possibly cancer-causing air may be more likely to smoke than people who work in offices or other professional settings) and whether one lives in an urban or rural environment (people in an urban area, who live amid more pollution, may also be more likely to smoke).

So the observational study by Doll and Hill was not by itself proof that smoking causes lung cancer. More evidence was needed. We shall get back to the link between smoking and lung cancer. But first we shall give a classical illustration of the most effective way to prove cause-and-effect.

Randomized Controlled Experiments: The Salk Polio Vaccine

A hundred years ago polio became a recurring epidemic health threat in the United States. It primarily struck children, often leaving its victims paralyzed. Throughout the first half of the 20th century, it afflicted hundreds of thousands of people, including the future President Franklin Delano Roosevelt. By the late 1940s and early 1950s, approximately 30,000 to 60,000 people per year were contracting polio. Research revealed that polio was caused

---

[3]See R. D. Cook, "Smoking and Lung Cancer," in *R. A. Fisher: An Appreciation,* ed. S. E. Fienberg and D. V. Hinkley (New York: Springer-Verlag, 1980), pp. 182–191; P. D. Stolley, "When Genius Errs: R. A. Fisher and the Lung Cancer Controversy," *American Journal of Epidemiology,* vol. 133, pp. 416–425; and B. W. Brown, Jr., "Statistics, Scientific Method, and Smoking," in *Statistics: A Guide to the Unknown,* ed. J. M. Tanur, F. Mosteller, W. H. Kruskal, R. F. Link, R. S. Pieters, and G. R. Rising (San Francisco: Holden-Day, 1972), pp. 40–51. These references discuss Fisher's role in the smoking-lung cancer controversy and provide many additional references.

by a virus, so scientists began to search for a vaccine.[4] In the early 1950s Dr. Jonas Salk developed a promising killed-virus vaccine against polio. At the same time Dr. Albert Sabin was developing a competing live-virus oral vaccine.

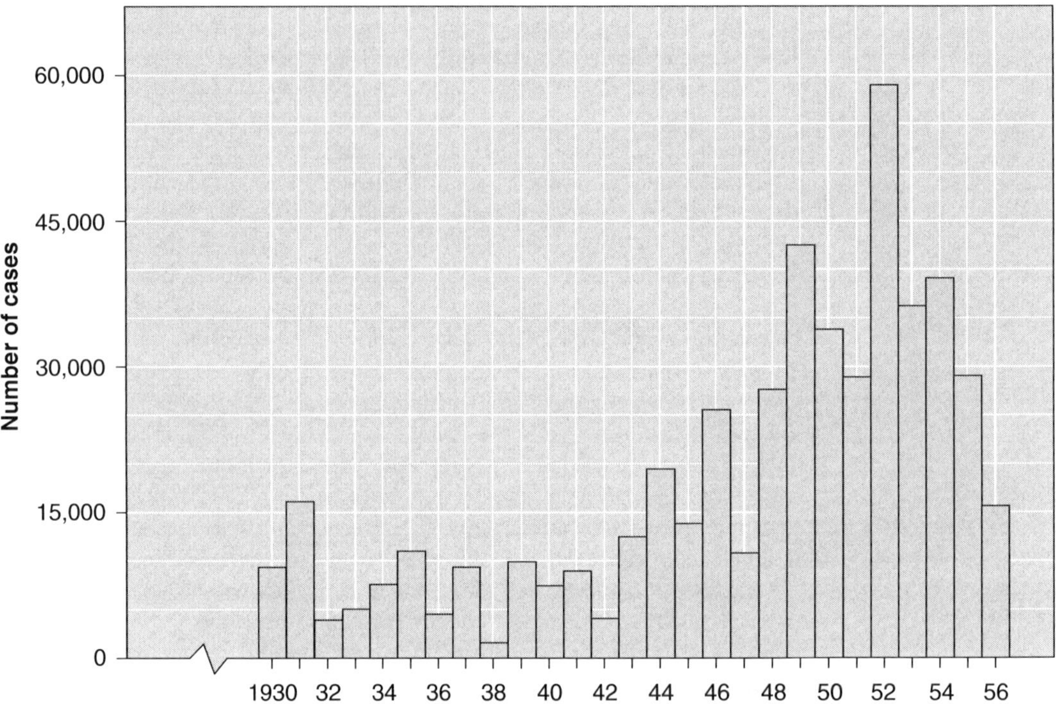

Figure 4.2 Annual occurrences of polio in the United States, 1930–1956.

In 1954 it was decided that a large-scale study should be conducted to assess the effectiveness of the Salk polio vaccine. Some observational studies were contemplated but rejected. One such proposal was to use the previous year, 1953, to provide controls (children not vaccinated) and use the next year, 1954, as the treatment year (children are vaccinated). In the treatment year, the vaccine would be widely disseminated, and whoever wished to receive it could do so. One could then compare the polio rates for the two years to see if the rate decreased. The problem with such an experimental design is that the severity of the polio outbreaks in 1953 and 1954 would be a confounding factor. Polio is an epidemic disease, like the flu. The number of polio cases varied widely from year to year. One year might have only a mild outbreak of polio; another year might have a very severe outbreak. For example, look at Figure 4.2. If the control year had been 1946 and the treatment year 1947, then it would have looked as if the vaccine were quite effective even if it were worthless. If the control year

[4]A vaccine is a substance that causes the body to believe it has contracted the virus. The body reacts by developing antibodies in the blood that recognize that virus. These antibodies remain in the body, so that if later the person does contract the viral disease, the antibodies will mark the invading viruses for destruction before they can cause harm. Vaccines can be made from small amounts of live virus, or of killed virus. The danger in using live virus is that the vaccine may cause the person to contract the disease before enough antibodies can be aroused. The danger in using killed virus is that it may not be able to fool the body into creating the antibodies. See P. Meier, "The Biggest Public Health Experiment Ever: The 1954 Field Trial of the Salk Poliomyelitis Vaccine," in *Statistics: A Guide to the Unknown*, ed. Tanur, Mosteller, Kruskal, Link, Pieters, and Rising (San Francisco: Holden-Day, 1978), pp. 2–13, for an account of the Salk experiment.

had been 1947 and the treatment year 1948, then even if the vaccine had cut the polio rate in half, it would have appeared ineffective.

Another possibility was to give the children in some regions of the country the vaccine, leaving the children in the rest of the country as the controls. The problem with this experimental design is again confounding: In particular, polio is prone to regional epidemics.

If the vaccine worked, it was important to have strong evidence that it worked, so that it could be widely used immediately to stop the dreaded disease. If the vaccine did not work, then it was just as important that researchers found out so that they could concentrate on more promising vaccines. Ambiguous results could have delayed the adoption of an effective vaccine for years. Because valid, and hence believable, statistical evidence was required, these observational approaches would not do.

controlled study
controls

Any study involving a group of units or subjects not receiving the treatment is called a **controlled study,** with the units or subjects not receiving the treatment called **controls.**

NFIP Study

In 1938 President Roosevelt established the National Foundation for Infantile Paralysis (NFIP), now known as the March of Dimes Birth Defects Foundation. (In honor of President Roosevelt and his association with the March of Dimes, the United States placed Roosevelt's image on the dime in 1946.) The NFIP planned a large field trial of the Salk vaccine in 1954 that would deal with the possible confounding effects of year and region. In participating school districts, the second-graders would be the treatment group getting the vaccine, and the first- and third-graders would be the control group. There could be some differences between the treatment and control groups, such as age and possibly more contagion within certain grades than in others, but because the control ages bracketed the treatment ages and all the subjects were within the same schools, those factors were expected to be minimal. However, an ethical problem with scientific implications arose: People could not be forced to take the vaccine. More precisely, children could not be given the vaccine without their parents' permission. For this reason the NFIP asked parents of second-graders to volunteer their kids for the study. About 64% did volunteer. Thus the NFIP plan was as follows:

- *Treatment group:* Second-graders, but *only* those with their parents' permission, are given the vaccine.
- *Control group:* The first-and third-graders at the same schools are not given the vaccine.

Although not randomized, the design for this statistical experiment is fairly good, but there are still some important ways in which the treatment and control groups may differ, three of which are stated next.

The volunteer effect The people in the treatment group receiving the vaccine were volunteers. The control group, by contrast, was made up of both some people who would have volunteered if they had been asked and some who would not. Volunteers are somewhat different from nonvolunteers in ways that matter medically. In this instance, volunteers tended to be more well-to-do, and, surprisingly, polio has been shown to be more likely to strike the more affluent[6] Thus this design is somewhat biased against the vaccine because at the outset of the vaccine trial, before anyone had been vaccinated, the treatment group would be at greater risk of polio than the control group.

---

[6]One usually associates affluence with better health, but in the case of polio it works the other way around. The explanation is that in less-affluent areas the polio virus is more likely to be present (because of poorer hygiene, higher population density, etc.). Hence, poor children are likely to be exposed to polio while they are still very young and protected by antibodies transmitted at birth from their mothers, allowing the children to develop their own polio antibodies to protect them from contracting the disease when they are older.

Change in children's behavior In the NFIP study the parents knew whether their child was vaccinated or not. This knowledge may have changed their behavior. For example, if the parents knew their child was vaccinated, they may have been more likely to let the child engage in more risky behavior, such as using a public swimming pool or attending summer camp with lots of other children, where exposure to the virus may have been more likely. This effect could also work against the vaccine.

Effect on diagnosis At the end of the study doctors had to evaluate the participants to determine who had been infected by polio. In the NFIP study the doctors evaluating the children usually knew who had been vaccinated and who had not. Depending on whether the doctor believed in the vaccine or not, this knowledge could sway the doctor's diagnosis in borderline cases, perhaps with the doctor being totally unaware of being so influenced.

The Randomized Controlled Experiment That Was Needed to Supplement the NFIP Study

Many public health experts involved in the study recognized the flaws in the NFIP's experimental design. It was too important to reach an unambiguous conclusion to allow these confounding effects to muddy the waters. Mindful of the ethical problems but aware that randomization was essential, a second design was proposed to supplement the NFIP observational experimental design with the following features:

1. a large number of experimental subjects in both treatment and control group
2. randomization
3. placebo
4. double-blind protocol

These features would ensure that at the outset of the trial, the treatment and control groups would be very similar in all respects, except for the treatment. If the polio rate for the treatment group turned out to be much lower than the rate for the control group during the trial, then the only possible explanation would be that the vaccine was at least partially effective. (The killed-virus Salk vaccine is not as effective as the riskier, more potent, live-virus Sabin vaccine adopted later.)

In order to achieve this design for this study, parents of children in grades 1 through 3 in selected school districts were asked to volunteer their children. The parents were explicitly told that their volunteered child might or might not receive the vaccine. Each eligible child was randomly assigned to either the vaccine group or the control group, essentially by the toss of a coin. Notice that this approach balances out the volunteer effect: Since both groups contained only volunteers, the difference between volunteers and nonvolunteers did not bias the comparison of the polio rates for the vaccine group and the control group. Randomization will also tend to equalize the effect of other confounding factors, such as overall health, age, sex, and level of affluence, especially when the number of replications is very large.

It is important to note that, being volunteers, the children undergoing the study were *not* a random sample from some well-defined population of school children. "Randomization" in a randomized experiment refers to the random assignment of units to treatments and not the mechanism for selecting the units, which is often nonrandom.

To eliminate differences in behavior based on knowing whether one received the vaccine or not, it was important that neither the children nor their parents knew who was in which group. There is no way a child would not notice being given an injection, so the plan was to give everyone in the study a seemingly identical injection. The treatment group had the vaccine in their injections, whereas the control group had plain saltwater. The saltwater injection

placebo

is an example of a **placebo.** A placebo is an inert treatment (that is, having no active medicinal ingredients) that, to the recipient, looks and feels the same as the real treatment (more about placebos soon in the section "Those Amazing Placebos"). The use of a placebo produces a **single-blind randomized controlled experiment;** that is, the subjects are blind as to whether they received the treatment or only the placebo. As a result, the risk-taking behavior and the psychological state (hopeful, anxious, etc.) of the two groups should be the same.

single-blind randomized controlled experiment

Finally, the experimenters who carry out the study (those giving the injections and the doctors making the diagnoses of the children at the end of the trial) are not told which children received the vaccine and which received the placebo. Hence, the single-blind experiment becomes a **double-blind experiment,** the second "blindness" being that of the people carrying out the experiment.

double-blind experiment

Now we have the plan for the randomized controlled experiment:

- *Treatment group:* Vaccine is given to a randomly selected half of the volunteers.
- *Control group:* Placebo is given to the other half of the volunteers.

The two studies (the nonrandom NFIP study and the randomized controlled double-blind experiment) were both implemented, each in about the same number of schools. The results of the two studies were announced on April 12, 1955, the 10th anniversary of President Roosevelt's death. Some of the data are in Table 4.1. The experiment was a success! The randomized controlled study showed that the polio rate among the nonvaccinated children was about $71/28 \approx 2.5$ times greater than for the vaccinated ones. The NFIP study also showed that the vaccine group did better, but not by quite as wide a margin: $54/25 \approx 2.1$. We see that confounding factors such as the volunteer effect, and possibly the other effects we discussed, did lessen the vaccine's apparent effectiveness in the NFIP study. Fortunately, we do not have to worry about confounding factors in the randomized controlled double-blind experiment.

Table 4.1 Polio Studies

Study	Number of subjects	Number contracting polio	Polio rate per 100,000
NFIP			
Vaccinated (2nd-grade volunteers)	221,998	56	25
Control (1st- and 3rd-graders)	725,173	391	54
Randomized control			
Vaccinated (volunteers)	200,745	57	28
Control (volunteers)	201,229	142	71

Here is how we can detect the volunteer effect. The polio rates in the vaccinated groups of the two studies were very close (25 versus 28 per 100,000), but the polio rates in the control groups were different: 54 per 100,000 in the NFIP study compared to 71 per 100,000 in the randomized control study (Table 4.1). Because neither control group received the vaccine, the difference in polio rates is likely due to the volunteer effect: The controls in the NFIP study consisted of both would-be volunteers and nonvolunteers, whereas the controls in the randomized study were all volunteers. Recall that volunteers were on average more affluent and, hence, more likely to contract polio.

The experiment proved that the Salk vaccine worked. Unfortunately, there were quality control problems with the Salk vaccine, and the vaccination program was temporarily halted when a bad batch of the vaccine caused 79 children to contract polio. However, the vaccination program was restarted, using new and improved vaccines. Subsequent widespread use of both the Salk and Sabin vaccines eventually led to the virtual eradication of polio from the United States. As a student of statistics, note the vital role that a randomized controlled experiment played in the fight against polio.

Back to Smoking

The observational studies showing association between smoking and lung cancer are not as convincing as the randomized controlled Salk vaccine study. Why not execute a similar study for smoking? Imagine such a study. A group of people are identified (volunteers would be fine) to be in the study, say, all about 18 years old. Half are randomly assigned to smoke for the rest of their lives; half are randomly assigned to never smoke. Obviously, we can not get volunteers for such an experiment. So studies involving cigarette smoking must be observational.

The collective force of many observational studies can, however, be convincing. The U.S. surgeon general appointed a blue-ribbon committee in 1962 to review the scientific (experiments on rats, tissue studies on deceased lung cancer victims, etc.), medical, and statistical evidence of the health effects of smoking and to arrive at a summary conclusion. The committee surveyed a large number of various kinds of studies, and in 1964 it issued a comprehensive report flatly stating that "cigarette smoking is causally related to lung cancer in men; the magnitude of the effect of smoking far outweighs all other factors."

Taking the studies as a whole, the idea that the relationship between smoking and lung cancer is causative is convincing because of a number of considerations:

1. The association appears in many types of studies using many types of subjects.
2. The association is so strong (the mortality due to lung cancer among smokers is 10 to 20 times that among nonsmokers) that it is unlikely that the effect could be totally explained by other factors.
3. The suspected cause (smoking) is shown to occur before the effect (lung cancer).
4. Tobacco smoke contains substances that are known to cause cancer in animals (established by doing randomized controlled experiments).
5. People are more likely to contract lung cancer the more they smoke per day, the earlier they start smoking, and the more they inhale (shown observationally).

The overwhelming and multifaceted evidence allows us to bypass the need for randomized controlled experimentation and yet still conclude causality. Successfully bypassing randomized experimentation is unusual, and this instance required an enormous investment of scientific and medical resources.

As an interesting aside, R. A. Fisher made the case that according to some of the Doll and Hill 1948–1949 data, noninhalers seemed more likely to contract lung cancer than inhalers. This is an example of an association that did not hold up under more careful examination. In fact, even Doll and Hill had additional data showing the difference was minimal. By contrast, the surgeon general's report cited a large body of evidence that showed that the more people inhaled, the higher the rate of lung cancer. It seems likely that some subtle third-variable influence was present in the earlier Doll and Hill data or, as happens occasionally in statistical studies, the statistical "gods of chance" conspired to fool us—that is, the natural randomness that produced the data yielded an unlikely result (such as 9 heads in 10 tosses of a fair coin).

The case against smoking was strong, but there were still skeptics, in particular the tobacco companies. In 1979 the surgeon general produced another (heftier) report, even more comprehensive and even more damning for smoking.[7] Still, it was not until January 1998 that even the tobacco companies had to admit, "We recognize that there is a substantial body of evidence which supports the judgment that cigarette smoking plays a causal role in the development of lung cancer and other diseases in smokers."[8] After so many years of statistical research, at least no one can argue with the pundit who said, "It is now proved beyond any doubt that smoking is one of the leading causes of statistics."

Observational studies can be of inferential value, but it takes many more of them, and a wide variety of types, to produce evidence as strong as evidence from randomized controlled experiments. Further, in observational studies, efforts to control for lurking variables like age, gender, etc., are vital, but are only possible to the extent that we are aware of the fact these variables affect the responses we are studying.

Those Amazing Placebos

The notion that an inert substance can heal because the patient believes it can is an old one. Such inert substances were dubbed placebos, from the Latin word for "I shall please," because a healer, lacking a truly curative substance, would prescribe a placebo to make the patient happy. Over the years placebos have included "usnea (moss from the skulls of victims of violent death), Gascoyne's powder (bezoar, amber, pearls, crabs' eyes, coral, and the black tops of crabs' claws), triangular Wormian bone from the juncture of the sagittal and lambdoid sutures of the skull of an executed criminal, wood lice, human placenta and perspiration,"[9] and many other strange and unpleasant substances. The placebos worked in the sense that people would often feel better after having taken them, even though these "medicines" had no active ingredients. The placebo effect is real; that is, true healing can sometimes occur if a person believes strongly enough in a treatment. In wartime front line medics have successfully given wounded soldiers fake painkillers when they ran out of real painkillers. About a third of those who received fake painkillers felt relief. Psychologists and medical researchers have studied the influence of mental functioning on the immune system and on other physiological characteristics related to disease and pain.

When a new drug or medical procedure is introduced, investigators often find that patients respond well to the new treatment. For example, one treatment for angina (suffocating chest pains associated with heart disease) was to tie off the mammary artery. Two studies *without any controls* reported 68% and 91% of the patients improving after the surgery. These rates seem very impressive, and this procedure was widely used from 1955 to 1960. Popularity plummeted, however, after two more experiments *with controls* showed 67% improvement in the treatment group, but 71% in the control group. (The controls received a placebo in the form of a skin incision that did not affect the artery.)[10] It appears that the improvements people felt were based on the placebo effect. The actual surgery did not provide the relief. It was the knowledge that their chest pain was being treated that gave the patients psychological and physical relief. Without carefully controlled experiments comparing the treatment to the placebo, surgeons would likely have continued performing a dangerous but useless procedure.

Chemonucleolysis is a treatment for alleviating the pain of a slipped disc in the spine. It

---

[7] *Smoking and Health: Report of the Advisory Committee to the Surgeon General of the Public Health Service,* U.S. Department of Health, Education, and Welfare, Public Health Service publication no. 1103, 1964, p. 106; *Smoking and Health: A Report of the Surgeon General,* U.S. Department of Health, Education, and Welfare publication no. (PHS) 79–50066, 1979.

[8] Statement before the U.S. House of Representatives Commerce Committee, January 19, 1998, by Geoffrey C. Bible, chairman and chief executive officer of Philip Morris Companies, Inc.

[9] A. K. Shapiro and E. Shapiro, "The Placebo: Is It Much Ado about Nothing?" in *The Placebo Effect,* ed. A. Harrington (Cambridge, Mass.: Harvard University Press, 1997), pp. 13–14.

[10] Ibid.

involves injecting an enzyme directly into the disc.[11] Between 1963 and 1975, almost 17,000 people had this treatment, with the studies reporting success rates from 50% to 80%. These studies did not use controls. Before the procedure could be approved for use in the United States, it needed to be evaluated in a randomized controlled experiment. In 1975 a double-blind randomized controlled experiment using 106 patients (53 in each group) was performed in which the placebo was an injection without the active enzyme. The treatment group had 60% success, and the placebo control group had 50% success. Thus the treatment appears to be a little better than the placebo. Because of the small number of participants in the study, the 10% difference was not statistically significant (i.e., the statistical evidence was weak; see the next chapter), and the treatment failed to be approved.

A later study, without controls, was reported in 1977 to have a 70% success rate with the procedure. Still, no approval for its use was forthcoming. Two more double-blind randomized controlled experiments (with placebos) were conducted. Combining the three placebo-control studies (this combining of statistical studies is called a **meta-analysis**), the success rate for the chemonucleolysis treatment was 70%, and that for the placebo was 47%, enough to produce strong statistical evidence in favor of the treatment. Finally, the treatment was approved. This appears to be a case in which the treatment really is effective, but again the message is clear: We found this out only by doing careful randomized double-blind controlled experiments, with enough replications to obtain strong statistical evidence.

meta-analysis

Notice that the uncontrolled studies for both the angina surgery and the chemonucleolysis injection showed success rates of 50% to 90%. The angina surgery actually did slightly worse than the placebo, suggesting that the surgery was useless at best and possibly even harmful. For the disc treatment, the placebo still did well, but not quite as well as chemonucleolysis. Chemonucleolysis was only about 23% better than a placebo. Both examples suggest a strong placebo effect. Thus, by not comparing the treatment of interest to a placebo, one often obtains overly optimistic, or even wrong, impressions of the effectiveness of a drug or medical treatment.

Drop-out and Non-Adherence

non-adherers

drop-outs

People who participate in an experiment but do not follow the rules are called **non-adherers.** People who initially participate in an experiment but do not complete the experiment are called **drop-outs.** Just as non-response can bias sample surveys, drop-out and non-adherence can bias carefully planned randomized controlled experiments. For example, suppose that we are running a 6-month trial of a new weight-loss pill in a randomized controlled double-blind experiment. Suppose that, in addition to receiving either treatment or placebo, all participants in the study are given information about lifestyle changes necessary for sustained weight loss. Suppose, for the sake of argument, that the weight-loss pill works and that during the first month of the trial the people in the treatment group lose more weight than the people in the control group. If the least successful individuals in the placebo group get discouraged and drop out, then the study would be biased against the new pill since at the end of the 6-month trial we would be comparing the treatment group to what is left of the control group, which would be a disproportionate number of individuals having the discipline to lose weight by lifestyle changes alone.

Clinical trials of experimental drugs to fight HIV are often affected by both non-adherence and drop-out. Some AIDS patients participating in such trials augment their prescribed medication with a concoction of their own, and some have broken the blind of the trial by having their medication tested by an outside lab. If their medication turned out to be the placebo, they

[11] R. L. Sanford, "The Wonders of Placebo," in *Statistics in the Pharmaceutical Industry,* ed. C. R. Buncher and J.-Y. Tsay (New York: Marcel Dekker, 1994).

would drop out of the trial. While all this is very understandable, it has hampered the search for effective treatments for AIDS.

Adherers are different from non-adherers in ways that can easily bias a statistical study. (We shall return to this fact in our discussion of the Key Problem.) That they are different was carefully documented by the investigators of the Coronary Drug Project, a randomized controlled double-blind experiment carried out to evaluate the safety and effectiveness of several drugs in the long-term treatment of heart disease in middle-aged men. The five-year mortality rate in 1103 men treated with one of the drugs, clofibrate, was 20.0%, compared to 20.9% in 2789 men given placebo. So clofibrate did not have the hoped-for ability to save lives. A closer inspection of the data revealed that good adherers to clofibrate (i.e., patients who took 80% or more of their prescribed clofibrate medication during the five-year follow-up period) had a substantially lower five-year mortality rate than the non-adherers to clofibrate (i.e., the patients in the clofibrate group who took less than 80% of their medication). The mortality rates were 15.0% versus 24.6%. So those who conscientiously took their clofibrate had a lower risk of dying than those who forgot to take their medication. That seems to imply that clofibrate did save lives, after all. However, similar findings were noted in the placebo group (i.e., 15.1% mortality rate for good adherers to placebo versus 28.3% mortality rate for non-adherers to placebo). The mortality rates are presented in Table 4.2.

As Table 4.2 shows, the big difference in mortality rates is not between the clofibrate and placebo groups (20.0% versus 20.9%). It is between adherers and non-adherers. Those heart patients who had the discipline to faithfully take their medications (whether clofibrate or placebo) had only a 15% five-year mortality rate. The explanation must be that the adherers also had the discipline to control their heart disease by making lifestyle changes such as exercising, dieting, reducing stress, and giving up smoking.

Table 4.2 Five-year Mortality Rates in the Clofibrate Trial

	Clofibrate	Placebo
Adherers	**15.0%**	**15.1%**
Non-adherers	**24.6%**	**28.3%**
Total group	**20.0%**	**20.9%**

Source: Influence of adherence to treatment and response of cholesterol on mortality in the Coronary Drug Project, *New England Journal of Medicine*, vol. 303 (1980) pp. 1038–41.

Example 4.15 Hormone Replacement Therapy

In the 1990s, the Women's Health Initiative recruited over 16,000 healthy post-menopausal women age 50 to 79 to participate in a double-blind randomized controlled experiment to determine once and for all the risks and benefits of hormone replacement therapy (HRT). Despite decades of observational evidence, it was still unclear whether health benefits outweighed risks.

Half of the volunteers were randomly selected to receive a daily tablet containing a combination of estrogen and progestin. The women in the control group were to receive an identical-looking placebo. Follow-up was by semiannual interviews and questionnaires and annual in-clinic visits. Monitoring began in the fall of 1997 and the study was expected to run until 2005. But after the 10th interim analysis in May 2002, overwhelming evidence had accumulated that risks outweighed benefits. So the clinical trial was stopped.

The Women's Health Initiative study gave conclusive proof that hormone replacement therapy causes a slight increase in a postmenopausal woman's risk of heart disease, stroke, and breast cancer, but reduces her risk of hip fractures and colorectal cancer. That raises the question: Why had all the earlier observational studies shown that women who took HRT under their doctor's supervision had a 35% to 50% lower risk of coronary heart disease than women who did not take HRT? What was wrong with the earlier observational studies?

A confounding factor in the observational studies was that the women receiving HRT took better care of their health than the women in the control groups. There are several reasons for this. Women who conscientiously took HRT for a long time were obviously good at following their doctor's orders. Not everybody is. And as we discussed in connection with the Coronary Drug Project, adherence (having the discipline to faithfully take a prescribed medication, even if it is a fake medication) is associated with a lower risk of heart disease. In other words, those who have the discipline to take their medications regularly tend to watch their diet, exercise, and live healthier lives. The observational studies also failed to pay enough attention to confounding socioeconomic factors. The women receiving HRT were on average wealthier and more educated than the women not receiving HRT. Middle-class and well-to-do women have lower rates of heart disease than poor women of comparable age. So lifestyle and socioeconomic background were confounding factors in the observational studies. The observational studies documented healthy women taking HRT, not HRT keeping women healthy!

Section 4.3 Summary

treatment
response
method of comparison
treatment group
control group
randomized controlled experiment
replicated
placebo
double-blind

To determine the effect of a **treatment** (like hormone replacement therapy) on a **response** (like getting heart disease), statisticians use the **method of comparison.** They compare the responses of a **treatment group** (receiving the treatment) to the responses of a **control group** (that does not receive the treatment). Without a control group it is not possible to judge the effects of a treatment. If the treatment and control groups are very similar in all respects, except for treatment, then any major difference in the responses of the two groups is likely to have been caused by the treatment. The best way to ensure that the treatment and control groups are very similar, except for the treatment, is to randomly assign the available subjects or units to treatment or control in a **randomized controlled experiment.** To reduce chance variation in the results, each treatment (including "no treatment") should be **replicated** (repeated) on many subjects or units. If a randomized controlled experiment has human subjects, then the control group should be given a **placebo,** whenever possible, to prevent the subjects from changing their behavior, based on knowledge of which group they were assigned to. In a **double-blind** experiment, neither the subjects nor the investigators interacting with the subjects know who is receiving which treatment. Double-blinding prevents the subjects from changing their behavior and prevents those who evaluate the responses from giving biased evaluations.

observational studies
association

Unfortunately, it is often not possible to conduct randomized designed experiments to investigate cause-and-effect. Investigators cannot tell people to smoke or to drop out of high school. So a randomized controlled experiment to determine the effects of smoking or the effects of dropping out of high school is not possible. Investigators have to rely on **observational studies.** In an observational study, the investigators passively observe the subjects or units being studied. The investigators do not assign the subjects to treatment or control. For example, in a study on smoking, the subjects who themselves choose to smoke form the treatment group and the subjects who themselves choose not to smoke form the control group. Observational studies can establish **association.** For example, smokers have a higher rate of lung cancer than nonsmokers. That links smoking to lung cancer. But association does not prove causation. To prove that the association is cause-and-effect we need additional evidence. The problem is that in observational studies there are often confounding factors whose effects are not easily separated from the effects of the treatment being studied. For example, women who smoke (the treatment) have twice the rate of cervical cancer (the response) of women who do not smoke. So smoking is associated with both lung cancer and cervical cancer. But whereas years of painstaking medical research has established that smoking does indeed cause lung cancer, the association between smoking and cervical cancer is not cause-and-effect. As we discussed in Example 4.13, lifestyle and socioeconomic background account for the association between smoking and cervical cancer.

4.3 Exercises

1. A university study found that children who lived in poverty during their first 5 years of life, on average, scored 9 points lower on IQ tests at age 5 than children who had not lived in poverty.
   a. Identify the treatment group and the control group.
   b. Identify the response variable.
   c. Was this an observational study or a designed experiment?

2. A magazine article reported that 95% of cold sufferers who treated their cold with a daily dose of at least 500 mg vitamin C recovered within a week. Explain why this is insufficient evidence to conclude that vitamin C is a cure for the common cold.

3. An Associated Press article from December 14, 1997, reported on a study to see whether Prozac helped people control their anger. The article said:

   > Forty people who considered their explosive tempers a problem enrolled in the study. In a week, each subject typically had fits of temper equivalent to fifteen verbal outbursts directed at others, eight physical outbursts directed at objects, or two physical assaults against others, the researchers said. The forty subjects were randomized to treatment with Prozac or placebo at an assignment ratio of 2:1. Twenty-seven participants were given Prozac and thirteen got a placebo. Neither group knew what they were getting. Prozac significantly reduced aggression for some subjects, although not everyone was helped, researchers reported in the December issue of the *Archives of General Psychiatry,* published by the American Medical Association.

   a. Was this study an observational study or a randomized controlled experiment?
   b. Which of the following statements is most accurate?
      i. There were no controls, but it was at least single-blind because there was a placebo.
      ii. There were no controls because it was an observational study.
      iii. There were controls, with a placebo.
      iv. One cannot determine whether there were controls or not.
      v. There were controls, but the people in the control group were volunteers, whereas the people who got the treatment were non-volunteers.

4. According to many observational studies, people who eat lots of fruits and vegetables rich in anti-oxidant vitamins have lower rates of colon cancer than people who don't. It was therefore hoped that a daily supplement of antioxidant vitamins would protect against colon cancer. In a randomized controlled clinical trial, the people in the treatment group were given a large daily dose of vitamin supplements, while the people in the control group were given a daily placebo. The experiment did not find any significant difference between the rates of colon cancer in the two groups.

   Determine which of the three statements below are true and which are false. Explain your answers.
   a. The experiment confirmed the findings of the observational studies.
   b. Since the lifestyles of people who eat lots of fruits and vegetables differ from the lifestyles of people who don't in many other ways, the observational studies could easily have reached the wrong conclusions, due to confounding.

c. Since the lifestyles of people who eat lots of fruits and vegetables differ from the lifestyles of people who don't in many other ways, the randomized experiment could easily have reached the wrong conclusions, due to confounding.

5. An article in the December 16, 1997, issue of the *New York Times* reports on a study of suicide and gambling. The article says,

> In the study, Dr. David Phillips, a professor of sociology at the University of California in San Diego, examined death certificates in major gaming cities in the United States—Atlantic City, Las Vegas, and Reno—and found that suicide rates were up to four times higher than in comparably sized cities where gambling is not legal.

a. Identify the treatment group and the control group.
b. Identify the response variable.
c. Was this an observational study or a designed experiment?

6. A university study found that the teen-age sons of women who smoked got into trouble with the law more often than the teen-age sons of women who did not smoke.
a. Identify the treatment group and the control group.
b. Identify the response variable.
c. Was this an observational study or a designed experiment?

7. The following is a quote about a recent study on breast cancer. Use this paragraph to answer the following questions.

> The cancer institute's study involved 13,338 women in the United States and Canada, making it one of the largest cancer prevention studies ever. Some women were given tamoxifen, others placebos. For those given tamoxifen over a 5-year period, one in 236 developed breast cancer. The placebo breast cancer rate was one in 130 women. There were significant reductions in the occurrence of both invasive and non-invasive breast cancers in every age group, from 35–45 to the over-60 group.

a. Identify the treatment group and the control group in this study.
b. How were placebos used in this study? Do you think it was necessary to use them?
c. Name factors that could influence the outcomes of this study.
d. Write down one question you would like to ask the researchers about their study.

8. How would you conduct an observational study to determine the effects of alcohol consumption on heart disease? Consider the following questions: How do you choose the control group and the treatment group? What information should you get from the subjects?

9. Design a randomized controlled experiment to determine the effects of alcohol consumption on heart disease. In the design of your experiment, consider the following questions: How do you choose the control group and the treatment group? How do you choose these groups to minimize the volunteer effect? Would a placebo be necessary for this kind of experiment? If so, what would you use as a placebo for this experiment? What limitations or obstacles would you face with this experiment?

10. A study to determine whether eating fish is associated with lower blood pressure compared the inhabitants' diets in two villages in Tanzania. One village was on the shore of a lake and had a diet consisting primarily of fish, and the other village was inland and had a diet consisting primarily of vegetables. Of the people in the fish-eating village, only 3% had high blood pressure, whereas in the vegetable-eating village, 16% of the people had high blood pressure.
a. Is there a control group in this study? If so, describe it.
b. Is this an observational study or a randomized controlled experiment?
c. Is there a placebo? If so, what is it?
d. How strong is the evidence in this study concerning the benefits of eating fish on high blood pressure?
e. What other factors could affect blood pressure?

11. A study was conducted to see whether a new drug, Copolymer 1, was effective in preventing relapses among people with multiple sclerosis. One group of patients was given the drug and another was given a placebo. Patients were randomly assigned to the two groups. After 2 years, 33.6% of the people taking the drug were relapse free, and 27% of the people taking the placebo were relapse free.
    a. Is this an observational study or a randomized controlled experiment?
    b. What other factors could be affecting the responses of the patients?

12. An article in the October 7, 1997, *Chicago Tribune* reported a study trying to determine whether children who are older than their classmates behave differently than children who are the same age as their classmates. The article said that a study in the October issue of the journal *Pediatrics* found that 12% of the children who started school when they were a year or more older than their classmates displayed extreme behavior problems, compared with 7% of children whose ages were normal for their grade. According to the study, the problems become more apparent as the children grow older.
    a. Was there a control group in this study?
    b. Is this study a randomized controlled experiment or an observational study?
    c. Does the study prove that being older than one's classmates tends to cause bad behavior? What other explanation could there be?

4.4 RANDOMIZED BLOCK DESIGNS, INCLUDING MATCHED-PAIRS DESIGNS

Randomized Block Design

In observational studies it is important that we control for confounding factors. For example, in a study on the link between smoking and heart disease it would be a bad idea to compare a group of middle-aged male smokers (the treatment group) with a group of young female nonsmokers (the control group). Because heart disease develops over time and women usually develop heart disease later in life than men do, the middle-aged men in the treatment group would on average have more heart disease than the young women in the control group even if the men didn't smoke. In other words, we wouldn't be able to tell to what extent the higher rate of heart disease in the treatment group was due to smoking (the treatment) and to what extent it was due to the confounding factors, age and gender. To control for age and gender we should compare middle-aged male smokers with middle-aged male nonsmokers; and we should compare middle-aged female smokers with middle-aged female nonsmokers. The same comparisons should be made for young and old subjects.

In randomized experiments it is also a good idea to actively control for potentially influential confounding factors rather than letting randomization alone attempt to accomplish the task. Suppose, for example, that we want to conduct a randomized controlled experiment to determine whether middle-aged people can lower their risk of a heart attack by taking a low-dose (81 mg) aspirin tablet every other day. In a **completely randomized design** we would randomly assign all the available experimental units (in this case, all the middle-aged subjects participating in the study) to the various treatment groups (in this case, the aspirin group and the placebo-based control group).

completely randomized design

In such a design both treatment and control group would contain men as well as women. But because we know that middle-aged women have less heart disease than middle-aged men, it would be better to use a randomized block design in which we actively control for gender by studying women and men separately. We will more clearly see the possible benefits from taking aspirin regularly if we don't mix the high-risk males and the low-risk females. Such a design would also help us find out whether men and women respond differently to the aspirin treatment. The point is that although a completely randomized design (with enough participants) tends to balance men and women in the treatment and control groups, and thus control for gender, a design that totally controls for gender by looking at blocks of men and women separately is more informative.

explanatory variable
factor
levels
response variable

The completely randomized experimental design in Figure 4.3 has one **explanatory variable** or **factor,** namely the aspirin dose. The factor has two **levels,** namely no aspirin (control or placebo) and 81 mg of aspirin (treatment) every other day. Whether or not the experimental subject has a heart attack is the **response variable.** The randomized block experimental design in Figure 4.4 has two explanatory variables or factors, namely, gender and aspirin dose. Gender has two levels (possible values), male and female.

randomized block design
blocks

In a **randomized block design,** the available experimental subjects (or units) are first divided into **blocks.** Each block consists of subjects (or units) that are known before the experiment to be similar in some way relevant to the experimental outcome. Within each block the experimental subjects (or units) are randomly assigned to the various treatments, perhaps one treatment versus a control.

In the aspirin trial we use a randomized block design, not just to control for the confounding caused by gender but also because we are interested in the confounding variable, gender. We want to know not just whether aspirin therapy could reduce the number of heart attacks in the middle-aged population as a whole, but also whether middle-aged women respond as well to aspirin therapy as middle-aged men do. Note that the resulting design has four groups, with two treatment groups and two control groups.

In the next example we are not really interested in the confounding factors themselves. We often use randomized block designs to prevent the effects of the confounding factors that are not of interest from obscuring the effect of the factor we are interested in.

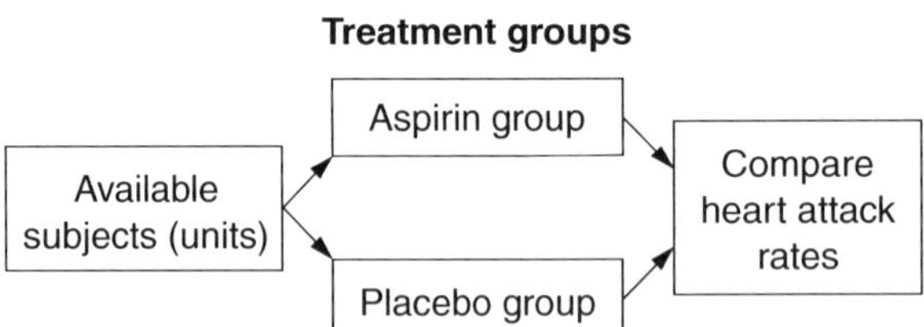

Figure 4.3 A completely randomized design.

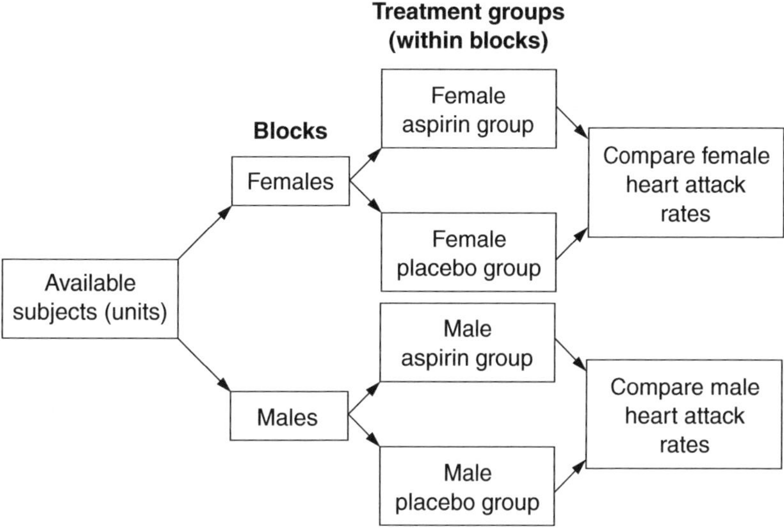

Figure 4.4 A randomized block design.

Example 4.16

Consider the problem of determining whether different brands of car tires suffer different amounts of tread loss after 30,000 miles of driving. A car rental fleet manager wants to find out which of four brands, A, B, C, and D, shows the least amount of tread wear after 30,000 miles. Although tire wear might be simulated in a laboratory, for authenticity the fleet manager wants to test the four brands under actual driving conditions. The variable to be measured for each tire tested is the tread loss between the time the new tire is mounted on the wheel of a car and the time the car has completed 30,000 miles.

Suppose that the fleet manager wants to test 4 tires of each brand, a total of 16 tires mounted on four test vehicles, which we shall denote I, II, III, and IV. One possible nonrandomized experimental design would be to put all four brand A tires on Car I, all four brand B tires on car II, and so forth, as shown in Table 4.3.

Table 4.3 A Completely Confounded Nonrandomized Design for Tire Brand Assignment

	Car			
Tire position	**I**	**II**	**III**	**IV**
Front left	A	B	C	D
Front right	A	B	C	D
Rear left	A	B	C	D
Rear right	A	B	C	D

This is a very bad design. The amount of tread wear depends not only on the quality of the tire brand (the factor we are interested in) but also on the terrain each car travels over, the weight of each car, how the driver operates each car, and the location of the tire on the car (the confounding factors). With the design in Table 4.3 we cannot distinguish between the brand effect and the confounding factors.

We might instead try a completely randomized design in which the 16 tires are randomly assigned to the 16 tire positions. One such resulting assignment is shown in Table 4.4.

Table 4.4 A Possible Tire Brand Assignment Resulting from a Completely Randomized Design

	Car			
Tire position	**I**	**II**	**III**	**IV**
Front left	D	B	A	B
Front right	B	B	D	A
Rear left	A	C	C	C
Rear right	A	D	C	D

This is a better design than the previous one, but it is not a good design. The problem with the assignment in Table 4.4 is that if, for example, Car I happens to put much more wear on its tires than the other three cars, then tire brand A will look bad even if it is just as good as the other three brands: Brand A will look bad because two brand A tires will be more severely worn. And if either Car II or Car III happens to put much more wear on its tires than the other three cars, then brand B or brand C, respectively, will look bad. Thus, variation between cars might obscure the variation between tire brands even though the design is a randomized one. Similarly, if front tires wear faster, Brand B will look bad and Brand C will look good.

To remedy this shortcoming, we could either test a larger number of tires of each brand, mounted on more cars (replication), using a completely randomized design, or we could test the 16 tires using a randomized block design in which each car receives one tire of each brand, randomly selected, as shown in Table 4.5. Here we have 4 blocks (cars) receiving 4 treatments (tire brands).

In this design, we randomly select one tire of each brand for each test car and randomly position the four tires on the car. Thus, we randomly assign treatments (tire brands) within each block (car), producing a randomized block design. Since tire wear depends on tire position, to improve the design still further, we balance out the effect of position by rotating the tires every 7500 miles, giving each tire 7500 miles in each of the four positions.

Table 4.5 A Randomized Block Design for Tire Brand Assignment

	Car			
Tire position	**I**	**II**	**III**	**IV**
Front left	C	D	B	C
Front right	B	C	C	D
Rear left	A	B	D	A
Rear right	D	A	A	B

Matched-Pairs Design

Matched-pairs designs

Matched-pairs designs compare two treatments using blocks of two experimental units (or subjects) that are matched as closely as possible. This special case of blocking (blocks of size 2) occurs so often that we consider it separately.

Example 4.17

Suppose a national pest control company wants to test two types of roach traps, brand A and brand B, to find out which is better. We select several dark, secluded locations appealing to roaches throughout a roach-infested high-rise building. The locations are far enough apart to ensure that the number of roaches visiting our different locations are independent.

We could use a completely randomized design in which we randomly select which of the locations to put brand A traps in and which locations to put brand B traps in. But since the number of roaches probably varies a lot from location to location, in a completely randomized design, the number of roaches trapped in a given location would reflect not only the effectiveness of the trap but also the number of roaches in that location. Thus, the variation between locations might obscure the variation between brands.

It is much better to use a matched-pairs design in which we place both a brand A trap and a brand B trap in each location. In such a design the variation between blocks (locations) will not obscure the variation between the two treatments (brands) because both treatments are compared within every block. Note that our knowing the advantage of a matched-pairs design made it easy to create such a design.

Example 4.18

A new growth hormone supplement for chicken feed is to be tested in an experiment involving 200 newly hatched chickens placed in eight large cages, with 25 chickens randomly assigned to each cage. We want to compare the weight gain over a 3-week period of 100 chickens receiving the hormone supplement (treatment) to the weight gain of 100 chickens that do not receive this hormone supplement (control group).

The trial takes place in winter, and the eight chicken cages are placed in a poorly insulated room that has considerable temperature variation. Specifically, cages on the south side of the room tend to be colder and get more drafts of cold air than those on the north side of the room because of the locations of the heater and the door. The heat loss due to each of the four windows is about the same. The cages have arbitrarily been labeled I, II, ..., VIII in Figure 4.5.

Because of the east-west symmetry of the room it is possible to group the eight cages into four matched pairs (four blocks of size 2 each) such that the two cages in each pair have very similar temperatures:

$$(\text{I}, \text{IV}), \quad (\text{II}, \text{III}), \quad (\text{V}, \text{VIII}), \quad \text{and} \quad (\text{VI}, \text{VII}).$$

From each pair we randomly select one cage to receive the hormone supplement and use the other cage as a control. This selection can be made by the toss of a fair coin. This matched-pairs design ensures that variation between locations does not obscure the studied variation between treatments. It does so much more effectively than a completely randomized design, which is likely to accidentally advantage or disadvantage the hormone treatment.

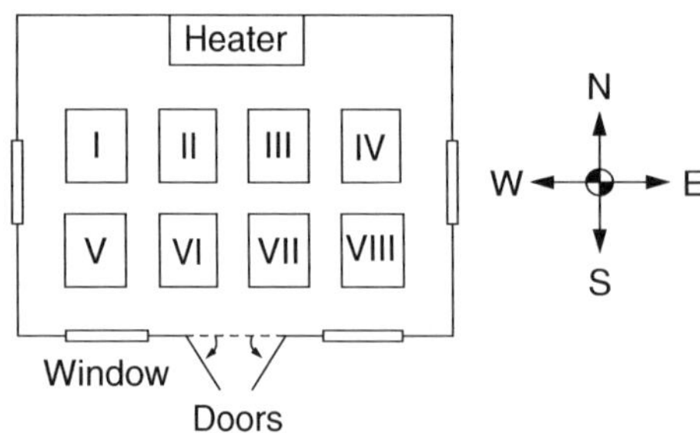

Figure 4.5 Arrangement of chicken cages.

❒ ❒

treatment factor(s)
blocking factor

The treatments of a designed experiment are the various combinations of levels of the explanatory variable(s), often called the **treatment factor(s),** together with the experimentally created blocks (the **blocking factor**)—for example, each car—tire brand combination in Example 4.16.

Section 4.4 Summary

experiment
treatments
experimental units, subjects, participants
factor
level
treatment
responses
completely randomized experimental design
randomized block design
blocks
Matched-pairs designs

In an **experiment** investigators apply various **treatments** to a collection of individuals or objects, called **experimental units,** or **subjects,** or **participants** (if they are human). A variable whose value is controlled by the experimenter is called a **factor.** The value chosen by the experimenter for a factor is called the factor's **level.** A **treatment** is a combination of levels of the factors in the experiment. The experimenter wants to determine how the different treatments affect the **responses** of the experimental units. In a **completely randomized experimental design,** all the available experimental units (or subjects) are assigned at random to the various treatments. In a **randomized block design,** the available experimental units (or subjects) are first divided into **blocks.** Each block consists of units that are known before the experiment to be similar in some way relevant to the experimental outcome. Within each block the experimental units are randomly assigned to the various treatments. **Matched-pairs designs** compare two treatments using blocks of two experimental units matched (paired) as closely as possible. Within each matched pair, one unit is randomly chosen to receive one treatment and the other unit receives the other treatment. Randomized block designs (and matched-pairs designs) are used to prevent variation between blocks (between pairs) from obscuring the effect of the factor(s) we are interested in.

Section 4.4 Exercises

1. Before final processing, steel (and other metals) is often cast into ingots for convenient storage and transport. Such steel ingots can weigh anywhere from a few pounds to 100 tons. A quality control supervisor conducts an experiment in which steel ingots are produced at three different temperatures, T1, T_2, and T_3, using four different deoxidation agents, A, B, C, and D. Five ingots are produced for each temperature-deoxidation combination. For each ingot the researcher determines the percentage of the ingot that must be discarded because of impurities in the top.

   Identify

   a. the factors of the experiment
   b. the experimental units
   c. the treatment combinations

d. the response variable
e. Is this a randomized block design?

2. A textile researcher wants to compare how two types of color dye affect the durability of fabrics. Since the effect of color dye may depend on cloth type, the researcher plans to apply the two dyes, A and B, to 2 specimens of each of five different types of fabric, I, II, III, IV and V, that is, a total of 10 fabric specimens.
   a. Describe a completely randomized design in which 5 specimens receive dye A and 5 specimens receive dye B.
   b. Describe a matched-pairs design in which 5 specimens receive dye A and 5 specimens receive dye B.
   c. Which design is better?

3. To compare the lifetimes of four brands of flashlight battery, A, B, C, and D, 20 flashlights will be used, 4 each of brands I, II, III, IV, and V. Each flashlight uses 2 batteries. We will test 10 batteries of each brand.
   a. Describe a completely randomized design in which 5 flashlights receive 2 brand A batteries each, 5 flashlights receive 2 brand B batteries each, and so forth.
   b. Describe a randomized block design in which 5 flashlights receive 2 brand A batteries each, 5 flashlights receive 2 brand B batteries each, and so forth.
   c. Explain why the part (b) design is better.

4. There is a strong correlation between a person's interest in something he or she sees and the person's pupil dilation. Some marketing organizations try to evaluate potential consumer interest in new products by measuring the pupil dilation of test subjects. Suppose that the pupil dilations of 25 test subjects were measured as they each were shown two different silverware choices, pattern A and pattern B, with a coin flip in each case deciding whether pattern A or B was shown first.
   a. What is the response variable?
   b. What is the explanatory variable?
   c. Is this a matched-pairs experiment or a completely randomized experiment? Explain your answer.

CHAPTER 4 SUMMARY

This chapter focused on two major ways to obtain real data: sampling from a population and conducting studies to determine cause-and-effect relationships.

randomization
randomized experiment
treatment
experimental unit
controlled experiment
treatment group
control group
randomized controlled experiment
replication

The key to successful data collection is to use an objective **randomization** mechanism. When the investigators use a randomization mechanism to assign the available test subjects or test units to the various treatments in an experiment, the experiment is called a **randomized experiment.** A randomized experiment is used to estimate the effect of a **treatment** when it is applied to an **experimental unit.** In a **controlled experiment** the effect of the treatment is measured by comparing the outcomes produced by units that received the treatment of interest to outcomes produced by units that did not receive the treatment of interest. The units that received the treatment are called the **treatment group,** and the units that did not receive the treatment are called the **control group.** Randomization is used to randomly assign the available experimental units to the treatment and control groups in a **randomized controlled experiment.** Random assignment of units to groups together with sufficient **replication** help ensure that the two groups are roughly similar in all respects except for whether they receive the treatment of interest (this by the law of large numbers).

response variables
factors
explanatory variables
levels
treatment

Outcome variables are also called **response variables.** The variables in a designed experiment that the experimenter controls are called **factors** or **explanatory variables.** The possible values of an explanatory variable are called the **levels** of that factor. A **treatment** is a particular combination of the levels of each factor.

completely randomized design
randomized block design
blocks
Matched-pairs designs

In a **completely randomized design,** the available experimental subjects (or units) are randomly allocated to all treatment combinations. In a **randomized block design,** the available experimental subjects (or units) are first divided into **blocks.** Each block consists of subjects (or units) that are known before the experiment to be similar in some way relevant to the experimental outcome. Within each block the experimental subjects (or units) are randomly assigned to the various treatments. **Matched-pairs designs** compare two treatments using blocks of two experimental units (or subjects) that are matched (paired) as closely as possible.

observational studies
confounding

In **observational studies** a randomization mechanism is not used to assign units to the treatment and control groups. As a result, the two groups might be different in more ways than just whether they received the treatment of interest. These other differences between the two groups are likely to cause differences in the outcomes in addition to differences caused by the treatment effect. This mixing up of the treatment effect with other factors or variables is called **confounding.**

volunteer effect

One of the more common confounding influences that occurs in observational studies of people is the **volunteer effect.** If the treatment group consists of volunteers and the control group does not, and volunteers differ from nonvolunteers in terms of the outcomes of the experiment, then the volunteer effect will be confounded (mixed up) with the effect of the treatment.

placebo

Another possible confounding influence is the change in behavior that can occur when participants know which group they are in. In experiments where this effect can occur, control group members should if at all possible be given a fake treatment, called a **placebo,** indistinguishable from the real treatment, so that they will be blind as to knowing which group they are in.

single-blind study
double-blind study

Another, related confounding influence is the effect on diagnosis that can occur when the measurement of the outcome on each unit is done by someone who knows which group the unit is in. This effect can be mitigated by ensuring that those conducting the experiment are blind as to which units are in which group. A **single-blind study** is one in which the experimental subjects are blind as to which group they are in. A **double-blind study** is one in which both the experimental subjects and the experimenters are blind as to which units are in which group.

survey sampling
probability sampling
Simple random sampling
representative

When a randomization mechanism is used in sampling from a population (**survey sampling**), the sampling is called **probability sampling.** The goal of probability sampling is to estimate the unknown value of some numerical population characteristic (a parameter) by randomly sampling units from the population. The randomization mechanism ensures that all members of the population have a known probability of being selected for the sample. **Simple random sampling** is impartial in the following sense: Each individual in the population has the same chance of being included in the sample and any combination of individuals is as likely to be included as any other combination of the same size. This means that the sample will probably be **representative** of the population, provided the sample is large.

non-probability sampling
selection bias

When a randomization mechanism is not used in sampling from a population, the sampling is called **non-probability sampling.** Non-probability sampling is likely to suffer from **selection bias,** which means that in the selection process some people are more likely to be included in the sample than other people. As a result the sample is not likely to be representative of the population in important ways with respect to the population characteristic that is being estimated.

nonresponse bias

Another error that can bias any survey sampling that requires a response from each sampled subject is **nonresponse bias,** which refers to the fact that the subjects who do not respond are different from those who do respond in important ways. As a result, those who respond do not represent the entire population.

Simple random sampling
stratified sampling
cluster samples

There are many techniques that can be used for probability sampling. **Simple random sampling** is the simplest technique, although **stratified sampling** and **cluster samples** are often used.

quota samples, samples of convenience
self-selected samples

Examples of non-probability samples abound. These include **quota samples, samples of convenience,** and **self-selected samples.** All of these techniques are inferior to probability sampling because they are all liable to unforeseen selection bias.

census

A **census** is an attempt to sample an entire population.

CHAPTER REVIEW EXERCISES

1. Define the following terms and give a short example of each: observational study, quota sampling, simple random sampling, placebo effect, randomized experiment, randomized controlled experiment, treatment group, control group, blocking, randomized block design, matched-pairs design.

2. There has been much emphasis in the past decade on finding a vaccine for HIV, the virus that causes AIDS. Suppose scientists are preparing to test a possible vaccine against the virus on humans. The vaccine is made up of a dead virus, so there is no possibility for a recipient to become infected. From the information in this chapter, design a randomized controlled experiment that would test the effectiveness of the vaccine. In your design, make sure to answer the following questions: How will you choose the sample? How will you reduce the effect of volunteerism in your sample? What problems will you face with your experiment? Are there certain problems unique to HIV transmission and infection that could affect your experiment? How can you control for these problems? (You might need to research characteristics of HIV transmission and infection.)

3. Find a research paper on a medical study and summarize the study and its results. Is this study valid? Which characteristics of the study make the results believable to the scientific community? Which characteristics of the study might make the results unbelievable to the scientific community? Is there anything about the study that you would change?

4. Design a survey of registered U.S. voters for a presidential election. How would you choose the people in the sample? How would you reduce selection bias and nonresponse bias?

5. Research the use of exit polls for predicting the presidential election. (An exit poll is a poll of voters taken on the day of an election after they have voted.) Report on its use in the past 30 years. Have the exit polls ever incorrectly predicted the winner of the presidential election?

6. According to a report by Dr. John Olney in the *Journal of Neuropathology and Experimental Neurology* in November 1998, data from the U.S. government show that brain cancer rates jumped 10% shortly after NutraSweet was approved by the Food and Drug Administration for widespread use in 1983.
   a. Is this a randomized controlled experiment or an observational study?
   b. Are there controls?
   c. Is there a placebo?
   d. Does this study give evidence that NutraSweet causes cancer? Why or why not?

7. Dr. Jose Manuel Silva and his colleagues at Coimbra University Hospital (in Portugal) conducted a study to see whether fish oil is effective in reducing triglyceride levels in the blood. (It is good for the heart to have low levels of triglycerides.) According to an article by Reuters on December 10, 1996:

   In the study, published in the *International Journal of Cardiology* this month, 40 patients, ages 18 to 70 years, were randomly selected to receive either 12 fish oil capsules a day (3.6 grams of omega 3), or 12 similar appearing soya oil capsules per day for two months. Prior to taking the capsules, all study participants avoided eating any fish for 1 month. After 8 weeks, triglycerides increased 19.9% among people taking soya oil and decreased 27.8% in the fish oil group (low levels of triglycerides are considered healthy). Total cholesterol levels were not affected significantly.

a. Is this a randomized controlled experiment or an observational study?
b. Is there a placebo? If so, what?
c. Why did the participants avoid eating fish for one month before the study?
d. Does fish oil appear effective in reducing triglycerides?

8. Eye-movement desensitization and reprocessing, abbreviated EMDR, is a technique developed by a California therapist, Dr. Shapiro, to help victims of severe psychological trauma. The treatment begins with a therapist passing two fingers rapidly back and forth near the patient's face; then the patient recalls the event that caused his or her trauma while following the fingers' movement with his or her eyes. Shapiro conducted an experiment on 22 patients, using her professional judgment to assign half the patients to a conventional treatment, and half to the conventional treatment plus EMDR. She reported dramatic improvement among those receiving EMDR, but little improvement in those receiving the conventional treatment without EMDR. Shapiro herself provided the treatments to the people in the experiment. Improvement was measured by the patient's own impression of the treatment.
a. Is this study a controlled experiment or an observational study?
b. Identify the treatment group and the control group.
c. Were the experimental subjects assigned to treatment or control at random?
d. Is there a placebo?
e. Is this study double-blind?
f. How convincing is this study? Why?

9. In a college statistics class, people who did well on the homework assignments tended to do about 5–10 points better on the exams than the people who did not do so well on the homework.
a. Is this study a randomized controlled experiment or an observational study?
b. Is there an association between performance on the homework and performance on the exams?
c. True or false: These data prove that doing well on the homework causes one to do well on the exams. Explain.

10. Gallup conducted a survey in March 1999 to determine support for legalization of marijuana, finding 29% for legalization, 69% against, and 2% don't know. According to Gallup, "The results are based on telephone interviews with a randomly selected national sample of 1018 adults, 18 years and older."

Section 4.2 describes the Scotty J. survey, which found that 89% support legalization and 7% oppose it. That survey is based on 5152 responses.
a. True or false: The Scotty J. survey is more reliable because it surveyed more people.
b. Which do you think is a better estimate of the percentage of people in the United States who support legalization: 29% or 89%? Why?

11. Suppose a newspaper is interested in including a new comic strip, but does not know which one of three possible ones to select. Discuss some possible factors that might be considered. Decide on one particular research question. Describe a possible data collection method you might use.

12. Teenage drivers pay enormous sums on automobile insurance compared to older drivers. Can you think of any ideas for what can be done to help deserving teenage drivers convince insurance companies that they deserve lower rates? Come up with a research question and describe a data collection method you might use.

13. Suppose you are a member of a club that is having difficulty attracting new members. Consider possible questions that you could discuss with the members to try to find out what is causing the problem. Decide on a question that can be answered with statistics and whose answer could help solve the problem. Describe a data collection method you might use.

14. Teaching methods are often a hot topic of debate. What kinds of questions might we ask about teaching methods that can be answered with either probability sampling or randomized controlled experiments? Decide on one question and describe how you might collect data.

II

Probability Modeling

PROFESSIONAL PROFILE

Thomas Bayes
(c. 1702–1761),
mathematician and
Presbyterian minister

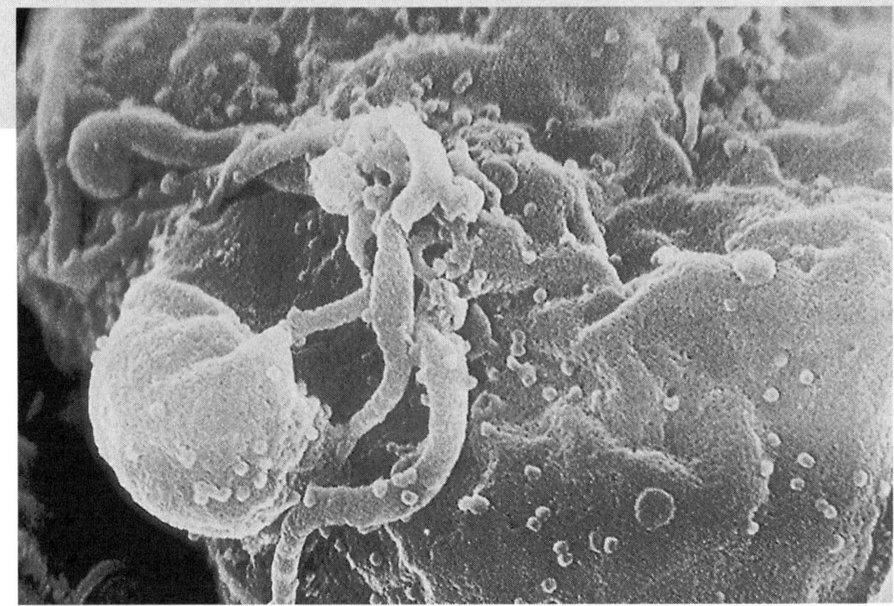

A scanning electron micrograph of HIV-1 budding from a cultured lymphocyte. From the CDC Public Health Image Library. Bayes' theorem can be used to determine the probability of actual HIV infection when a patient has tested positive for the virus.

This portrait is often described as being one of Thomas Bayes. The identification of the person shown has been questioned, however. No more reliable portrait is now available.

A nonconformist in many matters, including religion, Thomas Bayes is best known today for his innovative insights into probability.

By the middle of the eighteenth century, many problems in probability had been solved. One that remained was a related but more difficult problem: Assume a given situation is being studied probabilistically. Assume a particular outcome or result has occurred, one that has several possible causes. What is the probability of each possible cause of the observed result? Bayes offered a solution to this type of inverse probability problem—now called Bayes' Theorem—in his *Essay Towards Solving a Problem in the Doctrine of Chances,* published posthumously in 1764.

Let's look at a modern situation that can be addressed using Bayes' Theorem. Consider a test for AIDS infection. Suppose we know both the probability of testing positive if one has the AIDS virus and the probability of testing positive if one does not have the virus. For the tested patient, the important "inverse" probability question is the probability of actually having an AIDS infection when one has tested positive. Because we also know the incidence of AIDS in the population in question, we can use Bayes' Theorem to provide the probability of actual infection when one has tested positive.

Today, Bayes' Theorem has far-reaching applications in statistics. Bayesian statistics is revolutionizing many applied areas of statistics, including settings where useful probability models are very complex and hard to analyze using traditional statistical approaches.

Our modern understanding of Bayes' theorem owes a great deal to Pierre-Simon Laplace (1749–1827), who used this formula to study problems in celestial mechanics, medical statistics, reliability, and jurisprudence. Since Laplace's time, others (including Ronald A. Fisher) have built on the foundations laid by Thomas Bayes to develop areas of modern statistics.

There are even those who argue that the scientific method is an application of Bayesian theorem. This view supports the idea that scientific theories are proven or rejected based on the updating of probabilities as new observations or experiments are conducted.

5

Probabilities and Simulation

Objectives

After studying this chapter, you will understand the following:

- The experimental probability $\hat{P}(A)$ as an accurate estimator of the theoretical probability $P(A)$
- The five-step simulation method for computing experimental probabilities
- Estimating probabilities of events via five-step simulation
- How to simulate random phenomena with coins, dice, random digits, and especially with box models by using highspeed computing
- Rules for doing elementary theoretical probability computations involving "and," "or," "not"
- The theoretical probability model of equally likely outcomes
- Mutually exclusive events
- Independent events
- Introduction to conditional probability
- Probability trees
- Sampling from a population with and without replacement

KEY PROBLEM
Tired MDs

Program directors at teaching hospitals insist that the mental alertness of young doctors does not suffer when the doctors are on call and work for over 36 hours.

To test this claim, a group of 25 doctors was given a battery of mental tests before they started their 36-hour call and a similar battery of tests of the same difficulty at the end of their call period. Twenty of the doctors did better on the tests they took before their call than on the tests they took after their call; only 5 did better on the second set of tests. Does this prove that the program directors are wrong?

If sleep deprivation does not affect a doctor's mental abilities, as the program directors insist, then each doctor would be just as likely to do better on the second set of tests as on the first set. If the claim of no loss in mental abilities were true, how likely would it be that only 5 or fewer out of the 25 doctors would do better on the second set of tests? If it is highly unlikely, then we will find it difficult to accept the directors' claim.

5.1 INTRODUCTION

In Chapter 4 we discussed how statisticians use *randomization* to reduce bias is sample surveys and in comparative experiments: To avoid selection bias in a sample survey, the sample should be selected *at random* by an impartial, objective, chance mechanism. To minimize the bias caused by confounding in a comparative experiment, the available test subjects or units should be assigned *at random* to the various treatment groups by an impartial, objective, chance mechanism.

Once data have been obtained in a statistical study, the data have to be interpreted. The tools statisticians use are probability theory and probability modeling. Probability is the theory of randomness and uncertainty.

We live in a world of uncertainty. For example, no one knows how many hurricanes will threaten the Atlantic and Gulf coasts next hurricane season. But uncertainty is not the same as total unpredictability. A news item a couple of years ago predicted an increase in the annual number of hurricanes over the next decade. Supposedly, the coming hurricane seasons will resemble those between 1920 and 1970 more than those between 1970 and 2000.

Meteorologists (weather forecasters) make such forecasts using probability models and data on current weather patterns. Each year the National Oceanographic and Atmospheric Administration (NOAA) spends 100 million dollars trying to forecast the weather as far in advance as possible. NOAA has models for predicting next year's weather and models for predicting tomorrow's weather. When a hurricane approaches, NOAA uses its probability models and current weather data to assess where the hurricane might hit land so that the coastal residents can be warned of the risk. Unfortunately, the weather is fundamentally chaotic: No amount of data collection and computing power will ever give 100% accurate hurricane forecasts. Similarly, weather forecasters can tell the atmospheric conditions that make tornadoes *likely* in a certain region during a certain time period, but they will never be able to tell for sure when and where a tornado will strike.

chance phenomena
random phenomena
statistical inference
statistical inferences

Phenomena such as the weather, the stock market, and even urban automobile traffic congestion are affected by a large number of influences that are not all known and that interact in a complicated manner. The behavior of such phenomena cannot be predicted in detail. They are called **chance phenomena** or **random phenomena**. Statisticians look for patterns in the apparent chaos of raw data obtained from measuring such phenomena, and they make **statistical inferences** (draw conclusions) about such patterns. To make statistical inferences, the statistician must choose a probability model for how the data were generated. For example, data from repeatedly throwing an ordinary die is usually modeled as having been produced by a "fair" die, that is, a die for which all six faces are assumed equally probable.

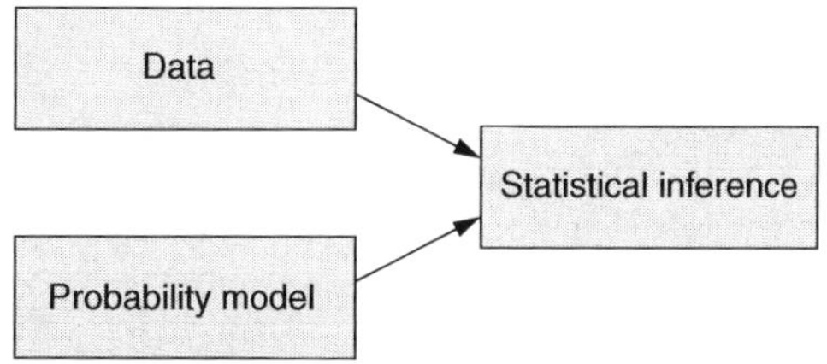

Figure 5.1 Model-based statistical inference.

five-step simulation method

A major goal of Chapter 5 is to introduce the **five-step simulation method**, which lets us build probability models and then observe data produced by the models we build. This experience is an important prelude to understanding and doing effective, probability-model-based statistical inference.

In summary, Chapter 5 first explains how one uses experimental probabilities obtained from data to estimate theoretical probabilities. In Section 5.3 the rules used to compute theoretical probabilities of events of interest are presented. Then the chapter presents the five-step simulation method for obtaining experimental probabilities that reliably estimate theoretical probabilities of interest.

5.2 EXPERIMENTAL PROBABILITY

Every day, people make decisions and act in the face of uncertainty. In the 1990s people invested in risky startup stocks hoping to make a lot of money. More recently, many investors have switched to mutual funds or U.S. Treasury bonds to reduce their risk of losing money. We get flu shots to reduce our risk of getting the flu. We send out numerous résumés with the hope of increasing our chances of getting a good job. We buy life insurance to protect our dependents in case we should die. To be able to quantify uncertainty, people talk about the "chance" or "probability" of an event taking place. The terms are almost but not quite interchangeable. *Probability* is almost always expressed as a number. *Chance* can be used in a qualitative sense as well ("a slim chance"), but it often means the same numeric concept as *probability* ("a 30% chance of rain"), and that is how it will be used in this book. Thus, *chance* and *probability* will be used interchangeably.

Consider questions such as the following:

- If the federal government introduces random drug testing for all its employees, what is the chance that someone who tests positive actually is drug free?
- If a man and a woman expecting a child are both carriers of Tay-Sachs disease, what is the probability that the child will be born with the deadly disease?
- If I show up at the airport with only carry-on luggage 30 minutes before departure, what is the chance that, because of security or overbooking, I will not be able to get a seat on the airplane?

- If a quality control inspector randomly chooses 10 of 500 DVD players for inspection, what is the probability that none of the inspected ones will be defective if there are 5 defective DVD players among the 500?
- If I have a kidney transplant, what are my chances of surviving 5 years?

Probability is important in a wide variety of fields, including medicine, manufacturing, public policy, and in daily life, as illustrated in the above questions. Probability assesses the likelihood of somethimg happening: The smaller the probability of something, the less likely it is to happen.

In this book we adopt the experimental, or relative frequency, interpretation of probability, which leads naturally to making statistical inferences. For example, data show that out of the hundreds of tornadoes that touch down every year, roughly 6 out of 10 strike between 2 P.M. and 8 P.M. We express this fact by saying that if a tornado strikes on a given day, the probability that it will happen between 2 P.M. and 8 P.M. is around 6/10, or 60%.

random experiment

Observed data often result from what we think of as a **random experiment**. A random experiment, such as observing whether or not a person with a kidney transplant survives 5 years, can be repeated over and over. For example, if 72 out of 100 kidney transplant recipients survive 5 years, then we would estimate the probability that a kidney transplant patient will survive 5 years to be around $72/100 = 0.72$. Of course, if another 100 kidney recipients are sampled, the observed survival rate would likely differ somewhat: We might get 76/100, say. Such estimates (72 / 100 and 76 / 100) vary from sample to sample.

Some of the simplest random experiments are found in games of chance. We shall often use examples from gambling to teach important probability concepts because in games of chance the nature of "chance variation" can be clearly understood. Gambling is an ancient human practice. Archaeologists have found artifacts used in games of chance in Egypt dating back to 3500 B.C. Egyptian tomb paintings show gods and people tossing *astragali*, small, irregular, six-sided sheep ankle bones. Cubical dice of wood and pottery have been found in Iraq and India, dating back to 3000 B.C. The Aryans, who invaded India around 1500 B.C., loved to gamble. The *Rigveda,* a collection of Aryan hymns, includes a poem, "The Gamester's Lament," in which the sun god Savitr scolds the gambler, telling him to stop playing with dice and go home to his wife and kids!

For thousands of years astragali, dice, and "throwing-sticks" of wood or ivory were used both in gambling and in divining the future. They were also used in the casting of lots. The Bible has numerous references to the casting of lots as a way to settle disputes or divide property. During the Middle Ages a popular game of dice in Western Europe was called *hazard*. The game was similar to modern-day craps, and its name was derived from the Arabic *al-zahr*, which means "the die." The game is believed to have been brought to Europe by soldiers returning from the Crusades.

It was in the seventeenth century that gamblers (and mathematicians and philosophers) began to think about probability in a more systematic way—with the obvious hope of improving their winnings! The serious study of probability apparently began in the seventeenth century when Galileo in Italy and Pascal and Fermat in France were asked by gamblers to help them find probabilities of certain events in various games of chance involving dice. In the process of helping the gamblers, Galileo, Pascal, and Fermat laid the groundwork for the formal rules of probability theory.

experimental probability of an outcome

relative frequency of the outcome

To model a random experiment, we first list all the possible outcomes, for example {1, 2, 3, 4, 5, 6} for the experiment of throwing a six-sided die once. It would take up more space, but it would be conceptually clearer to write this collection of all the possible outcomes as {1 or 2 or 3 or 4 or 5 or 6}. We will use commas throughout. If an experiment is repeated a number of times, then the **experimental probability of an outcome** of the experiment is the fraction of times (in the media, it is usually given as the percentage of the time) the outcome occurs. This fraction is also called the **relative frequency of the outcome**, namely the number of times the outcome occurs relative to the total number of repetitions of the experiment.

theoretical probability of an outcome

The **theoretical probability of an outcome** is the number that the experimental probability of the outcome gets increasingly closer to as the number of repetitions of the random experiment gets larger and larger. The larger the number of repetitions, the closer the experimental probability is likely to be to the theoretical probability. When you do simulations you will observe this. Thus, for practical purposes, when the number of repetitions is very large we can treat the experimental probability we obtain as if it were the theoretical probability. In doing so, we are doing statistical inference. Usually, we refer to a theoretical probability simply as a **probability**. We illustrate this notion of a probability being estimated by its experimental probability with some examples.

probability

Example 5.1

What is the probability that a pregnant woman will have a boy?

Solution: It is impossible to predict the sex of a child at the time of conception. However, if we look at all the children born each year in the United States, we see that the relative frequency (that is, the experimental probability) of male births is almost the same from year to year. This striking fact is illustrated by Table 5.1. (The relative frequency of male births is of course the number of male births divided by the total number of births.)

Table 5.1 Male Births in the United States, 1940–2005

Year	Live births in thousands	Male births in thousands	Relative frequency (experimental probability)
1940	2559	1313	0.513
1950	3632	1863	0.513
1955	4097	2099	0.512
1960	4258	2180	0.512
1965	3760	1927	0.513
1970	3731	1915	0.513
1975	3144	1613	0.513
1980	3612	1853	0.513
1985	3761	1928	0.513
1990	4158	2129	0.512
1995	3900	1996	0.512
2000	4059	2077	0.512
2005	4138	2119	0.512

Source: Statistical Abstract of the United States, 2009.

Based on these experimental probabilities, we estimate that the theoretical probability that a newborn baby will be a boy is between 0.512 and 0.513 or, for those who prefer their probabilities in percents, between 51.2% and 51.3%. Interestingly, we have discovered that the theoretical probability is not 0.5, as most people think.

Note that this does not tell us whether a particular woman will have a boy or a girl. Probabilities do not tell us what will happen in one specific case or even in the short run. Rather, they refer to what will happen in the long run, here when we look at thousands of births. For example, we can be sure that very close to 51.3% of all the babies born next year in New York City will be boys. This is an example of the **law of statistical regularity**, also called the **relative frequency interpretation of probability**. This law tells us that if a random experiment (here the birth of a child) is repeated over and over, then in the long run the proportion of times that the experiment has a particular outcome (e.g., a boy) will stabilize around a value we call the probability of the outcome (here 0.513).

law of statistical regularity

relative frequency interpretation of probability

Example 5.2

When we toss a coin, it can land either heads or tails. The outcome is unpredictable. What is the probability that it lands heads?

Solution: Two hundred fifty years ago, the French naturalist Count Buffon tossed a coin 4040 times, getting 2048 heads. His experimental probability of heads was $2048/4040 = 0.5069$. A hundred years ago, the English statistician Karl Pearson tossed a coin 24,000 times, getting 12,012 heads. His experimental probability was $12{,}012/24{,}000 = 0.5005$.

During World War II, the English mathematician John Kerrich tossed a coin 10,000 times while being held a prisoner of war. He got 5067 heads, corresponding to an experimental probability of 0.5067. Let us take a closer look at John Kerrich's data. To see what can happen as the number of coin tosses increases, refer to Table 5.2. The table cumulatively tracks Kerrich's results of tossing a coin 10,000 times. The first row gives the number of heads that appeared in the first 10 tosses, the second row gives the number appearing in the first 20 tosses, and so on. Notice the sequence of relative frequencies (each an experimental probability of heads). At first they show considerable variability as they dance up and down. However, after hundreds of coin tosses, the experimental probabilities, although continuing to vary a little, appear to approach a number near 0.500, possibly $1/2$ itself. Indeed, if we had the patience to toss an absolutely fair coin 1,000,000 times, the relative frequency of heads would be very close to $1/2$, say 0.5004 when rounded. This statistical regularity is a characteristic of random experiments: As the number of independent repetitions is increased (see Table 5.2), the continuously updated experimental probabilities of heads approach a fixed number, the theoretical probability of heads.

The approach of the experimental probability of heads to some number near 0.500 is shown graphically in Figure 5.2. Note that the horizontal scale is compressed for easy viewing by using a logarithmic scale.

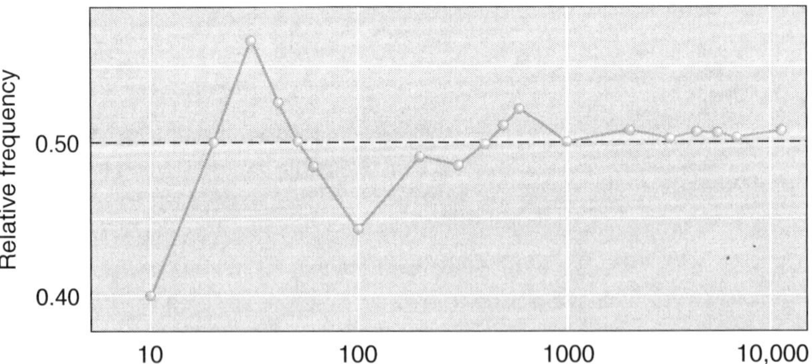

Figure 5.2 The relative frequency of heads as a function of the number of coin tosses in Kerrich's data.

Table 5.2 10,000 Tosses of a Coin

Number of tosses	Number of heads	Relative frequency (experimental probability)
10	4	0.400
20	10	0.500
30	17	0.567
40	21	0.525
50	25	0.500
100	44	0.440
200	98	0.490
300	146	0.487
400	199	0.498
500	255	0.510
600	312	0.520
700	368	0.526
800	413	0.516
900	458	0.509
1000	502	0.502
2000	1013	0.507
3000	1510	0.503
4000	2029	0.507
5000	2533	0.507
6000	3009	0.502
7000	3516	0.502
8000	4034	0.504
9000	4538	0.504
10,000	5067	0.507

Source: J.E. Kerrich, *An Experimental Introduction to the Theory of Probability,* J. Jorgenson, Copenhagen, 1946, p. 14.

❒ ❒

We are not surprised to find that when we toss an actual coin, the probability of heads seems to be close to $1/2$: Our intuition tells us that heads and tails should be equally likely outcomes of the coin toss. Unfortunately, we cannot always rely on our intuition. Only by looking at data (such as Kerrich's coin tosses and birth statistics of the United States) are we able to tell that heads and tails are indeed very close to equally likely outcomes of a coin toss, whereas boys are slightly more likely than girls among newborn babies, in spite of our intuition that they too should be equally likely outcomes. Probabilities often need to be determined or confirmed experimentally. This basic experimental principle is the foundation of statistical inference, the major focus of this textbook.

The next example is a game of chance that is so complicated that our uninformed intuition does not even tell us whether the chances (the theoretical probability) of winning are *favorable* (better than $1/2$) or *unfavorable* (less than $1/2$). Observing the experimental probability of winning for a large number of repetitions can, however, approximate and hence give us a rough idea of the unknown chance of winning!

Example 5.3

In the game of craps, the player repeatedly throws a pair of dice and adds up the scores (numbers of spots on the top faces) of the two dice until he or she wins or loses. Figure 5.3 shows the 36 possible outcomes for a pair of dice.

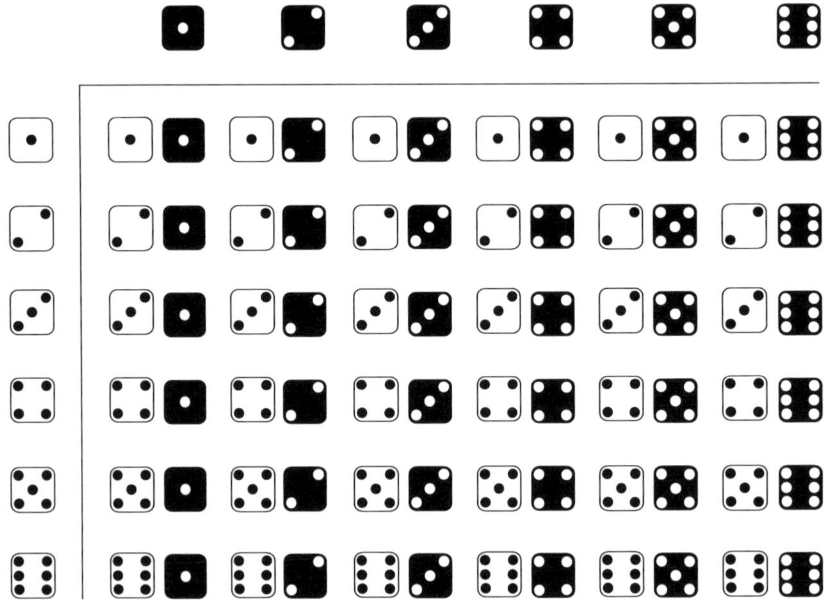

Figure 5.3 Throwing a pair of dice: chart of outcomes.

Here are the rules of the game: If the player's first sum (score) is 7 or 11, the player wins immediately. If the sum is 2, 3, or 12, the player loses immediately. If the sum is 4, 5, 6, 8, 9, or 10, then that number becomes the player's *point*. Once the point is established, the player has to keep throwing the pair of dice until he or she either wins by repeating the point or loses by throwing a 7. Following in the footsetps of the seventeenth-century gamblers, we can estimate our chance of winning in craps by playing a lot of games. Consider, for example, the 10 plays in Table 5.3.

Table 5.3 Estimating the Probability of a Player's Winning at Craps

Repetition	Scores	Outcome
1	11	Win
2	6, 4, 3, 7	Loss
3	4, 8, 5, 10, 12, 4	Win
4	10, 6, 4, 7	Loss
5	10, 4, 5, 9, 7	Loss
6	9, 3, 7	Loss
7	6, 7	Loss
8	7	Win
9	8, 5, 5, 5, 3, 8	Win
10	6, 8, 9, 9, 7	Loss

Based on the 10 repetitions listed in Table 5.3, we crudely estimate that the probability P of winning in craps is around 4/10 (the experimental probability). To improve the accuracy of our estimate, we would need many more repetitions.

Imagine that we play 100 times, winning 54 times. We estimate

$$P \approx \frac{54}{100} = 0.54 \approx 0.5.$$

With only 100 plays we can trust only the first decimal of our estimate.

Say we play 10,000 times, winning 4,852 times. We estimate

$$P \approx \frac{4852}{10{,}000} = 0.4852 \approx 0.49.$$

With only 10,000 plays, we can trust only the first two decimals of our estimate.

Say we play 1 million times, winning 493,483 times. We estimate

$$P \approx \frac{493{,}483}{1{,}000{,}000} = 0.493483 \approx 0.493.$$

With 1 million plays, we can trust only the first three decimals of our estimate.

Using the addition rule and the multiplication rule, we can show that if we use fair dice, then

$$P = 0.49293.$$

Our chance of winning in a real game of craps obviously depends on whether or not we are using fair dice, that is, dice for which, in the long run, all six faces will turn up with the same relative frequency of 1 / 6.

Section 5.2 Summary

The outcome of a **random experiment** (such as counting the number of Atlantic hurricanes next summer or observing whether or not a person with a kidney transplant survives 5 years), is unpredictable. However, in a very long series of **independent repetitions** of the same random experiment, a pattern emerges that can be described in terms of the stabilization of observed relative frequencies or experimental probabilities of the various possible outcomes.

For a particular outcome of a random experiment, the **theoretical probability of the outcome** is

$$P(\text{outcome}) = \textbf{fraction of times the outcome would occur in an infinitely long series of independent repetitions.}$$

To estimate this probability, one method is to consider a large number of repetitions of the random experiment and compute the **experimental probability** of the outcome, using

$$\hat{P}(\text{outcome}) = \frac{\textbf{number of successful repetitions}}{\textbf{total number of repetitions}} = \textbf{relative frequency of the outcome.}$$

Here, a "successful repetition" means a repetition of the random experiment in which the outcome of interest occurs. The more repetitions we observe, the closer the experimental probability will likely be to the unknown theoretical probability we seek, this fundamental fact called the law of statistical regularity. The symbol $\hat{P}$ is read "P hat"; the "hat" merely reminds us that this is not the theoretical probability P but an estimate of P computed from experimental data.

Section 5.2 Exercises

1. Look up the *Statistical Abstract of the United States* (in a library, or online at http://www.census.gov/ statab/www/) to determine the relative frequency of male births for each of the last 10 years.

2. Roll a die 60 times and observe how often you obtain each of the six values 1, 2, 3, 4, 5, and 6. Use your data to compute the following experimental probabilities:
   a. $\hat{P}$(rolling a 1); $\hat{P}$(rolling a 2); $\hat{P}$(rolling a 3); $\hat{P}$(rolling a 4); $\hat{P}$(rolling a 5); $\hat{P}$(rolling a 6)
   b. $\hat{P}$(number on die is even) *Hint for part* (b): The number on the die being even is an "event" that occurs when an outcome of 2, 4, or 6 is observed. We use the $\hat{P}$(outcome) formula with "event" replacing "outcome."
   c. Can you view the answers in parts (a) and (b) as accurate estimates of the corresponding theoretical probabilities?

3. Shuffle a standard deck of 52 cards and draw 1 card. Repeat this experiment at least 50 times. Estimate the probability of drawing an ace. Is your answer consistent with what you think is the theoretical probability of drawing an ace?

4. Toss three coins 50 times (50 repetitions) and keep a tally of the number of heads on each toss of the three. Then use your data to compute the following experimental probabilities:
   a. $\hat{P}$(all three are heads)
   b. $\hat{P}$(exactly two are heads)
   c. $\hat{P}$(exactly one is heads)
   d. $\hat{P}$(none of the three is heads)
   e. The probability from parts (a) and (d) should be roughly equal, and the probabilities from parts (b) and (c) should be roughly equal. Is this true for your 50 repetitions?
   Hint: Getting exactly one head means you got exactly two tails.

5. A building supplies manager wants to estimate the probability that a certain type of carpet tack will fall point-up when dropped on a smooth wood floor. A worker drops one of the tacks 100 times and finds that it falls point-up (its head flat on the floor) 35 times. What is the experimental probability that the tack will fall point-up? Do you think that this estimate will be fairly accurate?

6. Over his career a basketball player has made 1210 free throws and missed 214. What is his experimental probability of making a free throw? Do you think this experimental probability is a good estimate?

7. Toss a pair of dice 50 times, and each time record the sum of the dots on the two dice. Then compute the following experimental probabilities.
   a. $\hat{P}$(the sum of the two dice equals 2)
   b. $\hat{P}$(the sum of the two dice equals 7)
   c. $\hat{P}$(the sum of the two dice equals 10)
   d. $\hat{P}$(the sum of the two dice equals 6)

8. Define or explain the concepts of relative frequency, experimental probability, statistical regularity, and (theoretical) probability. Relate the concepts to each other.

9. Play craps 100 times and estimate the theoretical probability of winning (time permitting).

10. Toss a coin. If it lands heads up, then stop. Otherwise keep tossing the coin until you get heads. Count the number of tosses needed to get a head. Repeat this random experiment 50 times and keep a record of the number of tosses in each of the 50 repetitions. Use your data to compute the following experimental probabilities:
    a. $\hat{P}$(we need only 1 toss to get a head)
    b. $\hat{P}$(we need 2 tosses)
    c. $\hat{P}$(we need 3 tosses)
    d. $\hat{P}$(we need 4 tosses)
    e. $\hat{P}$(we need 5 tosses)
    f. $\hat{P}$(we need 6 tosses)
    g. Is the theoretical probability P(1000 tosses needed) greater than 0? Explain. Is the probability close to 0?

11. Hold a penny on its edge on a tabletop with your left index finger and flick it with your right index fingernail so that the penny spins upright for several seconds before it falls. Note whether the penny falls heads up or tails up. Repeat this experiment 100 times and estimate the probability that the penny falls heads up. Do heads and tails seem equally likely (that is, does each have theoretical probability 1/2) in this experiment? *Hint:* If you get many more or many fewer than 50 heads, the answer is "no."

12. An article in a medical journal reported on 492 patients who underwent a specific heart valve procedure. Of these patients, 33 died (a fatal complication) during or soon after the procedure, and 84 suffered nonfatal complications. Based on these data, estimate the probability that a patient who undergoes this procedure
    a. will suffer a complication
    b. will suffer a fatal complication

13. In April 1992, *USA Today* presented the data on over 17,000 tornadoes touching down over a 20-year period (see following table). Note that these experimental probabilities should be very good estimates of the theoretical probabilities, because 17,000 is a huge number of repetitions. Thus, you can assume that these are the true probabilities.
    a. What is the most likely 2-hour period for a tornado to touch down?
    b. If a tornado touches down on a given day, what is the chance that it happens between 2 P.M. and 8 P.M.? What is the chance that it happens between noon and 10 P.M.? *Hint:* Add the appropriate experimental probabilities.

Time of day tornadoes touch down	Experimental probability
Midnight to 2 A.M.	0.029
2 to 4 A.M.	0.021
4 to 6 A.M.	0.022
6 to 8 A.M.	0.022
8 to 10 A.M.	0.031
10 A.M. to noon	0.045
Noon to 2 P.M.	0.097
2 to 4 P.M.	0.179
4 to 6 P.M.	0.224
6 to 8 P.M.	0.191
8 to 10 P.M.	0.098
10 P.M. to midnight	0.041

5.3 PROBABILITY

Section 5.2 makes clear that one way to accurately estimate theoretical probabilities of interest is to compute corresponding experimental probabilities using data obtained from a large number of independent repetitions of the chance experiment in question. This approach will be expanded into a powerful method for accurately estimating unknown probabilities of interest in Section 5.4.

In this section we will develop the basic rules for manipulating theoretical probabilities. Understanding these rules will help us better understand what the probability of an event means. Second, these rules will sometimes allow us to bypass doing a large number of replications to obtain an experimental probability and directly obtain the desired theoretical probability using simple formulas.

probability model
outcomes
sample space
event
event's occurring

We describe, or *model*, a random phenomenon (called a *random experiment*) by listing all the possible outcomes and their assigned probabilities. Such a description is called a **probability model**. The **outcomes** are just all the different possible results of the experiment. The probabilities are assigned based on experimental probabilies or based on certain assumptions made about the nature of the phenomenon being modeled. For example, certain outcomes may be assumed to be equally likely. We call the set, or collection, of all possible outcomes of the random experiment the **outcome space** or the **sample space**, often denoted by S. An **event** is a set, or collection, of some of the outcomes. For example, if we toss a six-sided die, then the outcome space is $S = \{1, 2, 3, 4, 5, 6\}$. The event that the score on the die is even is the set of outcomes $\{2, 4, 6\}$. We usually use capital letters, such as A, B, C, and D, to denote events. We want to be able to compute the probability of an event's occurring. By an **event's occurring**, we mean that the random experiment results in one of the event's listed outcomes. For any event, A, say, its probability is the long-run relative frequency of the event in repeated trials. The probability of A is written $P(A)$. To illustrate, consider the random experiment of counting the number of boys in a three-child family. Then the collection of all possible outcomes is

$$S = \{0, 1, 2, 3\} = \{0 \text{ or } 1 \text{ or } 2 \text{ or } 3\}.$$

This is the **sample space**. Suppose the parents want at least one boy and at least one girl. So out of the three children they would like either one or two boys. Then, this hoped for event, denoted by A, say, is given by

$$A = \{1, 2\}.$$

We say that A has **occurred**, provided the three-child family has either one or two boys. As we shall see in Example 5.9, $P(A)$ is close to $3/4$ in this case. Of the millions of three-child families, roughly 3/4 have either one or two boys.

We have various natural ways of combining events to form new more complex events of interest.

Example 5.4

Let M be the event it rains in Champaign County next Monday and T be the event it rains next Tuesday. Often the sample space and events of interest are represented graphically as shown next.

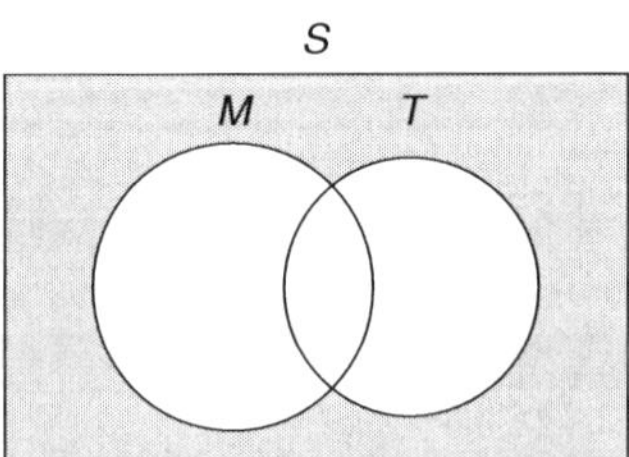

Here the rectangular region represents the sample space S of all possible outcomes and the circular region M represents the outcomes of S leading to M occurring, etc. Since, $P(S) = 1$, the area of S is always 1 in these diagrams. It is intuitively useful to view the area of M to be $P(M)$.

We can naturally form new events using M and T as building blocks. For example, a gardener wanting rain may be interested in the event A that rain occurs on at least one of the two days. In particular, he may hope that $P(A)$ is close to 1, meaning the event is likely to occur. A is the shaded area shown next.

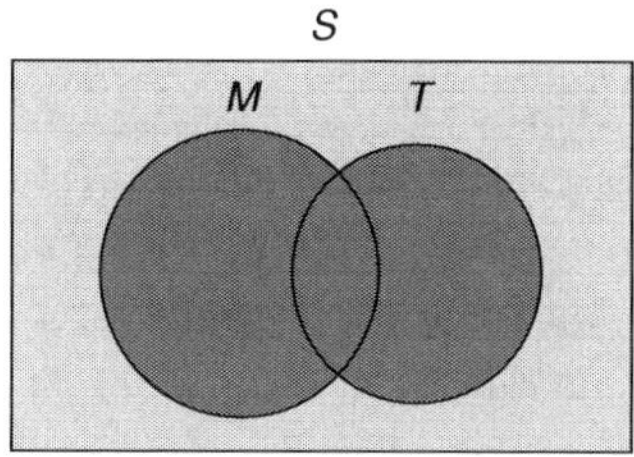

Here $A = M$ or T, meaning the event that at least one of M and T occurs, and that possibly both occur.

A person with guests staying both Monday and Tuesday, who hopes to have an outdoor picnic, may be worried about the occurrence of the event B that it rains on both days. She hopes that $P(B)$ is close to 0, meaning the event is unlikely to occur. B is the shaded area shown next.

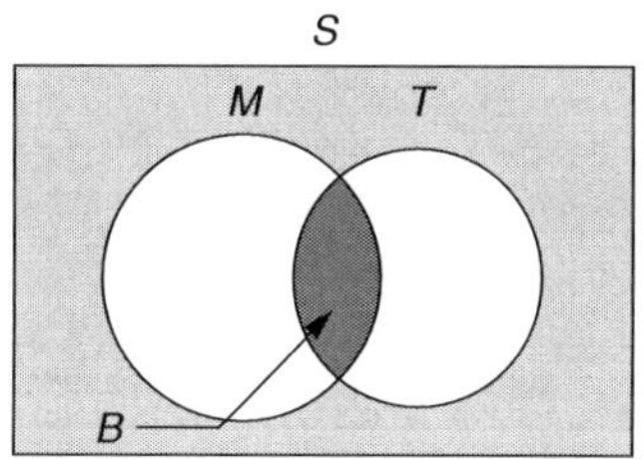

Here $B = M$ and T, meaning that M and T *both* occur.

A person in the local park district tennis tournament has a match scheduled Monday night. She is interested in the event C (shaded below), that it does not rain on Monday. She wants $P(C) \approx 1$. C is the shaded area shown next.

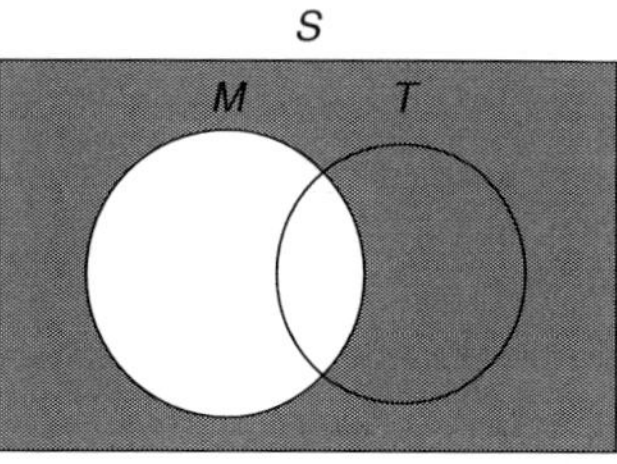

Here $C =$ not M, meaning that M does not occur.

❒ ❒

The above example illustrates verbally and visually how new events can be naturally built up from other events using "and," "or," and "not." These three ways of forming new events are both basic and important.

Next we study a type of random experiment where it is often not necessary to collect data and obtain experimental probabilities in order to estimate theoretical probabilities. These are random experiments where it is reasonable to assume that all of the possible outcomes are equally likely.

Equally Likely Outcomes

equally likely outcomes

If we expect all the outcomes of a random experiment to occur with the same relative frequency in the long run, then we say that the **outcomes** are **equally likely**. Many random phenomena are modeled as having a finite number of possible outcomes, all equally likely. The flip of a coin, the toss of a casino die, and the spin of a roulette wheel come to mind. As we have seen, among newborn babies, boys and girls are almost equally likely and for many purposes, can be modeled as equally likely. If all the listed outcomes are equally likely, then we have a simple formula for computing the probability that a particular event A will take place:

P(event) for equally likely outcomes

Fundamental rule for $P(A)$ if all outcomes are equally likely

$$P(A) = \frac{\textbf{number of possible outcomes in } A}{\textbf{total number of possible outcomes}}$$

Note: In this fundamental rule, we are computing a theoretical probability and not an experimental probability. Hence, there is no observed data; there is only a list of possible outcomes, a subset of which is the event A. Note that we may be able to compute the theoretical probability $P(A)$ without needing to collect data.

Example 5.5

The roulette wheel in an American casino has 38 pockets, numbered 1 through 36 plus 0 and 00. The croupier sets the wheel in motion and tosses a ball inside in the direction opposite the motion of the wheel. Bets are made on the number of the pocket where the ball will come to rest.

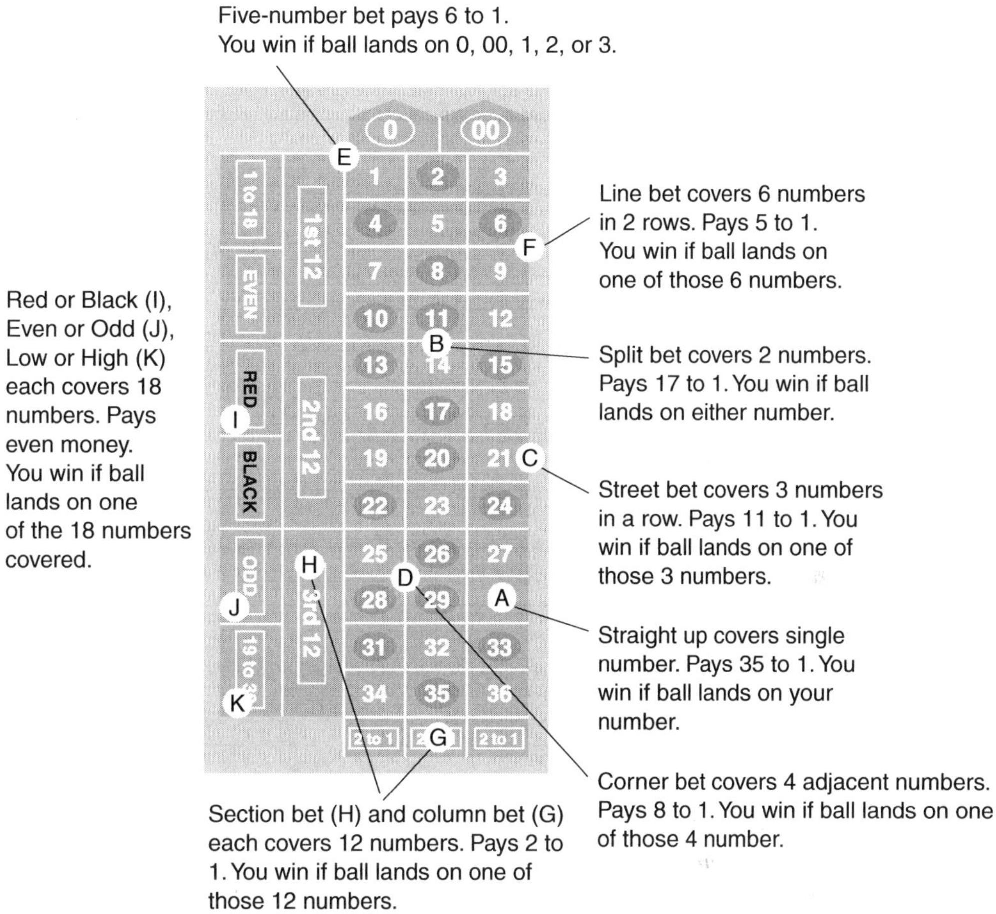
Five-number bet pays 6 to 1.
You win if ball lands on 0, 00, 1, 2, or 3.
Line bet covers 6 numbers in 2 rows. Pays 5 to 1. You win if ball lands on one of those 6 numbers.
Red or Black (I), Even or Odd (J), Low or High (K) each covers 18 numbers. Pays even money. You win if ball lands on one of the 18 numbers covered.
Split bet covers 2 numbers. Pays 17 to 1. You win if ball lands on either number.
Street bet covers 3 numbers in a row. Pays 11 to 1. You win if ball lands on one of those 3 numbers.
Straight up covers single number. Pays 35 to 1. You win if ball lands on your number.
Corner bet covers 4 adjacent numbers. Pays 8 to 1. You win if ball lands on one of those 4 number.
Section bet (H) and column bet (G) each covers 12 numbers. Pays 2 to 1. You win if ball lands on one of those 12 numbers.
0
00
1 to 18
EVEN
RED
BLACK
ODD
19 to 36
1st 12
2nd 12
3rd 12
2 to 1
2 to 1
A
B
C
D
E
F
G
H
I
J
K

Figure 5.4 The various bets in American roulette.

The sample space of this chance experiment is

$$S = \{0, 00, 1, 2, 3, \ldots, 36\}.$$

Because the casino makes every effort to ensure that the wheel is perfectly balanced, we assume that all 38 outcomes are equally likely, and we assign the probability $1/38$ to each of the 38 outcomes in S. One of the many ways you can place a bet is to bet on the block of numbers from 31 to 36. If any one of the six outcomes occurs, then you win the bet. Your chance of winning this bet is

$$P(\{31, 32, \ldots, 36\}) = 6/38$$

according to the fundamental rule for P(event) when all outcomes are equally likely. Remember to think of each comma as an "or." Thus, the event is really {31 or 32 or 33 or 34 or 35 or 36}.

Example 5.6

Consider the random experiment of having two children. If we focus on the children's sex, then the sample space S for this experiment has four outcomes,

$$S = \{GG, GB, BG, BB\}.$$

Here GB indicates the outcome that the first child born is a girl and the second is a boy. The event A that the two children are of the same sex is the collection of two outcomes:

$$A = \{GG, BB\}.$$

The event A will occur if the family has either two girls or two boys.

We usually assign the probability $1/4$ to each of the four outcomes in S because we presume that the four outcomes are equally likely, even though in reality they are only approximately so: As we saw in Example 5.1, boys are slightly more probable than girls. To get better estimates of the four outcome probabilities, we would need data from a very large number of two-child families. In fact, for most practical purposes this approximate model of equally likely outcomes will do, and it is the one people almost always use. If we use this model, then $P(A)$ = (number of outcomes in A) / (number of outcomes in S) $= 2/4$ by the formula for computing $P(A)$ for equally likely outcomes.

Example 5.7

If we toss a die, then the sample space is

$$S = \{1, 2, 3, 4, 5, 6\}.$$

If the die is carefully made to meet exacting casino standards, then we assign the probability $1/6$ to each of the six outcomes because we assume that the outcomes are equally likely. The list of all possible outcomes and their corresponding probabilities is called the **probability distribution** of the random experiment. Its probabilities add to 1, as must always be the case.

probability distribution

Outcome	1	2	3	4	5	6
Probability	1/6	1/6	1/6	1/6	1/6	1/6

Let A be the event the die is even. Thus $A = \{2, 4, 6\}$. Then $P(A) = 3/6$ by the equally likely outcome rule.

Example 5.8

On a Las Vegas roulette wheel, the numbers 0 and 00 are colored green, 18 of the other numbers are colored red, and 18 are black. If you bet on "red," you win if the ball comes to rest in a red slot. Assuming that the roulette wheel is perfectly balanced and all outcomes are equally likely, your chance of winning a bet on red is $18/38$, since there are 18 outcomes for which the color is red out of the 38 possible outcomes. That is, $P(\text{red}) = 18/38$ by the rule for equally likely outcomes.

Example 5.9

A family has three children. What can we say about the children's sexes? We can use a

tree diagram of a multistage experiment

so-called **tree diagram of a multistage experiment** to represent the eight possible outcomes (Figure 5.5). Here each stage represents a different child.

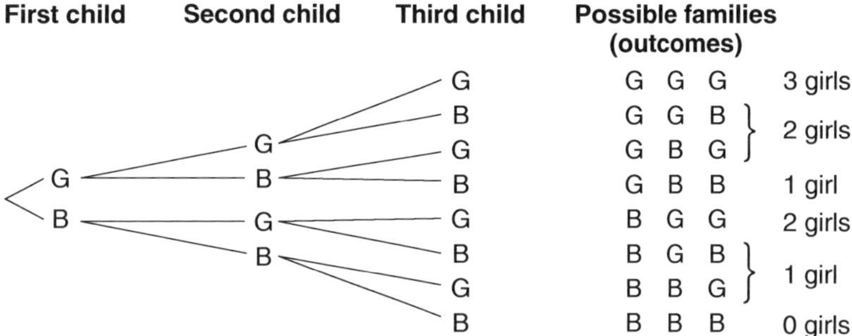

Figure 5.5 Tree diagram for possible three-child families.

Let us assume, as an approximation to reality, that the eight outcomes are equally likely. In that case, each must have probability $1/8$. Referring to Figure 5.5, for the four events involving the number of girls, these probabilities follow from the fundamental rule for equally likely outcomes.

$$
\begin{aligned}
P(\text{three girls}) &= P(\text{GGG}) = 1/8 \\
P(\text{two girls}) &= (\text{Number of outcomes with 2 girls, 1 boy})/8 \\
&= 3/8 \\
P(\text{one girl}) &= (\text{Number of outcomes with 1 girl, 2 boys})/8 \\
&= 3/8 \\
P(\text{no girls}) &= P(\text{BBB}s) = 1/8.
\end{aligned}
$$

We have thus found the probability distribution for the random number of girls in a three-child family:

Number of girls	0	1	2	3
P (number of girls)	$1/8$	$3/8$	$3/8$	$1/8$

Notice that the probabilities add up to 1, as they always do for a probability distribution.

❐ ❐

Tree diagrams are often used to represent the outcomes of experiments that are carried out in stages. In particular, as illustrated in Example 5.9, they help us count the number of outcomes in an event A resulting from multiple stages, and in the sample space S, as needed when outcomes are equally likely.

Recall that we construct events by combining other usually more elementary events, using "and," "or," and "not." Clearly, if one knows the probabilities of these building block events, we will want, if easily possible, to use them to find the probabilities of the combined events.

Complementary Events

Imagine that we play 1000 games of craps (see Example 5.3) and we win 474 times. It follows that we lost 1000 – 474 = 526 times. That gives us the experimental probabilities

$$\hat{P}(\text{we win}) = 474/1000 = 0.474, \quad \text{and} \quad \hat{P}(\text{we lose}) = 526/1000 = 0.526.$$

It is no accident that these two experimental probabilities add up to 1. Consider any event A (like winning), associated with any random phenomenon (like playing a game of craps). Clearly, if the random phenomenon is repeated, then

(No. of times A does not occur) = (No. of repetitions) – (No. of times A occurs).

Divide both sides of this equation by the number of repetitions. Then we get

$$\hat{P}(A \text{ does not occur}) = 1 - \hat{P}(A \text{ occurs}).$$

Because theoretical probabilities are (very) long-run relative frequencies, we conclude that it also must be true that

$$P(A \text{ does not occur}) = 1 - P(A \text{ occurs}).$$

The event "A does not occur," denoted "not A" for short, is called the complement of A. This leads to one of the basic rules useful for computing probabilities of events:

complementary events rule

Rule for probabilities of complementary events

$$P(\text{not } A) = 1 - P(A).$$

The Addition Rule (the "Or" Rule) for Probabilities

addition rule
multiplication rule

In probability theory there is both an **addition rule** and a **multiplication rule**. You sometimes will find these rules useful to solve what at first appear to be difficult probability problems. We shall first learn the addition rule.

We first state the addition rule in the simplest and most commonly occurring case of two events with no outcomes in common, namely two mutually exclusive events.

addition rule for mutually exclusive events

Addition rule for two mutually exclusive events

Suppose two events A and B have no outcomes in common; that is, A and B are **mutually exclusive**. Then the probability that one or the other event occurs, namely A or B occurs, is the sum of the probabilities of each event.

$$P(A \text{ or } B) = P(A) + P(B)$$

The addition rule holds for theoretical probabilities because it holds for experimental probabilities.

Diagramatically, below are two events A and B with no outcomes in common:

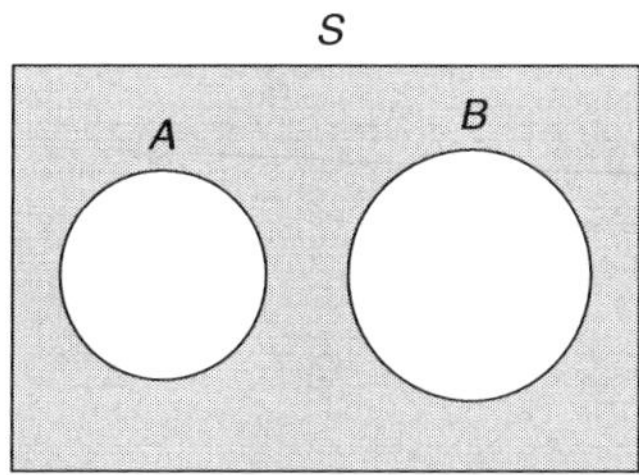

To illustrate the addition rule for two mutually exclusive events, consider once again the game of craps in Example 5.3. There are two mutually exclusive ways of winning in craps on the first toss:

$$A = (\text{you score 7}) \qquad \text{or} \qquad B = (\text{you score 11})$$

Applying the addition rule we get

$$P(\text{you win on first toss}) = P(A \text{ or } B) = P(A) + P(B) = 6/36 + 2/36 = 8/36.$$

(See Figure 5.3.) Applying the complement rule we get

$$\begin{aligned} P(\text{you do not win on the first toss}) &= 1 - P(\text{you win on the first toss}) \\ &= 1 - 8/36 = 28/36. \end{aligned}$$

The addition rule applies for more than two mutually exclusive events, as we shall see in Example 5.10. For example, for mutually exclusive events A, B, and C, we have

$$P(A \text{ or } B \text{ or } C) = P(A) + P(B) + P(C).$$

Example 5.10

Before a person can safely receive a blood transfusion, it is necessary to determine what is called his or her ABO blood group. There are four ABO blood groups: denoted as A, B, O, and AB (four outcomes). The possible donor groups for each recipient blood group are given in Table 5.4. A person of blood type O can receive blood only from a donor of type O, for example.

The distribution of the four ABO blood types varies with ethnic groups. For example, Peruvian Incas are all type O. The blood group distributions for African Americans and for European Americans are given in Table 5.5. What is the probability that a randomly chosen European American is of type O? What is the probability that he or she is not of type O?

Table 5.4 The ABO Group Transfusion Rules

Recipient	Permitted donors	Forbidden donors
O	O	A, B, and AB
A	O or A	B and AB
B	O or B	A and AB
AB	All groups	None

Table 5.5 ABO Blood Type Population Proportions for African Americans and European Americans

	Blood type probability			
Ethnicity	**O**	**A**	**B**	**AB**
African American	0.49	0.27	0.20	0.04
European American	0.45	0.42	0.10	0.03

Solution: The European American row of proportions in Table 5.5 provides the probability distribution for the blood type of a randomly sampled European American.

We have just applied a fundamental principle that is a consequence of the rule for $P(A)$ in the case of equally likely outcomes: When we randomly sample from a population in such a way that all members of the population are equally likely to be selected, then the population proportion for each population trait becomes the probability that one randomly sampled member of the population has that trait. That is,

population properties become probabilities for a randomly sampled member

Population proportion having trait = *P*(randomly sampled member has trait)

Note that the four outcomes (A, B, O, AB) are *not* equally likely. According to the above bolded rule, the probability that a randomly chosen European American is type O is 0.45, i.e., $P(\text{type O}) = 0.45$, because 45% of European Americans are type O. The probability of not getting a person of type O is, by the rule of complementary events,

$$P(\text{not type O}) = 1 - P(\text{type O}) = 1 - 0.45 = 0.55.$$

There is another way to compute the probability of not getting a person of type O: Every person who is not type O is either type A, type B, or type AB. That is, the event

$$(\text{not type O}) = (\text{type A}) \text{ or } (\text{type B}) \text{ or } (\text{type AB}).$$

The four ABO blood types are mutually exclusive; that is, no person has more than one ABO blood type.

Diagramatically,

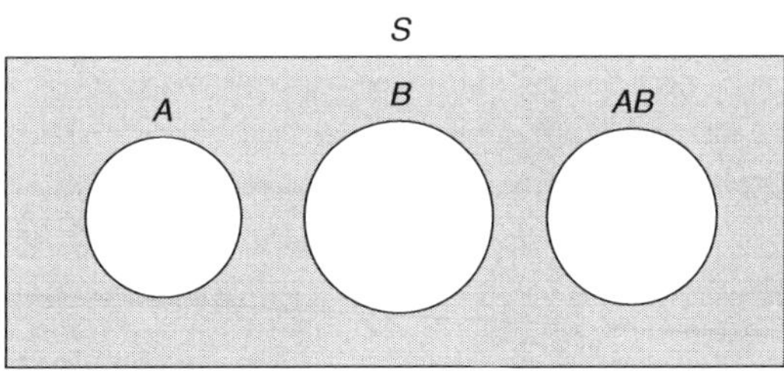

Because A, B, and AB have no outcomes in common, it follows that the population proportions must satisfy

$$\text{proportion not type O} = \text{proportion type A} + \text{proportion type B} + \text{proportion type AB}.$$

Because the population proportions are our sampling probabilities, for the European Americans, we get

$$\begin{aligned} P(\text{not type O}) &= P(\text{type A}) + P(\text{type B}) + P(\text{type AB}) \\ &= 0.42 + 0.10 + 0.03 = 0.55. \end{aligned}$$

❐ ❐

Of course, two arbitrary events A and B may not be mutually exclusive. Then the addition rule requires an extra term:

addition rule for events not mutually exclusive

Addition rule for events not mutually exclusive

If the events A and B are not mutually exclusive, then the addition rule becomes

$$P(A \text{ or } B) = P(A) + P(B) - P(A \text{ and } B)$$

Diagramatically, the event (A or B) is the entire darkly shaded area.

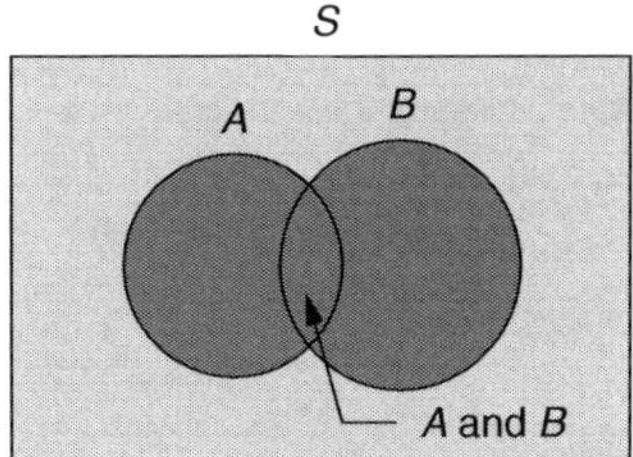

event (A or B)
event (A and B)

Note: Recall that the **event (*A* or *B*)** is the event that at least one of the events A or B occurs. Because A and B may not be mutually exclusive, it includes the possibility of an outcome occurring for which both A and B occur. The **event (*A* and *B*)**, whose probability is subtracted in the above addition rule, is the event that the outcome occurring is such that both A and B occur. This can happen only when A and B are not mutually exclusive.

Example 5.11

To illustrate the addition rule for events not mutually exclusive, imagine that we toss a pair of symmetric casino dice, one white and one black. See Figure 5.6 and consider the events

$$A = \text{(the white die shows five dots);}$$

$$B = \text{(the black die shows five dots).}$$

Since the 36 possible outcomes are equally likely,

$$P(A) = 6/36 = 1/6, P(B) = 6/36 = 1/6, \text{ and } P(\text{both A and B}) = 1/36.$$

Further, using our new addition rule for overlapping events,

$$\begin{aligned} P(\text{we get at least one 5}) &= P(A \text{ or } B) \\ &= P(A) + P(B) - P(A \text{ and } B) \\ &= 1/6 + 1/6 - 1/36 = 11/36. \end{aligned}$$

Notice that the incorrect computation $P(A) + P(B) = 6/36 + 6/36 = 12/36$ counts twice the outcome of both dice being 5, namely the event (A and B). This is the reason we must subtract $P(A \text{ and } B)$ from $P(A) + P(B)$, to eliminate the extra probability accumulated by counting $P(A \text{ and } B) = P(\text{both dice are 5})$ twice.

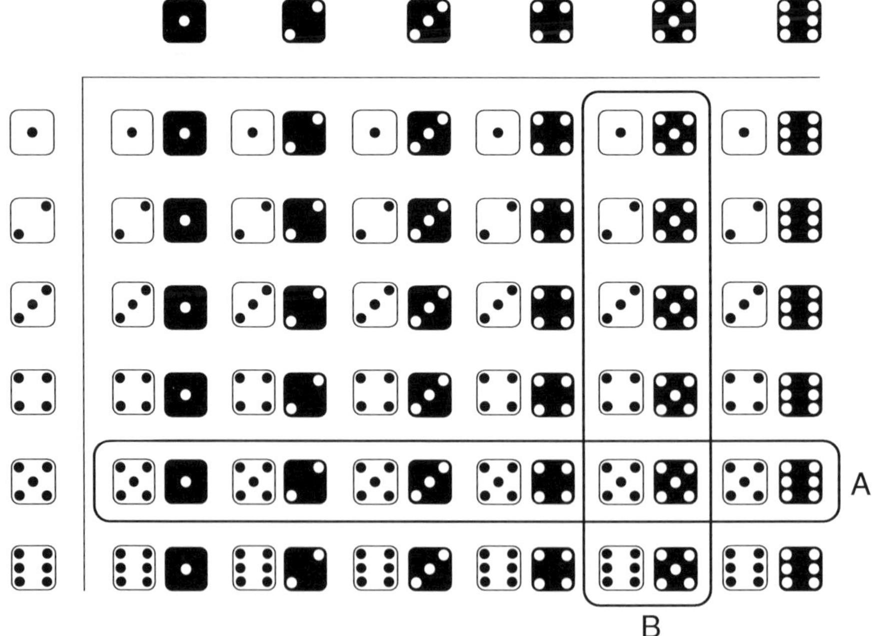

Figure 5.6 Possible dice rolls with at least one die showing five dots.

❒ ❒

Probability models play an important role in the genetic theory of heredity originated by Gregor Mendel (1822–1884). Many futurists predict that the twenty-first century will be the century of genetics, with breakthroughs in genetic engineering, gene therapy, cloning, and the like. In preparation for the next example that involves genetics, let us review some elementary genetics (and perhaps explain genetics for the first time!). Consider an individual (a person, an animal, or a plant) that is the result of sexual reproduction (as opposed to cloning). The individual's genetic makeup is inherited from its parents. For every genetic trait, each parent has a gene pair. Each parent contributes half of his or her gene pair, with equal probability for the two halves, to an offspring, who therefore gets a gene pair, with one gene from each parent, the combination called a *genotype*. The two parents make their contributions independently of which gene the other parent contributed.

What observed trait appears in the offspring depends on which of two or more forms (called *alleles*) its two genes have. For many observed traits, the genes have only two alleles, one *dominant* (often denoted A) and the other *recessive* (often denoted a). A recessive trait is observed in the offspring only when *both* halves of the offspring's inherited gene pair have the recessive form. Otherwise, the dominant trait appears.

Example 5.12

An important example of a recessive genetic disease caused by a single gene pair is sickle-cell anemia (SCA), a severe and potentially deadly disorder that, in the United States, affects roughly 1 in 500 African Americans. The gene responsible for sickle-cell anemia comes in two alleles: the dominant healthy allele A and the recessive defective allele a. A baby born with genotype aa will develop sickle-cell anemia and probably die young. A person of genotype Aa (A from mother, a from father) or aA (a from mother, A from father) will not develop the symptoms of sickle-cell anemia, but he or she may pass the disease on to the next generation. He or she is thus a *carrier* of the disease. (There is no difference between genotypes Aa and aA; it does not matter which parent contributed which form.) A person with genotype AA will neither develop sickle-cell anemia nor be able to pass the disease on to the next generation.

Data suggest that 1 in 10 healthy African Americans is of type Aa or aA and thus is a carrier of sickle-cell anemia. *If two healthy sickle-cell anemia carriers are expecting a child, what is the chance that the child will be healthy?*

Solution: Both parents have the genetic makeup either Aa or aA. As far as the sickle-cell anemia gene pair is concerned, we can view having a child as a random experiment carried out in two independent stages:

First stage: Inherit a gene from the mother

Second stage: Independently inherit a gene from the father

We can use a tree diagram to represent all possible genotypes for the child (see Figure 5.7). The dominant healthy gene is denoted by A. The recessive sickle-cell anemia gene is denoted by a.

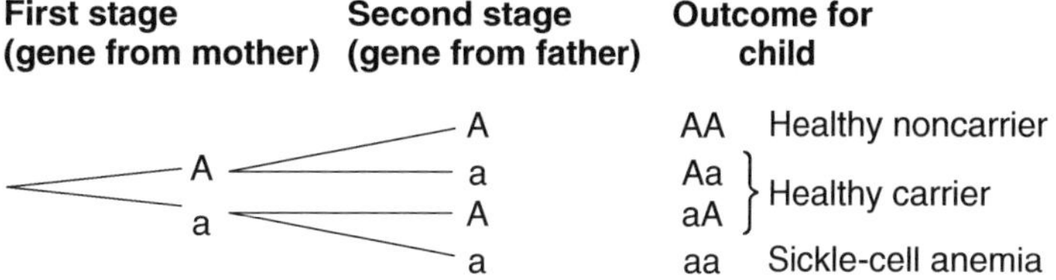

Figure 5.7 Possible sickle-cell anemia genotypes.

According to Mendel's theory the mother and father independently contribute their alleles. Hence, the four outcomes (genotypes) are equally likely. Note that there are three events of interest, as seen next. Therefore,

$$P(\text{child is healthy and not a carrier}) = P(\text{AA}) = 1/4.$$
$$P(\text{child has sickle-cell anemia}) = P(\text{aa}) = 1/4.$$

By the basic rule for P(event) where the outcomes are equally likely,

$$P(\text{child is a healthy carrier}) = P(\text{aA or Aa}) = 2/4 = 1/2.$$

By the addition rule for mutually exclusive events,

$$\begin{aligned} P(\text{child is healthy}) &= P((\text{healthy carrier}) \text{ or } (\text{healthy noncarrier})) \\ &= P(\text{healthy carrier}) + P(\text{healthy noncarrier}) \\ &= 1/2 + 1/4 = 3/4 \end{aligned}$$

The a allele is not an unmitigated curse. It turns out that healthy carriers (with genotype Aa or aA) have higher resistance to malaria than noncarriers (with genotype AA). If a person lives in a region with high incidence of malaria, such as central Africa, then that person is better off being Aa than AA.

Is there a cure for sickle-cell anemia? The following medical case offers hope. An 8-year-old girl in Boston suffering from both leukemia and sickle-cell anemia was treated by destroying her cancerous bone marrow and replacing it with healthy marrow from a donor. This cured both her leukemia *and* her sickle-cell anemia! For most people it is too risky to treat sickle-cell anemia with bone marrow transplants. In the future, gene therapy may offer a cure.

❐ ❐

Cystic fibrosis and Tay-Sachs disease are other examples of important recessive single-gene diseases. Cystic fibrosis affects mainly Caucasians, and Tay-Sachs disease mainly Jews of Eastern European ancestry. Huntington's disease is an example of a *dominant* single-gene disease. If A denotes the lethal Huntington's gene and a denotes the corresponding healthy gene, then everyone with genotype AA, Aa, or aA will develop this degenerative neurological disorder when they reach middle age. That is, the presence of even one A allele suffices to cause the disease.

Genetics is complex but fascinating, and our developing understanding of it is vital for finding medical cures for inherited diseases.

*Probability Trees

probability tree

It is often helpful to draw a tree diagram to represent the possible outcomes of an experiment that is carried out in multiple stages, as we have seen in Figure 5.5. A **probability tree** for a multistage experiment, which is especially useful when outcomes are not equally likely, displays on each branch of the tree the probability of the next stage outcome, assuming that ("conditional" on that) the present stage outcome has occurred.

sensitivity
specificity

Consider, for example, a diagnostic urine test for bladder cancer. Ideally, every person with bladder cancer should test positive if given this test, and every person without bladder cancer should test negative. But no diagnostic test is that good. The **sensitivity** of the test is the probability that a person with cancer will test positive. The **specificity** of the test is the probability that a person who does not have cancer will test negative. Clearly, a good test has both its sensitivity and specificity close to 1. Suppose the sensitivity of the test is 0.95 and the specificity of the test is 0.90.

Finally, suppose that 1 out of 5000 adult Americans has this rare bladder cancer, that is, P(cancer) $= 1/5000 = 0.0002$ for a randomly chosen adult. Suppose that we give a randomly chosen adult this screening test for bladder cancer. The probability tree is easy to construct and is shown in Figure 5.8. The terms commonly used for the four possible multistage outcomes, such as *true positive* and *false negative*, are also given in Figure 5.8.

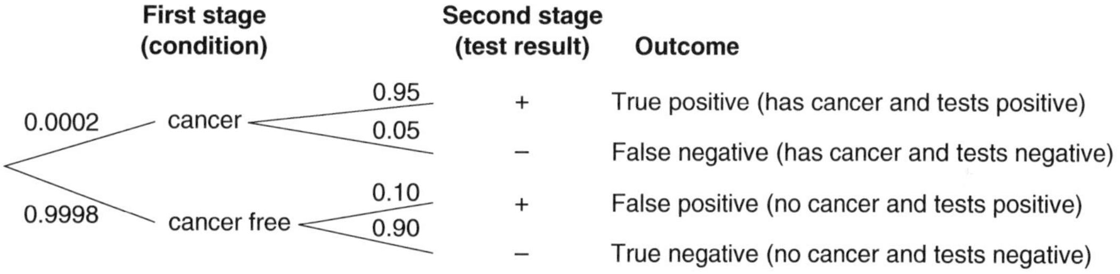

	First stage (condition)		Second stage (test result)	Outcome
0.0002	cancer	0.95	+	True positive (has cancer and tests positive)
		0.05	−	False negative (has cancer and tests negative)
0.9998	cancer free	0.10	+	False positive (no cancer and tests positive)
		0.90	−	True negative (no cancer and tests negative)

Figure 5.8 Testing a randomly chosen adult for bladder cancer.

Note that 0.9998, 0.05, and 0.1 in Figure 5.8 were all obtained from the above three given probabilities of 0.0002, 0.95, and 0.9, respectively, using the rule $P(\text{not } A) = 1 - P(A)$. For example, $P(\text{cancer free}) = 1 - P(\text{cancer}) = 1 - 0.0002 = 0.9998$. If we randomly test a very large number of adults, then about 95% of those who have bladder cancer will test positive. Since 1 in 5000 adults has bladder cancer, it follows that the proportion of adults in the population of tested adults who both have cancer and test positive will be around

$$95\% \text{ of } 0.0002 = 0.95 \times 0.0002 = 0.00019.$$

Moreover, because population proportions become probabilities for a randomly sampled person, we conclude that the probability that a randomly screened person is a true positive is 0.00019. That is,

$$P(\text{true positive}) = 0.00019.$$

Reasoning similarly, the probability of getting a false negative test result for a randomly screened person is

$$P(\text{false negative}) = 0.05 \times 0.0002 = 0.00001.$$

Similarly, the probability of getting a false positive test result is

$$P(\text{false positive}) = 0.10 \times 0.9998 = 0.09998.$$

We have discovered a new and very useful probability rule:

multiplication rule for probability trees

Multiplication rule for probability trees

We can compute the probability of any multistage outcome of a probability tree by multiplying the probabilities of the branches along the route that leads to the outcome. That is,

$$P(\textbf{outcome}) = \textbf{product of probabilities along the branches producing the outcome}$$

Looking at Figure 5.8 and applying the addition rule for mutually exclusive events, we see that the probability that a randomly chosen adult will test positive, namely the event ((true positive) or (false positive)), is

$$\begin{aligned} P(\text{positive}) &= P((\text{true positive}) \text{ or } (\text{false positive})) \\ &= P(\text{true positive}) + P(\text{false positive}) \\ &= 0.00019 + 0.09998 \approx 0.1, \end{aligned}$$

where 0.00019 and 0.09998 were obtained from the multiplication rule for probability trees, as shown previously. That is, if we randomly screen a large number of individuals for bladder cancer using this test, around 10% of those screened will test positive in spite of the fact that only 1 out of 5000 has bladder cancer. Most of the positive test results will be false alarms!

Let us return to genetics for another illustration. The gene for albinism (lack of skin coloration pigment) in humans is recessive, like the gene for sickle-cell anemia (see Example 5.12). Imagine that a husband and a wife both are carriers of the albinism gene and that they have two children. Assume that the children are not identical twins, so that they inherit their genotypes *independently* of each other according to Mendelian genetics. We therefore get the probability tree shown in Figure 5.9.

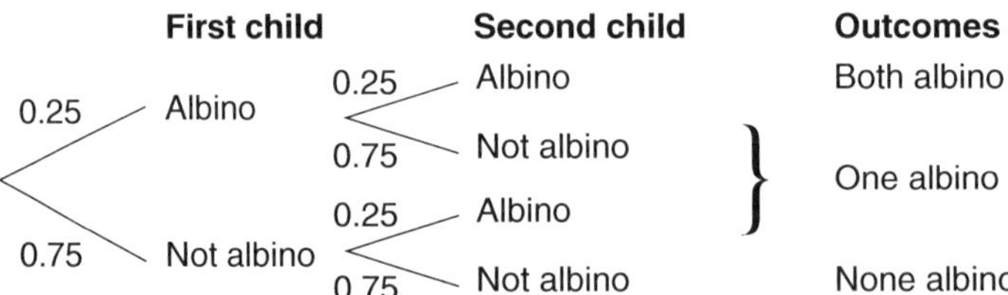

Figure 5.9 Albinism in a two-child family.

The probability that both children are albino is, by the multiplication rule for probability trees,

$$P(\text{first is albino and second is albino}) = 0.25 \times 0.25 = 0.0625.$$

The probability that the first child is not an albino and the second child is an albino equals

$$0.75 \times 0.25 = 0.1875.$$

We have discovered a special case of the multiplication rule for independent events:

multiplication rule for independent events

Multiplication rule for independent events

If two events A and B are independent, then

$$P(A \text{ and } B) = P(A) \times P(B)$$

The analogous rule holds for three or more independent events:
If the events A, B, and C are independent, then

$$P(A \text{ and } B \text{ and } C) = P(A) \times P(B) \times P(C).$$

independent events

By "**independent events**," we mean that the occurrence of one event has no effect on the likelihood that the other occurs or not. It is usually clear when two events are independent, allowing us to use this simple formula for events judged to be independent. A coin turning up heads and a die showing at least five dots are clearly independent events, but rain in San Francisco and rain in Oakland, across San Francisco Bay, will not be independent events. Rain in San Francisco and rain in Singapore will be, though.

In the albinism example, whether the second child will be an albino is not influenced by the first child's condition, since the children inherit their genes independently of each other. Whenever we know events A and B are independent, we are allowed to use the product rule to compute for $P(A \text{ and } B)$. We will learn more about the important concept of independence later. We now give an example where the multiplication rule for independent events is used to compute probabilities.

Example 5.13

It is well-known by medical experts that ABO blood type (see Example 5.10) is independent of the Rh factor. That is, the distribution of the Rh factor in the European American population (85% Rh+ and 15% Rh−) is the same within each of the four ABO blood groups (see Table 5.5). Converting a population proportion into a probability, we will use the fact that

$$P(\text{Rh+}) = 0.85.$$

Imagine that we select a European American at random. We use the product rule for independent events to compute the probabilities for the various combinations of Rh factor and ABO blood type, using independence

$$\begin{aligned} P((\text{type O}) \text{ and } (\text{Rh+})) &= P(\text{type O}) \times P(\text{Rh+}) \\ &= 0.45 \times 0.85 \times 0.3825. \end{aligned}$$

Similarly,

$$\begin{aligned} P((\text{type AB}) \text{ and } (\text{Rh}-)) &= P(\text{type AB}) \times P(\text{Rh}-) \\ &= 0.03 \times 0.15 \times 0.0045. \end{aligned}$$

Genetic theory tells us that ABO blood type, Rh factor, and gender are all independent. Thus, for a randomly chosen European American,

$$\begin{aligned} P((\text{type A}) \text{ and } (\text{Rh+}) \text{ and } (\text{male})) &= P(\text{type A}) \times P(\text{Rh+}) \times P(\text{male}) \\ &= 0.42 \times 0.85 \times 0.50 = 0.1785. \end{aligned}$$

Note that without the assumption of independence we could not compute these probabilities.

*Conditional Probability

Let us return to the bladder cancer screening example (Figure 5.8). If a randomly screened person tests positive, how likely is it that the test result is correct? That is, what is the (conditional) probability that the person has bladder cancer, given that he or she tested positive? We are asking for a probability of an event, conditional on knowing that a second event has occurred. For another example, imagine that you are waiting to board a plane from New York to San Fransisco. You are talking on your cell phone to a friend in Oakland California, who tells you that the rain is pouring down where she is. Then the probability that it is raining in San Francisco, given it is raining in Oakland, will be close to 1.

Returning to the screening example, recall that $P(\text{positive test}) \approx 0.1$. Imagine that we randomly screen a million adults. Because probabilities are population proportions, the number of positive test results will be around

$$P(\text{positive test}) \times 1,000,000 = 0.1 \times 1,000,000 = 100,000.$$

Out of these 1,000,000 screened, the number who both actually have cancer and test positively, that is, the number of true positive test results, will be around

$$P(\text{true positive}) \times 1,000,000 = 0.00019 \times 1,000,000 = 190.$$

So only around 190 out of 100,000 persons who test positive will actually have cancer! The rest will be false alarms. Thus, we estimate that the **conditional probability** that a randomly chosen adult has bladder cancer, given that he or she tested positive, denoted by $P(\text{cancer} \mid \text{positive test})$, will be around

conditional probability

$$P(\text{cancer} \mid \text{positive test}) = \frac{190}{100,000}.$$

Because this probability is so small, it should be noted that doctors would not use this bladder cancer test for random screening of the general population. Rather, they would focus on testing people known to be at greater risk of bladder cancer because of clinical indications, family history, ethnicity, and so forth.

Let us make rigorous the argument we just used and thus determine a general formula for the conditional probability. Consider a random experiment (such as screening a randomly chosen adult for bladder cancer). Consider two events, A (such as the person tests positive) and B (such as the person has cancer) in the random experiment. If we independently repeat the random experiment a number of times, the **experimental conditional probability of B given A**

experimental conditional probability of B given A

$$\hat{P}(B|A) = \frac{\text{number of times both } A \text{ and } B \text{ occur}}{\text{number of times } A \text{ occurs}} \quad \begin{array}{l}\text{(the proportion of times } B \text{ occurs for}\\ \text{those repetitions where } A \text{ has occured)}\end{array}$$

$$= \frac{(\text{number of times both } A \text{ and } B \text{ occur})/(\text{number of repetitions})}{(\text{number of times } A \text{ occurs})/(\text{number of repetitions})}$$

$$= \frac{\hat{P}(A \text{ and } B \text{ occur})}{\hat{P}(A)}.$$

Because experimental probabilities become theoretical probabilities when the number of repetitions gets very large, by imagining the number of repetitions getting very large we obtain the same formula for the **(theoretical) conditional probability of B, given A:**

(theoretical) conditional probability of B, given A:

formula for the conditional probability of B given A

Formula for the conditional probability of B given A

$$P(B|A) = \frac{P(A \text{ and } B \text{ occur})}{P(A)}$$

$P(A$ and B occur) is usually written $P(A$ and $B)$.

We shall consider an elementary illustration. Imagine that you deal a card face down to a friend from a well-shuffled deck of 52 cards. The deck of cards has four suits: spades, hearts, diamonds, and clubs. In each suit there are 13 cards: 2 through 10, jack, queen, king, and ace. Jacks, queens, and kings are face cards. Your friend picks up the card you dealt, but she does not show or tell you which card it is. Let B denote the event that she got a king. Since there are 4 kings in the deck and all 52 cards are equally likely,

$$P(B) = \frac{4}{52}.$$

Imagine now that after picking up the card, your friend exclaims, "I got a face card!" How likely is it now that she got a king? (Your friend obviously knows whether she got a king or not, but you don't. The question is: How likely is it *from your perspective* that she got a king, knowing that she got a face card?) Let A denote the event that she got a face card. There are 12 face cards. We seek $P(B|A)$. Since the 12 face cards (of which we know she got one) are equally likely, the conditional probability that she got a king (4 ways to get a king), given that she got a face card, is clearly

$$P(B|A) = \frac{4}{12} = \frac{1}{3}.$$

We could instead have used the formula in the box for computing this conditional probability. First, note that in this particular example the event (A and B) is the same event as the (more restrictive) event B. Hence, $P(A \text{ and } B) = P(B)$. The answer when using the formula for a conditional probability is

$$P(B|A) = \frac{P(A \text{ and } B)}{P(A)} = \frac{P(B)}{P(A)} = \frac{4/52}{12/52} = \frac{4}{12} = \frac{1}{3}.$$

Note that the two approaches give the same answer. We caution students that in complicated problems it can be dangerous to depend on intuition to solve for a conditional probability. But, as we shall see, in Example 5.14 all the conditional probabilities are easy to figure out intuitively.

*The Multiplication Rule (the "And" Rule) for Probabilities of Two (Possibly Dependent) Events

Earlier we learned the multiplication rule for independent events. There is a similar rule that works in general, that is, whether the events are independant or not. The rule follows from the formula for conditional probabilities, $P(B|A) = P(\text{both } A \text{ and } B)/P(A)$. If we multiply both sides by $P(A)$, then we get the general case of the multiplication rule for probabilities:

multiplication rule for two and three dependent events

General multiplication rule for two events and for three events

For any two events A and B,

$$\boldsymbol{P(A \text{ and } B) = P(A) \times P(B|A).}$$

Similarly, for any three events A, B, and C,

$$\boldsymbol{P(A \text{ and } B \text{ and } C \text{ all occur}) = P(A) \times P(B|A) \times P(C|A \text{ and } B \text{ occur}).}$$

Example 5.14

Suppose that the cards in a deck of 52 playing cards are shuffled well and that 3 cards are dealt 1 at a time. Let A denote the event that the first card dealt is a heart, B the event that the second card dealt is a heart, and C the event that the third card dealt is a heart. Using the multiplication rule for two events, we get,

$$\begin{aligned} P(\text{first 2 cards dealt are hearts}) &= P(A \text{ and } B) \\ &= P(A) \times P(B|A) \\ &= (13/52) \times (12/51). \end{aligned}$$

Here, $P(B|A) = 12/51$ follows because of the 51 remaining cards after the first draw 12 are hearts because the first card was a heart. Similarly,

$$\begin{aligned} P(\text{all 3 cards are hearts}) &= P(A \text{ and } B \text{ and } C \text{ all occur}) \\ &= P(A) \times P(B|A) \times P(C|A \text{ and } B) \\ &= (13/52) \times (12/51) \times (11/50). \end{aligned}$$

As another illustration,

$$P(\text{all 3 cards are aces}) = (4/52) \times (3/51) \times (2/50)$$

As still another illustration,

$$\begin{aligned} &P(\text{first card is a spade, second card a heart, and third card a diamond}) \\ &= (13/52) \times (13/51) \times (13/50). \end{aligned}$$

In the above applications of the multiplication rule, all the needed conditional probabilities can easily be found by careful reasoning about the remaining equally likely outcomes. The multiplication rule for probability trees we studied previously is a special case of the general multiplication rule.

Conditional probability allows us to formally define the concept of events A and B being statistically independent in a way that makes common sense. Recall that the informal idea is that the occurrence of one event—A, say—has no influence on the likelihood of the other's—B's, say—occurring:

definition of independent events

Definition of events A and B being statistically independent

Events A and B are independent if

$$\boldsymbol{P(B|A) = P(B)}$$

Equivalently, events A and B are independent if

$$\boldsymbol{P(A|B) = P(A)}$$

When one recalls that $P(B|A)$ is defined by

$$P(B|A) = \frac{P(\text{both } A \text{ and } B)}{P(A)}$$

it follows immediately that the definition of independence becomes equivalent to

$$P(\text{both } A \text{ and } B) = P(B) \times P(A),$$

multiplication rule for independent events

producing the **multiplication rule for independent events** that we studied previously. Thus, the nonintuitive multiplication rule for independent events is actually equivalent to the highly intuitive definition of independence in the above box.

Imagine again that we deal a card from a well-shuffled deck of 52 cards. The deck of cards has four suits: spades, hearts, diamonds, and clubs. In each suit there are 13 cards: 2 through 10, jack, queen, king, and ace. The events

$$A = (\text{we deal a heart}) \text{ and } B = (\text{we deal a king})$$

are statistically independent.

$$P(B|A) = \frac{1}{13} = \frac{4}{52} = P(B)$$

and

$$P(A|B) = \frac{1}{4} = \frac{13}{52} = P(A).$$

The likelihood that the card is a king is the same whether the card is a heart or not. The likelihood that the card is a heart is the same whether the card is a king or not.

*Not Equally Likely Outcomes

To guide us in assigning probabilities to events when the outcomes of the random experiment are not equally likely, we note two fundamental rules that always hold for probabilities: The probabilities we assign to each of the possible outcomes must be between 0 and 1, and they must add up to 1. That is,

$$\mathbf{0 \le P(\text{outcome}) \le 1,}$$
$$\sum_{\substack{\text{all outcomes}\\ \text{in } S}} \mathbf{P(\text{outcome}) = 1.}$$

Here $\sum$ is the summation symbol we introduced in Chapter 2, which merely tells us to add up what is to its right. A subscript, such as "all outcomes in S," tells us what values to sum over.

The reason the two fundamental rules must hold is that relative frequencies (experimental probabilities) obey these rules and that probabilities, being long-run relative frequencies, inherit all the properties of relative frequencies.

A fundamental principle for all random phenomena is that the probability that any particular event A of a random experiment will take place equals the sum of the probabilities of the outcomes in the event. That is, for any event A:

P(event) for unequally likely outcomes

$$\mathbf{P(A) = \sum_{\substack{\text{all outcomes}\\ \text{in } A}} P(\text{outcome}).}$$

This very important and totally general formula is a consequence of the fact that the corresponding formula holds for experimental probabilities. It holds whether the outcomes are equally likely or not. When the outcomes are equally likely, it reduces to the formula we learned for that case. We shall use the above formula in the next example.

Example 5.15

A dollar coin is flipped successively until the first head is observed. If we count the number of flips needed to get heads for the first time, then the sample space consists of all the counting numbers,

$$S = \{1, 2, 3, 4, 5, 6, 7, \ldots\}.$$

Either depending on one's intuition, or actually doing some random experimentation, one can conclude that these outcomes cannot be equally likely. In Exercise 10 of Section 5.2, we try to estimate the probability of some of the outcomes of this chance experiment. It turns out that the theoretical probability distribution for the number of tosses it takes to get the first head is actually given by

Number of Flips	1	2	3	4	5	6	...
Probability	1/2	1/4	1/8	1/16	1/32	1/64	...

As required, these probabilities add up to 1: it can be shown that $1/2 + 1/4 + 1/8 + 1/16 + 1/32 + \cdots = 1$. Using this probability distribution for the outcomes of the random experiment, we can compute the probability of any event A. For example,

$$P(\text{we need at most three tosses to get the first head}) =$$
$$P(1) + P(2) + P(3) = 1/2 + 1/4 + 1/8 = 7/8.$$

❐ ❐

Section 5.3 Summary

Equally likely outcomes If all outcomes are equally likely, then we can compute the probability of any event *A*:

$$P(A) = \frac{\textbf{number of different outcomes in } A}{\textbf{total number of outcomes in } S}$$

Complement rule for any events The probability that a given event *A* does not occur is 1 minus the probability that *A* does occur:

$$P(\text{not } A) = 1 - P(A)$$

Addition rule for two mutually exclusive events If two events *A* and *B* have no outcomes in common (are mutually exclusive), they cannot both occur at the same time. Then

$$P(A \text{ or } B) = P(A) + P(B)$$

Addition rule for two not mutually exclusive events If the events *A* and *B* are not mutually exclusive (i.e., they can both occur at the same time), then the chance that at least one of the events occurs is

$$P(A \text{ or } B) = P(A) + P(B) - P(A \text{ and } B)$$

***General multiplication rule for two events**

$$P(A \text{ and } B) = P(A) \times P(B|A)$$

This generalizes to three events:

$$P(A \text{ and } B \text{ and } C) = P(A) \times P(B|A) \times P(C|A \text{ and } B)$$

A caution: The addition rules apply in one situation (dealing with the event (*A* or *B*)) and the multiplication rules in another (dealing with the event (*A* and *B*)). To apply either of these rules correctly, one must make sure that the appropriate situation holds.

Rules for P(event) for a probability tree

- Multiply along the appropriate branches to find *P*(outcome) for each outcome in the event.
- Add P(outcome) across the appropriate end branches to find *P*(event).

***Conditional probability**

$$P(B|A) = \frac{P(A \text{ and } B)}{P(A)}$$

Independent events The events A and B are defined to be statistically independent if either

$$P(B|A) = P(B) \quad \text{or} \quad P(A|B) = P(A).$$

Either of these defining formulas becomes the multiplication rule for independent events: If the events A and B are independent, then

$$P(A \text{ and } B) = P(A) \times P(B).$$

Similarly, if the events A, B, and C are independent, then

$$P(A \text{ and } B \text{ and } C) = P(A) \times P(B) \times P(C)$$

*The probabilities assigned to the outcomes of a random phenomenon must be between 0 and 1 and must add up to 1:

$$0 \le P(\text{outcome}) \le 1,$$

$$\sum_{\text{all outcomes}} P(\text{outcome}) = 1.$$

*The probability of any event A equals the sum of the probabilities of the outcomes in A:

$$P(A) = \sum_{\substack{\text{all outcomes}\\ \text{in } A}} P(\text{outcome})$$

Section 5.3 Exercises

The probabilities required in these exercises are theoretical, not experimental.

1. Identify three probability experiments having multiple stages (such as tossing two dice) and equally likely outcomes. Draw a tree diagram for each to justify your choices. Compute a couple of probabilities of interest to you.

2. A four-sided die (what must it be shaped like?) is called a tetrahedral die. Assume that the possible outcomes of tossing a four-sided die are 1, 2, 3, and 4. Under the assumption of equally likely outcomes, find the following:
   a. $P(1)$
   b. $P(2)$
   c. $P(3)$
   d. $P(4)$
   e. $P(\text{even number})$
   f. $P(\text{number less than } 4)$

3. A 10-spinner is shown below. The arrow is free to spin randomly until it stops in 1 of the 10 sectors. Assume that the arrow is equally likely to stop in each of the 10 sectors.

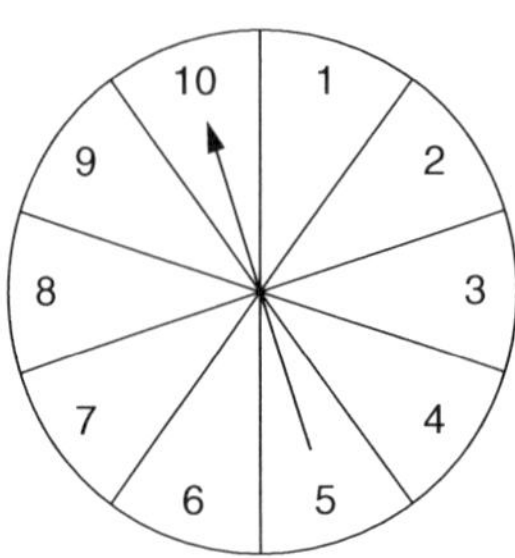

   a. What is the theoretical probability of obtaining each of the numbers 1 through 10?
   b. What is the theoretical probability of obtaining an odd digit?

c. What is the theoretical probability of obtaining a digit greater than 3?

4. Suppose you have 19 classmates, and the instructor is about to choose 2 members of your class to present problems on the board.
   a. What is the theoretical probability that you and your friend Tasha will both be chosen, assuming that each class member has an equal chance of being selected? (Hint: How many ways can two people be chosen by choosing one person and then a second from the remainder?)
   b. The class has four seniors. One person is randomly chosen to critique the solutions. What is the probability a senior will be chosen to critique the solutions?

5. A regular deck of playing cards has 52 cards. These cards consist of four suits, called spades, hearts, diamonds, and clubs. Each suit has 13 cards with values 2, 3, ..., 10, jack, queen, king, and ace. Hearts and diamonds are red, whereas spades and clubs are black.

   Suppose you deal the top card from a well-shuffled deck that is face down. Find the following theoretical probabilities:
   a. $P(\text{black card})$
   b. $P(\text{heart})$
   c. $P(\text{ace})$
   d. $P(\text{king of diamonds})$

6. Suppose you throw a pair of casino dice (see Example 5.3). Compute:
   a. $P(\text{the sum of the two dice equals 2})$
   b. $P(\text{the sum of the two dice equals 7})$
   c. $P(\text{the sum of the two dice equals 10})$
   d. $P(\text{the sum of the two dice equals 6})$
   e. $P(\text{the sum is at most 5})$
   f. $P(\text{the sum is at least 10})$

   Hint: Use the following addition table to see all of the possible outcomes of throwing a pair of dice and the corresponding sum in each case.

	Die 2					
Die 1	**1**	**2**	**3**	**4**	**5**	**6**
1	2	3	4	5	6	7
2	3	4	5	6	7	8
3	4	5	6	7	8	9
4	5	6	7	8	9	10
5	6	7	8	9	10	11
6	7	8	9	10	11	12

7. In roulette half of the numbers between 1 and 36 are odd, half are even. The remaining two numbers 0 and 00 are considered neither odd nor even. If you bet on "odd," what is your chance of winning? If you bet on the four numbers {17, 18, 20, 21}, what is your chance of winning? (You win if the ball comes to rest in any of the those four pockets.)

8. A coin is tossed four times, and the sequence of heads and tails is recorded.
   a. List each of the 16 sequences in the sample space S.
   b. Find the probability of getting two heads and two tails in any order.
   c. Find the probability of getting heads on the second toss.
   d. Find the probability of getting at most one tail.
   e. Find the probability of getting at least one tail.

   Hint: It may help to draw a tree diagram.

9. Consider the coin toss experiment in Example 5.15 (Section 5.3) and in Exercise 10 of Section 5.2. Someone suggests that we assign the following probabilities to the outcomes in the sample space $S = \{1, 2, 3, 4, 5, \ldots\}$:

Outcome	1	2	3	4	5	...
Probability	1/2	1/3	1/4	1/5	1/6	...

What is wrong with this assignment? Would 1/2, 1/4, 1/8, ... be a logically correct assignment? (Do these probabilities add up to 1?)

10. The gene responsible for eye color comes in two forms (alleles): dominant brown (B) or recessive blue (b). A person with genotype BB, Bb, or bB has brown eyes, whereas a person with genotype bb has blue eyes. (This model is obviously an oversimplification. Eyes come in many shades of color. But the model is a good first approximation.) Suppose that a blue-eyed man and a brown-eyed woman have a blue-eyed child.
    a. What is the genotype of the woman: BB or Bb? (By "Bb" here we really mean either Bb or bB.)
    b. What is the chance that the couple's second child will also have blue eyes?

    Now consider a different couple, a brown-eyed woman and a brown-eyed man, with a blue-eyed child.
    c. Identify their genotypes.
    d. What is the chance that the couple's second child will have brown eyes?

11. If two healthy cystic fibrosis carriers have a baby, there is a $3/4$ chance that the child will be healthy and a $1/4$ chance that the child will develop cystic fibrosis. Does this mean that if the couple has four children, then three will be healthy and one will get cystic fibrosis? Explain.

12. (Complex but interesting genetic problem) In humans there is a special chromosome pair that determines sex. The female pair is XX and the male pair is XY. A child inherits one of the mother's X chromosomes and, with equal probability, the father's X chromosome to become a female or the father's Y chromosome to become a male. (The different birth rates for boys and girls is explained by different prenatal survival rates.) Some genes are carried only by the X chromosome. These genes are said to be *sex-linked*. Females have a pair of such genes; males have only one. Females can have genotype AA, Aa, or aa, but males have only two genotypes: A or a. Many sex-linked genes, such as hemophilia (a condition in which blood cannot coagulate, so one easily suffers life-threatening bleeding) and color blindness, are recessive and cause harm.

    Let a be such a gene. Then all a males and all aa females show the defect. A males and AA, Aa, or aA females will be healthy.
    a. Can a boy inherit hemophilia from his father? Explain.
    b. Can a boy inherit hemophilia from a healthy mother?
    c. Suppose the mother and the maternal grandparents of a newborn baby boy are healthy, but that one of the boy's maternal uncles has hemophilia. What is the chance that the boy has hemophilia?

13. Explain why breast cancer screening for younger women will detect few cancers but will produce many false positive test results.

14. Recall that there are four ABO blood types: A, B, O, AB. The gene for ABO blood type has three forms (alleles): A, B, and O. O is recessive. A person with genotype AA or AO (in either order) will have blood type A. A person with genotype BB or BO (in either order) will have blood type B. A person with genotype AB (one parent contributed A, one contributed B) will have blood type AB. A person with genotype OO will have blood type O. Now suppose that two parents have blood type AB. Find the probability that their first child has
    a. blood type A
    b. blood type B
    c. blood type AB

15. A symmetric casino die is thrown four times. Compute the probability of getting
    a. four 6s
    b. no 6
    c. at least one 6

16. Assume that both male and female African Americans have the ABO blood type distribution shown on the "African American" line of Table 5.5. Also assume that the blood types of husbands and wives are independent. Find the probability that in a randomly chosen African-American married couple

a. The wife is type AB.
b. Both husband and wife are type O.
c. The wife is type A, and the husband is either type O or type A.
d. The wife is type B, and the husband is either type O or type B.
e. The wife can receive a blood transfusion from her husband (see Table 5.4).

17. A wildcatter (oil prospector not employed by a major oil company) is drilling an exploratory oil well in a region not known to contain oil or gas. His chance of drilling through shale is 0.6. If the well hits shale, the chance of striking oil is 0.3. If the well does not hit shale, the chance of striking oil is only 0.05. Draw a probability tree in which the first pair of branches represents hitting shale or not hitting shale and the second set of branches represents striking oil or not striking oil. What is the chance that the wildcatter strikes oil?

18. About 1 in 10 healthy African Americans is a carrier of sickle-cell anemia (see Example 5.12). This holds for both men and women.
a. What is the chance that in a randomly chosen, healthy African-American couple, both husband and wife are carriers of sickle-cell anemia?
b. Recall that if a couple of healthy carriers has a child, the risk that the child will have sickle-cell anemia is 0.25. Draw a probability tree in which the first branches represent the sickle-cell anemia status of a randomly chosen African-American couple. Let the last set of branches represent the sickle-cell anemia status of the couple's firstborn. What is the chance that the child gets sickle-cell anemia? *Hint:* Are both parents carriers or not?

19. Here are the number (in thousands) of college degrees awarded in 1996, classified by level and sex of the degree recipient.

	Bachelor's	Master's	Professional	PhD	Total
Male	522	179	45	27	773
Female	642	227	32	18	919
Total	1164	406	77	45	1692

a. If you select a degree recipient at random, what is the chance that you pick a woman?
b. What is the conditional probability that you picked a woman, given that you selected a PhD?
c. What is the conditional probability that you picked a bachelor's degree recipient, given that you selected a woman?

20. What is the probability that a family with three children has all boys, if you know that at least one of the children is a boy (see Figure 5.5)?

21. Compute the probability of winning in craps (see Example 5.3). Draw a probability tree in which the first set of branches represents your score on the first toss of the dice and the second set of branches represents winning and losing.

22. Out of four bottles of milk, two are spoiled. First John buys one of the four bottles, and then Jack buys one of the remaining bottles. Find
a. P(Jack gets spoiled milk | John got spoiled milk)
b. P(Jack gets spoiled milk | John got fresh milk)

23. A somewhat absentminded hiker forgets to bring his insect repellent on 30% of his hikes. The probability of being bitten is 0.9 if he forgets the repellent and 0.2 if he uses the repellent. Draw a probability tree and compute the theoretical probability that
a. He will forget the insect repellent and will get bitten.
b. He will get bitten.
c. Given that he gets bitten, he forgot the repellent.

24. ELISA (enzyme-linked immunosorbent assay) tests are used to screen donated blood for the AIDS virus. Imagine that the probability that an infected person tests positive is 0.997. The probability that a person who is not infected tests negative is 0.985. *Hint:* These last two probabilities really are conditional probabilities.
    a. What is the probability that a person who is not infected will test positive?

    Suppose that 0.1% of all blood donors are infected and that we test a randomly chosen donor. Draw a probability tree (see Figure 5.8) and compute the probability that we get
    b. a true positive
    c. a false positive
    d. a positive test result
    e. If the donor tests positive, how likely is it that he or she is infected?
25. Consider the AIDS test in the previous exercise. Imagine that we randomly test a person from a country where 10% of the population is infected. Find the probability that we get
    a. a true positive
    b. a false positive
    c. a positive test result
    d. If the person tests positive, what is the conditional probability that he or she is infected?
26. Two cards are dealt from a deck of 52 cards. Find the probability that
    a. Both are aces.
    b. Neither is an ace.
    c. At least 1 of them is an ace.

    Hint: Use $P(A \text{ and } B) = P(B|A)P(A)$.

5.4 SIMULATION: A POWERFUL TOOL FOR LEARNING AND DOING STATISTICS

Smallpox is a contagious, airborne viral disease that within a month kills around 30% of all unvaccinated victims. The disease was eradicated 30 years ago, and routine smallpox vaccination was discontinued because smallpox no longer posed a threat. If bioterrorists were now to unleash the smallpox virus on an unprotected American city, hundreds of thousands of people might die. To protect ourselves, we could restart the vaccination program. The government has a stockpile of half a billion doses of smallpox vaccine. But vaccination is not completely risk free. According to the Center for Disease Control, the risk of dying as the result of a smallpox vaccination is one in a million.

After 9/11 the government was considering a proposal to vaccinate up to 10 million people by early 2004. If 10 million individuals were vaccinated, how many would die as a result of the vaccination? One way to answer such a question is to run computer simulations. A fast computer can mimic the chance experiment of vaccinating 10 million people and counting the number of resulting deaths. Ten computer simulations of 10 million people being vaccinated might yield the following number of deaths:

10	8	11	11	9	9	14	10	7	13

By having the computer run 10,000 such simulations, we would get a very accurate approximation of the probability distribution of the number of deaths that would result from the vaccination of 10 million people.

As we often do when we introduce new probability concepts, we shall use gambling schemes to illustrate the concept of simulation. Imagine that we want to find the theoretical probability of winning a game of craps (see Example 5.3). To do so, we could follow in the footsteps of seventeenth-century gamblers and play a lot of craps games and use the experimental probability we get as our estimate of the theoretical probability. Unfortunately, we would need at least 400 craps games to be reasonably confident that our experimental probability is within 0.05 of the theoretical probability. If we want to be reasonably sure that we are within 0.01 of the theoretical probability, we need at least 10,000 craps games. This approach does not look very promising.

simulation

Fortunately, we can make a computer do it for us. In a few seconds a fast computer can mimic 10,000 craps games and report back to us the proportion of times it won. **Simulation** is useful for studying complex chance experiments that are built from simpler underlying chance experiments. (The game of craps is a complex chance experiment based on the simple, well-understood chance experiment of tossing a pair of dice.) This simulation approach is often called **Monte Carlo simulation**. It was first used under that name by physicists studying nuclear chain reactions. The five-step simulation method presented shortly will allow us to use simulation to estimate unknown probabilities.

Monte Carlo simulation

A computer uses a random number–generation program to produce digits to simulate the toss of a pair of dice or any other chance experiment. To show how the computer does this, we will use a table of random digits to do the simulations. The numbers in Table B.2 in Appendix B look just like numbers produced by 1800 independent tosses of a symmetric casino die. The random number generator in a computer can produce many more such numbers.

We can simulate the toss of a pair of casino dice by picking two digits from Table B.2. Here are the first two rows from this table:

Row									
1	66533	45332	24614	22231	26431	35541	12165	62116	16111
2	61261	22613	26252	14622	32262	33244	34614	13316	41136

To use a random number table, you may start with any row and at any place in that row. Once you pick a starting point, however, you should keep moving in a fixed way, such as from left to right, from that place through the rest of the table. You can move right or left or diagonally or in whatever direction you wish; the important thing is to move systematically so that you do not "cherry pick": Do not allow for any selective human behavior that could invalidate independent selections.

We use these digits to simulate 10 games of craps (see Example 5.3). Let us pick the digits from left to right in the first row and then continue from left to right in the second row, and so forth. The results are shown in Table 5.6. Based on these 10 simulated craps games, we crudely estimate that the probability of winning in craps is around 0.30. To get a more accurate estimate, we should make the computer simulate at least 10,000 craps games. (The theoretical probability of winning with casino dice (absolutely fair) is close to 0.493. If we use 10,000 simulations, we can be reasonably confident that the experimental probability will be within 0.01 of the true probability. Thus we will at least discover that the chance of winning a craps game is close to 0.5. If we did a very large number of repetitions, say 1,000,000, then we would discover the probability is close to 0.49 because the margin of error will be reduced from 0.01 to 0.001.)

Table 5.6 Simulating 10 Craps Games

Repetition	Scores	Outcome
1	(6,6)	Loss
2	(5,3) (3,4)	Loss
3	(5,3) (3,2) (2,4) (6,1)	Loss
4	(4,2) (2,2) (3,1) (2,6) (4,3)	Loss
5	(1,3) (5,5) (4,1) (1,2) (1,6)	Loss
6	(5,6)	Win
7	(2,1)	Loss
8	(1,6)	Win
9	(1,6)	Win
10	(1,1)	Loss

Example 5.16 Safety of Blood Transfusions

Suppose you are injured and lose so much blood that you need several blood transfusions, and you are in a poor, underdeveloped country where health authorities lack the resources to screen donated blood to make sure it is free of infectious diseases. Assume that the local blood supply is such that 1/6 (a made-up probability) of the available units of blood contain an infectious organism, such as hepatitis B virus or HIV. Assuming that you need five units of blood, what is the chance that all five units are free from the infectious agent?

Solution: Receiving one unit of blood is like rolling a casino die once. If the die shows 1 through 5, the blood is safe, and you are not exposed to the infectious agent; if the die shows a 6, you are exposed. We can therefore simulate one transfusion by selecting a digit from Table B.2 of Appendix B. Select a 6, and you are exposed; select 1 through 5, and you are not exposed. We can simulate receiving five units of blood by selecting five digits from Table B.2. If none of the digits is a 6, then all five units are safe, and you are not exposed at all. We shall use the first five digits of each row of Table B.2 to repeat the simulation. For the first simulation we use the first five digits in the first row, 66533. For the second simulation we use the first five digits in the second row, 61261; and so forth. The first seven simulations are shown in Table 5.7.

Table 5.7 Simulating 5 Blood Transfusions

Repetition	Digits	Outcome
1	66533	Exposed
2	61261	Exposed
3	61144	Exposed
4	21341	Not exposed
5	15365	Exposed
6	51612	Exposed
7	31244	Not exposed

If we use the first five digits in all 40 rows of Table B.2, we can repeat the simulation 40 times. We can simulate 40 persons each receiving five units of blood. It turns out that 15 persons escape exposure while 25 persons receive at least one contaminated unit of blood. We therefore estimate the probability of receiving five units of safe blood to be around $15/40 = 0.375$, a number we don't trust much because 40 is very few repetitions. (The theoretical probability is approximately 0.402.)

The Five-Step Simulation Method

five-step simulation method

The preceding blood transfusion example illustrates our **five-step simulation method** for doing a simulation study when using a random number table:

Step 1. Choose a (Probability) Model First we identify receiving a blood transfusion with selecting one of the numbers 1, 2, 3, 4, 5, 6 at random with equal probability. We identify getting a 6 with receiving contaminated blood and getting 1 through 5 with receiving safe blood.

Step 2. Define One Simulation Five transfusions corresponds to selecting five random digits as described in Step 1.

Step 3. Define the Event of Interest We get five safe transfusions (person not exposed) if none of the five digits chosen in Step 2 is a 6.

Step 4. Repeat the Simulation In Table 5.7 we listed seven simulations. The greater the number of simulations, the more accurate our estimate of the theoretical probability will be. If we use a computer, we should do at least 10,000 simulations.

Step 5. Find the Experimental Probability of the Event of Interest In Table 5.7 $\hat{P}$(five safe transfusions) $= 2/7$, a number we cannot even begin to trust as having any accuracy! With 40 simulations, $\hat{P}$ (5 safe transfusions) comes to 0.375, which begins to roughly approach the theoretical probability 0.402.

In the next example we will use simulation to estimate a probability distribution (that is, the probabilities for the various possible values of a statistic of interest).

Example 5.17 Multiple-Choice Quiz

A multiple-choice quiz has five questions. To each question there are six options to choose from, one of which is correct. Imagine that you have no clue whatsoever what the correct answers are and blindly (randomly) guess the answer to each of the five questions. How likely are you to get 0, 1, 2, 3, 4, or 5 correct answers? That is, find the probability distribution of the number of correct answers.

Solution: We will use the five-step method.

Step 1. Choose a Model Since there are six answers to choose from, the probability that we guess the correct answer to any particular question is 1/6. We can therefore identify trying to guess an answer with selecting one of the numbers 1, 2, 3, 4, 5, or 6 at random with equal probability. We identify getting a 6 with guessing the correct answer. (You might, of course, have selected one of the other digits, such as 1.)

Step 2. Define One Simulation Trying to guess the answers to five questions corresponds to selecting five random numbers as described in Step 1. Each 6 corresponds to a correct answer.

Step 3. Define the Statistic of Interest The statistic of interest is the number of correct answers. It corresponds to the number of 6s among the five digits chosen in Step 2.

experimental distribution of a statistic of interest

Note: Because we are looking at the entire **experimental distribution of a statistic of interest** rather than a particular event, the title for Step 3 is changed from "Event of Interest" to "Statistic of Interest."

Step 4. Repeat the Simulation We shall use the second five-digit column in Table B.2 of Appendix B to simulate 40 values of the statistic of interest, the number of correct answers. We list the first five simulations.

Repetition	Digits	Number of correct answers
1	45332	0
2	22613	1
3	46631	2
4	66222	2
5	46135	1

Table 5.8 Experimental and Theoretical Probability Distributions for the Number of Correct Answers to the Quiz

Number of correct answers	Experimental frequency	Experimental probability	Theoretical probability
0	15	0.375	0.402
1	14	0.350	0.402
2	7	0.175	0.161
3	3	0.075	0.032
4	1	0.025	0.003
5	0	0.000	0.000

Step 5. Find the Experimental Probability Distribution of the Statistic of Interest We put the results of the 40 simulations into Table 5.8. The title of Step 5 has changed because we now want the experimental probabilities for all possible values of the Step 3 "statistic of interest." We have also put the corresponding theoretical probabilities in Table 5.8 for comparison. We see that the experimental probabilities are not too far off, but in fact we were lucky to come as close as we have and cannot trust experimental probabilities when only 40 simulations have been done. If we had done 10,000 simulations, then the experimental probabilities would have been very close to the theoretical ones. (We do not expect students to derive the theoretical probabilities in Table 5.8 at this point. We only provide them for comparison. They could be derived from a probability tree or from the binomial formula in Chapter 7.) Note that the theoretical probabilities are rounded to a certain number of significant digits. For example,

$$P(5\text{ correct}) = (1/6)^5 \approx 0.00013 \approx 0.000.$$

experimental probability histogram
theoretical probability histogram
probability histogram

We can also compare the experimental and theoretical distributions by looking at the corresponding histograms. The **experimental probability histogram** (also called a density histogram as in Chapter 1; see Figure 5.10) shows the shape of the experimental probability distribution of the statistic of interest; the **theoretical probability histogram**, also called the **probability histogram** (see Figure 5.11) shows the shape of the theoretical probability distribution of the statistic of interest. Each bar graph rectangle in Figures 5.10 and 5.11 has an area equal to the corresponding probability from Table 5.8. When the possible values of the statistic of interest are all integers, as is often the case, then the rectangles, whose heights are now the probabilities because a rectangle of width one has its area equalling its height, provide a convenient visual impression of the probability distribution's center, spread, and shape, just as a density histogram does with data.

In a probability histogram the area of each rectangle equals the probability of obtaining the corresponding value. Because the sum of all outcome probabilities is 1, the sum of the areas of the rectangles of a probability histogram equals 1. We note that the experimental probability histogram of Figure 5.10 is fairly similar to the theoretical probability histogram of Figure 5.11. In fact, if the number of simulations had been large, (say, 10,000), the histograms would have been indistinguishable.

❐ ❐

Sometimes the outcome values of a probability distribution are not all integers. Then we can simply use the heights of lines at the values to graph the distribution. To illustrate, consider the distribution that is graphed in Figure 5.12.

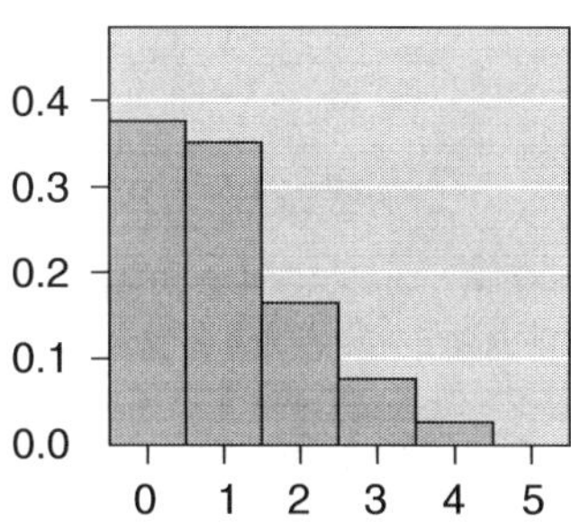

Figure 5.10 Experimental probability histogram derived from Table 5.8.

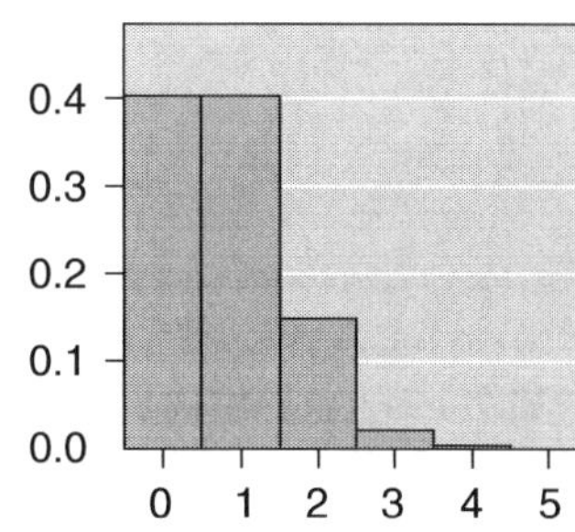

Figure 5.11 Theoretical probability histogram derived from Table 5.8.

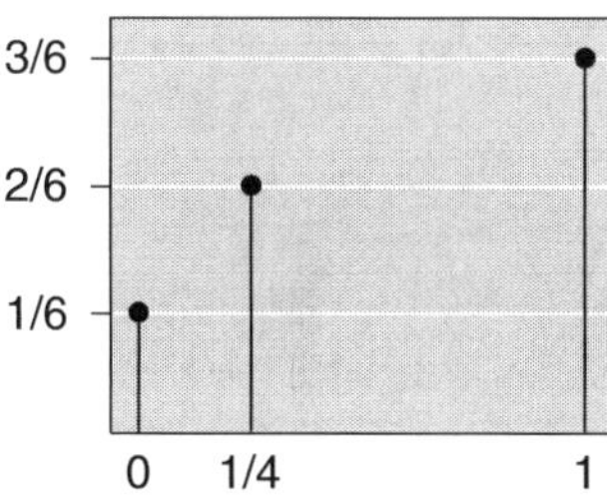

Figure 5.12 Graph of probability distribution whose outcome values are not integers.

random variable

The number of correct guesses on the quiz in Example 5.17 is an example of a random variable. That is, a **random variable** is a commonly used name for a statistic of interest. As such, a random variable is a numerical outcome affected by chance, such as the number of tornadoes to touch down in Kansas next spring or Alex Rodriguez's batting average next season.

There is a mathematical formula for computing the "binomial" theoretical probabilities in Table 5.8, but we will not learn it until Chapter 6. For now we can use the experimental probabilities obtained by simulation to estimate the theoretical probabilities. To get accurate estimates, we need a large number of simulations, but that is no problem if we have a computer program that can do the simulations for us (as the on-line instructional software accompanying the book does).

We now give an example that illustrates how the experimental probability distribution gets closer and closer to the theoretical probability distribution as we have the computer increase the number of simulations from 100 to 100,000. We shall focus on a random variable that is useful for studying the game of craps. In the game of craps (see Example 5.3), we repeatedly throw a pair of dice and add up the spots on the two dice. This sum of the scores on the two dice is a random variable. Let us assume that the dice are fair dice (e.g., symmetric casino dice) and that the 36 outcomes in Figure 5.3 are equally likely. In that case we can find the theoretical probability distribution for the sum of the two dice. In the standard notation for displaying such a probability distribution, the possible values of the random variable are denoted by x and the associated probabilities by $p(x)$. (Note that we used the notation $P(A)$ with a capital P for the probability of *an event* A, whereas we use the notation $p(x)$ with a lowercase p for the probability *at a value* x for a **probability distribution of a random variable.**) In this example the variable has a limited number of possible values x, so we can list all of them with their associated probabilities $p(x)$ (see Exercise 6, Section 5.3):

probability distribution of a random variable

Sum of dice (x)	P(sum = x) (denoted p(x))
2	1/36
3	2/36
4	3/36
5	4/36
6	5/36
7	6/36
8	5/36
9	4/36
10	3/36
11	2/36
12	1/36

probability density function (pdf)

Note that $\sum_{\text{all } x} p(x) = 1$. The function $p(x)$ is called a **probability density function (pdf)** of a random variable.

We can simulate one observation of the sum by picking a pair of digits from the random number table for a fair six-sided die (Table B.2 in Appendix B) and adding the two digits. We can simulate 100 observations of the sum by picking 100 such pairs of digits and computing the sum of each pair.

Our instructional computer software simulated 100,000 such sums of two tossed fair dice. Table 5.9 shows the experimental probabilities $\hat{p}(x)$ obtained for the first 100, 1000, 10,000, and the entire 100,000 simulations. The next-to-last column provides the theoretical probabilities $p(x)$, rounded to five significant figures for easy comparison. The last column, which computes the difference $\hat{p}(x) - p(x)$, shows the very high accuracy achieved with 100,000 simulations (the largest error being 0.00137 in size). As this example convincingly demonstrates, simulation is a powerful tool of great practical use for finding probability distributions accurately. Professional statisticians make heavy use of such simulation analyses in settings in which they are unable to derive the true probability distribution mathematically.

Table 5.9 Experimental and Theoretical Probability Distributions for the Sums of the Scores on Two Casino Dice

Sum	100 simulations	1000 simulations	10,000 simulations	100,000 simulations	Theoretical probability	Error (exp. – theor.)
2	0.02	0.025	0.0298	0.02789	0.02778	0.00011
3	0.07	0.053	0.0549	0.05610	0.05556	0.00054
4	0.05	0.093	0.0853	0.08318	0.08333	−0.00015
5	0.12	0.104	0.1112	0.10974	0.11111	−0.00137
6	0.15	0.127	0.1311	0.13834	0.13889	−0.00055
7	0.19	0.169	0.1626	0.16590	0.16667	−0.00077
8	0.11	0.128	0.1369	0.13901	0.13889	0.00012
9	0.14	0.129	0.1134	0.11118	0.11111	0.00007
10	0.07	0.081	0.0848	0.08319	0.08333	−0.00014
11	0.07	0.062	0.0581	0.05665	0.05556	0.00109
12	0.01	0.029	0.0319	0.02882	0.02778	0.00104

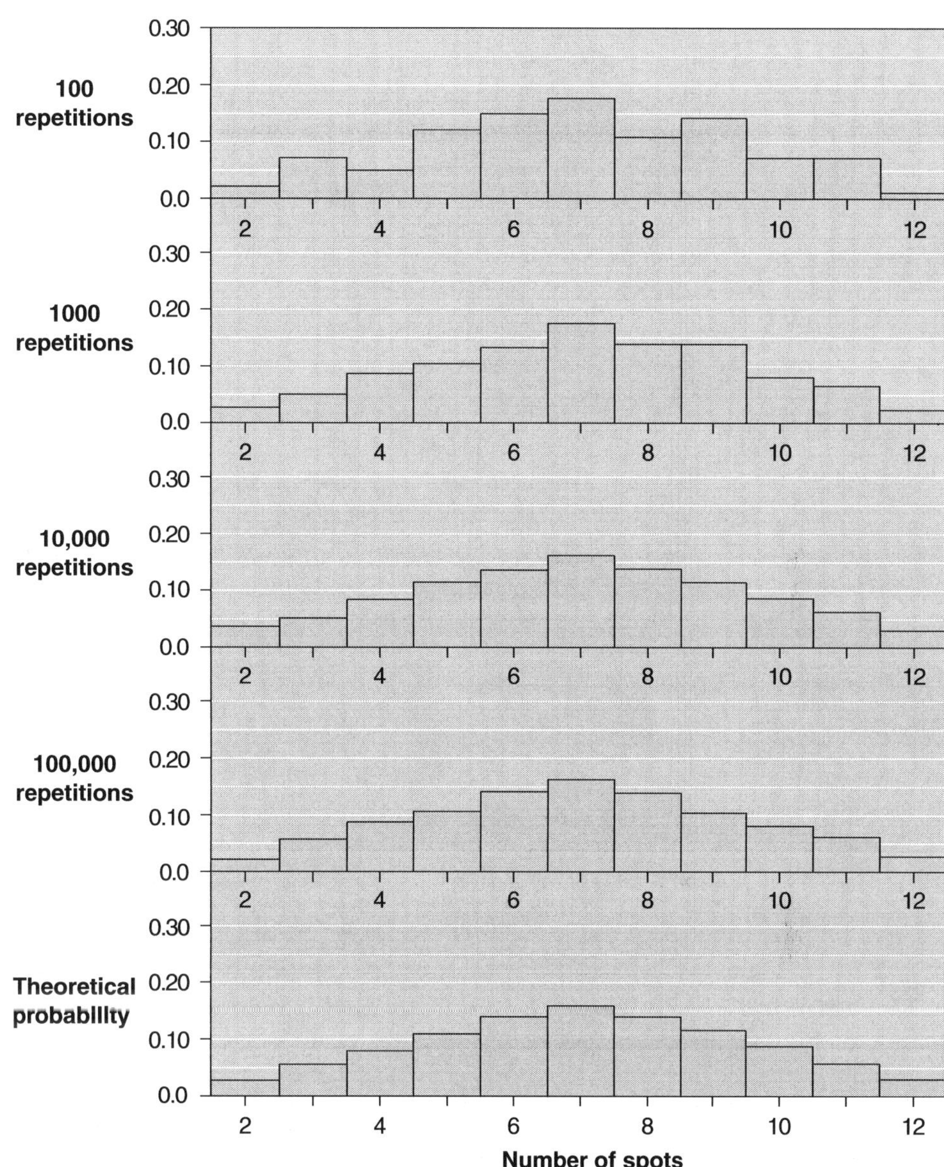

Figure 5.13 Experimental probability histograms (top four) and theoretical probability histrogram (bottom) devided from Table 5.9.

To the eye, the 100,000-simulation-based experimental probability histogram and the theoretical probability histogram are indistinguishable (see Figure 5.13). The last column of Table 5.9 makes this closeness numerically clear.

Random Experiments with Two Equally Likely Outcomes Per Trial

Many random experiments consist of repeated trials in which there are only two equally likely outcomes per trial. The numbers in Table B.1, the two-digit (0 or 1) random numbers in Appendix B, look just like numbers produced by 1800 independent tosses of a fair coin, if we identify heads with the digit 1 and tails with the digit 0. We can use the digits in Table B.1 to simulate the outcomes of independent repetitions of any multiple-trial random experiment with two equally likely outcomes per trial.

Example 5.18 True-False Quiz

Imagine that you blindly guess the answer to each of the five questions on a true-false quiz. How many correct answers will you likely get?

Solution: We use the five-step method to estimate the distribution of the random variable defined as the number of correct answers.

Step 1. Choose a Model The probability that we guess the correct answer to any particular question is 1/2. We can therefore simulate guessing an answer with selecting one of the digits 0 or 1, with equal probability. Let us identify the digit 0 with a wrong answer and the digit 1 with a correct answer.

Step 2. Define One Simulation Guessing the answers to five questions corresponds to selecting five digits according to the probability model described in Step 1. Each 1 occurring corresponds to a correct answer.

Step 3. Define the Statistic of Interest The statistic of interest is the number of correct answers. It corresponds to the number of 1s among the five digits chosen in Step 2.

Step 4. Repeat the Simulation We shall use the 40 rows of the second five-digit column in Table B.1 to simulate 40 values of the statistic of interest.

Repetition	Random digit	Number of correct answers
1	10000	1
2	00111	3
3	11000	2
4	00101	2
5	11011	4
.	.	.
.	.	.
.	.	.
40	10100	2

Step 5. Find the Experimental Probability Distribution of the Statistic of Interest We put the results of the 40 simulations in Table 5.10 (rounding the theoretical probabilities to three significant digits). The theoretical distribution is again a binomial distribution, to be studied in Chapter 6. Note that with the relatively small number of repetitions of simulations, some of the errors are fairly large, for example −0.088.

Table 5.10 Experimental and Theoretical Probabilities for the Number of Correct Answers to the Quiz

Number of correct answers	Experimental frequency	Experimental probability	Theoretical probability	Error (exp. – theor.)
0	1	0.025	0.031	−0.006
1	8	0.200	0.156	0.044
2	14	0.350	0.313	0.037
3	9	0.225	0.313	−0.088
4	6	0.150	0.156	−0.006
5	2	0.050	0.031	0.019

Example 5.19

We now consider a second example with two equally likely outcomes per trial, this time with the number of trials random.

Family Size

A young couple wants to have a daughter but does not want a large family. If their firstborn is a girl, they plan not to have any more children. If their firstborn is a boy, they will try again until they have a girl. How many children will they likely have?

Solution: We use the five-step method.

Step 1. Choose a Model We assume that each child is equally likely to be male or female. If the couple has more than one child, we assume that the children's sexes are independent (ignoring the slight possibility of identical twins). We identify having a child with selecting one of the digits 0 or 1, with equal probability. We identify the digit 0 with a boy and the digit 1 with a girl.

Step 2. Define One Simulation If the first digit we select is a 1, then we stop. Otherwise, we continue selecting digits until we first get a 1. Note that the number of trials per simulation is random here.

Step 3. Define the Statistic of Interest The statistic of interest is the number of children the couple has, which corresponds to the number of digits in the simulation sample.

Step 4. Repeat the Simulation We pick digits from left to right in the first row of Table B.1 (in Appendix B) to simulate 10 families. Table 5.11 shows the simulation of 10 families. Our computer software simulated 100,000 families. Table 5.12 shows the experimental probability distribution of the statistic of interest for the first 100, 1000, 10,000, and 100,000 simulated families, with rounding to three decimal places. In Table 5.12 we have combined 9, 10, ... into one category called "Over 8." We have to do this somewhere for distributions where the number of values for the random variable is high or even infinite. One should stop at a point when the experimental probabilities have begun to be very small; here we have stopped at 8.

Step 5. Find the Experimental Probability Distribution of the Statistic of Interest The results of the full 100,000 simulations provide accurate estimation of the theoretical probabilities, as one can see from the last three columns of Table 5.12.

Table 5.11 Simulating 10 Families

Repetitions	Digits	Number of children
1	01	2
2	1	1
3	1	1
4	01	2
5	00000001	8
6	01	2
7	01	2
8	1	1
9	1	1
10	0001	4

Table 5.12 Experimental Distributions ($\hat{p}(x)$ values) and Theoretical Distribution ($p(x)$ values) of the Number of Children

No. of children	100 simulations	1000 simulations	10,000 simulations	100,000 simulations	Theoretical probability p(x)	Error* ($\hat{p}(x) - p(x)$)
1	0.44	0.502	0.503	0.501	0.500	0.001
2	0.25	0.249	0.246	0.249	0.250	−0.001
3	0.13	0.123	0.127	0.124	0.125	−0.001
4	0.13	0.055	0.062	0.062	0.063	−0.001
5	0.03	0.034	0.029	0.032	0.031	0.001
6	0.02	0.019	0.017	0.015	0.016	−0.001
7	0.00	0.004	0.008	0.008	0.008	0.000
8	0.00	0.006	0.004	0.004	0.004	0.000
Over 8	0.00	0.008	0.003	0.004	0.004	0.000

* $\hat{p}(x)$ is from the 100,000 simulations column.

Some of the columns in Table 5.12 do not add up to 1 because of rounding (but are close to 1). With 1000 simulations, all of the experimental probabilities differ from the corresponding theoretical probabilities by less than 0.01. With 100,000 simulations, the average size of difference is slightly less than 0.001, certainly impressive accuracy. Figure 5.14 presents a visual comparison of the true and estimated probability distributions. Again, the density histogram of the statistic of interest for a large number of simulations is indistinguishable from the probability histogram of the statistic of interest. By contrast, note that with only 100 simulations there are a couple of rather large errors.

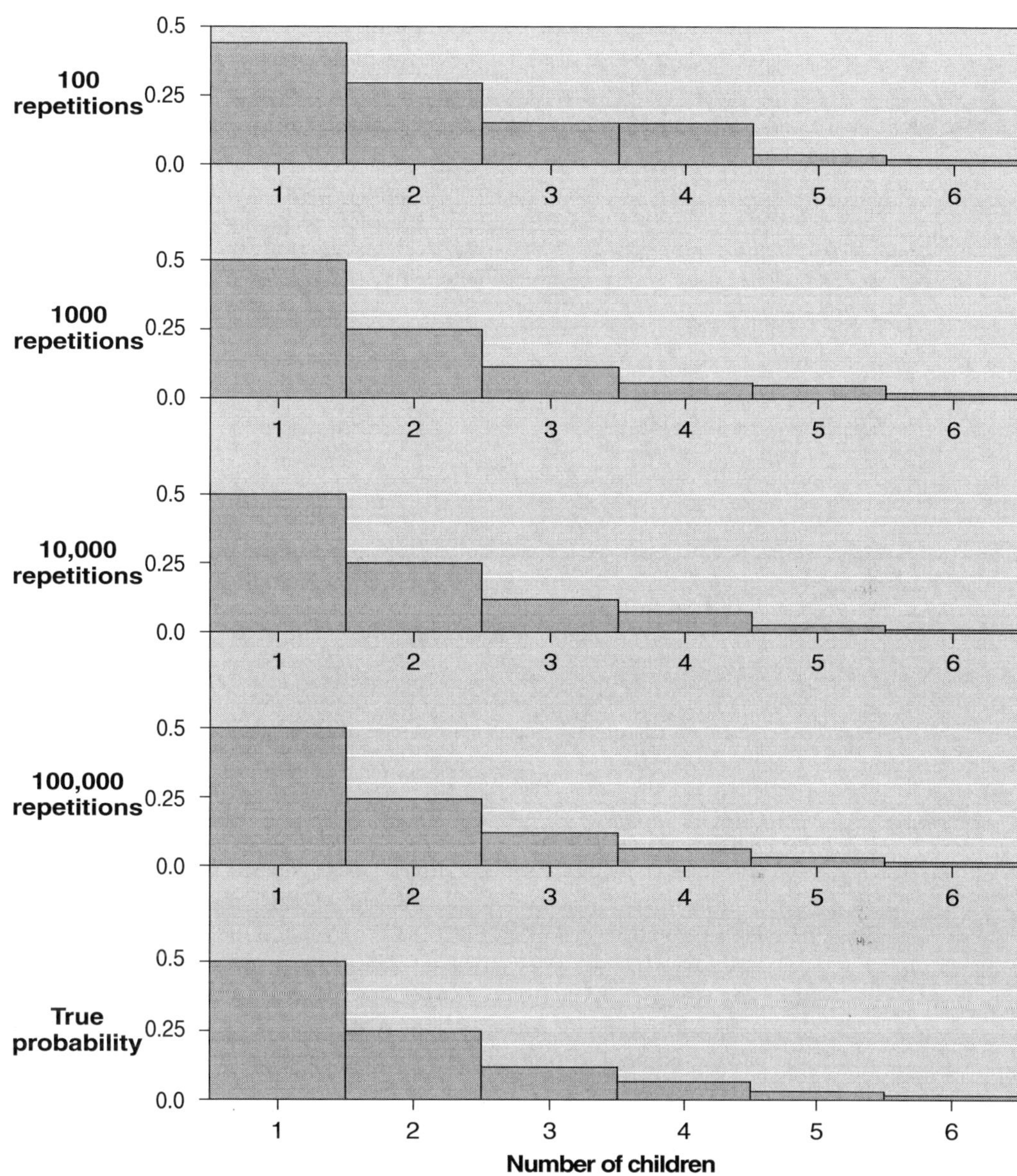

Figure 5.14 Experimental probability histrograms (top four) and theoretical probability histrogram (bottom) derived from Table 5.12.

❐ ❐

The point of the last two examples is that for all practical purposes, we can accurately discover the *entire* probability distribution of a statistic of interest (of a random variable) via simulation.

Example 5.20 What Looks Random?

If we toss a fair coin three times, there are $2 \times 2 \times 2 \times 8$ possible outcomes, shown in Figure 5.15. All of the outcomes are equally likely. The probability of getting HHH is the same as getting HTH, namely $1/8$.

First coin toss	Second coin toss	Third coin toss	Outcomes
H	H	H	H H H
		T	H H T
	T	H	H T H
		T	H T T
T	H	H	T H H
		T	T H T
	T	H	T T H
		T	T T T

Figure 5.15 Possible outcomes of three coin tosses.

If we toss the coin 10 times, there are $2 \times 2 \times 2 \times 2 \times 2 \times 2 \times 2 \times 2 \times 2 \times 2 = 1024$ equally likely possible outcomes. Each of the three outcomes

THTHHTHTHT HHHHHHHHHH HHHTTHHTHT

has the same probability, namely $1/1024 \approx 0.001$. Most people would think that the first outcome must be more likely than the second outcome because it looks "more random." But this impression is wrong, because each specific outcome (result of 10 tosses) has the same very small probability. Our intuition tells us that heads and tails should tend to alternate. This is "dead wrong," because the coin has no memory of its previous tosses! A head is just as likely to be followed by another head as by a tail. If the first five tosses happen to result in heads, then it is still just as likely that the sixth toss will result in another head as a tail. In real life a string of bad luck can in fact be followed by another bad luck event!

The third outcome contains a run of three heads. Is having at least one such run of at least three heads or at least three tails unusual? More precisely, what is the probability that in 10 tosses of a fair coin we get at least one run of three or more consecutive heads or three or more consecutive tails? Note here that there are many outcomes causing this event to happen in addition to the second and the third outcome above. This is actually a fairly challenging problem to solve theoretically.

Solution: We use the five-step method.

Step 1. Choose a Model We identify tossing the coin once with selecting one of the digits 0 or 1, with equal probability. We shall identify the digit 0 with tails and the digit 1 with heads.

Step 2. Define One Simulation Tossing the coin 10 times corresponds to selecting 10 digits as described in Step 1. Each 0 corresponds to a tail, and each 1 corresponds to a head.

Step 3. Define the Event of Interest We are looking for runs of three or more consecutive 0s or 1s.

Step 4. Repeat the Simulation We use the 40 rows of third and fourth five-digit columns in Table B.1 to do 40 simulations.

Repetition	Random digit	Run of 3 or more?
1	00010 10111	Yes
2	10111 10011	Yes
3	01101 01101	No
4	00001 01100	Yes
5	10100 00110	Yes
.	.	.
.	.	.
.	.	.
40	01101 00110	No

Step 5. Find the Experimental Probability of the Event of Interest In 35 out of the 40 repetitions, we get a run of three or more. We therefore estimate the probability of getting a run of three or more consecutive heads or tails in 10 tosses to be roughly $35/40 = 0.875$. Of course, doing 10,000 simulations would produce a much more accurate probability estimate. (Careful counting reveals that out of the 1024 possible equally likely outcomes, 846 outcomes have a run of three or more 0s or 1s, so the theoretical probability is $846/1024 = 0.826$.) That is, most of the time (over 80% of the time) we will see a run of three or more consecutive heads or tails in 10 tosses of a fair coin. Does this surprise you? Even longer runs are not that unusual. For example,

$$P(\text{a run of at least six heads or at least six tails}) = 96/1024 \approx 0.094.$$

In 10 tosses of a fair coin, runs of length six or longer occur about 10% of the time—a very counterintuitive result because such a long run seems so unlikely.

Psychologists have done studies that show that most people do not realize how likely such long runs are. People usually do not realize how often true randomness produces streaks, like a "streak of bad luck." For example, if people are asked to write down "random" sequences of 10 heads or tails, they invent fewer long runs (most people try to avoid runs of four or more) than will naturally occur with 10 independent tosses of a fair coin. Would you create anywhere near nine sequences with a run of at least six heads or six tails in a row if you made up 100 sets of 10 coin tosses? Probably not! You probably wouldn't make any! It just seems too "nonrandom."

More Simulations

Tables B.1 and B.2 in Appendix B are very convenient for simulating random phenomena by hand that have either two (like the toss of a fair coin) or six (like the roll of a fair die) equally likely outcomes. We will now learn how to simulate random phenomena with different numbers of outcomes and random phenomena whose outcomes are not equally likely. To simulate those experiments, we use the 10-digit random numbers in Table B.3 of Appendix B. This is the standard random number table appearing in so many textbooks and for so many calculators and numerical computer packages.

First we discuss how we might generate random numbers such as those in Table B.3. Seeing how a random number table can be constructed from a physical process helps us understand its basic character better.

Many state lotteries have a daily "pick three" game in which a new winning three-digit number is selected every day. The lottery uses a set of 10 ping-pong balls that are identical except for being numbered 0, 1, 2, 3, 4, 5, 6, 7, 8, 9. The balls are placed in a large glass bowl, where they are tossed about and mixed by an air jet until one is forced out at random; the number on that ball is the first digit of the winning number. The ball is then returned to the glass bowl, and the process is repeated, generating the second digit independently of the first. The second ball is replaced in the bowl, that is, put back in the bowl, and the process is repeated, generating the third digit independently of the first two. This is one example of **random sampling with replacement**, which will be discussed in greater detail in Section 5.6.

random sampling with replacement

For developing our intuition about random numbers, we may reason as if the digits in Table B.3 were generated as in the state lottery and as if the random number generator in a TI calculator or computer produces digits that way, as well. In other words, the random digits are independent of each other and are equally likely to be any of the numbers from 0 to 9, as the concrete reality of the random choosing of numbered ping-pong balls makes clear.

Example 5.21

Approximately 9 out of 10 Americans are right-handed, and 1 in 10 is left-handed. If we randomly select 15 Americans, how many will be left-handed?

Solution: We use the five-step method.

Step 1. Choose a Model We represent selecting one person at random by selecting one of the numbers 0, 1, 2, 3, 4, 5, 6, 7, 8, 9 at random with equal probability. We interpret getting 1 through 9 as selecting a right-handed person, and we interpret getting 0 as selecting a left-handed person. Therefore our chance of picking a right-handed person is 9 / 10, and our chance of picking a left-handed person is 1 / 10.

Step 2 Define One Simulation Selecting 15 persons corresponds to selecting 15 digits from Table B.3, interpreting the digits as explained in Step 1. Each 0 corresponds to a left-handed person.

Step 3. Define the Statistic of Interest The number of left-handed persons corresponds to the number of 0s among the 15 digits chosen in Step 2.

Step 4. Repeat the Simulation We shall use the 40 rows of fourth, fifth, and sixth five-digit columns in Table B.3 to simulate 40 samples of 15 persons each.

Repetition	Random digits	Number of left-handed persons
1	91807 57883 65394	1
2	78007 58644 73823	2
3	47004 48304 77410	4
4	78232 57097 01430	3
5	56211 85446 13656	0
.	.	.
.	.	.
.	.	.
40	03265 12748 77513	1

Step 5. Find the Experimental Probability Distribution of the Statistic of Interest The results of the 40 simulations are shown in Table 5.13.

Note: As remarked earlier, when the number of possible values of a random variable is large, we lump its values beyond a certain point in a single category. Thus, we create the "Over 5" category here.

Table 5.13 Experimental and Theoretical Probabilities for the Number of Left-Handed Persons in a Random Sample of 15 Persons

Number of left-handed persons	Experimental frequency	Experimental probability	Theoretical probability	Error $(\hat{p}(x) - p(x))$
0	9	0.225	0.206	0.019
1	14	0.350	0.343	0.007
2	11	0.275	0.267	0.008
3	5	0.125	0.129	−0.004
4	1	0.025	0.043	−0.018
5	0	0.000	0.010	−0.010
Over 5	0	0.000	0.002	−0.002

❐ ❐

We often have to select digits in pairs or triples, and so forth, to simulate correctly the probabilities of the various outcomes of a random phenomenon, as the next example shows.

Example 5.22

Imagine that two healthy cystic fibrosis (CF) carriers marry and have four children. CF is a recessive disease like sickle-cell anemia (see Example 5.12). How many of their children will get cystic fibrosis?

Solution: We use the five-step method.

Step 1. Choose a Model The children inherit their genotypes independently of each other (we ignore the possibility of identical twins). For each of the four children, the probability of getting cystic fibrosis is $1/4$. We represent determining a child's CF status by selecting a pair of random digits from Table B.3 (in Appendix B). There are 100 equally likely pairs: $00, 01, 02, \ldots, 98, 99$. If we pick 01 through 25, the child has cystic fibrosis. If we pick 00 or 26 through 99, the child does not have cystic fibrosis. Convince yourself that this yields $P(\text{CF}) = 0.25$ exactly. How would you model a probability of 0.38? Of 0.615?

Step 2. Define One Simulation Checking the CF status of four children corresponds to selecting four pairs of random digits, each as described in Step 1.

Step 3. Define the Statistic of Interest The number of children with CF corresponds to the number of two-digit numbers chosen in Step 2 that are between 01 and 25.

Step 4. Repeat the Simulation We shall use the first eight digits in each row of Table B.3 to form four pairs of digits. We shall simulate 40 families of four children each.

Repetition	Digits	Number of children with cystic fibrosis
1	32, 23, 61, 26	1
2	40, 33, 65, 06	1
3	88, 79, 59, 37	0
4	12, 80, 76, 51	1
5	65, 92, 99, 67	0
.	.	.
.	.	.
.	.	.
40	04, 86, 16, 38	2

Step 5. Find the Experimental Probability Distribution of the Statistic of Interest: The results of the 40 simulations are shown in Table 5.14. The experimental probability that at least one of the four children has cystic fibrosis is

$$1 - \hat{P}\,(\text{none have CF}) = 1 - 0.275 = 0.725$$

or

$$\hat{P} = (1 \text{ or } 2 \text{ or } 3 \text{ or } 4) = 0.475 + 0.200 + 0.025 + 0.025 = 0.725,$$

both methods providing the same answer, as they must. According to Table 5.14, the theoretical probability is

$$1 - 0.316 = 0.684$$

or

$$0.422 + 0.211 + 0.047 + 0.004 = 0.684.$$

Table 5.14 Experimental and Theoretical Probabilities for the Number of Children with Cystic Fibrosis

Number of children with cystic fibrosis	Experimental frequency	Experimental probability	Theoretical probability	Error $(\hat{p}(x) - p(x))$
0	11	0.275	0.316	−0.041
1	19	0.475	0.422	0.053
2	8	0.200	0.211	−0.011
3	1	0.025	0.047	−0.022
4	1	0.025	0.004	0.021

Note that we found the experimental probability of the event of interest in two ways: (1) using the complement rule and (2) breaking the event into mutually exclusive events of 1, 2, 3, or 4 children's having CF and then applying the addition rule.

If we want to simulate the random event that determines whether a child gets cystic fibrosis or not, we have several other possibilities. We could take single random digits from Table B.3, interpret 1 as the child having CF, interpret 2 through 4 as the child not having CF, and ignore 0 and 5 through 9. We could also take pairs of digits from the coin-tossing Table B.1, use 11 to represent the child having CF, and use the other three combinations—00, 01, and 10—to represent the child not having CF.

❒ ❒

Now we consider an example of disease transmission. Although it is a simplistic example, it will serve to introduce the use of statistics in the field of epidemiology. Probability models and statistical inference are vital to epidemiology. We shall only spell out Steps 1 through 3 of the five-step method.

Example 5.23

The chance of contracting strep throat when coming into contact with an infected person is estimated as 0.15. Suppose the four children of a family come into contact with an infected person. What is the chance of at least one of the four children's getting the disease?

Solution:

Step 1. Choose a Model Use pairs of random digits, sampling with replacement, with the following rules:

01 through 15	Child gets strep throat.
16 through 99, plus 00	Child does not get strep throat.

Step 2. Define One Simulation Read four pairs of random digits, one for each child in the family.

Step 3. Define the Event of Interest The event of interest occurs (some of the children in the family get sick with strep throat) if at least one of the four pairs of digits is in the range 01 through 15.

Step 4. Repeat the Simulation The more the better.

Step 5. Find the Experimental Probability of the Event of Interest Compute

$$\frac{\text{number of times event of interest occurs}}{\text{total number of repetitions}}$$

❒ ❒

Here is another model for the spread of an infectious disease.

Example 5.24

Consider one person infected by human papilloma virus (HPV). As our probability model, based on past experience, we assume that a person infected with HPV will infect one other person with probability of 0.20, will infect two other persons with probability 0.45, will infect three persons with probability 0.30, and will not infect anyone with probability 0.05. Using a pair of random digits, we can simulate the number of persons infected by this person:

00 through 04:	No one is infected.
05 through 24:	One person is infected.
25 through 69:	Two persons are infected.
70 through 99:	Three persons are infected.

If the first person infects anyone, we can simulate the number of persons infected by each of these, and so on. Will we have an epidemic that infects a lot of people, or will the disease die out quickly without infecting many people?

The disease could die out entirely by chance, especially early on. For example, $P(\text{nobody infected}) = 0.05$; that is, there is a 0.05 chance that the disease dies out immediately! If we simulate the spread of the disease many times, however, we will find that the first person will infect about two persons on average (this a topic for Chapter 6); each of these will on average infect about two additional persons; and so on. The disease will tend to spread at an exponential rate (tending to double the number infected at every stage). One could use the five-step simulation approach here to develop this into a nice group project.

Example 5.25

Data collected show that a secretary received telephone calls at a rate described by the following table:

Calls in one hour	Percent of the time
0	15%
1	35%
2	30%
3	15%
4	5%

Estimate the theoretical probability that he will receive four or more calls during a two-hour period.

Solution:

Step 1. Choose a Model Using a random number table, assign pairs of chosen digits in the following way:

Calls in an hour	Percentage of the time	Random digits
0	15%	00–14
1	35%	15–49
2	30%	50–79
3	15%	80–94
4	5%	95–99

Step 2. Define One Simulation Read two pairs of random digits, one for each of the two hours.

Step 3. Define the Event of Interest Using the table in Step 1, convert the random two-digit numbers into a number of calls in each of the two hours. Add these together, and if it is at least 4, then the simulation is a "success" in the sense that the event of interest occurs. Otherwise, it is called a "failure." (The "success/failure" terminology is convenient, and we will use it from time to time.)

Step 4. Repeat the Simulation We shall use the first four digits in each row of Table B.3 (in Appendix B) to form two pairs of digits. We simulate five repetitions.

Repetition	Random digits	Calls in Hour 1	Calls in Hour 2	Total number of calls	Success?
1	32 33	1	1	2	No
2	40 33	1	1	2	No
3	88 79	3	2	5	Yes
4	12 80	0	3	3	No
5	65 92	2	3	5	Yes

Step 5. Find the Experimental Probability of the Event of Interest Based on only five repetitions, we estimate the probability of getting four or more calls to be around $2/5 = 0.40$. By sheer coincidence this happens to be almost the theoretical answer. The theoretical probability is 0.405.

Section 5.4 Summary

We use simulation to estimate theoretical probabilities. We can simulate independent repetitions of a random phenomenon by using a random number table. Several random number tables are given in Appendix B.

The Five-Step Method

Step 1. Choose a Probability Model When using a random number table, we identify each possible outcome of the random phenomenon with the behavior of a block of digits in the random number table. This is done in such a way that the known probability of the outcome equals the probability of the corresponding behavior of the block of digits.

Step 2. Define One Simulation Take one block of digits at a time from the random number table, progressing from block to block in a consistent manner that rules out the influence of human preference or bias.

Step 3. Define the Event of Interest or Define the Statistic of Interest The event or statistic of interest is the main focus of the problem being solved.

Step 4. Repeat the Simulation Ten thousand or more simulations are desirable if using computer software.

Step 5. Find the Experimental Probability of the Event of Interest or Find the Experimental Probability Distribution of the Statistic of Interest These experimental probabilities can be used as estimates of the theoretical probabilities.

Section 5.4 Exercises

1. Estimate the probability of receiving 10 safe blood transfusions under the conditions described in Example 5.16. Make 40 simulations by using the last two five-digit columns in Table B.2 of Appendix B. For the first simulation use 62116 16111 in the first row. Identify 10 safe transfusions with not finding any sixes among the 10 digits. For the second simulation use 13316 41136 in the second row, and so on.

2. Estimate the probability distribution of the number of 6s in 10 tosses of a fair die (see Example 5.17). Make 40 simulations by using the third and the fourth five-digit columns in Table B.2. For the first simulation use 24614 22231 in the first row. Count the number of 6s. For the second simulation use 26252 14622 in the second row, and so on.

3. Estimate the probability distribution of the number of heads in 10 tosses of a fair coin. Make 40 simulations using the last two five-digit columns in Table B.1 of Appendix B. For the first simulation use 10001 10110 in the first row. Identify heads with 1 and tails with 0. For the second simulation use 00000 01111 in the second row, and so on.

4. Estimate the probability distribution of the number of girls in a family with four children. Make 40 simulations using the first four digits in each row in Table B.1. For the first simulation use 0111 in the first row. Identify 1 with a girl and 0 with a boy. For the second simulation use 1110 in the second row, and so on.

5. Estimate the probability that in 10 tosses of a fair coin, we get at least one run of four or more consecutive heads or four or more consecutive tails (see Example 5.20). Make 40 simulations using the fifth and sixth five-digit columns in Table B.1. For the first simulation use 00010 11001 in the first row. Identify 0 with tails and 1 with heads. For the second simulation use 11001 11011 in the second row, and so on. (The first simulation has a run of three 0s [tails] and the second simulation has a run of three 1s [heads], but neither has a run of four or more consecutive 0s or 1s.)

6. Estimate the probability distribution of the number of left-handed persons in a random sample of five persons (see Example 5.21). Make 40 simulations using the first five-digit column in Table B.3 of Appendix B. For the first simulation use 32236 in the first row. Identify each 0 with a left-handed person. Count the number of left-handed persons (zeros) in the sample. For the second simulation use 40336 in the second row, and so on.

For each of the following exercises, spell out Steps 1 through 3 of our five-step method. You do not need to carry out simulations. In fact, imagine the instructional software doing them for you, with the number of repetitions being 10,000 or more. In each exercise identify the statistic of interest and the event of interest.

7. A manufacturer of 60-second film advertises that 95 out of 100 prints will develop successfully. (By this it means that the probability of a print's developing successfully is 0.95.) Suppose you buy a pack of 12 and find that 2 prints do not develop. If the manufacturer's claim is true, what is the probability that 2 or even more prints do not develop in a pack of 12?

8. A pharmaceutical company knows from previous testing that a certain antibiotic capsule falls below prescribed strength 6% of the time, making the dosage ineffective. What is the probability that a prescription of 20 such capsules will contain 2 or more with ineffective dosages?

9. The following table gives the proportion of each type of hit Babe Ruth achieved during the 1928 season:

Type of hit	Percentage of total hits
Single	50
Double	15
Triple	4
Home run	31

Assume that these sample-based percentages represent the probability hitting distribution for Babe Ruth. If in a given game Ruth had three hits, estimate the probability that he had at least one home run.

10. Using the data from Exercise 9, estimate the probability that Ruth had no singles in a game in which he had exactly two hits.

5.5 SIMULATING RANDOM SAMPLING VIA A BOX MODEL

The box models presented in this section represent probability models for random phenomena (random experiments) and, as such, they help us understand what happens when some random phenomenon is repeatedly observed. But the box models also have another very important purpose: When we use the instructional software accompanying this book to simulate a random experiment on a computer, the first step in the simulation is to design a box model. This box model informs the computer what the possible outcomes of the random experiment are and what the corresponding probabilities are. With this information, the computer can then simulate the random experiment in question.

Recall the state lottery machine described in the preceding section: the glass bowl with the 10 hollow plastic balls (ping-pong balls) numbered from 0 to 9. The device is used to generate the digits of the winning "pick three" number, one digit at a time. Three times, one of the 10 balls is ejected from the bowl at random by an air jet, the number on the ball is noted, and the ball is returned to the bowl. This process generates three-place numbers in which each of the 10 digits has an equal probability of appearing in each place, independently of which digits appear in the other two places.

Imagine a simpler and cheaper version of the machine. Instead of a glass bowl, the balls are contained in a sturdy cardboard box. Rather than having an air jet shake up the balls until one is ejected, we shake the box, open it, reach in without looking, and randomly select a ball. We record the number on the ball and put the ball back. We could use this kind of arrangement to represent probability models of a wide variety of experiments, not just those involving dice, coins, or digits. Such a device is called a **box model**. A box model is really a probability model made concrete.

box model

Tossing a fair casino die, for example, can be modeled by drawing a ball at random from a box containing six balls numbered 1 through 6.

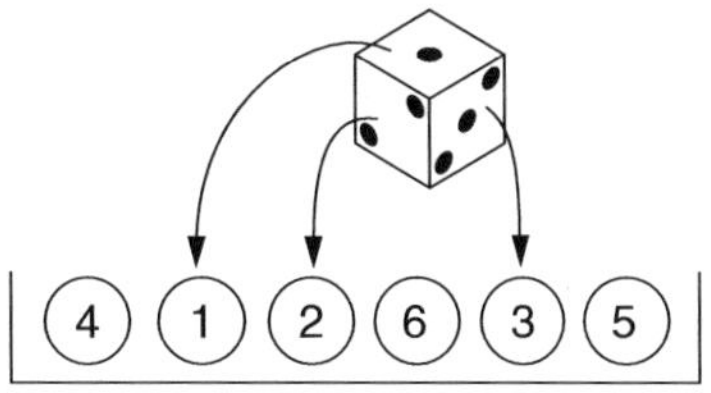

4 1 2 6 3 5

Many board games, such as Monopoly®, and many gambling games, such as craps (see Example 5.3), involve tossing two dice. Tossing two dice and recording both scores can be simulated by first drawing one ball at random from the box, noting the number on it, and then returning the ball to the box. After the ball has been returned, we again have six balls in the box. Then a second drawing is made at random from the box, and the number on that ball is recorded. We can simulate the sum of the scores on two fair dice by computing the sum of the two numbers drawn from the box.

Tossing a die 10 times (or equivalently tossing 10 different dice) and recording each die score can be modeled by 10 times drawing one of the six balls at random, noting the number on the ball, and replacing the ball in the box before the next drawing. This box model sampling is an example of **random sampling with replacement**. We can simulate the sum of the scores on 10 fair dice by drawing 10 numbers at random with replacement between draws from the box and adding up the 10 numbers.

random sampling with replacement

As the next example illustrates, if a gambler makes the same bet repeatedly, then we can simulate his or her net gain by drawing at random with replacement from a box and computing the sum of the numbers drawn.

Example 5.26

Keno is a popular game in casinos. A player bets on one or more numbers from 1 to 80. Each bet costs \$1. The casino (the "house") randomly picks 20 of the numbers from 1 to 80. If the player has placed a bet on a single number, say 33, the player wins if 33 is among the 20 numbers chosen by the house. The probability of winning is $20/80 = 1/4$. A bet on a single number pays 2 to 1. If the player wins on a \$1 bet, he gains \$2: his dollar back plus \$2 more. If he loses, the house keeps his dollar. We can simulate a player's gain or loss on a \$1 bet by drawing from the box:

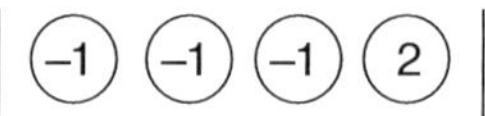

The amount he gains on the bet is the number on the chosen ball. (A "gain" of –\$1 is of course a loss of \$1.) Notice that the chance of gaining \$2 is the same whether he draws from the box or plays keno, namely $p(2) = 1/4$. And the chance of gaining –\$1 (losing a dollar) is the same whether he draws from the box or plays keno, namely $p(-1) = 3/4$. Thus, if you bet \$1 on number 33 ten times, then we can simulate your net gain or loss by randomly drawing from the box ten times with replacement between draws and adding the resulting ten numbers.

A keno player who bets on a pair of numbers wins if both her numbers are chosen by the house. Her theoretical probability of winning is $(20/80) \times (19/79)$, which is very close to 0.06. We will build the box model by assuming this probability is exactly 0.06. The bet pays 11 to 1: If she wins on a \$1 bet, she gains \$11: her stake back plus \$11 more. If she loses, the house keeps her dollar and she "wins" –\$1. We can simulate her gain or loss on a \$1 bet on a pair of numbers by drawing one ball at random from the box:

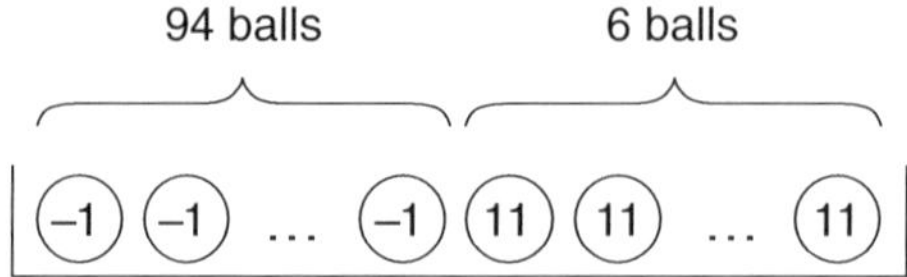

With this box, her probability of gaining \$11 is 0.06, and her probability of losing \$1 is 0.94. If she bets \$1 on a pair of numbers 30 times, we can simulate her gain or loss by randomly drawing from the box 30 times with replacement and adding up the resulting 30 numbers. It may help to emphasize that what is being simulated here is not the keno game itself but rather the probability distribution of one's chances of winning or losing, the essence of the situation, actually. That is, we are simulating 30 draws from the distribution with pdf given by:

x	-1	11
$p(x)$	0.94	0.06

Building a Box Model and Running a Five-Step Simulation

When we make a box model to simulate a random experiment, we have to decide:

What distinct numbers should be on the balls?

How many balls of each kind should there be?

Answering these two questions allows us to specify the box model completely, thus performing Step 1 of the five-step method. To carry out one simulation, we have to decide:

How many draws from the box?

Do we draw with or without replacement between draws?

This completes Step 2.

In the second keno example, the numbers on the balls were the different amounts of money that could be gained or lost on each play, namely -1 and 11. To figure out how many balls of each kind we should have, we note that the probability of drawing a ball with any particular number, say 11, should be the same as the probability of winning that amount on one play: 6/100 in the keno example. Thus, 6 out of 100 balls are 11s, and 94 of them are -1s. If we place a $1 bet on the same pair of numbers in several *independent* keno games, then we can simulate our net gain by drawing repeatedly *with replacement* from the box and adding up the resulting numbers. Finally, the number of draws from the box must equal the number of keno plays.

counting random variable

The next examples illustrate that when we simulate a **counting random variable**, we draw from a box in which the number on each ball is either 0 or 1.

Box Models for Counting Random Variables

Let us revisit Example 5.21, in which we used random digits to simulate the number of left-handed persons in a random sample of 15 Americans. Recall that 9 out of 10 Americans are right-handed and 1 in 10 is left-handed. Imagine that the 15 persons are selected one at a time. As we select the persons, we keep count of the number of left-handed persons in the sample. With each person selected, our count either goes up by 1 or stays the same:

- Zero is added to the count if the person is right-handed.
- One is added to the count if the person is left-handed.

The count has 9 chances in 10 to stay the same and 1 chance in 10 to go up by 1. It follows that with each person selected, the count changes as if we had added the number on a randomly chosen ball from the box:

We can therefore simulate the number of left-handed persons in a random sample of 15 Americans by drawing 15 times at random with replacement from this box and adding the 15 numbers we get.

Note that if a box consists of all 0s and 1s, with the proportion of 1s being equal to the proportion in a large, real-world population with a certain trait (such as left-handedness), or the proportion of 1s being the chance of a random experiment resulting in a certain outcome (such as winning at keno), then sampling from the box at random, with replacement after each draw, and adding the numbers we get simulates counting the number of times the trait or outcome of interest occurs in the sample.

Example 5.27

Blindly guessing the answers to 15 true-false questions on a quiz can be simulated by performing 15 random draws with replacement from this box:

Each guess corresponds to the random draw of one ball. If ball number 1 is chosen, the guessed answer is correct. If ball number 0 is chosen, the guessed answer is wrong. We record the number that is on the ball and return the ball to the box before the next drawing. Then the sum of the 15 numbers drawn counts the number of correct answers on the simulated 15-question quiz. It is important to realize that there is no difference in the data produced whether a real student guesses on 15 true-false questions or we do 15 random draws with replacement from the box model. The simulation is a perfect imitation of the randomly guessing student.

We can use the same box model as in Example 5.27 for simulating tossing of a coin if we identify heads with ball number 1 and tails with ball number 0. To simulate the number of heads in 50 tosses of a fair coin, we can look at the sum of the numbers on 50 balls drawn at random with replacement from the box. Similar to the above example of the guessing student, a box model–based simulation is a perfect stand-in for actual coin tossing.

We can also use the same simple (0,1) box model of Example 5.27 to solve the chapter's Key Problem.

Example 5.28 **The Tired MDs**

Assume for the moment that the program director at the teaching hospital is correct when he insists that the young doctors are just as alert at the end of their 36-hour call as at the beginning. Then, if we gave 25 doctors the same type of mental tests at the beginning of their call and again at the end, each doctor would be just as likely to do better on the second battery of tests as on the first set of tests.

With these two equally likely outcomes assumed, we can use the (0,1) box model of Example 5.27 to simulate the situation. We draw randomly with replacement from the box 25 times, once for each of the 25 doctors. If we draw ball number 0, the doctor does better on the first set of tests. If we draw ball number 1, the doctor does better on the second set of tests. The number of doctors who do better on the second set of tests will equal the number of ones in these 25 draws. Even more simply, the sum of the numbers on the 25 balls drawn equals the number of doctors who do better on the second set of tests, because summing these 25 zeros and ones is the same as just counting the ones.

Recall that only 5 of the 25 doctors did better on the second set of tests in the actual data collected by the program director. That is surprisingly few, if we were to believe the program director's claim that the young doctors were just as mentally alert at the end of their 36-hour call as at the beginning of the call. Could this low result have happened just by chance? This leads us as statisticians to ask how likely it is to observe behavior by the 25 doctors that is this extreme, or even more extreme, if the program director's claim is true. That is, if we assume that the program director is correct, what is the probability that only 5 or fewer than 5 of the 25 doctors do better on the second set of tests?

Based on 50,000 computer simulations, the experimental probability estimate is

$$\hat{P}(\text{5 or fewer do better on the second battery of tests}) = \hat{P}(\text{sum} \leq 5) = \frac{101}{50,000} \approx 0.002.$$

Since this probability is so small, we conclude that more than 5 of the 25 doctors would have done better on the second test if there had been no decline in mental alertness. Thus, the observed data of only 5 doctors doing better on the second battery of tests contradicts the program director's claim! As we will see in Chapter 9 and beyond, this line of reasoning is called **hypothesis testing** and is one of the central ideas of statistical inference.

hypothesis testing

Example 5.29 Boy or Girl?

The probability that a newborn baby is a girl is 0.487 (see Example 5.1). What is the probability that out of 100 newborn babies, at least half are girls? What is the probability that out of 2500 newborn babies, at least half are girls? How about out of 10,000? Note that we are here interested in a subtle effect: As more and more children are born, it becomes more and more likely that the proportion of girls will be close to 0.487—this is the law of statistical regularity. It thus becomes less and less likely that at least half ($\geq$0.5) of the children will be girls.

Solution: In the long run more boys than girls will be born, but in the short run we may beat the odds and have at least as many girls as boys. In order to model $P(\text{girl}) = 0.487$, we construct a box with 487 ones and 513 zeros. To simulate counting the number of girls among 100 newborns, we randomly draw with replacement 100 times from the box and add up the 100 numbers we get. (Remember that this adding counts just the 1s.) Here each 0 corresponds to a boy and each 1 to a girl. Note that the proportion of 1s in the box was set equal to the probability of a girl. We draw *with replacement* to guarantee that the simulated births are *independent* of each other. That is, the box is the same on each draw regardless of the result of the previous draw. The event of interest is getting at least 50 girls. Simulating 50,000 times reveals that the probability that at least half of the 100 newborns are girls is very close to 0.44.

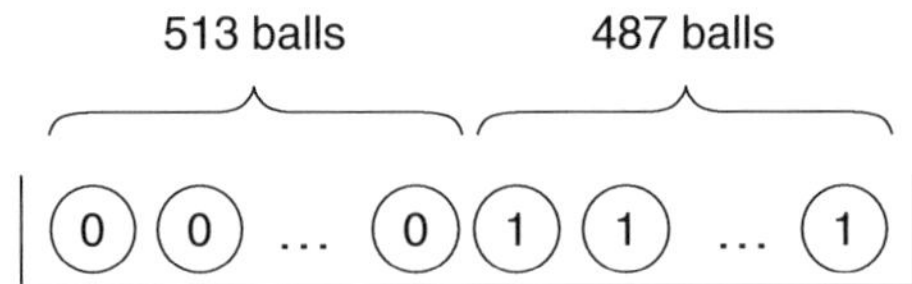

To simulate counting the number of girls among 2500 newborns, we draw 2500 times with replacement from the box. The event of interest is getting at least 1250 girls. The probability is very close to 0.10, as 50,000 simulations reveals to us.

To simulate counting the number of girls among 10,000 newborns, we draw 10,000 times with replacement from the box. The event of interest is getting at least 5000 girls. The probability is very close to 0.005.

Clearly, "beating the odds" and having at least as many girls as boys gets progressively harder as the number of draws increases from 100 to 2500 to 10,000.

Although the balls in all the box models we have seen so far have been marked with numbers, they do not have to be, even when we do computer simulation (although the textbook's accompanying instructional software does require numbering the balls).

In the following situation, numbering the balls is artificial and distracting: Four-o'clocks are ornamental plants. If we cross red-flowered ones (genotype RR) with white-flowered ones (genotype WW), all the offspring will be pink-flowered (genotype RW or WR). If we cross two pink-flowered four-o'clocks (both parents of genotype RW or WR), a tree diagram analysis like the one in Example 5.12 tells us that the offspring will have white flowers with probability 0.25, pink flowers with probability 0.5, and red flowers with probability 0.25. (Do you see why?) To simulate growing 100 offspring having two pink-flowered four-o'clock parents, we can sample 100 times at random with replacement from the box.

We then count the number of times we get a white ball, a pink ball, and a red ball. The point is that these color values are most naturally represented not as numbers but rather as nonnumerical categories (three different colors).

Section 5.5 Summary

A sampling **box model** is a boxful of balls (for concreteness, they can be viewed as ping-pong balls) that are all equally likely to be chosen. The number of balls can be large. The balls are marked to represent the different possible outcomes (usually, but not necessarily, with numbers), and the numbers of balls with the different marks are in proportion to the probabilities of the different outcomes.

If we know the possible outcomes of a random experiment and the corresponding probabilities, then we can make a box model for simulating independent repetitions of the random experiment. Each repetition (one simulation) corresponds to randomly drawing a certain number of times from the box. Specifying the box really amounts to a concrete representation of a probability distribution. It allows us to use the simulation software accompanying the textbook.

When we make a Step 1 box model for a random phenomenon, we have to decide:

What numbers should be on the balls?

How many balls should there be of each kind?

When we specify one simulation in Step 2, we need to decide:

How many draws will be made from the box?

Do we draw with or without replacement between draws?

Drawing with replacement simulates independent trials. For, clearly, the results of two such trials are unrelated. The number of draws in Step 2 always equals the number of trials or draws that constitute one simulation of the random phenomenon, such as the number of children in a family or the number of persons interviewed in a population survey.

For gambling problems in which the player makes the same bet in several independent plays:

- The numbers on the balls are the amounts that can be won or lost on a single play.
- The fraction of balls with any particular number equals the probability of winning that amount on a single play.
- The number of draws equals the number of plays.
- We draw with replacement.
- The gain or loss is the sum of the numbers on the balls drawn.

Gambling problems are very helpful in our learning about probability modeling and simulation.

We can simulate the number of times an event of interest occurs (this a counting random variable) in n independent trials by drawing n times at random with replacement from a box of 0s and 1s, where the proportion of 1s equals the probability that the event of interest occurs in one trial. The sum of the n numbers drawn will represent the number of times the event of interest occurred in the n trials.

Section 5.5 Exercises

Note: In the gambling problems below, losing an amount, say losing \$2, is the same as "winning" –\$2, so the ball to be used has the number –2 on it.

1. In the casino game high-low, a pair of fair six-sided dice is rolled, and the sum of the two scores computed (see Figure 5.3). If you bet \$1 on low, you win \$1 (get your dollar and 1 more back, with a net gain of \$1) if the sum is 2, 3, 4, 5, or 6. If you bet \$1 on high, you win \$1 if the sum is 8, 9, 10, 11, or 12. If you bet \$1 on 7, you win \$4 if the sum is 7. For each of the three \$1 bets, "low," "high," and "7," design a box model (that is, specify which numbers should be on the balls and how many balls there are with each number) for your gain or loss.

2. A roulette wheel has 38 slots, of which 18 are red, 18 are black, and 2 are green (see Examples 5.5 and 5.8). A bet on "red" pays *even money*: If you bet \$1 on red and a red number comes up, you win \$1. If a black or a green number comes up, you lose \$1. Design a box model for a \$1 bet on red. Specify which numbers should be on the balls and how many balls of each kind there should be. How many times do we draw from the box to simulate our net gain on 100 independent \$1 bets on "red"? What is the statistic of interest? Do 10,000 computer simulations to estimate your chance of making money on 100 independent \$1 bets on "red."

3. A bet on a single number at roulette (38 slots, recall) pays "35 to 1": If you bet \$1 on 3, say, and the ball ends up in slot number 3, you get your \$1 back plus \$35 in winnings. If the ball ends up in any other slot, you lose the dollar. Design a box model for a \$1 bet on a single number. Specify which numbers are on the balls an how many balls of each kind there are. How many times do we draw from the box to simulate our net gain on 35 independent \$1 bets on a single number? Do we draw with or without replacement? What is the statistic of interests? Do 10,000 computer simulations to estimate your chance of making money on 35 independent \$1 bets.

4. A "split" is two adjacent numbers on a roulette (38 slots, recall) table, such as 2 and 3. If you bet on a split and either number comes up, you get your \$1 back plus \$17 in winnings. Design a box model for a \$1 bet on a split. Do 10,000 computer simulations to estimate your chance of making money on 100 independent \$1 bets.

5. A quiz has 10 multiple-choice questions. To each question there are five suggested answers, one of which is correct. Imagine that you blindly guess the answer to each of the 10 questions. Design a box model for simulating the number of correct answers you get on the quiz. Do 10,000 computer simulations to estimate your chance of getting at least five correct answers.

6. A husband and a wife are both carriers of the albinism gene. Any child of theirs has a 0.25 chance of being an albino. Suppose the couple has five children. Design a box model for simulating the number of albinos among the five children.

7. A traffic light at an intersection is green 58% of the time for people traveling on the main street. Imagine that you pass this light twice a day: once in the morning and once in the afternoon. How would you set up a box model and simulate how many times in the 5-day workweek the light is green?

8. Consider the following probability distribution:

x	1	6	7	12
$p(x)$	0.33	0.17	0.49	0.01

How would you construct a box model and simulate drawing from the corresponding population until the first 7 is obtained?

9. Consider the following box model:

Find the probability distribution it is simulating by finding the probability $p(x)$ for each x occurring in the box.

10. Consider the following box model:

Find the probability distribution it is simulating by finding the probability $p(x)$ for each x occurring in the box.

5.6 RANDOM SAMPLING WITH OR WITHOUT REPLACEMENT

In Section 5.5 we simulated independent trials of a multistage random experiment (for example, tossing a dice twice) by drawing at random *with replacement* from a box of numbered balls. Box models can also be used to study population surveys in which a representative cross section of the population is interviewed (see Example 1.9). In such surveys the people who will be interviewed are selected at random but *without replacement* between draws, since we do not want to contact the same person twice. This is **random sampling without replacement**.

random sampling without replacement

To see the difference, consider a box containing three balls: one red, one black, and one white.

Drawing two balls at random *with* replacement means first drawing one ball at random, noting which ball was selected, replacing the ball in the box, and then drawing a second ball. Because the first ball is put back into the box before the second draw, the result of the first draw in no way influences the result of the second draw: The two draws are independent! Figure 5.16 shows a tree diagram that represents the nine possible outcomes, each with probability 1/9.

Drawing two balls at random without replacement means that the first ball is not replaced in the box before the second drawing is made. There are six possible outcomes, all equally likely, with probability 1/6 (see Figure 5.17). When we draw without replacement, the draws are not independent!

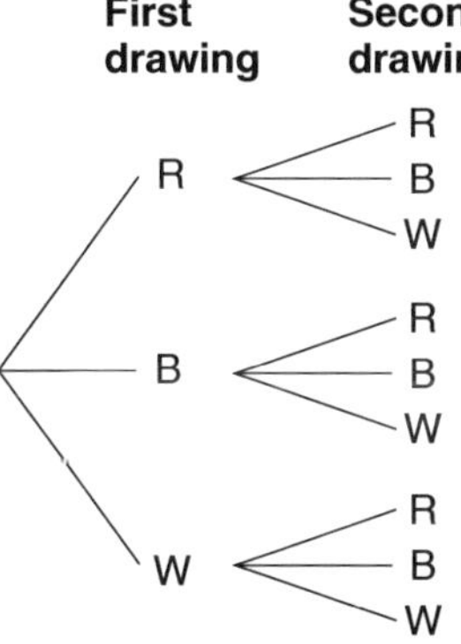

Figure 5.16 Drawing two balls at random with replacement.

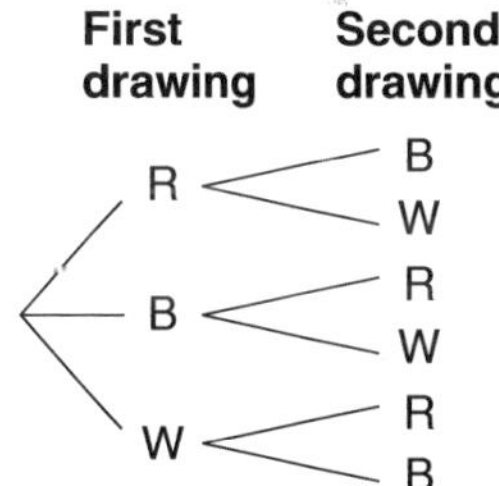

Figure 5.17 Drawing two balls at random without replacement.

Sample Surveys

Each month the Current Population Survey interviews a random sample of over 100,000 Americans to help policy makers in the government judge the economic health of the nation. The size of the labor force, the unemployment rate, the distribution of income, and educational level are all estimated from these survey results.

Market surveys use random samples to discover consumer preferences and usage of products. Public opinion polls ask people's candidate preferences before elections; their satisfaction with prominent elected officials between elections; and people's views on social controversies like same-sex marraige, legalization of undocumented aliens, stem cell research, gun control, abortion, and the death penalty.

All sample surveys use random samples drawn without replacement. By sampling without replacement, the survey organizations avoid, as they certainly should, selecting any person more than once. Unfortunately, the needed probability computations for doing statistical inference are much easier if the sampling is with replacement.

Simulation is also simpler and faster if we sample with replacement. Fortunately, whenever the sample size is less than 5% of the population size, it makes little difference whether we sample with or without replacement: Even if we draw with replacement, not too many persons will be drawn more than once. So, when we analyze sample survey data from samples of size less than 5% of the population, we can pretend that the sample was drawn with replacement. In particular, with good accuracy we can simulate the sample survey by random sampling from a box model with replacement.

Example 5.30

There are about 4 million people in Kentucky. One in eight, or about 1/2 million, are 65 years or older, whereas seven out of eight, or about 3.5 million, are under the age of 65. What is the chance that in a random sample of 448 residents of Kentucky, at least 15% are 65 years old or older?

Solution: We can simulate the number of persons 65 years old or older in the sample by computing the sum of 448 numbers drawn at random without replacement from Box A

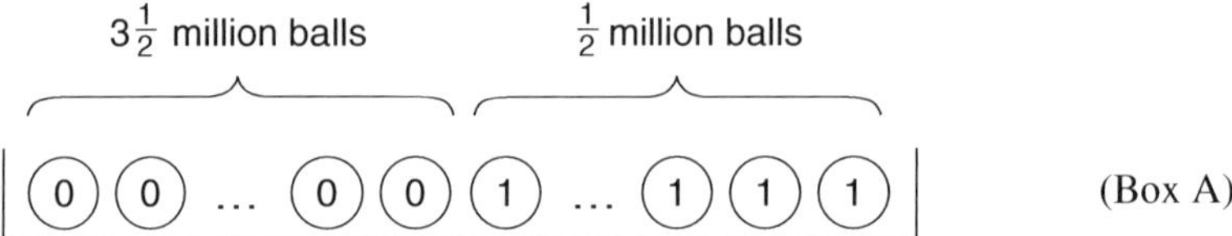

There is a 0-ball for each person under the age of 65 and a 1-ball for each person 65 or older. We want to estimate

$$P(\text{sum} \geq 68),$$

since 15% of 448 is 67.2.

Since the sample size $n = 448$ is small compared to the population size $N = 4$ million, it makes very little difference whether we sample with or without replacement from the box. So when we simulate, there is no harm in drawing the 448 balls *with* replacement.

If we draw at random *with* replacement, then it makes no difference at all whether we draw from Box A above or from Box B below

0 0 0 0 0 0 0 1 (Box B)

In both Box A and Box B, 7/8 of the balls are 0-balls and 1/8 of the balls are 1-balls. Each time we draw from either box, the chance that we get a 0-ball is 7/8, and the chance that we get a 1-ball is 1/8. When we sample *with* replacement, it is the *proportion* of balls with each particular value that matters, not the number of balls. In fact, Box A is too big for our accompanying simulation software, but Box B is easy and fast to simulate sampling from. Ten thousand simulations yielded

$$\hat{P}(\text{sum} \geq 68) = \frac{503}{10,000} = 0.0503,$$

a value that will be very close to the theoretical probability.

Example 5.31 Flag Signals

How many different signals can be made using four flags of different colors lined up on a vertical flagpole if exactly three flags are used for each signal? If one of the flags is yellow and we pick the three flags at random, what is the chance that the signal uses the yellow flag?

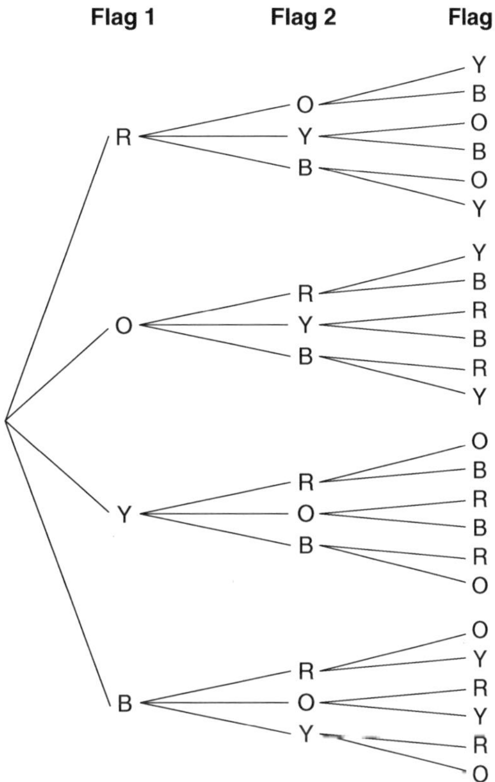

Figure 5.18 Possible three-flag signals without replacement.

Solution: Let the four colors be R (red), O (orange), Y (yellow), and B (blue). To make a flag signal, we first pick the flag to go on top; then we pick the flag to go in the middle; finally, we pick the flag to go at the bottom. In other words, we sample three flags, one at a time without replacement. A total of $4 \times 3 \times 2 = 24$ different signals are possible; they are shown in Figure 5.18. The number of signals with at least one yellow, using Figure 5.18, is $4+4+6+4 = 18$. Thus, using the equally likely outcomes rule for computing a probability,

$$P(\text{signal uses the yellow flag}) = \frac{18}{24} = \frac{3}{4}.$$

By contrast, one can show that when sampling with replacement (that is, if we hoist and lower three signal flags, one at a time, with repetitions of the same flag allowed)

$$P(\text{at least one yellow flag}) = \frac{37}{64},$$

a much smaller number than 3/4.

Whether we sample with or without replacement, there are usually too many possible outcomes to draw a tree diagram. Instead, we can use the multiplication counting rule to compute the number of possible outcomes:

multiplication counting rule

Multiplication counting rule for multiple-stage experiments

Suppose an experiment is conducted in k stages. Suppose that Stage 1 can be performed in N_1 ways; for each of these, Stage 2 can be performed in N_2 ways; for each of the ways of performing Stages 1 and 2, Stage 3 can be performed in N_3 ways; and so forth. Then there are

$$N_1 \times N_2 \times N_3 \times \cdots N_k$$

ways of performing the Stage 1 through k experiment.

To illustrate the rule, if we toss a die five times, then there are six possible outcomes for the first toss, six possible outcomes for the second toss, and so on. If we record the five scores in order, then there are

$$6 \times 6 \times 6 \times 6 \times 6 = 7776$$

possible ordered outcomes. Nobody wants to draw the tree diagram for this situation!

A regular deck of playing cards has 52 cards. Each card has a suit and a face value. The four suits are spades, hearts, diamonds, and clubs. There are 13 cards in each suit, with face values $2, 3, \ldots, 10$, jack, queen, king, and ace. Imagine that you deal five cards from a well-shuffled deck. The number of ways of dealing 5 cards, one at a time without replacement, from a deck of 52 cards is

$$52 \times 51 \times 50 \times 49 \times 48 = 311{,}875{,}200$$

(the total number of possible outcomes for the experiment of dealing a 5-card hand) if we keep track of the order in which the cards are dealt. The first card can be chosen in 52 ways, the second card in 51 ways, the third card in 50 ways, and so on. Thirteen of the 52 cards are hearts. The number of ways of dealing 5 hearts is

$$13 \times 12 \times 11 \times 10 \times 9 = 154{,}440.$$

Using

$$P(A) = (\text{number in } A)/(\text{number in } S),$$

permitted because all 5-card hands are equally likely,

$$\begin{aligned} P(\text{all five cards are hearts}) &= \frac{\text{number of 5-card hands with 5 hearts}}{\text{number of 5-card hands}} \\ &= \frac{154{,}440}{311{,}875{,}200} \approx 0.0005. \end{aligned}$$

Here is another probability problem with cards involving simple counting: Four of the 52 cards are aces. It follows that 48 cards are not aces. The number of ways that we can pick 5 cards and not get any aces is

$$48 \times 47 \times 46 \times 45 \times 44 = 205,476,480,$$

if we keep track of the order in which we pick the cards. (The first card can be chosen in 48 ways, the second card in 47 ways, and so on.) Thus

$$P(\text{no aces among 5 cards}) = 205,476,480/311,875,200 \approx 0.659.$$

Thus, about $2/3$ of the time, we will be dealt a hand with no aces in it.

Traditional statistics courses often go much further in developing counting principles useful for solving such equally likely outcome problems. But we shall not do so, because it is a distraction. Moreover, we can always use simulation to estimate probabilities relating to sampling without replacement.

Example 5.32

What is the probability of getting at least 1 ace in a hand of 5 cards dealt from an ordinary deck of 52 playing cards?

Solution: The drawing involved here is without replacement, because each card can occur only once in a hand of 5 cards. We could estimate the probability of getting at least 1 ace in a hand of 5 cards by repeatedly dealing out 5-card hands to find out the number of times 1 or more aces occur, but this would take a huge amount of time. Another approach would be to use a table of random numbers and follow the five-step procedure.

Step 1. Choose a Model Use two-digit random numbers from 01 to 52, inclusive. Ignore all others.

01–04:	ace
05–52:	remaining cards in deck

If the first six digits are 09 75 48, we treat this as 09 48 because 75 is greater than 52, so it is ignored. We now have 52 equally likely outcomes by this trick of ignoring 53–99 and 00. This is a powerful tool for obtaining equally likely probabilities when the number of outcomes is not 2, 6, 10, 100, or 1000, say.

Table 5.15 Estimating the Probability of at Least One Ace in Five Cards

Repetition	Random number	Success?
1	49 29 25 02 52	Yes
2	42 45 40 49 07	No
3	37 20 30 38 21	No
4	48 32 07 30 22	No
5	43 49 04 26 09	Yes
6	09 38 44 22 36	No
7	39 16 51 19 06	No
8	10 09 49 50 24	No
9	01 21 03 26 02	Yes
10	10 02 16 47 13	Yes

Step 2. Define a Simulation A simulation consists of reading off five random numbers between 01 and 52, ignoring duplicates. That is, if the first six digits in the table are 03 03 27, then we treat this as 03 27 because the second 03 is a duplicate.

Step 3. Define the Event of Interest A successful simulation occurs when at least one of the five two-digit random numbers is between 01 and 04, inclusive (i.e., when at least one of the numbers obtained represents an ace).

Step 4. Repeat the Simulation Do at least 100 simulations. Suppose 10 repetitions produced the results listed in Table 5.15. Here we have removed all pairs larger than 52. In 4 of the 10 repetitions, at least one of the random numbers is less than or equal to 4 (repetitions 1, 5, 9, and 10). Therefore, in 4 of the repetitions we have drawn at least one ace.

Step 5. Find the Experimental Probability of the Event of Interest Based on only 10 simulations, we estimate that the probability of getting at least one ace is around 0.4. Since we theoretically computed

$$P(\text{no aces}) = 0.659,$$

it follows from $P(\text{not } A) = 1 - P(A)$ that

$$P(\text{at least one ace}) = 1 - P(\text{no aces}) \approx 0.341.$$

Note: We could take a counting variable box model approach in Step 1, using this box:

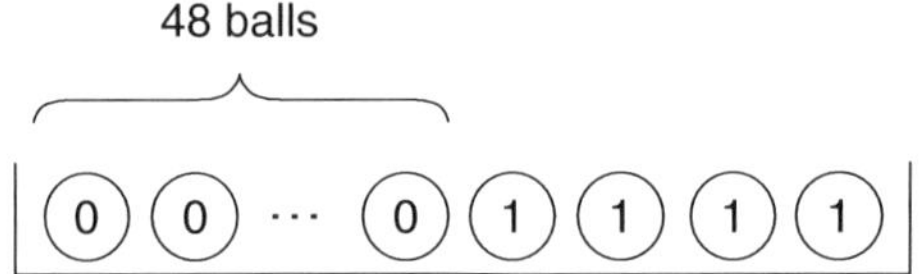

where 1 represents an ace and 0 represents a non-ace. Then Step 2 amounts to randomly sampling five times *without* replacement. The rest of the steps are the same. Moreover, by using box model–based simulation software, we can now do 10,000 or more repetitions of the basic Step 2 simulation in Step 4.

Section 5.6 Summary

The multiplication counting rule is used to count the number of different outcomes in a multistage experiment. The rule applies to sampling both with and without replacement. This rule helps us compute probabilities in multistage equally likely outcome random experiments.

Sample surveys use random sampling without replacement. If the sample size is no more than 5% of the population size, then we can simulate the sample by drawing with replacement and the box model can be built using population proportions instead of actual numbers of population members. This allows us to use a box model with a much smaller number of balls than the actual population size—which speeds up the computer simulation.

Section 5.6 Exercises

You may find some of these probability problems difficult to solve theoretically, especially when they involve sampling without replacement. Try to solve them theoretically first, but you may need to resort to the five-step method. Using the instructional software, you can obtain high accuracy by doing a larger number of simulations, 10,000 or more being ideal.

1. A bag contains five black marbles, four red marbles, and three white marbles. Three marbles are drawn in succession, producing a sample space of $12 \times 12 \times 12$ outcomes if we replace, and $12 \times 11 \times 10$ outcomes if we do not replace.
   a. If each marble is replaced before the next one is drawn, what is the probability that at least one of three is white? *Hint:* What is the complement of at least one being white?
   b. If each marble is not replaced before the next one is drawn, what is the probability that at least one of the three is white?
   c. If each marble is not replaced before the next one is drawn, what is the probability that all three are of the same color? *Hint:* The event that all three are the same color is the event (all are black) or (all are red) or (all are white).

2. A student has to match three terms that she has never seen before with their definitions. If she guesses, find her probability of
   a. getting all three correct
   b. getting none correct
   c. getting exactly one correct
   d. getting exactly two correct

   Hint: Label the terms 1, 2, 3 for convenience. Write the student's answer as a sequence such as 2, 3, 1, meaning the student assigns definition 2 to term 1, definition 3 to term 2, and so on. List all the different possible (equally likely) orders of 1, 2, 3 with which the student can respond. For example, the response 3, 2, 1 produces one match: that for term 2 and its correct definition.

3. A man has 10 keys, and exactly 1 of them fits the lock on his office door. He tries the keys 1 at a time, never trying any key more than once.
   a. Is this sampling with or without replacement? Why?
   b. What is the probability that he will obtain the correct key within his first three choices? *Hint:* Find the probability of the complement first.

4. In a small town with only 100 registered voters, two candidates are running for mayor: Mr. Jones and Mrs. Smith. You know that out of the 100 registered voters, 40 would vote for Mr. Jones and 60 for Mrs. Smith, but not everyone will actually vote. In fact, because of a terrible snowstorm, only 10 of the voters are able to vote. For each of the following questions, assume that the people who show up to vote are a random sample from the population of 100.

   Simulate the election, using at least 10,000 simulations, recalling that this is sampling without replacement (since each person can vote only once).
   a. What is the probability that Mr. Jones will win the election (that is, receive 6 or more votes)?
   b. What is the probability that Mrs. Smith will win?
   c. What is the probability that they each receive 5 votes?
   d. Do you think that the low voter turnout helped Mr. Jones's chances or hurt them? (*Hint:* What would his chances have been if all 100 had voted?)

5. Assume the setting of Exercise 4.
   a. Could you legitimately have done the simulation by sampling with replacement, assuming that your goal is to obtain answers that are approximately correct?
   b. If your answer in part (a) is "yes," build a simple a box model to simulate the 10 voters. In particular, use as few balls as possible (let 0 denote a vote for Jones and 1 a vote for Smith).
   c. Find the theoretical probability that Mrs. Smith gets all 10 votes. *Hint:* Can you assume sampling with replacement to do your computation? If so, recall that the 10 trials are independent, allowing use of the multiplication rule for independence of the different voters.

6. A group of 10 men are choosing teams to play a game of basketball, and they decide to assign people randomly to either team, with 5 players on each team. Two of the players are taller than the rest, both of them being 6 feet 5 inches tall. What is the probability that they will end up on the same team? *Hint:* This is hard to do theoretically. Simulate it with a 0,1 box of size 10, where 1 indicates being of height 6 feet 5 inches!

7. A "population" of 10 consists of 5 in favor and 5 opposed to raising the drinking age.
   a. Two are sampled randomly with replacement. What is the probability the sample is evenly split?
   b. Answer part (a) for sampling without replacement. *Hint:* Sample the 2 people one at a time.
8. Four cards are dealt one at a time. Find P(all are aces).
9. John has three pairs of shoes, five shirts, and four pairs of pants. How many ways can John dress?

CHAPTER 5 SUMMARY

The outcome of a **random experiment** (such as counting the number of Atlantic hurricanes next summer or observing whether or not a person with a kidney transplant survives 5 years), is unpredictable. However, in a very long series of **independent repetitions** of the same random experiment, a pattern emerges that can be described in terms of the stabilization of observed relative frequencies or experimental probabilities of the various possible outcomes.

For a particular outcome of a random experiment, the **theoretical probability of the outcome** is

$$\textbf{\textit{P}(outcome)} = \textbf{fraction of times the outcome would occur in an infinitely long series of independent repetitions.}$$

To estimate this probability, one method is to consider a large number of repetitions of the random experiment and compute the **experimental probability** of the outcome, using

$$\hat{P}(\text{outcome}) = \frac{\text{number of successful repetitions}}{\text{total number of repetitions}} = \text{relative frequency of the outcome.}$$

Here, a "successful repetition" means a repetition of the random experiment in which the outcome of interest occurs. The more repetitions we observe, the closer the experimental probability will likely be to the unknown theoretical probability we seek, this fundamental fact called the law of statistical regularity. The symbol $\hat{P}$ is read "P hat"; the "hat" merely reminds us that this is not the theoretical probability P but an estimate of P computed from experimental data.

The *five-step method* uses simulation—often called *Monte Carlo simulation*—to repeat the random experiment thousands of times so that we can estimate the theoretical possibilities of the various possible outcomes of the random experiment with reasonable accuracy. Our instructional software uses *box models* to model the random experiment.

The Five-Step Method

Step 1. Choose a Probability Model

Step 2. Define One Simulation

Step 3. Define the Event of Interest or Define the Statistic of Interest The event or statistic of interest is the main focus of the problem being solved.

Step 4. Repeat the Simulation Ten thousand or more simulations are desirable if using computer software.

Step 5. Find the Experimental Probability of the Event of Interest or Find the Experimental Probability Distribution of the Statistic of Interest These experimental probabilities can be used as estimates of the theoretical probabilities.

A sampling **box model** is a boxful of balls (for concreteness, they can be viewed as ping-pong balls) that are all equally likely to be chosen. (Think of the state lottery's "Pick 3" game.) The number of balls can be large. The balls are marked to represent the different possible outcomes (usually, but not necessarily, with numbers), and the numbers of balls with the different marks are in proportion to the probabilities of the different outcomes.

If we know the possible outcomes of a random experiment and the corresponding probabilities, then we can make a box model for simulating independent repetitions of the random experiment. Each repetition (one simulation) corresponds to randomly drawing a certain number of times from the box. Specifying the box really amounts to a concrete representation of a probability distribution. It allows us to use the simulation software accompanying the textbook.

When we make a Step 1 box model for a random phenomenon, we have to decide:

What numbers should be on the balls?

How many balls should there be of each kind?

When we specify one simulation in Step 2, we need to decide:

How many draws will be made from the box?

Do we draw with or without replacement between draws?

Drawing with replacement simulates independent trials. For, clearly, the results of two such trials are unrelated. The number of draws in Step 2 always equals the number of trials or draws that constitute one simulation of the random phenomenon, such as the number of children in a family or the number of persons interviewed in a population survey.

For gambling problems in which the player makes the same bet in several independent plays:

- The numbers on the balls are the amounts that can be won or lost on a single play.
- The fraction of balls with any particular number equals the probability of winning that amount on a single play.
- The number of draws equals the number of plays.
- We draw with replacement.
- The gain or loss is the sum of the numbers on the balls drawn.

Gambling problems are very helpful in our learning about probability modeling and simulation.

We can simulate the number of times an event of interest occurs (this a counting random variable) in n independent trials by drawing n times at random with replacement from a box of 0s and 1s, where the proportion of 1s equals the probability that the event of interest occurs in one trial. The sum of the n numbers drawn will represent the number of times the event of interest occurred in the n trials.

We present the *rules for theoretical probabilities* in Section 5.3:

Equally likely outcomes If all outcomes are equally likely, then we can compute the probability of any event A:

$$P(A) = \frac{\textbf{number of different outcomes in } A}{\textbf{total number of outcomes in } S}$$

Complement rule for any events The probability that a given event A does not occur is 1 minus the probability that A does occur:

$$P(\text{not } A) = 1 - P(A)$$

Addition rule for two mutually exclusive events If two events A and B have no outcomes in common (are mutually exclusive), they cannot both occur at the same time. Then

$$P(A \text{ or } B) = P(A) + P(B)$$

Addition rule for two not mutually exclusive events If the events A and B are not mutually exclusive (i.e., they can both occur at the same time), then the chance that at least one of the events occurs is

$$P(A \text{ or } B) = P(A) + P(B) - P(A \text{ and } B)$$

***General multiplication rule for two events**

$$P(A \text{ and } B) = P(A) \times P(B|A)$$

This generalizes to three events:

$$P(A \text{ and } B \text{ and } C) = P(A) \times P(B|A) \times P(C|A \text{ and } B)$$

A caution: The addition rules apply in one situation (dealing with the event (A or B)) and the multiplication rules in another (dealing with the event (A and B)). To apply either of these rules correctly, one must make sure that the appropriate situation holds.

Rules for P(event) for a probability tree

- Multiply along the appropriate branches to find P(outcome) for each outcome in the event.
- Add P(outcome) across the appropriate end branches to find P(event).

***Conditional probability**

$$P(B|A) = \frac{P(A \text{ and } B)}{P(A)}$$

Independent events The events A and B are defined to be statistcally independent if either

$$P(B|A) = P(B) \quad \text{or} \quad P(A|B) = P(A).$$

Either of these defining formulas becomes the multiplication rule for independent events: If the events A and B are independent, then

$$P(A \text{ and } B) = P(A) \times P(B).$$

Similarly, if the events A, B, and C are independent, then

$$P(A \text{ and } B \text{ and } C) = P(A) \times P(B) \times P(C)$$

*The probabilities assigned to the outcomes of a random phenomenon must be between 0 and 1 and must add up to 1:

$$0 \le P(\text{outcome}) \le 1,$$

$$\sum_{\text{all outcomes}} P(\text{outcome}) = 1.$$

*The probability of any event A equals the sum of the probabilities of the outcomes in A:

$$P(A) = \sum_{\substack{\text{all outcomes} \\ \text{in } A}} P(\text{outcome})$$

CHAPTER REVIEW EXERCISES

If a problem calls for simulation to estimate a probability, design a box model (specify the values of the balls and how many balls there are of each value in the box), and solve the problem by doing at least 10,000 computer simulations. State how many times you draw from the box, whether you draw with or without replacement, what the statistic of interest is, and what the event of interest is.

1. Assume you know that 90% of the entire student population at a large university is right-handed and 10% is left-handed. Your class has 10 students, and the room the class is assigned to has 10 desks, all designed for right-handed people.
   a. What is the probability that all the students in your class will be right-handed?
   b. What is the probability that in your class of 10 students, there will be at least 1 left-handed person?
   c. Assume now that there is 1 left-handed desk in the room. Estimate the probability that there will be a perfect match between the students and their preferred types of desk. Do 10,000 computer simulations.
   d. The probabilities in parts (b) and (c) should be different. Why?
2. State each of the following probability rules:
   a. Complement rule
   b. Addition rule—in general and for mutually exclusive events
   c. Multiplication rule—in general and for independent events
   d. P(event) when outcomes are equally likely

3. You are interested in estimating the percentage of married couples in which both husband and wife are Democrats and the percentage in which both husband and wife are Republicans. You know that in the population in general, 50% are Democrats.
   a. You should not assume that the political party of the wife is independent of the party of the husband. Why not?
   b. Since the husband's and wife's political parties are not independent, this situation cannot be simulated by simply flipping two coins, one for the husband and one for the wife. Assume that 75% of women married to Republican men and 75% of women married to Democrats share their husband's party affiliation. So, to estimate the desired percentages, begin as follows: Flip a coin 25 times, letting heads indicate that the husband is a Democrat and tails indicate that the husband is a Republican. Now, for each of the husbands, simulate the political party of the wife. (*Hint:* This is equivalent to performing 25 simulations, with each trial having a 75% chance of the outcome "same party" and a 25% chance of "different party.")
   c. Using your results, estimate the probabilities of couples in which husband and wife are both Democrats, then both Republicans. What is the proportion in which both are of the same party?
   d. Solve part (c) theoretically by using a probability tree argument.
4. A friend brings you a coin and claims it is not fair (not equally likely to give heads or tails). He claims that he flipped it five times and saw only one head.
   a. Use the five-step method with 10,000 simulations to estimate the probability of seeing at most one head in five flips of a fair coin.
   b. Does your friend have a legitimate claim? Why or why not?
   c. Solve part (a) using a theoretical probability argument.
5. Explain the difference between sampling with replacement and sampling without replacement. Give an example of where each method is used.
6. The following table presents the results of a survey asking people whether they have volunteered during the past 12 months:

By gender	
Men	45% volunteered
Women	52% volunteered
By employment status	
Full-time	50% volunteered
Part-time	58% volunteered
Not employed	46% volunteered
Retired	40% volunteered

Source: Champaign-Urbana News-Gazette, March 6, 1997, B1.

   a. Assume that these percentages represent true probabilities. If you ask a woman at random, what is the probability that she has not volunteered during the past 12 months?
   b. Estimate the probability that if you ask five retirees (men and women) whether they have volunteered in the past year, at least one of them will tell you that he or she has. Perform at least 50 simulations using a random number table, or design a box model and do a large number of computer-based simulations.

7. Now we want to calculate the probability in part (b) of Exercise 6 theoretically.
   a. Write out the complement of the event "at least one of the five retirees has volunteered during the past 12 months."
   b. Assume that the retirees are independent of each other with respect to whether they have volunteered. Using the product law of independent outcomes, write the probability of the event you found in part (a) as the product of five separate probabilities.
   c. Finally, use the property of complementary events for theoretical probability to calculate *P*(at least one of the five retirees has volunteered).

8. Once again, look at the data in Exercise 6. In a certain neighborhood, to be more specific, the breakdown of employment status (for both men and women) is as follows:

Full-time	65%
Part-time	10%
Not employed	5%
Retired	20%

   a. Using a random number table, estimate the probability that if you draw an individual at random from the neighborhood, that person has volunteered during the past 12 months.
   (*Hint:* Use four consecutive digits from the random number table. The first two tell you what employment status the person has, and the second two tell you whether the person has volunteered, noting that how one interprets the second two digits depends on the first two.)
   b. Using the results from part (a), estimate the probability that, given that a person volunteers, he or she has a full-time job.
   c. Suppose you walk around the neighborhood asking people whether they have volunteered. Are you sampling with or without replacement?
   d. Use a probability tree to solve part (a) theoretically.

9. Five friends go to a restaurant where there are 10 choices on the menu. Estimate the probability that none of the five chooses the same item, assuming each chooses 1 item and that they all make their choices independently of each other.

10. You and some friends begin to draw straws to see who has to go pick up the pizza you ordered. There are a total of 10 people in the group, and there are 10 straws. The loser is whoever draws the 1 long straw.
   a. Before any straws are drawn, what is the theoretical probability that you will be the one who has to go pick up the pizza? Assume that each straw is equally likely to be drawn.
   b. Three of your friends draw first, and none of them draws the long straw. What is your probability of being the loser now? In other words, what is *P*(drawing long straw | first 3 drew short straws)?

11. A bus is scheduled to arrive at a bus stop at exactly 8:24 A.M., but, rounding to the nearest minute, it actually arrives according to the following probability distribution at the left. You arrive at the stop to catch the bus according to the probability distribution at the right, rounding to the nearest minute:

x	$p(x)$
8:21	0.05
8:22	0.10
8:23	0.15
8:24	0.35
8:25	0.20
8:26	0.10
8:27	0.05

x	$p(x)$
8:21	0.05
8:22	0.15
8:23	0.30
8:24	0.20
8:25	0.20
8:26	0.05
8:27	0.05

a. Estimate your probability of catching the bus. (If you arrive at the same time the bus does, assume that you catch the bus.)
b. Estimate the probability that you will have to wait 3 minutes or more for the bus. (If you arrive at 8:21 and the bus comes at 8:24, you have waited exactly 3 minutes.)
c. Solve parts (a) and (b) theoretically, using a probability tree argument.

12. Return to the bus data in Exercise 11. Estimate the following conditional probabilities:
    a. *P*(you catch the bus | the bus arrives at 8:24)
    b. *P*(you miss the bus | you arrive at the bus stop at 8:25)
    c. *P*(you are forced to wait 3 minutes or more | the bus arrives at 8:26)
    d. Solve parts (a) through (c) theoretically, using the definition of conditional probability.

13. Four football teams are in a league together. Based on the strengths and weaknesses of each team, you deduce the following probabilities:

P(Team A beats Team B) = 0.75
P(Team A beats Team C) = 0.40
P(Team A beats Team D) = 0.50
P(Team B beats Team C) = 0.60
P(Team B beats Team D) = 0.30
P(Team C beats Team D) = 0.40

    a. Compute *P*(Team C beats Team A).
    b. Assume that each game is independent of the others and that each team plays each of the other teams twice during the season. Estimate the probability that Team A will finish the season with two or fewer losses (that is, a 4-2 record or better). Do at least 50 trials, with each trial being the six-game season that Team A has to play.

14. Return to the data in Exercise 13. The four teams are to play a single-elimination tournament with the following form. The teams are to be randomly assigned to the brackets.

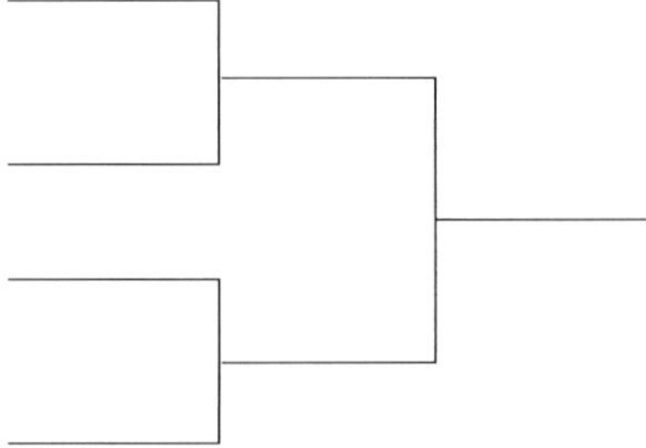

    a. Draw out 25 bracket diagrams such as the one shown above. In each case, assign the teams randomly to the four starting positions. Is this assignment of four teams an example of sampling with replacement or without replacement?
    b. Simulate each of the 25 tournaments of part (a) using the probabilities given in Exercise 13. What is your estimate for the probability that Team C wins the tournament?
    c. Using the 25 simulations you just performed, what is your estimate for the conditional probability *P*(Team C wins the tournament | Team C plays Team A in the first round)?

15. Jan plays a game in which she spins a wheel with 10 equally likely outcomes, numbered 0 through 9. Whatever number she spins is the amount of money she wins. After her first spin, she gets to decide whether she wants to spin again and try to improve on the amount she has won. If she spins a second time, she wins only the amount she receives on the second spin, not what she would have received from the first spin.
    a. Assume that Jan always decides to spin the second time (clearly a poor strategy if she got a 9 on the first spin). Estimate the probability that she will reduce the amount of money she wins on her second spin. Do at least 50 simulations. This can be done very easily by means of a random number table.
    b. Find the following theoretical conditional probabilities:

$$P(\text{on the second spin she increases her winnings} \mid \text{on first spin she won \$3})$$
$$P(\text{on the second spin she increases her winnings} \mid \text{on first spin she won \$4})$$
$$P(\text{on the second spin she increases her winnings} \mid \text{on first spin she won \$5})$$

    c. Jan decides to continue if the probability of increasing her winnings is $> 1/2$. On the basis of the results of part (b), what should Jan's rule be for deciding whether to spin the wheel a second time?

16. Design box models and, if appropriate, describe the sampling scheme for
    a. Exercise 8, part (a) (*Hint:* This requires two boxes, with the second box dependent on the result of the first box draw).
    b. Exercise 9
    c. Exercise 10, part (a)
    d. Exercise 11 (*Hint:* This requires two independent boxes.)

17. A box contains six balls, labeled 1 through 6. What is the probability that if you draw out two of the balls, the sum of the numbers on the two is equal to or greater than 6 if you
    a. draw with replacement
    b. draw without replacement

    Perform simulations for each part to estimate the probabilities, or solve the problem theoretically.

18. A study of 100 married couples is conducted as to whether they exercise at least once a week. The results are as follows:

$$\hat{P}(\text{wife exercised}) = 0.56$$
$$\hat{P}(\text{husband exercised}) = 0.49$$

    a. Is it reasonable to assume that the behaviors of the husband and wife in one random couple are independent?
    b. Suppose that within each couple, the husband's and wife's exercise behaviors are independent.
    i. Estimate *P*(both exercise).
    ii. Estimate *P*(neither exercises).

19. Write out the complement of each of the following events, but do not use the word *not*.
    a. "There was at least one game in which he scored a point."
    b. "At least five of the houses on the block are painted white."
    c. "The first card dealt from the deck of cards is a diamond."
    d. "Exactly 7 of the 10 members of the class passed the final exam."
    e. "There is at least one boy among the three children."

20. The ABO blood type distribution for African Americans is shown in Table 5.5. What is the probability that a randomly chosen African American is
    a. either type O or type A
    b. not type O

    If we randomly choose two unrelated African Americans, what is the probability that
    c. both are type O
    d. neither is type O

21. Deal one card from a well-shuffled deck of 52 cards. What is the probability that you get
    a. the queen of hearts
    b. an ace
    c. a face card
    d. a club
    e. a red card

    (Spades and clubs are black; hearts and diamonds are red; jacks, queens, and kings are face cards.)

22. Deal two cards, one at a time without replacement. What is the probability that
    a. The first card is the queen of hearts and second card is the ace of spades?
    b. Both cards are aces?
    c. Both cards are hearts?
    d. Neither card is a heart?

23. One out of 10 Americans is left-handed, and 9 out of 10 are right-handed. Draw a probability tree for selecting two unrelated persons at random and determining whether they are left-handed or right-handed. What is the probability that
    a. Both are left-handed?
    b. Both are right-handed?
    c. One of them is left-handed and the other right-handed?

24. Draw a probability tree for selecting three unrelated persons at random and determining whether they are left-handed or right-handed. What is the probability that
    a. All three are left-handed?
    b. All three are right-handed?
    c. One of three is left-handed and the other two right-handed?
    d. Two are left-handed and one is right-handed?

25. On a slot machine there are three reels, with digits 0, 1, 2, 3, 4, and 5 and a flower on each reel. When a coin is inserted and the lever pulled, the three reels spin independently, and each comes to rest at one of the seven positions mentioned with equal probability. What is the probability that
    a. Flowers show on all three reels?
    b. No flower shows?
    c. A flower shows on one of the three reels (either the first, the second, or the third reel) but not on the other two reels?
    d. Flowers show on two of the three reels but not on the third? *Hint:* To answer the last two questions, you will need to draw a probability tree.

26. The probability that a newborn baby is a boy is 0.51. If a family has five children (none are twins), what is the chance that
    a. The firstborn is a girl?
    b. The oldest and the youngest are girls?
    c. All of the children are boys?
    d. All of the children are girls?
    e. All five children have the same sex?

27. Three out of ten adults in Kentucky smoke. A telephone interview is conducted with 450 randomly chosen adults from Kentucky. Design a box model for simulating on a computer the selection of the sample and the determination of the percentage of the people in the sample who smoke. Use as few balls as possible. Define one simulation. Define the statistic of interest. Do 10,000 computer simulations to estimate the probability that the sample percentage will be between 27% and 33%, inclusive.

28. Repeat the previous exercise with a random sample of 720 adults from Utah. In Utah only one out of ten of all adults smoke. Do 10,000 computer simulations to estimate the probability that over 12% of the people in the sample smoke.

29. Hospital patients face a 1 in 20 risk of acquiring an infection in the hospital. Each year over 2 million infections are acquired in hospitals in the United States, leading directly or indirectly to over 80,000 deaths. Design a box model for randomly selecting the records of 50 former hospital patients and determining how many of the 50 patients got infected in the hospital. Define one simulation. Define the statistic of interest. Do 10,000 computer simulations to estimate the probability that at most 5 of the 50 patients got infected.

30. Repeat the previous exercise with a random sample of 30 intensive care unit patients, who face a 1 in 10 risk of getting infected in the hospital. Estimate the probability that at least 5 of the 30 intensive care patients got infected.

31. It is estimated that 40% of the 13,000 women in the United States who were diagnosed with AIDS in 1997 got infected through heterosexual contact. Imagine that we randomly select the medical records of 80 women who were diagnosed with AIDS in 1997 and check how many of these 80 women got infected through heterosexual contact. Design a box model. Define one simulation. Define the statistic of interest. Do 10,000 computer simulations to estimate the probability that at most half of the 80 women got infected through heterosexual contact.

32. In the 1980s the percentage of preschool children in the United States who were immunized for measles decreased to 70%, apparently because the public became less fearful of the disease. As a result, the number of reported cases of measles increased, reaching 30,000 in 1990. Imagine that we randomly select the 1990 medical records of 180 preschoolers and check how many of the children were immunized for measles. Design a box model. Define one simulation. Define the statistic of interest. Do 10,000 computer simulations to estimate the probability that at least 120 of the preschoolers were immunized.

PROFESSIONAL PROFILE

Blaise Pascal (1623–1662), mathematician, scholar, philosopher, inventor, and pioneer in statistics.

"The Cardsharps" by Michelangelo Merisi da Caravaggio, circa 1594. The painting is currently in the Kimbell Art Museum, Fort Worth, Texas.

Blaise Pascal

As was true for both Thomas Bayes and Christiaan Huygens, much of Blaise Pascal's interest in probability was prompted by questions arising from gambling. In 1654, his friend, the writer Antoine Gombaud (Chevalier de Méré), turned to Pascal and others for help in solving a gambling problem that had been around for centuries.

The problem centered on a case in which two individuals agree to play a certain number of games of chance (for example, a best of seven series). However, these individuals decide part way through the series to stop early, or they are interrupted before the series is complete. How should the winnings be divided fairly between them if, for example, one has won three games, and the other has won only one?

Pascal took up the challenge with his friend and fellow mathematician, Pierre de Fermat, who, in turn, exchanged letters on this question with Christiaan Huygens. From this collaboration came the notion of "expected value" and the foundation for today's understanding of probability. In addition, the collaboration helped lay the groundwork for the development of calculus, as later famously developed by Leibniz and Newton.

This breakthrough in understanding probability was just one of Pascal's contributions. He made many others. As a teenager he wrote a significant paper on the geometry of conical sections. He made major contributions to the philosophy of mathematics, including a theory of definition and a convenient tabular presentation of binomial coefficients now known as Pascal's triangle. He made significant discoveries on the principles of hydraulic fluids, and his theory is today known as Pascal's Law. He is regarded as an important author of the French Classical Period. In other areas, he invented the syringe, was the first to understand a vacuum, and, at age nineteen, invented what was one of the earliest mechanical calculating machines. Pascal's calculating machine, known today as the pascaline, was developed to help his father in his work as a tax collector. Eventually some fifty pascalines were built. Few individuals have made so many widely varying contributions to human knowledge and society.

6

Expected Value and Simulation

Measure, measure, measure.
Measure again and again.
Galileo

Objectives

After studying this chapter, you will understand the following:

- The law of large numbers for repeated, independent observations of a random variable
- The expected value (theoretical mean) $\mu_X = E(X)$ of a random variable X as a measure of the center of its distribution and as approximated by the sample mean $\overline{X}$
- The five-step simulation method for estimating the mean of a random variable
- Three theoretical formulas for computing the mean of a random variable
- The standard deviation of a random variable X, denoted σ_X or SD (X), as a measure of its spread and as approximated by the sample standard deviation S
- The five-step method for estimating the SD of a random variable
- Two theoretical formulas for computing the SD of a random variable
- The theoretical mean and SD for a sum of random draws from a box (population)

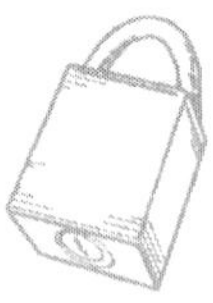

KEY PROBLEM

The Hermits' Epidemic

Six hermits live on an otherwise deserted island. An infectious disease strikes the island. None of the hermits has ever contracted the disease before. No one is immune. Once a person is exposed and contracts the disease, he becomes infectious for a short time until he recovers, becoming immune to the disease in the future.

Suppose one of the hermits gets the disease. While infectious, he randomly visits one (and only one) of the other five hermits. The visited hermit contracts the disease and, while infectious, visits one (and only one) of the other five hermits at random. If the visited hermit has not yet had the disease, he gets it and becomes infectious. The disease is transmitted in this manner until an infectious hermit visits an immune hermit (who has already had the disease), at which time the disease dies out. Assuming this pattern of visits, how many hermits can be expected to get the disease?

Probability modeling is very important in epidemiology, the study of the spread of diseases. In April 2009 the so-called swine flu broke out in Mexico and started spreading to other countries. Many people feared that it would soon become a dangerous worldwide pandemic. But computer simulations run by epidemic-modeling teams at Northwestern University and at Indiana University indicated that the actual number of flu cases would (at least in the short run) be much smaller than feared. Each four-week computer projection made by the university teams required about 10 hours of supercomputing time and used the latest knowledge of the disease parameters (how contagious the flu was and how lethal it was) plus detailed models of human mobility. Our interesting but unrealistic example is a model of an epidemic. Try imagining the kinds of probability modeling assumptions you might make to turn the Hermit problem into a realistic model for the spread of a contagious disease.

Scientists like to count and measure things. Gregor Mendel, the founder of the science of genetics, discovered the laws of heredity by counting pea plants. Counts and measurements are easy to analyze and easy to communicate to other people. Johannes Kepler deduced his three planetary laws from Tycho Brahe's careful astronomical measurements.

Alternatively, subjective information, such as a burn patient's pain level, is not as easy to interpret. Is it low, moderate, or severe? How will a doctor interpret the patient's description of his or her pain level? One person's "moderate" pain might be another person's "severe" pain. We can assign an integer indicating the pain's level or rank. For example, a burn victim's pain in a study on pain reduction might be scaled as so:

None	0
Low	1
Moderate	2
Severe	3

But the use of such a scale does not eliminate the subjectivity of the pain assessment. In the social sciences, such a rank-based scale is often called a *Likert scale*. If we use such a scale, each patient is assigned an integer to characterize his or her level of pain. Similarly, a university course evaluation form might use rankings from 1 (lowest) to 5 (highest). In a study on urbanization, families might be classified as living in a rural, suburban, or urban setting. We can assign these categories numbers, such as 1, 2, and 3 for rural, suburban, and urban. Such data are called **ranked** or **ordinal** (meaning ordered) **data**. Just as we can compare measurements and counts, we can compare ranked data. Pain level 2 is more severe than pain level 1; residential category 2 is less urban than residential category 3. We can easily handle such data statistically, as well; we would use the median and interquartile range as measures of center and spread.

ranked or ordinal data

In this manner, almost any random phenomenon being observed can be viewed as producing numerical data, whether it is measurements, counts, or rankings. This leads us to the importance of the concept of a random variable.

random variable

As stated in Chapter 5, a **random variable** is a numerical description of the outcome of a random experiment. As such, *random variable* is just another name for a statistic of interest. We usually denote random variables by capital letters, such as X, Y, and Z. As we learned in Chapter 5, the **probability distribution of a random variable** X lists the possible values X can take on (the x's) and the corresponding probabilities (the $p(x)$'s). The function $p(x)$ is called a **probability density function (pdf)**.

probability distribution of a random variable

probability density function (pdf)

In Chapter 5 we considered several random variables and their probability distributions. The examples included

- the sum of the scores on two dice
- the number of correct guesses on a five-question true-false quiz
- the number of left-handed persons in a random sample of 15 persons
- the number of children with cystic fibrosis in a four-child family with carrier parents

In this chapter we take a closer look at random variables. In particular, we study two quantities that describe the center and spread of a random variable, or more accurately, of the probability distribution (law) of the random variable. We shall define the mean and standard deviation of a random variable X and thereby of its probability distribution. The mean is usually denoted by μ_X ("mu sub X") and the standard deviation by σ_X ("sigma sub X"), with the subscript often dropped to produce μ and σ if it is obvious which random variable X we are referring to.

The mean of X is also called its "expected value." The expected value can be interpreted as the long-run average score of X. Consider, for example, the hundreds of thousands of bets made on roulette in a casino. (See Figure 5.4 in Section 5.3.) The outcome of an individual bet is unpredictable. But, as we shall see, in the long run the casino will make around 5.26 cents on average per \$1 bet. We express this fact by saying that the casino's expected gain on a \$1 bet is 5.26 cents.

In Chapter 5 we used either real data or simulated data to estimate event probabilities and probability distributions of random variables: We used the experimental probability $\hat{P}(A)$ of an event A of interest to estimate its theoretical probability $P(A)$.

In this chapter we shall use either real data or simulated data obtained by the five-step method to estimate the theoretical mean μ_X and the theoretical standard deviation σ_X of a random variable X of interest. We shall use the sample mean $\overline{X}$ and the sample standard deviation S_X of repeated observations of X as our estimates. In short, we use the experimental values to estimate the theoretical values:

$$\hat{P}(A) \approx P(A), \overline{X} \approx \mu_X, S_X \approx \sigma_X$$

6.1 THE EXPECTED VALUE (THEORETICAL MEAN) OF A RANDOM VARIABLE

On a pack of GE 40-watt Soft White lightbulbs, you can find the statement "Avg. life 1000 hrs." General Electric is not promising that every one of its 40-watt Soft White bulbs will give you 1000 hours of lighting. Some may last for fewer than 1000 hours, and some will last for more than 1000 hours. The exact number of hours may depend on many factors—slight variations in the manufacturing process, how gently the lightbulbs were handled in transport, how oily your fingers were when you inserted the bulb in the socket, or how often you turn the light on and off, for example. What General Electric is promising is that *in the long run* people will on average get (about) 1000 hours of lighting per lightbulb.

Statisticians imagine finding the average (mean) life of all 40-watt GE lightbulbs, past and future. They call this mean the *expected value* of the life of the lightbulbs. The intuitive notion is that the number of hours of life we will get from an individual lightbulb will be somewhere near the "expected life." In this sense, the expected value indicates the "typical" value of the random variable.

theoretical mean of X
expected value of X
μ_X, $E(X)$

Consider now any random variable X determined by the outcome of some random phenomenon. We define the **theoretical mean of** $\boldsymbol{X}$ (equivalently the **expected value of** $\boldsymbol{X}$) as the average value of X in a very, very long series of independent repetitions of the random phenomenon. We usually use the notation $\boldsymbol{\mu_X}$ or $\boldsymbol{E(X)}$ to denote the theoretical mean or expected value of X. Note that μ_X and $E(X)$ denote the same quantity.

The following example illustrates how we can compute μ_X when X is a counting random variable. The example also illustrates the interpretation of μ_X as the "typical" value of X, when X is observed repeatedly.

Example 6.1

In the card game of bridge, a well-shuffled deck of 52 cards is dealt, 1 card at a time, to four players, who receive 13 cards each. A bridge hand is thus a random sample of 13 cards drawn without replacement. The number of hearts, X, among the 13 cards varies from bridge hand to bridge hand. Sometimes we get no hearts. At other times we get 6 hearts. We are interested in two related questions. First, if we play a lot of bridge hands, how many hearts will we get per hand on average? That is, what is $E(X)$? Secondly, how does the distribution of the random number of hearts in a bridge hand compare to the central value $E(X)$? In particular, how close does X tend to be to its "typical value" $E(X)$?

Solution: The first question is easy to answer: A quarter of the cards in the deck of cards are hearts. Thus, as we are dealt hundreds of bridge hands, we know from Chapter 5 that very close to 1/4 of the cards we are dealt will be hearts; that is, (number of hearts)/(number of cards dealt) $\approx$1/4. There are 13 cards in a bridge hand. Thus, in the long run, the average number of hearts per bridge hand, namely,

$$\begin{aligned}\frac{\text{number of hearts}}{\text{number of bridge hands}} &= \frac{\text{number of cards dealt}}{\text{number of bridge hands}} \cdot \frac{\text{number of hearts}}{\text{number of cards dealt}} \\ &= 13 \cdot \frac{\text{number of hearts}}{\text{number of cards dealt}} \\ &\approx 13 \cdot \frac{1}{4} = 3.25.\end{aligned}$$

So $E(X)$, the long-run average number of hearts per bridge hand, is 3.25.

Now, the second question is trickier. Since one cannot be dealt 3.25 hearts, the mean is not even a possible value of X. How likely is it that the number of hearts dealt will be close to 3.25? For example, is being dealt 0 hearts or 8 hearts very unusual? To help answer the second question, we will simulate a large number of bridge hands, focusing on the experimental probability distribution of the number of hearts obtained. To simulate drawing a bridge hand and counting the number of hearts, we can use a box model with a 0-ball for each of the 39 nonhearts and a 1-ball for each of the 13 hearts in the deck of cards.

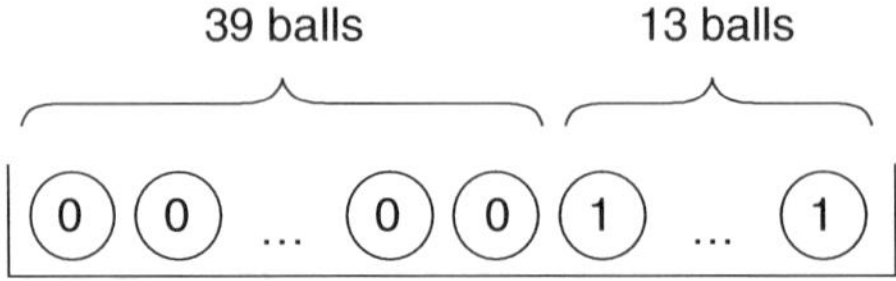

Note that we do not need to have the balls in the box indicate irrelevant aspects, such as the face value of the card and whether a nonheart is a spade, a diamond, or a club. The box models only those aspects of the physical reality that are important to us!

We can simulate the number of hearts in a bridge hand by drawing 13 balls at random from the box *without replacement*. The number of hearts corresponds to the sum of the numbers on the 13 balls drawn. In fact, we simulated 100,000 bridge hands, getting an average of $\bar{x} = 3.245$ hearts per hand and obtaining the experimental probability distribution of X given in Table 6.1 (rounded to three digits to the right of the decimal point) and Figure 6.1.

Table 6.1 Number of Hearts X in a Bridge Hand

Number of hearts, x	Probability $\hat{p}(x)$
0	0.014
1	0.080
2	0.207
3	0.290
4	0.236
5	0,124
6	0.041
7	0.008
Over 7	0.000
Total	1.000

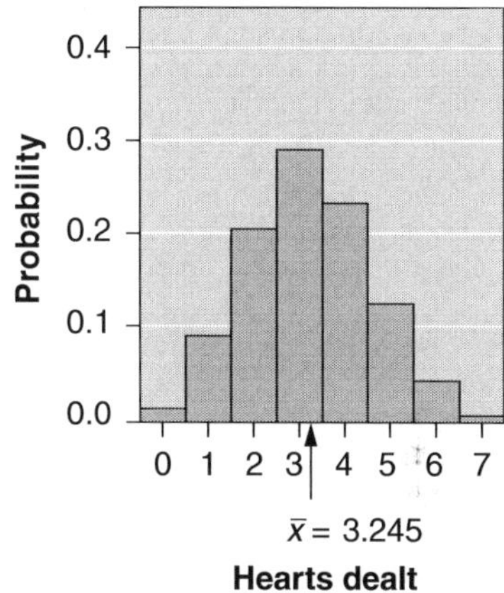

Figure 6.1 Experimental probability distribution and mean of the number of hearts in a bridge hand.

We see that $\hat{p}(2) + \hat{p}(3) + \hat{p}(4) + \hat{p}(5) = 0.857$. This tells us that in the long run, we will be dealt between two and five hearts about 86% of the time. Thus, the number of hearts is usually close to the mean (or expected value), 3.25.

law of large numbers

The fact that, in the long run, the average number of hearts per bridge hand (we observed 3.245 in the above example) will be close to $\mu_X = E(X) = 3.25$ is an example of what statisticians call the **law of large numbers**. This law holds for any sequence of independent repetitions of a numerically valued random phenomenon, that is, independent replications of a random variable.

❒ ❒

Example 6.1 illustrated the law of large numbers:

law of large numbers for independent repetitions

Law of Large Numbers for Independent Repetitions

Consider a random variable X determined by the outcome of some random phenomenon (experiment). In a very long sequence of independent repetitions of the random experiment, the sample mean $\overline{X}$ of the observed values of X will be very close to the theoretical mean μ_X. That is,

$$\overline{X} \approx \mu_X.$$

random sample

Note: A sequence of independent repetitions of the same random variable is usually called a **random sample**.

Important Expected Value Formulas

The argument we used in Example 6.1 to compute the expected number of hearts in a bridge hand can be used in many situations. Consider the following examples, which will lead us to an important formula for $E(X)$.

Example 6.2

A multiple-choice exam has 40 questions. To each question there are five options to choose from, one of which is correct. Imagine that you have no clue whatsoever what the correct answers are and blindly (randomly) guess the answer to each of the 40 questions. How many correct answers should you expect?

Solution: $40 \times (1/5) = 8$

Reason: If each one of thousands of people guesses at the answers to all 40 questions, then very close to 1 / 5 of all the tens of thousands of guesses will be correct; that is,

$$\frac{\text{number of correct guesses}}{\text{number of guesses}} \approx \frac{1}{5}.$$

It follows that the average number of correct answers per person is

$$\begin{aligned}\frac{\text{number of correct guesses}}{\text{number of persons}} &= \frac{\text{number of guesses}}{\text{number of persons}} \times \frac{\text{number of correct guesses}}{\text{number of guesses}}\\ &= 40 \times \frac{\text{number of correct guesses}}{\text{number of guesses}} \approx 40 \times \frac{1}{5} = 8.\end{aligned}$$

This shows that the expected number of correct answers on the exam is $40 \times (1/5) = 8$.

Example 6.3

A fair coin is tossed 25 times. What is the expected number of heads?

Solution: $25 \times (1/2) = 12.5$. Do not round 12.5 off to 12 or 13.

Reason: If thousands of persons each tossed a fair coin 25 times, then very close to 1/2 of the tens of thousands of coin tosses would result in heads. It follows that the average number of heads per person would be very close to $25 \times (1/2)$.

Example 6.4

A fair die is tossed 45 times. What is the expected number of 6s?

Solution: $45 \times (1/6) = 7.5$

Reason: In the long run, very close to 1/6 of the tosses of a fair die will result in a 6. If thousands of persons each tossed a fair die 45 times, their average number of 6s per person would be very close to $45 \times 1/6$.

Example 6.5

Ten percent of all Americans are left-handed. What is the expected number of left-handed persons in a random sample of 85 Americans?

Solution: $85 \times 0.1 = 8.5$

Example 6.6

Thirty-five percent of all American women age 25 to 29 are single. What is the expected number of single women in a random sample of 60 American women age 25 to 29?

Solution: $60 \times 0.35 = 21$

❐ ❐

counting random variable

The solutions to Examples 6.1 through 6.6 all have a common form. Each random variable is a **counting variable**. That is, it counts the number of times something happens in repeated trials, like the number of times we draw a heart in 13 draws from a deck of cards.

expected value of a counting random variable

The n Times p Formula for the Expected Value of a Counting Random Variable

Consider an event that has probability p of occurring in each of n trials. Let the counting random variable X denote the number of times that the event of interest occurs in the n trials. Then the expected value $E(X)$ (theoretical mean μ_X) is given by

$$E(X) = \mu_X = np.$$

In Example 6.1 we counted the number of times we got a heart in $n = 13$ drawings from a well-shuffled deck of cards. In each drawing the chance of getting a heart was $p = 1/4$. Thus, $E(X) = np = 13 \cdot (1/4) = 3.25$. In Example 6.5 we counted the number of times we selected a left-handed person in $n = 85$ drawings from the American population. In each drawing the chance of getting a left-handed person is $p = 0.1$. Thus, $E(X) = np = 85 \cdot 0.1 = 8.5$.

The n times p formula for μ_X only applies when we are counting the number of times that something happens in n repeated trials. We next learn a formula for computing the theoretical mean in general.

distribution formula for the expected value of a random variable

Distribution Formula for the Expected Value of a Random Variable

If the possible values of X are $x_1, x_2, \ldots, x_k$, with corresponding probabilities $p(x_1), p(x_2), \ldots, p(x_k)$, we can compute the expected value of X by multiplying each possible value by its probability and adding up all the products. That is,

$$\begin{aligned} E(X) = \mu_X &= x_1 p(x_1) + x_2 p(x_2) + \cdots + x_k p(x_k) \\ &= \sum xp(x). \end{aligned}$$

Example 6.7

If X denotes the score on a fair die, then each of the possible values $1, 2, \ldots, 6$ has probability 1/6, producing the probability density function (pdf):

$$p(1) = p(2) = \cdots = p(6) = 1/6.$$

Imagine that the die is tossed thousands of times. In the long run each of the values $1, 2, \ldots, 6$ will occur about 1/6 of the time, and the average score $\overline{X}$ will approach

$$E(X) = \sum xp(x) = \left(1 \cdot \frac{1}{6}\right) + \left(2 \cdot \frac{1}{6}\right) + \left(3 \cdot \frac{1}{6}\right) + \left(4 \cdot \frac{1}{6}\right) + \left(5 \cdot \frac{1}{6}\right) + \left(6 \cdot \frac{1}{6}\right)$$
$$= \frac{21}{6} = 3.5.$$

❒ ❒

Recall the box models in Section 5.5 and in Example 6.1. We used box models to simulate the number of hearts in a bridge hand, the number of correct guesses on a 15-question true-false quiz, our net gain on thirty \$1 bets on a pair of numbers in keno, and the number of girls among 100 newborns. In each case we computed the sum Y of n random draws from the box. It turns out we can compute $E(Y)$ directly and simply from the contents of the box.

The following formula applies to sums of repeated scores.

box model formulas for expected value of the sum of random draws from a box

Box Model Formulas for the Expected Value of the Sum Y of n Random Draws from a Box

If Y is the sum of n repeated random draws from a box (made with or without replacement between draws), then

$$E(Y) = \mu_Y = n \cdot (\textbf{box mean}).$$

Here the box mean is just the mean of all the numbers in the box.

In particular, if we have just one random draw ($n = 1$), producing the random variable X, then

$$\mu_X = \textbf{box mean.}$$

We also have a box model–based formula for the expected value of the sample mean of n random draws from a box.

box model formula for $E(\overline{X})$

Box Model Formula for Expected Value for the Sample Mean of n Random Draws from a Box

The expected value of the sample mean $\overline{X}$ of n repeated random draws (either with or without replacement) from a box is the box mean:

$$E(\overline{X}) = \mu_{\overline{X}} = \textbf{box mean.}$$

Note that the formula for $E(\overline{X})$ is the same regardless of the sample size n.

Clearly, the above expected value formulas apply whenever we use a box model to model sampling from a real-world population, in which case the box mean is the population mean.

Population Surveys and Random Variables

Many national surveys are carried out via random sampling from various real-world populations. These surveys have great influence on public policy, political campaigns, and marketing.

Example 6.8

A "household" is either a single person living alone or a group of people living together, regardless of their relationship to each other. There are over 100 million households in the United States. What is the average size of American households?

Solution: Here is the distribution (pdf) of American households by size, according to the U.S. Census Bureau:

Household size, x	1	2	3	4	5	6	7
Fraction of households, $p(x)$	0.26	0.32	0.17	0.15	0.07	0.02	0.01

Suppose we select a household at random. That selection is equivalent to drawing a ball from a box containing over 100 million balls, one for each American household, each ball numbered with the size of the corresponding household. The size of the household drawn is a random variable X, with probability distribution given by the preceding table. The mean μ_X equals the average size of all American households according to the box model formula. But, by our distribution formula, μ_X can be computed

$$\begin{aligned}\mu_X &= \sum xp(x) \\ &= 1(0.26) + 2(0.32) + 3(0.17) + 4(0.15) + 5(0.07) + 6(0.02) + 7(0.01) \\ &= 2.55.\end{aligned}$$

So that is the average size of all American households.

We have ignored households with eight or more members, because less than 0.5% of all households are that large. Figure 6.2 locates the mean on the probability histogram.

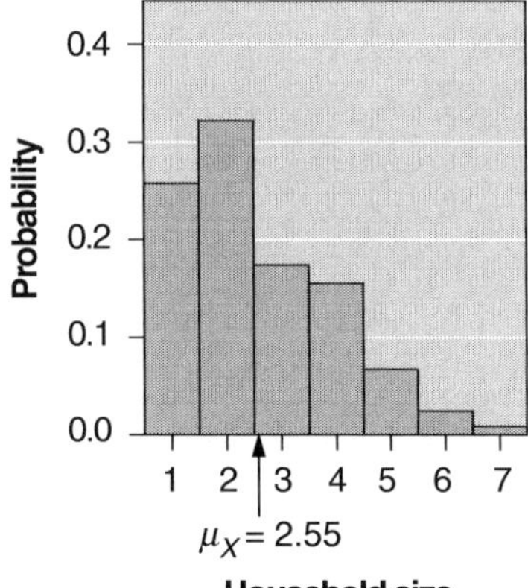

Figure 6.2 Probability distribution of U.S. households by size.

❐ ❐

How are incomes distributed in the United States? Are household incomes keeping up with inflation? How do the incomes of men and women compare? How do the incomes of various minority groups compare with those of the rest of the population? What is the unemployment rate for various population groups? The monthly Current Population Survey attempts to answer such questions, using a nationwide random sample of about 57,000 households. A large amount of economic and social data is published in the annual *Statistical Abstract of the United States* (see the U.S. Census Bureau Web site, www.census.gov).

Suppose we select one of the over 100 million U.S. households at random. The income of the chosen household is a random variable X, whose mean μ_X equals the average income of all American households (the box model mean). As in Example 6.8, selecting one household corresponds to drawing from a box model with one ball for each of the over 100 million households. In this case the number on each ball is the previous year's income for the corresponding household. If we sample many households at random without replacement (as the Current Population Survey does), then the average income for the households in the sample, $\overline{X}$, will be close to the population mean, μ_X. This is an application of the law of large numbers for human population surveys and other real-world population random sampling.

law of large numbers for real-world population random sampling

Law of Large Numbers for Real-World Population Random Sampling

Draw observations $X_1, X_2, \ldots, X_n$ at random *without replacement* from a real-world population with mean μ_X. If the sample size n is large, then the sample mean $\overline{X} = \sum X/n$ of the observed values will be close to μ_X. That is,

$$\overline{X} \approx \mu_X.$$

Box Models and Expected Values

Consider again the game of keno, described in Section 5.5. We can simulate a player's gain on a $1 bet on a single number by drawing one ball at random from the following box:

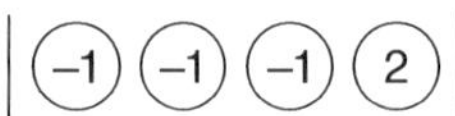

The corresponding probability distribution (pdf) for the player's gain X on one bet is

x	-1	2
$p(x)$	3/4	1/4

We can simulate a player's average gain (or loss) on 100 independent $1 bets on a single number by drawing 100 times at random with replacement from the box and computing the mean $\overline{X}$ of the resulting winnings. What can we say about this sample mean? It is obviously a random variable. So we can not tell the exact value of $\overline{X}$ ahead of time. But although $\overline{X}$ depends on chance, it will obey the law of large numbers. That is, the average gain $\overline{X}$ on the 100 bets will be close to the theoretical mean,

$$\mu_X = \text{box mean} = \frac{(-1-1-1+2)}{4} = -\frac{1}{4}$$

Here we could just as easily have computed the theoretical mean of one bet μ_X using the distribution formula:

$$\mu_X = (-1)p(-1) + 2p(2) = \left(-1 \cdot \frac{3}{4}\right) + \left(2 \cdot \frac{1}{4}\right) = -\frac{1}{4}$$

We conclude that in a very long series of independent \$1 bets on a single number in keno, your average winning per bet will approach -1/4 dollar. In other words, in the long run you will, on average, lose a quarter per \$1 bet. If you make 100 one-dollar bets on a single number in keno, then by the box model formula for the sum of n draws from a box, your expected gain is $100 \cdot (-\$0.25) = -\$25$, that is, you can expect to lose around \$25. If you make 1000 one-dollar bets, your expected loss is \$250.

Example 6.9

If you repeatedly bet \$1 on a pair of numbers in keno, what will your average loss per bet be in the long run?

Solution: Recall the discussion of keno in Section 5.5. We learned that we can simulate a gambler's gain on a \$1 bet on a pair of numbers by drawing one ball at random from the following box:

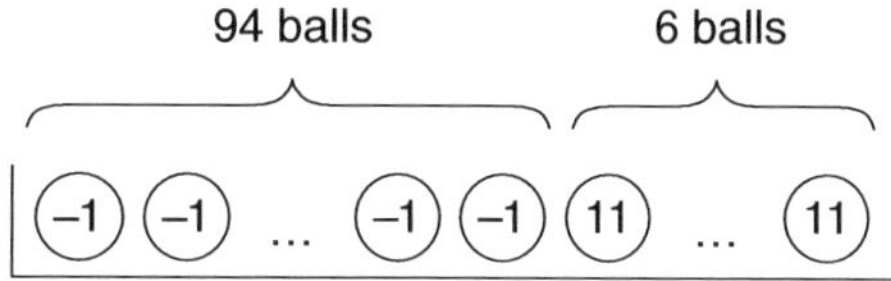

The gambler's expected gain on a single \$1 bet therefore equals the box mean:

$$\mu_X = \frac{(94 \cdot (-1)) + (6 \cdot 11)}{100} = \frac{-94+66}{100} = -\frac{28}{100} = -\$0.28$$

In the long run the gambler will on average lose \$0.28 per \$1 bet.

In the short run it is possible that the gambler beats the odds and wins more money than he loses and comes out ahead. But the more bets the gambler makes, the less likely it is that he comes out ahead. For example, as we shall see in Chapter 7 (or as we can show using simulation), the gambler has a 46% chance (i.e., a probability of 0.46) of making money on 10 independent \$1 bets on a pair of numbers, a 15% chance of making money on 100 bets, but only 1 chance in 1000 of making money on 1000 bets. If the gambler bets on a single number, then his chances of coming out ahead are even less: The gambler has a 22% chance of making money on 10 independent \$1 bets on a single number, a 2.5% chance of making money on 100 bets, and no practical chance at all of making money on 1000 bets. You cannot beat the odds in the long run.

This explains why gambling is a can't-lose business for the casinos; it is not gambling for them at all. Imagine, for example, a thousand gamblers each playing keno 10 times, making \$1 bets on either a single number or on a pair of numbers independently of each other. Each of the gamblers has a reasonable chance of making a little money, but most will not. On 10,000 independent \$1 bets (1000 gamblers betting 10 times), it can be shown that the house is for all practical purposes guaranteed to make over \$2000. (If you do not believe this, simulate 5000 one-number plays and 5000 two-number plays using the course software twice.)

As thousands of gamblers independently make bets, the *law of large numbers* guarantees the casino an average gain of around \$0.25 per \$1 bet on a single number in keno and around \$0.28 per \$1 bet on a pair of numbers. However, the law of large numbers does not apply to the individual gambler who makes only a few bets. A few gamblers do make money. That is why gamblers flock to casinos. Most gamblers don't make money, of course.

Life insurance companies work much the same way as casinos. The company is betting that not too many of its customers will collect on their policies. Using life tables, which list the probabilities of persons of various ages and conditions in life dying within the next year, the life insurance company computes the expected payout on each policy and sets the insurance rates high enough to cover the total of all the expected payouts, company expenses, plus dividends to shareholders. With thousands of policyholders, the insurance company "plays" often enough to rely on the law of large numbers: The company's average payout per policy will be close to its expected value. The probabilities in the life tables are experimental probabilities obtained from careful analysis of large numbers of lifetimes; these experimental probabilities are thus very close to their associated theoretical probabilities. They are developed by professionals known as *actuaries,* who are trained in statistics.

Finally, we shall illustrate the *law of large numbers* by simulating 100,000 tosses of a symmetric casino die. The box for simulating the score X on a fair die is

(1) (2) (3) (4) (5) (6)

Since μ_X = box average, this yields

$$\mu_X = \frac{1+2+\cdots+6}{6} = \frac{21}{6} = 3.5.$$

This is, of course, the same value we got in Example 6.7 using the distribution formula for expected values $(E(X) = \sum xp(x))$.

In the following we have tabulated the sample mean $\overline{x}$ and the difference between the sample mean and the theoretical mean $\mu_X = 3.5$, namely, $(\overline{x} - 3.5)$, for 10, 50, 100 simulations, up to 100,000 simulations of a fair die. We see that $\overline{x}$ dances around the theoretical mean 3.5, tending to get closer as the number of trials increases.

Number of throws	$\overline{x}$	$\overline{x} - 3.5$
10	2.800	−0.700
50	3.460	−0.040
100	3.230	−0.270
500	3.450	−0.050
1000	3.493	−0.007
5000	3.540	0.040
10,000	3.496	−0.004
50,000	3.498	−0.002
100,000	3.503	0.003

Figure 6.3 displays the average score $\overline{x}$ as a function of the number of tosses, using a logarithmic scale for n.

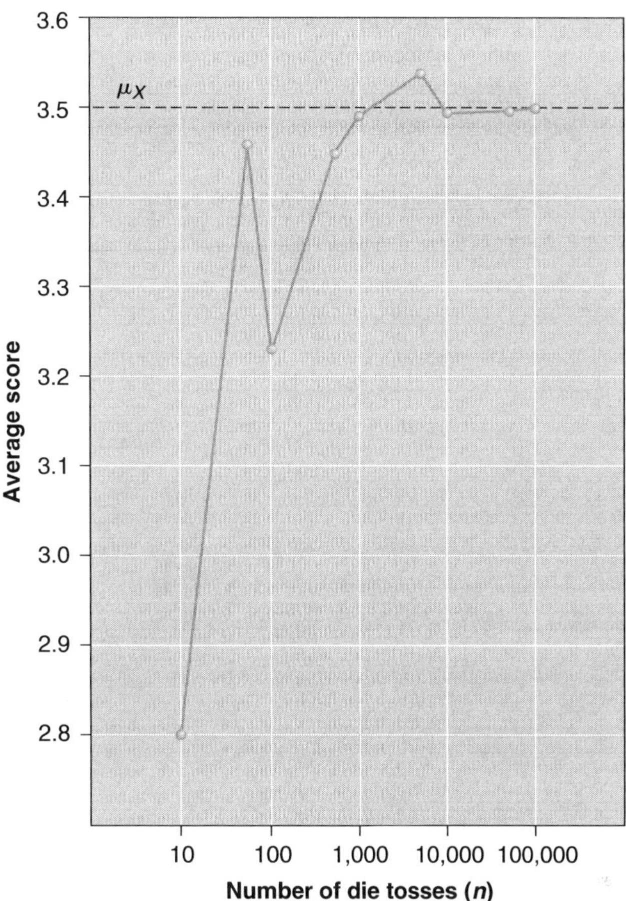

Figure 6.3 Average score $\bar{x}$ as a function of the number of tosses.

Section 6.1 Summary

Consider a random variable, X determined by the outcome of some random phenomenon. The mean of X, denoted by μ_X, or equivalently the expected value of X, denoted $E(X)$, is the average value of X in a very long series of independent repetitions of the random phenomenon.

The *n* Times *p* Formula for the Expected Value of a Counting Variable Consider an event that has probability p of occurring in each of n trials. If the counting variable X is the number of times that the event of interest occurs in the n trials, then

$$\mu_X = np.$$

The Distribution Formula for the Expected Value of a Random Variable If the possible values of X are $x_1, x_2, \ldots, x_k$ with corresponding probabilities $p(x_1), p(x_2), \ldots, p(x_k)$, then

$$\begin{aligned}\mu_x &= x_1 p(x_1) + x_2 p(x_2) + \cdots + x_k p(x_k) \\ &= \sum xp(x)\end{aligned}$$

The set of $(x, p(x))$ values is called the **probability distribution** of X.

The Box Model Formula for the Expected Value of a Random Variable If Y is the sum of n repeated random draws from a box (made with or without replacement between draws), then

$$E(Y) = \mu_Y = n \cdot (\text{box mean})$$

The expected value of X, a single number drawn from a box, and the expected value of the sample mean $\overline{X}$ of repeated random draws from the box both equal the box mean. That is,

$$\mu_X = E(X) = E(\overline{X}) = \mu_{\overline{X}} = \text{box mean}$$

Section 6.1 Exercises

1. What is the expected number of aces in a bridge hand? (See Example 6.1.)
2. A fair die is rolled once. What is the expected value of the score?
3. Compute the expected gain for each of the three \$1 bets in the game of high-low, described in Exercise 1 of Section 5.5.
4. Compute the expected gain for a \$1 bet on red in roulette. (See Exercise 2 of Section 5.5.)
5. Compute the expected gain for a \$1 bet on a single number in roulette. (See Exercise 3 in Section 5.5.)
6. Compute the expected gain for a \$1 bet on a split in roulette. (See Exercise 4 in Section 5.5.)
7. A true-false quiz has 15 questions. Imagine that you blindly guess the answer to each of the 15 questions. How many correct answers can you expect to get? (Compute the expected value. Do not round off.)
8. Simulate your net gain on 10 independent \$1 bets on a pair of numbers in keno. (See Example 6.9.) Use a computer to repeat the simulation 10,000 times. Estimate the probability distribution of your net gain. Estimate your chances of coming out ahead. What is your expected gain?
9. Using a computer, simulate the casino's net gain on 10,000 independent \$1 bets on a pair of numbers in keno (this is one simulation). Run 10,000 simulations to estimate the probability that the casino makes at least \$2000 on the 10,000 bets. What is the casino's expected gain?
10. Simulate your net gain on 10 independent \$1 bets on a single number in keno. Use a computer to repeat this 10-play simulation 10,000 times. Estimate the probability distribution of your net gain. Estimate your probability of coming out ahead. What is your expected gain?
11. Simulate the casino's net gain on 10,000 independent \$1 bets on a single number in keno. Run 10,000 simulations to estimate the probability that the casino makes at least \$2000 on the 10,000 bets. What is the casino's expected gain?
12. Five symmetric casino dice are tossed and the five scores added up. What is the expected value of the sum?
13. A quiz has 15 multiple-choice questions. Each question has five answers to choose from, one of which is correct. A correct answer is worth four points, but a point is taken off for each wrong answer. What is your expected score if you blindly (randomly) guess the answer to each of the 15 questions?
14. On the day before her 50th birthday, a businesswoman buys a life insurance policy that pays \$1 million if she dies before her 55th birthday. The policy costs \$8000 per year for up to 5 years, payable on her 50th, 51st, 52nd, 53rd, and 54th birthdays. The amount X that the insurance company earns on this policy is \$8000 per year for up to 5

years as long as the woman is alive, minus the $1 million the company has to pay if she dies before her 55th birthday. Following is the distribution of X = amount insurance company earns, computed using data from the U.S. National Center for Health Statistics. For example, 0.00686 is the probability that she will die at age 51. Fill in the missing probability that she lives to be 55 or older, and compute the insurance company's expected earnings (μ_X) on this policy.

Table 6.1

Age at death	Earnings	Probability
50	−992,000	0.00638
51	−984,000	0.00686
52	−976,000	0.00732
53	−968,000	0.00777
54	−960,000	0.00822
55 or older	40,000	

15. In a state lottery's Pick 3 game, a three-digit integer is selected at random. All three-digit numbers from 000 to 999 are equally likely. If a player bets $1 on a particular number, and if that number is selected, the payoff is $500 (minus the $1 paid for the lottery ticket). Let X equal the payoff to the bettor, namely, −$1 or $499. Compute the expected payoff μ_X.

16. If a gambler bets one dollar on "fives" in the casino game "chuck-a-luck," he loses his dollar with probability 125/216, and gains 1, 2, or 3 dollars with probabilities 75/216, 15/216, and 1/216, respectively. So his net gain is either −1, 1, 2, or 3 dollars. Use the distribution formula to compute the gambler's expected gain on a one-dollar bet.

6.2 USING FIVE-STEP SIMULATION TO ESTIMATE MEAN VALUES

Consider a random variable X determined by the outcome of some random experiment. The mean μ_X is the average score of X in a very long series of independent repetitions of the random experiment. Thus, we can in principle estimate μ_X if we repeat the random experiment many times or, more practically, use five-step simulation to produce a large number of replications of X via the textbook software.

Example 6.10 The Cereal Box Problem

Suppose Tripl Crisp Cereal is running a promotion by including one of six different colored pens in each box of cereal. The pens come in the following colors:

Fire orange	Brilliant yellow	Baby blue
Shell pink	Rosie red	Grassy green

Assume that when you buy a box of cereal, your chances of getting any one of the six colored pens are equal. About how many boxes of cereal would you expect to have to buy to get a complete set of all six colored pens? We let the random variable X equal the number of cereal boxes needed.

The first thing to notice is that this problem involves chance. What is the *smallest* number of boxes of cereal that you would have to buy? Clearly, 6. It is possible (but not very likely!) that each box purchased would have a different-colored pen. In that case, 6 boxes would do it. On the other hand, what can we say about the largest number of boxes that you might have to buy? It is possible (but again, not likely) that you could buy 1000 or even 10,000 boxes and not get that last color of pen! So, for any one person the answer is at least 6 boxes, but it could be an arbitrarily large number of boxes.

How can we find the expected number of boxes needed in our cereal box problem? There is a theoretical answer, but it requires sophisticated mathematical reasoning. We prefer an experimental data-collecting approach, because it is actually far more helpful in learning to think statistically.

We could go on a shopping trip and start repeatedly buying cereal boxes until we get pens of every color. We could record our purchases as started in Table 6.2. Here is how Table 6.2 works: For each cereal box, a tally mark (/) shows what color of pen was obtained. Table 6.2 shows that so far in shopping trip 1, three boxes have been purchased, with one red and two green pens. A shopping trip ends when we have a complete set of all six colors. Table 6.3 shows the results of the completed shopping trip 1. We see that we bought 21 boxes of cereal before we had a complete set of six colors. We also see, for example, that we have three orange pens and two blue pens.

We now have one estimate of how many boxes of cereal are needed to get a complete set of six pens: 21 boxes. But we could have been lucky, or unlucky. With this single shopping trip, we have no way of knowing whether 21 is unusually large, unusually small, or fairly typical. In order to get a more trustworthy estimate of the typical number of boxes needed, we can go on several more shopping trips and compute the average number of boxes per shopping trip. If we carry out enough trips, the average will get as close to the theoretical expected value as we wish. *Collecting a large amount of data can therefore substitute for a mathematical derivation of the unknown expected value!* That is, we substitute an average of data for a theoretically derived mathematical average.

Table 6.2 Partial Results of One Cereal-Buying Shopping Trip

Shopping trip	Outcomes (pens obtained)						Number of boxes of cereal
	Orange	Yellow	Blue	Pink	Red	Green	
1					/	//	3

Table 6.3 Results of One Completed Cereal-Buying Shopping Trip

Shopping trip	Outcomes (pens obtained)						Number of boxes of cereal
	Orange	Yellow	Blue	Pink	Red	Green	
1	///	/	//	////	~~////~~ //	////	21

It is not practical to solve this problem in this way (actually going out and buying cereal). It is too time-consuming, too expensive, and not sensible (unless, of course, we *really* like Tripl Crisp cereal). Another approach to solving this problem is to do a simulation as a substitute for the real experiment of actually going shopping for cereal. In particular, we can use a physical simulation model—a fair, six-sided die— to substitute for the process of purchasing boxes of cereal.

We use the five-step method.

Step 1. Choose a Model There are six different pens to collect. Assume that when we buy a box of Tripl Crisp, our chances of getting any one of the six colors of pen are equal. We can imagine a box model with six balls, each marked with a color, but we can also easily use a familiar six-sided die as a physical model for buying cereal boxes with their pens. Let each side of the die correspond to one of the six pen colors:

1 = orange	2 = yellow	3 = blue
4 = pink	5 = red	6 = green

One toss of the die will correspond to the purchase of one box of cereal. An equivalent way of simulating the purchase of boxes of cereal is to use the random digits (1–6) in Table B.2 in Appendix B. Each digit will correspond to a color.

Step 2. Define One Simulation A completed shopping trip corresponds to rolling the die until all six sides are obtained.

Step 3. Define the Statistic of Interest Record the number of rolls of the die (number of cereal boxes purchased) in Step 2. This is the statistic of interest in the cereal box problem.

Step 4. Repeat the Simulation It took 16 rolls to obtain all six sides in our first simulated shopping trip. We let this count as shopping trip 2, shown in Table 6.4.

Table 6.4 Simulation of a Complete Shopping Trip

Shopping trip	Outcomes (pens obtained)						Number of boxes of cereal
	Orange	Yellow	Blue	Pink	Red	Green	
2	//	///	///	/	~~////~~	//	16

The next three shopping trips we simulate by using the random digits in the first two rows of Table B.2 (fair die random number table):

Row									
1	66533	45332	2461 4	22231	26431	35541	12165	62116	16111
2	61261	22613	26252	14622	32262	33244	34614	13316	41136

We regard the first five numbers in the first row, 66533, as indicating the first five pen colors obtained. Since we are using the rules 1 = orange, 2 = yellow, 3 = blue, 4 = pink, 5 = red, and 6 = green, the pens corresponding to (6, 6, 5, 3, 3) are (green, green, red, blue, blue). This can be the beginning of shopping trip 3.

We now go to the next group of five numbers, 45332, and record the outcomes in Table 6.5. Then we go to the next group, 24614. In this group we need only the first four numbers, since the 1 gives us the last color needed to complete the set. We can draw a loop around these 14 numbers in the table to show that this is a completed shopping trip, which is shown in Table 6.5.

We next begin shopping trip 4 with the number 4 (in the random number table, following the trip 3 loop) and continue in the same way to complete trips 4 and 5. The results are recorded in Table 6.5. The first set of all six colors (trip 3) was obtained in 14 rolls of the die, the second set was obtained in 13 rolls, and the third set was obtained in 28 rolls.

Step 5. Find the Mean of the Statistic of Interest In the cereal box problem, the statistic of interest is the number of boxes that have to be obtained to get a pen of every color. Divide the total number of boxes obtained in all the simulations (shopping trips) by the number of simulations (shopping trips), thereby obtaining the mean number of cereal boxes per shopping trip.

In Table 6.5 three shopping trips are listed, in which 14, 13, and 28 boxes were purchased. This gives a sample mean of $\overline{X} = \hat{E}(X) = (14 + 13 + 28)/3 = 18.3$ boxes. If we include the first two shopping trips, in Table 6.3 and Table 6.4, we get a sample mean of $\hat{E}(X) = (21 + 16 + 14 + 13 + 28)/5 = 18.4$ boxes. Here $\hat{E}(X)$ denotes the experimental expected value of the random variable X, just as $\hat{P}(A)$ denotes the experimental probability of the event A.

The theoretical expected value can be shown by advanced probability techniques to be equal to 14.7. Our simulated average of 18.4 is not that close. A run of 10,000 simulations obtained by using the textbook software, which includes the cereal box problem, produced an average of

$$\hat{E}(X) = 14.73,$$

which is much closer to the theoretical 14.7. Thus, if we didn't know the value of the theoretical mean, $E(X) = 14.7$, we could estimate it using the experimental sample mean, $\hat{E}(X)$, computed from a large number of simulations, knowing that our estimate would be reasonably accurate.

Table 6.5 Using a Random Number Table to Simulate Complete Shopping Trips

Shopping trip	Outcomes from table digits: Orange	Yellow	Blue	Pink	Red	Green	Number of boxes of cereal
3	/	//	////	//	//	///	14
4	//	////	///	//	/	/	13
5	~~////~~ ~~////~~ //	~~////~~	/	/	//	~~////~~ ///	28

It is interesting to note the variations that occur from simulation to simulation. Only 142 lucky shoppers out of 10,000 got all their pens in only 6 boxes. Thus,

$$\hat{P}(\text{6 different pens in 6 boxes}) = \frac{142}{10,000} = 0.0142,$$

which is close to the theoretical probability (an equally likely outcome situation with $6^6 = 46,656$ ways of choosing 6 pens).

$$P(\text{different pens in 6 boxes}) = \frac{6 \times 5 \times 4 \times 3 \times 2 \times 1}{46656} = 0.0154.$$

In our simulations (not shown), one unlucky shopper needed to buy 70 boxes!

One could ask what happens when one color of pen is very rare. A situation was simulated where the orange pen, say, was 100 times as rare as each of the other five colors of pens which were equally likely to be found in a randomly chosen box. We again let X denote the number of boxes needed to get a complete set. Then, doing 10,000 simulations, we obtained $\overline{X} = \hat{E}(X) = 503.2$, which must be close to the theoretical mean $E(X)$. Thus, as expected, many more cereal boxes are required on average to get all six colors, with one unlucky shopper requiring 5271 boxes purchased. However, a few incredibly lucky shoppers required only 6 boxes, the minimum possible.

Example 6.11 Tickets, Please!

Airline flights are sometimes overbooked. Airlines expect that a certain percentage (the no-show rate) of ticketed passengers will not show up to claim their seats. So, to ensure that as many seats as possible are filled with paying passengers, the airline issues more tickets for the flight than there are seats on the airplane. Thus, there is a slight chance that a seat will not be available on a flight for which a person has a ticket reservation. Often, a person with a reservation who is denied boarding gets a compensatory payment, such as a free ticket.

The no-show rate for each regularly scheduled flight is an experimental probability, $\hat{P}$(no show), that the airline computes from its daily records for that flight. Suppose the airplane has 40 seats and the airline's records indicate a no-show rate for a certain flight of 10%. The airline accepts 43 reservations for this flight. We will estimate via box model simulation for this flight (a) the expected number of empty seats and (b) the expected number of ticketed passengers who do not get a seat. Both (a) and (b) are of interest, because the airline expects to lose the amount in (a) times the average price of a ticket if there are empty seats and the amount in (b) times the penalty paid to each passenger denied a seat if more passengers show up than there are seats.

Solution: As usual, we use the five-step method:

Step 1. Choose a Model We can use a table of random digits. One of several appropriate assignments if we use Table B.3 (Appendix B) is

$$0 = \text{person with reservation is a no-show}$$
$$1\text{–}9 = \text{person with ticket reservation shows up}$$

If we prefer a box model, (as required when using the textbook's instructional software) and let 0 denote a no-show, and 1 a person with a reservation showing up, an appropriate box model is

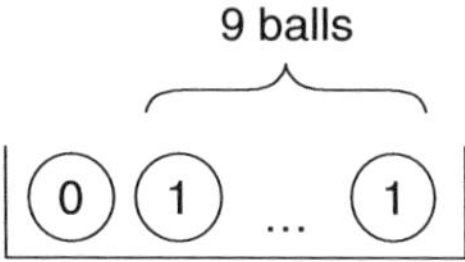

Note that for both models, P(no-show) $= 0.1$, as required.

Step 2. Define One Simulation If we are using an ordinary random number table (Table B.3), a simulation consists of reading 43 digits, one for each reservation made. In the case of the box model, 43 balls are sampled, with replacement, from the box above.

Step 3. Define the Statistics of Interest We are interested in the number of no-shows, that is, the number of 0s among the 43 digits or balls. (With a no-show rate of 10%, the expected number of no-shows is $43 \times 0.1 = 4.3$, by our basic $n \times p$ formula.) If there are more than three no-shows, then there will be empty seats on the flight, the number of which is the first statistic of interest. If there are fewer than three no-shows, then there will be people with reservations who will not get a seat, the number of which is the second statistic of interest. If we use a computer to draw 43 times at random, with replacement, from the box, then the sum of the 43 numbers drawn corresponds to the number of persons with reservations who show up for the flight.

Step 4. Repeat the Simulation We'll use the random number table approach. Our instructional software is not set up to do large numbers of simulations for this specialized problem. However, it would be easy to program the five-step method for this example with the help of a decent computer programmer. For the first simulation we use the first 43 digits in the first row in Table B.3 in Appendix B. There is only one 0 in the first row, corresponding to only one no-show. This means that 42 persons show up. Out of these, 40 will be seated and 2 will not be seated. There will be no empty seats.

For the second simulation we use the first 43 digits in the second row. There are five 0s, corresponding to five no-shows. So, 38 persons show up, everyone gets seated, and there are 2 empty seats.

For the third simulation we use the first 43 digits in the third row, and so forth. Table 6.6 shows the results of 10 simulations using the first 10 rows.

Step 5. Find the Mean of the Statistic(s) of Interest The average number of no-shows in the 10 simulations is 4.1, which is close to the expected number of no-shows, 4.3. The average number of empty seats per flight is 1.6 (the theoretical expected value is 1.56), this being the experimental expected value for the first statistic of interest. The average number of persons not seated per flight is 0.5. (The theoretical expected value is found to be 0.26 by advanced probability techniques.) It is instructive to note how much the number of no-shows in column 2 varies. That is why it is called a random variable.

To gain somewhat more accuracy, 100 simulations were done via computer. Let X denote the number of empty seats and Y the number of persons not seated. The result of the 100 simulations was

$$\hat{E}(X) = 1.5 \text{ (theoretical } E(X) = 1.56)$$
$$\hat{E}(Y) = 0.31 \text{ (theoretical } E(Y) = 0.26)$$

Table 6.6 Airline Seating with Overbooking Strategy

Repetition		Number of 0s (no-shows)	Persons seated	Empty seats	Persons not seated
1		1	40	0	2
2		5	38	2	0
3		4	39	1	0
4		8	35	5	0
5		0	40	0	3
6		5	38	2	0
7		4	39	1	0
8		6	37	3	0
9		4	39	1	0
10		4	39	1	0
	Sum	41	384	16	5
	Average	4.1	38.4	1.6	0.5

There was one case of 10 no-shows, producing 7 empty seats.

Example 6.12 The Hermits' Epidemic (Revisited)

This problem is somewhat similar to the cereal box problem. It too cannot be simulated via computer without modifying our instructional software. Again, we can instead design a five-step simulation using dice or a random number table. We can use a fair die or Table B.2 in Appendix B to simulate the spread of the disease. Number the hermits $1, 2, \ldots, 6$. For the first simulation we can use the numbers in the first row of Table B.2:

$$66533 \quad 45332 \cdots$$

The first hermit to get the disease is number 6. He then visits hermit number 5. (We ignore the second 6, because hermit number 6 cannot visit himself.) Hermit number 5 visits hermit number 3, who visits hermit number 4. (We ignore the second 3.) Finally, hermit number 4 visits hermit number 5, who already had the disease and by now is immune, and the disease dies out. In this simulation, four hermits got infected and two did not. That is, $X = 4$, if we let the random variable X denote the number of infected hermits.

For the second simulation we use the numbers in the second row of Table B.2:

$$61261 \quad 22613$$

In this simulation, hermits 6, 1, and 2 get infected (before hermit 2 visits the immune hermit 6) and hermits 3, 4, and 5 do not, so three hermits get infected. If we program a computer to do 50,000 simulations (or use a theoretical approach for finding the expected value), we will find that the expected number of infected hermits $E(X) \approx 3.51$.

The Key Problem could be changed in a variety of ways to make it more realistic as a model for the spread of disease. For example, the number of hermits on the island could be increased. Suppose there were 100 hermits. We could use the digits of a random number table in pairs (00, 01, 02, ..., 99) to designate which hermit is visited each time. With 100 hermits, how many do you think will get infected? Thirty? Forty? Or fifty? Actually since each infected hermit only gets one chance to pass the disease on, the disease will die out quickly. There is a probability of 0.50 that at most 12 hermits will be infected, and the expected number of infected hermits is around 13.

If we increase the number of hermits to 1000, then only between 1% and 8% of the hermits will get infected before the disease dies out. If we consider one million hermits, then at most 0.2% (this is $0.2 \times 0.01 = 0.002 =$ one out of every 500) of the hermits will get infected before the disease dies out.

This model for the spread of a disease is realistic only for hermits! To make our model more realistic, we have to allow for the possibility that an infected person passes the disease on to more than one person.

Other possible modifications include making the number of days a hermit is infectious be random, assuming that every member of the hermit society makes a random number of visits each day (that is, comes into contact with a random number of other hermits), assuming that the disease is initiated with several infectious hermits, and assuming that the disease is transmitted by a visit with probability less than 1. Now the model begins to seem fairly realistic in its attempt to simulate the way a disease actually spreads.

Section 6.2 Summary

Consider a random variable X determined by the outcome of some random experiment. We can estimate the theoretical mean $E(X)$ by the average score of X, namely $\overline{X}$, denoted as $\hat{E}(X)$, in a long sequence of independent repetitions (e.g., independent simulations using the five-step method) of the random experiment.

Often we can use the textbook's simulation software and achieve high accuracy by doing a large number of simulations. Otherwise, we can design a random number-based, five-step simulation, which, if high accuracy were needed, could be programmed by a computer programmer.

Section 6.2 Exercises

For all exercises, use the five-step method. In most cases the simulation software will be needed.

1. The following table gives the results of five simulated shopping trips in the cereal box problem:

Trip	1. Orange	2. Yellow	3. Blue	4. Pink	5. Red	6. Green	Number of boxes
1	///	/	//	////	~~////~~ //	////	21
2	//	///	///	/	~~////~~	//	16
3	/	//	////	//	//	///	14
4	//	////	///	//	/	/	13
5	~~////~~ ~~////~~ //	~~////~~	/	/	//	~~////~~ //	28

a. Using these five simulations, estimate the expected number of boxes of cereal needed to get the complete set of six pens.
b. Using Table B.2, simulate another 20 trips. To simulate the sixth shopping trip, use the numbers in row number 6, from left to right, starting with 51612. To simulate the seventh shopping trip,use the numbers in row number 7, and so forth. Extend the table to provide the information for the additional trips.
c. Find the (experimental) expected number of boxes of cereal needed to be purchased, based on the 25 trials.
d. It can be shown by advanced mathematics (probability theory) that the experimental expected number of boxes required to obtain all six colors will get closer and closer to 14.7, the theoretical expected value, as more and more trials (shopping trips) are carried out. Compare your answer in part (c) to the theoretical result of 14.7.
e. Repeat parts (c) and (d) doing 10,000 computer simulations.

2. Suppose that instead of six different colors of pens, Tripl Crisp Cereal decides to use *four* different colors of pens in its boxes. What is the average number of boxes of cereal one would expect to have to buy in order to get the complete set of four pens?
a. Make 40 simulations by using the digits 1 through 4 and ignoring 5 and 6 in Table B.2 of Appendix B. For the first simulation use the digits in the first row, from left to right, starting with 66533. For the second simulation, use the digits in the second row, and so forth.
b. Do 10,000 simulations using the textbook's software.

3. For the airline reservation problem in Example 6.11, make the following changes but still assume that the aircraft has 40 seats. For each part below, set up the first three steps to find the expected number of empty seats and the expected number of persons not getting a seat. Do 10 simulations.
a. Suppose the airline accepts 45 (instead of 43) reservations for a flight. Assume a no-show rate of 10%. For the first simulation use the 45 digits in the 11th row of Table B.3 (in Appendix B). Identify each 0 with a no-show. For the second simulation use the 45 digits in the 12th row in Table B.3, and so forth.

b. Suppose the airline accepts 45 reservations but that the no-show rate is 1 in 6 (instead of 10%). For the first simulation use the 45 digits in the 11th row of Table B.2. Identify each 6 with a no- show. For the second simulation use the 45 digits in the 12th row in Table B.2, and so forth.

4. In Example 6.11 assume that the airline accepts 43 reservations for the 40 seats but that the no-show rate is only 5% (instead of 10%). Make a box model for the number of no-shows. Use a computer to simulate the number of no-shows. Repeat 20 times. For each simulation, find the number of empty seats and the number of persons not getting a seat.
   a. Estimate the expected number of empty seats.
   b. Estimate the expected number of persons not getting a seat.

5. A newly married couple agree that they want to have children. The husband wants to have at least one son, and the wife wants to have at least one daughter. They are interested in finding how many children they can expect to have before they will have at least one boy and at least one girl. Assume that the probability of having a boy is 0.5. For the first simulation use the digits in the first row of Table B.1 (in Appendix B) from left to right. Identify 0 with a boy and 1 with a girl. For the second simulation use the digits in the second row of Table B.1. Do 20 simulations. Estimate the expected number of children.

6. Imagine a place where the chance of rain is 1 in 6 every day of the year, independent of the weather the previous days. A newcomer to the place fails to bring an umbrella. How many days of dry weather can he or she expect before the first rainy day? For the first simulation use the digits in the first row of Table B.2, going from left to right. Identify 3 with "rain" and 1, 2, 4, 5, and 6 with "no rain." For the second simulation use the digits in the second row. Do 20 (or more) simulations. Each time count the number of dry days before the first rainy day.

7. Simulate the Hermits' Epidemic (Example 6.12) with 10 hermits labeled 0, 1, . . . , 9. For the first simulation use the digits in the 21st row of Table B.3, going from left to right. For the second simulation use the digits in the 22nd row of Table B.3. Do 10 simulations. Each time count the number of hermits who get infected. Estimate the expected number.

8. It seems intuitive that the more different colors the pens come in, the more cereal boxes you need to open before having one of every color. How many more? Does it take, on average, one more box if there are six colors than if there are five colors? Or is it proportional—that is, would it take twice as many boxes to get all six colors as to get all three colors? This problem asks you to use the five-step method to try to discover the general character of the relationship between the number of different colors and the expected number of boxes needed to obtain one of each color.
   a. If there is only one color, then you only have to buy one box of cereal to get it. (Why?) For other numbers, you need to use the five-step method. If the pens come in two colors in equal numbers, then we can simulate the number of cereal boxes it takes to get one of each color by drawing at random with replacement from the box

   until each value occurs. The statistic of interest is the sample size.

   If the pens come in three colors in equal numbers, then we simulate the number of cereal boxes it takes by drawing from the box

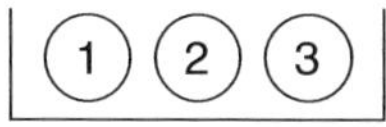

   until all three values occur. Again, the statistic of interest is the sample size. If the pens come in four colors, then we use the box

and so forth.

Run 10,000 computer simulations to estimate the expected number of cereal boxes needed if there are 2, 3, ..., 10 different colors in equal numbers. (This is nine separate five-step simulations, with a different box model in each and with 10,000 repetitions in each.) Fill in the following table with the average number of cereal boxes needed in each of the 10 cases.

Number of colors	Average number of boxes	(Average number of boxes)/ (number of colors)
1		
2		
3		
⋮		
10		

b. Does the average number of cereal boxes needed in part (a) increase by approximately 1 when the number of colors increases by 1?

c. Fill in the third column in the table above by dividing the second column by the first column. This column has the number of cereal boxes needed per color. Are the numbers in this column approximately the same? Are they increasing? Are they decreasing?

d. Make a scatterplot, with X being the number of colors and Y being the average number of cereal boxes needed. Are the points approximately on a straight line, or do you notice some curvature in their pattern?

9. Repeat Exercise 5, the couple wanting both a son and a daughter. Run 10,000 computer simulations to estimate the expected number of children. Explain what box you use, define one simulation, and define the statistic of interest. *Hint:* Is this related to Exercise 8?

10. Repeat Exercise 6, the man without an umbrella. Run 10,000 computer simulations to estimate the expected number of days without rain. Explain what box you use, define one simulation, and define the statistic of interest. *Hint:* Use a counting box of only 0s and 1s.

11. Consider the cereal box problem with six colors but with the orange pen being 100 times as rare as each of the other five colors.

a. Describe the box model.

b. Estimate the number of cereal boxes needed to get one pen of every color.

6.3 THE STANDARD DEVIATION OF A RANDOM VARIABLE

Consider a random variable X determined by the outcome of some random phenomenon. We define the theoretical standard deviation of X, denoted σ_X, or SD(X), as the number that the sample standard deviation of a very, very long sequence of independent observations of X approaches. We shall learn a distribution-based and a box model-based formula to compute σ_X.

distribution formula for SD(X)

> **The Distribution Formula for the Standard Deviation of a Random Variable**
>
> Recall that if the possible values of X are $x_1, x_2, \ldots, x_k$, with corresponding probabilities $p(x_1), p(x_2), \ldots, p(x_k)$, then we can compute the mean of X by
>
> $$\mu_X = x_1 p(x_1) + x_2 p(x_2) + \cdots + x_k p(x_k) = \sum x p(x).$$
>
> The theoretical variance of X, denoted either by Var (X) or σ_X^2, is computed by
>
> $$\begin{aligned}\sigma_X^2 &= (x_1 - \mu_X)^2 p(x_1) + (x_2 - \mu_X)^2 p(x_2) + \cdots + (x_k - \mu_X)^2 p(x_k) \\ &= \sum (x - \mu_X)^2 p(x)\end{aligned}$$
>
> The (theoretical) standard deviation of X, denoted either by SD(X) or σ_X, is the square root of the variance:
>
> $$SD(X) = \sigma_X = \sqrt{\sigma_X^2} = \sqrt{\sum (x - \mu_X)^2 p(x)}.$$

As an illustration, let X denote the value on a symmetric casino die. In this case, $\mu_X = 3.5$, as we showed in Example 6.7. The possible values of X are 1, 2, 3, 4, 5, and 6, each with probability 1/6. The variance of X is, by the distribution formula,

$$\begin{aligned}\sigma_X^2 &= \left[(1-3.5)^2 \cdot \frac{1}{6}\right] + \left[(2-3.5)^2 \cdot \frac{1}{6}\right] + \left[(3-3.5)^2 \cdot \frac{1}{6}\right] \\ &\quad + \left[(4-3.5)^2 \cdot \frac{1}{6}\right] + \left[(5-3.5)^2 \cdot \frac{1}{6}\right] + \left[(6-3.5)^2 \cdot \frac{1}{6}\right] \\ &= \left[(-2.5)^2 \cdot \frac{1}{6}\right] + \left[(-1.5)^2 \cdot \frac{1}{6}\right] + \left[(-0.5)^2 \cdot \frac{1}{6}\right] \\ &\quad + \left[(0.5)^2 \cdot \frac{1}{6}\right] + \left[(1.5)^2 \cdot \frac{1}{6}\right] + \left[(2.5)^2 \cdot \frac{1}{6}\right] \\ &= 35/12.\end{aligned}$$

Thus, the standard deviation of X is $\sigma_X = \sqrt{35/12} = 1.708$.

At the end of Section 6.1, we showed that in a sequence of tosses of a fair die, the sample mean $\overline{X}$ gets closer and closer to the theoretical mean μ_X as the number of tosses increases. In the same way, the sample standard deviation S_X gets closer and closer to the theoretical standard deviation σ_X as the number of tosses increases. This fact is illustrated by the following data obtained by five-step simulations. Briefly, the five steps are

Step 1. Choose a Model Box with balls $1, 2, \ldots, 6$.

Step 2. Define One Simulation Draw one ball.

Step 3. Define the Statistic of Interest Record the number on the chosen ball.

Step 4. Repeat the Simulation We shall let N, the number of simulations increase from 10 to 100,000.

Step 5. Find the Sample Standard Deviation of the Statistic of Interest For each value of N the computer calculates the sample standard deviation of the N simulated scores,

$$S_X = \sqrt{S_X^2},$$

where

$$S_X^2 = \sum (X - \overline{X})^2 / (N - 1).$$

The results of these five-step simulations are as follows:

Number of tosses	S_X	$S_X - \sqrt{35/12}$ (error)
10	1.989	0.281
50	1.695	−0.013
100	1.799	0.091
500	1.761	0.053
1,000	1.733	0.025
5,000	1.721	0.013
10,000	1.708	0.000
50,000	1.705	−0.003
100,000	1.704	−0.004

Recall the Empirical Rule from Section 2.3, which allows us to interpret the sample mean $\overline{x}$ and the sample standard deviation s of a data set without outliers when the histogram of the data is not too far from being "bell-shaped":

- About 68% of the data are in the range $\overline{x} \pm s$.

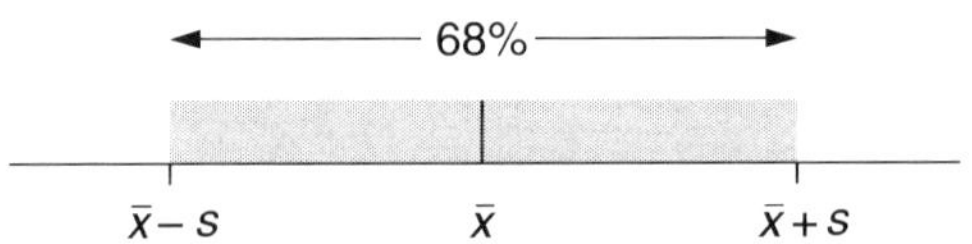

- About 95% of the data are in the range $\overline{x} \pm 2s$.

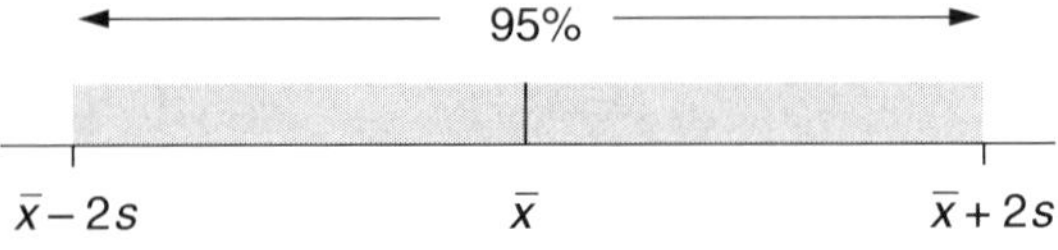

In summary, the mean of the data locates the center of the data, and the standard deviation of the data measures the spread around the sample mean.

Now consider any random variable X. We saw in Chapter 5 that if we have thousands of independent observations of X, then the experimental probability histogram for the data will be very close to the theoretical probability histogram of X. So we can think of the theoretical distribution of X as the experimental distribution of a very large number of independent observations of X. Think of μ_X as the mean of these thousands of observations ($\mu_X \approx \bar{x}$) and σ_X as the standard deviation of these thousands of observations ($\sigma_X \approx s_x$). The theoretical 68-95-99.7% rule for roughly bell-shaped distributions then follows from the Empirical Rule.

theoretical 68-95-99.7% rule for a random variable with a bell-shaped pdf

Theoretical 68%-95%-99.7% Rule for a Random Variable X with a Roughly Bell-Shaped Probability Density Function (pdf)

Suppose the distribution of X is roughly bell-shaped. Then:

1. There is about a 68% chance that X will fall in the range $\mu_X \pm \sigma_X$.
2. There is about a 95% chance that X will fall in the range $\mu_X \pm 2\sigma_X$.
3. There is about a 99.7% chance that X will fall in the range $\mu_X \pm 3\sigma_X$.

In other words, μ_X locates the "center" of the distribution (pdf) of X, and σ_X measures the "spread" of the distribution (pdf) of X around μ_X. We express this loosely by saying that "X will be around μ_X," with the typical size of the distance between X and μ_X being σ_X. We sometimes write, "give or take σ_X or so."

For the number X rolled on a fair die, the theoretical mean μ_X and standard deviation σ_X are

$$\mu_X = 3.5 \text{ and } \sigma_X = \sqrt{35/12} = 1.708, \text{ respectively.}$$

We note that even though the probability histogram is not bell-shaped, the probability that X falls in the range $3.5 - 1.708$ to $3.5 + 1.708$ is 4/6, which is very close to 0.68. And the probability that X falls in the range $\mu_X - 2\sigma_X$ to $\mu_X + 2\sigma_X$, or $3.5 - 3.4$ to $3.5 + 3.4$, is 1, which is not too far from 0.95. This probability histogram is illustrated in Figure 6.4. In fact, the 68-95% portion of the rule (either empirical or theoretical) holds surprisingly well even when the distribution is not bell-shaped. But one should *never* presume that the rule holds for all non-bell-shaped distribution

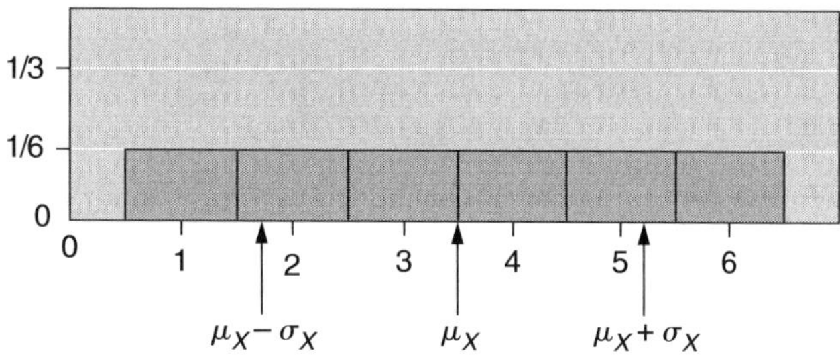

Figure 6.4 The probability histogram of the score on a fair die.

We have often considered a random variable X that we model as a random draw from a box. It is therefore useful to have formulas for the standard deviation of such a random variable.

box model formulas for μ_X, σ_X, and σ_X

Box Model Formulas for Box Means, Box Variance, and Box Standard Deviation

box mean

Consider a box with N numbered balls. Let X be the result of one random draw. Then the expected value of X equals the **box mean**:

$$\mu_X = \mu_{\text{box}} = \frac{\sum x}{N},$$

box variance

where we sum over all the balls in the box. The variance of X equals the **box variance**:

$$\sigma_X^2 = \sigma^2{}_{\text{box}} = \frac{\sum (x - \mu_{\text{box}})^2}{N}.$$

box standard deviation

The standard deviation of X equals the **box standard deviation**, which is the square root of the box variance.

$$\sigma_X = \sigma_{\text{box}} = \sqrt{\sigma_{\text{box}}^2}$$

Important: Notice that we are dividing by N, not $N - 1$. In the special case where the box contains balls of only two different numbers, there is a convenient simple formula for computing σ_{box},

$$\sigma_{\text{box}} = (\text{largest value} - \text{smallest value})\sqrt{p(1-p)},$$

where p is the proportion of balls with the larger value. In particular, if it is a box of 0s and 1s, then

$$\sigma_{\text{box}} = \sqrt{p(1-p)}$$

Consider again the box we use to simulate the toss of a fair die:

| ① ② ③ ④ ⑤ ⑥ |

With $N = 6$ balls in the box, we get

$$\begin{aligned}
\mu_{\text{box}} &= (1+2+3+4+5+6)/6 = 21/6 = 3.5 \\
\sigma_{\text{box}}^2 &= [(1-3.5)^2 + (2-3.5)^2 + \cdots + (6-3.5)^2]/6 = 35/12 \\
\sigma_{\text{box}} &= \sqrt{35/(12)} \approx 1.708.
\end{aligned}$$

We already know that if Y is the sum of the numbers on n randomly selected balls from a box (drawn with or without replacement between draws), then

$$\mu_Y = n \cdot \mu_{\text{box}}$$

The formula for σ_Y depends on whether we sample with replacement or not.

box model formula for σ_Y, with Y the sum of n box draws

Box Model Formulas for σ_Y where Y is a Sum of n Random Draws from a Box

If Y is the sum of the numbers on n randomly selected balls from a box, drawn *with replacement* between draws, then

$$\sigma_Y = \sqrt{n} \cdot \sigma_{\text{box}}$$

If Y is the sum of the numbers on n randomly selected balls from a box, drawn *without replacement* between draws, then

$$\sigma_Y = \sqrt{n} \cdot \sigma_{\text{box}} \cdot \sqrt{\frac{N-n}{N-1}},$$

where N = number of balls in the box. If we draw without replacement, then the sample size n cannot be greater than the total number of balls N. Here

$$\sqrt{(N-n)/(N-1)}$$

finite population factor

is called the **finite population (adjustment) factor** to the with-replacement formula for σ_Y due to the finite "population size" N when sampling without replacement.

Note: Recall that any distribution, given by its set of $p(x)$'s (its pdf), can be used to build a box model. Thus, the above formulas for σ_Y for random draws from a box actually apply to the sum of repeated observations from any distribution. All one has to do is find σ_{box} by converting the $p(x)$'s into a box model, and then compute the appropriate box model formula. A similar remark applies for μ_Y where Y is a sum of random draws from a box.

To simulate the sum of the scores on a pair of symmetric casino dice, we can draw at random twice with replacement from the box we just looked at, and compute the sum Y of the two numbers drawn from the box. We get

$$\mu_Y = 2 \cdot \mu_{\text{box}} = 2 \cdot 3.5 = 7$$

$$\sigma_Y = \sqrt{2} \cdot \sigma_{\text{box}} = \sqrt{2} \cdot \sqrt{35/12} \approx \sqrt{35/6} \approx 2.4.$$

The (theoretical) probability histogram for Y was given in Figure 5.12 and is shown here as Figure 6.5. Clearly, this distribution is roughly bell-shaped. We therefore expect that the theoretical 68-95-99.7% Rule applies well.

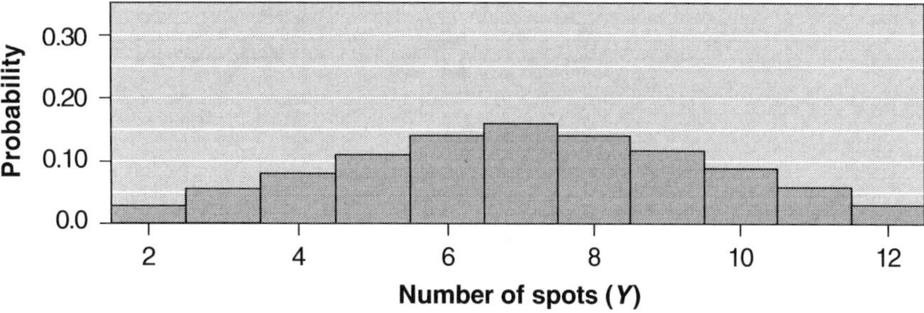

Figure 6.5 Probability distribution of the sum of two fair dice.

According to the rule, there should be around a 68% chance that Y falls in the interval $\mu_Y \pm \sigma_Y$, or between $7 - 2.4 = 4.6$ and $7 + 2.4 = 9.4$. Similarly, there should be around a 95% chance that Y falls in the interval $\mu_Y \pm 2\sigma_Y$, or between $7 - 4.8 = 2.2$ and $7 + 4.8 = 11.8$.

This turns out to be almost exactly right. In fact, if we look at Figure 5.3, we see that the probability that Y takes one of the values 5, 6, 7, 8, or 9 is 24 / 36 = 2 /3, or 66.7%. And the chance that Y takes one of the values from 3 to 11 is 34 /36, or 94.4%. The accuracy of the rule is excellent.

Example 6.13

A fair die is tossed 100 times. The sum of the scores most likely will be around ________ give or take ________ or so.

Solution: The sum Y of the 100 scores will be around its expected value μ_Y, give or take one standard deviation σ_Y or so (the "typical size" of the deviation of Y from μ_Y). We can simulate Y by drawing 100 times at random with replacement from the box

1	2	3	4	5	6

and adding up the 100 numbers we get. The theoretical mean and standard deviation of Y are

$$\begin{aligned} \mu_Y &= 100 \cdot \mu_{\text{box}} = 100 \cdot 3.5 = 350, \\ \sigma_Y &= \sqrt{100} \cdot \sigma_{\text{box}} = \sqrt{100} \cdot \sqrt{35/12} \approx 17.08 \end{aligned}$$

The sum Y of the 100 scores most likely will be around 350, give or take 17 or so. There is about a 95% chance that the sum will be between $\mu_Y - 2\sigma_Y \approx 350 - 34 = 316$ and $\mu_Y + 2\sigma_Y \approx 350 + 34 = 384$.

It is interesting to note that the textbook's instructional software can be used to confirm these results. Indeed, 10,000 simulations produced

$$\bar{y} = \hat{\mu}_Y = 349.89, \quad s_Y = \hat{\sigma}_Y = 16.86, \quad \hat{P}(316 \le Y \le 384) = 0.9571,$$

all quite close to their theoretical complements.

Section 6.3 Summary

If the possible values of X are $x_1, x_2, \ldots, x_k$, with corresponding probabilities $p(x_1), p(x_2), \ldots, p(x_k)$, then we can compute the mean of X by

$$\mu_X = x_1 p(x_1) + x_2 p(x_2) + \cdots + x_k p(x_k) = \sum x p(x).$$

The theoretical variance of X, denoted either by Var (X) or σ_x^2, is computed by

$$\begin{aligned} \sigma_x^2 &= (x_1 - \mu_X)^2 p(x_1) + (x_2 - \mu_X)^2 p(x_2) + \cdots + (x_k - \mu_X)^2 p(x_k) \\ &= \sum (x - \mu_X)^2 p(x) \end{aligned}$$

The (theoretical) standard deviation of X, denoted either by SD(X) or σ_X, is the square root of the variance:

$$\mathrm{SD}(X) = \sigma_X = \sqrt{\sigma_X^2} = \sqrt{\sum (x - \mu_X)^2 p(x)}.$$

Just as we can estimate the theoretical mean μ_X of a random variable X by the simulated average score of X, namely $\overline{X}$, in a long series of independent repetitions of the random phenomenon (using the five-step method), we can also estimate the standard deviation σ_X by the simulated S_X. This estimate of σ_X is accurate if the number of independent repetitions n is large.

If the probability histogram of a random variable X is roughly bell-shaped, then the theoretical 68-95-99.7% Rule applies using μ_X and σ_X.

If Y is the sum of the numbers on n randomly selected balls from a box, drawn *with* replacement between draws, then

$$\sigma_Y = \sqrt{n} \cdot \sigma_{\text{box}}$$

If Y is the sum of the numbers on n randomly selected balls from a box, drawn *without* replacement between draws, then

$$\sigma_Y = \sqrt{n} \cdot \sigma_{\text{box}} \cdot \sqrt{\frac{N - n}{N - 1}},$$

where N = number of balls in the box. In either case (with or without replacement), Y is likely to be around μ_Y, give or take σ_Y or so.

We can approximately confirm the theoretical values for μ_Y and σ_Y by simulating 10,000 values of the sum Y. The experimental values $\overline{Y}$ and S_Y will be close to μ_Y and σ_Y.

Section 6.3 Exercises

1. A professor has a key ring with 10 keys, one of which opens his office door. Without his reading glasses he cannot tell the keys apart, so he tries to use them, one at a time, to open the door. He is equally likely to try the correct key first, second, third, and so on. How many wrong keys will he try before he finds the correct one? We can simulate the number of wrong keys by picking one random number from Table B.3 in Appendix B. Do 45 simulations using the digits in the last row of Table B.3, going from left to right. Estimate the expected number of wrong keys (the theoretical expected value is 4.5). Estimate the standard deviation for the number of wrong keys.

2. A student blindly guesses all the answers to a 15-question, true-false quiz. Use Table B.1 (in Appendix B) to simulate the number of correct answers Y he or she gets. Estimate the expected number, the variance, and the standard deviation of Y. Do 25 simulations. For the first simulation use the first 15 digits in the first row of Table B.1. For the second simulation use the first 15 digits in the second row, and so forth. Identify each 0 with a wrong answer and each 1 with a correct answer.

3. Let Y be the sum of the numbers on 25 balls drawn at random with replacement from this box:

   a. What is the smallest Y can be?
   b. What is the largest Y can be?

c. Most likely, Y will be around __________, give or take __________ or so.
d. Use the textbook software to confirm the theoretical values for μ_Y and σ_Y.

4. A true-false quiz has 36 questions. Imagine that you blindly guess the answer to each question. The number of correct answers you get most likely will be around __________, give or take __________ or so.

5. A fair die is tossed 250 times. The sum of the 250 scores is likely to be around __________, give or take __________ or so.

6. A fair die is tossed 180 times. The number of 6s is likely to be around __________, give or take __________ or so.

CHAPTER 6 SUMMARY

Consider a random variable, X determined by the outcome of some random phenomenon. The mean of X, denoted by μ_X, or equivalently the expected value of X, denoted $E(X)$, is the average value of X in a very long series of independent repetitions of the random phenomenon. If the possible values of X are $x_1, x_2, \ldots, x_k$, with corresponding probabilities $p(x_1), p(x_2), \ldots, p(x_k)$, then we can compute the mean of X by

$$\mu_X = x_1 p(x_1) + x_2 p(x_2) + \cdots + x_k p(x_k) = \sum x p(x).$$

The theoretical variance of X, denoted either by Var (X) or σ_X^2, is computed by

$$\begin{aligned}\sigma_X^2 &= (x_1 - \mu_X)^2 p(x_1) + (x_2 - \mu_X)^2 p(x_2) + \cdots + (x_k - \mu_X)^2 p(x_k) \\ &= \sum (x - \mu_X)^2 p(x)\end{aligned}$$

The (theoretical) standard deviation of X, denoted either by SD(X) or σ_X, is the square root of the variance:

$$\mathrm{SD}(X) = \sigma_X = \sqrt{\sigma_X^2} = \sqrt{\sum (x - \mu_X)^2 p(x)}.$$

Just as we can estimate the theoretical mean μ_X of a random variable X by the simulated average score of X, namely $\overline{X}$, in a long series of independent repetitions of the random phenomenon (using the five-step method), we can also estimate the standard deviation σ_X by the simulated S_X. This estimate of σ_X is accurate if the number of independent repetitions n is large.

If the probability histogram of a random variable X is roughly bell-shaped, then the theoretical 68-95-99.7% Rule applies using μ_X and σ_X.

Four counting variables we have an easy-to-use formula for the expected value.

The n Times p Formula for the Expected Value of a Counting Variable Consider an event that has probability p of occurring in each of n trials. If the counting variable X is the number of times that the event of interest occurs in the n trials, then

$$\mu_X = np.$$

For sums and averages we also have easy-to-use formulas.

The Box Model Formula for the Expected Value of a Random Variable If Y is the sum of n repeated random draws from a box (made with or without replacement between draws), then

$$E(Y) = \mu_Y = n \cdot (\text{box mean})$$

The expected value of X, a single number drawn from a box, and the expected value of the sample mean $\overline{X}$ of repeated random draws from the box both equal the box mean. That is,

$$\mu_X = E(X) = E(\overline{X}) = \mu_{\overline{X}} = \text{box mean}$$

If Y is the sum of the numbers on n randomly selected balls from a box, drawn *with* replacement between draws, then

$$\sigma_Y = \sqrt{n} \cdot \sigma_{\text{box}}$$

If Y is the sum of the numbers on n randomly selected balls from a box, drawn *without* replacement between draws, then

$$\sigma_Y = \sqrt{n} \cdot \sigma_{\text{box}} \cdot \sqrt{\frac{N-n}{N-1}},$$

where N = number of balls in the box. In either case (with or without replacement), Y is likely to be around μ_Y, give or take σ_Y or so.

CHAPTER REVIEW EXERCISES

1. A fair coin is tossed 45 times. What is the expected number of heads?
2. A fair die is tossed 60 times. What is the expected number of 6s?
3. Recall from Table 5.5 that the ABO blood type distribution for African Americans is

Blood type	O	A	B	AB
Probability	0.49	0.27	0.20	0.04

For a random sample of 150 African Americans, compute the expected number of persons of each of the four ABO blood types.

4. Assume that both male and female African Americans have the blood type distribution in Exercise 3.
   a. If we randomly choose an African-American couple, what is the chance that both are type O?
   b. In a random sample of 50 African-American couples, what is the expected number of couples where both are type O?
5. A study on heart attacks examined the records of 2254 heart attack victims admitted to a coronary care unit over a 10-year period. Assume that heart attacks are equally likely to occur on any day of the week. How many of the 2254 heart attacks would we expect to have occurred
   a. on Mondays
   b. on weekends (Saturdays or Sundays)

6. According to the U.S. Census Bureau, the distribution of American families by size is

Family size, x	2	3	4	5	6	7
Fraction of families, $p(x)$	0.43	0.23	0.21	0.09	0.03	0.02

(A family is defined as a group of two or more related persons living together. We have ignored the few families with eight or more members.) Compute the average size of all American families.

7. Each kernel on a corncob is the result of an independent event. A certain genetic experiment with corn produces four types of kernels, when we consider both color and shape: (purple, normal), (purple, shrunken), (yellow, normal), and (yellow, shrunken), with probabilities 9/16, 3/16, 3/16, and 1/16, respectively. If there are 624 kernels on a cob, compute the expected number of each of the four types of kernels.

8. Estimate the expected number of times you have to roll a fair die before one of the six sides appears a second time. (That is, how many times do you have to roll before a side that you have already seen comes up again?)
   a. What is the smallest number possible?
   b. What is the largest number possible?
   c. Do 10 simulations (or more, if time permits). For the first simulation use the digits in the first row in Table B.2 (in Appendix B). For the second simulation use the digits in the second row. Estimate the mean, the variance, and the standard deviation.

9. John is taking a test consisting of 10 multiple-choice questions with four choices for each question. For each question he gets right, he receives 4 points, but for every one he answers incorrectly, he loses 1 point.
   a. If John does not know the answer to a question and blindly guesses the answer, what is his expected score for that question?
   b. If he does not answer a question, then he neither loses nor gains any points for that question. Which is the better strategy if he does not know the answer to a question: guessing the answer or not answering the question?

10. If you bet \$1 on a three-digit number in the Illinois state lottery Pick 3 game and the lottery matches your three digits in exact order, you win \$500. You do not get your \$1 purchase price back.
    a. What is your chance of winning a single bet?
    b. Imagine that every day, you make five \$1 bets on five different three-digit numbers. What is your chance of winning on a given day?
    c. Make a box model for your gain on a given day. (You gain either \$495 or −\$5.)
    d. What is your expected gain on a given day?
    e. Imagine that you bet on five numbers on 300 different days over a year. If you have simulation software, do 10,000 simulations on a computer to estimate the distribution and the expected value of the number of times you win over 300 days.

11. One in five automobile accidents involves a rear-end collision. Use the third and the fourth five-digit columns in Table B.3 (in Appendix B) to simulate 10 automobile accidents. Identify a 0 and a 1 with a rear-end collision. Do 20 simulations. For the first simulation use the digits in the first row (of the third and fourth five-digit columns.) Count the number of rear-end collisions. For the second simulation use the digits in the second row. Estimate the distribution and the expected value of the number of rear-end collisions.

12. Are fatal automobile accidents involving drunk drivers caused by weekend binge drinkers or by everyday drunks (equally likely to be drunk on any day of the week)? A study looked at a random sample of 133 fatal-accident reports involving drunk drivers. Assuming that all the accidents were caused by everyday drunks (not true, of course), how many of the accidents would we expect to have occurred
    a. on Wednesdays?
    b. on weekends (Fridays, Saturdays, and Sundays)?

13. If a gambler bets one dollar on "fives" in the casino game "chuck-a-luck", the casino gains 1 dollar with probability 125/216, and loses 1, 2, or 3 dollars with probabilities 75/216, 15/216, and 1/216, respectively. In other words, the pdf for the casino's gain on a \$1 bet is

$$p(-3) = \frac{1}{216}, \quad p(-2) = \frac{15}{216}, \quad p(-1) = \frac{75}{216}, \quad p(1) = \frac{125}{216}.$$

a. Create a box model for the casino's gain on a single one-dollar bet on "fives." State what distinct numbers should be on the balls and state how many balls of each kind there should be.
b. What is the casino's expected net gain (in dollars) on 10 one-dollar bets?
c. Run 10,000 computer simulations to estimate the probability that the casino makes money on 10 independent one-dollar bets.
d. What is the casino's expected net gain (in dollars) on 10,000 one-dollar bets?
e. Run 10,000 computer simulations to estimate the probability that the casino makes at least \$500 on 10,000 independent one-dollar bets.

PROFESSIONAL PROFILE

View of Saturn compiled by the National Aeronautics and Space Administration (NASA) from images captured in early 2004 by the Cassini spacecraft. Christiaan Huygens was first to find the rings of Saturn.

Christiaan Huygens (1629–1695), mathematician, astronomer, and writer on our solar system.

Christiaan Huygens. From an engraving by G. Edelinck, based on a painting by Caspar Netscher.

Like other outstanding men of his time, Dutchman Christiaan Huygens had interests that spanned many areas of knowledge. He was a mathematician, astronomer, and influential writer on probability. He made contributions to physics, including his wave theory of light, and experimented with double refraction in Icelandic crystal (calcite). He worked on the design of accurate clocks. He invented the pendulum clock (the patent is dated 1657), published a mathematical analysis of pendulums, and patented a pocket watch. He even experimented with internal combustion (using gun powder).

A dedicated astronomer, in 1655 Huygens studied the sky using a 50-power telescope of his own design. From the observations he made with his telescope, he theorized that Saturn was surrounded by a thin, flat ring. He made the first discovery of a moon circling Saturn, and made the first known sketch of the Orion nebula (published in *Systema Saturnium,* 1659). A believer in extraterrestrial life, Huygens wrote a book, *Cosmotheros,* in which he discussed his belief that life on other planets was very similar to life on Earth.

Huygens's importance for statisticians is based on his authorship of the first book on probability theory, *De ratiociniis in loudo aleae* ["On Reasoning in Games of Chance"]. The book, published in 1657, is result of Huygens's collaborative activities with Blaise Pascal, Pierre de Fermat, and a number of other individuals including Giles Roberval, and Pierre de Carcavi. The book has had unusual influence on mathematics and statistics because it provided, for the first time, a readily available source of information on probability to a wide audience. Also, given the widespread interest in gambling in his day, the book sparked interest in anyone who wanted to do well at the gambling tables.

For all of his other interests, Huygens's world view was strongly influenced by his work on probability. Anticipating the statistical hypothesis testing insights we use today, Huygens wrote to Pierre Perrault, "I do believe that we do not know anything for certain, but everything probably."

7

Probability Distributions: The Essentials

You have a very serious disease. Of ten persons who get the disease only one survives. But do not worry, it is lucky you came to me for I've recently had nine patients with the disease and they all died of it.

G. Polya

Objectives

After studying this chapter, you will understand the following:

- Binomial distribution of the number of successes in independent trials
- Random sampling from a large two-attribute population and the binomial distribution
- Geometric distribution of the waiting time until a first success in independent trials
- Poisson distribution of the number of totally random occurrences
- The Poisson approximation to binomial probabilities
- The normal distribution (bell-shaped curve)
- The computation of normal probabilities using a standard normal table
- Practical applications of the normal distribution
- How to use a normal probability plot to assess whether a data set looks like a sample from a normally distributed population.
- The normal approximation to binomial probabilities

KEY PROBLEM

In the 1898 book *Das Gesetz der kleinen Zahlen (The Law of Small Numbers)*, Dr. L. von Bortkiewicz summarized some data on 14 Prussian cavalry units over the 20 years from 1875 to 1894, producing 280 observations. For each unit and year, he counted the number of unfortunate soldiers in that unit who were kicked to death by horses that year. In most instances, no one was kicked to death by horses. The worst instances were two units that had four such fatalities (unit 11 in 1880 and unit 14 in 1882). The table shows the number of units that had 0,1,2,3, or 4 of these deaths. A unit is counted each year. Thus, if a specific unit had 0 fatalities in 17 of the 20 years, it contributes 17 to the 144 cases with 0 fatalities.

Number of soldiers kicked to death	Number of units
0	144
1	91
2	32
3	11
4	2
5 or more	0
	280

According to the table, in 144 observations there were no deaths, 91 observations had one, and so on.

Could Dr. Bortkiewicz have used this data to estimate the probability of a unit in the next year having at least one soldier kicked to death? Do these frequencies show a predictable pattern? Can we fit a probability model to these data?

Certain distributions of random variables are used so often that these probability distributions have been given names. In this chapter we introduce the binomial, hypergeometric, and Poisson distributions. These distributions are used to model the number of times an outcome of interest occurs in certain chance experiments.

discrete distributions

discrete random variables

We also study the geometric distribution, used to model the waiting time until an outcome of interest occurs for the first time. All these distributions are said to be **discrete distributions,** and their associated random variables are called **discrete random variables** because they take on only certain discrete values, in this case, the integers 0,1,2, ... Later in the chapter, we resume our discussion of the normal distribution that we used to model physiological variables and repeated measurements (see Chapter 2). In particular, the notion of a random variable having a normal distribution is introduced. The normal distribution is an example of a **continuous distribution** of a **continuous random variable,** so called because the possible values of the random variable flow together continuously.

continuous distribution

continuous random variable

factorial

The **factorial** notation is convenient for representing certain discrete probability distributions. For $k = 1, 2, 3, \ldots$, the product of the first k positive integers is called "k factorial" and is written $k!$ For example,

$$1! = 1.$$
$$2! = 1 \times 2 = 2.$$
$$3! = 1 \times 2 \times 3 = 6.$$
$$4! = 1 \times 2 \times 3 \times 4 = 24.$$

Finally, "zero-factorial" is defined as 1:

$$0! = 1.$$

binomial coefficient

For two non-negative integers $0 \leq x \leq n$, the **binomial coefficient**

$$\binom{n}{x} = \frac{n!}{x!(n-x)!}$$

gives the number of different ways of selecting x objects from a collection of n distinct objects, when the order in which the x objects are chosen does not matter. (This can be proved using the multiplication rule for a multistage experiment of Section 5.6.) The binomial coefficient is read as "n choose x." For example, there are "3 choose 2" $= 3!/(2! \times 1!) = 3$ ways of selecting 2 objects from (a, b, c), namely, (a, b), (a, c), (b, c). As an example of computing a binomial coefficient,

$$\binom{5}{2} = \frac{5!}{2! \times 3!} = \frac{5 \times 4 \times 3 \times 2 \times 1}{2 \times 1 \times 3 \times 2 \times 1} = \frac{5 \times 4}{2} = 10.$$

7.1 THE BINOMIAL DISTRIBUTION

two-outcome trial

A **two-outcome trial** is a chance experiment in which we classify the outcome in one of two mutually exclusive and exhaustive ways. We often call the two outcomes "success" and "failure," and we call the two-outcome trial a "success-failure" trial.

repeated-trials random experiment
independent trials

In statistics we often consider repeated independent two-outcome trials, which we also call a **repeated-trials random experiment. Independent trials** means that the outcomes of previous trials have no influence on which outcomes will occur in future trials. We are often interested in the number of successes in a predetermined number of independent, repeated success-failure trials.

Imagine, for example, that you make three independent \$1 bets on "red" in roulette (see Example 5.5). Each time you bet, your chance of winning a dollar is $p = 18/38$, and your chance of losing a dollar is $1 - p = 20/38$. You have to win at least two out of the three times to make money. If we list the outcomes of the three bets, then the three-bet (three-trial) chance experiment has eight possible outcomes:

$$\{\text{LLL, WLL, LWL, LLW, WWL, WLW, LWW, WWW}\},$$

where L stands for "loss" and W stands for "win," and WLL stands for a win on first bet and losses on second and third bets, for example.

Let X denote the number of times you win. Since the three bets are independent, we can use the multiplication rule for independent events (Section 5.3). We will also use the addition rule for mutually exclusive events (Section 5.3). For example, the event $(X = 2)$ occurs if any of the three mutually exclusive events WWL, WLW, and LWW occurs. Therefore, by the addition rule, remembering that an outcome is also an event,

$$p(2) = P(X = 2) = P(\text{WWL}) + P(\text{WLW}) + P(\text{LWW}).$$

Since the outcomes of the three bets are independent, P(WWL) $= pp(1-p)$, and so on. Hence,

$$P(X=2) = p^2(1-p) + p(1-p)p + (1-p)p^2$$
$$= 3p^2(1-p) = 3 \cdot \left(\frac{18}{38}\right)^2 \frac{20}{38} = 0.3543.$$

Also, since

$$p(3) = P(X=3) = P(\text{WWW}) = p \cdot p \cdot p = \left(\frac{18}{38}\right)^3 = 0.1063,$$

your chance of making money is

$$P(X \geq 2) = P(X = 2 \text{ or } X = 3) = p(2) + p(3) = 0.3543 + 0.1063 = 0.4606$$

That is, you have about a 46% chance of making money. We could also have obtained this result by simulating our net gain on three \$1 roulette bets on a computer 10,000 times, using a box model of eighteen 1s and twenty −1s. (Try it!)

The Binomial Probability Distribution

Consider n independent trials, each resulting in one of two possible outcomes, which for convenience we call "success" and "failure." Assume that the probability of success, p, is the same for all n trials, for example, $p = 18/38$ for repeated bets on "red" in roulette. Let the random variable X denote the total number of successes in the n trials. Then X has the

binomial probability density function (pdf)

binomial probability density function (pdf):

$$p(x) = P(X = x)$$
$$= P(x \text{ successes and } n - x \text{ failures})$$
$$= \binom{n}{x} p^x (1-p)^{n-x},$$

for $x = 0, 1, \ldots, n$.

binomially distributed

We say that X is **binomially distributed** with parameters n and p. The expected value of X and the standard deviation of X are, respectively,

$$\mu_X = np \quad \text{and} \quad \sigma_X = \sqrt{np(1-p)}.$$

In our roulette example there are three independent trials (bets). A win is a success and a loss is a failure. The probability of a success is 18 / 38 for each trial. It follows that X, the number of times you win, is binomially distributed, with paramaters $n = 3$ and $p = 18/38$.

Now imagine that you make 35 one-dollar bets on a pair of numbers in keno (Section 5.5), losing \$1 or winning \$11 on each bet. The number of wins, X, is binomial, with $n = 35$ and $p = 0.06$. The probability of losing all 35 keno bets is

$$p(0) = \binom{n}{0} p^0 (1-p)^n = (0.94)^{35} = 0.1147.$$

The chance of winning once and losing 34 times is

$$p(1) = \binom{n}{1}p(1-p)^{n-1} = 35(0.06)(0.94)^{34} = 0.2562.$$

The chance of winning twice and losing 33 times is

$$p(2) = \binom{n}{2}p^2(1-p)^{n-2} = \frac{35 \cdot 34}{2}(0.06)^2(0.94)^{33} = 0.2780.$$

To come out ahead and make money on the 35 bets, you have to win at least 3 times. You lose money as a result of the 35 bets if $X \leq 2$. We can compute the probability that you lose money by using the addition rule:

$$\begin{aligned} P(X \leq 2) &= p(0) + p(1) + p(2) \\ &= 0.1147 + 0.2562 + 0.2780 = 0.6489. \end{aligned}$$

Therefore, your chance of making money on the 35 bets, noting that $(X \geq 3)$ is the complement of $(X \leq 2)$, is

$$P(X \geq 3) = 1 - P(X \leq 2) = 1 - 0.6489 = 0.3511.$$

Thus, you have only about a 35% chance of coming out ahead, as we have established theoretically, using the binomial distribution. Again, we could have obtained this result with great approximation accuracy by simulating our net gain on the 35 one-dollar bets on a computer 10,000 times, using the appropriate box model. That is, it is easy to simulate Bernoulli trials and a binomial random variable by using the textbook's five-step simulation software.

binomial experiment

Binomial Experiment Conditions

- Each trial has two possible outcomes, generically denoted "success" and "failure."
- The number of trials, n, is fixed in advance.
- The success-probability p is the same from trial to trial.
- The trials are independent
  (Trials satisfying the above binomial experiment conditions are called **Bernoulli trials.**)
- The random variable X of interest is the total number of successes in the n trials.

Bernoulli trials

If these five conditions hold for our real-world setting, then X is binomially distributed, and we compute $P(X = x)$ using the binomial pdf given above.

Example 7.1

Four-o'clocks are ornamental plants. If we cross two pink-flowered four-o'clocks, the chance that an offspring will have red flowers is $1/4$, pink flowers $1/2$, and white flowers $1/4$. (See the discussion at the end of Section 5.5.) Different offspring inherit their genetic makeup independently. Imagine that 150 offspring are produced. Then, the 150 trials are independent, and the number X of red-flowered offspring is binomial, with $n = 150$ and $p = 1/4$: Denote each red-flowered offspring a success and each pink-flowered or white-flowered offspring a failure. Similarly, the number Y of white-flowered offspring is also binomial, with $n = 150$ and $p = 1/4$. Finally, the number Z of pink-flowered offspring is binomial, with $n = 150$

and $p = 1/2$.

Example 7.2

A symmetric casino die is tossed 15 times, and we count the number X of sixes. If we can assume that the outcomes of successive tosses are independent, and for each toss identify success with getting a 6 and failure with not getting a 6, then it follows that X is binomial, with $n = 15$ and $p = 1/6$.

Example 7.3

A true-false quiz has 10 questions. Imagine that you do not know any of the answers and blindly (and hence independently) guess each of the 10 answers. Then your number X of correct answers will be binomial with $n = 10$ and $p = 1/2$.

❐ ❐

Binomial random variables are "counting variables." We count the number of times an event of interest occurs in repeated trials. But not all counting random variables are binomial. It is important to be able to tell whether a counting variable is binomial or not. A counting random variable X is binomial only when the real-world situation is such that the five bulleted conditions of a binomial experiment hold.

Example 7.4

You have \$5 and you make repeated \$1 bets on "red" in roulette until you either double your money (gain \$5) or go broke. Here the success-failure trials (the individual bets) are independent, with the same success probability $p = 18/38$ in each trial. But *the number of trials has not been fixed in advance*, and hence X, the number of bets you win, is *not* binomial.

Example 7.5

A bridge hand of 13 cards is dealt at random, 1 card at a time, from a well-shuffled deck of 52 cards. Let X denote the number of hearts among the 13 cards. Each card dealt is either a heart or not a heart. If we identify getting a heart with success and not getting a heart with failure, then we have 13 success-failure trials, with success probability $p = 1/4$ on each trial (the suits are equally likely). But because we do not replace cards between draws, successive draws from the deck are *not independent*, and X is *not* binomial. In fact, X has the hypergeometric distribution that was estimated through simulation in Table 6.1 in Section 6.1.

Binomial Distribution Histograms

Figures 7.1, 7.2, and 7.3 display the probability histograms (the graphs of the pdfs $p(x) = P(X = x)$) of three different binomial distributions. The heights of the various rectangles equal the corresponding binomial probabilities. A missing rectangle means that the corresponding probability is less than 0.005. Notice that the histogram in Figure 7.1 is symmetric, almost bell-shaped, whereas the shapes of the histograms in Figures 7.2 and 7.3 are skewed.

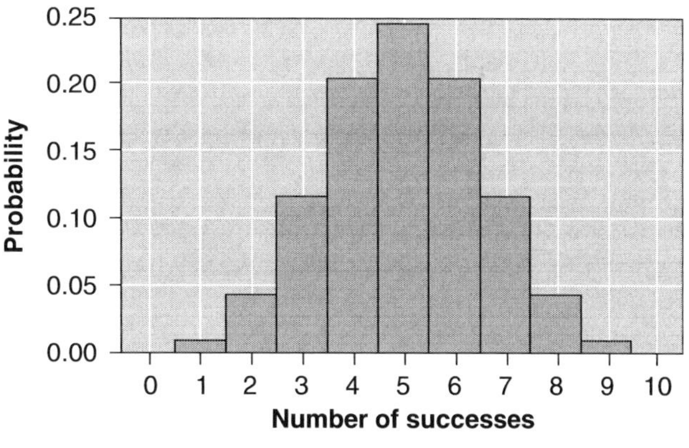

Figure 7.1 Binomial distribution histogram for $p = 0.5$, $n = 10$.

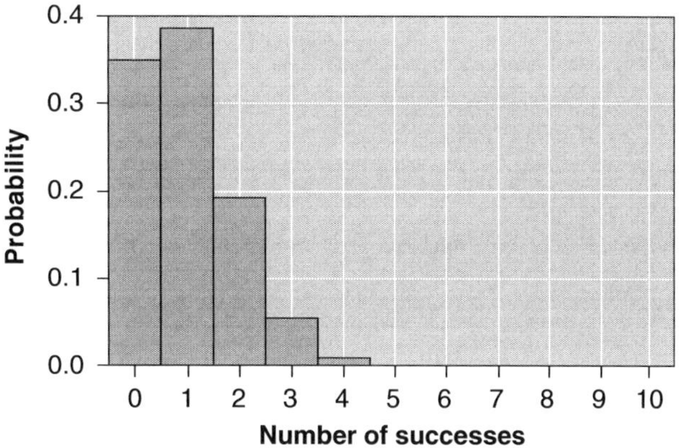

Figure 7.2 Binomial distribution histogram for $p = 0.1$, $n = 10$.

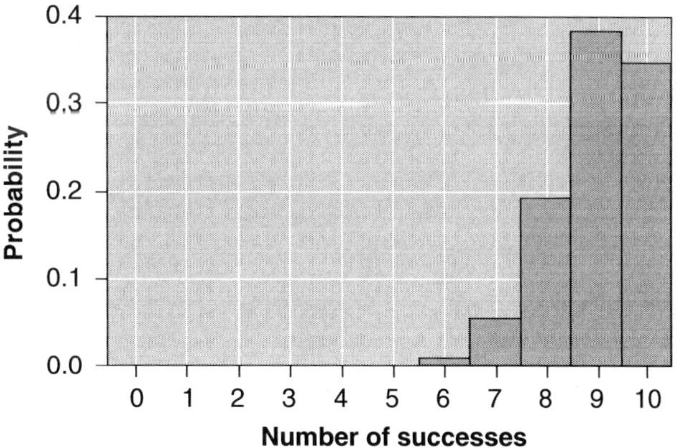

Figure 7.3 Binomial distribution histogram for $p = 0.9$, $n = 10$.

Using the Tables for Cumulative Binomial Probabilities

If the binomial parameter n is not too large, then it is easy to compute binomial probabilities, using a calculator, but it may be possible to save time by using the binomial tables in Appendix G.

Example 7.6

Consider 25 married couples where both husband and wife are carriers of cystic fibrosis (see Example 5.12). If each of the couples has a child (25 independent trials), what is the probability that at most 8 of the 25 children develop the deadly disease? What is the chance that at least 4 of the children get cystic fibrosis?

Solution: Based on an analogous argument in Example 5.12 for sickle-cell anemia, each child has a 25% risk of getting cystic fibrosis. If X denotes the number of the 25 children who get cystic fibrosis, then X is binomial, with $n = 25$ and $p = 0.25$. The pdf of X is

$$p(x) = P(X = x) = \binom{25}{x}\left(\frac{1}{4}\right)^x\left(\frac{3}{4}\right)^{25-x}, \text{for } x = 0,\ 1, \ldots, 25.$$

We are interested in $P(X \leq 8)$ and $P(X \geq 4)$. These probabilities can easily be obtained from the binomial tables. Table 7.1 provides the relevant binomial table from Appendix G. Appendix G provides the cumulative binomial probability $P(X \leq x)$ for $n = 2, 3, \ldots, 25$ and $p = 0.05, 0.1, \ldots, 0.5$.

To determine $P(X \leq 8)$, locate the binomial table for $n = 25$ and find the number in the row for $x = 8$ and the column under $p = 0.25$. We get

$$P(X \leq 8) = 0.8506.$$

To determine $P(X \geq 4)$, find the number in the row for $x = 3$ and in the column under $p = 0.25$ to get

$$P(X \geq 4) = 1 - P(X \leq 3) = 1 - 0.0962 = 0.9038.$$

Note that $P(4 \leq X \leq 8) = P(X \leq 8) - P(X \leq 3) = 0.8506 - 0.0962 = 0.7544.$

❐ ❐

The binomial distribution can also be used to approximately model random sampling *without* replacement from a finite (real) population if the population is much larger than the sample (our rule of thumb is: sample size $n \leq 5\%$ of population size N) and if each member of the population either has or lacks an attribute of interest.

Simple Random Sampling and the Binomial Distribution

Consider an actual real-world population of size N, of which N_1 members of the population have a certain attribute of interest and the remaining $N_2 = N - N_1$ members do not. For example, we might consider a population of $N = N_1 + N_2$ persons, of which N_1 smoke and N_2 do not smoke. Or we might consider a shipment of $N = N_1 + N_2$ identical electronic devices, of which N_1 devices are defective and the remaining N_2 devices are not defective.

simple random sampling

random sampling without replacement

Suppose we randomly draw a sample of n members from the population, one at a time, *without* replacement between draws. This method of sampling, discussed in Section 5.6, is called **simple random sampling** or **random sampling without replacement.** When we sample in this way, then all possible samples of size n are equally likely to be obtained.

Table 7.1 Cumulative Binomial Probabilities for 25 Independent Trials[1]

	p									
x	**0.05**	**0.10**	**0.15**	**0.20**	**0.25**	**0.30**	**0.35**	**0.40**	**0.45**	**0.50**
0	0.2774	0.0718	0.0172	0.0038	0.0008	0.0001	0.0000	0.0000	0.0000	0.0000
1	0.6424	0.2712	0.0931	0.0274	0.0070	0.0016	0.0003	0.0001	0.0000	0.0000
2	0.8729	0.5371	0.2537	0.0982	0.0321	0.0090	0.0021	0.0004	0.0001	0.0000
3	0.9659	0.7636	0.4711	0.2340	0.0962	0.0332	0.0097	0.0024	0.0005	0.0001
4	0.9928	0.9020	0.6821	0.4207	0.2137	0.0905	0.0320	0.0095	0.0023	0.0005
5	0.9988	0.9666	0.8385	0.6167	0.3783	0.1935	0.0826	0.0294	0.0086	0.0020
6	0.9988	0.9905	0.9305	0.7800	0.5611	0.3407	0.1734	0.0736	0.0258	0.0073
7	1.0000	0.9977	0.9745	0.8909	0.7265	0.5118	0.3061	0.1536	0.0639	0.0216
8	1.0000	0.9995	0.9920	0.9532	0.8506	0.6769	0.4668	0.2735	0.1340	0.0539
9	1.0000	0.9999	0.9979	0.9827	0.9287	0.8106	0.6303	0.4246	0.2424	0.1148
10	1.0000	1.0000	0.9995	0.9944	0.9703	0.9022	0.7712	0.5858	0.3843	0.2122
11	1.0000	1.0000	0.9999	0.9985	0.9893	0.9558	0.8746	0.7323	0.5426	0.3450
12	1.0000	1.0000	1.0000	0.9996	0.9966	0.9825	0.9396	0.8462	0.6937	0.5000
13	1.0000	1.0000	1.0000	0.9999	0.9991	0.9940	0.9745	0.9222	0.8173	0.6550
14	1.0000	1.0000	1.0000	1.0000	0.9998	0.9982	0.9907	0.9656	0.9040	0.7878
15	1.0000	1.0000	1.0000	1.0000	1.0000	0.9995	0.9971	0.9868	0.9560	0.8852
16	1.0000	1.0000	1.0000	1.0000	1.0000	0.9999	0.9992	0.9957	0.9826	0.9461
17	1.0000	1.0000	1.0000	1.0000	1.0000	1.0000	0.9998	0.9988	0.9942	0.9784
18	1.0000	1.0000	1.0000	1.0000	1.0000	1.0000	1.0000	0.9997	0.9984	0.9927
19	1.0000	1.0000	1.0000	1.0000	1.0000	1.0000	1.0000	0.9999	0.9996	0.9980
20	1.0000	1.0000	1.0000	1.0000	1.0000	1.0000	1.0000	1.0000	0.9999	0.9995
21	1.0000	1.0000	1.0000	1.0000	1.0000	1.0000	1.0000	1.0000	1.0000	0.9999
22	1.0000	1.0000	1.0000	1.0000	1.0000	1.0000	1.0000	1.0000	1.0000	1.0000
23	1.0000	1.0000	1.0000	1.0000	1.0000	1.0000	1.0000	1.0000	1.0000	1.0000
24	1.0000	1.0000	1.0000	1.0000	1.0000	1.0000	1.0000	1.0000	1.0000	1.0000
25	1.0000	1.0000	1.0000	1.0000	1.0000	1.0000	1.0000	1.0000	1.0000	1.0000

[1] The table gives the probability of obtaining x or fewer successes in $n = 25$ independent trials, namely, $P(X \le x)$, where $p =$ probability of success in a single trial.

hypergeometric distribution

Let X denote the number of population members in the sample who have the attribute in question. For example, X might be the number of smokers in the sample or the number of defective electronic devices in the sample. X is said to have a **hypergeometric distribution.** It has pdf

$$p(x) = P(X = x) = \binom{N_1}{x}\binom{N_2}{n-x} \Big/ \binom{N_1 + N_2}{n}$$

for $x = 0, 1, \ldots n$. For large values of N, n, and x, the distribution is tedious to compute. Note that if we randomly sample from the population, one at a time, *with replacement* between draws, then the draws will be *independent,* and we instead have a binomial experiment. Then, X, the number of times we select a population member with the attribute, will be binomial, with parameters n and $p = N_1/(N_1 + N_2)$.

If the sample size n is less than 5% of the population size N, it makes very little difference in computing $P(X = x)$ whether we sample with or without replacement. So even if we sample without replacement, X is approximately binomial, with parameters n and $p = N_1/(N_1 + N_2)$.

We state this basic principle allowing finite population sampling to sometimes be modeled by the binomial distribution:

binomial approximation of the hypergeometric distribution

5% rule

Binomial Approximation of the Hypergeometric Distribution

Suppose that a simple random sample of size n is taken (without replacement) from a finite population in which the proportion of members that have a specified attribute is p. If the **sample size n is less than 5% of the population size,** then the number X of members in the sample that have the attribute is approximately binomial, with parameters n and p.

Example 7.7

We mentioned at the end of Section 5.3 that 85% of all Americans have Rh+ blood type. If we take a simple random sample of 25 Americans, what is the chance that all 25 are Rh+? What is the chance that at least 20 of the persons in the sample are Rh+? What is the chance that at most 20 of the persons in the sample are Rh+?

Solution: If X denotes the number of persons in the sample who are Rh+, then, since the population size N is huge, X, which has a hypergeometric distribution, is approximately binomial by the 5% rule, with $n = 25$ and $p = 0.85$. Since the binomial tables of Appendix G only cover binomial distributions with $p \leq 0.50$, we shall consider instead the number Y of persons in the sample who are not Rh+ (and hence are Rh−). Clearly, the probability of a person being Rh− is $1 - 0.85 = 0.15$. Thus, Y is (almost) binomial, with $n = 25$ and $p = 0.15$. Because $X + Y = 25$, any event of interest involving X can be rewritten as an event involving Y, whose probability we can solve for. For example, using the binomial Table 7.1,

$$\begin{aligned} P(X = 25) &= P(Y = 0) \approx 0.0172 \\ P(X \geq 20) &= P(Y \leq 5) \approx 0.8385, \text{ and} \\ P(X \leq 20) &= P(Y \geq 5) = 1 - P(Y \leq 4) \\ &= 1 - 0.6821 = 0.3179. \end{aligned}$$

Example 7.8

A bridge hand of 13 cards is dealt at random, 1 card at a time, from a well-shuffled deck of 52 cards. Let X denote the number of hearts among the 13 cards. X has a hypergeometric distribution, with $N_1 = 13, N_2 = 39$, and $n = 13$.

$$\begin{aligned} p(x) &= P(X = x) = p(\text{we get } x \text{ hearts and } n - x \text{ nonhearts}) \\ &= \binom{13}{x}\binom{39}{13-x} \Big/ \binom{52}{13} \end{aligned}$$

for $x = 0, 1, 2, \ldots, 13$. This distribution is tabulated in Table 6.1 in Section 6.1. Here the sample size, n, is 25% of the population size (52), and we *cannot* use the binomial approximation.

Section 7.1 Summary

Let X denote the number of times an event of interest occurs in a series of trials. Assume that

- Each trial has two possible outcomes: Either the event of interest occurs or it does not occur.
- The number of trials n is fixed in advance.
- The trials are independent.
- For each trial, the probability that the event of interest will occur equals the same value p.

Then X has the **binomial probability density function**:

$$p(x) = P(X = x) = P(\text{the event occurs exactly } x \text{ times.})$$
$$= \binom{n}{x} p^x (1-p)^{n-x}, \quad \text{for } x = 0, 1, \ldots, n.$$

We say that X is binomially distributed, with parameters n and p. The mean and the standard deviation of X are

$$\mu_X = np \quad \text{and} \quad \sigma_X = \sqrt{np(1-p)}.$$

Suppose we randomly draw a sample of n members from a population, one at a time, without replacement between draws. As stated, this method of sampling is called *random sampling without replacement* or *simple random sampling*. Let p denote the proportion of population members that have a certain attribute of interest. If the sample size n is less than 5% of the population size, then the number X of members in the sample that have the attribute of interest is approximately binomial, with parameters n and p. Regardless of whether the 5% rule holds or not, the true probability density function of X is:

$$p(x) = P(X = x) = \binom{N_1}{x}\binom{N_2}{n-x} \Big/ \binom{N_1 + N_2}{n}$$

where N_1 is the number of population members with the attribute and N_2 is the number of population members without the attribute.

Section 7.1 Exercises

1. Compute the binomial coefficients $\binom{4}{0}, \binom{4}{1}, \binom{4}{2}, \binom{4}{3}$.
2. A fair coin is tossed five times. Let X denote the number of times the coin lands heads up. For $x = 0, 1, 2, 3, 4, 5$, compute

$$p(x) = P(X = x) = P(\text{we get } x \text{ heads and } 5 - x \text{ tails in any order}).$$

3. A fair coin is tossed 20 times. Let X denote the number of times the coin lands heads up.
   a. Determine the expected number of heads μ_X.
   b. Use the binomial tables to determine

$$P(X \leq 9),\ P(X \leq 10), \text{ and } P(X = 10).$$

   Hint: $P(X = 10) = P(X \leq 10) - P(X \leq 9)$.
   c. Use five-step simulation to approximate these binomial probabilities. Describe the box model and define the statistic of interest. Run 10,000 simulations to approximate the three binomial probabilities.
4. For families with four children, what are the separate probabilities that a randomly selected family will have 0,1, 2, 3, or 4 girls, assuming that boys and girls are equally likely at each birth and ignoring the possibility of identical twins? Given 2000 families, each with four children, how many families would you expect to have no girls? One girl? Two girls? Three girls? Four girls?

5. Your chance of winning a bet on a single number in keno is $p = 1/4$ (see Section 5.5). The bet pays 2 to 1: If you bet \$1 and you win, then you gain \$2. (You get your dollar back plus 2 additional dollars.) Of course, if you lose, your gain is -1. Suppose you make 25 independent, \$1 bets on a single number in keno, \$1 at a time.
   a. What is your net gain if you win 8 times and lose 17 times?
   b. What is your net gain if you win 9 times and lose 16 times?
   c. What is your net gain if you win 10 times and lose 15 times?
   d. How many times do you have to win to make money?
   e. What is your chance of making money on the 25 bets?

6. Ninety percent of all Americans are right-handed. Imagine that we take a random sample of 25 Americans. Use the tables to determine the probability that
   a. All 25 are right-handed.
   b. At least 20 of the 25 are right-handed.
   (Hint: Consider the number of persons in the sample who are *not* right-handed.)
   c. At most 20 of the 25 are right-handed.
   d. Exactly 20 of the 25 are right-handed.
   e. Explain why you were allowed to use the binomial probability distribution.
   f. Approximate the binomial probabilities in parts (a)–(d) using five-step simulation.

7. A multiple-choice test has 25 questions. Each question has five answers, one of which is correct. Suppose you blindly guess the answer to each of the 25 questions.
   a. What is the chance that you guess the answer to the first question correctly?
   b. What is your expected number of correct answers on the test?
   c. What is the probability that you get at most 8 correct answers on the test?

8. There is about an 80% chance that a randomly chosen person age 20 will still be alive at age 65, according to the U.S. National Center for Health Statistics. Suppose that we randomly select fifteen 20-year-olds.
   a. How many do you expect to still be alive at age 65?
   b. What is the chance that all 15 are alive at age 65?
   c. What is the chance that at least 12 of the 15 are alive at age 65?
   d. What is the chance that at most 12 of the 15 are alive at age 65?
   e. Why is it legitimate to use the binomial probability distribution?

9. It is estimated that 40% of all U.S. females get pregnant at least once before reaching the age of 20. Consider a random sample of 16 U.S. females who are at least 20 years old. Find the probability that the number of women in the sample who got pregnant at least once before reaching the age of 20 will be
   a. at most 8.
   b. at most 4.
   c. between 5 and 8, inclusive.

10. In the casino game of chuck-a-luck, three fair, six-sided dice are rolled. If you bet a dollar on 5s, then you lose your dollar if no 5s are rolled. If at least one 5 is rolled, you get your dollar back plus \$1 for each 5 shown by the three dice. If X denotes your net gain on a \$1 bet on 5s, then X can equal -1, 1, 2, or 3.
    a. Compute the corresponding probabilities, $p(-1)$, $p(1)$, $p(2)$, and $p(3)$. *Hint:* What is the probability that you get no 5s? One 5? Two 5s? Three 5s?
    b. Compute your expected net gain μ_X using the distribution-based formula ($\mu_X = \Sigma x p(x)$) as discussed in Chapter 6.

11. A labor dispute has arisen concerning the allegedly discriminatory way that 20 laborers at a construction site were given their job assignments. Six of the 20 job assignments were considered highly undesirable, whereas the remaining 14 jobs were considered desirable. The dispute was triggered by the fact that all 4 African-American laborers working on the site were given undesirable job assignments. The grievance states that "if the jobs were assigned without regard to race or color, then it is highly unlikely that all 4 African-American workers would have received undesirable assignments."

Imagine that the 20 jobs were randomly assigned to the 20 workers. Let X denote the number of African-American workers who were assigned undesirable jobs.

a. What is the distribution of X?
b. What is the result of computing $P(X = 4)$?
c. Do we have reason to doubt that the jobs were randomly assigned? Why do you think so?

7.2 THE GEOMETRIC DISTRIBUTION

If you buy a Pick 3 lottery ticket and bet on one of the numbers from 000 to 999, your chance of winning is $p = 1/1000$. Imagine that you buy a lottery ticket 5 days a week, every week of the year, year in and year out, until you win a bet. How many Pick 3 lottery tickets do you have to buy to win once? We can list the possible outcomes of this chance experiment:

$$S = \{\text{W, LW, LLW, LLLW, LLLLW}, \ldots\}.$$

Here LW means that you lose your first bet but win the second bet; LLLW means that you lose your first three bets but win the fourth bet; and so forth. Let X denote the number of bets you have to make to win once. Since the daily lottery drawings (trials) are independent, we can use the product rule for independent events to get

$$\begin{aligned} P(X = 1) &= P(\text{W}) = P(\text{win first bet}) = p \\ P(X = 2) &= P(\text{LW}) = P(\text{lose first bet}) \cdot P(\text{win second bet}) = (1 - p)p \\ P(X = 3) &= P(\text{LLW}) \\ &= P(\text{lose first bet}) \cdot P(\text{lose second bet}) \cdot P(\text{win third bet}) \\ &= (1 - p)(1 - p)p. \end{aligned}$$

Generalizing this, we see that X has the probability density function (pdf)

$$p(x) = P(X = x) = P\text{ (you lose first } x - 1 \text{ bets and win bet number } x) = (1 - p)^{x-1}p,$$
for $x = 1, 2, 3, 4, \ldots$

geometrically distributed

We say that X is **geometrically distributed** with success-probability $p = 1/1000$. Our analysis did not depend on the specific value of the success-probability p in the lottery. We can introduce the geometric distribution with parameter p as follows:

conditions producing a geometric distribution

geometric distribution

Conditions Producing a Geometric Distribution

Consider a sequence of independent success-failure trials with the same success-probability p in each trial (Bernoulli trials). Let X denote the number of trials it takes to observe the first success. Then X has a **geometric distribution** with the following pdf:

$$p(x) = P(X = x) = (1 - p)^{x-1}p \text{ for } x = 1, 2, 3, \ldots$$

Furthermore, the mean and standard deviation of X are

$$\mu_X = 1/p \quad \text{and} \quad \sigma_X = \sqrt{(1 - p)/p^2}.$$

Let X be a geometric random variable. Using the multiplication rule for independent events, note that for $x = 1, 2, 3, \ldots,$

$$\begin{aligned} P(X > x) &= P[(\text{1st trial a failure}) \text{ and } (\text{2nd trial a failure}) \text{ and } \cdots (x\text{th trial a failure})] \\ &= P(\text{1st trial a failure}) \times P(\text{2nd trial a failure}) \times \cdots \times P(x\text{th trial a failure}) \\ &= (1-p)^x. \end{aligned}$$

This is a useful formula.

As Figures 7.4 and 7.5 show, geometric distributions are very skewed (and not bell-shaped). For that reason we cannot use the 68-95-99.7% Rule of Section 6.3 to interpret the value of the mean and the standard deviation. We will still say, however, that a geometric variable is likely to be around its expected value, give or take a standard deviation or so.

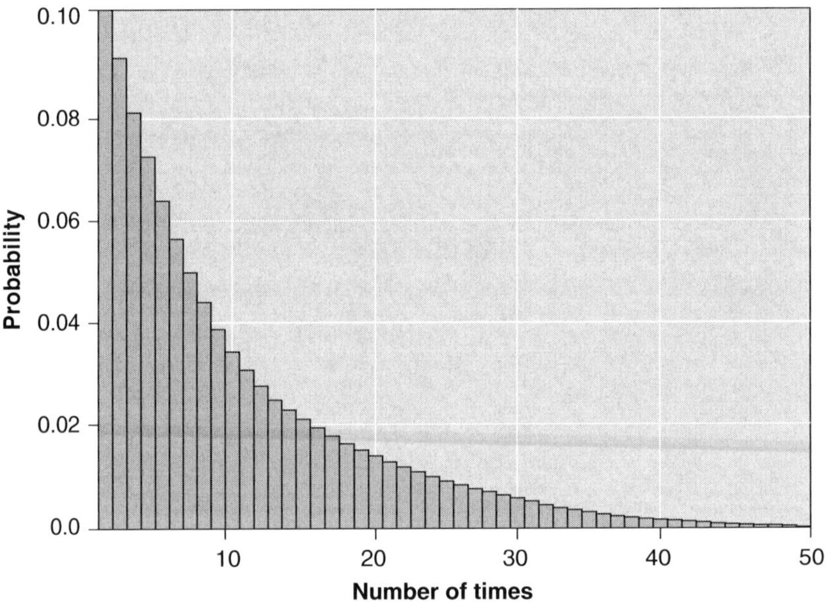

Figure 7.4 Geometric probability histogram, $p = 0.1$.

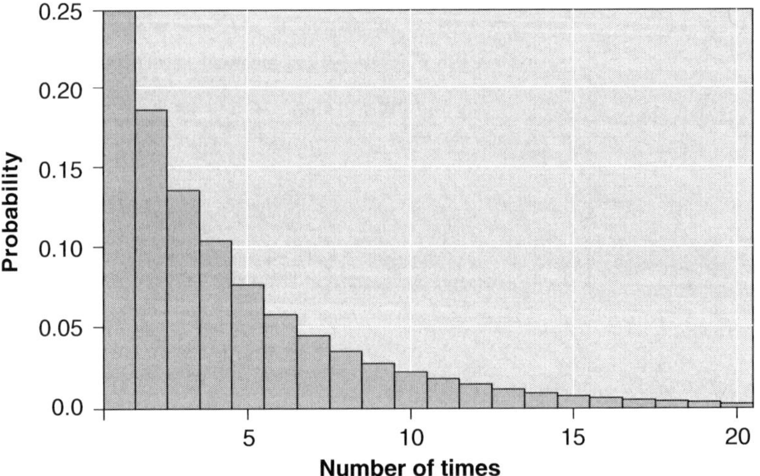

Figure 7.5 Geometric probability histogram, $p = 0.25$.

Example 7.9

Toss a pair of casino dice and add up the scores on the two dice. How many times do we have to toss the pair to get a sum of 7?

Solution: As Figure 5.3 in Section 5.2 helps show, when we toss a pair of fair dice, the chance of getting a sum of 7 is $p = 6/36 = 1/6$. The number X of such tosses of pairs of dice needed to get a sum of 7 is geometric with parameter $p = 1/6$. Thus, the pdf of X is

$$p(x) = \frac{1}{6}\left(\frac{5}{6}\right)^{x-1}, \text{ for } x = 1, 2, 3, \ldots.$$

The mean and standard deviation of X, respectively, are

$$\mu_X = 1/p = 6 \quad \text{and} \quad \sigma_X = \sqrt{(1-p)/p^2} = \sqrt{30} \approx 5.5.$$

To get a sum of 7, we will most likely need around 6 tosses, give or take 5.5 tosses or so. That is, requiring between 1 and 11 tosses is typical, using 6 ± 5.5. It is unlikely that we will need exactly 6 tosses. In fact, the chance that we need exactly 6 tosses is only

$$p(6) = \frac{1}{6}\left(\frac{5}{6}\right)^{5} = 0.0670.$$

It is quite possible that we will need more than 11 tosses to get a sum of 7:

$$P(X > 11) = \left(\frac{5}{6}\right)^{11} = 0.1346.$$

But it is unlikely that we will need more than 20 tosses:

$$P(X > 20) = \left(\frac{5}{6}\right)^{20} = 0.0261.$$

Example 7.10

Recall the game of craps, discussed in Example 5.3 in Section 5.2. A player (called the shooter) rolls a pair of fair, six-sided dice. A first roll of 7 or 11 means that the shooter wins, whereas 2, 3, or 12 on the first roll means that the shooter loses. Anything else becomes the shooter's point, and in that case the shooter continues to roll the dice until either a 7 turns up, in which case the shooter loses, or the point turns up again, and the shooter wins (this is called making the point). For example, if the first roll is a 5, then the shooter continues to roll until he or she either makes another 5 and wins, or makes a 7 and loses. Assume that 5 has just been established as the shooter's point. How many more rolls X will the shooter have to make to finish the game?

Solution: As Figure 5.3 in Section 5.2 shows, on each roll of the two dice, the chance of making either a 5 (shooter wins) or 7 (shooter loses) is $p = (4/36) + (6/36) = 10/36$. Therefore, the number X of rolls needed to finish the game is geometric, with parameter $p = 10/36$. The expected value of X is

$$\mu_X = 1/p = 36/10 = 3.6,$$

and the standard deviation is $\sigma_X = \sqrt{(1-p)/p^2} = \sqrt{9.36} \approx 3.06$. The shooter will most likely need around 3.6 rolls, give or take 3 rolls or so. It is unlikely that he will need more than 10 rolls:

$$P(X > 10) = \left(1 - \frac{10}{36}\right)^{10} = \left(\frac{26}{36}\right)^{10} = 0.0386.$$

Because Bernoulli trials are easy to simulate, it is convenient to use the textbook's software to generate a geometric random variable and to accurately approximate geometric probabilities.

Section 7.2 Summary

Let X denote the number of trials it takes for an outcome of interest to occur in a series of trials. Assume the following:

- Each trial has two possible outcomes.
- The trials are independent.
- In each trial the chance that the outcome of interest occurs equals the same probability p.

Then X has the **geometric probability density function**

$$p(x) = P(X = x) = p(1-p)^{x-1}, \text{ for } x = 1, 2, 3, \ldots .$$

Section 7.2 Exercises

1. How many days until the next day of measurable rain? Should this be modeled by a geometric probability distribution? Explain.
2. A deck of cards is dealt one card at a time. Each card is either an ace or not an ace. We are interested in the number of cards until the first ace. Should this situation be modeled by a geometric probability distribution? Explain.
3. A couple is determined to have children until they have a girl. (See Example 5.19.) We are interested in the number of children they will have. Should this situation be modeled by a geometric probability distribution? Explain. Assuming that the chance of having a girl is 0.49, simulate 10,000 such families. Describe the box model and the statistic of interest. Compare the experimental distribution you obtained from your simulations and the theoretical distribution of the number of children. (See Table 5.12.) Compare the expected number of children and the average number of children for the 10,000 simulated families. Discuss your results.
4. To reduce population growth, China implemented a "one-child" per family policy that has been extremely unpopular in parts of the country. Some rural communities have suggested changing the policy to a "one-son" policy. This would mean that families could have children until they had a son. Suppose that the probability of having a boy is 0.51 (boys are slightly more likely than girls) and that the sex of a child is independent from birth to birth. Carry out

a five-step simulation that gives the geometric random variable experimental probabilities and its expected value for the number of children in families that follow the one-son policy. Assume that families have children until they have a boy and there are no twins, triplets, and so forth. Simulate 10,000 such families. Discuss your results.

5. a. Write the mathematical function for the theoretical pdf for the number of children in a family that follows the one-son policy mentioned in Exercise 4. Assume that families have children until they have a boy, and there are no twins, triplets, and so forth.
   b. Determine the theoretical expected number of children in such a family. Give the theoretical standard deviation.
   c. Make a pdf table of the probabilities for 1, 2, 3, 4, and "5 or more" children.
   d. Make a graph of the histogram of the theoretical pdf (the probability histogram).

6. In a game of craps, bystanders can also place bets at almost any point of the game. They can bet on the shooter or against the shooter. Many people prefer to wait to see the result of the shooter's first shot before placing a bet, because some points (first rolls of 4, 5, 6, 8, 9, or 10) are easier to make than others.
   a. Suppose the shooter's first roll is a 4. What is the probability of shooting a 4 on any given roll? *(Hint:* See Figure 5.3. There are 36 possible outcomes. How many of these are equal to 4?)
   b. Theoretically, how many rolls should we expect before the shooter rolls another 4? (Do not count the first roll.) Remember that earlier we figured out that it typically takes around six rolls to get the first 7.
   c. Suppose the shooter's first roll is a 6. Theoretically, how many rolls should we expect it to take the shooter to roll another 6? Which is the better situation to be in: first roll a 4, or first roll a 6?
   d. Graph the theoretical pdfs for the number of rolls until the first 4, the number of rolls until the first 6, and the number of rolls until the first 7, side by side.

7. Carry out a five-step simulation to determine experimentally the geometric distribution and its expected number of the rolls of two dice before the first 7. Run 10,000 simulations.

8. Suppose you are tossing a fair coin until the first heads. You have tossed it twice and got tails both times. How many more times should you expect to toss the coin until a heads appears? Explain.

7.3 THE POISSON DISTRIBUTION

We have seen that the binomial and geometric distributions have very important probability modeling roles. For example, the number of successes in a fixed number of trials is surprisingly often well modeled as a binomial random variable. The Poisson distribution also applies in a wide variety of real-world settings, as disparate as radioactive decay, occurrence times of natural disasters, locations of galaxies, locations of the "homes" of a biological species in a homogeneous environment, and times of arrival of customers or messages for processing at a server. Indeed, the wide applicability of the Poisson distribution as a probability modeling tool can seem almost magical.

The reason the Poisson distribution is so widely applicable is that, as for the binomial distribution, the assumptions leading to it are so basic and hence widely applicable. Informally, the Poisson distribution is an appropriate model whenever we are counting isolated events that are occurring "totally randomly," either over time or in space. This vague statement needs detailed explanation, which will be given later in this section.

Poisson distribution

Poisson intensity parameter

Poisson Distribution (pdf)

A random variable X is said to have a **Poisson distribution** if its **pdf** is given by

$$p(x) = P(X = x) = \frac{\lambda^x}{x!}e^{-\lambda}$$

for $x = 0, 1, 2, \ldots$, where $\lambda > 0$ is the rate of occurrence,or **Poisson intensity parameter.** Then

$$\mu_X = \lambda \text{ and } \sigma_X = \sqrt{\lambda}.$$

Binomial probabilities are easy to compute or look up in a binomial table if n is small. But when n is large, it is time-consuming to evaluate binomial probabilities and especially a sum of binomial probabilities, even with a calculator. Imagine the task facing scientists 200 years ago when they had to evaluate such binomial probabilities without calculators!

For example, consider the problem of computing the probability that out of 2500 randomly chosen newborn babies, at least half are girls. If we ignore the possibility of identical twins, then the number X of girls among the 2500 newborns is binomial, with $n = 2500$ and $p = 0.487$. (See Example 5.1, Section 5.2.) The pdf of X is

$$p(x) = \binom{2500}{x}(0.487)^x(0.513)^{2500-x}, \ x = 0, 1, \ldots, 2500.$$

The probability that at least half of the children are girls is

$$P(X \geq 1250) = p(1250) + p(1251) + p(1252) + p(1253) + p(2500) + \cdots + p(2500).$$

It would be very time-consuming to compute this probability on a hand-held calculator. Our textbook software can estimate this probability (a value close to 0.1) very accurately using five-step simulation. A theoretical result allowing quick approximation of binomial probabilities for large n was worked out early in the eighteenth century by the French Protestant Abraham de Moivre, who had fled to England to avoid religious persecution. De Moivre's normal distribution-based approximation, which works well for large n provided p is not too close to 0 or 1, will be presented in Section 7.4. In this section, though, we shall use a different approximation found in the early nineteenth century by the French mathematician and physicist Siméon Poisson. Poisson's approximation of the binomial by the Poisson distribution works only for large n and small p. It provides easy-to-compute approximations for hard-to-compute binomial probabilities if n is large and p is small.

the Poisson approximation to the binomial distribution

The Poisson Approximation to the Binomial Distribution

Let X be binomial, with the n and p parameters satisfying $n \geq 25$ and $p \leq 0.1$. If, also, $np \leq 15$ (i.e., the larger n is, the smaller p must be!), we can approximate binomial probabilities by Poisson probabilities. That is, in terms of probability density functions,

$$p(x) = P(X = x) = \binom{n}{x}p^x(l - p)^{n-x} \approx \frac{\lambda^x}{x!}e^{-\lambda}, \text{ for } x = 0, 1, \ldots,$$

where the Poisson intensity parameter λ satisfies $\lambda = \mu_X = np$.

Example 7.11

Hospital patients face a 1 in 20 risk of acquiring an infection in the hospital. Suppose we randomly select the medical records of 30 former hospital patients and determine the number X of these patients who got infected during their hospital stay. X is approximately binomial, with $n = 30$ and $p = 1/20$. Because the criteria are $(p \leq 0.1, n \geq 25, np \leq 15)$ and the criteria are met, the Poisson approximation should be accurate with $\lambda = np = 30 \cdot (1/20) = 1.5$. In Table 7.2 we present the corresponding binomial and Poisson probabilities. In this case it is not difficult to compute the binomial probabilities. We are just illustrating how accurate the Poisson approximation is. Notice that even though n is only 30, the approximation is quite good, as seen by subtracting the two probabilities for fixed x. In each row the approximation error is less than 0.01.

Table 7.2 Binomial Probabilities with $n = 30$ and $p = 1/20$ and Corresponding Approximating Poisson Probabilities with $\lambda = np = 1.5$

x	$\binom{n}{x} p^x (1-p)^{n-x}$	$\frac{\lambda^x}{x!} e^{-\lambda}$
0	0.215	0.223
1	0.339	0.335
2	0.259	0.251
3	0.127	0.126
4	0.045	0.047
5	0.012	0.014
6	0.003	0.004
7	0.000	0.001

Example 7.12

According to the *Canadian Immunization Guide*, about 1 in 24,000 people develop transient thrombocytopenia from a measles shot. This medical condition involves a temporary suppression of the production of clot-forming blood platelets, leading to increased bruising and bleeding. If 36,000 children are given measles shots, what is the chance that at most 4 of the vaccinated children will develop transient thrombocytopenia? What is the probability that at least 2 of the children develop this condition?

Solution: If X denotes the number of vaccinated children who develop this side effect, then X is binomial, with $n = 36{,}000$ and $p = 1/24{,}000$. Clearly, the criteria of $p \leq 0.1$, $n \geq 25$, $np \leq 15$, are easily met. The approximating Poisson distribution has intensity parameter $\lambda = np = (36{,}000)(1/24{,}000) = 1.5$. In Table 7.3 we present the corresponding binomial and Poisson probabilities. We present both with six decimal places to show how close they are. Notice that with such a large value of n and such a small p, the difference between the binomial probabilities and the corresponding Poisson probabilities is very small, in fact, less than 0.00001 in every case.

Table 7.3 Binomial Probabilities with $n = 36,000$ and $p = 1/24,000$ and Corresponding Approximating Poisson Probabilities with $\lambda = np = 1.5$

x	$\binom{n}{x} p^x (1-p)^{n-x}$	$\frac{\lambda^x}{x!} e^{-\lambda}$
0	0.223123	0.223130
1	0.334699	0.334695
2	0.251028	0.251021
3	0.125512	0.125511
4	0.047065	0.047067
5	0.014119	0.014120
6	0.003529	0.003530
7	0.000756	0.000756
8	0.000142	0.000142

If we use the probabilities in the last column of Table 7.3, we find that the approximating Poisson probability that at most 4 of the vaccinated children suffer side effects is

$$P(X \leq 4) = p(0) + p(1) + p(2) + p(3) + p(4) \approx 0.9814,$$

whereas the approximating probability that at least 2 children suffer side effects is

$$P(X \geq 2) = 1 - P(X \leq 1) = 1 - (p(0) + p(1)) \approx 0.4422.$$

These approximating Poisson probabilities differ less than 0.00001 from the corresponding binomial probabilities. But the usual and quickest way to estimate $P(X \leq 4)$ and $P(X \geq 2)$ is to use the Poisson tables (see Appendix J), a part of which is displayed in Table 7.4.

Table 7.4 Cumulative Poisson Probabilities[1]

	$\lambda = E(X)$									
x	**1.1**	**1.2**	**1.3**	**1.4**	**1.5**	**1.6**	**1.7**	**1.8**	**1.9**	**2.0**
0	0.333	0.301	0.273	0.247	0.223	0.202	0.183	0.165	0.150	0.135
1	0.699	0.663	0.627	0.592	0.558	0.525	0.493	0.463	0.434	0.406
2	0.900	0.879	0.857	0.833	0.809	0.783	0.757	0.731	0.704	0.677
3	0.974	0.966	0.957	0.946	0.934	0.921	0.907	0.891	0.875	0.857
4	0.995	0.992	0.989	0.986	0.981	0.976	0.970	0.964	0.956	0.947
5	0.999	0.998	0.998	0.997	0.996	0.994	0.992	0.990	0.987	0.983
6	1.000	1.000	1.000	0.999	0.999	0.999	0.998	0.997	0.997	0.995
7	1.000	1.000	1.000	1.000	1.000	1.000	1.000	0.999	0.999	0.999
8	1.000	1.000	1.000	1.000	1.000	1.000	1.000	1.000	1.000	1.000

[1]The table gives the probability $P(X \leq x)$ of x or fewer events when the expected number of such events is λ.

To determine $P(X \leq 4)$, find the number in the row with $x = 4$ in the column under $\lambda = 1.5$ to get $P(X \leq 4) \approx 0.981$. To determine $P(X \geq 2) = 1 - P(X \leq 1)$, find the number in the row with $x = 1$ in the column under $\lambda = 1.5$ to get $P(X \leq 1) = 0.558$. Therefore, $P(X \geq 2) \approx 1 - 0.588 = 0.442$. As you can see, the $P(X \leq 4)$ and $P(X \geq 2)$ approximations are extremely accurate.

Example 7.13

In the United States the annual death rate due to leukemia is 7.5 per 100,000 population. What is the expected number of leukemia deaths in a simple random sample of 10,000 Americans over a 4-year period? Determine the probability that a simple random sample of 10,000 Americans over a 4-year period has at most 5 leukemia deaths.

Solution: Over a 4-year period the death rate due to leukemia is $4(7.5) = 30$ per 100,000 population; thus the probability of a randomly chosen person dying of leukemia over a 4-year period is $30/100{,}000$. Let X denote the number of leukemia deaths in a simple random sample of 10,000 Americans over a 4-year period. Because the population of all Americans is large, the conditions of a binomial experiment are met even though the sampling is without replacement. Hence X is binomial, with $n = 10{,}000$ and $p = 30/100{,}000$. The expected number of deaths thus satisfies $E(X) = np = 3 \leq 15$. Because $n = 10{,}000 \geq 25$ and $p = 30/100{,}000 < 0.1$, X is approximately Poisson with $\lambda = np = 3$. Using the Poisson table in Appendix J,

$$P(X \leq 5) \approx 0.916.$$

Poisson Modeling of Total Randomness

isolated events occurring totally at random

As stated at the beginning of the section, Poisson distributions are used to model the number of times an event of interest occurs during a predetermined amount of time, or over a predetermined area or volume, when **isolated events** are **occurring totally at random.** Typical examples include

- the number of α-particles emitted by a tiny radioactive source during a specified time interval
- the number of traffic accidents at a busy intersection over a year
- the number of an insurance company's policyholders who die on a given day
- the number of telephone calls received at a switchboard between 9 A.M. and 10:30 A.M.
- the number of flaws in a square meter of cloth
- the number of nematodes in a cubic centimeter of earth

All of these real-world phenomena, and so many others, are well-modeled by a Poisson distribution because they satisfy three assumptions that make the notion of occurring "totally at random" precise.

We have just seen that a binomial distribution with large n and small p is well approximated by a Poisson distribution. In fact, the larger that n is and the smaller that p is, the better the approximation is. The next example shows us that this mathematical fact discovered by Poisson is precisely the reason that a random variable X satisfying the conditions of a Poisson experiment (loosely, events occur totally at random and are isolated from each other) in fact has a Poisson distribution (pdf),

$$P(X = x) = \lambda^x e^{-\lambda}/x!.$$

We now give a precise description of the real-world conditions that lead to a random variable X having a Poisson distribution.

conditions producing a Poisson distribution

events occur totally at random and are isolated

The Poisson Experiment Conditions Producing a Poisson Distribution

The random number of occurrences X of an event of interest during a preset amount of time (or over a preset area or volume) will have a Poisson distribution if the following real-world conditions (briefly described as **events occur totally at random and are isolated**) are judged to be satisfied:

1. The probability of an occurrence within a small time interval (small relative to the preset amount of time) of fixed duration is the *same regardless of when the time interval begins*. That is, the rate of occurrence is constant over time.
2. The probability of an occurrence during a given interval of time is independent of whether there is an occurrence in another, nonoverlapping interval (even if the intervals are close to each other). That is, past occurrences do not influence future occurrences.
   Note: Conditions 1 and 2 amount to events occurring totally at random.
3. It is much less likely that there will be more than one occurrence during a very small interval of time than that there will be exactly one occurrence in that interval (the real-world setting described by this condition is that nearly simultaneous occurrences are highly unlikely). That is, events are isolated in time (or in space).

Example 7.14

One milligram of uranium 238, containing around 2.5×10^{18} atoms, undergoes spontaneous radioactive decay at an average rate of 13 disintegrations per second. Individual atoms disintegrate independently of each other at different times. Each *isolated* disintegration results in the emission of an α-particle, which can be detected by a Geiger counter. Let X denote the number of α-particles emitted during a specified 1-second time period. As stated, $E(X) = 13$. What is the probability distribution of X?

Solution: Break the 1-second time period into one million consecutive 1-microsecond periods (a very large n). The average expected number of emissions per microsecond is obviously $13/1{,}000{,}000$. We make the following assumptions (justified by the physical reality of uranium decay described immediately above):

- The probability of a single emission during a 1-microsecond time period is the same for all 1-microsecond time periods; call it p. (That is, the probability of a single emission is constant over time.)
- The numbers of emissions during nonoverlapping time intervals are independent.

These two assumptions are a careful statement that the emissions occur totally at random and at a constant rate. We also conclude the following assumptions that result from a careful study of the physics:

- It is much less likely that there will be greater than one emission during any 1-microsecond time period than that there will be exactly one emission (emissions are isolated in time from one another).

Call each emission a "success." Under these Poisson experiment assumptions, and ignoring the much less likely possibility of 2 or more emissions in any of the one million microsecond intervals, the total number of emissions, X, during the specified 1-second time period becomes the number of successes in $n = 1{,}000{,}000$ independent success-failure Bernoulli trials, with a very small success-probability p in each trial. Thus, X has a binomial pdf with

$n = 1{,}000{,}000$. Recall for a binomial pdf that $E(X) = np$. We already know that $E(X) = 13$. Solving for p, we get, $p = 13/1{,}000{,}000$. Clearly, the criteria for X being well approximated by a Poisson distribution are amply met ($n \geq 25$, $p \leq 0.1$, $np \leq 15$). It follows that by the Poisson approximation to the binomial that X approximately has the Poisson pdf, with intensity parameter $\lambda = np = 13$. In fact, radioactive decay is usually modeled using a Poisson distribution.

In summary, the fact that the Poisson experiment conditions hold for X allows X to be viewed as a binomial pdf with very large n and very small p. Then, X being binomial allows us to apply the Poisson approximation to the binomial pdf of X to infer that X has a Poisson pdf.

Example 7.15

Imagine standing on a bridge over a rural interstate and counting the number of cars, X, that pass under it in an hour. Compared with the preset 1-hour interval, a 1-second interval can be considered small. Let's examine the three conditions of a Poisson experiment that, if met, justify X being modeled by a Poisson distribution. Condition 1 of a Poisson experiment means that the chance that a car passes under the bridge during any 1-second period is the same for every 1-second period in the hour. This condition would be violated if, for example, halfway through the hour an athletic event at a nearby high school ended. As long as the flow of traffic over the hour is steady (constant), condition 1 will be met.

Condition 2 of a Poisson experiment means that a car passing during a particular second has no effect on the chance that one passes during the next (or any other) second. That would not be the case if you are counting regularly scheduled buses stopping at a bus stop. If a bus has just passed, then for the next several minutes, it would be very unlikely that another one would come. If we are counting cars, however, condition 2 seems very reasonable to assume.

Condition 3 of a Poisson experiment is clearly valid for a small time interval because cars passing under the bridge are isolated events. If instead of counting cars, you counted the number of people in cars, condition 3 would not hold because there could be several people in one car.

Since we are confident that the three conditions of a Poisson experiment hold, we can use the Poisson probability distribution as the pdf of X.

Let us now consider the Key Problem, involving the Prussian cavalry soldiers.

Example 7.16

Let X be the number of fatalities due to horse kicks in a given cavalry corps in a given year. We first confirm that X obeys the conditions of a Poisson experiment. Assume that each of the 14 Prussian cavalry corps had approximately the same number of men and horses, so the risk of someone's being kicked to death was the same each day of the 20 years for each of the 14 cavalry corps. In other words, assume that condition 1 is satisfied, with the small time interval, relative to the preset 1-year time interval, being a single day. Condition 2 also follows because it is reasonable to assume that one soldier being kicked to death is independent of another soldier being kicked to death. Clearly, condition 3 holds because soldiers being kicked to death by horses are isolated events. X is therefore Poisson.

We can consider the data in the Key Problem as 280 independent observations of X. The theoretical mean μ_X, of a Poisson variable is the parameter λ, and it can be estimated by the sample mean $\bar{x}$. There were a total of $(1 \cdot 91) + (2 \cdot 32) + (3 \cdot 11) + (4 \cdot 2) = 196$ fatalities in 14 cavalry corps over 20 years ($14 \cdot 20 = 280$ units, recalling that each unit is counted in each of the 20 years). Therefore, $\bar{x} = 196/280 = 0.7$ per unit per year.

In Table 7.5 we have listed the experimental and theoretical Poisson probabilities corresponding to $\lambda = 0.7$. We have also computed the expected number of units with x fatalities by multiplying each Poisson probability (second from last column) by the number of units, 280, according to our np formula for the mean of a counting random variable, with $n = 280$ and $p = P(X = x)$. The Poisson model fits the data reasonably well, as illustrated by Figure 7.6. Note that each experimental probability $\hat{P}(X = x)$ is simply the number of units with x fatalities divided by the total number of units, 280.

Table 7.5 Prussian Soldier Data

Number of soldiers x kicked to death	Number of units	Experimental probability $\hat{P}(X = x)$	Theoretical Poisson probability $P(X = x)$ when $\lambda = 0.7$	Expected number of units with x fatalities
0	144	0.514	0.497	139.0
1	91	0.325	0.348	97.3
2	32	0.114	0.122	34.1
3	11	0.039	0.028	7.9
4	2	0.007	0.005	1.4
5 or more	0	0.000	0.001	0.2
Total	280	0.999	1.001	279.9

Note: The expected number of units with x fatalities is $nP(X = x)$.

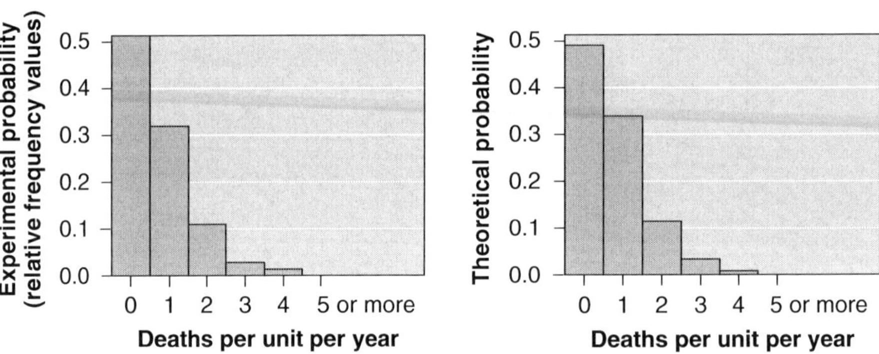

Figure 7.6 Prussian soldier data density histogram (left) and theoretical Poisson probability histogram for $\lambda = 0.7$ (right).

Example 7.17

It is tempting to use the Poisson distribution any time the description of the problem sounds superficially like the horse-kick example, in that isolated things are occurring randomly over time or randomly in space. But consider the number of runs the Oakland Athletics baseball team scored in each of the 144 games they played in 1995. "How many runs were scored per game?" sounds like "How many soldiers were kicked to death per unit per year?" Because the Poisson distribution worked so well in the latter case, one might be inclined to assume that it applies to the former. The question is whether the number of runs per game is reasonably modeled as a Poisson random variable.

The average number of runs per game scored by the Athletics was 5.07; let this be our estimate of λ in the Poisson pdf formula. In Table 7.6 we have, for each possible number of runs, the number of games in which the Athletics scored that number of runs as well as the observed proportion of games (that is, the experimental probability of that number of runs per game), and the theoretical Poisson probability of that number of runs, assuming a theoretical mean of $\lambda = 5.07$. Figure 7.7 has the corresponding density histogram constructed from these data and the theoretical Poisson probability histogram for $\lambda = 5.07$. They do not look close at all! Even after ignoring the small random upward and downward fluctuations of the experimental probability (as one should), the two shapes are still quite different.

Table 7.6 Oakland Athletics Runs per Game, 1995

Number of runs x	Number of games	Experimental probability $\hat{P}(X = x)$	Theoretical Poisson probability $P(X = x)$ when $\lambda = 5.07$
0	8	0.0556	0.0063
1	14	0.0972	0.0319
2	14	0.0972	0.0808
3	14	0.0972	0.1365
4	18	0.1250	0.1730
5	11	0.0764	0.1754
6	20	0.1389	0.1482
7	13	0.0903	0.1073
8	14	0.0972	0.0680
9	5	0.0347	0.0383
10	2	0.0139	0.0194
11	6	0.0417	0.0089
12	1	0.0069	0.0038
13	3	0.0208	0.0015
14	1	0.0069	0.0005
15 and over	0	0.0000	0.0003

Figure 7.7 shows that the number of runs for the Oakland A's (top graph) are more spread out than the Poisson model (bottom graph) predicts. There were more bad days (few runs) for the A's than the Poisson model predicts. There were more really good days (lots of runs) for the A's than the Poisson model predicts. You should try to explain how this situation might violate some of the three conditions for the Poisson distribution, and to explain why the Poisson distribution therefore does not model these data well. (See Exercise 9 of this section.)

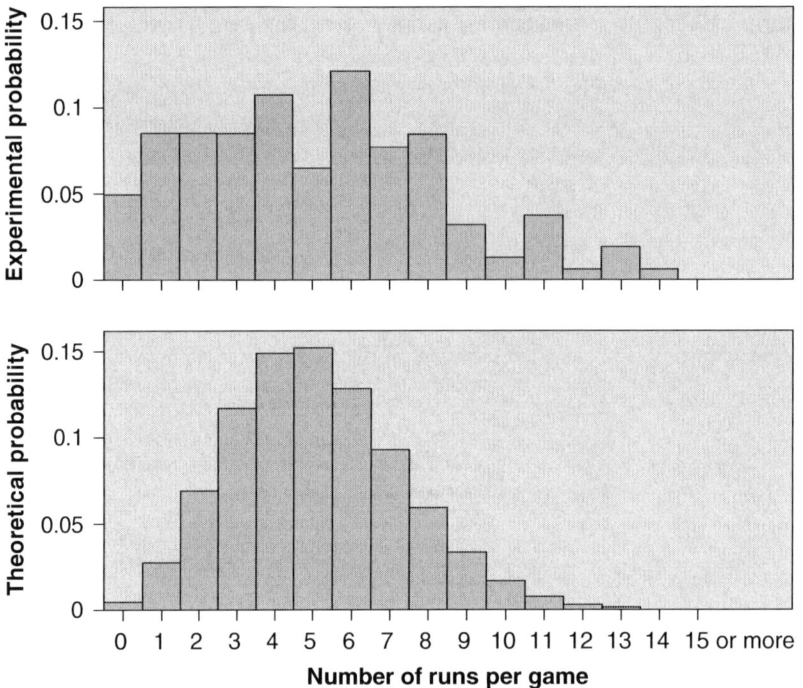

Figure 7.7 Oakland Athletics data density histogram (top) of experimental probabilities and theoretical Poisson probability histogram for $\lambda = 5.07$ (bottom).

Section 7.3 Summary

The pdf for a Poisson random variable X with mean $\lambda > 0$ is

$$p(x) = P(X = x) = \frac{\lambda^x}{x!}e^{-\lambda}, \qquad \text{for } x = 0, 1, 2, \ldots.$$

The mean of X is $\mu_X = \lambda$, and the standard deviation of X is $\sigma_X = \sqrt{\lambda}$.

Poisson distributions are often used to model the number of times an event of interest occurs during a predetermined time interval (or over a predetermined area or volume). The number of occurrences will be well modeled by a Poisson distribution if the following real-world conditions are judged to be approximately satisfied (called a Poisson experiment): The rate of occurrences is constant over time; past occurrences do not influence the chances of future occurrences; and near-simultaneous occurrences are highly unlikely.

Sometimes these three assumptions are loosely characterized as saying that the occurrences are isolated and totally at random. Poisson distributions can be used to accurately model the number of misprints in a book, the number of bacterial colonies on a petri dish, the number of diseased trees on an acre of woodland, the number of earthworms per square meter of soil, the number of stars in a region of the Milky Way Galaxy, and so forth.

Finally, Poisson distributions provide easy-to-compute approximations for hard-to-compute binomial probabilities if n is large and p is small: If X is binomial with $n \geq 25$, $p \leq 0.1$, and $np \leq 15$, then

$$P(X = x) = \binom{n}{x} p^x (1-p)^{n-x} \approx \frac{\lambda^x}{x!}e^{-\lambda},$$

where $\lambda = \mu_X = np$. Moreover, the fact that this mathematically verifiable Poisson approximation holds justifies X having a Poisson distribution when X in fact satisfies the conditions of a Poisson experiment.

Section 7.3 Exercises

1. According to the *Statistical Abstract of the United States,* the infant mortality rate of Australia is 5.0 deaths of under 1 year of age per 1000 live births. Use the Poisson approximation to determine the probability that among 2000 live-born infants selected at random there will be
   a. at most 15 infant deaths.
   b. at least 5 infant deaths.
   c. between 5 and 15 infant deaths, 5 and 15 included.

2. You play the Pick 3 lottery game by betting on one of the numbers from 000 to 999. Each number has the probability $p = 1/1000$ of being the winning number. Each bet costs you \$1. If you win, you receive \$500. (You do not get the purchase price of your winning lottery ticket back.) A man bets on 3 different numbers, 4 days a week, 50 weeks a year, for 2 years. On a given day his gain is either \$497 or –\$3 (–\$3 gain means a \$3 loss).
   a. What is the chance that he wins on a given day?
   b. What is his expected net gain over the 2 years?
   c. What is the probability that he does not win a single bet over the 2 years?
   d. What is the probability that he makes money over the 2 years? (*Hint:* How many times does he have to win to make money?)

3. In America routine smallpox vaccination was discontinued 30 years ago because smallpox no longer posed a threat. If bioterrorists now unleash the smallpox virus on an American city, hundreds of thousands of people might die. We could restart the vaccination program, but vaccination is not completely risk-free. According to the *CDC Morbidity and Mortality Weekly Report,* June 22, 2001, the risk of dying from a smallpox vaccination is one in a million. Imagine that 10 million individuals are vaccinated against smallpox. Determine the probability that among the vaccinated individuals
   a. At most 12 die from the vaccination.
   b. At least 10 die from the vaccination.
   c. Between 4 and 17, inclusive, die from the vaccination.

4. In the United States the annual death rate due to diabetes is 24 per 100,000 population.
   a. What is the expected number of deaths due to diabetes over a 2-year period in a simple random sample of 10,000 Americans?
   b. What is the probability that over a 2-year period a simple random sample of 10,000 Americans has no deaths due to diabetes? At most eight deaths due to diabetes? At least five deaths due to diabetes?

5. During the twentieth century an average of 1.6 hurricanes per year struck the U.S. mainland. Estimate the probability that in a given year
   a. No hurricane will strike the U.S. mainland.
   b. At most one hurricane will strike the U.S. mainland.
   c. At least two hurricanes will strike the U.S. mainland.

6. The world's most active volcano above sea level is Kilauea on the island of Hawaii, which since the mid-1980s has erupted an average of 3.5 times per year. (An eruption indicates vigorous activity either from a new volcanic vent or from an established volcanic vent after a prior slowdown.)
   a. What is the expected number of eruptions during the next 24 months?
   b. What is the probability that over the next 24 months, Kilauea will have between 4 and 10 eruptions, inclusive?

7. A Nevada roulette wheel has 38 pockets; one is numbered 0, another is numbered 00, and the rest are numbered from 1 through 36. The croupier spins the wheel and drops a ball inside. The ball is equally likely to land in any one of the 38 pockets. Before it lands, bets can be placed. If you bet a dollar on a single number at roulette and the number comes up, you get the \$1 back together with winnings of \$35. If any other number comes up, you lose the dollar. Gamblers say that a single number pays 35 to 1. Suppose you play roulette 152 times, betting a dollar on the number 17 each time.
   a. How many times can you expect to win? (Compute the expected value.)
   b. What is the chance that you don't win any bets?

c. How many times do you have to win to come out ahead (to win more money than you lose)?
d. What is your chance of coming out ahead? *Hint:* If you are feeling energetic, do this problem using formulas and simulation.
e. What is your expected gain (positive or negative)?

8. Thirty-five out of every 10,000 births result in a pair of identical twins, on average. Determine the probability that out of 3000 births selected at random there will be
   a. at most 5 pairs of identical twins.
   b. between 5 and 16 pairs of identical twins, inclusive.

9. Go through the three conditions for a Poisson random variable and give possible reasons why, in the example of the Oakland Athletics, the data do not seem to follow the Poisson distribution. *Hint:* As one of many reasons, are the number of runs scored by the Oakland A's during the last three innings of a game independent of the number of runs they scored during the first three innings? Explain.

10. Rebecca works as a greeter at a grocery store. She theorizes that during the 2 P.M. to 4 P.M. time period, customers arrive at a rate of two per minute. She is willing to make the three Poisson assumptions. She keeps track of the number of customers who arrive each morning during a 15-minute period. Her results are shown in the following table:

Minute	Number of customers	Minute	Number of customers
1	2	9	1
2	0	10	2
3	4	11	3
4	5	12	4
5	1	13	2
6	0	14	1
7	2	15	0
8	3		

   a. In how many 1-minute periods did no customer arrive? In how many 1-minute periods did one customer arrive? Two customers? Three customers? Four customers? Five customers? What is the average number of customers per minute?
   b. Using the answers from part (a), find the experimental probability that no customer arrives in a 1-minute period. Calculate such a probability for every number of customers up to five.
   c. According to the Poisson distribution with a theoretical mean of 2 (our estimate of A in part (a), you'll recall), the theoretical probabilities of the six outcomes are as follows:

Number of customers in a minute	Theoretical probability
0	0.135
1	0.271
2	0.271
3	0.180
4	0.090
5	0.036

Create two histograms, one showing the experimental probabilities and the other showing the theoretical probabilities. Compare the two histograms you created. Does this evidence seem to support Rebecca's hypothesis concerning the arrival of customers?

11. The following table shows experimental probabilities based on 1000 observations. The table also has the corresponding theoretical probabilities under the Poisson distribution with a mean of 3.4.

Number of occurrences x	Experimental probability $\hat{P}(X = x)$	Theoretical probability $P(X = x)$
0	0.053	0.033
1	0.153	0.113
2	0.138	0.193
3	0.103	0.219
4	0.225	0.186
5	0.214	0.126
6	0.068	0.072
7	0.047	0.035

a. Make a density histogram of the experimental probabilities and a histogram of the theoretical probabilities.
b. Compare the two histograms. Do the data seem to follow the Poisson distribution with mean $\lambda = 3.4$?

12. Since seismologists started using the Richter scale in 1935 to measure the magnitude of earth quakes, the most powerful earthquake ever recorded was a 1960 quake off the coast of Chile that measured 9.5. The quake on December 26, 2004, that triggered the deadly tsunami in the Indian Ocean that killed 300,000 people measured 9.0. The aftershock on March 28, 2005, measured 8.7. On average, there is an earthquake of magnitude over 8.0 somewhere in the world four times every 5 years. Estimate the probability that in a given year there are
a. no earthquakes of magnitude over 8.0.
b. at least two earthquakes of magnitude over 8.0.
c. at most two earthquakes of magnitude over 8.0.

7.4 THE NORMAL PROBABILITY DISTRIBUTION

In Sections 2.3 and 2.4, we looked at different data sets whose density histograms roughly followed a normal (bell-shaped) curve. Examples included the batting averages of Major League Baseball players, IQ scores, and the heights of men and women, considered separately. Many other variables are also approximately normally distributed. For example, cholesterol level, systolic blood pressure, reaction time, and lung capacity are all approximately normally distributed. On the other hand, weight distributions are only crudely bell-shaped, being usually moderately skewed to the right.

Suppose we have the heights of a random sample of $n = 100$ women. If we group the data into class intervals 2 inches wide, a typical density histogram would look like Figure 7.8(a). If we had a sample of 1000 women, and if we group their heights into class intervals of 1 inch width, a typical density histogram is given in Figure 7.8(b). (See, for example, Pearson and Lee's histogram in Figure 1.13 for the heights of 1052 women.) If we had a sample of 50,000 women and we grouped their heights into class intervals of 1/2 inch width, then a typical density histogram is given in Figure 7.8(c). Notice that as n gets large and the class width gets small, the typical density histogram approximates a normal curve increasingly well. Finally, if it were possible to measure the exact height (with no round-off error) of the over 100 million adult women in the United States, then as the class interval width approaches 0 we would get closer and closer to the limiting normal density histogram (pdf) in Figure 7.8(d).

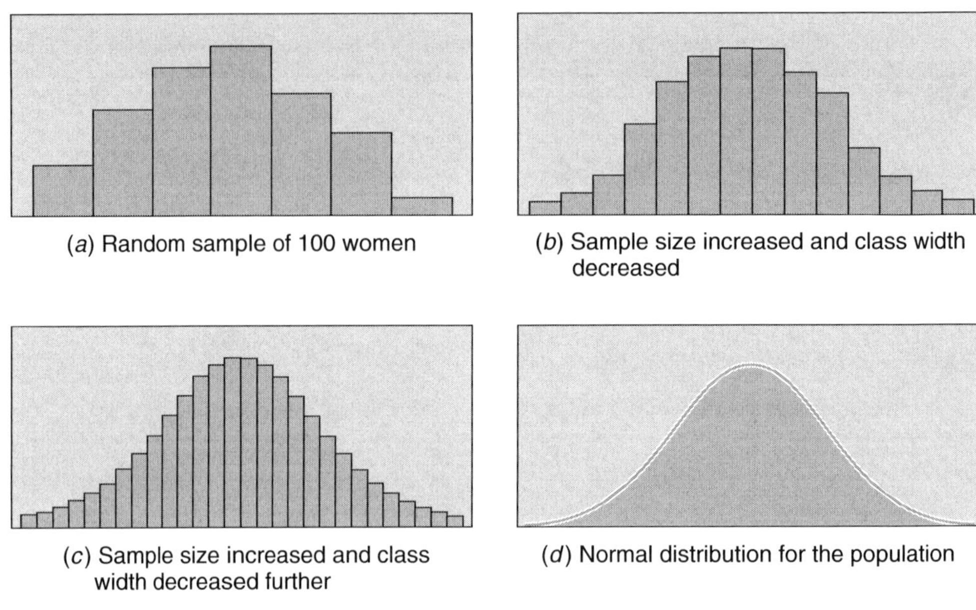

Figure 7.8 Histograms for the heights of women.

Let μ denote the mean height of the over 100 million women in the United States and σ the standard deviation of the heights. The normal curve in Figure 7.8(d), scaled so that the total area under the curve equals 1, is given by the equation

$$f(x) = \frac{1}{\sigma\sqrt{2\pi}} e^{-(x-\mu)^2/(2\sigma^2)},$$

normal probability density (pdf)
normal curve

where $e \approx 2.71828$ and $\pi \approx 3.14159$. This **normal probability density (pdf)** or **normal curve,** with mean μ and standard deviation σ, is displayed in Figure 7.9. This curve

- is bell-shaped and symmetric around the mean μ,

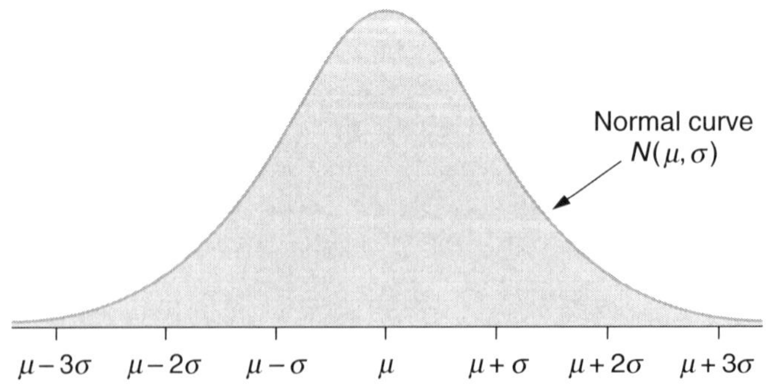

Figure 7.9 The normal probability density curve with mean μ and standard deviation σ.

- extends indefinitely in both directions, always above the horizontal axis, rapidly approaching, but never touching the axis,

- has a total area under the curve of 1,
- obeys the 68-95-99.7% Rule almost perfectly, for example, having 99.7% of the area under the curve between $\mu - 3\sigma$ and $\mu + 3\sigma$.

standard normal density

The **standard normal density** (discussed in Section 2.4) is the special normal curve (pdf) with mean $\mu = 0$ and standard deviation $\sigma = 1$. Its probabilities are tabulated in Appendix E.

Recall that for density histograms of data, which have total area 1, that the *area of the bars* over a given interval equals the *proportion of observations* in that interval. If we apply this to the probability density curve in Figure 7.8(*d*), which we view as a histogram with class interval width close to 0 for the exact heights of the more than 100 million U.S. women, we conclude that the proportion of women who have heights between two given limits, a and b, equals the area from a to b under the probability density curve, as shown in Figure 7.10.

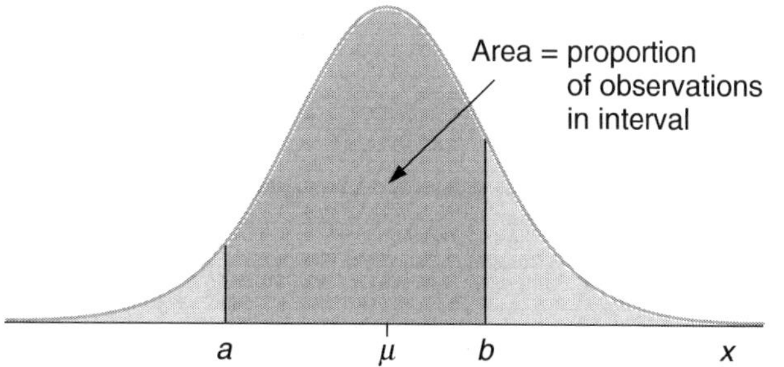

Figure 7.10 The proportion of observations in an interval.

As we explained in Section 2.4, we can determine the interval-based area under any normal curve by converting the interval endpoints into z-scores. Then, we find the area over the z-score interval under the standard normal curve (via tables in Appendix E), as learned in Section 2.4. This method is illustrated by Figure 7.11.

Now let X denote the exact height (measured without round-off error) of a randomly selected woman. We have seen (recall Figure 7.8(d)) that X is normally distributed with mean μ and standard deviation σ, written "X is $N(\mu, \sigma)$". (Refer to Figure 7.11.) The probability that the woman's height is between a and b is

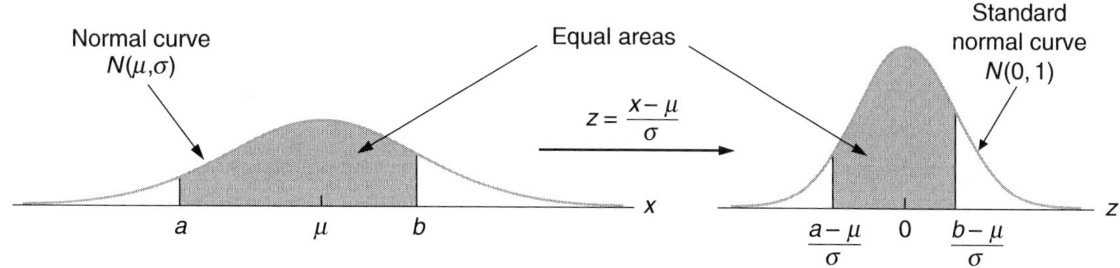

Figure 7.11 Finding areas under normal curves using the standard normal curve.

$$\begin{aligned} P(a \le X \le b) &= \text{proportion of women in the population with height between } a \text{ and } b \\ &= \text{area from } a \text{ to } b \text{ under the normal curve with mean } \mu \\ &\quad \text{and standard deviation } \sigma \\ &= \text{area from } a - \mu/\sigma \text{ to } (b - \mu)/\sigma \text{ under the standard normal curve.} \end{aligned}$$

This is illustrated in the next example.

Example 7.18

Assume that the mean height of U.S. women is $\mu = 63.7$ inches with a standard deviation of $\sigma = 2.6$ inches. Let X denote the exact height (measured without round-off error) of a randomly selected woman. Find

$$P(60 \le X \le 66) = \text{the proportion of women with height between 60 inches and 66 inches.}$$

Solution: We shall proceed in three steps. (Learn these steps for the many similar problems to be solved.)

steps for solving a normal probability problem

Step 1. Draw a number line and shade the interval of interest.

Step 2. Convert the interval endpoints into z-scores.

$$(60 - \mu)/\sigma = (60 - 63.7)/2.6 = 1.42$$
$$(66 - \mu)/\sigma = (66 - 63.7)/72.6 = 0.88$$

−1.42 0.88 z

Step 3. Find the area under the standard normal curve over the z-score interval.

$$\begin{aligned} P(60 \le X \le 66) &= \text{area under } N(0, 1) \text{ pdf over the interval}(-1.42, 0.88) \\ &= 0.8106 - 0.0778 = 0.7328, \end{aligned}$$

noting that $P(Z \le 0.88) = 0.8106$ and $P(Z \le -1.42) = 0.0778$ from Appendix E.

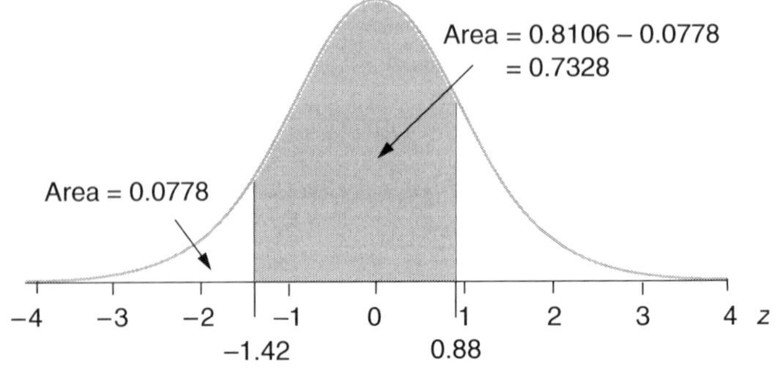

❐ ❐

68-95-99.7% Rule for a $N(\mu, \sigma)$ and a standard $N(0, 1)$ distribution.

As another example of converting from a $N(\mu, \sigma)$ setting to a $N(0, 1)$ setting, Figure 7.12 illustrates that the **68-95-99.7% Rule for a $N(\mu, \sigma)$ distribution** is derived from the **68-95-99.7% Rule for the standard normal ($N(0, 1)$) distribution.**

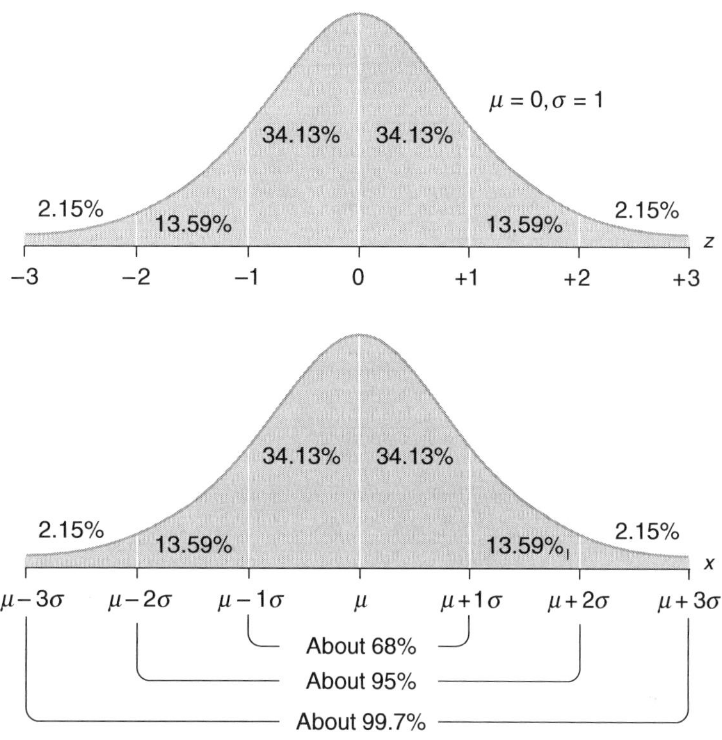

Figure 7.12 Areas under the standard normal curve derived from the normal table and the corresponding areas under the normal curve with mean μ and standard deviation σ.

Normally Distributed Random Variables

continuous random variables

Counting random variables (binomial, Poisson, etc.) are discrete random variables. Measurement variables, like exact height, are **continuous random variables.**

A continuous random variable is defined as a random variable all of whose probabilities $P(a \le X \le b)$ are found by finding the area under its probability density function (pdf) from a to b. A random variable modeled by a normal pdf thus is one example of a continuous random variable. In later chapters we shall discuss additional continuous distributions, including the t-distributions and the χ^2 distributions.

Consider a continuous random variable, X, determined by the outcome of some random phenomenon. If the density histogram of tens of thousands of independent observations of X follows the normal probability density curve depicted in Figure 7.9, then we say that X is **normally distributed with mean μ and standard deviation σ,** and as remarked above, use the shorthand notation $N(\mu, \sigma)$ to denote the normal distribution with mean μ and standard deviation σ.

normally distributed with mean μ and standard deviation σ

standardized normal random variable is $N(0, 1)$

Important fact: If a random variable X is $N(\mu, \sigma)$, then the standardized variable

$$Z = \frac{X - \mu}{\sigma}$$

has the standard normal distribution, $N(0, 1)$. It is this fact that allows us to determine any normal probability from areas under the standard normal density curve, as the next example shows.

Example 7.19

According to the National Center for Health Statistics, the systolic blood pressure of adult women is (approximately) normally distributed, with mean $\mu = 123$ mmHg and standard deviation $\sigma = 21$ mmHg. If we randomly select an adult woman and measure her systolic blood pressure X, find

$$P(X \geq 150) = \text{proportion of adult women with systolic blood pressure over 150.}$$

Solution: The random variable X is (approximately) $N(123,21)$. We shall proceed in three steps.

Step 1. Draw a number line and shade the interval of interest.

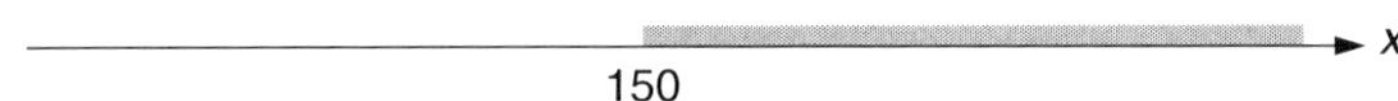

Step 2. Convert the interval endpoint into z-score.

$$(150 - \mu)/\sigma = (150 - 123)/21 = 1.29$$

z
1.29

Step 3. Find the area under the standard normal curve over the z-score interval. Consulting Appendix E,

$$P(X \geq 150) = 1 - P(X \leq 150) = 1 - P(Z \leq 1.29) = 1 - 0.9015 = 0.0985.$$

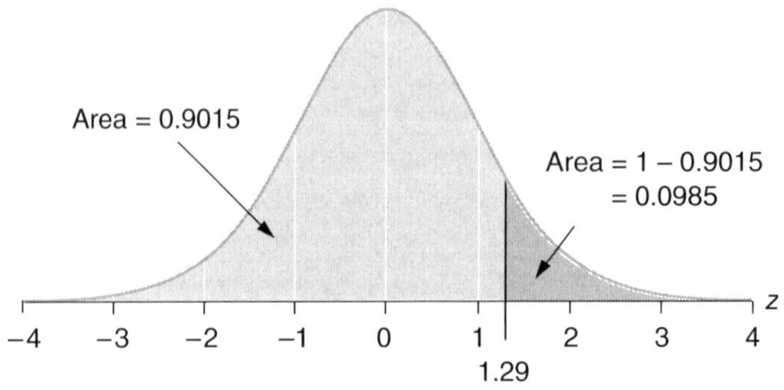

Assessing Normality: Normal Probability Plots

The normal distributions give us good models for many data sets. Then we can easily compute probabilities (that is, population proportions of interest). For any fairly homogeneous group of people (such as the population of all six-year-old boys or the population of all female undergraduate students), many test scores (such as IQ scores) and many physiological variables (such as height and systolic blood pressure) will be approximately normally distributed. So will repeated measurements like Newcomb's data (if we drop the occasional outlier). However, not every data set is approximately normally distributed. Age distributions (See Section 1.5) and income distributions (see Section 2.2) do not follow a normal curve.

In statistics we often use sample data to draw conclusions about a population whose distribution is unknown. The population can be real, as in Example 1.9, or conceptual, as in the repeated measurement setup. Many important statistical procedures for drawing inferences from sample to population are based on the assumption that the unknown population distribution is (very close to) normal. To justify the use of such procedures we have to make sure that the normality assumption is reasonable. If the sample is large, then the sample histogram can tell us whether the population distribution is approximately normal or not. The sample histogram should be roughly bell-shaped without outliers. Nonnormal features such as skewness or gaps and clusters would cast doubt on the normality of the population distribution. If we only have a small sample, however, then it is not always easy to tell from the sample histogram whether the unknown population distribution is approximately normal or not. There is more chance variation in small samples.

normal probability plot

A **normal probability plot** is a graphic tool for checking whether a data set looks like a sample from a normally distributed population. Although such plots can be constructed by hand, using statistical tables, we shall only consider computer-generated normal probability plots. Such plots come in several slightly different versions. But the basic idea is the same for all of these plots: We plot the observed data against the z-scores expected if the data came from a normal population. If we have a sample from a normal population, then the points of the probability plot will follow a straight line. If, on the other hand, we have a sample from a skewed distribution, then the plot will be curved. Outliers in the data set will show up as outliers in the normal probability plot.

Guidelines for Assessing Normality Using a Normal Probability Plot

- If the plotted points lie close to a straight line, the data set looks like a sample from a normal population; i.e., the data set does not seem to be a set of observations of a normally distributed random variable.
- If the plot has a nonlinear trend (such as curvature) or gaps and clusters, then the data set seems to be a sample from a normal population; i.e., the data set does not look like a set of observations of a normally distributed random variable.

Figures 7.13 to 7.16 are normal probability plots for data sets we have looked at before. The data are plotted vertically against the corresponding standard normal z-score along the horizontal axis. (Minitab and TI-83 PLUS switch the axes.)

Example 7.20

Figure 7.13 is a normal probability plot for the age at death of American presidents, listed in Example 2.5. The plotted points are tightly clustered around a straight line. We conclude that the data set seems to be a sample from a normal population.

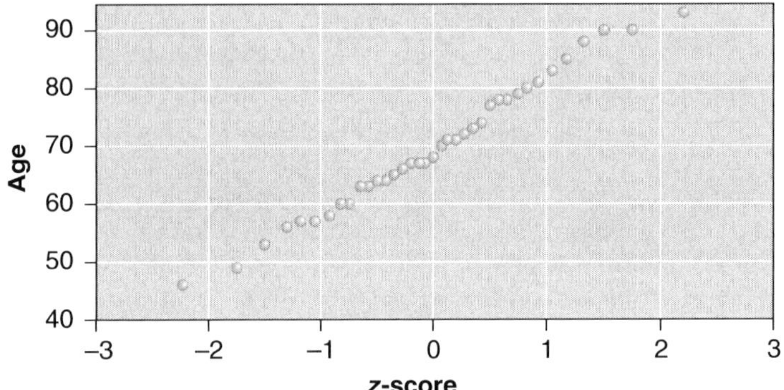

Figure 7.13 Normal probability plot for the age at death of American presidents from George Washington to Ronald Reagan. The straight linear pattern shows that the data set looks like a sample from a normal population.

Example 7.21

Figure 7.14 is a normal probability plot for the salaries of the Miami Heat for the 2005-2006 season, listed in Example 2.6. As the histogram in Figure 2.4 shows, the salary distribution is right-skewed and Shaq's salary is an outlier. The normal probability plot shows significant departure from linearity, reflecting the fact that the data set does not seem to be a sample from a normal population.

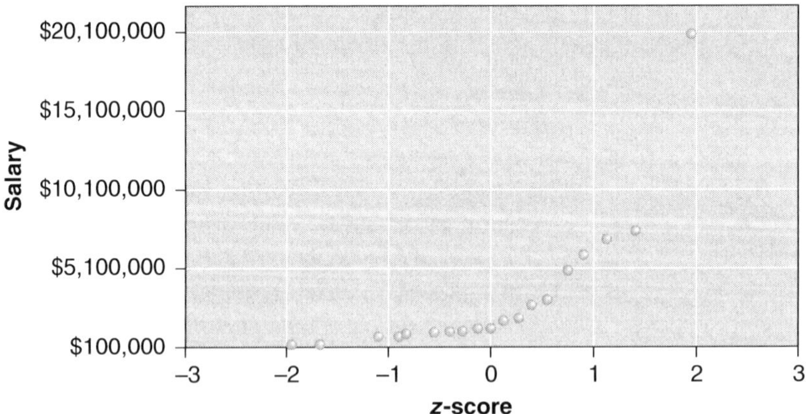

Figure 7.14 Normal probability plot for the salaries of the Miami Heat for the 2005-2006 season. The upward curvature and the outlier in the upper right corner shows that this data set is not a sample from a normal population.

Example 7.22

Figure 7.15 is a normal probability plot for Simon Newcomb's passage times listed in Table 2.2 in Example 2.8. Except for the two outliers, the plotted points are tightly clustered around a straight line. This indicates that a normal model fits well if we drop the two outliers. Figure 7.16 confirms this. Figure 7.16 is a normal probability plot for Newcomb's passage times with the two outliers omitted. The only significant departure from normality is caused by Newcomb's rounding of the passage times to the nearest nanosecond. Because of the rounding, many values are repeated several times. For example, six measurements were rounded to 27 and seven measurements were rounded to 28. Each of the values 24, 25, 26, 29, and 32 were repeated five times. As a result, the normal probability plot has a nonnormal "stairstep" appearance. This phenomenon is called **granularity.** This nonnormal feature would have disappeared if Newcomb had been able to measure the passage times to the nearest tenth of a nanosecond.

granularity

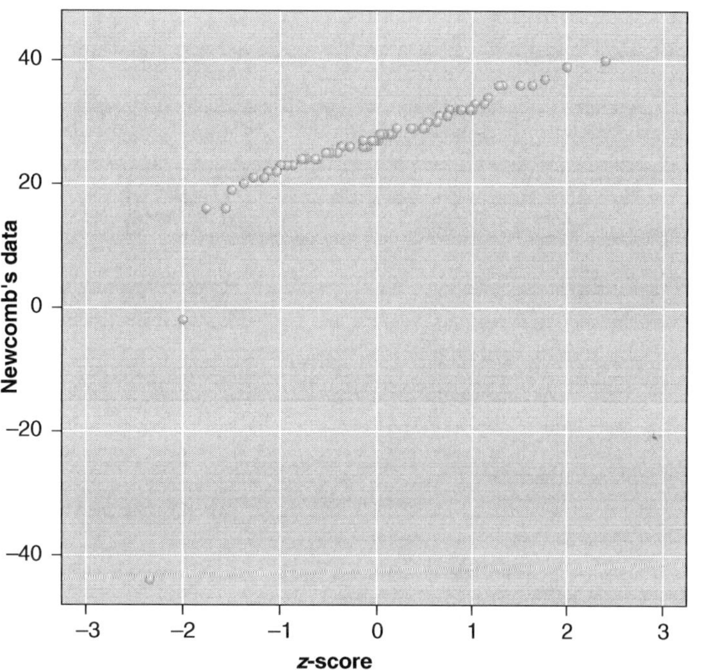

Figure 7.15 Normal probability plot for Newcomb's 66 measurements of the passage time of light. Except for the two low outliers, the plotted points follow a straight line.

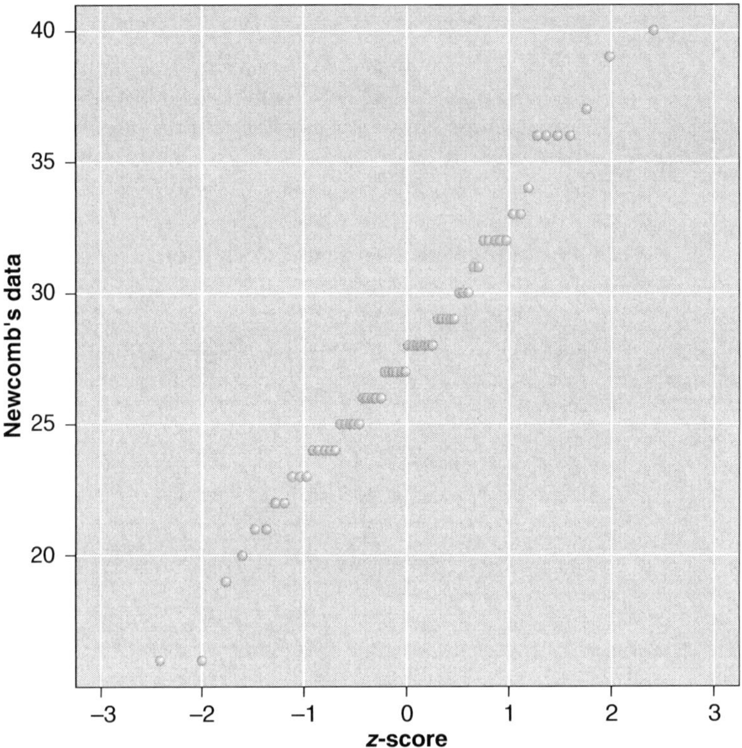

Figure 7.16 Normal probability plot for Newcomb's measurements without the two low outliers. Except for the granularity caused by Newcomb's rounding of his measurements, the data set looks like a set of observations of a normally distributed random variable.

❐ ❐

You should view normal probability plots as showing how you decide to decide whether the data was sampled from a normal population.

The Normal Approximation to a Binomial Distribution

A binomial random variable X (See Section 7.1) with parameters n and p can be thought of as the number of successes in n independent success-failure trials with success-probability p in each trial (Bernoulli trials). The pdf of X is

$$p(x) = P(X = x) = \binom{n}{x} p^x (1-p)^{n-x},$$

for $x = 0, 1, \ldots, n$. As we discussed at the beginning of Section 7.3, this formula becomes awkward to use as the number of trials n gets large. In Section 7.3 we presented the Poisson approximation to binomial probabilities, an approximation that works well for large n when p is sufficiently small. Here we present the normal approximation to binomial probabilities, discovered 300 years ago by Isaac Newton's contemporary Abraham de Moivre. His approximation works for any fixed value of p provided n is large enough.

Recall that the mean and standard deviation of a binomial variable are

$$\mu = np \quad \text{and} \quad \sigma = \sqrt{np(1-p)},$$

respectively. As exhibited in Figure 7.17, the basic idea of the normal approximation is that we can replace the probability histogram of a binomial variable X with the normal density curve that has the same mean and standard deviation as X.

The bars of a binomial probability histogram are centered at the values $0, 1, \ldots, n$, and each bar has a width of 1 unit. The bar centered at the integer c covers the interval from $c - 0.5$ to $c + 0.5$ and has an area equal to $P(X = c)$. If the corresponding continuous normal curve follows the histogram closely, then the area of the bar centered at the integer c will be close to the area under the normal curve from $c - 0.5$ to $c + 0.5$. We can therefore estimate $P(X = c)$ by the area under the corresponding normal curve from $c - 0.5$ to $c + 0.5$. The same reasoning applies to a binomial probability of the form $P(a \leq X \leq b)$, where a and b are integers: We need the normal curve areas corresponding to the rectangles from a to b, with the left-hand rectangle starting at $a - 0.5$ and the right-hand rectangle ending at $b + 0.5$.

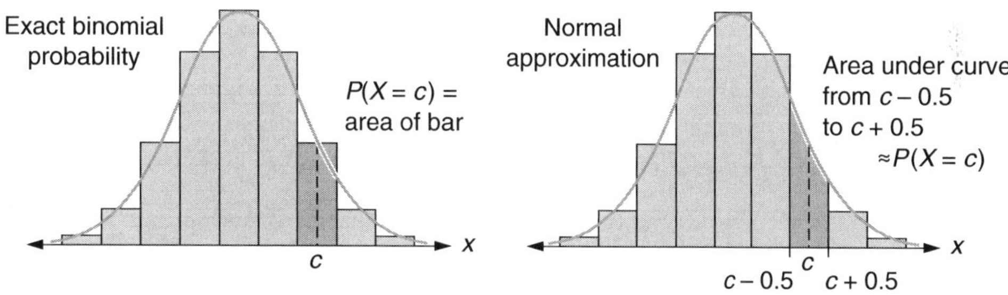

Figure 7.17 The normal approximation to a binomial probability.

continuity correction

Hence, $P(a \leq X \leq b) \approx$ Area under normal curve with an interval from $a - 0.5$ to $b + 0.5$. These adjustments of a and b by 0.5 are often referred to as the **continuity correction.** Note that we do not want strict inequalities. Thus, $P(2 < X < 9)$ becomes $P(3 \leq X \leq 8)$, producing the interval $(3 - 0.5, 8 + 0.5) = (2.5, 8.5)$.

Example 7.23

Let X denote the number of heads in 100 tosses of a fair coin. Estimate $P(45 \leq X \leq 55)$.

Solution: X is binomial with $n = 100$ and $p = 1/2$. Therefore, X has mean $\mu = np = 50$ and standard deviation $\sigma = \sqrt{np(1-p)} = \sqrt{25} = 5$. As we can see in Figure 7.18, the normal curve $N(50, 5)$ follows the probability histogram of X very closely.

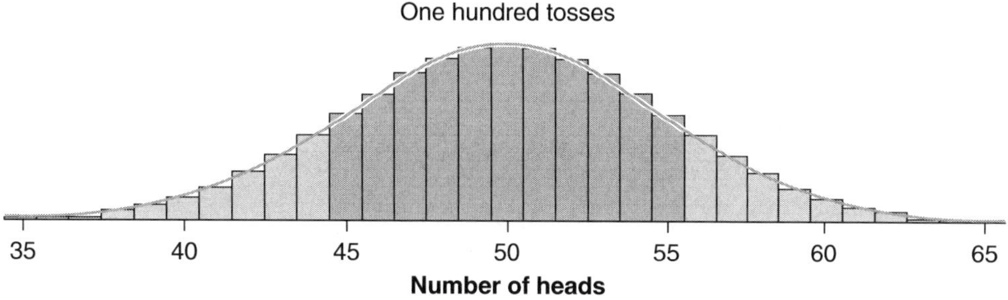

Figure 7.18 The probability histogram for the number of heads in 100 tosses, with the normal curve $N(50, 5)$ superimposed.

The cumulative binomial probability $P(45 \leq X \leq 55)$ equals the combined area of the 11 histogram bars centered over the integer values 45 through 55 in Figure 7.18. These 11 bars cover the interval from $45 - 0.5 = 44.5$ to $55 + 0.5 = 55.5$. We can therefore estimate the binomial probability $P(45 \leq X \leq 55)$ by the area under the normal curve $N(50, 5)$ from 44.5 to 55.5. To determine this area we proceed in three steps:

Step 1. Draw a number line and shade the interval of interest.

Step 2. Convert the original endpoints into z-scores.

$$(44.5 - \mu)/\sigma = (44.5 - 50)/5 = -1.1$$
$$(55.5 - \mu)/\sigma = (55.5 - 50)/5 = 1.1$$

−1.1 1.1

Step 3. Find the area under the standard normal curve over the z-score interval.

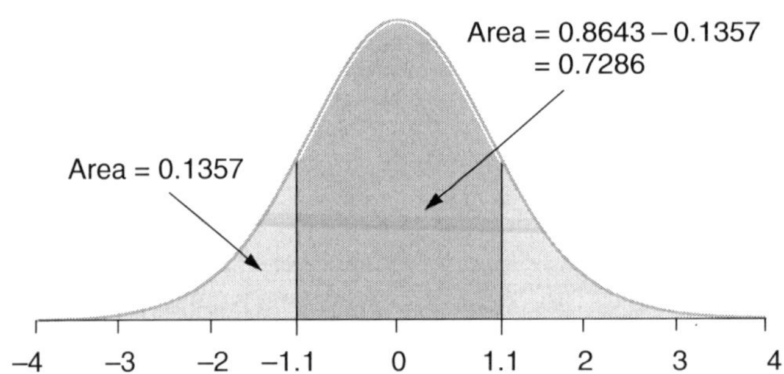

The normal approximation gives us $P(45 \leq X \leq 55) = P(-1.1 \leq Z \leq 1.1) = P(Z \leq 1.1) - P(Z \leq -1.1) = 0.8643 - 0.1357 = 0.7286$ using Appendix E, where Z is $N(0, 1)$. The actual binomial probability is 0.7287. The normal approximation is extremely accurate in this case.

The normal approximation does not work equally well for all binomial distributions. If the binomial probability histogram is skewed because of a small p close to 0 or a large p close to 1, then the normal approximation will be poor. Figure 7.19 displays three binomial probability histograms with superimposed normal curves for $p = 0.1$. They are all skewed to the right, but less so for the larger n values. The highly skewed binomial probability histogram for $p = 0.1$ and $n = 10$ can be found in Figure 7.2 in Section 7.1. In both cases when $n = 10$ and $n = 25$, the normal approximation is poor. The normal approximation is not too bad when $n = 100$, but the Poisson approximation, which works well for large n and small p, is actually better. In the case when $n = 400$, the normal approximation is quite good.

For fixed p, the normal approximation gets better as n gets larger. How can we tell whether n is large enough to guarantee that the normal approximation will be reasonably accurate?

How large n should be depends on how small p is (or how close p is to 1). The following is a useful rule of thumb.

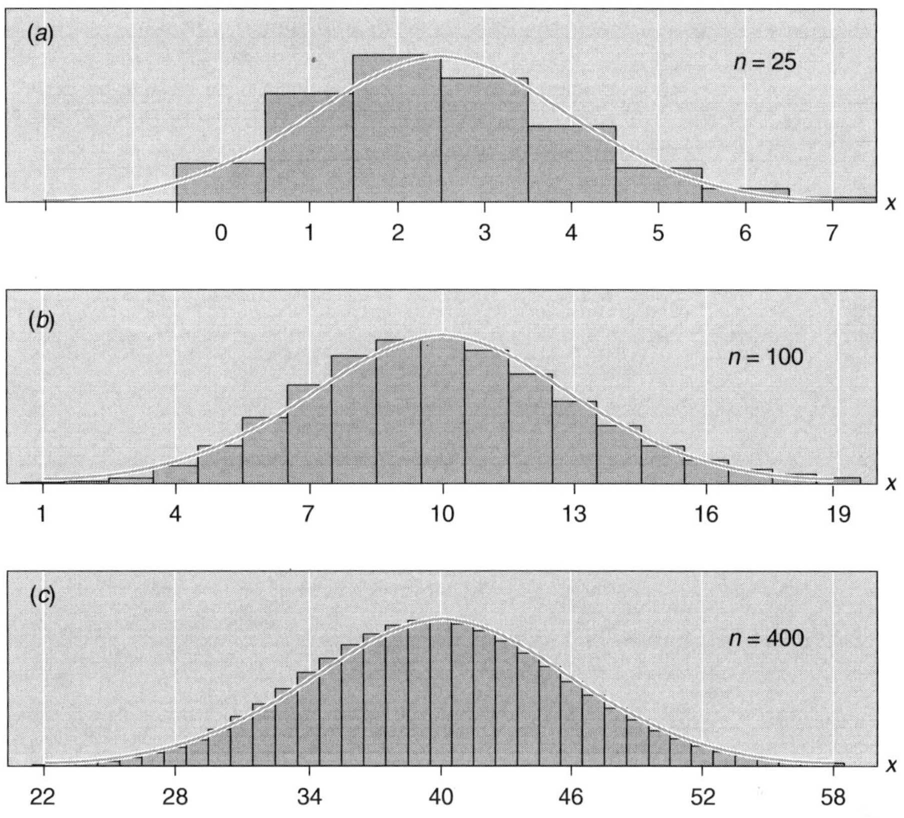

Figure 7.19 The normal approximation to binomial distributions with $p = 0.1$ for (a) $n = 25$, (b) $n = 100$, and (c) $n = 400$.

Recall that a normal variable is 99.7% certain to fall in the range $\mu \pm 3\sigma$ (see Figure 7.12). A binomial random variable has values in the range 0 to n; it satisfies $\mu = np$ and $\sigma = \sqrt{np(1-p)}$. If the 99.7% normal range $\mu \pm 3\sigma$ is contained in the binomial range 0 to n, namely

$$0 \le \mu - 3\sigma \le \mu + 3\sigma \le n,$$

which, substituting for μ and σ becomes

$$0 \le np - 3\sqrt{np(1-p)} \le np + 3\sqrt{np(1-p)} \le n,$$

rule of thumb for using the normal approximation to the binomial

then the normal approximation will be reasonably accurate. But if this normal range $\mu \pm 3\sigma$ extends below $0 - 1/2 = -1/2$ or above $n + 1/2$ the approximation can be poor. The condition $0 \le np - 3\sqrt{np(1-p)} \le np + 3\sqrt{np(1-p)} \le n$ can be restated, using algebra, as

$$np \ge 9(1-p) \text{ and } n(1-p) \ge 9p.$$

This pair of inequalities will be this textbook's rule of thumb for when to use the normal pdf to accurately approximate the binomial distribution.

Remark Most textbooks instead give the following easy-to-use requirement for when to use the normal approximation to binomial probabilities: $np \geq 5$ and $n(l - p) \geq 5$. However, if $p \leq 0.1$ and $5 \leq np \leq 10$, then the Poisson approximation is slightly more accurate than the normal approximation, suggesting that the rule should be modified, as has been done here.

Figure 7.20 illustrates our rule of thumb. In Figure 7.20(a), we consider the binomial distribution with $n = 20$ and $p = 0.6$. The interval $\mu \pm 3\sigma = 12 \pm 3\sqrt{4.8}$ or (5.43,18.57) lies within the binomial range 0 to 20, as shown by the fact that $np \geq 9(1 - p)$, namely, $12 \geq 3.6$, and $n(1 - p) \geq 9p$, namely, $8 \geq 5.4$, both hold. So the normal approximation to the binomial distribution is good. In Figure 7.20(b) we try to use the normal approximation to the binomial with $n = 10$ and $p = 0.1$. Here $\mu \pm 3\sigma = 1 \pm 3\sqrt{0.9}$ or $(-1.85, 3.85)$ goes below the binomial range 0 to 10. The requirement, $np \geq 9(1 - p)$, namely, $1 \geq 8.1$, fails badly. Thus, the normal approximation is not good, as we see. For example, $P(X = 0) = 0.349$, while the normal approximation gives 0.2421, a serious error.

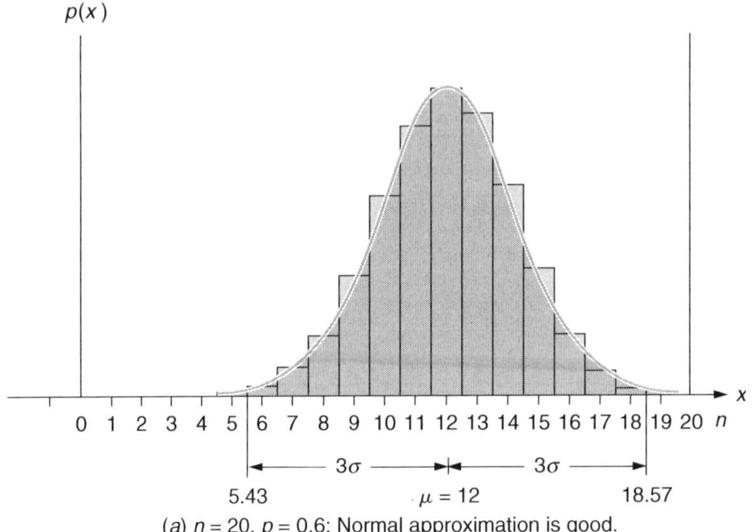

(a) $n = 20$, $p = 0.6$: Normal approximation is good.

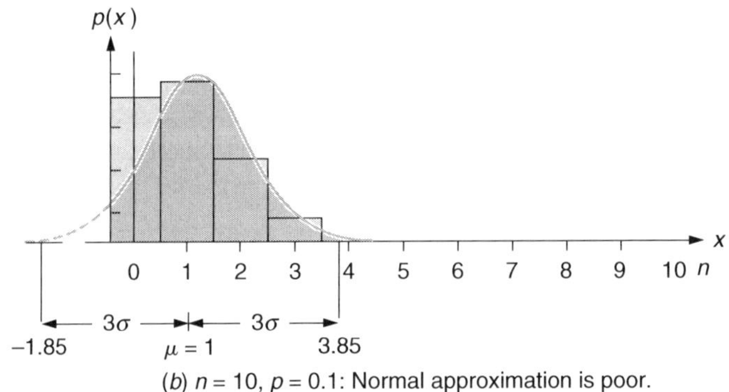

(b) $n = 10$, $p = 0.1$: Normal approximation is poor.

Figure 7.20 Rule of thumb for normal approximation to binomial probabilities.

Section 7.4 Summary

Consider a continuous random variable X determined by the outcome of some random experiment. If the density histogram of tens of thousands of independent observations of X closely follows the normal pdf $f(x)$ with mean μ and standard deviation σ,

$$f(x) = \frac{1}{\sigma\sqrt{2\pi}} e^{-(x-\mu)^2/(2\sigma^2)} \qquad \text{for } -\infty < x < \infty,$$

then we say that X is **normally distributed** with mean μ and standard deviation σ. We use the shorthand notation X is $N(\mu, \sigma)$.

If the random variable X is $N(\mu, \sigma)$, then the standardized variable

$$Z = \frac{X - \mu}{\sigma}$$

has the standard normal distribution $N(0, 1)$. If X is $N(\mu, \sigma)$, then

$$\begin{aligned} P(a \le X \le b) &= \text{area from } a \text{ to } b \text{ under the normal curve with mean } \mu \text{ and standard deviation } \sigma \\ &= \text{area from } (a-\mu)/\sigma \text{ to } (b-\mu)/\sigma \text{ under the standard normal curve.} \end{aligned}$$

In particular, if X is $N(\mu, \sigma)$, then there is about a 68% chance that X will be in the range $\mu - \sigma$ to $\mu + \sigma$, a 95% chance that X will be in the range $\mu - 2\sigma$ to $\mu + 2\sigma$, and a 99.7% chance that X will be in the range $\mu - 3\sigma$ to $\mu + 3\sigma$. This is the 68-95-99.7% Rule for normal random variables. It is the theoretical companion to the 68-95-99.7% Rule for approximately bell-shaped data sets.

Guidelines for Assessing Normality Using a Normal Probability Plot. To assess the normality of a variable using sample data, construct a normal probability plot (sometimes called a normal quantile plot).

- If the plotted points lie close to a straight line, the data set looks like a sample from a normal population.
- If the plot has a nonlinear trend (such as curvature) or gaps and clusters, then the data set does not look like a sample from a normal population.

normal approximation to a binomial random variable

The Normal Approximation to a Binomial Distribution. Let X be binomial with parameters n and p. Then X has mean and standard deviation

$$\mu = np \text{ and } \sigma = \sqrt{np(1-p)}.$$

If $0 \le \mu - 3\sigma \le \mu + 3\sigma \le n$, which is the same as $np \ge 9(1-p)$ and $n(1-p) \ge 9p$, then we can accurately estimate $P(a \le X \le b)$ for integers $a \le b$ by areas under the normal density curve with the same mean and standard deviation as X. We do this by applying the continuity correction to a and b.

Section 7.4 Exercises

1. A random variable X is assumed to have the normal distribution with mean 5.0 and standard deviation 1.5 (i.e., X is N (5, 1.5)).
   a. State a range in which 68% of the observations generated by this random variable will lie.
   b. State a range in which 95% of the observations generated by this random variable will lie.
2. Another random variable is assumed to have the normal distribution, this time with mean −3.5 and standard deviation 0.5.
   a. State a range in which 68% of the observations of this random variable will lie.
   b. State a range in which 95% of the observations of this random variable will lie.
3. A random variable was observed 100 times. Before gathering the data, the researcher hypothesized that the data would follow the normal distribution with mean 45 and standard deviation 3. The observations are given in the following table:

43.50	45.44	45.82	45.75	44.01
42.17	41.50	41.13	44.42	41.98
47.52	48.27	42.18	44.54	47.94
44.47	47.51	51.13	41.00	42.79
49.31	47.30	41.50	39.93	48.67
44.51	42.23	45.59	48.92	44.49
41.59	46.89	44.59	40.75	40.48
44.31	48.07	45.38	47.00	43.51
53.50	44.97	42.97	45.50	41.09
41.18	43.53	49.56	49.58	40.68
43.73	45.10	39.82	50.41	48.66
48.56	47.53	41.79	47.78	44.95
40.82	41.22	42.99	43.04	39.42
46.58	42.12	50.90	47.80	42.93
46.52	44.73	43.30	42.31	45.72
41.16	43.92	43.77	47.39	41.55
45.63	40.97	45.72	43.56	45.28
44.88	52.84	41.55	43.80	43.87
41.70	42.43	44.60	46.89	44.25
48.43	49.38	44.26	40.80	43.61

   a. If the hypothesis is correct, what percentage of observations should lie between 42 and 48?
   b. What percentage of the values really lies between 42 and 48?
   c. If the hypothesis is correct, what percentage of the observations should lie between 39 and 51?
   d. What percentage of the values really lies between 39 and 51?
   e. Does the researcher's hypothesis seem reasonable based on a–d?
   f. Construct a normal probability plot. What do you conclude?
4. A random variable X has a normal distribution with mean 10 and standard deviation 2 (i.e., X is N (10,2)). Calculate the following probabilities:
   a. $P(X < 10)$
   b. $P(X > 10)$
   c. $P(8 < X < 10)$
   d. $P(6 < X < 12)$
5. The 50 observations in the following table were made of the weights (in kilograms) of 12-year-old boys. The sample mean of this set of data is 40, and the sample standard deviation is 6.43.

40.96	33.54	37.64	26.82
42.99	44.39	32.04	31.02
38.53	34.20	31.54	34.66
44.81	39.69	40.73	38.73
37.06	38.15	34.85	33.01
44.57	42.96	49.89	50.05
42.11	49.99	54.18	35.13
41.06	37.14	39.58	41.42
48.64	32.82	41.62	38.07
44.61	40.61	38.93	44.10
36.13	52.28	29.71	35.03
50.87	45.82	34.87	49.15
31.00	41.86		

a. What percentage of the observations falls within one standard deviation of the mean?
b. What percentage of the observations falls within two standard deviations of the mean?
c. Do these data seem to follow the normal distribution in the sense of obeying the 68% and 95% rules?
d. Construct a normal probability plot. What do you conclude?

6. A random variable Z has the standard normal distribution.
   a. What is the mean of Z?
   b. What is the standard deviation of Z?
   c. What is the probability of Z's being between -1 and 1?

7. A random variable has the normal distribution with mean 3 and standard deviation 2. Convert the following observations of this random variable into standard units (z-scores):
   a. 3.234
   b. 5.193
   c. 1.401
   d. -0.0184

8. Repeat Exercise 7, this time for a normal random variable having mean 6 and standard deviation 3.
   a. 5.290
   b. 2.816
   c. 8.791
   d. 10.271

9. Repeat Exercise 7, this time for a normal random variable having mean -5 and standard deviation 4.
   a. 4.823
   b. -5.972
   c. -11.732
   d. 1.672

Note. Assume that Z is standard normal in Exercises **10–16.** In every case, first draw a simple sketch of the area (probability) or of the z value you are able to find.

10. Use the normal table to find each probability.
    a. $P(Z < 1.96)$
    b. $P(Z < -1.96)$
    c. $P(Z < 1.0)$
    d. $P(Z < -1.0)$
    e. $P(Z < 0.5)$
    f. $P(Z < -0.5)$
    g. $P(Z < 0)$

11. Use the normal table to find each probability.
    a. $P(Z > 1.96)$
    b. $P(Z > -1.96)$
    c. $P(Z > 1.0)$
    d. $P(Z > -1.0)$
    e. $P(Z > 0.5)$
    f. $P(Z > 0)$

12. Use the normal table to find each probability.

a. $P(Z < 0.68)$
b. $P(Z > 0.68)$
c. $P(Z < -0.68)$
d. $P(Z > -0.68)$

13. Use the normal table to find z.
a. $P(Z < z)0.95$
b. $P(Z < z)0.90$
c. $P(Z < z) = 0.98$
d. $P(Z < z) = 0.66$
e. $P(Z < z) = 0.50$

14. Use the normal table to find z.
a. $P(Z < z) = 0.05$
b. $P(Z < z) = 0.25$
c. $P(Z < z) = 0.01$
d. $P(Z < z) = 0.10$
e. $P(Z < z) = 0.5$

15. Use the normal table to find z.
a. $P(Z > z) = 0.95$
b. $P(Z > z) = 0.90$
c. $P(Z > z) = 0.99$
d. $P(Z > z) = 0.05$
e. $P(Z > z) = 0.01$
f. $P(Z > z) = 0.10$

16. Use the normal table to solve for z. But first decide whether z is positive, negative, or 0.
a. $P(Z < z) = 0.68$
b. $P(Z < z) = 0.16$

Note. For the remaining problems (**17–34**), if the exact value is not available in the Appendix E normal table, you may use the closest one (you do not need to interpolate).

17. The heights of a group of male students follow the normal distribution with mean 70 inches and standard deviation 3.1 inches.
a. What percentage of the students would you expect to be shorter than 68 inches?
b. What percentage of the students would you expect to be taller than 73.5 inches?
c. What is the height such that 31% of the students are shorter than that height?

18. To help ensure that boxes of bananas weigh at least 40 pounds upon arrival at their destination, a packing plant might adopt this rule: Pack boxes to have a weight of 41.5 pounds of bananas, with a maximum permissible range of 3 ounces above or below 41 pounds 8 ounces. (That is, pack boxes to have at least 41 pounds 5 ounces but no more than 41 pounds 11 ounces of bananas.) With this rule and the shrinkage in travel (mostly due to evaporation of water), the distribution of box weights upon arrival may be assumed to be approximately normal with a mean of 41 pounds and a standard deviation of 4 ounces.

Suppose 30 million boxes of bananas are packed each year. Given this packing plant rule and the assumption about the distribution of box weights upon arrival, how many boxes would be expected to weigh less than 40 pounds upon arrival? (*Note:* A more complete table tells us that for a standard normal Z, $P(Z < -4.00) = 0.000032$. Also, recall that the expected number of objects with a certain property is given by np, where $n =$ total number of objects, and p is a probability to be solved for.)

19. A car manufacturer is producing a piston for its engines. The piston is supposed to have a diameter of 5.300 inches, but, because of variability, the diameter of a piston actually follows a normal distribution with a mean of 5.300 and a standard deviation of 0.010. If a piston is more than 0.025 inch away from the needed size of 5.300 inches, the piston is rejected.
a. What percentage of the pistons would you expect to be too large?
b. What percentage of the pistons would you expect to be too small? *Hint:* Recall that a probability is converted to a percentage by moving the decimal point two places right.

20. Reread the situation presented in Exercise 19. Suppose instead that, because of a mechanical problem with the machine producing the pistons, the pistons have diameters that follow a normal distribution with a mean of 5.290 inches. The standard deviation is still 0.010. The specifications still require the diameter of the piston to be 5.300 inches plus or minus 0.025 inch.
a. What percentage of the pistons would you expect to be too large in this situation?
b. What percentage of the pistons would you expect to be too small?

21. A set of final exam scores has a sample mean of 52 and a sample standard deviation of 6. The scores are normally

distributed.

a. If a teacher wants to assign a grade of A to the top 15% of the scores, what test score should be the lowest A?
b. If the bottom 15% are to be F's, what test score should be the highest F?

22. Pipettes are used in chemical and biological laboratories to dispense controlled amounts of liquid. Each pipette is carefully calibrated, but, because it is a physical measuring instrument, it will not dispense the exact amount set by the user. The error in amount is well described by the normal distribution.

Suppose a scientist is using a pipette known to dispense amounts with a standard deviation of 0.05 microliter (a microliter is a millionth of a liter). The scientist is able to set the pipette to dispense varying amounts of liquid.

a. If the pipette is set to dispense 200 microliters, what percentage of the time will it dispense greater than 200.075 microliters?
b. If the pipette is set to dispense 175 microliters, what percentage of the time will it dispense less than 174.90 microliters of liquid?
c. The scientist wanted to set the pipette to dispense 200 microliters, but she accidentally set it to 199. Is there any chance that the pipette will dispense as much as 200 microliters?

23. Suppose diastolic blood pressures (the lowest pressures between heartbeats) for women age 18 to 74 are normally distributed with mean 76 and standard deviation 12.

a. Suppose high blood pressure is defined to be a diastolic pressure of 90 or above. What is the probability that a randomly selected woman has high blood pressure?
b. Suppose low blood pressure is defined to be diastolic pressure at or below 60. What proportion of the population has low blood pressure?
c. Suppose "typical" diastolic blood pressure is defined to be between 60 and 90. What proportion of the population has typical blood pressure?

24. Suppose a random variable X is normally distributed with mean 50 and standard deviation 8. Find the following:

a. $P(X < 58)$
b. $P(X < 46)$
c. $P(46 < X < 56)$
d. $P(40 < X < 46)$
e. $P(X < 54)$
f. $P(X > 40)$

25. Let X be binomial with parameters n and p. Use the correction for continuity and match each of the four binomial probabilities below with the corresponding normal probability with $\mu = np$ and $\sigma = \sqrt{np(1-p)}$.

Binomial probability	**Normal probability**
$P(X > 78)$	$P(X > 77.5)$
$P(X \geq 78)$	$P(X < 77.5)$
$P(X \leq 78)$	$P(X > 78.5)$
$P(X < 78)$	$P(X < 78.5)$

26. Let X be binomial with parameters n and p, as specified in parts (a) through (j). For which cases would it be appropriate to use the normal approximation?

a. $n = 20, p = 0.6$
b. $n = 10, p = 0.1$
c. $n = 10, p = 0.5$
d. $n = 25, p = 0.1$
e. $n = 100, p = 0.5$
f. $n = 45, p = 0.1$
g. $n = 100, p = 0.1$
h. $n = 10, p = 0.9$
i. $n = 100, p = 0.01$
j. $n = 400, p = 0.1$

27. A fair coin is tossed 25 times. Compute the chance that you get at least 15 heads

a. using the binomial tables.
b. using the normal approximation.

28. A multiple-choice test has 18 questions. To each question there are three possible answers, one of which is correct. Imagine that you do not know any of the answers and blindly guess the answer to each of the 18 questions. Compute the probability that you get at most 8 correct answers
    a. computing the exact binomial probability.
    b. using the normal approximation.

29. Ten percent of all Americans are left-handed. Imagine that we pick a simple random sample of 100 Americans. Let X denote the number of left-handed persons in the sample. Compute and compare various ways of computing $P(12 \leq X \leq 14)$ using
    a. the exact binomial probabilities.
    b. the Poisson approximation.
    c. the normal approximation.

30. The probability that a newborn baby will be a girl is $p = 0.487$. If 2500 babies are born in a city, compute the chance that at least half of them will be girls.

31. Repeat Exercise 30, this time for a city in which 10,000 babies are born. What is the chance that at least half of them will be girls? Repeat for 40,000 babies.

32. A manufacturer receives a shipment of 50,000 parts from an outside subcontractor. When the shipment arrives, the manufacturer wants to have assurance that the parts meet agreed-to specifications. A simple random sample of $n = 400$ parts is inspected. Let X denote the number of defective parts among the 400 inspected. The whole shipment is accepted if $X \leq 5$. If $X \geq 6$, then the whole shipment is rejected and returned to the outside subcontractor. (This is an example of an important area of statistics known as acceptance sampling.) Let p denote the (unknown) proportion of defective parts among the 50,000 in the shipment. Compute the probability that the shipment is accepted if
    a. $p = 0.005$
    b. $p = 0.010$
    c. $p = 0.015$
    d. $p = 0.020$
    e. $p = 0.025$
    f. $p = 0.030$
    g. $p = 0.035$
    h. $p = 0.040$

    Hint: For the smallest p use the Poisson approximation. For larger p use the normal approximation.

33. The risk of dying from a smallpox vaccination is one in a million. If 225 million individuals are vaccinated against smallpox, determine the probability that among the vaccinated individuals
    a. at least 180 die from the vaccination.
    b. between 200 and 250, inclusive, die from the vaccination.

34. In the United States the annual death rate due to leukemia is 7.5 per 100,000 population.
    a. What is the expected number of leukemia deaths in a simple random sample of 100,000 Americans over a 4-year period?
    b. What is the probability that a simple random sample of 100,000 Americans over a 4-year period has fewer than 20 leukemia deaths? Between 20 and 40 leukemia deaths, inclusive?

CHAPTER 7 SUMMARY

Binomial Experiment Conditions

- Each trial has two possible outcomes, generically denoted "success" and "failure."
- The number of trials, n, is fixed in advance.
- The success-probability p is the same from trial to trial.
- The trials are independent.
  (Trials satisfying the above binomial experiment conditions are called **Bernoulli trials**.)

- The random variable X of interest is the total number of successes in the n trials.

Then X has the **binomial probability density function**:

$$p(x) = P(X = x) = P(\text{the } n \text{ trials result in } x \text{ successes and } n - x \text{ failures})$$
$$= \binom{n}{x} p^x (1-p)^{n-x}, \quad \text{for } x = 0, 1, \ldots, n.$$

We say that X is binomially distributed, with parameters n and p. The mean and the standard deviation of X are

$$\mu_X = np \quad \text{and} \quad \sigma_X = \sqrt{np(1-p)}.$$

Conditions Producing a Geometric Distribution

Consider a sequence of independent success-failure trials with the same success-probability p in each trial (Bernoulli trials). Let X denote the number of trials it takes to observe the first success. Then X has a geometric distribution with the following pdf:

$$p(x) = P(X = x) = (1-p)^{x-1} p \quad \text{for } x = 1, 2, 3, \ldots$$

Furthermore, the mean and standard deviation of X are

$$\mu_X = 1/p \quad \text{and} \quad \sigma_X = \sqrt{(1-p)/p^2}.$$

Simple Random Sampling and the Binomial Distribution

Consider an actual, real-world population of size N, of which N_1 members of the population have a certain attribute of interest and the remaining $N_2 = N - N_1$ members do not. For example, we might consider a population of $N = N_1 + N_2$ persons, of which N_1 smoke and N_2 do not smoke. Or we might consider a shipment of $N = N_1 + N_2$ identical electronic devices, of which N_1 devices are defective and the remaining N_2 devices are not defective.

Suppose we randomly draw a sample of n members from the population, one at a time, *without* replacement between draws. This method of sampling, discussed in Section 5.6, is called **simple random sampling** or **random sampling without replacement.** When we sample in this way, then all possible samples of size n are equally likely to be obtained.

Let X denote the number of population members in the sample with the attribute in question. For example, X might be the number of smokers in the sample or the number of defective electronic devices in the sample. X is said to have a **hypergeometric distribution.** It has pdf

$$p(x) = P(X = x) = \binom{N_1}{x}\binom{N_2}{n-x} / \binom{N_1 + N_2}{n}$$

for $x = 0, 1, \ldots n$. For large values of N, n, and x, the distribution is tedious to compute. Note that if we randomly sample from the population, one at a time, *with replacement* between draws, then the draws will be *independent,* and we instead have a binomial experiment. Then, X, the number of times we select a population member with the attribute, will be binomial, with parameters n and $p = N_1 / (N_1 + N_2)$.

If the sample size n is less than 5% of the population size N, it makes very little difference in computing $P(X = x)$ whether we sample with or without replacement. So even if we sample without replacement, X is approximately binomial, with parameters n and $p = N_1/(N_1 + N_2)$.

Poisson Distribution (pdf)

A random variable X is said to have a Poisson distribution if its **pdf** is given by

$$p(x) = P(X = x) = \frac{\lambda^x}{x!}e^{-\lambda}$$

Poisson intensity parameter

for $x = 0, 1, 2, \ldots$, where $\lambda > 0$ is the rate of occurrence,or **Poisson intensity parameter.** Then

$$\mu_X = \lambda \text{ and } \sigma_X = \sqrt{\lambda}.$$

Poisson distributions are often used to model the number of times an event of interest occurs during a predetermined time interval (or over a predetermined area or volume). The number of occurrences can be modeled by a Poisson distribution if the following real-world conditions are judged to be approximately satisfied: The rate of occurrences is constant over time; past occurrences do not influence the chances of future occurrences; and near-simultaneous occurrences are highly unlikely.

The Poisson Approximation to the Binomial Distribution

Let X be binomial, with the n and p parameters satisfying $n \geq 25$ and $p \leq 0.1$. If, also, $np \leq 15$ (i.e., the larger n is, the smaller p must be!), we can approximate binomial probabilities by Poisson probabilities. That is, in terms of probability density functions,

$$p(x) = P(X = x) = \binom{n}{x} p^x (l - p)^{n-x} \approx \frac{\lambda^x}{x!}e^{-\lambda}, \text{ for } x = 0, 1, \ldots,$$

where the Poisson intensity parameter λ satisfies $\lambda = \mu_X = np$.

Normal Distributions

Consider a continuous random variable X determined by the outcome of some random experiment. If the density histogram of tens of thousands of independent observations of X closely follows the normal pdf $f(x)$ with mean μ and standard deviation σ,

$$f(x) = \frac{1}{\sigma\sqrt{2\pi}}e^{-(x-\mu)^2/(2\sigma^2)} \qquad \text{for } -\infty < x < \infty,$$

then we say that X is **normally distributed** with mean μ and standard deviation σ. We use the shorthand notation X is $N(\mu, \sigma)$.

If the random variable X is $N(\mu, \sigma)$, then the standardized variable

$$Z = \frac{X - \mu}{\sigma}$$

has the standard normal distribution $N(0,1)$. If X is $N(\mu,\sigma)$, then

$$P(a \le X \le b) = \text{area from } a \text{ to } b \text{ under the normal curve with mean } \mu \text{ and standard deviation } \sigma$$

$$= \text{area from } (a-\mu)/\sigma \text{ to } (b-\mu)/\sigma \text{ under the standard normal curve.}$$

In particular, if X is $N(\mu,\sigma)$, then there is about a 68% chance that X will be in the range $\mu - \sigma$ to $\mu + \sigma$, a 95% chance that X will be in the range $\mu - 2\sigma$ to $\mu + 2\sigma$, and a 99.7% chance that X will be in the range $\mu - 3\sigma$ to $\mu + 3\sigma$. This is the 68-95-99.7% Rule for normal random variables. It is the theoretical companion to the 68-95-99.7% Rule for approximately bell-shaped data sets.

Guidelines for Assessing Normality Using a Normal Probability Plot. To assess the normality of a variable using sample data, construct a normal probability plot (sometimes called a normal quantile plot).

- If the plotted points lie close to a straight line, the data set looks like a sample from a normal population.
- If the plot has a nonlinear trend (such as curvature) or gaps and clusters, then the data set does not look like a sample from a normal population.

The Normal Approximation to Binomial Distributions. Let X be binomial with parameters n and p. Then X has mean and standard deviation

$$\mu = np \text{ and } \sigma = \sqrt{np(1-p)}.$$

If $0 \le \mu - 3\sigma \le \mu + 3\sigma \le n$, which is the same as $np \ge 9(1-p)$ and $n(1-p) \ge 9p$, then we can accurately estimate $P(a \le X \le b)$ for integers $a \le b$ by areas under the normal density curve with the same mean and standard deviation as X. We do this by applying the continuity correction to a and b.

CHAPTER REVIEW EXERCISES

1. For each of the following distributions, give the value of the theoretical mean and standard deviation:
   a. binomial distribution with $n = 15$ and $p = 0.35$
   b. Poisson distribution with parameter $\lambda = 4.3$
   c. standard normal distribution.

2. Convert the following sampled values from a normal distribution with mean 5 and standard deviation 2 into standardized units (z-scores):
   a. 4.392
   b. 6.921
   c. 8.936
   d. 0.0638

3. Calculate the following probabilities, where Z has the standard normal distribution:
   a. $P(Z < -1.43)$
   b. $P(Z > 0.97)$
   c. $P(Z < -2.10)$
   d. $P(Z > 1.75)$

4. Solve for z:
   a. $P(Z > z) = 0.1131$
   b. $P(Z > z) = 0.3557$
   c. $P(Z > z) = 0.0250$
   d. $P(Z > z) = 0.0104$

5. Consider the following situation: An engineer keeps track of how often an unacceptable part comes off the production line. He hypothesizes that there are 2.5 bad parts per hour.
   a. What distribution might the number of bad parts in an hour be assumed to have? Explain.
   b. What are the theoretical mean and standard deviation, based on the engineer's hypothesis?
   c. Suppose, instead, we addressed the problem this way: The engineer knows that 1000 parts come through an hour, and he suspects that 0.0025 of them are defective. He is interested in finding out the probability of no defectives being produced in a given hour. What distribution would he use in this case? Explain.

6. When Jill leaves home in the morning, there is a 40% $(p = 0.4)$ chance that the traffic light she comes to first will be green. Based on a 5-day workweek, calculate the following information:
   a. What is the mean number of times she will come to the light when it is green? What is the standard deviation?
   b. What is the probability that on all 5 days she will reach the light when it is green?
   c. What is the probability that she will reach the light when it is green on exactly 4 out of the 5 days?
   d. What is the probability that she will reach the light when it is green on exactly 3 out of the 5 days?

7. A random variable is hypothesized to have the Poisson distribution with mean 1.3. The variable is observed 100 times, with the following distribution of results, along with the theoretical proportions:

Value	Observed proportion	Theoretical proportion
0	0.23	0.27
1	0.41	0.35
2	0.21	0.23
3	0.08	0.10
4	0.05	0.03
≥ 5	0.02	0.01

   Draw a density histogram for both the observed and the theoretical proportions. Do they seem to be similar?

8. For each of the following random variables, tell whether it is continuous or discrete:
   a. the amount of time you have to wait in line
   b. the number of people who are in front of you when you get into line
   c. the total number of cars that pass by in an hour
   d. the average speed of the cars that pass by in an hour

9. A random variable is suspected to follow the normal distribution with mean 6 and standard deviation 0.5. Once the random variable is observed many times, it is calculated that 53% of the observations lie between 5.5 and 6.5. Furthermore, 83% of the observations lie between 5 and 7. Do these observations of the random variable seem to give evidence for or against the case that the random variable has the normal distribution? Explain.

10. Convert the following observed values sampled from the normal distribution with mean −6.5 and standard deviation 2.4 into standard units (z-scores):
    a. −2.381
    b. −12.905
    c. −0.0964

11. Calculate the following probabilities from the normal distribution specified:
    a. $P(X < 4.0)$, where X has mean 5.0 and standard deviation 1.0
    b. $P(X > 6.5)$, where X has mean 5.5 and standard deviation 0.75
    c. $P(X > -5.44)$, where X has mean −7 and standard deviation 0.6
    d. $P(4 < X < 7)$ where X has mean 6 and standard deviation 1.5

12. An engineer is designing a pedestrian walkway over an expressway. He knows that the heights of semis passing

underneath follow the normal distribution with a mean of 14.5 feet and a standard deviation of 0.5 feet.

a. If he builds the walkway with a clearance of 15.5 feet, what percentage of the trucks will not fit?

b. How much clearance should the engineer build under the walkway if he wants to be sure that 99.9% of the trucks will fit?

13. Recall the situation in Exercise 12. The engineer determined how high the walkway should be, but now he wants to figure out how much weight the walkway should be able to hold. The walkway must be large enough that 50 people can be on it at once. The engineer figures that the weight of 50 people follows the normal distribution with a mean of 7500 pounds and a standard deviation of 175 pounds.

a. If the walkway is built to carry a weight of 7815 pounds, what percentage of the time will 50 people be able to overload it?

b. Suppose the engineer makes some mistakes in his calculations and has the bridge built to support a weight of only 7400 pounds. What percentage of the time will 50 people on the walkway overload it now?

14. A gambler is playing a game in which he has a 20% ($p = 0.2$) chance of winning. Perhaps he is willing to play because when he does win, he wins big. He decides to play the game seven times.

a. What is the probability that he will lose all seven games?

b. What is the probability that he will win exactly two games?

c. What is the probability that he will win exactly one game?

d. How many games will he win, on average?

15. Calculate the following probabilities. The variable X has the specified normal distribution:

a. $P(X > 9.5)$, where X has mean 8.4 and standard deviation 0.5

b. $P(X < -6.7)$, where X has mean -6.0 and standard deviation 0.4

c. $P(X > 0)$, where X has mean -0.4 and standard deviation 2

16. Suppose a drug is tested on 10 people, with probability of 0.75 of a cure.

a. Find the probability that exactly 8 people are cured.

b. Find the probability that at least 7 people are cured.

c. Find the probability that 6, 7, or 8 people are cured.

17. How would you simulate the drug trials of Exercise 16?

18. Consider a Poisson random variable with mean $\lambda = 5$. Find

a. P(at least three occurrences)

b. P(two, three, or four occurrences)

c. P(one occurrence)

d. P(less than three occurrences)

19. Let Z be standard normal. Find

a. $P(0.5 < Z < 1.5)$

b. $P(-0.5 < Z < 1)$

c. $P(Z < -1.5 \text{ or } Z > 0.5)$. *Hint:* What is the complement?

20. Let Z be standard normal. Let $P(-a < Z < a) = 0.8$. Find a.

21. Let X be normal with mean 10 and standard deviation 75. In each of the following cases, find a :

a. $P(X < a) = 0.9$

b. $P(X > a) = 0.05$

c. $P(-a < X - 10 < a) = 0.95$

22. On average, six major hurricanes strike the U.S. mainland every 10 years. Compute the probability that in a given year

a. no major hurricane will strike the U.S. mainland

b. at most two major hurricanes will strike the U.S. mainland

c. at least one major hurricane will strike the U.S. mainland

23. A leading genetic cause of mental retardation is Fragile *X* Syndrome, which affects 1 in 1500 males worldwide. Consider a simple random sample of 9000 males. Compute the probability that at least 8 of the men in the sample suffer from Fragile *X* Syndrome
   a. by computing binomial probabilities (if you can)
   b. by using the Poisson approximation
   c. by using the normal approximation
   d. Which approximation is slightly more accurate in this case, the Poisson approximation or the normal approximation?

24. Over the past century and a half, the Hawaiian volcano Mauna Loa has erupted, on average, once every 5 years. Compute the probability that in a given year Mauna Loa
   a. will not erupt
   b. will erupt exactly once

25. The normal probability plots for the two data sets in Table 2.11 are displayed below. Which plot corresponds to the first data set and which plot corresponds to the second data set? For each data set, discuss whether there are outliers or gaps and clusters and whether the data set is approximately normally distributed.

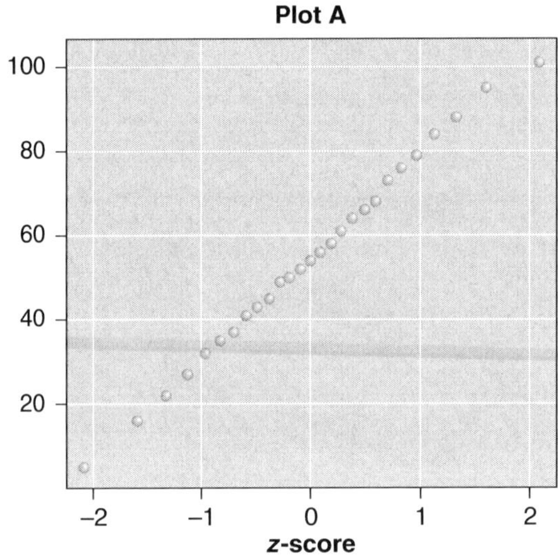

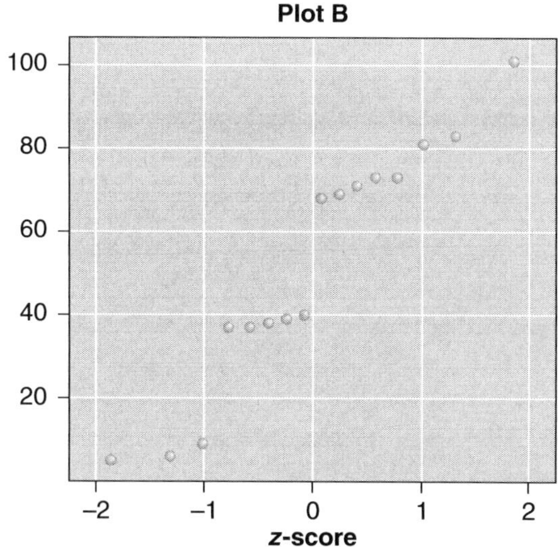

26. The normal probability plot for the salaries (in millions of dollars) of the San Diego Padres for the 2006 season is displayed below. Is the salary distribution bell-shaped, left-skewed, or right-skewed?

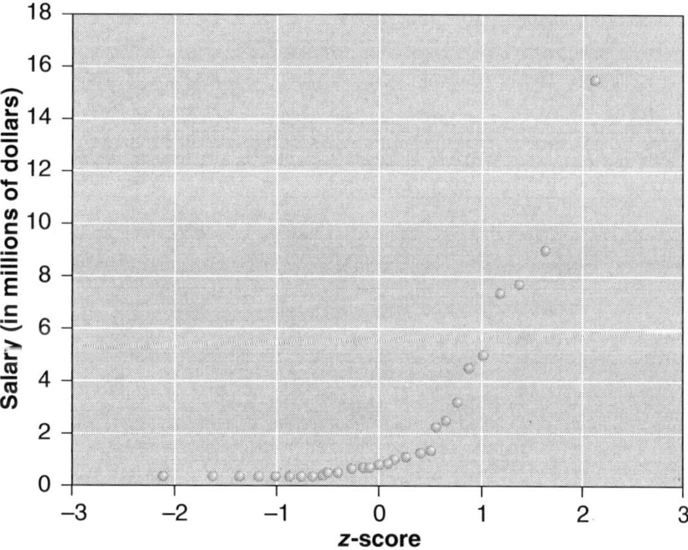

Statistical Inference Estimation and Hypothesis Testing

PROFESSIONAL PROFILE

Sir Ronald Aylmer Fisher (1890–1962), British statistician, biologist, and geneticist

The Morrow Plots on the University of Illinois at Urbana-Champaign campus. This agricultural research plot was started in 1879 and is the oldest in the Western Hemisphere. The world's oldest agricultural research center is Great Britain's Rothamsted (founded in 1856). Ronald Fisher's work at Rothamsted also benefited the Morrow Plots studies. Photo by author William F. Stout.

Sir Ronald Aylmer Fisher in 1912.

Following his graduation from Gonville and Caius College, Cambridge, in 1912, R. A. Fisher spent six years working as a statistician for the City of London. A precocious student, he studied the newly rediscovered theory of Mendelian genetics during his college years. This was a time when Charles Darwin's theory of evolution was hotly debated, and there was scientific concern that Mendelian genetics and Darwin's theory could not be reconciled. Fisher would later prove that these ideas are compatible, a major achievement.

In 1919, Fisher began work at the Rothamsted Agricultural Research Station at Harpenden, Hertfordshire, England. His job was to analyze and reinterpret the station's 76 years worth of data. Fisher stayed at Rothamsted until 1933, completing a highly productive period. Much of his pioneering work was first published in a series of reports under the general title Studies in Crop Variation. The results included his principles for the design of statistical experiments, elaborations of his universally used analysis of variance, and important studies of the statistics of small samples. Fisher's systematic approach to the analysis of real data became the springboard for his development of new statistical methods. He paid special attention to the work involved in computations and took care to recommend methods that were both rigorous and practical. The result was his 1925 book, Statistical Methods for Research Workers, a work that went through many editions and translations and became an international standard for researchers in many fields. Ten years later came The Design of Experiments, which received similar international acceptance.

If R. A. Fisher had contributed nothing more to the world of science, he would be recognized as a major contributor to statistics and scientific investigations. He did more, however. He is credited with the invention of maximum likelihood and the concepts of sufficiency, ancillarity, Fisher's linear discriminator, and Fisher information. Further, his own analysis of variance distribution z (commonly known today as the F distribution) is still widely used. R. A. Fisher is clearly one of the giants in statistics.

8

An Introduction to Estimation

Everybody believes in the central limit theorem, the experimenters because they think it is a mathematical theorem and the mathematicians because they think it is an experimental fact.

G. Lippmann (1845–1921)

Objectives

After studying this chapter, you will understand the following:

- ❐ The statistical estimation problem
- ❐ Bias and chance error of an estimator
- ❐ The sampling distribution of a sample statistic
- ❐ The standard error (SE) as the typical estimation error size
- ❐ Mean and SD formulas for $\overline{X}$ and $\hat{p}$
- ❐ The central limit theorem for $\overline{X}$ and $\hat{p}$
- ❐ The bootstrap method for estimating the SD of an estimator
- ❐ Estimation of the population SD
- ❐ Unbiasedness of an estimator

KEY PROBLEM

On May 22, 2007, the Pew Research Center in Washington, D.C., released the results of the first-ever random sample survey of America's estimated 2.4 million Muslims. One of the findings was that most Muslims "do not see a conflict between being a devout Muslim and living in a modern society." The poll interviewed a random sample of 1050 Muslim American adults between January 24 and April 30, 2007. Of the people interviewed, 71% said that they believed you can get ahead in the United States if you work hard—a higher percentage than the 64% of the general population who shared that belief. How are we to interpret this last finding?

a. Exactly 71% of all adult Muslim Americans share this belief.

b. Around 71% of all adult Muslim Americans share this belief, but this could be off by a few percentage points.

c. Roughly 71% of all adult Muslim Americans share this belief, but this could be off by 10 to 20 percentage points.

The issue here is how much does the sample percent usually differ from the population percent? Very little or a lot? How accurate can we expect the sample percent to be as an estimator of the population percent?

8.1 BIAS AND CHANCE ERROR

Measurement as Estimation

Our lives are full of measurements: We constantly check our watches to see the time; when we buy gasoline at a gas station the pump measures the amount (volume) it dispenses and charges us accordingly; at the supermarket the fruits and vegetables we purchase are often priced according to their measured weight; as we get older we check our pulse, blood pressure, and cholesterol level to watch out for heart disease.

Normally we pretend that all these measurements are 100% accurate. In fact, no measurement is 100% accurate. When the nurse measures your height and weight at the doctor's office the height measurement is only accurate to within 1/4 to 1/2 inch, depending on how carefully the nurse measures. The weight measurement for a digital scale is only accurate to within 1/2 pound to 1 pound. To see that this is so, go to a local drug store or medical clinic and try a machine that measures blood pressure, pulse, and weight for free. You will find that *repeated measurements of the same thing always differ slightly from each other*. Changes in your blood pressure readings could, of course, be explained by actual changes in your blood pressure: as you relax, your blood pressure goes down. But the way repeated successive measurements of your weight fluctuate randomly up and down clearly is not due to actual weight changes.

Example 8.1

Nineteen-year-old Svetlana measured her weight eight times at a medical clinic while waiting for her mother's drug prescription to be filled. The readings (in pounds) were

132.5	131.8	132.4	132.6	132.7	132.7	131.6	131.5

She may have shifted her position on the scale slightly between readings, but she did not make any changes in her clothing. In an ideal world, all eight readings should have been the same. They should have equaled her actual weight at the time. But since the readings differ slightly from each other, they must be off by a measurement error that changes from measurement to measurement, sometimes up and sometimes down:

$$\text{individual measurement} = \text{true value} + \text{measurement error}.$$

systematic error
bias

We distinguish between two types of measurement error: systematic error and chance error. The **systematic error** is usually called the **bias.** It affects all measurements in the same way, pushing them all in the same direction by the same amount. For example, consider one of those bathroom scales that displays your weight on a rotating disc seen through a small window, such as the one depicted in Figure 8.1.

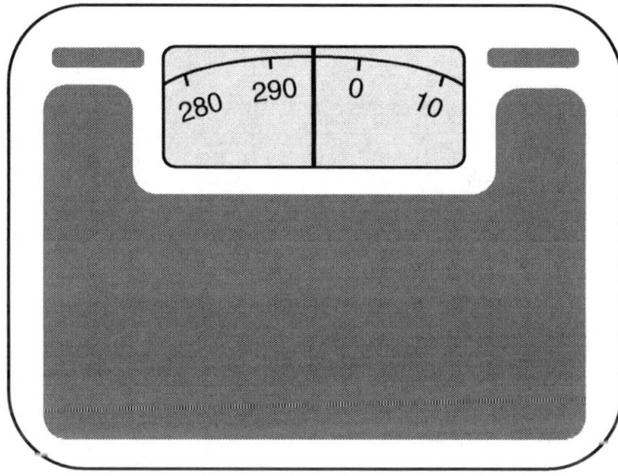

Figure 8.1 A bathroom scale with systematic error.

Imagine that you adjust the scale so that it reads −5 pounds when no one is standing on it. Then everyone who uses the scale will get a reading that is around 5 pounds below his or her true weight. The measurements will have a bias, or systematic error, of −5 pounds.

chance error

We refer to the apparently random variability of repeated measurements of the same thing (such as Svetlana's eight weight measurements) as **chance error.** Formally, our mathematical model for measurement error is

$$\text{measurement error} = \text{bias} + \text{chance error}.$$

In contrast to the bias, or systematic error, which affects each repeated measurement in the same way, the chance error varies from measurement to measurement, sometimes up and sometimes down. In the long run, the chance errors of repeated measurements will average out to zero.

model for repeated measurements of some quantity

Model for Repeated Measurements of Some Quantity

individual measurement = true value + bias + chance error.

If the measurement procedure has no bias, then the long-run average of repeated measurements will approach the true value because the chance errors will average out to zero.

Assuming the scale at the medical clinic to be unbiased, Svetlana estimates her weight to be around

$$\begin{aligned}\bar{x} &= (132.5 + 131.8 + 132.4 + 132.6 + 132.7 + 132.7 + 131.6 + 131.5)/8 \text{ lb}\\ &= 132.225 \text{ lb}.\end{aligned}$$

As we shall see in this chapter, if in fact the scale is unbiased, then the average of eight repeated measurements is likely to be two to three times more accurate than a single measurement. That is why students are told in laboratory courses to measure things repeatedly.

Unfortunately, we can not tell from Svetlana's eight readings whether the scale is biased or not. To check a scale for bias you have to weigh a so-called standard weight, an object whose true weight μ is known. Since the average $\overline{X}$ of many repeated measurements will be around

true weight + bias,

we can estimate the bias by $\overline{X} - \mu$ (μ being known). The scale can then be adjusted to reduce (if not completely eliminate) bias. You may have noticed official stickers on supermarket scales and on gas pumps certifying that they have been checked for accuracy.

We finally note that the sample standard deviation s of Svetlana's eight readings is around 0.5 pounds. The standard deviation s tells us the typical size of past and future chance errors for that scale.

The fact that even the most careful scientific measurements are distorted by measurement error was illustrated by Simon Newcomb's data in Example 2.8 (see Table 2.2). Newcomb knew that the average of repeated measurements would be much more accurate than a single measurement—assuming that the measurement procedure has no bias. He therefore made 66 repeated measurements in a famous scientific experiment designed to accurately estimate the true (actual) speed of light, μ_{speed}, one of the "fundamental constants" of our physical universe. Each such measurement consisted of the time it took a beam of light to travel 7442 meters. After eliminating one clear outlier (see Figure 2.9), the true time, μ_{time}, for light to travel 7442 meters or 7.442 kilometers (about 4.6 miles) was estimated by $\bar{t} = 24{,}827.29 \times 10^{-9}$ seconds (using the fact that 1 nanosecond = 10^{-9} seconds). Then, since speed = distance/time, we get the estimate of the speed of light:

$$\hat{\mu}_{\text{speed}} = \frac{7.442}{24{,}827.29 \times 10^{-9}} = 299{,}750.80 \text{ kilometers/second.}$$

Here we have introduced some commonly used statistical notation. For an unknown parameter μ, we often denote its estimated value by $\hat{\mu}$ ("mu hat"). Often, but not always, the estimated value of a parameter is the sample average of a set of repeated measurements, such as the sample average time $\bar{t}$ being used to estimate μ_{time}.

error of estimation

Modern and more sophisticated experiments have essentially established that the true speed of light in a vacuum is $\mu_{\text{speed}} = 299{,}792.5$ km/sec. Newcomb's **error of estimation** was

$$\hat{\mu}_{\text{speed}} - \mu_{\text{speed}} = -41.7 \text{ km/sec.}$$

The percentage of error is $(-41.7/299{,}792.5) \times 100\% = -0.014\%$, a magnitude of only around one-seventieth of 1%. Thus, Simon Newcomb's estimate of the speed of light made more than 100 years ago was very close to the true value. Part of his estimation error is due to the fact that light travels a tiny bit more slowly through air than through a vacuum.

Example 8.2

Eleven-year-olds Carolina and Marisol measure the height of their classmate Pilar six times, getting the following readings (in centimeters):

148.5	146.5	147.5	148.0	149.5	148.0

Estimate Pilar's actual height, assuming that the six measurements have no systematic error (bias). If the girls make a seventh measurement, roughly by how much might the seventh measurement be off from Pilar's actual height?

Solution: The average of the six measurements is $\overline{x} = 148.0$ cm. The standard deviation is $s = 1.0$ cm. We estimate Pilar's actual height to be 148 cm. If the girls make a seventh measurement it will more likely than not differ from Pilar's actual height by less than one standard deviation, or 1 cm. It could be off by a little more than one standard deviation, but the measurement is unlikely to be off by more than two standard deviations, or 2 cm. Note that four out of the six measurements differed by less than one standard deviation from the mean. The remaining two measurements differed by less than two standard deviations from the mean.

Box Models for Repeated Measurements

In Section 5.5 we learned to model repeated trials of a chance experiment by drawing balls from a box. For example, we can simulate the average score on five independent tosses of a fair die by drawing five times at random with replacement between draws from the box below:

The average of the five numbers drawn from the box then corresponds to the average of the five scores of the die.

We can also use box models for repeated measurements. For example, imagine that Newcomb had made 1 million independent measurements, and not just 66. We could then create a box with 1 million balls, one ball for each measurement. The number on a given ball should be the corresponding measurement value. (See Table 2.2 and surrounding text to understand what the numbers on the balls stand for.)

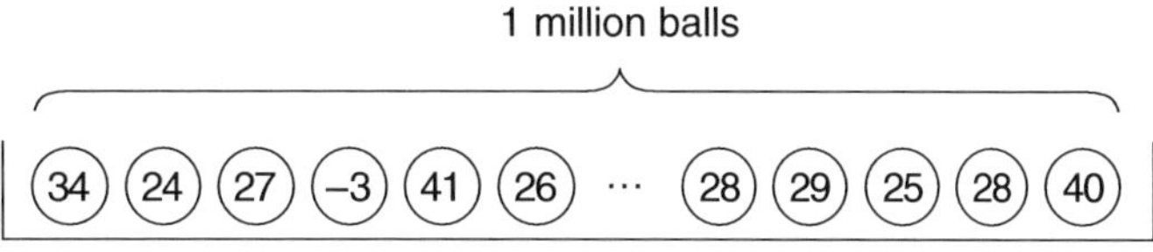

Based on 1 million measurements, the distribution of the numbers in the box would extremely accurately reflect the theoretical distribution of a single random measurement. We could therefore simulate 66 *independent* measurements (Newcomb's experiment) by drawing 66 times at random *with replacement* from the box.

Since Newcomb made only 66 measurements, we can only imagine the one million–measurement box model just described. But we shall use such imagined box models to explain why the average of repeated, independent, unbiased measurements is likely to be more accurate than a single measurement.

Estimation of a Characteristic of a Real Population

conceptual population
real population *sample survey*

The process of obtaining repeated measurements can be viewed as random sampling from a **conceptual population**. By contrast, many estimation problems involve random sampling from a **real population,** often a population of people such as in a **sample survey.**

population parameters

In Section 5.6 we learned that the Current Population Survey each month interviews a random sample of over 100,000 Americans. (See also Examples 1.9 and 2.7.) The survey helps policy makers judge the economic health of the nation by estimating **population parameters,** such as the size of the labor force, the median household income, and the unemployment rates for men and women, controlling for marital status, race, and age. The findings of the survey influence the fiscal policies of the government and the interest rates set by the Federal Reserve Board.

The goal of market surveys is also to estimate population parameters. Marketing organizations want to tailor their advertising to the demographics of the viewers of a given TV program or the readers of a given magazine. So they need to estimate the size of the audience, its age and gender distribution, and its average disposable income. As a result of market surveys, you see ads for beer and pickup trucks during football games, ads for Viagra on the History Channel, and ads for diapers and cleaning products during soap operas. (How do you think soap operas got their name?)

Public opinion polls play a major role in election campaigns. Their purpose is usually to estimate the population proportion p (a parameter) agreeing with a certain political message or favoring a particular candidate. Campaign messages are fine-tuned and campaign strategies are adjusted in response to the findings of such opinion surveys. As a result of polling, Democratic candidates often embrace traditionally Republican positions on tax cuts and national defense, and Republicans embrace traditionally Democratic positions on education and health care. Unfortunately, many candidates avoid any real policy debate before elections and resort to "negative campaigning" instead.

Some Terminology of Sample Surveys (Real Populations)

In Section 4.1 we introduced the following language:

population

- The **population** is the collection of all units under study.

units

- The **units** in the population can be individuals, groups of individuals (such as households or families), or objects.

sample

- The **sample** (often a simple random sample or a stratified random sample) is the part of the population that is actually used to get information about the population.

parameter

- A **parameter** is a number that describes some characteristic of the population. The exact value of a population parameter is usually unknown.

sample statistic
statistic

- A **sample statistic** or simply **statistic** is a number that can be computed from the sample. We use appropriately defined statistics to estimate parameters.

Example 8.3

Based on a random sample of around 50,000 households, the Census Bureau estimated that the median household income dropped 3.6% in 2008 to $50,303. Here the population of interest consists of all 112 million U.S. households in 2008. The unknown parameter is the median income of all U.S. households in 2008. The sample is the 50,000 households interviewed. The statistic of interest is the median income of the 50,000 households in the sample, that is, $50,303.

Example 8.4

Ten 1-milliliter containers are filled from different locations of the city's public swimming pool and tested for the density of *Escherichia coli* bacteria. The sample is the 10 containers of water, and the population is all the water in the swimming pool.

Suppose that the sample average $\overline{x}$ of the 10 observed *E. coli* bacteria densities is $1500/\text{cm}^3$ (that is, 1500 cells per cubic centimeter, or 1500 cells per milliliter). The estimate of the density μ of *E. coli* bacteria in the swimming pool is $\hat{\mu} = \overline{x} = 1500/\text{cm}^3$. The symbol $\hat{\mu}$ is read as "mu hat" or as the estimate of "μ."

❐ ❐

Example 8.5

A manufacturer receives a shipment of 50,000 parts from an outside subcontractor. When the shipment arrives, the manufacturer inspects 400 randomly chosen parts, finding 6 defective parts. Here the population of interest is the 50,000 parts in the shipment. The parameter of interest is the proportion p of defective parts among the 50,000 parts. The sample is the 400 parts inspected. Our estimate of p is the proportion of defective parts in the sample, $\hat{p} = 6/400 = 0.015$.

❐ ❐

It is straightforward to use the median income of the 50,000 households interviewed by the Current Population Survey to estimate the median income of all 112 million American households. It is also straightforward to use the mean size of the sample households to estimate the mean household size of the population of all 112 million households, and to use the unemployment rate in the sample to estimate the unemployment rate in the population. In this chapter we shall discuss the *accuracy* of such estimates, as in the following example.

Example 8.6

In order to reduce traffic congestion on campus a large university imposes a mandatory annual bus fee on every registered student. In return for paying this fee the students can use their student IDs as bus passes on the local buses. However, in a simple random sample of 356 students taken a year later, 89 students, or 25%, said that they still parked on campus during classes. How should we interpret the findings of this sample survey? Choose an option:

a. Exactly 25% of all students at the university park on campus.

b. Around 25% of all students park on campus, but this could be off by a few percentage points.

c. Around 25% of all students park on campus, but this could be off by 10 to 20 percentage points.

The issue here is how much does the sample percent usually differ from the population percent? Very little or a lot? Or do we expect the sample percent not to differ from the population percent at all? How accurate can we expect the sample percent to be as an estimator of the population percent?

Box Models for Simulating Sample Surveys

We simulate the behavior of the sample proportion, often used in sample surveys, by drawing from "counting" boxes of 0s and 1s.

Example 8.7

There are around 1.5 million adults in Utah. Suppose that 1 in 6, or around 0.25 million, smoke, whereas 5 out of 6, or around 1.25 million, do not smoke. Imagine that we take a simple random sample of 500 adults from Utah. How much will the proportion of smokers in the sample differ from the proportion of smokers in the population of all adults in Utah? (Note that in an actual estimation problem the proportion of smokers in the population would be unknown.)

population proportion
sample proportion

Solution: Following standard statistical notation we shall use the symbol p to denote the **population proportion** and the symbol $\hat{p}$ to denote the **sample proportion.** In other words, p denotes the proportion of all adults in Utah who smoke, and $\hat{p}$ denotes the proportion of smokers in the sample. Note that p is a population parameter, whereas $\hat{p}$ is a sample statistic that varies from one random sample to the next. Also note that the only way that we could learn the exact value of the population parameter p would be to interview every single adult resident of Utah, something that not even the U.S. Census has ever succeeded in doing. All we have are estimates of p based on random samples. But for the purpose of this example, let us assume that $p = 0.25/1.5 = 1/6$ is known to us.

We can simulate the process of estimating the proportion of adult smokers in Utah by simulating the proportion of smokers in a simple random sample of 500 adult residents of Utah. We compute the sum of 500 numbers drawn at random *without* replacement from the box:

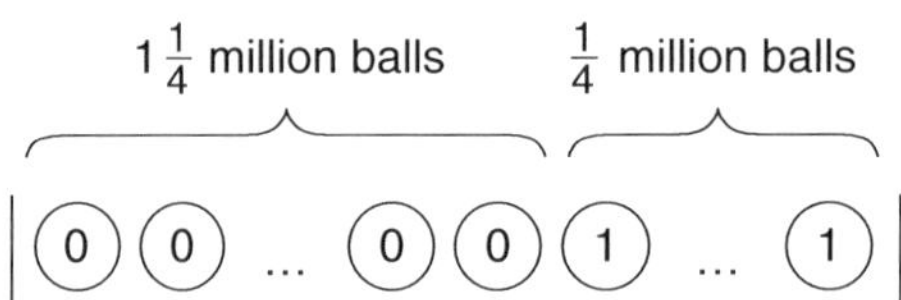

(Recall that sampling is done without replacement when one is obtaining a simple random sample from a real population.) There is a 0-ball for each nonsmoking adult and a 1-ball for each smoking adult in Utah. (See also Example 5.30.) Then

$$\text{sum of numbers drawn} = \text{simulated number of smokers in the sample.}$$

And $\hat{p}$, the proportion of smokers in the simulated sample, equals the proportion of 1-balls in the sample, which in turn equals the average of the numbers drawn. Thus,

$$\begin{aligned}\hat{p} &= \text{simulated proportion of smokers in the sample}\\ &= \text{mean of numbers drawn from box.}\end{aligned}$$

Because the sample size $n = 500$ is very small compared with the population size $N = 1.5$ million, it makes very little difference in the value of the number we obtain whether we sample with or without replacement from the box. So when we simulate, there is no harm in drawing the 500 balls *with* replacement. If we draw at random with replacement, then it makes no difference at all whether we draw from our previous box model, with 1.5 million balls, or from the following, much smaller box, with only 6 balls:

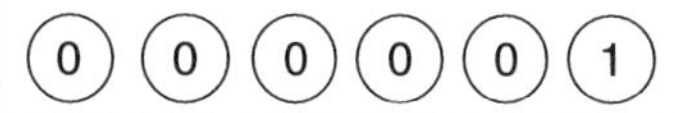

In *both* box models, 5/6 of the balls are 0-balls and 1/6 of the balls are 1-balls. Each time we draw from either box, the chance that we get a 0-ball is 5/6 and the chance that we get a 1-ball is 1/6. When we sample *with* replacement, it is the *proportion* of balls with each particular value that matters, not the number of balls. This fact is often very important when one is setting up the random sampling from a box model. Our simulations run faster the fewer balls there are in the box.

To investigate the *variability* of $\hat{p}$ and to see how much $\hat{p}$ typically differs from p, we simulated 10,000 values of $\hat{p}$, getting an average value of 0.1667 with a standard deviation of 0.0166. As Figure 8.2 shows, the experimental histogram for the 10,000 simulated $\hat{p}$-values follows a bell curve centered at p. The $\hat{p}$-values are neither systematically smaller than p nor systematically greater than p. They vary randomly around p. For some samples $\hat{p}$ is very close to p, whereas for other samples $\hat{p}$ is not so close to p. About two-thirds of the $\hat{p}$-values differ by less than one standard deviation from p, and about 95% of the $\hat{p}$-values differ by less than two standard deviations from p. These two facts are often noted in statistical studies.

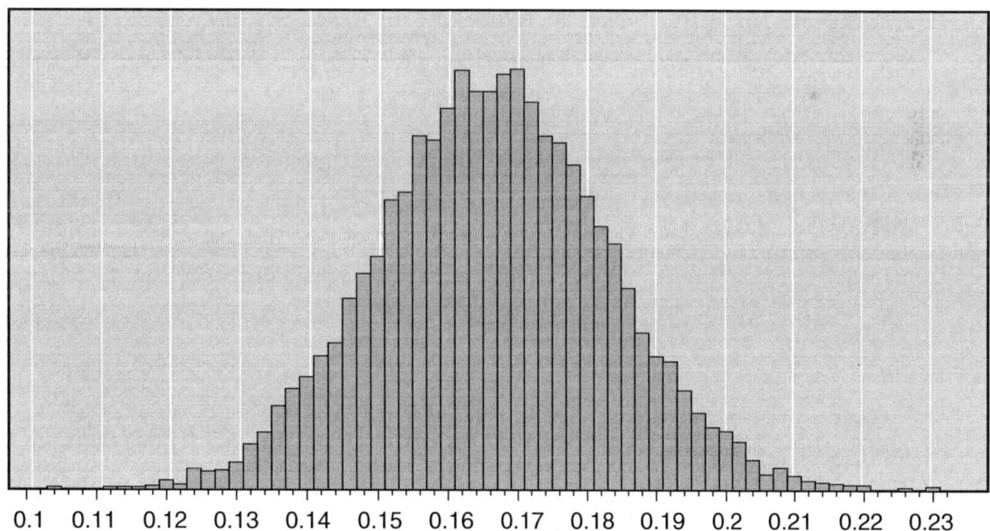

Figure 8.2 The experimental histogram for 10,000 simulated $\hat{p}$-values with $p = 1/6$ and $n = 500$.

❐ ❐

In a couple of Section 8.1 exercises you will be asked to judge how close $\hat{p}$ typically is to

p by simulating 10,000 values of $\hat{p}$ and examining the distribution of the simulated values for different values of p and n.

In many sample surveys we are interested in a quantitative variable, such as household income or household size.

Example 8.8

There are over 100 million households in the United States. Imagine that we take a simple random sample of 1000 households, observing household size. How much will $\overline{X}$, the mean size of the sample households, differ from μ, the mean size of all U.S. households?

Solution: Consider a box with one ball for each of the 100 million U.S. households. Let the number on each ball be the size of the corresponding household.

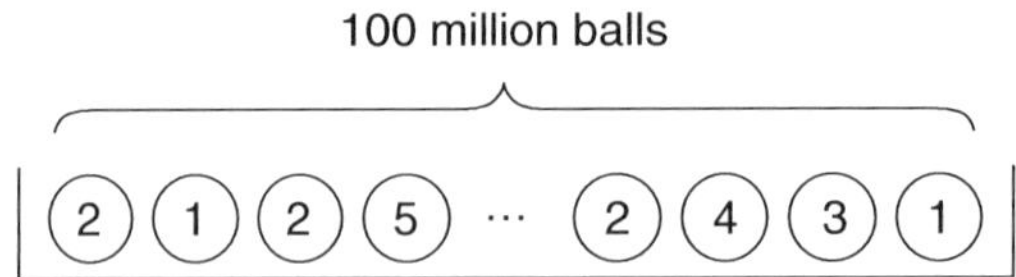

We can simulate $\overline{X}$ by computing the average of 1000 numbers drawn at random from the box. Because the sample size $n = 1000$ is much smaller than 5% of the population size $N = 100$ million, it does not matter whether we draw the 1000 numbers with or without replacement. And if we sample *with* replacement, then we can exchange this box for a smaller box, provided the two boxes have the same proportion of 1-balls, the same proportion of 2-balls, the same proportion of 3-balls, and so forth. The box model that follows, which contains 100 balls, reflects the distribution for household size given in Example 6.8.

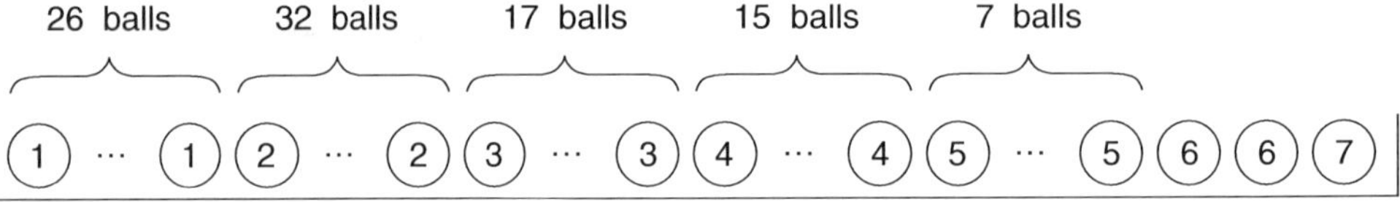

In an exercise you will be asked to judge how close $\overline{X}$ typically is to μ by simulating 10,000 values of $\overline{X}$ and examining the distribution of the simulated values. Note that each such $\overline{X}$ is the result of a random sample with replacement of size 1000.

❐ ❐

random sampling from a population

estimator of a parameter

In summary of estimation problems, whether the population is conceptual as in measurement or real as in surveying, the data used in an estimation problem is viewed as resulting from **random sampling from a population.** The quantity being estimated is a parameter, sometimes called a population parameter. The **estimator of a parameter** is computed from the random sample according to some formula or algorithm; for example, $\overline{X}$ is a common estimator of the population mean μ.

Recall that a random variable is a future or potential observation not yet made. A random variable is usually indicated by an uppercase letter and an observed value of that random variable by a lowercase letter. For example, the future score X on a die is a random variable. We will not know the value of X until we toss the die. We only know that the score could be 1, 2, 3, 4, 5, or 6. Once we toss the die we observe a score, which we denote by a lowercase letter, say $x = 5$. This notation is carried over to statistics (estimators) obtained from random sampling. For example, $\overline{X}$ and S denote two random variables (or statistics) that can be computed once the random sampling is done. By contrast, the lowercase $\overline{x}$ and s denote the observed numerical values of $\overline{X}$ and S that result after sampling. This notational convention for estimators in particular and random variables in general will be used throughout the remainder of the book.

For estimators of parameters we also use a slightly different notation. Let θ denote an unknown parameter. For example, θ could be the speed of light in vacuum, the mass of the proton, last year's median household income in the United States, the average household size in Los Angeles, the number of children in America without health insurance, or this month's unemployment rate in the United States. Consider some experiment or sample survey designed to estimate θ. We let $\hat{\theta}$ denote the resulting random estimate, sometimes called a

point estimate

point estimate because, once observed, it is a single number and hence, from the geometric perspective, a point on the number line. Our mathematical model for the estimation procedure is

$$\hat{\theta} = \theta + \text{bias} + \text{chance error}.$$

Here the bias is the systematic error associated with the estimation procedure. If we computed additional $\hat{\theta}$-values, using the same procedure and the same number of observations each time, then the bias would remain unchanged whereas the chance error would vary from $\hat{\theta}$-value to $\hat{\theta}$-value. In the long run, if we generated a very large number of $\hat{\theta}$-values, then the chance errors would average out to 0.

If the estimation procedure has no systematic error, or bias, then we say that $\hat{\theta}$ is unbiased. If $\hat{\theta}$ is unbiased, then the long-run average of simulated $\hat{\theta}$-values will approach the true parameter value θ, since the chance errors will average out to 0. Formally, this means that the expected value of the random variable $\hat{\theta}$ equals the parameter θ. (See Chapter 6.) An

unbiased estimator

estimator $\hat{\theta}$ of a parameter θ is said to be **unbiased** if its expectation, or expected value, $E(\hat{\theta})$, equals θ.

We learned in Section 6.1 that the sample mean $\overline{X}$, derived either from a simple random sample or from repeated independent observations, is an unbiased estimator of the theoretical mean μ:

$$E(\overline{X}) - \mu.$$

The sample proportion $\hat{p}$, derived from a simple random sample from a finite population, is an unbiased estimator of the corresponding population proportion p. The experimental success rate $\hat{p}$ in repeated independent trials with the same (unknown) success probability p is an unbiased estimator of p:

$$E(\hat{p}) = p.$$

If $\hat{\theta}$ is unbiased, then the standard deviation of a large number of (simulated) $\hat{\theta}$-values tells us the typical size of past and future estimation errors associated with this estimation procedure.

Unbiasedness means that if many statisticians were to carry out the same estimation procedure, the average of all their individual efforts would be close to the population parameter. As an example, we simulated 10,000 statisticians each deciding to estimate the mean μ of a fair die (of course we know the answer: $\mu = 3.5$) using five tosses each. Although each statistician using his or her average score $\overline{X} = \sum_1^5 X_i/5$ will be a little bit off from μ (the typical error being approximately of size $\sigma_{\text{die}}/\sqrt{5} = 1.708/\sqrt{5} = 0.764$), the average of our 10,000 simulated $\overline{X}$-values was 3.5023, very close to the theoretical value, $\mu = 3.5$. The fact that

$$\text{average of 10,000 simulated } \overline{X}\text{-values} = 3.5023 \approx 3.5 = \mu$$

is a practical illustration of the unbiasedness of the sample mean $\overline{X}$ as an estimator of the theoretical mean μ, in this case $E(\overline{X}) = \mu = 3.5$.

To reiterate, statistics—that is, estimators, in the context of this chapter—come from samples and can be calculated directly from the sample data. Parameters come from populations and are usually unknown and, hence, need to be estimated by statistics. Statistics are random variables, whereas parameters are nonrandom constants.

Table 8.1 contains a list of some important population parameters and the corresponding statistics usually used to estimate them.

Table 8.1 Some Important Parameters and the Statistics Used to Estimate Them

Parameter	Symbol	Statistic	Symbol
Population mean	μ ("mu")	Sample mean	$\overline{X}$ ("X bar")
Difference of two population means	$\mu_X - \mu_Y$	Difference of two sample means	$\overline{X} - \overline{Y}$
Population median*		Sample median*	
Population variance	σ^2 ("sigma squared")	Sample variance	S^2
Population standard deviation	σ ("sigma")	Sample standard deviation	S
Population correlation coefficient	ρ ("rho")	Sample correlation coefficient	r
Population proportion	p	Sample proportion	$\hat{p}$ ("p hat")
Difference of two population proportions	$p_1 - p_2$	Difference of two sample proportions	$\hat{p}_1 - \hat{p}_2$

*There is no standard notation in use for the population median or the sample median.

Section 8.1 Summary

Repeated measurements of the same thing always differ slightly from each other. Our model is:

$$\text{individual measurement} = \text{exact value} + \text{bias} + \text{chance error}.$$

The *bias*, or *systematic error*, affects all measurements the same way, pushing them all in the same direction by the same amount. The *chance error* varies from measurement to measurement, sometimes up and sometimes down. In the long run, the chance errors of repeated, independent measurements will average out to zero.

More generally, let θ denote any unknown *parameter* and let $\hat{\theta}$ denote a *sample statistic* used to estimate θ. Then our mathematical model is:

$$\hat{\theta} = \theta + \text{bias} + \text{chance error}.$$

If different statisticians independently were to compute estimates $\hat{\theta}$ in exactly the same way, using the same number of observations derived in exactly the same way, then the bias would affect all the different $\hat{\theta}$-values in the same way, pushing them all in the same direction by the same amount. By contrast, the chance error would vary from $\hat{\theta}$-value to $\hat{\theta}$-value, sometimes up and sometimes down. In the long run, if we generated more and more $\hat{\theta}$-values, then the chance errors would average out to zero.

Ideally, we want an estimator $\hat{\theta}$ to have both *small bias* and *small chance errors*. The smaller the bias is and the chance errors are, the more *accurate* the estimator will be.

If an estimator $\hat{\theta}$ has no bias, or systematic error, then we say that $\hat{\theta}$ is *unbiased*. The estimator $\hat{\theta}$ is unbiased if $E(\hat{\theta}) = \theta$. Note, however, that unbiasedness does not guarantee that the chance errors will be small. An estimator $\hat{\theta}$ can be unbiased and still have large chance errors.

Section 8.1 Exercises

1. For each of the following, identify the sample, estimator (statistic), population, and parameter involved. Use the standard notation when appropriate.
   a. In a poll of 64 freshmen at a large college, it was found that 41 of the students polled were in favor of lowering the legal drinking age from 21 to 18.
   b. Several ears of corn were randomly selected from a truck unloading corn at a grain elevator. They were found to have an average moisture content of 35%.
   c. Of 10 frogs randomly taken from a large forest pond, 2 were found to be deformed. (An increase in deformities in amphibians has been an unresolved environmental quality issue in recent years.)
   d. Twenty-five "one-pound" loaves of bread were randomly sampled from a day's output at a bakery. The average weight of these was 1.07 pounds.
   e. The average cholesterol level of 10 randomly selected long-distance runners from Boston was 157.

2. A poll of 20 randomly sampled people at a shopping center was taken to rate the taste of a new cola drink. The results (1 = "terrible" to 5 = "terrific") were as follows:

4	3	3	4	4	3	2	3	5	3
4	2	4	3	3	2	3	4	3	3

What is the mean preference score of the 20 people in the poll? What is the range of preference scores in the poll? What is the sample SD? What parameter could the mean preference score of the poll estimate? What parameter could the sample SD estimate?

3. A random sample of 25 college students at a midwestern university are surveyed about their political views. The students are asked to categorize themselves as liberal, conservative, or moderate. Ten of them consider themselves liberal. What parameter could this statistic be used to estimate?

4. In the newspaper account below, what sample and population are involved? What factors could affect how well the sample represents the population?

   By measuring annual tree rings in groves scattered across the western states, scientists here are seeking to define the likelihood of prolonged and extreme drought for each of nine river basins in that region. At least 10 trees are sampled at each site. A hollow drill extracts a pencil-sized core of wood without seriously harming the tree. The core is then studied under magnification to record climate-induced variations in tree ring width back to the time when the tree began growing. Trees on well-drained slopes are preferred since they respond quickly to a drought.

   An objective of the project, funded in part by $286,000 from the National Science Foundation, is to determine whether there is evidence for what some hydrologists call a "Noah effect." This would be a weather extreme beyond known precedent, like the 40 days of rain that, according to Genesis, flooded the world in the days of Noah. A variant, known as the "Joseph effect," would be a condition —such as a drought—far more prolonged than any on record. The reference, again from Genesis, is to the seven years of famine predicted by Joseph. It is assumed that, if radical departures from normal behavior have taken place in the past, they may occur again and the frequency of occurrence may be estimated.

   (W. Sullivan, "Ancient Tree Rings Tell a Tale of Past and Future Droughts," *New York Times*, September 2, 1980)

5. Give examples from newspaper or magazine articles in which a sample statistic is used (or assumed to be used) to estimate a population parameter.

6. Give the names and symbols for the parameters and statistics used in the estimation of a population mean, SD, and proportion.

7. The diameter of a human hair (in micrometers) was repeatedly measured in a laboratory as follows:

49	51	48	48	49	47	50	48	49
49	51	49	48	47	51	50	51	49
48	49	52	49	52	52	53	47	53
47	51	52						

Find the mean and standard deviation of the measurements. Draw a histogram of the data. Do the data appear to be normally distributed? Assuming that the measurements have no systematic error, estimate the diameter of the hair. If a 31st measurement was made, by roughly how much do you expect the 31st measurement to differ from the actual diameter of the hair?

8. A state inspection office tested 50 gasoline pumps around the state for measurement errors. Five gallons of gasoline (according to the meter) were pumped, and the error was measured. The results, in cubic inches, are as follows (minus signs mean less than five gallons):

2	−4	−6	−3	0
−4	−1	0	0	−2
−1	−1	−4	2	−4
−1	−2	−4	−3	1
−2	−4	−2	−1	−4
−2	−6	−2	−1	−5
0	−3	−4	−3	−7
−5	−1	−4	−4	2
−2	−3	−3	1	1
−4	−3	1	−3	−3

Construct a histogram of the errors. Do the data appear to be approximately normally distributed? Find $\overline{x}$. Do you suspect bias in the sense that pumps on average deliver less gas than the meter indicates?

9. The compressive strength of cement is the amount of squeezing pressure it can withstand before crumbling. The following data are values of the compressive strength (in pounds) of the same batch of cement as measured by 50 laboratories. Estimate the true compressive strength of the cement. Do the data appear to be normally distributed?

4466	3914	4084	4135	4084
4154	4123	4030	4120	3626
3771	3889	4180	4141	4524
3810	3777	4072	3871	4181
3948	4081	4334	4046	4130
3676	3888	4126	4128	4058
3589	4135	4018	3675	4251
3842	4409	4134	4017	3888
3922	4300	4447	4228	3907
3825	4172	4005	4241	4339

10. Weigh or measure something repeatedly, say, 25 times; compute the mean (your estimate of the true value) and the standard deviation (your estimate of measurement variability). Do the data appear to be normally distributed? Does using $\overline{X}$ to estimate the unknown true value provide a reasonably accurate measurement, assuming we know there is no bias?

11. In this exercise you have to judge how close the sample proportion $\hat{p}$ typically is to the population proportion p. You will see that the accuracy of $\hat{p}$ as an estimator of p depends on the sample size n.

    Assume that the proportion p of Kentucky residents who are 65 years old or older equals 1/8. (See Example 5.30.) Let $\hat{p}$ denote the proportion of persons 65 years old or older in a simple random sample of n Kentucky residents. For each of the specified sample sizes, n, use the box model in Example 5.30 to simulate 10,000 $\hat{p}$-values. Note the shape of the histogram for the 10,000 simulated $\hat{p}$-values. Note whether the mean of the 10,000 simulated $\hat{p}$-values is close to $p = 1/8$.

    a. $n = 50$
    b. $n = 200$
    c. $n = 800$
    d. $n = 3200$
    e. Refer to parts (a)–(d). Does $\hat{p}$ appear to be systematically smaller than p, appear to be systematically greater than p, or appear to vary randomly around p? Explain.
    f. Refer to parts (a)–(d). Judging from the standard deviations of the simulated $\hat{p}$-values, what happens to the variability of $\hat{p}$ if we quadruple the sample size n?

12. In this exercise you have to compare the variability of the sample proportion $\hat{p}$ for different values of p, but with the same sample size n.

    For each of the box models below, let p denote the proportion of 1-balls in the box. Let $\hat{p}$ denote the proportion of 1-balls in a random sample of 100 balls drawn with replacement from the box. Simulate 10,000 $\hat{p}$-values and record the mean and the standard deviation of the simulated $\hat{p}$-values. Note whether $\hat{p}$ appears to be systematically smaller than p, appears to be systematically greater than p, or appears to vary randomly around p.

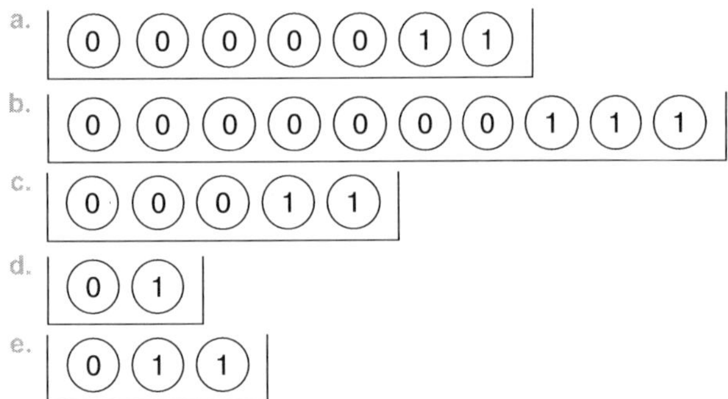

f. Judging from the five standard deviations of the simulated $\hat{p}$-values in parts (a)–(e), is the variability of $\hat{p}$ about the same for all five boxes? If not, which box produces the greatest variability of $\hat{p}$?

13. Use the second box model in Example 8.8 to simulate the average household size $\overline{X}$ of 100 sample households. Simulate 10,000 values of $\overline{X}$.
   a. Describe the shape of the histogram of the 10,000 simulated values.
   b. Does $\overline{X}$ appear to be systematically smaller than the box mean μ, appear to be systematically greater than μ, or appear to vary randomly around μ?
   c. What is the standard deviation of the 10,000 simulated values of $\overline{X}$?

14. We shall investigate whether the sample standard deviation S is an unbiased estimator of the theoretical standard deviation σ.

   Consider the box below, which contains a total of 50 balls:

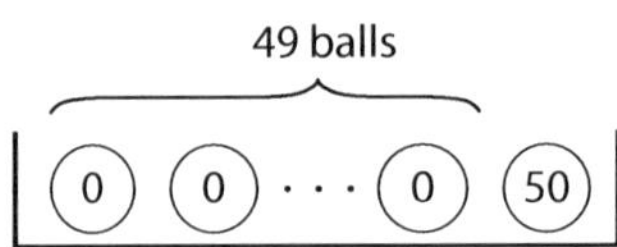

   a. Verify that $\sigma_{\text{box}} = 7$. (See Section 6.3.)
   b. Let S denote the sample standard deviation of $n = 10$ numbers drawn at random with replacement from the box. To investigate whether

$$E(S) = \sigma_{\text{box}} = 7,$$

   simulate 10,000 values of S. What is the mean of the simulated values? Is the mean close to 7? If not, is the mean smaller than 7 or greater than 7?
   c. Next, let S denote the sample standard deviation of $n = 1000$ numbers drawn with replacement from the box. Simulate 10,000 values of S. What is the mean of the simulated values? Is the mean close to 7?
   d. Does S seem to get more biased or less biased as we increase the sample size n?

8.2 The Large-Sample Distributions of $\overline{X}$ and $\hat{p}$

It is straightforward to use a sample mean $\overline{X}$ to estimate an unknown population mean μ, and to use a sample proportion $\hat{p}$ to estimate an unknown population proportion p. The central limit theorem (CLT) helps us evaluate the accuracy of the estimators $\overline{X}$ and $\hat{p}$.

sampling distribution

For large samples the CLT describes the probability distributions of $\overline{X}$ and $\hat{p}$. The probability distribution of a sample statistic is often called the statistic's **sampling distribution.**

The **sampling distribution** of a sample statistic is the statistic's probability distribution. It is the distribution of the values the statistic takes when we consider all possible samples of the same size from the same population.

As we did in Figure 8.2 in Example 8.7, we can approximate the sampling distribution of a sample statistic well by the experimental distribution of 10,000 or more simulated values of the statistic.

To put the CLT in perspective, we shall begin with the following theoretical result:

The Distribution of $\overline{X}$ When We Sample from a Normal Population

Suppose that we draw n times at random *with* replacement from a population that is *exactly normal* $N(\mu, \sigma)$. Then the theoretical probability distribution of the sample mean $\overline{X}$ will be *exactly* normal, too, with mean

$$\mu_{\overline{X}} = E(\overline{X}) = \mu$$

and standard deviation

$$\sigma_{\overline{X}} = \text{SD}(\overline{X}) = \sigma/\sqrt{n}.$$

The reason that we call this a theoretical result is that no real-world population is exactly normal. Some population distributions are close to normal, but they are never exactly normal.

The CLT tells us that even if the population distribution is not normal, the theoretical probability distribution of $\overline{X}$ will still be approximately normal if the sample size n is large enough. How large n has to be depends on how close the population distribution is to being normal. If the population distribution is approximately normal, then $\overline{X}$ is approximately normal for all n. But the further from normal the population is, the larger n has to be before $\overline{X}$ becomes approximately normal.

CLT for $\overline{X}$ for random sampling with replacement

The Central Limit Theorem (CLT) for $\overline{X}$

Suppose that we draw n times at random *with* replacement from a box of numbered balls. Assume either that the distribution of the numbers in the box is approximately normal or that the number of draws is reasonably large. Then the theoretical probability histogram for the sample mean $\overline{X}$ will approximately follow a normal curve with mean

$$\mu_{\overline{X}} = E(\overline{X}) = \mu_{\text{box}}$$

and standard deviation

$$\sigma_{\overline{X}} = \text{SD}(\overline{X}) = \sigma_{\text{box}}/\sqrt{n}.$$

In particular, $\overline{X}$ is likely to be around μ_{box}, give or take $\sigma_{\text{box}}/\sqrt{n}$ or so.

Note: The larger n is, the closer $\overline{X}$ is to being normal. The rule of thumb for "reasonably large" is $n \geq 20$, if the distribution of the numbers in the box is not too skewed and does not have extreme values (outliers). However, the more skewed the distribution of the numbers in the box is, the larger n has to be.

Recall that μ_{box} is the average of the numbers in the box. Similarly, σ_{box} is the standard deviation of the numbers in the box, as discussed in Section 6.3.

The CLT for $\overline{X}$ is an extremely important and useful result. It tells us that if the sample size n is large, then we can use the standard normal distribution table (Z-table) in Appendix E to estimate $P(a \leq \overline{X} \leq b)$ for *any* choice of a and b.

Recall that when a random variable X is $N(\mu, \sigma)$ we have to standardize in order to be able to use Table E to compute a probability involving X. For example, finding $P(X \leq 5)$ for X being $N(1,3)$ requires us to find $P(Z \leq (5-1)/3)$ using Appendix E. Thus, an important part of the statement of the CLT for $\overline{X}$ is to tell us the mean $\mu_{\overline{X}}$ (namely, μ_{box}) and standard deviation $\sigma_{\overline{X}}$ (namely, $\sigma_{\text{box}}/\sqrt{n}$) that we have to use to standardize $\overline{X}$ so that we can use Appendix E.

As we discussed in Section 5.5, we can model (simulate) the observations in n independent trials, repeated under identical conditions, by drawing n times at random with replacement from a box of numbered balls. It follows that the CLT applies to the average observation $\overline{X}$, computed from *independent trials*, repeated under identical conditions.

Example 8.9

Consider n repeated, independent, unbiased, approximately normally distributed measurements $X_1, X_2, \ldots, X_n$ of an unknown quantity μ. Because the measurements are unbiased, $E(X_i) = \mu$ for each measurement. We model the measurement process by drawing n times at random with replacement from a box of numbered balls with the box mean

$$\mu_{\text{box}} = \mu.$$

The theoretical standard deviation of each single measurement X_i is

$$\sigma_{X_i} = \sigma_{\text{box}},$$

where σ_{box} is large or small depending on whether the typical measurement error for a single measurement X_i is large or small. Each measurement X is likely to be around μ, give or take $\sigma_X = \sigma_{\text{box}}$ or so. Thus σ_X really tells us how accurate a single measurement is likely to be.

The formula for $\sigma_{\overline{X}}$ given as part of the statement of the CLT shows, assuming no bias, that the average of four measurements is likely to be about twice as accurate as a single measurement because

$$\sigma_{\overline{X}} = \sigma_{\text{box}}/\sqrt{4} = \sigma_{\text{box}}/2 = \sigma_X/2.$$

And the average of 100 measurements is likely to be about 10 times more accurate than a single measurement because

$$\sigma_{\overline{X}} = \sigma_{\text{box}}/\sqrt{100} = \sigma_{\text{box}}/10 = \sigma_X/10.$$

Assuming no bias, we improve measurement accuracy in two ways: making σ_{box} as small as possible (by using careful laboratory techniques) and making the number of measurements n as large as possible.

Example 8.10

We usually apply the CLT approximation only if $n \geq 20$. But in a few special cases the approximation is not too bad even for smaller n.

Consider again the box we use to simulate the toss of a fair die

As we computed in Section 6.3,

$$\begin{aligned}\mu_{\text{box}} &= (1+2+3+4+5+6)/6 = 21/6 = 3.5,\\ \sigma^2_{\text{box}} &= [(1-3.5)^2+(2-3.5)^2+\cdots+(6-3.5)^2]/6\\ &= 35/12,\end{aligned}$$

and hence,

$$\sigma_{\text{box}} = \sqrt{35/12} = 1.708.$$

The probability histogram for the score on a fair die is shown in Figure 8.3. It is symmetric and does not have extreme values. Recall from Chapter 6 the 68-95% rule for random variables whose probability histogram is approximately "bell-shaped." For such a variable X there is

- about a 68% chance that X will fall in the range $\mu_X \pm \sigma_X$
- about a 95% chance that X will fall in the range $\mu_X \pm 2\sigma_X$

In other words, μ_X locates the "center" of the distribution of X, and σ_X measures the "spread" of the distribution of X around μ_X. We express this loosely by saying that "X will be around μ_X," with the typical size of the distance between X and μ_X being σ_X. We often write "give or take σ_X or so."

For the number X rolled on a fair die, the theoretical mean μ_X and standard deviation σ_X are $\mu_{\text{box}} = 3.5$ and $\sigma_{\text{box}} = 1.708$, respectively.

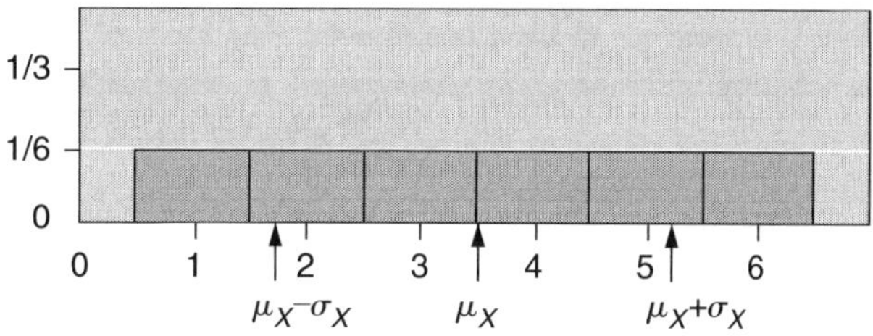

Figure 8.3 The probability histogram of the score on a fair die.

Although we saw in Section 6.3 that the 68 95% rule happened to work well for the distribution of X, this was a lucky accident because, as Figure 8.3 shows, X does *not* have a bell-shaped distribution.

Next, let's consider the mean of two die throws. To simulate the mean of the scores on a pair of symmetric casino dice we can draw at random twice with replacement from the box we just looked at and compute the average $\overline{X}$ of the two numbers drawn from the box. We get, by the statement of the CLT,

$$\begin{aligned}\mu_{\overline{X}} &= \mu_{\text{box}} = 3.5,\\ \sigma_{\overline{X}} &= \sigma_{\text{box}}/\sqrt{2} = 1.208.\end{aligned}$$

The theoretical probability distribution of $\overline{X}$ can easily be derived from Figure 5.3 in Section 5.2:

x	1	1.5	2	2.5	3	3.5	4	4.5	5	5.5	6
$p(x)$	1/36	2/36	3/36	4/36	5/36	6/36	5/36	4/36	3/36	2/36	1/36

The probability distribution of $\overline{X}$ is displayed graphically in Figure 8.4. The area of each bar (width = 1/2) equals the corresponding probability. Even though n is only equal to 2, the probability histogram of $\overline{X}$ is already roughly bell-shaped and the 68-95% rule applies.

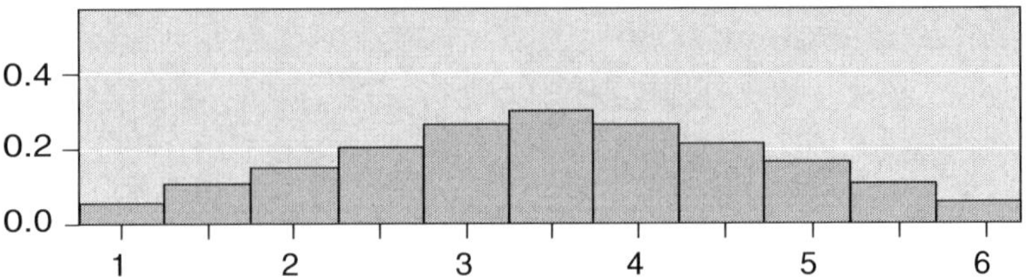

Figure 8.4 Probability distribution (pdf) of the mean score $\overline{X}$ on two fair dice.

The chance that $\overline{X}$ falls in the range $\mu_{\overline{X}} \pm \sigma_{\overline{X}}$, or between $3.5 - 1.21 = 2.29$ and $3.5 + 1.21 = 4.71$, is $24/36 = 2/3$, or $66.7\% \approx 68\%$. And the chance that $\overline{X}$ falls in the range $\mu_{\overline{X}} \pm 2\sigma_{\overline{X}}$, or between $3.5 - 2.42 = 1.08$ and $3.5 + 2.42 = 5.92$, is 34/36, or $94.4\% \approx 95\%$. Note, however, that even though the CLT normal approximation for $\overline{X}$ worked well for these two intervals, it would be incorrect to assume that the CLT normal approximation for $\overline{X}$ will work well for all intervals, because $n = 2$ is not large enough.

Finally, assume that we toss a fair die 25 times. We can simulate the average of the 25 scores by drawing 25 times at random with replacement from the box and computing the average $\overline{X}$ of the 25 numbers drawn. Hence

$$\mu_{\overline{X}} = \mu_{\text{box}} = 3.5$$

$$\sigma_{\overline{X}} = \text{SD}(\overline{X}) = \frac{\sigma_{\text{box}}}{\sqrt{25}} = 0.342.$$

We shall not attempt to derive the exact theoretical distribution of $\overline{X}$. The mathematical calculations involved are much too tedious in this case. Instead we shall use simulation to examine the experimental distribution of 100,000 simulated $\bar{x}$-values. Because 100,000 is so large, this experimental distribution of values will be very close to the theoretical distribution of $\overline{X}$. According to the CLT, noting that $n = 25$ is reasonably large and considering that even when $n = 2$ the distribution of $\overline{X}$ is roughly bell-shaped, we expect this experimental distribution to be close to a normal distribution. Figure 8.5 displays the experimental histogram of 100,000 simulated $\bar{x}$-values. It follows a bell curve centered at $\mu = 3.5$ extremely well.

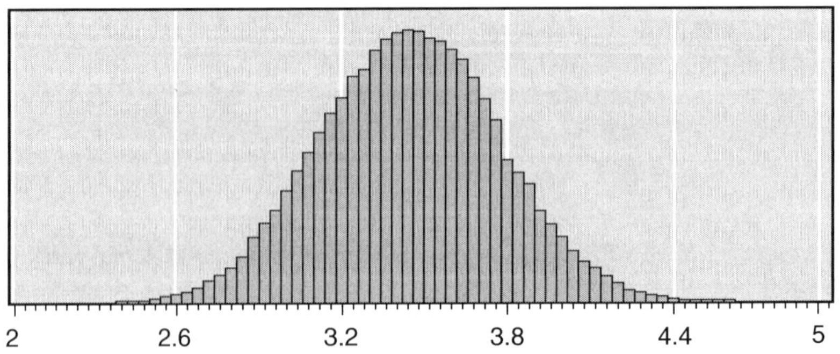

Figure 8.5 Experimental histogram for 100,000 simulated $\bar{x}$-values. Each $\bar{x}$-value is the average of $n = 25$ scores of a fair die.

The mean of the simulated $\overline{x}$-values is 3.499, which is very close to the theoretical mean, 3.500. The sample standard deviation of the simulated $\overline{x}$-values is $\widehat{\text{SD}}(\overline{X}) = 0.342$, which equals the theoretical standard deviation $\sigma_{\overline{X}} = 0.342$. Around 70.6% of the simulated $\overline{x}$-values are between $\mu_{\overline{X}} - \sigma_{\overline{X}} = 3.5 - 0.342 = 3.158$ and $\mu_{\overline{X}} + \sigma_{\overline{X}} = 3.5 + 0.342 = 3.842$. This is close to what the 68-95% rule predicts.

But the CLT allows us to estimate the probability of $\overline{X}$ falling in any interval, as the next example illustrates.

Example 8.11

Imagine that we toss five perfectly symmetric casino dice and let $\overline{X}$ denote the average of the five scores. Because the fair die distribution is symmetric (rather than skewed) and does not have any extreme values, we shall use the CLT to approximate $P(3.1 \leq \overline{X} \leq 3.9)$. As already mentioned, we usually use the CLT approximation only if $n \geq 20$. But in a few special cases the approximation is also good for smaller n.

Solution: We could simulate $\overline{X}$ by computing the mean of five numbers drawn at random with replacement from the box

It follows by the statement of the CLT that

$$\mu_{\overline{X}} = \mu_{\text{box}} = 3.5,$$

$$\sigma_{\overline{X}} = \text{SD}(\overline{X}) = \frac{\sigma_{\text{box}}}{\sqrt{5}} = 0.764.$$

Instead of estimating the probability of interest via five-step simulation, we will use the CLT to approximate $P(3.1 \leq \overline{X} \leq 3.9)$, proceeding in three steps.

Step 1. Draw a number line and shade the interval of interest.

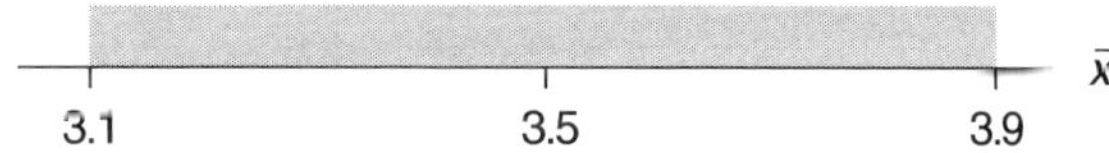

Step 2. Convert the interval endpoints into z-scores by first subtracting $\mu_{\overline{X}}$ and then dividing by $\sigma_{\overline{X}} = 0.764$.

$$(3.1 - \mu)/\sigma = (3.1 - 3.5)/0.764 = -0.52,$$

$$(3.9 - \mu)/\sigma = (3.9 - 3.5)/0.764 = 0.52.$$

−0.52 0 0.52 z

Step 3. Find the area under the standard normal curve over the z-score interval. (Use Appendix E.)

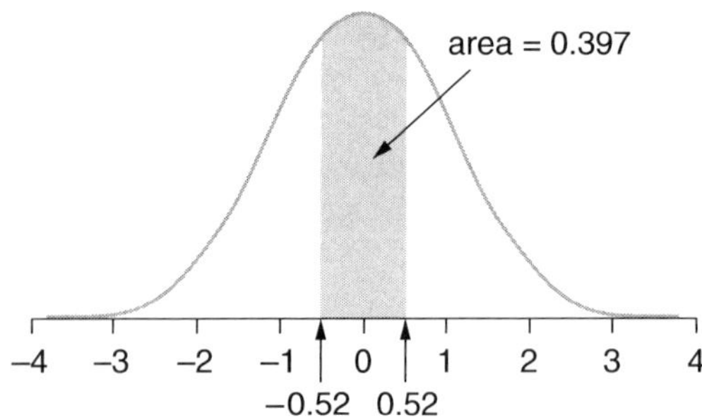

By Appendix E,

$$P(Z \le 0.52) - P(Z \le -0.52) = 0.6985 - 0.3015 = 0.397$$

Thus, by the CLT we approximate $P(3.1 \le X \le 3.9) \approx 0.397$. We check this answer by simulating 10,000 $\overline{X}$-values and noting the resulting experimental probability $\hat{P}(3.1 \le \overline{X} \le 3.9) = 3875/10{,}000 = 0.388$, quite close to 0.397.

Comment: If the sample size n is smaller than 500, then five-step simulation often gives slightly more accurate probability estimates than the CLT, provided we simulate at least 10,000 $\overline{x}$-values. Note, however, that in order to apply the CLT we need to know *only* μ_{box} and σ_{box}, whereas we need to know the entire contents of the box in order to run five-step simulations. That is, we must know the probability law for the box.

As we have discussed, the CLT applies to the average observation $\overline{X}$, computed from *repeated, independent trials*. We shall now see that the CLT also applies to the sample mean $\overline{X}$, computed from a *simple random sample* drawn from a large finite population, provided the population distribution is not too skewed and the sample size n is reasonably large, but still reasonably small compared to the population size N.

The CLT for $\overline{X}$ for Simple Random Samples

As we illustrated in Examples 8.7 and 8.8, we can model (simulate) simple random sampling from a finite population by drawing at random *without* replacement from a box with one ball for each member of the population. The expected value of the sample mean $\overline{X}$ is

$$\mu_{\overline{X}} = E(\overline{X}) = \mu_{\text{box}} = \mu_{\text{population}}.$$

The theoretical standard deviation of $\overline{X}$ is

$$\sigma_{\overline{X}} = \text{SD}(\overline{X}) = \frac{\sigma_{\text{box}}}{\sqrt{n}} \cdot \sqrt{\frac{N-n}{N-1}},$$

where n denotes the sample size, N the population size, and

$$\sigma_{\text{box}} = \sigma_{\text{population}}.$$

Here, $\sqrt{(N-n)/(N-1)}$ is called the finite population correction factor.

CLT for the Sample Mean $\overline{X}$ Computed from a Simple Random Sample Drawn from a Large Real Population

Consider a simple random sample of size n drawn without replacement from a large real population of size N with mean μ_{pop} and standard deviation σ_{pop}. Assume either that (i) the population distribution is approximately normal or that (ii) both n and $N - n$ are reasonably large and the population distribution is not too skewed and does not have outliers. Then the sampling distribution of $\overline{X}$ will be approximately normal with parameters

$$\mu_{\overline{X}} = \mu_{\text{pop}}$$

and

$$\sigma_{\overline{X}} = \frac{\sigma_{\text{pop}}}{\sqrt{n}} \cdot \sqrt{\frac{N-n}{N-1}}.$$

Example 8.12

Consider the mean $\overline{X}$ of $n = 10$ numbers drawn at random one at a time, without replacement, from the box

(1) (2) (3) . . . (16) (17) (18) (19)

Here $N = 19$, $\mu_{\text{pop}} = 10$, and $\sigma_{\text{pop}} = \sqrt{30}$.

$$\sigma_{\overline{X}} = \frac{\sqrt{30}}{\sqrt{10}} \cdot \sqrt{\frac{19-10}{19-1}} = \sqrt{1.5} = 1.225.$$

Figure 8.6 displays the experimental histogram of 50,000 simulated $\overline{x}$-values. The mean of the simulated $\overline{x}$-values is 9.998, which is very close to the theoretical mean $\mu_{\overline{x}} = 10{,}000$. The sample standard deviation is 1.221, which is very close to the theoretical standard deviation $\sigma_{\overline{x}} = 1.225$. The experimental histogram appears as a bell curve centered at $\mu = 10$.

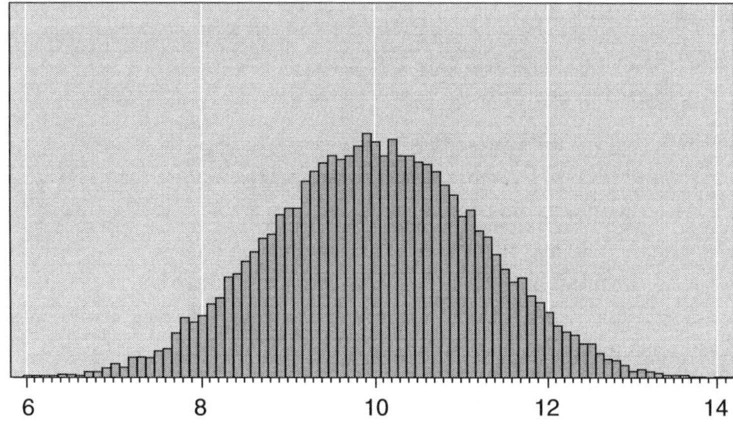

Figure 8.6 Experimental histogram for 50,000 simulated $\overline{x}$-values. Each $\overline{x}$-value is the mean of a simple random sample of $n = 10$ numbers drawn from the population $\{1, 2, 3, 4, \ldots, 17, 18, 19\}$.

The larger the population size N is compared to the sample size n, the closer "the finite population correction factor"

$$\sqrt{\frac{N-n}{N-1}}$$

is to 1. This is illustrated in Table 8.2.

Table 8.2 The Correction Factor as a Function of the Population Size N, for Fixed Sample Size $n = 500$

Population size N	Correction factor
5,000	0.9488
10,000	0.9747
25,000	0.9900
250,000	0.9990
2,500,000	0.9999

If the sample size n is less than 5% of the population size N, then the *finite population correction factor* is practically 1:

$$\sqrt{\frac{N-n}{N-1}} \approx 1.$$

It follows that the theoretical standard deviation of $\overline{X}$ is practically the same as if we had sampled *with* replacement:

$$\sigma_{\overline{X}} = \mathrm{SD}(\overline{X}) \approx \frac{\sigma_{\mathrm{pop}}}{\sqrt{n}}.$$

In fact, if the sample size n is less than 5% of the population size N, then it makes little difference whether the sampling is without or with replacement. We already know that $\overline{X}$ will be approximately

$$N(\mu_{\mathrm{pop}}, \sigma_{\mathrm{pop}}/\sqrt{n})$$

if we sample with replacement and n is reasonably large. It follows that if we sample without replacement and n is reasonably large, but less than 5% of N, then the sample mean $\overline{X}$ is still approximately

$$N(\mu_{\mathrm{pop}}, \sigma_{\mathrm{pop}}/\sqrt{n}).$$

This is the most commonly used version of the CLT for the sample mean $\overline{X}$, computed from a simple random sample drawn from a large finite population.

CLT for $\overline{X}$ Computed from a Simple Random Sample Drawn from a Many Times Larger Population

Consider a simple random sample drawn from a population at least 20 times larger than the sample. Let μ_{pop} and σ_{pop} denote the population mean and population standard deviation. Assume either that (i) the population distribution is approximately normal or that (ii) the population distribution is not too skewed and does not have outliers, and the sample size n is reasonably large ($n \geq 20$, say). Then the sampling distribution of the sample mean $\overline{X}$ is close to the normal distribution $N(\mu_{\text{pop}}, \sigma_{\text{pop}}/\sqrt{n})$.

Under the stated assumptions, $\overline{X}$ is likely to be around μ_{pop}, give or take $\sigma_{\text{pop}}/\sqrt{n}$ or so. In particular, if μ_{pop} is unknown, then the accuracy of $\overline{X}$ as an estimator of μ_{pop} will depend on σ_{pop} and the sample size n, but not on the population size N.

Example 8.13

One year the average Math SAT score of 25 high school students in an AP calculus class was 620. That year the Math SAT scores nationwide followed a normal curve with mean 500 and standard deviation 100. Estimate the probability of getting a mean ($\overline{X}$) Math SAT score of 620 or higher for 25 randomly selected students from the population of all high school students nationwide who took the Math SAT that year.

Solution: We know

$$\mu_{\text{pop}} = 500 \quad \text{and} \quad \sigma_{\text{pop}} = 100.$$

For the average $\overline{X}$ of 25 randomly selected students we get

$$\mu_{\overline{X}} = \mu_{\text{pop}} = 500 \quad \text{and} \quad \sigma_{\overline{X}} \approx \sigma_{\text{pop}}/\sqrt{25} = 20.$$

To approximate $P(\overline{X} \geq 620)$ we use the CLT for a large population sample survey, proceeding in three steps.

Step 1. Draw a number line and shade the interval of interest.

Step 2. Convert the interval endpoint into a z-score by first subtracting $\mu = 500$ and then dividing by $\sigma = 20$.

$$(620 - \mu)/\sigma = (620 - 500)/20 = 6.$$

0 6

Step 3. We need $P(Z \geq 6)$ for a $N(0, 1)$ random variable Z. Use Appendix E to find the area under the standard normal curve over the z-score interval. As noted in the footnote to Table E.2, for any $z \geq 4$, the area to the left of z is 1 to four decimal places. The area to the right of z must therefore be 0.0000 (rounded).

There is thus no chance that the average Math SAT score of 25 randomly selected test takers would be 620 or higher. But that should not surprise us since the students in a high school AP calculus class are not randomly selected from the population of U.S. high school students.

❐ ❐

The next example illustrates that even if the population distribution is skewed (and hence decidedly non-bell-shaped), the normal approximation to the distribution of $\overline{X}$ will be quite accurate if the sample size n is large enough.

Example 8.14

The proportion of left-handed persons in the U.S. population is about $p = 0.1$. We can simulate the number of left-handed persons in a random sample of n Americans by computing the sum of n numbers drawn at random with replacement from the box below. (Sampling with replacement is allowed because the U.S. population is huge, so the $n \leq 0.05N$ rule is amply satisfied.)

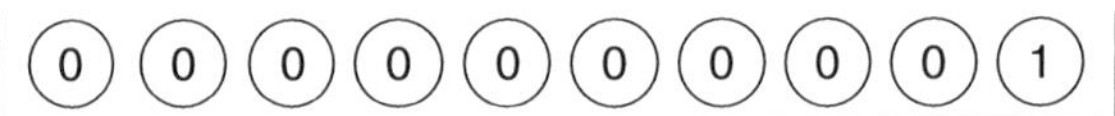

The simulated proportion $\hat{p}$ of left-handed persons in the sample is the average of the numbers drawn:

$$\hat{p} = \overline{X}.$$

The fact that $\hat{p}$ is a sample mean (an $\overline{X}$) is very important because we then know that the CLT applies to $\hat{p}$.

For a box of 0s and 1s,

$$\mu_{\text{box}} = \frac{\sum x}{N} = \text{proportion of 1-balls in box} = p,$$

$$\sigma_{\text{box}} = \sqrt{p(1-p)},$$

where the formula for σ_{box} was presented in Section 6.3 and will be derived later in the chapter.

In this case, $\mu_{\text{box}} = 0.1$ and $\sigma_{\text{box}} = 0.3$. We write μ_{box} or $\mu_{\text{population}}$ depending on whether the sampling is simulation from a box or actual random sampling from a real population, and similarly for σ_{box} and $\sigma_{\text{population}}$. Figure 8.7 displays the sampling distribution of $\hat{p}$ in the case $n = 10$. The histogram does not follow a bell curve.

Figure 8.8 displays the sampling distribution of $\hat{p}$ in each of the cases $n = 25$, 100, and 400.

Each of the histograms in Figures 8.7 and 8.8 is skewed to the right, but becomes less so the larger n is, with very little skew when $n = 400$. Because of the extreme skewness of the box model, the normal approximation is not good in the cases $n = 10$ and $n = 25$. The approximation is reasonable for $n = 100$ and is quite good for $n = 400$. The larger n is, the better the approximation is.

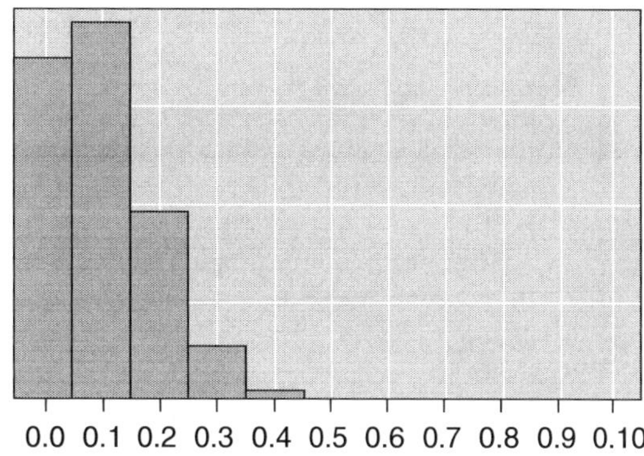

Figure 8.7 The sampling distribution of $\hat{p}$, when $p = 0.1$ and $n = 10$.

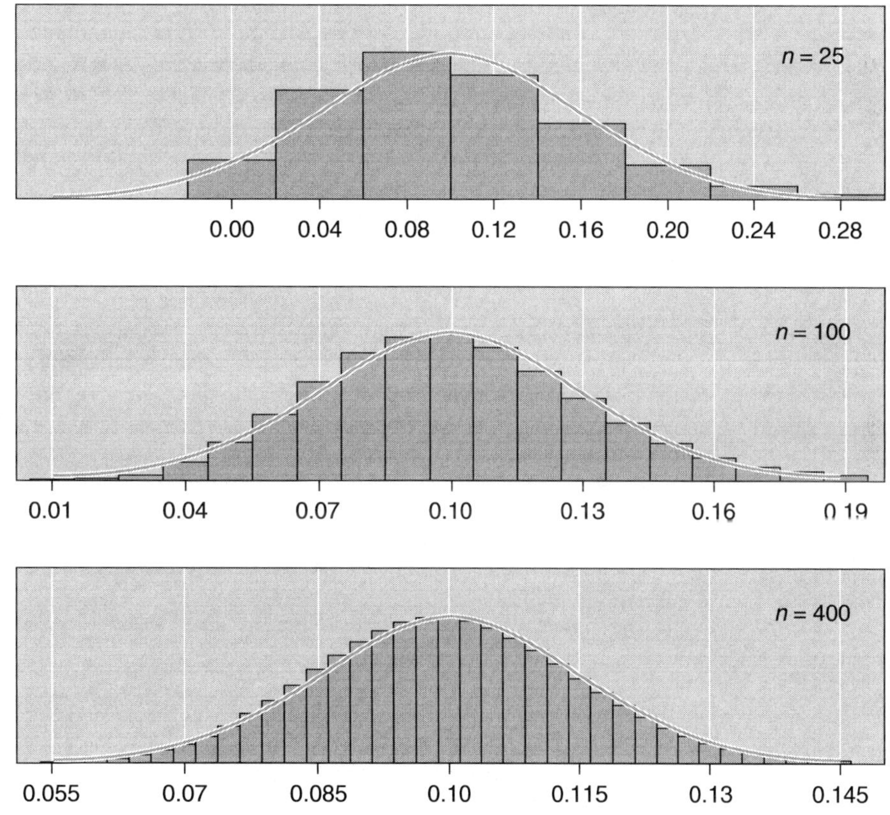

Figure 8.8 The normal approximation to the distribution of $\hat{p}$ when $p = 0.1$ and $n = 25, 100$, and 400.

❒ ❒

As Example 8.14 illustrates, the CLT for $\hat{p}$ is a special case of the CLT for $\overline{X}$. This is because $\hat{p}$ is an $\overline{X}$.

CLT for $\hat{p}$ for a random sample from a box

The Central Limit Theorem for $\hat{p}$ Resulting from a Random Sample with Replacement

Suppose that we draw n times at random *with* replacement from a box of 0s and 1s. Let p denote the proportion of 1-balls in the box and $\hat{p}$ the proportion of 1-balls in the sample. If n is so large that $np \geq 10$ and $n(1-p) \geq 10$, then the theoretical histogram for $\hat{p}$ will approximately follow a normal curve with mean

$$\mu_{\hat{p}} = E(\hat{p}) = p$$

$SD(\hat{p})$ for a random sample

and standard deviation

$$\sigma_{\hat{p}} = \mathrm{SD}(\hat{p}) = \sqrt{p(1-p)}/\sqrt{n}.$$

CLT for $\hat{p}$ for any random sample

Note: The **CLT for $\hat{p}$,** although stated for box model sampling with replacement, **applies to *any* random sampling with replacement** from a population of 0 and 1 values, that is, to any repeated-trials random experiment whose outcomes are classified into two categories, for convenience labeled 0 or 1. For example, if $\hat{p}$ is the proportion of times in 60 tosses that a fair die comes up a 6, then the CLT applies to $\hat{p}$ $(np = 60 \cdot \frac{1}{6} = 10 \geq 10$ and $n(1-p) = 50 \geq 10)$.

The formulas for $\mu_{\hat{p}}$ and $\sigma_{\hat{p}}$ follow from the general formulas for $\mu_{\overline{X}}$ and $\sigma_{\overline{X}}$ for a random sample. Recall that

$$\mu_{\overline{X}} = \mu_{\text{box}}, \quad \sigma_{\overline{X}} = \sigma_{\text{box}}/\sqrt{n}.$$

Here the box mean is the proportion of 1s in the box (or population), namely, p. Thus $\mu_{\hat{p}} = \mu_{\text{box}} = p$ because p is the proportion of 1s in the population.

Further, $\sigma_{\text{box}} = \sqrt{p(1-p)}$, as claimed earlier in the chapter. For the curious student, this holds because the number of 0 values in the box (or population) is $N(1-p)$ and the number of 1 values is Np. Hence, with $x = 0$ or 1 for each x,

$$\sigma^2_{\text{box}} = \frac{\Sigma(x-p)^2}{N} = \frac{[N(1-p)](0-p)^2}{N} + \frac{[Np](1-p)^2}{N} = p - p^2 = p(1-p).$$

Thus,

$$\sigma_{\hat{p}} = \sigma_{\text{box}}/\sqrt{n} = \sqrt{p(1-p)}/\sqrt{n},$$

as claimed.

Our rule of thumb: Requiring $np \geq 10$ and $n(1-p) \geq 10$ in the CLT for $\hat{p}$ tells us that if p is either close to 0 or close to 1, then the sample size n has to be really large if we want to use the normal curve to estimate probabilities involving $\hat{p}$. For example, if $p = 0.1$, as in Example 8.14, then n should be at least 100. By contrast, Figure 8.9, where $n = 30$, illustrates that if p is around 0.5, then $\hat{p}$ is approximately normally distributed for much smaller values of n.

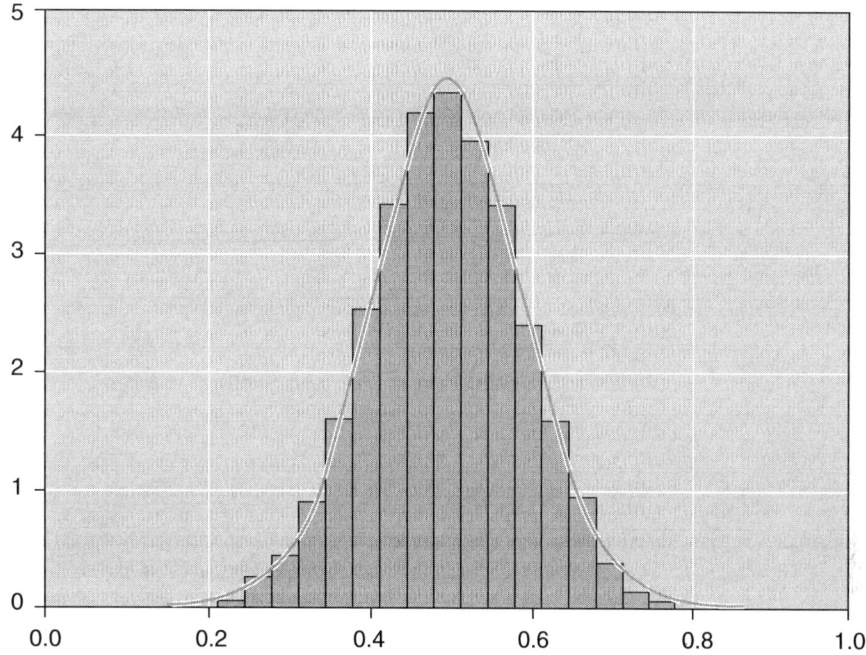

Figure 8.9 Experimental histogram of 10,000 simulated $\hat{p}$-values when $p = 0.5$, $n = 30$.

Next we illustrate that as the sample size n gets larger and larger, $\hat{p}$ gets closer and closer to p, a very useful fact when one wants to accurately estimate p by $\hat{p}$.

Example 8.15

We can simulate the proportion $\hat{p}$ of heads in n tosses of a fair coin by drawing n times at random with replacement from the box

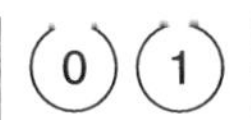

and computing the average of the numbers drawn.

$$E(\hat{p}) = p = 0.5 \quad \text{and} \quad \text{SD}(\hat{p}) = \sqrt{p(1-p)}/\sqrt{n} = 0.5/\sqrt{n}.$$

According to an application of the CLT, the proportion of heads $\hat{p}$ will most likely be around p, give or take $\text{SD}(\hat{p})$ or so:

- In 25 tosses, $\hat{p}$ will be around 0.50, give or take $0.5/5 = 0.10$ or so, since $\sqrt{n} = 5$.
- In 100 tosses, $\hat{p}$ will be around 0.50, give or take $0.5/10 = 0.05$ or so, since $\sqrt{n} = 10$.
- In 2500 tosses, $\hat{p}$ will be around 0.50, give or take $0.5/50 = 0.01$ or so, since $\sqrt{n} = 50$.

For example, we obtained in our box model simulations:

n	25	100	2500
$\hat{p}$	0.64	0.46	0.492

Thus, as the number of tosses n increases, $\hat{p}$ is more and more likely to be close to $p = 0.50$.

❐ ❐

Next we state the CLT for a sample proportion $\hat{p}$ derived from a *simple* random sample, that is, from sampling *without* replacement from a real population.

CLT for $\hat{p}$ for a large real population (without-replacement sampling)

The CLT for $\hat{p}$ Computed from a Simple Random Sample Drawn from a Large Real Population

Consider a *large* population of size N of which a proportion p has a certain characteristic of interest. Consider a simple random sample of size n drawn without replacement from the population. Let $\hat{p}$ denote the proportion of the sample having the characteristic of interest.

If p is unknown, then the sample proportion $\hat{p}$ is an unbiased estimator of p:

$$\mu_{\hat{p}} = E(\hat{p}) = p.$$

Using the fact that $\hat{p}$ is an $\overline{X}$ and that $\sigma_{\text{population}} = \sqrt{p(1-p)}$, we find that the theoretical standard deviation of $\hat{p}$ is

SD of $\hat{p}$ for a large population

$$\sigma_{\hat{p}} = \text{SD}(\hat{p}) = \frac{\sqrt{p(1-p)}}{\sqrt{n}} \cdot \sqrt{\frac{N-n}{N-1}}.$$

If n is so large that $np \geq 10$ and $n(1-p) \geq 10$, and $N - n$ is so large that $(N-n)p \geq 10$ and $(N-n)(1-p) \geq 10$, then $\hat{p}$ will be approximately normal, $N(p, \sigma_{\hat{p}})$.

SD of $\hat{p}$ for a finite population

The larger the population size N is compared to the sample size n, the closer "the finite population correction factor"

$$\sqrt{\frac{N-n}{N-1}}$$

is to 1. If $N \geq 20n$, then the factor is so close to 1 that we can ignore it and use

$$\sigma_{\hat{p}} \approx \frac{\sqrt{p(1-p)}}{\sqrt{n}}.$$

The Accuracy of $\hat{p}$ as an Estimator of p

Alaska has around 700,000 residents. Of these, around 32% are 45 years old or older. By contrast, Florida has around 17 million residents, of whom around 41% are 45 years old or older. Imagine that we take a simple random sample of 500 residents from each state. For which state, Alaska or Florida, is the sample proportion $\hat{p}$ of residents 45 years old or older likely to be closer to the population proportion p of residents 45 years old or older?

For both samples, $\hat{p}$ is likely to differ from p by around one standard deviation,

$$\sigma_{\hat{p}} = \frac{\sqrt{p(1-p)}}{\sqrt{n}} \cdot \sqrt{\frac{N-n}{N-1}}.$$

For Alaska,

$$\sqrt{p(1-p)} = \sqrt{0.32 \times 0.68} = 0.466.$$

For Florida,

$$\sqrt{p(1-p)} = \sqrt{0.41 \times 0.59} = 0.492.$$

We note that the two different values of p result in almost the same value of $\sqrt{p(1-p)}$.

For the Alaska sample, the finite population correction factor is practically 1:

$$\sqrt{\frac{N-n}{N-1}} = \sqrt{\frac{700{,}000-500}{700{,}000-1}} = 0.9996.$$

For the Florida sample, the finite population correction factor is even closer to 1:

$$\sqrt{\frac{N-n}{N-1}} = \sqrt{\frac{17{,}000{,}000-500}{17{,}000{,}000-1}} = 0.999985.$$

For the Alaska sample, the standard deviation is

$$\sigma_{\hat{p}} = \frac{\sqrt{p(1-p)}}{\sqrt{n}} \cdot \sqrt{\frac{N-n}{N-1}} \cdot \frac{0.466}{22.36} \cdot 0.9996 = 0.021.$$

For the Florida sample, the standard deviation is

$$\sigma_{\hat{p}} = \frac{\sqrt{p(1-p)}}{\sqrt{n}} \cdot \sqrt{\frac{N-n}{N-1}} = \frac{0.492}{22.36} \cdot 1.0000 = 0.022.$$

We note that the two standard deviations are almost equal. Thus, answering our question above, the expected accuracy of $\hat{p}$ is practically the same for the two samples. Given that $N \geq 20n$ for both samples, the different population sizes make no difference. And the different values of p have little effect on the expected accuracy of $\hat{p}$ (as long as p is not very close to 0 or 1). Hence, for the two states the expected accuracy is about the same, even though Florida has a much larger population.

What happens to the expected accuracy of $\hat{p}$ if we quadruple the sample size to $n = 2000$ in both states? Since we divide by $\sqrt{n}$ in the formula for $\sigma_{\hat{p}}$ (and the finite population correction factor will remain practically constant), multiplying the sample size by 4 will divide $\sigma_{\hat{p}}$ by 2. Therefore, quadrupling the sample size n will double the expected accuracy of $\hat{p}$ in both states. This illustrates that the expected accuracy of $\hat{p}$ as an estimator of p depends on n, but not on N, as long as $N \geq 20n$.

If the sampling distribution of $\hat{p}$ is approximately normal, then it follows from the 68-95-99.7% rule that there is about a 95% chance that $\hat{p}$ differs from p by less than $2\sigma_{\hat{p}}$. Since

$$2\sigma_{\hat{p}} \leq 1/\sqrt{n}$$

for all values of p, we derive the *margin of error* estimate from Section 4.2:

Margin of Error for $\hat{p}$ as an Estimator

Suppose we use the sample proportion $\hat{p}$ derived from a simple random sample of size n, drawn from a much larger population, to estimate an unknown population proportion p. Then we have about a 95% chance that our estimate will be off by less than $1/\sqrt{n}$ (in either direction). We call $1/\sqrt{n}$ the *margin of error*. The accuracy of $\hat{p}$ as an estimator of p does not depend on the size of the population—as long as the population size is at least 20 times greater than the sample size.

Section 8.2 Summary

The *sampling distribution* of a sample statistic is the statistic's probability distribution. It is the distribution of the values the statistic takes when we consider all possible samples of the same size from the same population.

Consider the sample mean $\overline{X}$ of n repeated random draws (made with or without replacement between draws) from a population with mean μ_{pop} and standard deviation σ_{pop}. The sampling distribution of $\overline{X}$ has mean

$$\mu_{\overline{X}} = E(\overline{X}) = \mu_{\text{pop}}.$$

If we sample with replacement, then the standard deviation of $\overline{X}$ is

$$\sigma_{\overline{X}} = \text{SD}(\overline{X}) = \sigma_{\text{pop}}/\sqrt{n}.$$

If we sample without replacement, then

$$\sigma_{\overline{X}} = \text{SD}(\overline{X}) = \frac{\sigma_{\text{pop}}}{\sqrt{n}} \cdot \sqrt{\frac{N-n}{N-1}},$$

where N denotes the population size.

The *central limit theorem* (CLT) for $\overline{X}$ tells us that in either case (whether we sample with or without replacement) the sampling distribution of $\overline{X}$ will be close to the normal distribution

$$N(\mu_{\overline{X}}, \sigma_{\overline{X}}),$$

provided n is reasonably large (and $N - n$ is reasonably large as well, if we sample without replacement) and the population distribution is not too skewed and does not have outliers.

The CLT for $\hat{p}$ is a special case of the CLT for $\overline{X}$. The experimental "success" rate $\hat{p}$ in n repeated independent trials with the same success probability p has mean

$$\mu_{\hat{p}} = E(\hat{p}) = p$$

and standard deviation

$$\sigma_{\hat{p}} = \text{SD}(\hat{p}) = \sqrt{p(1-p)}/\sqrt{n}.$$

If n is so large that $np \geq 10$ and $n(1-p) \geq 10$, then the sampling distribution of $\hat{p}$ will be close to the normal distribution

$$N(p, \sigma_{\hat{p}}).$$

Finally, consider a large population of size N of which a proportion p has a certain characteristic of interest. Consider a simple random sample of size n drawn without replacement from the population. Let $\hat{p}$ denote the proportion of the sample having the characteristic of interest. The sample proportion $\hat{p}$ has mean

$$\mu_{\hat{p}} = E(\hat{p}) = p,$$

and standard deviation

$$\sigma_{\hat{p}} = \mathrm{SD}(\hat{p}) = \frac{\sqrt{p(1-p)}}{\sqrt{n}} \cdot \sqrt{\frac{N-n}{N-1}}.$$

If n is so large that $np \geq 10$ and $n(1-p) \geq 10$ and $N-n$ is so large that $(N-n)p \geq 10$, and $(N-n)(1-p) \geq 10$, then the sampling distribution of $\hat{p}$ will be close to the normal distribution

$$N(p, \sigma_{\hat{p}}).$$

If we take a simple random sample (sampling without replacement) from a population at least 20 times larger than the sample, then we can ignore "the finite population correction factor"

$$\sqrt{\frac{N-n}{N-1}}$$

in the formulas for $\sigma_{\overline{X}}$ and $\sigma_{\hat{p}}$ and use

$$\sigma_{\overline{X}} \approx \sigma_{\text{pop}}/\sqrt{n}$$

and

$$\sigma_{\hat{p}} \approx \sqrt{p(1-p)}/\sqrt{n}.$$

It follows that if the population mean μ_{pop} is unknown, then the accuracy of the sample mean $\overline{X}$ as an estimator of μ_{pop} will depend on σ_{pop} and the sample size n, but not on the population size N (as long as $N \geq 20n$). If the population proportion p is unknown, then the accuracy of the sample proportion $\hat{p}$ as an estimator of p will depend on p and n, but not on N (as long as $N \geq 20n$). In fact, we have about a 95% chance that $\hat{p}$ will be off by less than $1/\sqrt{n}$ (in either direction). Note that $N \geq 20n$ allows us to ignore the finite population correction factor and the requirements $(N-n)p \geq 10$ and $(N-n)(1-p) \geq 10$. To apply the CLT for $\hat{p}$ we need only make sure that $np \geq 10$ and $n(1-p) \geq 10$.

Section 8.2 Exercises

1. One in six adults in Utah smokes. (See Example 8.7.) Suppose that we take a simple random sample of 500 adults from Utah. Let $\hat{p}$ denote the proportion of smokers in the sample. Estimate the probability $P(0.149 < \hat{p} < 0.183)$
   a. using the CLT
   b. simulating 10,000 $\hat{p}$-values
2. Suppose that the average household size in a large city is $\mu = 2.60$ persons per household with a standard deviation of $\sigma = 1.50$ persons. Let $\overline{X}$ denote the average size of $n = 400$ randomly chosen households.

a. Compute $E(\overline{X})$ and $\mathrm{SD}(\overline{X})$.
b. Use the CLT to estimate $P(2.50 < \overline{X} < 2.70)$.

3. Let $\hat{p}$ denote the proportion of heads in n tosses of a fair coin. Use the CLT to estimate $P(0.475 < \hat{p} < 0.525)$
   a. for $n = 100$
   b. for $n = 500$
   c. for $n = 1000$
   d. Check each of your estimates in (a) through (c) by simulating 10,000 $\hat{p}$-values.
   e. Compare your estimates in (a) through (c). As n gets larger, does $P(0.475 < \hat{p} < 0.525)$ get larger or smaller?

4. Let $\overline{X}$ denote the average of the scores on five symmetric casino dice. (See Example 8.11.) Estimate $P(3.1 \le \overline{X} \le 3.9)$ by simulating 10,000 $\overline{x}$-values. Compare your answer to the estimate derived in Example 8.11 using the CLT.

5. The score X on a fair casino die has mean $\mu = 3.5$ and standard deviation $\sigma = 1.708$. Let $\overline{X}$ denote the average score on n tosses of a fair die. Use the CLT to estimate $P(3.25 < \overline{X} < 3.75)$
   a. for $n = 10$
   b. for $n = 100$
   c. for $n = 1000$
   d. Check each of your answers in (a) through (c) by simulating 10,000 $\overline{x}$-values.
   e. What do you notice when comparing parts (a) through (c)? As n gets larger, does $P(3.25 < \overline{X} < 3.75)$ get larger or smaller?

6. We want to simulate 40 observations of $\overline{X}$, the average score on five fair dice. We shall use the first five-digit column in Table B.2. For the first simulation we use the five digits in the first line, 66533, yielding the average 4.6, and so forth. We get the 40 averages shown here.

4.6	3.2	3.2	2.2	4.0	3.0	2.8	3.2	3.6	3.6
3.8	4.0	3.8	3.8	3.6	3.4	3.0	4.2	4.8	4.6
4.0	3.2	3.4	4.0	4.2	3.6	4.0	3.0	2.6	2.6
5.6	4.0	1.6	3.0	3.4	3.8	4.6	3.6	3.2	3.4

   a. Compute the experimental probability $\hat{P}(3.3 < \overline{X} < 3.7)$.
   b. Estimate $P(3.3 < \overline{X} < 3.7)$ using the CLT.
   c. Compare the answers in (a) and (b).

7. Compute $\mathrm{SD}(\overline{X})$ where $\overline{X}$ is the average score on five fair dice. Use the data in the previous exercise to compute the experimental probabilities in parts (a) and (b).
   a. $\hat{P}(3.5 - \mathrm{SD}(\overline{X}) \le \overline{X} \le 3.5 + \mathrm{SD}(\overline{X}))$
   b. $\hat{P}(3.5 - 2\mathrm{SD}(\overline{X}) \le \overline{X} \le 3.5 + 2\mathrm{SD}(\overline{X}))$
   c. Compare the answers in (a) and (b) to 0.68 and 0.95, as given by our basic rule.

8. Let X denote the number of heads in 10 tosses of a fair coin. As we have learned, we can write $X = Y_1 + \cdots + Y_{10}$, where $Y_i = 1$ if toss number i results in heads, and 0 if toss number i results in tails. Each Y_i has $\mu = 1/2$ and $\sigma = 1/2$. The proportion of heads in the 10 tosses is

$$X/10 = (Y_1 + Y_2 + \cdots + Y_{10})/10 = \overline{Y}.$$

   Use the CLT to estimate:
   a. $P(0.25 < \overline{Y} < 0.85)$
   b. $P(0.35 \le \overline{Y} \le 0.65)$

   Since X is binomial with $n = 10$ and $p = 1/2$, use Appendix G to find the exact probabilities in (c) and (d).
   c. $P(0.25 < \overline{Y} < 0.85) = P(3 \le X \le 8) = P(X \le 8) - P(X \le 2)$

d. $P(0.35 \le \overline{Y} \le 0.65) = P(3.5 < X < 6.5) = P(X \le 6) - P(X \le 3)$

9. We can simulate $Y_1 + Y_2 + \cdots + Y_{10}$ in Exercise 8 by picking 10 random digits from Table B.1. We shall use the last 2 five-digit columns to simulate 40 observations of $\overline{Y}$. For the first simulation we shall use 10001 10110, yielding $X = 5$ and $\overline{Y} = X/10 = 0.5$. We get the following 40 observations of $\overline{Y}$:

0.5	0.4	0.6	0.7	0.6	0.5	0.5	0.7	0.6	0.8
0.2	0.4	0.8	0.4	0.4	0.4	0.5	0.5	0.6	0.5
0.3	0.4	0.5	0.5	0.3	0.4	0.6	0.2	0.7	0.7
0.5	0.2	0.6	0.4	0.3	0.8	0.5	0.4	0.6	0.4

Compute the experimental probabilities:
a. $\hat{P}(0.25 < \overline{Y} < 0.85)$
b. $\hat{P}(0.35 \le \overline{Y} \le 0.65)$
c. Compare with the results in Exercise 8.

10. If you make a \$1 bet in the casino game chuck-a-luck, then your net gain X can equal -1, 1, 2, or 3 dollars with probabilities

$$p(-1) = \frac{125}{216},\ p(1) = \frac{75}{216},\ p(2) = \frac{15}{216},\ p(3) = \frac{1}{216}.$$

(See Exercise 10 in Section 7.1.)
a. Describe a box model for simulating your gain on a \$1 bet. (Specify what numbers are on the balls and how many balls there are of each kind.)
b. Run 10,000 computer simulations to estimate your chance of making money on five independent \$1 bets. Does the experimental probability histogram for your net gain on the five bets follow a bell curve?
c. Repeat for your net gain on 25 independent \$1 bets. What is your chance of making money? Does the experimental probability histogram for your gain follow a bell curve?

11. Let $\overline{X}$ denote the sample mean of a simple random sample of $n = 10$ numbers drawn without replacement from the population $\{1, 2, 3, 4, \ldots, 27, 28, 29\}$.
a. Determine $\mu_{\overline{X}}$ and $\sigma_{\overline{X}}$, using the facts that $\mu_{\text{pop}} = 15$ and $\sigma_{\text{pop}} = \sqrt{70}$.
b. Use the CLT to estimate $P(12.25 < \overline{X} < 19.25)$.
c. Simulate 10,000 values of $\overline{X}$ to estimate $P(12.25 < \overline{X} < 19.25)$. Determine the mean and the standard deviation of the 10,000 simulated $\overline{x}$-values.
d. Describe the shape of the experimental distribution of the simulated $\overline{x}$-values. (You may have to rescale the axis. Start at 6. Use increments of size 3 and nine minor ticks.)

8.3 THE STANDARD ERROR

Let $X_1, X_2, \ldots, X_n$ be independent observations from a distribution with unknown mean μ and standard deviation σ. The CLT tells us that for reasonably large n, the sampling distribution of the sample mean $\overline{X}$ roughly follows a normal curve centered at μ with standard deviation $\text{SD}(\overline{X}) = \sigma/\sqrt{n}$ (see Figure 8.10).

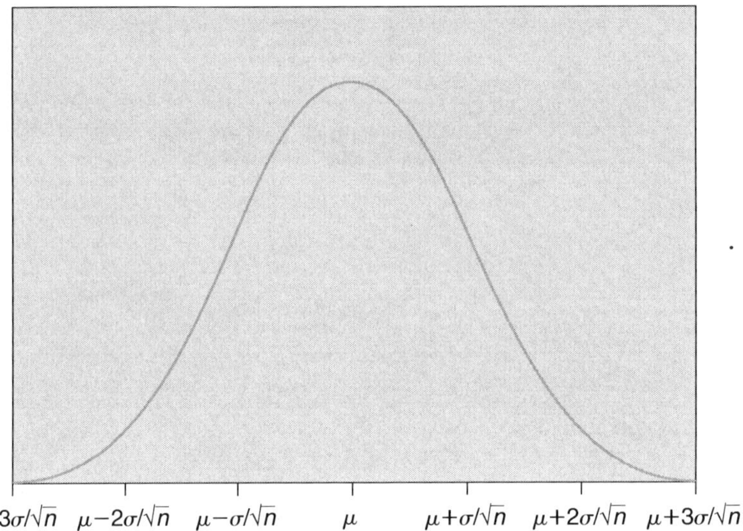

Figure 8.10 The normal density curve with mean μ and standard deviation $\sigma/\sqrt{n}$.

The sample mean $\overline{X}$ is likely to be around μ, give or take $\sigma/\sqrt{n}$ or so. Specifically, there is approximately

- a 68% chance that $\overline{X}$ differs from μ by less than $\sigma/\sqrt{n}$,
- a 95% chance that $\overline{X}$ differs from μ by less than $2\sigma/\sqrt{n}$,
- a 99.7% chance that $\overline{X}$ differs from μ by less than $3\sigma/\sqrt{n}$.

Since the sample mean $\overline{X}$ is likely to be around μ, give or take $\mathrm{SD}(\overline{X}) = \sigma/\sqrt{n}$ or so, it follows that the unknown mean μ is likely to be around $\overline{X}$, give or take $\sigma/\sqrt{n}$ or so.

Example 8.16

A laboratory makes 25 measurements to determine the mass μ of a certain object. Assume that the measurements are unbiased (good laboratory procedures are in place) and independent and that the typical accuracy of the measurements is known from past measurements. Assume that it is known that the typical measurement error is given by $\sigma = 300$ micrograms. If the sample mean is $\overline{x} = 1644$ micrograms, we estimate μ to be around 1644 micrograms, give or take

$$\sigma/\sqrt{n} = 300/\sqrt{25} \text{ micrograms} = 60 \text{ micrograms}$$

or so. Note the key role of dividing by $\sqrt{n} = \sqrt{25}$.

In Example 8.16 we assumed that σ was known. More often it is not only the distribution mean μ that is unknown, but also the distribution standard deviation σ. Since we can estimate σ by the sample standard deviation S, we can estimate the true

$$\mathrm{SD}(\overline{X}) = \sigma/\sqrt{n}$$

by the so-called **standard error** of $\overline{X}$,

$$\mathrm{SE}(\overline{X}) = S/\sqrt{n}.$$

> A standard error for a sample statistic Y (usually denoted SE(Y) and computed from the sampled data) is an estimate of SD(Y), the standard deviation of Y.

Example 8.17

The quality control manager at a food processing plant wants to make sure that most of the thousands of 14 oz Toasted Oats cereal boxes filled at the plant contain close to 14 ounces, or 397 grams, of cereal. If the boxes are overfilled, then the company will lose money. If the boxes are underfilled, then the consumers will be unhappy and government regulators may fine the company. To estimate the mean weight μ of all the 14 oz boxes filled at the plant and to estimate the variation in weight, the quality control manager takes a random sample of $n = 25$ boxes and weighs the contents of the sampled boxes. Suppose that the sample mean is $\overline{x} = 407$ grams and the sample standard deviation is $s = 12$ grams. She estimates μ to be around 407 grams, give or take the standard error of $\overline{X}$, given by

$$s/\sqrt{n} = 12/\sqrt{25}\text{ grams} = 2.4\text{ grams}$$

or so.

Example 8.18

Recall Newcomb's experiment to measure the speed of light. Excluding the two outliers, the computations in the remaining $n = 64$ measurements show that the average measured time Newcomb's beam of light took to travel 7442 meters was $\overline{t} = 24,827.75$ nanoseconds and that the sample SD was $s_T = 5.08$ nanoseconds, where a nanosecond is 1×10^{-9} seconds. Thus, the standard error of $\overline{T}$ is

$$\mathrm{SE}(\overline{T}) = \frac{s_T}{\sqrt{n}} = \frac{5.08}{8} = 0.635\text{ nanoseconds.}$$

Therefore, the estimated time for a beam of light to travel 7442 meters together with its standard error–based typical error is given by

$$\overline{t} \pm \mathrm{SE}(\overline{T}),$$

namely

$$24{,}827.75 \pm 0.635\text{ nanoseconds.}$$

Because the standard error is so small, any serious measurement error occurring is likely due to bias (systematic error), which cannot be detected by statistical methods. Recalling that the true speed of light in a vacuum is now known to be extremely close to 299,792.5 kilometers/second, as contrasted with Newcomb's estimate of 299,745 km/sec, we see that the bias in the estimated time $\overline{T}$ (around 3 to 4 nanoseconds) is the major source of the error. The systematic error in Newcomb's measurements can partly be explained by the fact that light travels a tiny, tiny bit more slowly through air than through a vacuum. Thus, even if Newcomb had the patience to make 1000 observations, his estimate of the time would still be high by 3 to 4 nanoseconds (due to bias) and hence, by using $v = d/\overline{T}$, the estimated speed of light in a vacuum would still have been too slow by 35 to 55 km/sec.

❐ ❐

Estimating an unknown success probability p in a chance experiment is a special case of estimating an unknown distribution mean. If $\hat{p}$ denotes the experimental success rate in n repeated independent trials, then $\hat{p}$ is an unbiased estimator of p with standard deviation

$$\mathrm{SD}(\hat{p}) = \sqrt{p(1-p)}/\sqrt{n}.$$

The standard error of p is the estimate

$$\mathrm{SE}(\hat{p}) = \sqrt{\hat{p}(1-\hat{p})}/\sqrt{n}.$$

Example 8.19

Suppose that we are using computer simulation to estimate the probability p of making money on five independent \$1 bets in the casino game chuck-a-luck. (See Exercise 10 in Section 7.1 and Exercise 10 in Section 8.2.) Suppose that our experimental success rate $\hat{p}$ in $n = 10{,}000$ independent simulations is

$$\hat{p} = 3{,}732/10{,}000 = 0.3732$$

The standard error of $\hat{p}$ is

$$\mathrm{SE} = \sqrt{\hat{p}(1-\hat{p})}/\sqrt{n} = \sqrt{0.3732 \cdot 0.6268}/\sqrt{10{,}000} = 0.0048$$

We estimate p to be around 0.373, give or take 0.005 or so.

❐ ❐

Suppose that we are interested in the proportion p of a large population that has a certain characteristic. Suppose that we take a simple random sample of size n from the population. Then the proportion $\hat{p}$ of the sample that has the characteristic is an unbiased estimator of p with standard deviation

$$\mathrm{SD}(\hat{p}) = \frac{\sqrt{p(1-p)}}{\sqrt{n}} \cdot \sqrt{\frac{N-n}{N-1}} \approx \frac{\sqrt{p(1-p)}}{\sqrt{n}},$$

assuming that the population size N is at least 20 times greater than the sample size n. The standard error of $\hat{p}$ is simplified to

$$\mathrm{SE}(\hat{p}) = \frac{\sqrt{\hat{p}(1-\hat{p})}}{\sqrt{n}} \cdot \sqrt{\frac{N-n}{N-1}} \approx \frac{\sqrt{\hat{p}(1-\hat{p})}}{\sqrt{n}},$$

thus allowing us to ignore the population size N.

Example 8.20

In Example 8.6 we were interested in the proportion p of all university students who parked on campus during classes. In a simple random sample of $n = 356$ students, 89 students said that they parked on campus during classes. We therefore estimate p to be around the sample proportion

$$\hat{p} = 89/356 = 0.25$$

give or take

$$\mathrm{SE}(\hat{p}) \approx \frac{\sqrt{\hat{p}(1-\hat{p})}}{\sqrt{n}} = \frac{\sqrt{0.25 \cdot 0.75}}{\sqrt{356}} = 0.02$$

or so.

❒ ❒

Consider any estimator $\hat{\theta}$ whose sampling distribution roughly follows a normal curve centered at the parameter θ we are trying to estimate. (The CLT tells us that $\overline{X}$ and $\hat{p}$ have this property.) Since normally distributed variables are likely to differ from their mean by one SD or so, it follows that $\hat{\theta}$ is likely to differ from θ by $\mathrm{SD}(\hat{\theta})$ or so. Assuming that the standard error $\mathrm{SE}(\hat{\theta})$ is close to $\mathrm{SD}(\hat{\theta})$, we can therefore estimate θ to be around $\hat{\theta}$, give or take $\mathrm{SE}(\hat{\theta})$ or so.

We shall give a couple of examples (in addition to $\overline{X}$ and $\hat{p}$) of sample statistics whose sampling distributions roughly follow normal curves. In the next section we shall discuss how we can estimate $\mathrm{SD}(\hat{\theta})$ for an estimator $\hat{\theta}$ in those situations where we do not have a formula for $\mathrm{SD}(\hat{\theta})$.

In many situations the sampling distribution of the sample median follows a bell curve centered at the population median. Unfortunately, we usually do not have a formula for the SD of the sample median, so computing the standard error of the sample median seems impossible. But we can use computer simulation to estimate the SD of the sample median and thus obtain a standard error for the sample median.

Example 8.21

Let Y denote the sample median of 25 numbers drawn at random with replacement from the following box, which contains a total of 29 balls:

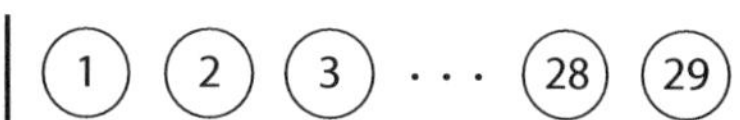

Figure 8.11 displays the experimental histogram of 50,000 simulated values of Y. The histogram follows a normal curve centered at the box median, 15.

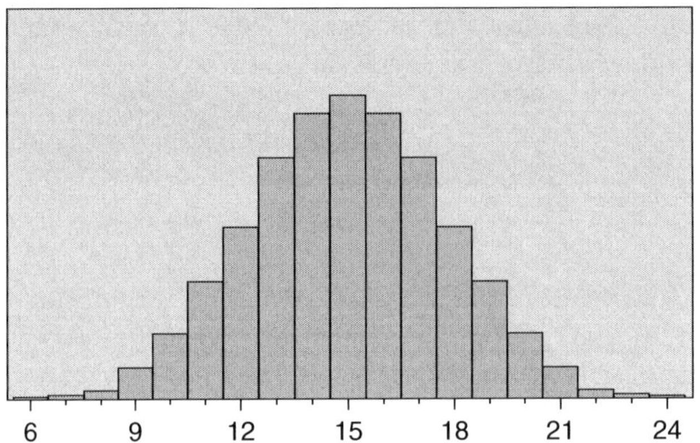

Figure 8.11 Experimental histogram for 50,000 simulated sample medians. Each simulated value is the median of 25 numbers drawn at random with replacement from the population $\{1, 2, 3, \ldots, 28, 29\}$.

The sample standard deviation of the 50,000 simulated values is 2.80. Our estimate of SD(Y) is therefore

$$\text{SE}(Y) = 2.80.$$

This example illustrates how we can use simulation to obtain standard errors of sample statistics.

❐ ❐

If the sample size n is large enough, then the sampling distribution of the sample standard deviation S will follow a bell curve centered at σ.

Example 8.22

Consider the sample standard deviation S of $n = 50$ numbers drawn at random with replacement from the box

Here, $\mu_{\text{box}} = 4$ and $\sigma_{\text{box}} = 2$. Figure 8.12 displays the experimental histogram of 50,000 simulated values of S. The histogram follows a bell curve centered at $\sigma = 2$.

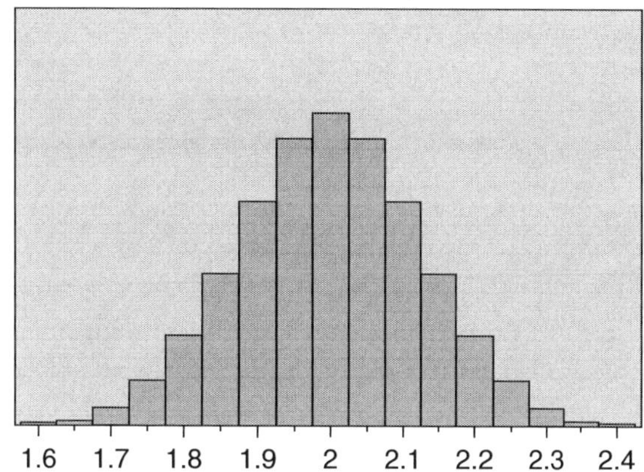

Figure 8.12 Experimental histogram for 50,000 simulated values of S. Each simulated value is the sample standard deviation of 50 numbers drawn at random with replacement from the population $\{1, 2, 3, 4, 5, 6, 7\}$.

The sample mean and sample standard deviation of the 50,000 simulated values are 1.9957 and 0.1264, respectively. Our estimate of σ_S, the standard deviation of S, is therefore

$$\text{SE}(S) = 0.1264.$$

We estimate μ_S, the mean of S, to be around 1.9957, give or take

$$0.1264/\sqrt{50{,}000} = 0.0006$$

or so. Notice that this shows that $E(S) < \sigma$, so S is a biased estimator of σ.

Section 8.3 Summary

The standard error of a sample statistic Y, which we shall denote SE(Y), is our estimate of SD(Y), the standard deviation of Y. If we have a formula for SD(Y), then we obtain SE(Y) by estimating the unknown parameters in the formula for SD(Y). For example, if the statistic of interest is $\overline{X}$, the mean of n independent observations from a distribution with unknown mean μ and the unknown standard deviation σ, then our estimate of

$$\text{SD}(\overline{X}) = \sigma/\sqrt{n}$$

is

$$\text{SE}(\overline{X}) = S/\sqrt{n},$$

where S denotes the sample standard deviation. If the statistic of interest is $\hat{p}$, the experimental success rate in n repeated independent success/failure trials with unknown success probability p, then our estimate of

$$\mathrm{SD}(\hat{p}) = \sqrt{p(1-p)}/\sqrt{n}$$

is

$$\mathrm{SE}(\hat{p}) = \sqrt{\hat{p}(1-\hat{p})}/\sqrt{n}.$$

Another way to estimate SD(Y) is to simulate a large number of values of Y. Then the sample standard deviation of the simulated values of Y is a good estimator of SD(Y).

If the sampling distribution of an estimator $\hat{\theta}$ roughly follows a normal curve centered at the parameter θ we are trying to estimate, then SE($\hat{\theta}$) gives the likely size of the estimation error. We estimate θ to be around $\hat{\theta}$, give or take SE($\hat{\theta}$) or so.

Section 8.3 Exercises

1. Describe a box model for simulating the number of heads in 100 tosses of a fair coin. Run 10,000 computer simulations to estimate the chance of getting 50 heads in 100 tosses. What is your estimate? Put a give-or-take number on your estimate.

2. Consider the box in Example 8.21. Both the box mean and box median equal 15. Let Y and $\overline{X}$ denote the sample median and the sample mean, respectively, of $n = 25$ numbers drawn at random with replacement from the box.
   a. Use the fact that $\sigma_{\text{box}} = \sqrt{70}$ to compute SD($\overline{X}$).
   b. Which is smaller, SE(Y) or SD($\overline{X}$)?
   c. Which statistic is likely to be closer to 15, the sample median Y or the sample mean $\overline{X}$? Explain.

3. Consider the sum of the scores on a pair of symmetric casino dice. (See Figure 5.3.) The possible values of the sum are $2, 3, \ldots, 11, 12$. (Note that these values are not equally likely.) Let X equal the number of times you have to toss the pair of dice to observe each of the sums $2, 3, \ldots, 11, 12$.
   a. Describe a box model for simulating X. (What are the numbers on the balls? How many balls are there with each value?)
   b. Simulate 10,000 values of X to estimate the mean μ_X of X. (See Section 6.2)
   c. Put a give-or-take number on your estimate.

4. An auditor for an insurance company took a simple random sample of 200 dental insurance claims and found that the average length of time it took a claim to be processed and paid was 63 days. If the sample standard deviation was 28 days, put a give-or-take number on the estimated average processing time.

5. In a random sample of 292 hospitalizations with a diagnosis of staph infection, 126 cases were caused by drug-resistant staph.
   a. Estimate the proportion of all staph hospitalizations that are caused by drug-resistant staph.
   b. Put a give-or-take number on your estimate.

6. An auditor wants to estimate the proportion of sales invoices at a large warehouse that contain errors. Suppose that in a simple random sample of 400 sales invoices, 38 contain errors.
   a. Estimate the proportion of all sales invoices that contain errors.
   b. Put a give-or-take number on your estimate.

7. Consider the sample standard deviation S of $n = 50$ numbers drawn at random with replacement from the box

Here $\mu_{\text{box}} = 3$ and $\sigma_{\text{box}} = 2$.

a. Simulate 10,000 values of S. Record the mean and the standard deviation of the simulated values. Describe the shape of the experimental histogram for the simulated values.
b. Estimate σ_S, the standard deviation of S.
c. Estimate μ_S, the mean of S.
d. Put a give-or-take number on your estimate of μ_S.

8. Consider the box

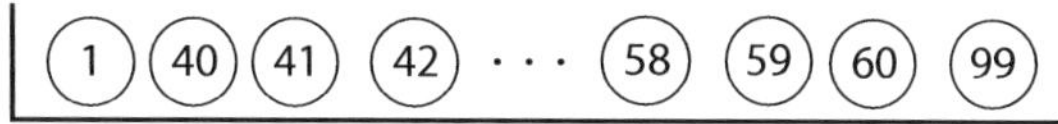

containing all the integers between 40 and 60 plus the outliers 1 and 99. Both the box mean and box median equal 50. Let M and $\overline{X}$ denote the sample median and the sample mean, respectively, of $n = 25$ numbers drawn at random with replacement from the box. We shall investigate which variable, M or $\overline{X}$, is likely to be closer to 50.
a. Simulate 10,000 values of M. Record the mean and the standard deviation of the simulated values. Describe the shape of the experimental histogram for the simulated values.
b. Simulate 10,000 values of $\overline{X}$. Record the mean and the standard deviation of the simulated values. Describe the shape of the experimental histogram for the simulated values.
c. Which is larger, SE(M), the sample standard deviation of the simulated values of M, or SE($\overline{X}$), the sample standard deviation of the simulated values of $\overline{X}$?
d. Which statistic appears to vary more, M or $\overline{X}$? Explain.
e. Which statistic is likely to be closer to 50? Explain.

8.4 BOOTSTRAPPING

Recall Simon Newcomb's experiment to measure the speed of light described in Example 2.8. He made 66 independent measurements of the time it took a beam of light to travel 7442 meters.

Imagine that Newcomb had made 1 million independent measurements, and not just 66. We could then create a box with 1 million balls, one ball for each measurement. The number on a given ball should be the corresponding measurement value. (See Table 2.2 and surrounding text to understand what the numbers on the balls stand for.)

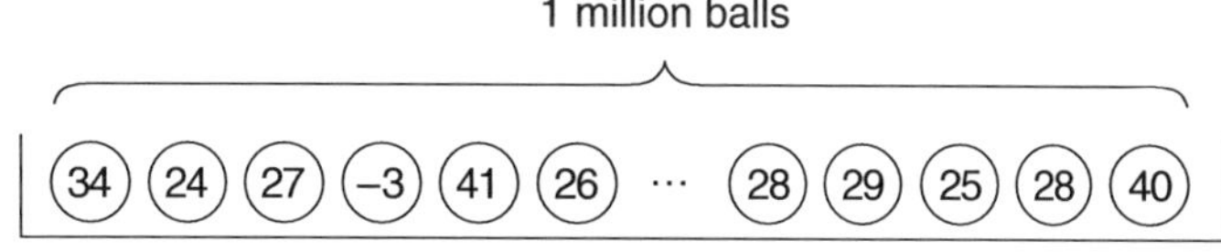

Based on 1 million measurements, the distribution of the numbers in the box would extremely accurately reflect the theoretical distribution of a single random measurement. We could therefore simulate 66 *independent* measurements (Newcomb's experiment) by drawing 66 times at random *with replacement* from the box.

Since Newcomb made only 66 measurements, we can only imagine the box model just described. But the imagined million-ball box model suggests how we could simulate Newcomb's 66 measurements and compare the variability of sample mean and sample median: Newcomb's 66 observed values in Example 2.8 is the closest thing we have to the imagined 1 million observations. If we cannot do five-step simulation using the million-ball box, the next best thing would be to do five-step simulation using a box containing Newcomb's 66 observed values, knowing that these 66 balls produce an experimental distribution shaped a lot like the distribution of the 1 million observations we wish we had. We could simulate 66 independent measurements by drawing 66 times at random with replacement from the box containing Newcomb's 66 observed values:

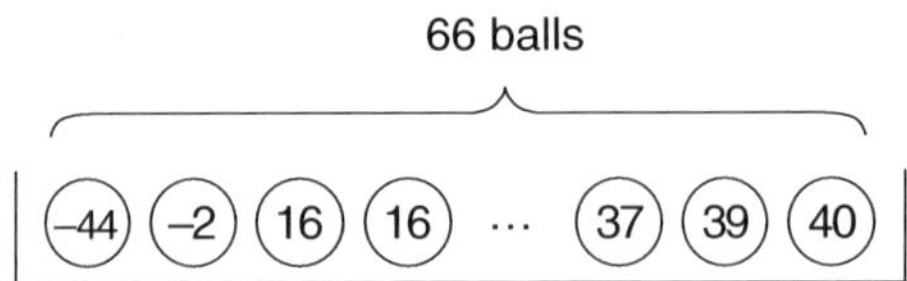

If, for example, the median of the 66 measurements is our statistic of interest used to compute an estimate of the speed of light, then we could study the variability of the sample median by repeating the simulation 10,000 times, producing 10,000 simulated sample medians. We could then study the variability of these simulated sample medians in order to understand how well the sample median works if we use it to estimate the true speed of light.

bootstrapping

The statistical approach just described is called **bootstrapping**. Bootstrapping is a modern statistical approach based on computer simulations—as opposed to the classical mathematical models and mathematical analysis that statisticians had to rely on for most of the twentieth century, before fast computers became readily available.

Example 8.23

The tensile strength of a material is the maximum stress it can be subjected to before failure. The data in Table 8.3 represent 25 independent measurements of the tensile strength of a composite material.

Table 8.3 Tensile Strength (in Thousands of Pounds per Square Inch) of a Composite Material

174.63	160.44	184.41	203.41	185.97
185.34	180.95	200.73	209.58	190.67
205.59	201.95	206.51	213.35	207.88
212.00	204.60	209.84	218.56	210.80
225.25	219.51	212.15	242.76	231.46

Source: Vangel, Mark G., "New Methods for One-Sided Tolerance Limits for a One-Way Balanced Random-Effects ANOVA Model," *Technometrics*; May, 1992, Vol. 34, Issue 2, pp. 176–185.

We are interested in both the mean tensile strength μ and the variability in tensile strength, as measured by the standard deviation σ. The sample mean and the sample standard deviation are $\bar{x} = 203.9336$ and $s = 18.320$, respectively. We estimate μ to be around 203.9, give or take $18.320/\sqrt{25} \approx 3.7$ or so.

To investigate the accuracy of S as an estimator of σ we imagine that we had 1 million independent tensile strength measurements, and not just 25. We could then create a box with 1 million balls, one ball for each measurement. The number on each ball would be the corresponding measurement value. We could then simulate S by drawing 25 balls at random with replacement from the million-ball box and compute the sample standard deviation of the 25 sampled values. If we simulated 50,000 values of S, then the SD of the simulated values would be very close to SD(S), the theoretical standard deviation of S.

The sample standard deviation S is always a slightly biased estimator of σ. We always have

$$E(S) < \sigma$$

The larger the sample size n is, however, the smaller the bias $E(S) - \sigma$ will be. If $\overline{S}$ denoted the sample mean of the 50,000 simulated values of S, then we could estimate the bias of S by

$$\overline{S} - \sigma_{\text{box}}.$$

We could then estimate σ to be around

$$S - (\text{estimated bias}),$$

give or take SE(S) or so, where SE(S) denotes the sample standard deviation of the 50,000 simulated values of S.

As in the case of Newcomb's data we do not have 1 million measurements. The 25 observations is all we have. But if we cannot do five-step simulation using the million-ball box, the next best thing is to do five-step simulation using a box containing the 25 observations in Table 8.3. Figure 8.13 displays the experimental histogram of 50,000 bootstrapped values of S. Each simulated value is the sample standard deviation S^* (note the notation S^* to distinguish this bootstrapped sample standard deviation from the standard deviation computed from real data) of 25 numbers drawn at random with replacement from the 25 observations in Table 8.3.

The mean and the standard deviation of the 50,000 bootstrapped values are 17.75 and 2.70, respectively. Our estimate of SD(S) is therefore

$$\text{SE}(S) = 2.70.$$

If $\overline{s}^*$ denotes the mean of the 50,000 bootstrapped s^* values, then our estimate of the bias of S is

$$\overline{s}^* - \sigma_{\text{bootstrap box}} = 17.75 - 17.95 = -0.20.$$

We estimate σ to be around

$$s - (\text{estimated bias}) = 18.3 + 0.2 = 18.5,$$

give or take $\text{SE}(S) = 2.7$ or so.

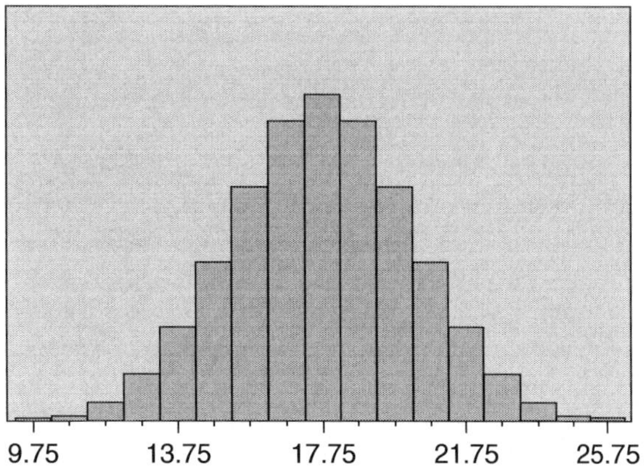

Figure 8.13 Experimental histogram for 50,000 bootstrapped values of S. Each simulated value is the sample standard deviation of 25 numbers drawn at random with replacement from the 25 observations in Table 8.3.

❐ ❐

Consider an estimator $\hat{\theta}$ (like S) of some unknown parameter of interest (like σ). If $\hat{\theta}$ is approximately normally distributed, then we want to estimate the standard deviation of $\hat{\theta}$, $\text{SD}(\hat{\theta})$, and the bias of $\hat{\theta}$, $E(\hat{\theta}) - \theta$. We can then estimate θ to be around $\hat{\theta}$ – (estimated bias), give or take $\text{SE}(\hat{\theta})$ or so.

The great insight of Brad Effron, of Stanford, and others who developed the bootstrap method of estimating the bias of $\hat{\theta}$ and $\text{SD}(\hat{\theta})$, was to realize that although we can never know the population box model in a statistical estimation problem, we can substitute a box model with the numbers of the observed sample. If the sample size n is moderate to large, this sample-determined box model will approximate the population box model and can be used to repeatedly simulate the statistic $\hat{\theta}$. In particular, for a large sample size n, the error in estimating $\text{SD}(\hat{\theta})$ by the sample standard deviation of a large number of simulated $\hat{\theta}$-values from the sample-determined box model will be small. This solves the problem of finding the unknown SE for a statistic $\hat{\theta}$ known to be approximately normal.

Using the sample as the box model, we can simulate $\hat{\theta}$ values, which we denote by $\hat{\theta}*$, the * being a reminder that they are simulated using the sample-determined box rather than a box representing the true population distribution. This approach is a way of "lifting ourselves by our bootstraps," that is, using only the sample data to obtain useful estimates for both the parameter that we are estimating and the standard error of our estimator of the parameter. Hence this simulation-based method is called bootstrapping. Bootstrap sampling is one of the most important and useful contributions to statistics made in the last quarter of the twentieth century.

In Example 8.23 we discussed bootstrap sampling using 25 independent measurements. We shall now discuss bootstrap sampling using a simple random sample drawn from a real population. Let us see in detail how the bootstrap approach to estimating the standard error of an estimator works. Consider a large real population from which we have taken a simple random sample (without replacement) of size n and have formed an estimator of some population parameter. We seek a bootstrap estimate of the standard error of this estimator. We first need to set up the Step 1 box model, from which we will obtain our bootstrap samples. We want this box model to be a stand-in for the unknown population. To create a box model that has almost the same distribution as the population producing the random sample and that is the same size as the population, we replicate the observed sample many times to build up a population-sized bootstrap box model. That is, we make a large number of identical copies of the sample to constitute the box model. This large, replicated box model is our best estimate of what the unknown large population is like in both size and distributional shape. We can now draw "bootstrap" random samples *without replacement* in the usual way we randomly sample from a real population; drawing one such bootstrap simulated sample is Step 2. From this simulated sample we compute the statistic of interest (Step 3) $\hat{\theta}*$ that is our estimator (such as $\overline{X}$). Step 4 consists of drawing many such bootstrap-simulated samples, each time computing our bootstrapped statistic $\hat{\theta}*$ of interest. This bootstrapping process replaces repeated sampling from the actual population to gauge the variability of $\hat{\theta}$, which is not an option in estimation problems.

Because the distribution created by a sample-determined box model will be much like the unknown theoretical population distribution, the experimental distribution of the simulated statistic of interest will be close to the statistic's unknown theoretical distribution. Thus, the experimental histogram of the statistic of interest $\hat{\theta}*$ obtained from the bootstrap-simulated samples should be close to the true sampling distribution of the statistic.

bootstrap estimate of $SD(\overline{X})$

Example 8.24

Suppose a random sample of size 5 is taken from a population of 200 elderly men and their diastolic (lower) blood pressures are recorded.

104	83	99	88	110

We note that, for the bootstrap approach to work, most practitioners would want at least a moderate-size sample, say $n \geq 10$, so that the shape of the sample distribution (the box model of the sampled observations) really does approximate the population distribution shape, at least roughly.

The sample mean is $\overline{x} = 96.8$. Suppose we wish to use $\overline{X}$ as the estimator of the population mean μ. How close is our estimator $\overline{X}$ likely to be to the true parameter μ? Even though we could use $\text{SE}(\overline{X}) = S/\sqrt{n}$ to estimate $\text{SD}(\overline{X})$, we will instead, for instructional purposes, use our just-learned bootstrap simulation approach to estimate $\text{SD}(\overline{X})$. We perform the five steps:

Step 1. Choose a Box Model. To form our box model, we "clone" the sample $200/5 = 40$ times to create a bootstrap box model of the same size (200) as the population.

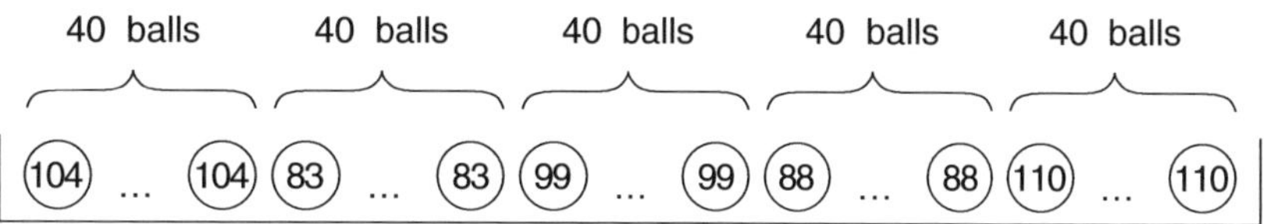

This is our best estimate of what the distribution of diastolic blood pressures in the population of 200 looks like. This box model stands in for the unknown population box model, that is, the unknown population distribution.

Step 2. Define One (Bootstrap Sample) Simulation. We choose our first bootstrap sample by drawing five balls at random *without replacement* from the bootstrap box model, just as we formed the original real data sample by taking the diastolic blood pressures of five men chosen randomly from the real population without replacement. Suppose the results were

$$(x_1^*, x_2^*, \ldots, x_5^*) = (83, 99, 83, 110, 104).$$

The number 83 (or any number in the sample) can occur two or more times in the simulated sample because of the "cloning" of the sample data.

Step 3. Define the (Bootstrapped) Statistic of Interest. Our bootstrapped statistic of interest in this sampling process is the mean $\overline{X}^*$ of the values in the simulated bootstrap sample. The asterisk ($*$) in the notation $\overline{X}^*$ indicates that this statistic is a bootstrap-simulated statistic. Do not confuse the actual "real-life" observed sample $(X_1, X_2, \ldots, X_5)$ and its estimator $\overline{X} = \sum X_i/5$ of μ with a bootstrap-simulated sample $(X_1^*, X_2^*, \ldots, X_5^*)$ and its corresponding bootstrap-estimator $\overline{X}^* = \sum X_i^*/5$.

Step 4. Repeat the Simulations. To illustrate the process, we perform three simulations (10,000–50,000 simulations is preferable). The results are shown in Table 8.4. The observed sample mean of these bootstrap-simulated sample means is $\overline{\overline{x}}{}^* = (95.8 + 96.8 + 94.6)/3 = 95.73$.

Table 8.4 Results of Three Bootstrap Samples

Sample number	Bootstrap samples					$\overline{x}^*$
1	83	99	83	110	104	95.8
2	99	104	83	110	88	96.8
3	104	104	83	83	99	94.6

Step 5. Estimate the Standard Deviation of the Statistic of Interest.

$$\text{SE}(\overline{X}) = \widehat{\text{SD}}(\overline{X}^*) = \sqrt{\frac{\sum_1^3(\overline{x}^* - \overline{\overline{x}}^*)^2}{2}} = \sqrt{1.2135} \approx 1.10.$$

$\widehat{\text{SD}}(\overline{X}^*) = 1.10$ is our bootstrapped estimate of $\text{SD}(\overline{X})$. Practically speaking, this will be a very bad estimate because the sample size $n = 5$ is small and the number of bootstrap samples, 3, is totally inadequate. The purpose of the example was only to instruct.

❐ ❐

Our proposed bootstrap box model is the same size as the real population. But that real population is many times larger than the sample. Therefore, as we learned in Chapter 5, we can sample *with replacement* when we do bootstrap sampling, and expect very similar results. When sampling with replacement, we know it is only the box model *proportions* that matter; for example, a box with 2 zeros and 1 one provides the same bootstrap sampling model as one with 2000 zeros and 1000 ones. The proportions in the bootstrap box model made by "cloning" the original sample are just those of the original sample itself. So the replication of the sample that we did to form a realistic population-sized box model is irrelevant and was a waste of time! We might as well have just drawn random samples *with replacement from a box consisting exactly of the original sample (uncloned).* Thus we can obtain bootstrap-simulated random samples by drawing samples of size n with replacement from a box model that consists only of the original sample (of size n) itself.

This sampling with replacement from the sample is what is done in practice when bootstrap sampling is carried out. Note that in forming the box model we no longer need to know the exact size of the original population. In summary, the foregoing strategy of cloning the sample to form the box model and then sampling without replacement was presented only to help us understand why bootstrapping is a sensible thing to do. **In practice, bootstrap sampling is always done with the original sample alone (without cloning) and with replacement.**

actual bootstrap sampling is done with replacement

When we bootstrap sample, we draw random samples of the same size n as our original sample *with replacement,* from our original observations or data. These samples with replacement from a box model of the actual sample are called **bootstrap samples** and are the same size as the number of observations, n, originally sampled from the real population. We calculate the bootstrapped statistic of interest from each of the bootstrap samples.

bootstrap samples

In Section 2.2 we saw that sometimes the median is a better parameter for characterizing the center of a distribution (population) than the mean because it is less affected by skewing and by extreme values. The sample median M of a sample drawn at random from that population is a reasonable estimate of the population median m. (The population median m is defined for the density (pdf) of a continuous random variable as the point where exactly half the area under the probability density function lies on each side, as shown in Figure 8.14.)

As we illustrated in Example 8.21, in many situations the sampling distribution of the sample median follows a bell curve centered at the population median. We do not have a formula for its standard error, which is essential in assessing the accuracy of M as an estimator of m. We can, however, use the bootstrap method to calculate a standard error by computing a bootstrap estimate of the standard deviation of the sample median.

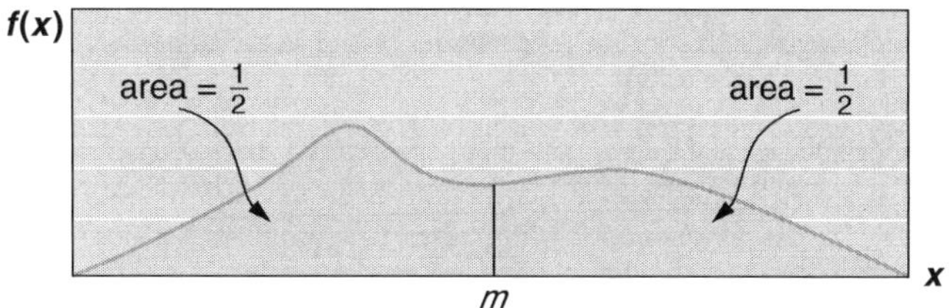

Figure 8.14 Theoretical continuous pdf; m denotes its theoretical median.

Example 8.25

The incomes of a random sample of 400 graduate students are determined. Because the income distribution for all graduate students is believed to be skewed, we decide to estimate the distribution center, the population median m, using the sample median $M = \$9.1$ K. We let our observed random sample of size 400 (not tabulated) stand in for the unknown population distribution. That is, the sample determines our 400-ball bootstrap box model for Step 1 of the five-step simulation method.

Next we draw samples of size 400 with replacement from our sample-based box model. Thus, Step 2 consists of drawing randomly with replacement one sample of size 400.

The statistic of interest for Step 3 is the bootstrapped sample median M^*. Here, as usual, the asterisk in M^* serves as a reminder that these sample medians are bootstrapped, as contrasted with the original M computed from the original real-world random sample.

In Step 4 we repeat 100 times this process of selecting a bootstrap sample of size 400, each time obtaining a sample median M^*. This is bootstrap sampling.

Then in Step 5 we compute the sample standard deviation of the statistic of interest, namely, the sample SD of the 100 bootstrapped sample medians (M^*-values), and this sample standard deviation will be our standard error, SE(M). The stem-and-leaf plot in Table 8.5 displays the 100 sample medians we obtained via bootstrap sampling.

That is, an estimate of the standard deviation of the sample median, SD(M), is obtained by calculating the sample SD of the 100 bootstrap sample medians, which in this case is equal to SE(M) = 1.005.

Table 8.5 Bootstrap Sample Medians (M^* values) for Example 8.25

Stem	Leaf
6	8
7	11444
7	66666
8	00011111111111111122233333
8	6666666668888888888888
9	000033
9	666666666888899
10	000000000111133
10	5577
11	1
11	
12	
12	5

Key: "12 5" stands for 12.5

The choice of 100 bootstrap samples is a bit low. In actual practice, 1000 is considered enough to produce decent results, although more is better in today's fast computing environment, including the textbook's software. Thus, we expect our sample median $M = 9.1$ to be roughly in error by around ± 1.005 as an estimate of the population median m.

Section 8.4 Summary

Let $x_1, x_2, x_3, \ldots, x_n$ denote data obtained either from a *repeated-trials random experiment* (repeated independent observations from a conceptual population) or from a *simple random sample* (from a real population at least 20 times larger than the sample). Let θ denote a parameter of interest that describes the population (conceptual or real). Let $\hat{\theta}$ denote an estimate of θ computed from the data $x_1, x_2, x_3, \ldots, x_n$.

To judge the likely accuracy of our estimate $\hat{\theta}$ we would like to know the *sampling distribution* of $\hat{\theta}$. The sampling distribution of $\hat{\theta}$ would tell us the *bias* of $\hat{\theta}$ (the systematic error) and the likely size of the *chance error* of $\hat{\theta}$ (the unsystematic error). The sampling distribution would tell us the likely accuracy of $\hat{\theta}$ as an estimator of θ. For example, since the large-sample sampling distribution of $\overline{X}$ is almost normal with mean μ and standard deviation $\sigma/\sqrt{n}$, we know that if both μ and σ are unknown, then we can estimate μ to be around $\overline{X}$, give or take $S/\sqrt{n}$ or so.

Three properties of $\overline{X}$ allow us to make this claim about the likely accuracy of $\overline{X}$ as an estimator of μ:

- The sampling distribution of $\overline{X}$ is almost normal.
- $\overline{X}$ is an unbiased estimator of μ.
- We have a simple formula for the standard error of $\overline{X}$, namely,

$$\text{SE}(\overline{X}) = S/\sqrt{n}.$$

Many estimators $\hat{\theta}$ have sampling distributions that are approximately normal. But often we do not know the bias of $\hat{\theta}$ and we do not have a simple formula for the standard error of $\hat{\theta}$. Consider, for example, the sample standard deviation S as an estimator of the theoretical standard deviation σ. For large n the sampling distribution of S is approximately normal. But S is slightly biased and we do not have a simple formula for the bias of S or the standard error SE(S).

We can approximate the sampling distribution of an estimator $\hat{\theta}$ and estimate the bias and the SD of $\hat{\theta}$ by *bootstrapping*. We create a bootstrap box containing the original data set

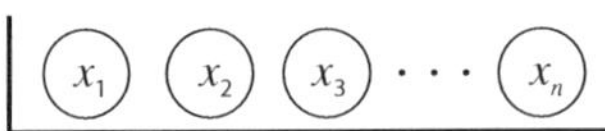

We draw n times at random with replacement from the bootstrap box to get the simulated observations

$$x_1^*, x_2^*, x_3^*, \ldots, x_n^*$$

We compute the simulated estimate $\hat{\theta}*$ from the simulated observations using the same formula or algorithm that gave us our original estimate $\hat{\theta}$ from the original data set. If we bootstrap 10,000 $\hat{\theta}*$-values, then the experimental distribution of the $\hat{\theta}*$-values will resemble the sampling distribution of $\hat{\theta}$. (We hope.) We use the sample standard deviation of the simulated $\hat{\theta}*$-values to estimate SD($\hat{\theta}$).

Let θ_{box} denote the value of the parameter for the bootstrap box. For example, if $\theta = \sigma$, then $\theta_{\text{box}} = \sigma_{\text{box}}$. If $\overline{\theta}*$ denotes the sample mean of the simulated $\hat{\theta}*$-values, then

$$\overline{\theta}^* - \theta_{\text{box}}$$

is an estimate of the bias of $\hat{\theta}$:

$$E(\hat{\theta}) - \theta.$$

Section 8.4 Exercises

1. When is it necessary to use the bootstrap method to find the standard error of an estimate?
2. Explain how to find the standard error of a sample mean for a sample of size 25 using the bootstrap method.
3. Ten people were sampled from a statistics class with 165 people. The weights of the 10 are

150	185	162	150	117	110	135	110	160	125

Weight is sometimes skewed to the right, leading a statistician to use the sample median M to estimate the population median m. Twenty-five bootstrapped samples of size 10 (with replacement) from these data were taken. The following are the sample medians (M^*s) of these bootstrapped samples:

133.5	155.0	142.5	150.0	150.0
150.0	126.0	155.0	150.0	130.0
160.0	150.0	137.5	150.0	142.5
150.0	117.0	137.5	137.5	155.0
150.0	147.5	150.0	160.0	150.0

a. What is the sample median of the original sample?
b. Find the bootstrapped SE of the sample median M using the 25 bootstrapped sample medians (M^*s).

4. Three dogs were randomly sampled from a local animal shelter. Their ages in years were given as 1, 2, and 7.
a. What is the median of the sample?
b. Ten bootstrap samples of size 3 ($n = 3$ is too small to produce a good SE(M), of course) from the three ages are listed in the following table. Compute the sample standard deviation of the sample median using the 10 bootstrap sample medians, thus finding the bootstrapped SE(M).

1.	1	2	1
2.	2	1	1
3.	1	7	2
4.	1	7	7
5.	1	7	7
6.	2	2	7
7.	2	7	2
8.	7	2	2
9.	1	1	2
10.	7	7	7

5. Imagine that we have 10 observations from a roughly bell-shaped distribution:

$x_0 = 91$	$x_1 = 151$	$x_2 = 110$	$x_3 = 117$	$x_4 = 101$
$x_5 = 97$	$x_6 = 107$	$x_7 = 95$	$x_8 = 76$	$x_9 = 90$

a. Compute the mean $\overline{X}$ of the sample.
b. Find the sample median M of the sample. (Remember to order the 10 observations first.) For bell-shaped (and hence symmetric) distributions, both the sample average and the sample median are reasonable estimates of the center of the distribution (population mean μ = population median, recall). To see which estimate is more accurate for this data set, use bootstrapping to evaluate which estimate has the smaller standard error, generating 50 bootstrap samples.
c. List each of the 50 bootstrap sample mean $\overline{x}^*$-values and median $M*$-values.
d. Find the bootstrap standard error $(\text{SE}(\overline{X}) = \widehat{\text{SD}}(\overline{X}^*))$ of the sample average.
e. Find the standard error $(\widehat{\text{SD}}(M^*) = \text{SE}(M))$ of the sample median.
f. Which standard error is smaller?

6. Consider the following six observations:

$x_1 = 10$	$x_2 = 7$	$x_3 = 12$
$x_4 = 8$	$x_5 = 5$	$x_6 = 9$

a. Find the sample median M. Use the first six digits in each of the first 10 lines in Table B.2 to generate 10 bootstrap samples of size 6 each. For the first bootstrap sample, use 665334 in the first line corresponding to the bootstrap sample $x_6, x_6, x_5, x_3, x_3, x_4$. For the second bootstrap sample, use 612612 in the second line, corresponding to the bootstrap sample $x_6, x_1, x_2, x_6, x_1, x_2$, and so forth.
b. List each of the 10 bootstrap samples and the corresponding 10 medians M_i^*.
c. Find the standard error of the sample median.

7. On your computer create a 66-ball box model containing Newcomb's 66 observed values in Example 2.8. (*Hint:* First organize the observations in a dotplot.)
a. Let the statistic of interest be $\overline{X}$, the mean of 66 values drawn at random with replacement from the box. Simulate 10,000 values of $\overline{X}$. Record the mean and the standard deviation of the 10,000 simulated values.
b. Let the statistic of interest be M, the median of 66 values drawn at random with replacement from the box. Simu-
late

10,000 values of M. Record the mean and the standard deviation of the 10,000 simulated values.

c. Which is larger, the standard deviation (of the 10,000 simulated values) of $\overline{X}$ or the standard deviation (of the 10,000 simulated values) of M? Which variable appears to have greater variability in Newcomb's experiment, $\overline{X}$ or M?

CHAPTER 8 SUMMARY

Repeated measurements of the same thing always differ slightly from each other. Our model is

$$\text{individual measurement} = \text{exact value} + \text{bias} + \text{chance error}.$$

The *bias*, or *systematic error*, affects all measurements the same way, pushing them all in the same direction by the same amount. The *chance error* varies from measurement to measurement, sometimes up and sometimes down. In the long run, the chance errors of repeated, independent measurements will average out to zero.

More generally, let θ denote any unknown *parameter* and let $\hat{\theta}$ denote a *sample statistic* used to estimate θ. Then our mathematical model is:

$$\hat{\theta} = \theta + \text{bias} + \text{chance error}.$$

If different statisticians independently were to compute estimates $\hat{\theta}$ in exactly the same way, using the same number of observations derived in exactly the same way, then the bias would affect all the different $\hat{\theta}$-values in the same way, pushing them all in the same direction by the same amount. By contrast, the chance error would vary from $\hat{\theta}$-value to $\hat{\theta}$-value, sometimes up and sometimes down. In the long run, if we generated more and more $\hat{\theta}$-values, then the chance errors would average out to zero.

Ideally, we want an estimator $\hat{\theta}$ to have both *small bias* and *small chance errors*. The smaller the bias is and the chance errors are, the more *accurate* the estimator will be.

If an estimator $\hat{\theta}$ has no bias, or systematic error, then we say that $\hat{\theta}$ is *unbiased*. The estimator $\hat{\theta}$ is unbiased if $E(\hat{\theta}) = \theta$. Note, however, that unbiasedness does not guarantee that the chance errors will be small. An estimator $\hat{\theta}$ can be unbiased and still have large chance errors.

The *sampling distribution* of a sample statistic is the statistic's probability distribution. It is the distribution of the values the statistic takes when we consider all possible samples of the same size from the same population.

Consider the sample mean $\overline{X}$ of n repeated random draws (made with or without replacement between draws) from a population with mean μ_{pop} and standard deviation σ_{pop}. The sampling distribution of $\overline{X}$ has mean

$$\mu_{\overline{X}} = E(\overline{X}) = \mu_{\text{pop}}.$$

This shows that if μ_{pop} is unknown, then $\overline{X}$ is an unbiased estimator of μ_{pop}. The sample mean $\overline{X}$ is likely to be around μ_{pop}, give or take $\text{SD}(\overline{X})$ or so.

If we sample with replacement, then the standard deviation of $\overline{X}$ is

$$\sigma_{\overline{X}} = \text{SD}(\overline{X}) = \sigma_{\text{pop}}/\sqrt{n}.$$

If we sample without replacement, then

$$\sigma_{\overline{X}} = \mathrm{SD}(\overline{X}) = \frac{\sigma_{\text{pop}}}{\sqrt{n}} \cdot \sqrt{\frac{N-n}{N-1}},$$

where N denotes the population size.

The *central limit theorem* (CLT) for $\overline{X}$ tells us that in either case (whether we sample with or without replacement) the sampling distribution of $\overline{X}$ will be close to the normal distribution

$$N(\mu_{\overline{X}}, \sigma_{\overline{X}}),$$

provided n is reasonably large (and $N - n$ is reasonably large as well, if we sample without replacement) and the population distribution is not too skewed and does not have outliers.

The CLT for $\hat{p}$ is a special case of the CLT for $\overline{X}$. The experimental "success" rate $\hat{p}$ in n repeated independent trials with the same success probability p has mean

$$\mu_{\hat{p}} = E(\hat{p}) = p$$

This shows that if p is unknown, then $\hat{p}$ is an unbiased estimator of p. The sample proportion $\hat{p}$ is likely to be around p, give or take $\mathrm{SD}(\hat{p})$ or so. The standard deviation of $\hat{p}$ is

$$\sigma_{\hat{p}} = \mathrm{SD}(\hat{p}) = \sqrt{p(1-p)}/\sqrt{n}.$$

If n is so large that $np \geq 10$ and $n(1-p) \geq 10$, then the sampling distribution of $\hat{p}$ will be close to the normal distribution

$$N(p, \sigma_{\hat{p}}).$$

Finally, consider a large population of size N of which a proportion p has a certain characteristic of interest. Consider a simple random sample of size n drawn without replacement from the population. Let $\hat{p}$ denote the proportion of the sample having the characteristic of interest. The sample proportion $\hat{p}$ has mean

$$\mu_{\hat{p}} = E(\hat{p}) = p$$

and standard deviation

$$\sigma_{\hat{p}} = \mathrm{SD}(\hat{p}) = \frac{\sqrt{p(1-p)}}{\sqrt{n}} \cdot \sqrt{\frac{N-n}{N-1}}.$$

If n is so large that $np \geq 10$ and $n(1-p) \geq 10$ and $N - n$ is so large that $(N-n)p \geq 10$, and $(N-n)(1-p) \geq 10$, then the sampling distribution of $\hat{p}$ will be close to the normal distribution

$$N(p, \sigma_{\hat{p}}).$$

If we take a simple random sample (sampling without replacement) from a population at least 20 times larger than the sample, then we can ignore "the finite population correction factor"

$$\sqrt{\frac{N-n}{N-1}}$$

in the formulas for $\sigma_{\overline{X}}$ and $\sigma_{\hat{p}}$ and use

$$\sigma_{\overline{X}} \approx \sigma_{\text{pop}}/\sqrt{n}$$

and

$$\sigma_{\hat{p}} \approx \sqrt{p(1-p)}/\sqrt{n}.$$

The standard error of a sample statistic Y, which we shall denote SE(Y), is our estimate of SD(Y), the standard deviation of Y. If we have a formula for SD(Y), then we obtain SE(Y) by estimating the unknown parameters in the formula for SD(Y). For example, if the statistic of interest is $\overline{X}$, the mean of n independent observations from a distribution with unknown mean μ and the unknown standard deviation σ, then our estimate of

$$\text{SD}(\overline{X}) = \sigma/\sqrt{n}$$

is

$$\text{SE}(\overline{X}) = S/\sqrt{n},$$

where S denotes the sample standard deviation. If the statistic of interest is $\hat{p}$, the experimental success rate in n repeated independent success/failure trials with unknown success probability p, then our estimate of

$$\text{SD}(\hat{p}) = \sqrt{p(1-p)}/\sqrt{n}$$

is

$$\text{SE}(\hat{p}) = \sqrt{\hat{p}(1-\hat{p})}/\sqrt{n}.$$

Another way to estimate SD(Y) is to simulate a large number of values of Y. Then the sample standard deviation of the simulated values of Y is a good estimator of SD(Y).

If the sampling distribution of an estimator $\hat{\theta}$ roughly follows a normal curve centered at the parameter θ we are trying to estimate, then SE($\hat{\theta}$) gives the likely size of the estimation error. We estimate θ to be around $\hat{\theta}$, give or take SE($\hat{\theta}$) or so.

Let $x_1, x_2, x_3, \ldots, x_n$ denote data obtained either from a *repeated-trials random experiment* (repeated independent observations from a conceptual population) or from a *simple random sample* (from a real population at least 20 times larger than the sample). Let θ denote a parameter of interest that describes the population (conceptual or real). Let $\hat{\theta}$ denote an estimate of θ computed from the data $x_1, x_2, x_3, \ldots, x_n$.

To judge the likely accuracy of our estimate $\hat{\theta}$ we would like to know the *sampling distribution* of $\hat{\theta}$. The sampling distribution of $\hat{\theta}$ would tell us the *bias* of $\hat{\theta}$ (the systematic error) and the likely size of the *chance error* of $\hat{\theta}$ (the unsystematic error). The sampling distribution would tell us the likely accuracy of $\hat{\theta}$ as an estimator of θ.

If the sampling distribution of $\hat{\theta}$ is approximately normal, then we estimate θ to be around

$$\hat{\theta} - (\text{estimated bias}),$$

give or take SE($\hat{\theta}$) or so.

Many estimators $\hat{\theta}$ have sampling distributions that are approximately normal. But often we do not know the bias of $\hat{\theta}$ and we do not have a simple formula for the standard error of $\hat{\theta}$. Consider, for example, the sample standard deviation S as an estimator of the theoretical standard deviation σ. For large n the sampling distribution of S is approximately normal. But S is slightly biased and we do not have a simple formula for the bias of S or the standard error SE(S).

We can approximate the sampling distribution of an estimator $\hat{\theta}$ and estimate the bias and the SD of $\hat{\theta}$ by *bootstrapping*. We create a bootstrap box containing the original data set

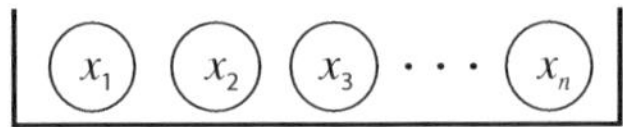

We draw n times at random with replacement from the bootstrap box to get the simulated observations

$$x_1^*, x_2^*, x_3^*, \ldots, x_n^*$$

We compute the simulated estimate $\hat{\theta}^*$ from the simulated observations using the same formula or algorithm that gave us our original estimate $\hat{\theta}$ from the original data set. If we bootstrap 10,000 $\hat{\theta}^*$-values, then the experimental distribution of the $\hat{\theta}^*$-values will resemble the sampling distribution of $\hat{\theta}$. (We hope.) We use the sample standard deviation of the simulated $\hat{\theta}^*$-values to estimate SD($\hat{\theta}$).

Let θ_{box} denote the value of the parameter for the bootstrap box. For example, if $\theta = \sigma$, then $\theta_{\text{box}} = \sigma_{\text{box}}$. If $\overline{\theta}^*$ denotes the sample mean of the simulated $\hat{\theta}^*$-values, then

$$\overline{\theta}^* - \theta_{\text{box}}$$

is an estimate of the bias of $\hat{\theta}$:

$$E(\hat{\theta}) - \theta.$$

CHAPTER REVIEW EXERCISES

1. In this exercise you have to judge how close the sample proportion $\hat{p}$ typically is to the population proportion p. You will see that the accuracy of $\hat{p}$ as an estimator of p depends on the sample size n.

   Assume that the proportion p of Kentucky residents who are 65 years old or older equals 1/8. (See Example 5.30.) Let $\hat{p}$ denote the proportion of persons 65 years old or older in a simple random sample of n Kentucky residents. For each of the specified sample sizes, n, use the box model in Example 5.30 to simulate 10,000 $\hat{p}$-values. Note the shape of the histogram for the 10,000 simulated $\hat{p}$-values. Note whether the mean of the 10,000 simulated $\hat{p}$-values is close to $p = 1/8$.

   a. $n = 25$
   b. $n = 100$
   c. $n = 400$
   d. $n = 1600$
   e. Refer to parts (a)–(d). Does $\hat{p}$ appear to be systematically smaller than p, appear to be systematically greater than p, or appear to vary randomly around p? Explain.
   f. Refer to parts (a)–(d). Judging from the standard deviations of the simulated $\hat{p}$-values, what happens to the variability of $\hat{p}$ if we quadruple the sample size n?

2. We shall investigate whether the sample standard deviation S is an unbiased estimator of the theoretical standard deviation σ.

Consider the box below, which contains a total of 26 balls:

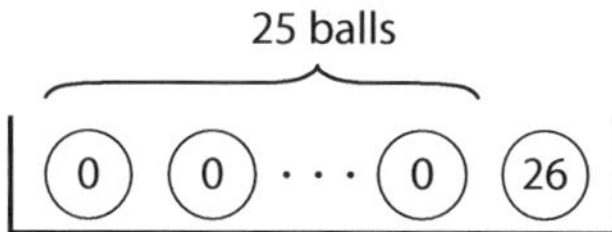

a. Verify that $\sigma_{\text{box}} = 5$. (See Section 6.3.)
b. Let S denote the sample standard deviation of $n = 10$ numbers drawn at random with replacement from the box. To investigate whether

$$E(S) = \sigma_{\text{box}} = 5,$$

simulate 10,000 values of S. What is the mean of the simulated values? What is the SD of the simulated values?
c. Estimate $E(S)$.
d. Put a give-or-take number on your estimate. (Compute the standard error of your estimate.)

3. We can simulate a gambler's gain on a \$1 bet on "red" in roulette by drawing once from the box

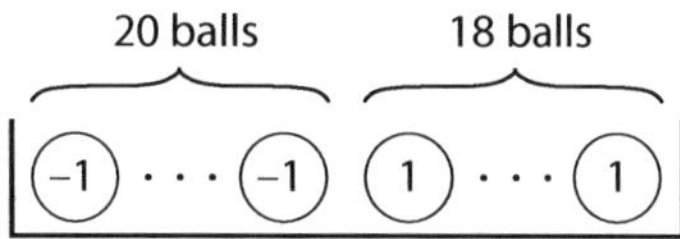

(See Exercise 2 in Section 5.5). Here

$$\mu_{\text{box}} = -0.0526 \text{ and } \sigma_{\text{box}} = 0.9986.$$

a. Use the CLT to estimate the probability that the gambler makes money on 25 independent \$1 bets on "red." (Estimate $P(\overline{X} > 0)$, where $\overline{X}$ denotes the gambler's average gain on the 25 bets. See Example 8.11.)
b. Simulate 10,000 values of $\overline{X}$ to estimate $P(\overline{X} > 0)$. Put a give-or-take number on your estimate. (Compute the standard error of your estimate.)

4. Let $\overline{X}$ denote the sample mean of a simple random sample of $n = 10$ numbers drawn at random one at a time without replacement from the population $\{1, 2, 3, 4, \ldots, 24, 25\}$.
a. Determine $\mu_{\overline{X}}$ and $\sigma_{\overline{X}}$, using the fact that $\mu_{\text{pop}} = 13$ and $\sigma_{\text{pop}} = \sqrt{52}$.
b. Use the CLT to estimate $P(12.25 < \overline{X} < 19.25)$.
c. Simulate 10,000 values of $\overline{X}$ to estimate $P(12.25 < \overline{X} < 19.25)$. Determine the mean and the standard deviation of the 10,000 simulated $\overline{X}$-values.
d. Describe the shape of the experimental histogram for the simulated $\overline{X}$-values. (You may have to rescale the axis. Start at 6. Use increments of size 3 and nine minor ticks.)

5. We can use the box in Example 5.29 to simulate the proportion $\hat{p}$ of girls among $n = 625$ newborn babies.
a. Use the CLT to estimate the probability that more than half of the babies are girls. (Estimate $P(\hat{p} > 0.5)$.)
b. Simulate 10,000 values of $\hat{p}$ to estimate $P(\hat{p} > 0.5)$. Put a give-or-take number on your estimate. (Compute the standard error of your estimate.)

6. Nine independent measurements were made on the tensile strength of a certain metal alloy. In increasing order, the measurements were

$$72,\ 88,\ 97,\ 105,\ 111,\ 118,\ 129,\ 139,\ 153.$$

Here $\overline{x} = 112.4$ and $s = 25.5$.
a. Estimate the mean tensile strength of the alloy.
b. Put a give-or-take number on your estimate. (Compute the standard error of your estimate.)

7. a. Estimate the median tensile strength of the alloy in Exercise 6.
   b. Bootstrap 10,000 values of the sample median to put a give-or-take number on your estimate of the median tensile strength.

8. Use the data in Exercise 6 to bootstrap 10,000 values of the sample standard deviation S.
   a. Estimate $\mathrm{SD}(S)$.
   b. Estimate the bias of S. (*Hint:* $\sigma_{\text{box}} = 24.00$.)

9. An office supply store receives a large shipment of inexpensive ball-point pens from a new supplier. To estimate the proportion of pens that are defective, the store manager tests a random sample of 225 pens and finds 25 to be defective.
   a. Estimate the proportion of all pens that are defective.
   b. Put a give-or-take number on your estimate. (Compute the standard error of your estimate.)

10. A company manufactures car windows that are attached by studs. A pull-out test is performed to determine the force required to pull a stud out of a window. Thirty measurements were made of the required force. In increasing order, the measurements were

120.3	128.2	132.5	135.4	139.2
123.6	128.7	132.9	135.8	139.9
124.9	129.4	133.4	136.3	140.7
126.0	130.8	133.9	137.1	142.5
127.1	131.3	134.4	138.0	143.8
127.9	131.9	135.0	138.9	148.1

    Here $\overline{x} = 133.60$ and $s = 6.34$.
    a. Estimate the mean force required.
    b. Put a give-or-take number on your estimate. (Compute the standard error of your estimate.)

11. a. Estimate the median force required to pull out one of the studs discussed in Exercise 10.
    b. Bootstrap 10,000 sample medians to put a give-or-take number on your estimate of the median force required. Describe the shape of the experimental histogram for the 10,000 simulated values. List the mean and the standard deviation of the simulated values.

12. Use the data in Exercise 10 to bootstrap 10,000 values of the sample standard deviation S.
    a. Describe the shape of the experimental histogram for the 10,000 simulated values.
    b. Estimate $\mathrm{SD}(S)$.
    c. Estimate the bias of S. (*Hint:* $\sigma_{\text{box}} = 6.24$.)

PROFESSIONAL PROFILE

Dr. John Snow
1813–1858
Physician,
Pioneer Epidemiologist

Map drawn by Dr. John Snow. The map shows how clusters of cholera cases were centered near a particular water pump on Broad Street during the 1854 epidemic in the Soho section of London.

Dr. John Snow

A mid-nineteenth-century British physician, Dr. John Snow was an early student of anesthesia and one of the first physicians to set doses of ether and chloroform for surgical anesthesia. He personally administered chloroform to Queen Victoria during the births of her last two children.

Throughout his professional life, Snow sought to collect data and analyze the results. His work made him a skeptic of the most popular theory of disease of his day, the miasma theory. This theory held that illness was caused by polluted "bad air." In 1849 Snow published an essay suggesting that water might be the carrier for cholera.

In 1854 cholera broke out in Soho. With the help of the Rev. Henry Whitehead, Snow interviewed victims and family members of the deceased. He soon discovered that a very high number of the victims had used a public water pump on Broad Street (today Broadwick Street). He drew a spots map to clearly show the clusters of cholera victims and their proximity to the suspect well. Snow believed that cholera was spread when people consumed water contaminated by sewage. He also used statistics to show a connection between water sources and cholera cases. The results were convincing enough that the local council disabled the well by removing the pump handle. Although, Snow acknowledged, the cholera cases were going down by the time the pump handle was removed, the cholera epidemic truly subsided once the well was no longer in use.

One of Snow's initial opponents was William Farr, a member of the General Register Office, responsible for collecting official medical statistics in England and Wales. As Snow collected his data in 1854, Farr collected his own. Farr hoped to prove the miasma theory, only to find that the data supported Snow's water contamination theory. Farr then became a supporter of Snow's ideas and applied that knowledge during another outbreak of cholera in 1866.

Today Snow's 1854 investigation is usually regarded as the founding event of epidemiology.

9

An Introduction to Hypothesis Testing

You don't have to eat the whole ox to know the meat is tough.
Samuel Johnson, quoted by James Boswell

Objectives

After studying this chapter, you will understand the following:

- Formulating a null hypothesis and its alternative
- Using a test statistic to assess the strength of evidence supporting a real-world claim
- Understanding level of significance and P-value
- Testing the hypothesis of a treatment effect in a randomized controlled experiment
- Testing hypotheses about one population proportion
- Testing hypotheses via the central limit theorem and five-step simulation methods

KEY PROBLEM

Does Aspirin Prevent Heart Attacks?

An article in the *New York Times* of January 27, 1988, described the results of a study on aspirin's ability to prevent heart attacks. The study focused on a group of 22,071 healthy males at least 40 years old who were followed over 5 years. The men were randomly divided into two groups. The 11,037 men in the treatment group took one children's aspirin tablet (81 mg) every other day for nearly 5 years. The 11,034 men in the control group received placebos that looked and tasted like children's aspirin.

Heart attacks often occur when blood clots form somewhere in the body and travel in the blood stream to the heart, where they choke off the heart's blood supply, thereby killing a portion of the heart muscle. Aspirin works as a blood thinner, making the blood less likely to clot. Aspirin may also reduce arterial inflammation, a suspected cause of clogged arteries.

Over the 5 years, 104 of the men in the aspirin group had a heart attack, compared to 189 heart attacks in the placebo group. Could this difference be explained as nothing more than chance variation, or is the difference too big to be explained by chance variation alone? Can we conclude that (at least for healthy middle-aged males) a children's aspirin every other day reduces the risk of a heart attack?

There were 80 strokes in the aspirin group and 70 in the control group. Could this difference be explained as chance variation, or is the difference too big?

INTRODUCTION

In Chapter 5 we learned how to model a random phenomenon by listing all possible outcomes and assigning probabilities to each of them. We learned that this list of all possible outcomes and their corresponding probabilities is called a probability distribution. We also learned how to set up a theoretical box model for any specified discrete probability distribution. With such box models we study chance variation by using the computer to simulate sampling from the distribution.

In Chapter 8 we studied the chance variation of estimators. Questions such as "Roughly by how much does an estimator $\hat{p}$ typically differ from the parameter p that we are trying to estimate?" can be answered by simulating a large number of values of $\hat{p}$ and computing the sample SD of the simulated $\hat{p}$-values.

test of significance

hypothesis testing

In this chapter we shall study chance variation in the context of a **test of significance**. For example, the data in the Key Problem suggest that aspirin reduces a person's risk of a heart attack. We shall use **hypothesis testing** to decide whether the evidence provided by the data is so strong that it "proves" beyond reasonable doubt that aspirin reduces the risk of heart attacks. We shall use both box model simulation and classical distribution theory (in particular, the central limit theorem for $\hat{p}$) to perform such tests.

Example 9.1

In the congressional election of November 8, 1994, 39% of all eligible Americans voted. During the next 2 days Yankelovich Partners, Inc., conducted a telephone poll for Time/CNN.

They interviewed a random sample of 800 Americans of voting age to find out why people had voted the way they had or why they had not voted. A surprising 56% of the persons interviewed told the pollster that they had voted. Can this difference—only 39% of the population voting, but 56% of the sample claiming to have voted—be explained by chance variation, or can we conclude that some of the persons interviewed lied when they claimed that they had voted? Assume the poll was a simple random sample.

Solution: Let p denote the proportion of eligible Americans who voted on November 8, 1994. This is a population parameter, with known value $p = 0.39$. We shall simulate a simple random sample of 800 Americans of voting age and let $\hat{p}$ denote the proportion of people in the sample who voted. On November 8, 1994, around 74 million Americans voted and around 116 million Americans of voting age did not vote. This suggests simulating $\hat{p}$ by drawing 800 times at random without replacement from the following box model:

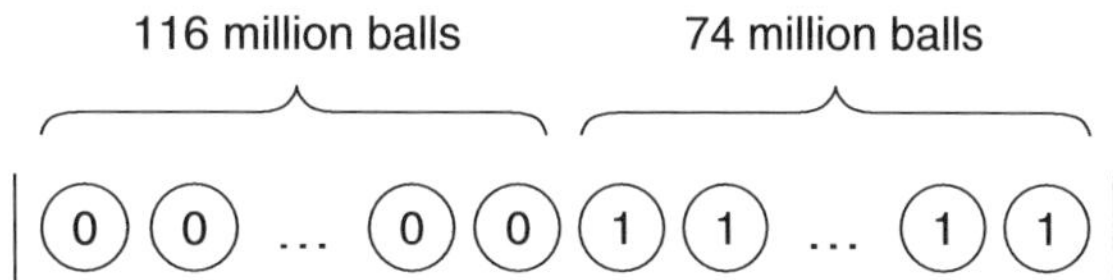

Each American of voting age who did not vote corresponds to a ball with a 0 on it, whereas each of those who voted corresponds to a ball with a 1 on it. The sum of the 800 numbers drawn corresponds to the number of people in the sample who voted. The average of the numbers on the sampled balls is the proportion $\hat{p}$ of people in the sample who voted.

Because the sample is so small compared to the population, it makes virtually no difference whether we draw the 800 balls with or without replacement. We shall therefore draw 800 times from the box *with* replacement to carry out one simulation. In that case it makes no difference if we replace our original box of 190 million balls with a much smaller box with the same proportions of 0s and 1s. Thus, noting that (116/74) = (58/37), we shall draw 800 times at random with replacement from the following box:

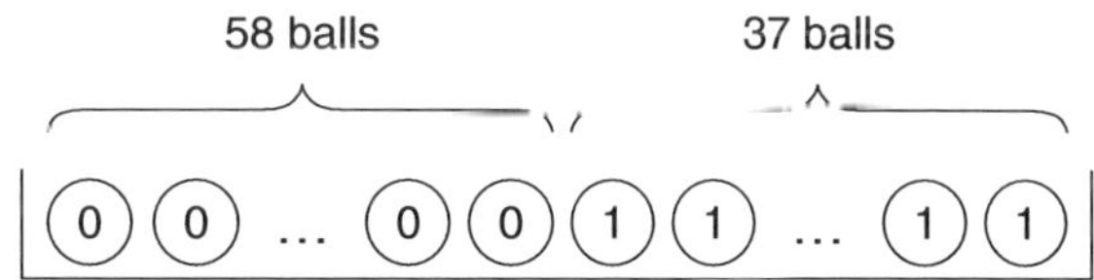

Note that this sampling scheme involving 95 balls is more practical, whether drawing by hand from an actual box or by computer, than working with 190 million balls. In fact, as a rule, we prefer to draw from a box with as few balls as possible. For each computer simulation of 800 draws we shall compute:

$$\hat{p} = \text{mean of the 800 numbers drawn.}$$

Ten thousand simulations yielded the experimental distribution of $\hat{p}$ given in Table 9.1.

Table 9.1 Proportion of People in Sample Who Voted

Simulated $\hat{p}$	Experimental frequency	Simulated $\hat{p}$	Experimental frequency
0.32	0	0.40	1905
0.33	7	0.41	1222
0.34	32	0.42	537
0.35	180	0.43	165
0.36	513	0.44	42
0.37	1165	0.45	6
0.38	1951	0.46	1
0.39	2274		

Notice that 0.56 is not among the simulated $\hat{p}$-values. Moreover, not even one of the simulated sample proportions came close to the observed $\hat{p} = 0.56$. In fact, one can show through theoretical calculations that the chance of getting a sample proportion $\hat{p}$ as high as the one Yankelovich got, that is, the probability

$$P(\text{simulated } \hat{p} \geq 0.56),$$

is incredibly small. There is less than one chance in a thousand billion billions that $\hat{p} \geq 0.56$ in a simple random sample of 800 balls from a box with its proportion of 1s being 0.39.

In other words, it would take a sampling miracle—the equivalent of a monkey randomly typing Lincoln's Gettysburg Address! So, we conclude that it is out of the question that 56% of the people interviewed by Yankelovich actually voted. Some of the people interviewed may have been too embarrassed to admit that they had failed to do their civic duty to vote. Others may have wanted to pretend that they were part of what was then called the "Republican Revolution"—even though they had not bothered to go to the polls. All we can say for sure is that our *statistical test proves* beyond any doubt that less than 56% of the people interviewed by Yankelovich did in fact vote.

❐ ❐

Statisticians have invented the *test of significance* to answer the question: *Was it possibly due to chance alone, or clearly due to more than chance?* In Example 9.1 we provided statistical proof that chance variation alone could not explain the difference between the 39% of the population who voted and the 56% of the sample who claimed to have voted. In the next example the question is again: Could what happened be explained by *chance alone*, or was it clearly due to *more than chance?*

Example 9.2

A labor dispute has arisen concerning the allegedly discriminatory way that 20 laborers at a construction site were given their job assignments. Six of the 20 job assignments were considered highly undesirable, whereas the remaining 14 jobs were considered desirable. The dispute was triggered by the fact that all four minority laborers working on the site were given undesirable job assignments. Could this have happened just by chance? Were the jobs assigned randomly without regard to race, or do we have statistical evidence of an injustice?

Solution: Imagine that we randomly assign the 20 jobs to the 20 laborers. Let X denote the number of undesirable jobs assigned to the 4 minority workers in this random assignment. We can simulate X by drawing four times at random without replacement from the box:

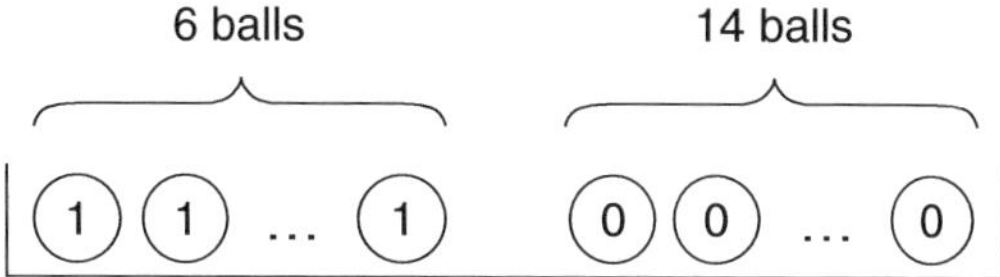

The logic behind how the box was built is as follows: Corresponding to each undesirable job assignment there is a ball with a 1 on it; corresponding to each desirable job assignment there is a ball with a 0 on it. Each of the four minority workers draws a ball (a job assignment) from the box. This model presumes that job assignments are done totally at random. The labor dispute is over whether this model is correct or incorrect! The sum of the four numbers drawn corresponds to the number of undesirable jobs assigned to the four minority workers. Ten thousand simulations (each consisting of job assignments to the four minority laborers) yielded the data in Table 9.2.

Our experimental probability of the event of interest is

$$\hat{P}(\text{all four minority workers get undesirable jobs}) = \hat{P}(X = 4) = 32/10{,}000 = 0.0032$$

Thus, although it is not impossible that a random assignment of jobs would give each of the four minority workers an undesirable job, it is *highly unlikely*. The odds are very much against it. So we conclude statistically that the jobs were not assigned at random. Note that our statistical test does not tell us how the jobs were nonrandomly assigned. Did seniority play a role? Were the 20 laborers equally qualified? Was discrimination indeed involved? The statistical test does not answer those questions. The test only tells us that the jobs were not assigned at random. It would be up to the courts or a mediation panel to decide whether the nonrandom assignment of the jobs represents an injustice.

Table 9.2 Number of Undesirable Jobs Assigned to the Four Minority Workers

Number of undesirable jobs assigned	Experimental frequency
0	2055
1	4523
2	2802
3	588
4	32

❐ ❐ ❐ ❐ ❐ ❐ ❐ ❐ ❐ ❐ ❐ ❐ ❐ ❐ ❐ ❐ ❐ ❐ ❐ ❐

In the next sections we shall introduce the rather technical language used in tests of significance: *null hypothesis, alternative hypothesis, test statistic, level of significance, and P-value.*

9.1 THE NULL HYPOTHESIS AND THE ALTERNATIVE HYPOTHESIS

research hypothesis

In the Key Problem the medical **research hypothesis,** also referred to as the claim, is that taking a children's aspirin every other day reduces a middle-aged man's risk of a heart attack. The data seem to support this hypothesis: The men in the treatment group had a lower rate of heart attacks than the men in the control group. Because we are looking at a randomized controlled experiment (see Section 4.1), there are only two possible explanations for these observed different rates of heart attacks:

Research Hypothesis: Aspirin reduces the risk of a heart attack.

Chance Explanation: The observed difference in heart attack rates is due to chance variation alone.

null hypothesis
H_0
alternative hypothesis
H_A

In this setup, statisticians call the chance explanation the **null hypothesis.** It nullifies the research hypothesis and is denoted by $\boldsymbol{H_0}$, the subscript suggesting "null." To prove the research hypothesis, we have to rule out the chance explanation. Following tradition, we call the research hypothesis **the alternative hypothesis** and denote it by $\boldsymbol{H_A}$, the subscript A suggesting "alternative." In summary, in the Key Problem we hope to establish the research claim

$$H_A : \text{Aspirin reduces the risk of a heart attack}$$

by ruling out the null hypothesis

$$H_0 : \text{The observed difference in heart attack rates is due to chance alone.}$$

In stating these two hypotheses, we use the standard statistical notation of writing "H_A :" and "H_0 :" followed by a statement of the corresponding hypothesis.

In statistical inference, we use information from random samples to draw conclusions about populations, as we did in Chapter 8 in the context of estimation. Therefore, it is important to distinguish between numerical indices that describe a characteristic of a population and those that describe a characteristic of a sample.

parameter
population
statistic
sample

Recall that a **parameter** is a number that describes some numerical characteristic of a **population.** It is a fixed number, but we do not know its value unless we examine the entire population. A **statistic** is a number that describes a numerical characteristic of a **sample.** Its value is computed from the sample using an appropriate formula, but its value varies from sample to sample. That is, a statistic is a random variable, such as $\bar{X}$ or $\hat{p}$. In Chapter 8, we learned that we can use sample statistics to estimate corresponding unknown population parameters. In this chapter, we will learn how to use a statistic to test a statistical hypothesis.

Example 9.3

Consider the population of all undergraduates at a large university. The proportion p of the undergraduates who smoke is an example of a population parameter. The mean family income μ of all undergraduates is another example of a population parameter. If we take a simple random sample of 100 undergraduates, then the proportion $\hat{p}$ of undergraduates in the sample who smoke is a statistic, as is the mean family income $\overline{X}$ of the students in the sample.

❒ ❒ ❒ ❒ ❒ ❒ ❒ ❒ ❒ ❒ ❒ ❒ ❒ ❒ ❒ ❒ ❒ ❒ ❒ ❒

test statistic

In hypothesis testing we often express the null hypothesis and the alternative hypothesis in terms of a population parameter. Further, we usually carry out the hypothesis test using a statistic to make our decision. Such a statistic is called a **test statistic.**

Example 9.4

Breast cancer rates are much higher in the affluent West than in the Third World. (See Example 3.1 in Chapter 3.) Heredity no doubt plays a role. But migrant studies suggest that lifestyle and environmental factors also contribute to a person's cancer risk. For example, in the 1980s there were over 200,000 Puerto Rican–born women living in New York City. Most arrived between 1945 and 1960. Between 1980 and 1986, 62 breast cancer cases were diagnosed in a random sample of 10,500 Puerto Rican–born women who lived in New York City. The breast cancer rate observed in the New York sample, $\hat{p} = 62/10{,}500 = 0.0059$, was a lot higher than the corresponding breast cancer rate for women living in Puerto Rico, whose population value was accurately known from government statistics to be 0.0035. Can this difference be explained by chance variation alone, or can we conclude that women of Puerto Rican descent living in New York City have a higher breast cancer risk than the 0.0035 rate for women living in Puerto Rico?

Solution: Let p denote the proportion of the *entire population* of Puerto Rican–born women living in New York City during 1980–1986 who were diagnosed with breast cancer at the time of the study. Using standard statistical notation, our *research hypothesis* is written as

$$H_A : p > 0.0035.$$

The null hypothesis, which nullifies our research hypothesis, is

$$H_0 : p \leq 0.0035.$$

As often happens, the null hypothesis involves an inequality; thus more than one value of the parameter of interest will satisfy the inequality. For instance, both $p = 0.0025$ and $p = 0.0035$ satisfy the null hypothesis. To carry out such a hypothesis test, this null hypothesis must be replaced by a null hypothesis that specifies a single value of the parameter.

replacing a many-valued null hypothesis by a single-valued one

How do we choose this single null-hypothesis parameter value? Our guiding principle is simple. To prove the research hypothesis made about the population parameter we must rule out the null hypothesis. So, if there are multiple possible values of the parameter for which the null hypothesis can be true, we must have strong enough evidence to reject *every one of these values* in order to have strong enough evidence to reject the null hypothesis. As it turns out, there will always be one value of the parameter under the null hypothesis that is more difficult to reject (that is, rule out) than all the other values. If the evidence is strong enough to reject this most-difficult-to-reject null-hypothesis value, then the evidence is *even stronger* that all the other null-hypothesis parameter values should be rejected. This most difficult value is used as the null-hypothesis value.

In our example, the single parameter value under the null hypothesis that will be most difficult to reject in favor of the alternative research hypothesis is the one whose value is closest to the alternative hypothesis parameter values, namely

$$H_0 : p = 0.0035.$$

❐ ❐

In general, whenever the null hypothesis is written as an inequality with respect to a certain parameter value, we will simply rewrite it as an equality, noting that this specified value is always chosen to be the one closest to the alternative hypothesis set of parameter values and hence the most difficult to reject.

Section 9.1 Exercises

1. For each of the following examples, identify the sample and the population.
   a. A poll of 40 college freshmen is taken to find out the views of the freshman class on lowering the drinking age from 21 to 18.
   b. A few small pieces of rock from a river valley region are analyzed at a laboratory to see how much gold they contain.
   c. Ecologists are worried about a certain parasite infecting tree frogs in northern Minnesota. Twenty frogs are examined for the presence of the parasite.
   d. Scientists are concerned about the salinity (saltiness) of the Gulf of Mexico. Small amounts of water from 30 locations are analyzed.
   e. A building inspector wants to see whether the concrete being used is of sufficient quality. She takes a small amount of mix from a truckload every morning for 5 days and sends the samples to the laboratory for analysis.
   f. Lisa is believed to have high blood pressure. Her doctor instructs her to measure her blood pressure at 20 random times during the week.

2. What are the different symbols we use to distinguish a population proportion from a sample proportion?

3. A state's Department of Transportation claims, "More than three out of four drivers in our state wear seat belts." To test this claim, a statistician interviews a random sample of 300 drivers.
   a. Identify the population about which the claim is made.
   b. What is the population parameter that is referred to in the claim?
   c. Rephrase the claim and the null hypothesis in terms of the population parameter.
   d. What is the sample statistic that the statistician will calculate?

4. For each of the following examples, identify the population and the parameter involved, state the claim in terms of the population parameter (i.e., in terms of the population proportion, p), state the null hypothesis in words, and state the null hypothesis in standard statistical notation.
   a. An organic farming group claims that more than 75% of consumers purchase organic produce on a regular basis.
   b. A polling company claims that more than two-thirds of the public favor reducing the size of the military budget.
   c. A social services organization claims that over 20% of American children are living in poverty.
   d. A public transport advocacy group claims that the majority of Americans living in cities would use public transport if it were available.

5. For each situation in Exercise 4, give an example of a statistic that could be computed from a sample to help determine whether the null hypothesis should be rejected or not.

6. A newspaper claims that over 40% of college undergraduates pull at least one "all-nighter" before they obtain their bachelor's degree.
   a. State the claim in symbolic form, state the null hypothesis in words, and, finally, state the null hypothesis in symbolic form using standard statistical notation.
   b. A statistician performs a statistical study on a randomly selected sample of recent graduates. In this sample, the proportion who pulled at least one all-nighter during the time they were undergraduates turned out to be 0.45. The statistician determines that the probability of observing a value of 0.45 or more in such a sample is 0.20 when assuming the null hypothesis is true. Discuss whether this constitutes strong evidence for rejecting the null hypothesis.

7. Find two or more examples from newspapers or magazines that involve claims or hypotheses that might be testable by observing a real data sample and testing the null hypothesis. Name the null hypothesis in each case and give an example of a statistic that could be used in testing it.

9.2 TESTS FOR A POPULATION PROPORTION

As mentioned above, the kind of statistical decision making we are describing is called hypothesis testing. Many statistical problems are hypothesis-testing problems. In this section we will learn how to test research hypotheses (claims) about the proportion of units in a population (either real or conceptual) that exhibit a particular property or quality.

Claims about population proportions underlie many of the most hotly debated issues that confront our society. Questions that involve single population proportions include

- What proportion of children in America have no health insurance?
- What proportion of "partial-birth" abortions are performed to save the mother's life?
- What is the breast cancer rate for women who have had an abortion? (Abortion foes claim that women increase their risk of breast cancer if they have an abortion.)
- What proportion of drugs illegally imported into America is intercepted by drug enforcement officers?

In testing hypotheses about population proportions, we formulate the null hypothesis by assuming that the research claim is false. Recall that these claims are more formally referred to as alternative hypotheses. The null hypotheses considered in this section will be of the form

$$H_0 : p = p_0,$$

where p_0 is the value of the population proportion assumed by the null hypothesis. (See Example 9.4; keep in mind that H_0 may have initially been stated as an inequality.) We will then evaluate the probability that a sample proportion $\hat{p}$ generated under the null hypothesis is at least as extreme in the direction of the alternative hypothesis as the sample proportion calculated from the real data sample.

In the first part of this section, we will use five-step simulation to test H_0. In the second part of this section, we shall use classical central limit theorem–based large-sample theory to test H_0.

box model–based simulation test of $H_0 : p = p_0$

Population Proportion Hypothesis Testing Box Model–Based Five-Step Simulation

We shall illustrate the box model–based simulation approach to hypothesis testing by completing the analysis of Example 9.4. We shall assess whether a sample proportion $\hat{p}$ as high as 0.0059 would be likely if the null hypothesis $H_0 : p = 0.0035 = 35/10{,}000 = 700/200{,}000$ were true for the 200,000 women of Puerto Rican descent.

Imagine a population of 200,000 women out of which 0.35%, or 700 women, are diagnosed with breast cancer. Consider a box with a 1-ball for each breast cancer victim and a 0-ball for each woman who is not diagnosed with breast cancer:

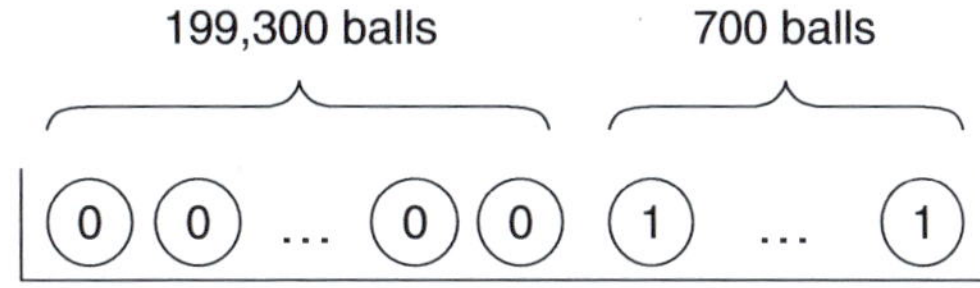

We can simulate the proportion $\hat{p}$ of breast cancer victims in a simple random sample of 10,500 women drawn from this population by drawing 10,500 times at random *without* replacement from the box and computing the proportion of 1-balls in the sample.

Because we are only sampling slightly more than 5% of the population, we will get almost the same results if we draw 10,500 times at random *with* replacement, in which case we can use the smaller box with the same proportions of 1-balls and 0-balls:

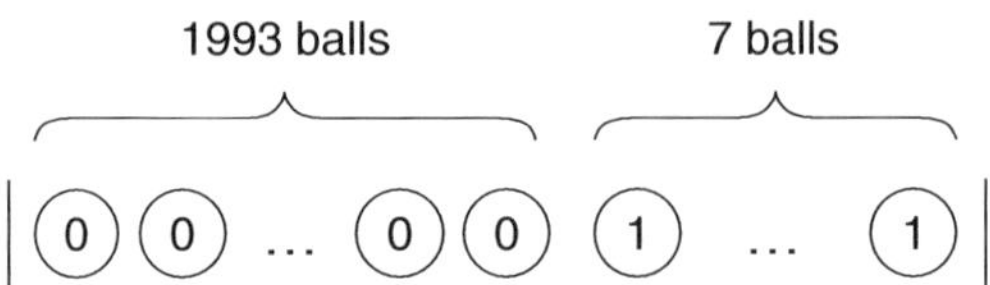

In fact, if we sample *with* replacement, then we can use any box of 0s and 1s for which the proportion of 1-balls equals the population proportion $p_0 = 0.0035$, specified by H_0.

To test H_0 we must answer the question: How likely is a sample proportion $\hat{p}$ as high as or higher than 0.0059 if the population proportion p is only 0.0035? If we had drawn a simple random sample of 10,500 women living in Puerto Rico, where the cancer rate is known to be $p = 0.0035$, could we have gotten a cancer rate in the sample as high as or higher than $\hat{p} = 0.0059$? We ran 100,000 simulations and got

$$\hat{P}(\text{simulated } \hat{p} \geq 0.0059 \mid p = 0.0035) = \frac{8}{100{,}000} = 0.00008.$$

In other words, only 8 of the 100,000 simulations yielded a simulated cancer rate as high as or higher than the actual cancer rate of 0.0059. So it would be *extremely* unlikely that we got a sample proportion $\hat{p}$ as high as 0.0059 if the population proportion p were only 0.0035, as the null hypothesis claims. Because the experimental probability is so small (0.00008), it would take a minor sampling miracle to get a sample rate as high as $\hat{p} = 0.0059$ if the population rate were only $p = 0.0035$ (H_0). In other words, the New York City population cancer rate must be higher than 0.0035, as claimed by our research hypothesis H_A.

The high breast cancer rate in the sample of Puerto Rican women living in New York City cannot be accounted for by chance variation alone. The evidence against H_0 is so strong that we conclude that the population of all Puerto Rican women living in New York City must have a higher risk than those living in Puerto Rico. We *reject* H_0 (the chance explanation) in favor of the research hypothesis H_A.

Note that we measured the strength of the evidence against H_0 and in favor of H_A by estimating the theoretical probability

$$P(\text{simulated } \hat{p} \geq \text{observed } \hat{p} \mid p = 0.0035).$$

P-value, P

The smaller this probability is, the stronger the evidence against H_0 is. This probability, which determines whether we can rule out the chance explanation H_0 or not, is known as the ***P*-value** of the test and is often denoted by ***P***.

Also note that the inequality in the event of interest (EOI),

$$\text{EOI} = (\text{simulated } \hat{p} \geq 0.0059),$$

goes in the same direction as the inequality in the alternative hypothesis (the research hypothesis),

$$H_A : p > 0.0035.$$

> The P-value of a statistical test tells us whether the observed data contradict H_0 or not. If the P-value is very small, then the data contradict H_0. If the P-value is not too small, then the data do not contradict H_0. The P-value (also called the observed significance level) is the probability of getting evidence against H_0 as strong or stronger than the evidence at hand—if H_0 were true.

The data in Example 9.4 proved that women of Puerto Rican descent living in New York City have a higher risk of breast cancer than women living in Puerto Rico. In the next example we shall compare the same New York City women of Puerto Rican descent with the rest of the female population of the United States.

Example 9.5

The breast cancer rate for the random sample of New York City women of Puerto Rican descent in Example 9.4, $\hat{p} = 62/10{,}500 = 0.0059$, was a bit lower than the corresponding breast cancer rate for the entire female population of the United States, which was accurately known to be 0.0073 at the time. Could this difference be the result of chance variation alone, or can we conclude that 20 years ago New York City women of Puerto Rican descent had a lower breast cancer rate than the rest of the female population of the United States?

Solution: As in Example 9.4, we let p denote the proportion of all Puerto Rican–born female residents of New York City who were diagnosed with breast cancer between 1980 and 1986. Our research hypothesis is

$$H_A : p < 0.0073.$$

The null hypothesis, which nullifies H_A, is

$$H_0 : p \geq 0.0073.$$

The most-difficult-to-reject null-hypothesis parameter value is the one p closest to H_A, namely $p = 0.0073$. We therefore change the null hypothesis to

$$H_0 : p = 0.0073.$$

If the breast cancer rate for all Puerto Rican–born women living in New York City were $p = 0.0073$, then we could simulate the proportion of breast cancer victims in a random sample of size 10,500 by drawing 10,500 times at random with replacement from the box

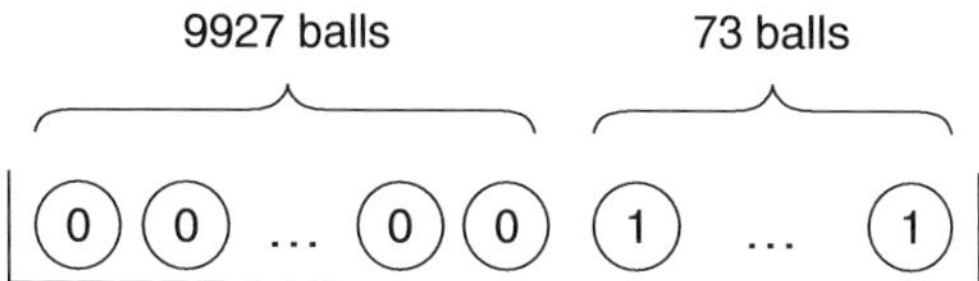

and computing the proportion $\hat{p}$ of 1-balls in the sample. The event of interest is

$$(\text{simulated } \hat{p} \leq \text{observed } \hat{p}) \text{ or } (\text{simulated } \hat{p} \leq 0.0059).$$

Note that the inequality in the event of interest goes in the same direction as the inequality in the alternative hypothesis (the research hypothesis claim):

$$H_A : p < 0.0073.$$

We ran 10,000 simulations and got

$$\hat{P}(\text{simulated } \hat{p} \leq 0.0059 \mid p = 0.0073) = \frac{527}{10{,}000} = 0.0527.$$

In other words, 527 of the 10,000 simulations yielded a simulated cancer rate as low as or even lower than the observed cancer rate of 0.0059. If we had drawn a simple random sample of 10,500 women from the entire female population of the United States, then there would have been a 5% chance of getting a cancer rate as low as or lower than the rate observed in the sample of Puerto Rican–born women. So while the cancer rate observed in the sample of Puerto Rican–born New York City women, $\hat{p} = 0.0059$, was lower than the national rate, we cannot completely rule out that this was due to chance alone. The cancer rate p for all Puerto Rican–born New York City women could have been as high as 0.0073. We cannot completely rule out (reject) H_0 (the chance explanation), and we therefore lack sufficient evidence to prove the research hypothesis beyond any doubt.

Of course, we do have some evidence against H_0—just not enough. Maybe the population of Puerto Rican–born women in New York City had a slightly lower cancer rate than the entire female population of the United States, and maybe the observed lower rate for the sample was just due to sampling variation and nothing more. We cannot tell which. Later in this section, we will discuss the issue of how strong the evidence against H_0 must be before we decide that we have "proof beyond a reasonable doubt" that H_A is true.

When we *fail to reject* H_0 we often say that we *accept* H_0. Note, however, that this does not mean that H_0 has been proven. It only means that we have *insufficient evidence* to rule out H_0. In summary,

- If we reject H_0, then we have proved H_A.
- If we fail to reject H_0, then we have proved nothing. We have insufficient evidence to tell for sure whether H_0 or H_A is correct. Although we often describe this as accepting H_0, it would be more precise to say we fail to reject H_0.

We can summarize our statistical analysis in Examples 9.4 and 9.5 as follows: Consider a population of size N, out of which an unknown proportion p has a characteristic of interest. In Examples 9.4 and 9.5 the parameter p of interest was the proportion of all Puerto Rican–born women living in New York City who were diagnosed with breast cancer between 1980 and 1986. In other examples p could be the proportion of American adults who favor a total ban on "partial-birth" abortions, or the proportion of American children who lack health insurance.

Assume that a simple random sample of size n has been drawn from the population, and let $\hat{p}$ denote the observed proportion of the sample that has the characteristic of interest. We shall view a statistical test for the population parameter p as a four-stage decision procedure, noting that these four stages are just the natural progression.

four stages of a hypothesis test

Stage I: Choose the null hypothesis and its alternative.

Stage II: Choose the null hypothesis box model and specify the number of draws from the box.

Stage III: Choose the test statistic and specify the event of interest. Either use simulation or distribution theory to estimate the P-value of the hypothesis test.

Stage IV: Make the decision to either reject or not reject the null hypothesis.

Stage I: Formulate the Null Hypothesis H_0 and the Alternative Hypothesis H_A

As we discussed in Section 9.1, the "alternative hypothesis" is often the research hypothesis (claim) that someone wants to prove. It could be $H_A : p < p_0$ or $H_A : p > p_0$. The null hypothesis nullifies the research hypothesis and should have the form

$$H_0 : p = p_0,$$

a single specified value for p, this single value allowing us to create an H_0 box model for the sample data.

Stage II: Choose a Null-Hypothesis-Based Sampling Box Model

The obvious choice of box model has N balls, one for each member of the population, with a 1-ball for each population member with the characteristic of interest and a 0-ball for each member without the characteristic. Under H_0, the proportion of 1-balls equals p_0:

H_0 box model

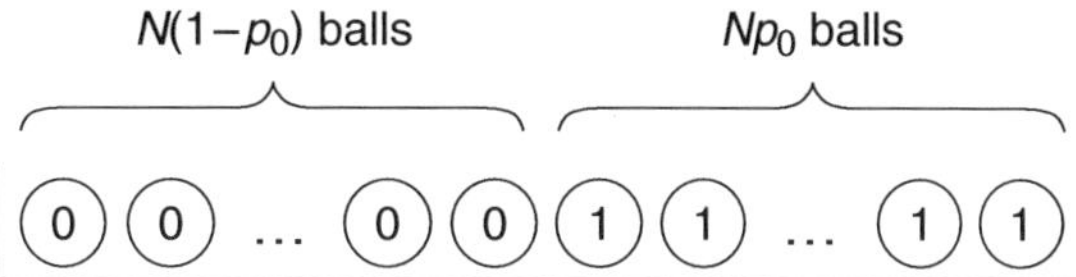

We can simulate the sample proportion by drawing n times (n denoting the sample size) at random without replacement from the box and computing the proportion of 1-balls in the sample.

If the sample size n is less than 5% of the population size N, then we will get almost the same results if we use sampling *with* replacement instead. If we sample with replacement, then we can use any box of 0s and 1s for which the proportion of 1-balls equals p_0, as specified by H_0. Note, however, that in Section 9.3 we shall consider several examples where the sample size is greater than 5% of the population size N. In this case, in order to authentically mimic the actual simple random sampling process, we should sample *without* replacement from the size N box shown above.

Stage III: Estimate the P-Value of the Hypothesis Test

observed significance level

We measure the strength of the evidence against H_0 by computing the P-value (also called the **observed significance level**) of the test. The P-value is the chance, when we are doing simulations that presume H_0 true, of getting evidence against H_0 as strong as or even stronger than the real data evidence at hand. As stated, we run the simulations assuming that H_0 is true, that is, using the H_0 box model of stage II. If

$$H_A : p > p_0,$$

then the event of interest is (simulated $\hat{p} \geq$ observed $\hat{p}$) and

$$P\text{-value} = P(\text{simulated } \hat{p} \geq \text{observed } \hat{p} \mid p = p_0).$$

If the H_A inequality is in the other direction, namely

$$H_A : p < p_0,$$

then the event of interest is (simulated $\hat{p} \leq$ observed $\hat{p}$) and

$$P\text{-value} = P(\text{simulated } \hat{p} \leq \text{observed } \hat{p} \mid p = p_0).$$

Notice that the direction of the H_A inequality is the same as the direction of the inequality for the P-value event. The smaller the P-value is, the stronger the evidence against H_0. We often denote the P-value simply as P.

Stage IV: Decide Whether to Reject H_0

How small should P be before we reject H_0? The standard procedure is to compare the P-value with some fixed threshold called the "level of significance." In this book we shall use a 0.05 threshold:

statistically significant

- If $P \leq 0.05$, the result is called **statistically significant,** and we reject H_0.
- If $P > 0.05$, the result is *not statistically significant*, and we do *not reject* H_0.

The smaller P is, the stronger the evidence against H_0.

highly significant
statistically significant at level 0.01
very highly significant
statistically significant at level 0.001

- If $P \leq 0.01$, the result is called **highly significant**, or **statistically significant at level 0.01.**
- If $P \leq 0.001$, the result is **very highly significant**, or **statistically significant at level 0.001.**

Note: In Example 9.4, the experimental P-value was 0.00008, *very highly significant.* In Example 9.5, the experimental P-value was 0.0527, *not quite statistically significant.*

large sample Z-test of $H_0 : p = p_0$

Population Proportion Hypothesis Testing Using the Normal Distribution (Large-Sample Case)

In Examples 9.4 and 9.5 we used simulation to estimate the P-value. Because the sample size n was so large in those examples, we could also have used the normal distribution and hence Appendix E to estimate the P-value.

central limit theorem for $\hat{p}$

Recall the **central limit theorem (CLT) for** $\hat{p}$ in Section 8.2: Suppose we draw n times at random from the stage II null hypothesis box of 0s and 1s. Let p_0 denote the proportion of 1-balls in the null-hypothesis box and $\hat{p}$ the proportion of 1-balls in the sample. Assume that $np_0 \geq 10$ and $n(1 - p_0) \geq 10$, as required for accurate application of the CLT. Also assume one of the following settings that allow use of the CLT:

- We are sampling with replacement.
- We are sampling without replacement, but the sample size n is less than 5% of the total number of balls, N, in the box.

standardized test statistic or Z-statistic

Then the sampling distribution of $\hat{p}$ is approximately normal with expected value $E(\hat{p}) = p_0$ and standard deviation $\text{SD}(\hat{p}) = \sqrt{p_0(1 - p_0)}/\sqrt{n}$. That is, the **standardized test statistic** or Z-statistic

$$Z = \frac{\text{statistic} - \text{parameter}}{\text{SD of statistic}} = \frac{\hat{p} - p_0}{\sqrt{p_0(1 - p_0)}/\sqrt{n}}$$

is approximately standard normal.

Let z_{OBS} denote the observed real data value of the Z-statistic,

$$z_{\text{OBS}} = \frac{\text{observed } \hat{p} - p_0}{\sqrt{p_0(1 - p_0)}/\sqrt{n}}.$$

To test H_0, we can simulate $\hat{p}$ by drawing n times from the null-hypothesis box described in stage II. We can estimate the P-value of the hypothesis test as follows. If the alternative hypothesis is $H_A : p > p_0$, then, as Figure 9.1 shows,

$$\begin{aligned} P\text{-value} &= P(\text{simulated } \hat{p} \geq \text{observed } \hat{p} \mid p = p_0) \\ &= P(\text{simulated } Z \geq z_{\text{OBS}}) \\ &\approx \text{area under the standard normal curve to the right of } z_{\text{OBS}}. \end{aligned}$$

Here the $\hat{p}$ random variable is viewed as being simulated. However, there is no simulation done because the CLT calculation of the P-value stands in place of doing simulations. That is, the CLT approach, when applicable, bypasses box model simulation.

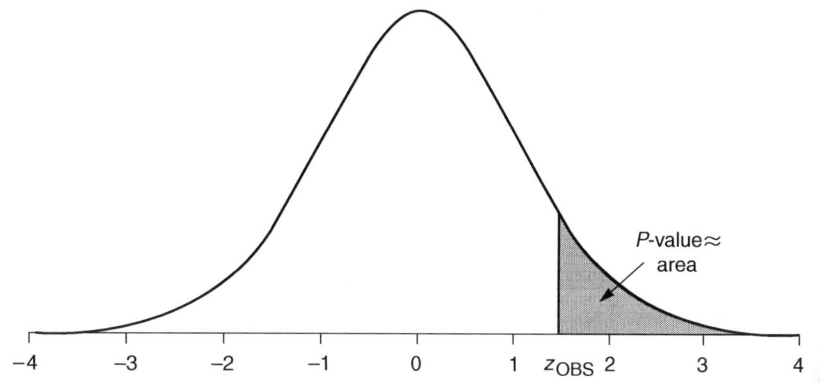

Figure 9.1 Estimating the P-value for $H_A : p > p_0$.

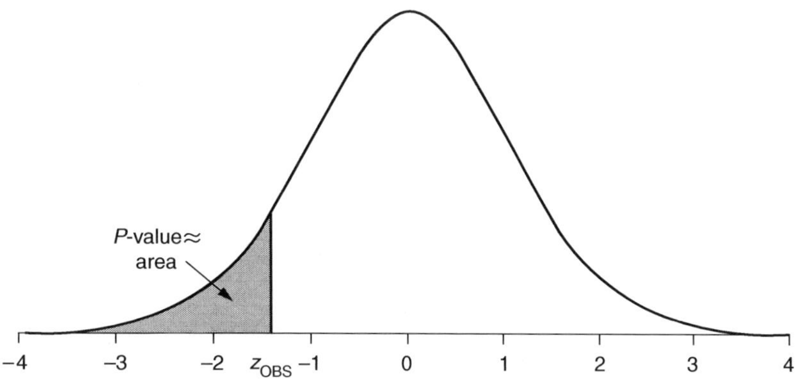

Figure 9.2 Estimating the P-value for $H_A : p < p_0$.

If the alternative hypothesis is $H_A : p < p_0$, then, as Figure 9.2 shows,

$$\begin{aligned} P\text{-value} &= P(\text{simulated } \hat{p} \leq \text{observed } \hat{p} \mid p = p_0) \\ &= P(\text{simulated } Z \leq z_{\text{OBS}}) \\ &\approx \text{area under the standard normal curve to the left of } z_{\text{OBS}}. \end{aligned}$$

In either case, if $P \leq 0.05$, then we reject H_0 in favor of H_A.

Comment: We thus have two methods for estimating the P-value for a test of the null hypothesis

$$H_0 : p = p_0.$$

- We can always estimate P by simulating 10,000 or more $\hat{p}$-values, drawing with or without replacement from the H_0 box, depending on the phenomenon we are modeling.
- Under the stated conditions ($np_0 \geq 10$, $n(1 - p_0) \geq 10$, and $n \leq 0.05N$ in the finite-population case), we can also apply the CLT and hence use the normal curve to estimate the P-value.

Note, however, that the classical CLT method often *underestimates* the actual P-value somewhat, especially if the sample size n is smaller than 200. The reason that we underestimate P is that we (and everyone else) are ignoring the so-called continuity correction discussed in Section 7.4 in connection with the normal approximation to a binomial distribution. Thus the simulation method actually estimates P more accurately when we perform a large number of simulations, such as 10,000, and the sample size is small (<200), as is sometimes the case.

So if the CLT approach produces a P-value that is significant (e.g., $P = 0.039$) but not highly significant, it would be wise, especially if $n < 100$, to use five-step simulation to get a more accurate P-value.

Example 9.6

box model
simulation-based test of $H_0 : p = p_0$

In a sample survey of 506 high school students carried out by the Roper Organization, 268 of the students did not know that the term *Holocaust* refers to the Nazi killing of over 6 million Jews during World War II. Does this prove that a majority (i.e., more than 50%) of American high school students don't know what *Holocaust* refers to?

Solution: We shall apply our four-stage decision procedure.

Stage I: Formulate the Null Hypothesis H_0 and Alternative Hypothesis H_A

Let p denote the proportion of high school students who don't know what *Holocaust* refers to. Our research hypothesis is

$$H_A : p > 0.5.$$

The corresponding null hypothesis is

$$H_0 : p = 0.5.$$

Stage II: Choose a Null-Hypothesis-Based Sampling Box Model

Clearly, the sample size $n = 506$ is less than 5% of the total number of high school students in the United States. We can therefore sample with replacement from any box of 0s and 1s for which the proportion of 1-balls equals $p_0 = 0.5$. We shall use the box:

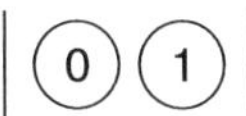

Stage III: Estimate the P-Value of the Hypothesis Test

We could simulate the proportion $\hat{p}$ of students who don't know what *Holocaust* refers to in a simple random sample of 506 high school students by drawing 506 times at random with replacement from the sampling box above and computing the proportion of 1-balls in the sample. The observed real data values are

$$\text{observed } \hat{p} = 268/506 = 0.5296,$$

$$z_{\text{OBS}} = \frac{\text{observed } \hat{p} - p_0}{\sqrt{p_0(1-p_0)}/\sqrt{n}} = \frac{0.5296 - 0.5000}{\sqrt{(0.5)(0.5)}/\sqrt{506}} = 1.33.$$

Because

$$np_0 = 506(0.5) = 253 \geq 10$$

and

$$n(1-p_0) = 506(1-0.5) = 253 \geq 10,$$

as Figure 9.3 illustrates, we can use the bell curve and Appendix E to estimate the P-value instead of actually carrying out the proposed simulations. Thus,

$$\begin{aligned} P\text{-value} &\approx \text{area under the standard normal curve to the right of } 1.33 \\ &= 0.0918. \end{aligned}$$

For comparison we ran 10,000 box model simulations and got

$$\hat{P}(\text{simulated } \hat{p} \geq 0.5296 \mid p = 0.5) = 0.0987.$$

The exact value is $P = 0.0986$. We note that, as explained earlier, the CLT-based normal curve estimate of P, that is, 0.0918, is a bit low.

Stage IV: Decide Whether to Reject H_0

Because P is greater than our 0.05 threshold, the result is *not statistically significant,* and we cannot reject H_0 in favor of H_A. We conclude that *roughly* 50% of American high school students don't know what *Holocaust* refers to, but our sample was too small to provide statistically conclusive evidence that it is more than 50%. Of course, for $H_0 : p_0 = 0.45$ versus $H_A : p_0 > 0.45$, we do get a very small P-value around 0.0002, which is very highly significant. Hence there is strong evidencee that over 45% of high school students at the time of the survey didn't know what *Holocaust* refers to.

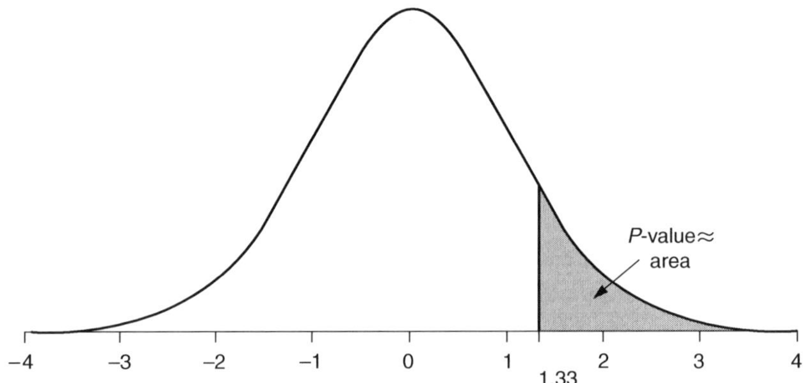

Figure 9.3 The P-value approximated by the CLT.

Section 9.2 Exercises

1. Several states are considering banning the use of cellular phones while driving because of the increased risk of traffic accidents associated with such use. State legislators who support such laws might claim that a large majority of their constituents are in agreement with them. Assume that a "large majority" is interpreted as "over 60%."

   Suppose one of these legislators surveys 20 randomly sampled constituents asking whether they think their state government should or should not pass a law making it illegal to use a cellular phone while driving, and that the results of the survey showed that 75% (15) of those surveyed did indeed think that their state should pass such a law.

   a. In words only, what is the population proportion under consideration in this problem?
   b. What is the claim being made about this population proportion?
   c. In terms of the population proportion, what is the null hypothesis?
   d. Why is the box model simulation approach preferred over the Z-statistic approach for testing the null hypothesis?
   e. To test the null hypothesis, we need to generate simulated samples that conform to the null hypothesis. On each simulation, we will generate a simulated sample having the same size as the original real data sample. What sample statistic will we measure on each simulated sample that we generate?
   f. The EOI for testing the null hypothesis will be based on comparing the sample statistic from the simulated sample to something. To what will we compare it? Why do we make this comparison?
   g. The results from 1000 simulations under the null hypothesis are shown in the following table. What are the most frequently occurring values of $\hat{p}$ in the simulated samples? What are the least frequently occurring values in the direction of rejecting the null hypothesis?

$\hat{p}$ observed in simulated sample	Frequency of occurrence	$\hat{p}$ observed in simulated sample	Frequency of occurrence
.00	0	.55	168
.05	0	.60	178
.10	0	.65	182
.15	0	.70	126
.20	1	.75	70
.25	1	.80	24
.30	5	.85	12
.35	7	.90	0
.40	26	.95	0
.45	75	1.00	0
.50	125		

h. Is the value of $\hat{p}$ observed in the real data sample among the least frequently observed simulated $\hat{p}$-values in the direction of rejecting the null hypothesis?
i. Decide whether to reject or not reject the null hypothesis and explain why you chose that option.
j. What would your answer have been if 80% (16) of the sample had indicated their state should pass such a law?

2. In Exercise 1, the simulated samples are generated from a box model. The choice of the box model is necessary to model the population from which the null hypothesis says the real data sample was drawn.
   Describe the choice for the box model and the type of sampling that should be done. Explain how this results in simulated samples that correspond to the null hypothesis.

3. In a simple random sample of 64 adults living in a large city, 25 persons were in favor of lowering the drunk-driving blood alcohol level from 0.10 to 0.08. Let p denote the proportion of all adults in that city who favor lowering the level.
   a. Test the null hypothesis $H_0 : p = 0.5$, where prior to data collection it was claimed that $p < 0.5$. Simulate 10,000 $\hat{p}$-values to estimate P. Do you reject H_0?
   b. Use the bell curve and Appendix E to estimate P. Is the bell curve estimate smaller or larger than the simulation estimate? Did you expect this? Why?

4. It is important to travelers that mass transit vehicles exhibit a high percentage of on-time performance. Suppose a regional train system boasts that its trains run on time over 80% of the time. To check out this claim, an investigative reporter might collect a sample of real data. Suppose a random sample of 40 train arrivals is observed, and 37 of them, that is, 92.5%, are observed to be on time.
   a. In words, what is the population proportion in this problem?
   b. What claim is being made about this population proportion?
   c. Why is a box model simulation approach preferred to a CLT approach in this problem?
   d. What is the null hypothesis?
   e. What choice of the box model should be used to generate simulated samples corresponding to the null hypothesis?
   f. What is the definition of one simulation from the box?
   g. Define the EOI and explain why this is of interest in regard to testing the null hypothesis.
   h. In 1000 simulations, 19 of them had a $\hat{p}$ as large as or larger than $37/40 = 0.925$. Explain what this means and state whether the null hypothesis should be rejected.

5. The president's press secretary claims that a majority of people approve of the president's performance. In a random sample of 2000 U.S. citizens over the age of 18, 53.5% approve of the president's performance. Test the null hypothesis that the population proportion is equal to 0.5. Estimate the P-value in two different ways.

6. In a random sample of 2000 residents of Illinois, 75% believe in the fairness of the jury system. Test whether the population proportion is equal to 0.8, as opposed to < 0.8, as has been claimed.

7. It is claimed that over 60% of voters favor a proposed constitutional amendment. A telephone poll of 20 voters results in 15 being in favor of the amendment. Should it be concluded, on the basis of this poll, that over 60% of the community favor the amendment? Use the simulated $\hat{p}$-values from Exercise 1.

8. Records from several years ago show that in a certain community 60% of the people owned their homes. An advocacy group for affordable housing claims that home ownership has declined since that time. A survey of 20 families, just taken, shows that 9 own their homes. Is this conclusive evidence that there has been a decrease in home ownership in the community as claimed by the advocacy group? Use the simulated $\hat{p}$-values from Exercise 1.

9. In a random sample of 60 college freshmen, 75% are in favor of lowering the drinking age from 21 to 18. Test the null hypothesis that the population proportion is equal to 0.7 or more.

10. In a random sample of 100 people, 70 are regular viewers of the television show *The Simpsons*. Test whether the population proportion is equal to 0.75 or less.

11. It is claimed that over 60% of the voters of Oregon favor a constitutional amendment. A random sample of 36 voters yields 30 favoring the amendment. Find the P-value. What is your conclusion?

9.3 COMPARING RATES FOR TREATMENT AND CONTROL GROUPS IN A RANDOMIZED CONTROLLED EXPERIMENT

Recall from Chapter 4 that in a randomized controlled experiment we assess the effect of a treatment (often a medical treatment) by contrasting the responses of units or subjects receiving the treatment with the responses of units or subjects not receiving the treatment. The difference in the responses is used to estimate the effect of the treatment. For this method of comparison to be successful, the only difference between the experimental units that receive the treatment (the treatment group) and the units that do not receive the treatment (the control group) must be the presence or absence of the treatment itself. The only way to ensure that this principle holds is to randomly assign the available units to the treatment and control groups. When a sufficiently large number of units is used in the study, random assignment will ensure that the two groups will be approximately alike in all respects except for whether or not they receive the treatment. Usually the set of units obtained for a medical or other randomized controlled experiment are not randomly selected from the population of interest. For example, physicians might recruit 200 volunteers (hence not a random sample) for a clinical trial of a new drug. The randomization needed occurs in the random assignment of 100 people to the treatment group and 100 people to the control group.

The standard method of analyzing data from a randomized controlled experiment is to assume that the treatment group and the control group were drawn independently and at random with replacement from two separate populations. But that is clearly not the way such experiments are done. First of all, the test subjects are volunteers, often from high-risk groups, and hence not at all a random sample from the general population. For example, the children who participated in the 1954 Salk vaccine field trial were volunteers from experimental areas with higher polio rates than the rest of the country. Second, when the available subjects are randomly assigned to either the treatment or control group, we do not sample with replacement. The treatment group is a random sample drawn without replacement from the available test subjects, and the control group consists of the remaining units (not chosen). So the treatment and control groups are not independent samples. Mathematically it can be shown that these two errors (pretending the two samples are independent and pretending that they are drawn with replacement) cancel each other out. But that seems like magic and is hard to explain! (See the discussion in Chapter 27, Section 3, of *Statistics* by Freedman, Pisani, and Purves, 4th edition.)

In this section we shall take a lesser-known but more intuitive approach to analyzing data from randomized controlled experiments. We shall consider only response variables with two possible values. For example, the child either gets polio or does not get polio, and the heart patient either survives 3 years or does not survive 3 years. But the box model approach presented here works for any response variable. For example, we might use this approach to analyze data from a randomized controlled experiment that compared how much two different blood pressure medications, on average, lowered the diastolic blood pressure of heart patients.

Consider a randomized controlled experiment with N test subjects (or experimental units) in which each test subject (or experimental unit) at the end of the experiment either will have or will not have a characteristic of interest (such as being a heart attack victim or a polio victim). Keep in mind that these subjects are probably not a random sample. Assume that N_T of the test subjects are randomly selected to make up the treatment group, and thus the remaining $N_C = N - N_T$ make up the control group.

For example, in the Key Problem there were $N = 22{,}071$ test subjects, out of which $N_T = 11{,}037$ were randomly assigned to the aspirin group and the remaining $N_C = 11{,}034$ were assigned to the placebo group. The proportion of the test subjects (experimental units) assigned to the treatment group is

$$p_T = \frac{N_T}{N}$$

This proportion is determined by the investigator and often equals 1/2 or close to it. For example, in the Key Problem, where the goal probably was to choose $p_T = 1/2$, in actuality (perhaps a few dropped out or were removed),

$$p_T = \frac{11{,}037}{22{,}071} = 0.5001.$$

Assume that at the end of the experiment we have observed a total of n test subjects with the characteristic of interest, x_T in the treatment group and x_C in the control group. That is,

$$n = x_T + x_C.$$

For example, in the Key Problem there were a total of $n = 293$ heart attacks, out of which $x_T = 104$ occurred in the aspirin group and $x_C = 189$ occurred in the placebo group, noting that $293 = 104 + 189$. The proportion with the characteristic of interest that occurred in the treatment group is

$$\hat{p} = \frac{x_T}{n}.$$

If the observed $\hat{p} < p_T$, then we have evidence (maybe strong, maybe weak) suggesting that the treatment reduces the incidence of the characteristic of interest. In the Key Problem, the observed value of $\hat{p}$ was

$$\hat{p}_{OBS} = \frac{104}{293} = 0.355,$$

which clearly is less than $p_T = 0.5001$, suggesting (perhaps strongly) that aspirin reduces the risk of heart attacks. This is because even though half of the subjects were assigned to the treatment group, only about one-third of the heart attacks occurred in the treatment group.

Let the parameter p denote the proportion of the cases expected to occur in the treatment group. The research hypothesis that the treatment reduces the incidence of the characteristic of interest can be expressed as

$$H_A : p < p_T.$$

The corresponding null hypothesis is

$$H_0 : p \geq p_T.$$

To prove H_A we only have to rule out the most-difficult-to-reject parameter value, $p = p_T$.

The null hypothesis in a randomized controlled experiment is that there is *no treatment effect*—that is, that the chances of a unit displaying the characteristic of interest are the same regardless of its random assignment to treatment group or control group.

There is a clever and simple way to simulate the random process of acquiring or not acquiring the characteristic of interest in a randomized controlled experiment *when the null hypothesis of no treatment effect is true* using a box model approach. To build the null-hypothesis box model, represent all the N test subjects or units of the experiment by balls in a box, labeling each of the N balls either T or C depending on whether it was assigned to the treatment (T) group or the control (C) group. For example, for the Key Problem the null-hypothesis box of $N = 22{,}071$ balls is

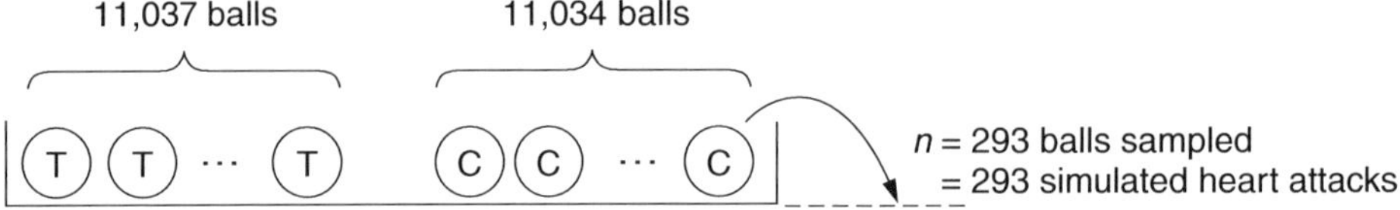

We can simulate $\hat{p}$, the proportion of the n heart attacks that occur in the treatment group, under the assumption that H_0 is true, by drawing n balls randomly from the above box *without replacement*. Each drawn ball is assigned the characteristic of interest (for example, being a heart attack victim) and each of the $N - n$ balls remaining in the box is assigned to not have the characteristic of interest.

The simulated proportion of the n cases in the treatment group is then the proportion of T-balls in the sample. Clearly, whether a unit is assigned the characteristic of interest in the simulation experiment is not influenced by whether the unit is a T or a C. Thus the null-hypothesis model of no treatment effect is being simulated. For example, the $n = 293$ heart attacks in the Key Problem can be simulated by taking 293 random draws without replacement from the box displayed above (step 2 of the five-step model). T and C units are equally likely to have a heart attack (by being drawn), so clearly H_0 (no effect for the aspirin) is true. Thus we do have a probability model of the experiment when there is no treatment effect.

We shall now complete our analysis of the Key Problem.

four-stage simulation-based hypothesis test for a randomized controlled experiment

Stage I: Formulate the Null Hypothesis H_0 and Alternative Hypothesis H_A

The research hypothesis or claim is

$$H_A : \text{Aspirin reduces the risk of a heart attack,}$$

which we symbolically rephrase as

$$H_A : p < p_T,$$

where p_T is the proportion of the test subjects assigned to the treatment (aspirin) group and p is the proportion of the heart attacks expected to occur in the aspirin group. The corresponding null hypothesis negating the research claim is

$$H_0 : \text{Aspirin has no effect on the risk of a heart attack,}$$

which we rephrase as

$$H_0 : p = p_T.$$

Stage II: Choose a Null-Hypothesis-Based Sampling Box Model

Given that the test subjects had a total of $n = 293$ heart attacks, we simulate the proportion of the heart attacks that occurred in the aspirin treatment group, under H_0, by drawing $n = 293$ times at random *without* replacement from the box

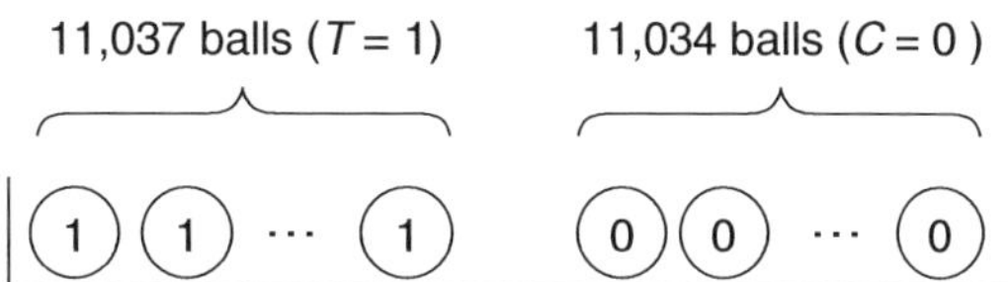

and computing the proportion $\hat{p}$ of 1-balls in the sample. Note that the box has a 1-ball for each person in the aspirin group (T) and a 0-ball for each person in the nonaspirin control group (C). Of course the nonaspirin control group will be given a placebo so that we have a double-blind experiment where neither the medical subjects nor the doctors know who is receiving the treatment.

Stage III: Estimate the *P*-Value

large-sample (and large-population) Z-test for randomized controlled experiment

Note that the total number of heart attacks, $n = 293$, was less than 5% of the number of test subjects, $N = 22,071$. Thus we have a very large population size compared to the sample size and can pretend that we are sampling with replacement. We have

$$n \cdot p_T = 293 \cdot \frac{11{,}037}{22{,}071} = 146.5 \geq 10$$

and

$$n(1 - p_T) = 293 \cdot \frac{11{,}034}{22{,}071} = 146.5 \geq 10$$

Thus by the CLT we can use the standard normal distribution and Appendix E to estimate the P-value of the test for

$$H_0 : p = p_T \text{ versus } H_A : p < p_T.$$

The observed value of the Z-statistic is

$$z_{\text{OBS}} = \frac{\hat{p}_{\text{OBS}} - p_T}{\sqrt{p_T(1 - p_T)}/\sqrt{n}} = \frac{0.355 - 0.5001}{\sqrt{(0.5001)(0.4999)}/\sqrt{293}}$$
$$= -4.96.$$

We get (instead of simulating the $\hat{p}$-values under H_0)

$$P = P(\text{simulated } \hat{p} \leq \hat{p}_{\text{OBS}} \mid H_0) \approx P(Z \leq -4.96) = 0.0000,$$

according to the Appendix E tables, rounded off to four places to the right of the decimal point.

Stage IV: Decide Whether to Reject H_0

Because P is less than our 0.001 threshold—indeed, P is actually about 5 in 10 million—the result is *very highly significant*, and we reject the null hypothesis that there is no treatment effect. In fact, since P is so small we have very strong statistical evidence that aspirin reduces the risk of a heart attack for middle-aged men.

Next we consider two examples where the sample size n is greater than 5% of the population size N.

box model
simulation-based test for randomized controlled experiment (population not large relative to sample size)

Example 9.7

Prevnar is a recently developed vaccine meant to protect infants against bacterial meningitis (a potentially fatal disease for babies that is also of great concern on many college campuses) and other pneumococcal infections. It is typically administered in several doses, starting when the infant is around 2 months of age. In preliminary tests for adverse reactions, 1321 infants were randomly assigned to either a treatment group receiving Prevnar or a control group receiving the standard vaccine. (There was no placebo group for ethical reasons.) After the first dose, 107 of the 710 infants in the Prevnar group and 67 of the 611 infants in the control group experienced fever as a side effect.

Could this difference be explained as nothing more than chance variation, or can we conclude that infants who receive the Prevnar vaccine are more likely to experience fever as a side effect than infants receiving the control vaccine?

Solution: The proportion p_T of test subjects assigned to the Prevnar treatment group was

$$p_T = \frac{710}{1321} = 0.5375.$$

The characteristic of interest is fever as a side effect of the first vaccination. The proportion of the total of 174 fevers that occurred in the Prevnar treatment group was

$$\hat{p}_{\text{OBS}} = \frac{107}{174} = 0.6149.$$

Because $\hat{p}_{\text{OBS}} > p_T$, Prevnar may cause more fevers than the control vaccine. The question is: Do we have enough statistical evidence to prove it? We shall use our four-stage method of hypothesis testing.

Stage I: Formulation of Null Hypothesis H_0 and Alternative Hypothesis H_A

Let p denote the proportion of the fevers expected to occur in the Prevnar group. Our conjecture (research hypothesis) is

$$H_A : p > p_T.$$

The corresponding null hypothesis of no difference in the risk of fever as a side effect is

$$H_0 : p = p_T.$$

Stage II: Choose a Null-Hypothesis-Based Sampling Box Model

Given that the test subjects had a total of $n = 174$ cases of fever after the first shot, we can simulate the proportion of the fever cases to occur in the Prevnar group. We perform the simulation under H_0, the assumption of no treatment effect, by drawing $n = 174$ times at random *without* replacement from the null-hypothesis box

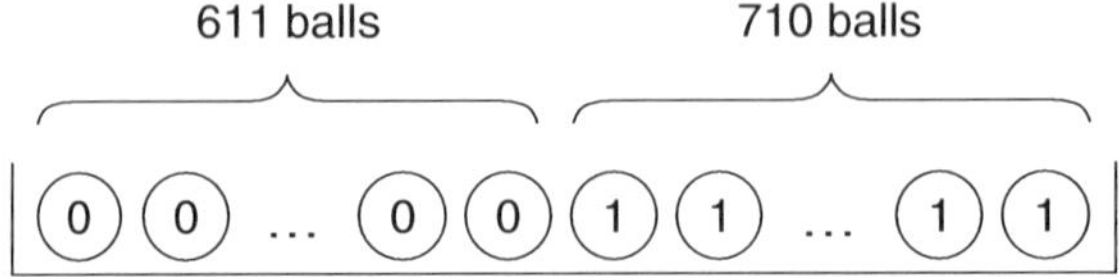

and computing the proportion $\hat{p}$ of 1-balls in the sample. Note that the box has a 0-ball for each infant in the control group and a 1-ball for each infant in the Prevnar treatment group.

Stage III: Estimate the P-Value

Because the number of fever cases, $n = 174$, is greater than 5% of the number of test subjects, $N = 1321$, we *cannot* draw upon the CLT for sampling *with* replacement to estimate the P-value. We could use the slightly more complicated CLT for sampling without replacement to estimate P. Instead, we ran 10,000 computer simulations of randomly sampling 174 balls without replacement from the box of 1321 balls. We got

$$\begin{aligned}\hat{P} &= \hat{P}(\text{simulated } \hat{p} \geq \hat{p}_{\text{OBS}} \mid H_0) \\ &= \hat{P}(\text{simulated } \hat{p} \geq 0.6149 \mid H_0) = \frac{173}{10{,}000} = 0.0173.\end{aligned}$$

Stage IV: Decide Whether to Reject H_0

Because the experimental P-value, $\hat{P}$, is less than our 0.05 threshold, we reject the null hypothesis that Prevnar is no worse in terms of side effects (of fever) than the control vaccine. The result is *statistically significant*, but not *highly significant* (although almost so).

Example 9.8

Viagra has become a heavily advertised prescription drug in America. One side effect studied is that Viagra seems to give some men headaches. In preliminary tests for adverse reactions, 1459 men were randomly assigned to either a treatment group receiving Viagra or a control group receiving a placebo. Out of the 734 men assigned to the Viagra group, 117 experienced headaches after taking the pill. Out of the 725 men in the placebo group, 29 experienced headaches. Could this difference be explained as nothing more than chance variation, or can we conclude that Viagra increases a man's risk of headaches?

Solution: The proportion p_T of test subjects assigned to the treatment group was

$$p_T = \frac{734}{1459} = 0.503.$$

The characteristic of interest is headaches after taking the medication. The proportion of the 146 clinically occurring headaches that occurred in the treatment group was

$$\hat{p}_{\text{OBS}} = \frac{117}{146} = 0.801.$$

Because $\hat{p}_{\text{OBS}} > p_T$, Viagra may increase a man's risk of a headache. The question is: Can we prove it statistically? We shall use our four-stage method of hypothesis testing.

I. Formulate the Null Hypothesis H_0 and Alternative Hypothesis H_A

Let p denote the proportion of the headaches expected to occur in the Viagra group. Our conjecture is

$$H_A : p > p_T.$$

The corresponding null hypothesis, the assumption of no treatment effect (the assumption that a man's risk of headaches is not influenced by Viagra), is

$$H_0 : p = p_T.$$

II. Choose a Null-Hypothesis-Based Sampling Box Model

The H_0 box has a 0-ball for each man in the control group and a 1-ball for each man in the Viagra treatment group:

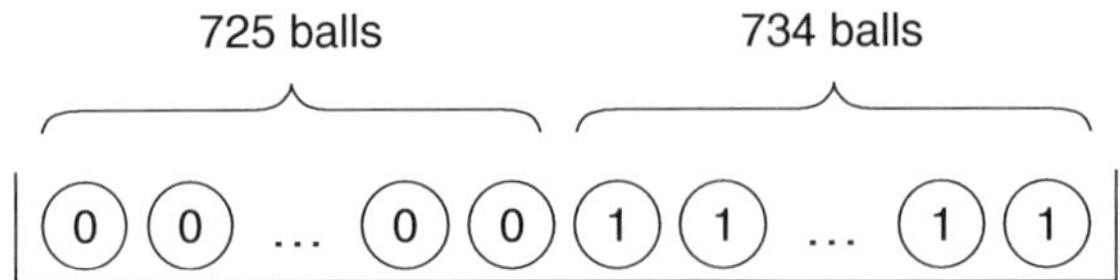

Given that a total of $n = 146$ test subjects suffered headaches, we can simulate the proportion of the 146 headaches occurring in the Viagra group under H_0, the assumption of no treatment effect, by drawing $n = 146$ times at random *without* replacement from the box and computing the simulated proportion $\hat{p}$ of 1-balls in the sample.

III. Estimate the *P*-Value

Instead of using the CLT to estimate P we ran 10,000 computer simulations, and we got

$$\hat{P} = \hat{P}(\text{simulated } \hat{p} \geq \hat{p}_{\text{OBS}} \mid H_0) = \hat{P}(\text{simulated } \hat{p} \geq 0.801 \mid H_0) = \frac{0}{10,000} = 0.0000.$$

IV. Decide Whether to Reject H_0

Because $\hat{P}$ is less than our 0.05 threshold, we reject the null hypothesis that Viagra is no more likely to cause headaches than a placebo. Indeed, the result is *very highly significant*, that is, statistically significant at level 0.001.

Section 9.3 Summary

summary of hypothesis testing for a randomized controlled experiment

Consider a randomized controlled experiment with N test subjects (or experimental units) in which each test subject (or experimental unit) at the end of the experiment either will have or will not have a characteristic of interest (such as being a heart attack victim or suffering a drug side effect). Assume that N_T of the test subjects are randomly selected to make up the treatment group and the remaining $N_C = N - N_T$ make up the control group. The proportion of the test subjects that are assigned to the treatment group is

$$p_T = \frac{N_T}{N}.$$

Assume that by the end of the experiment we have observed a total of n test subjects with the characteristic of interest, x_T cases in the treatment group and x_C cases in the control group $(n = x_T + x_C)$. The proportion of the cases that occurred in the treatment group is

$$\hat{p} = \frac{x_T}{n}.$$

If $\hat{p} < p_T$, then we have evidence suggesting that the treatment reduces the incidence of the characteristic of interest. If $\hat{p} > p_T$, then we have evidence suggesting that the treatment increases the incidence of the characteristic of interest. Is the evidence statistically strong?

Let p denote the proportion of the cases expected to occur in the treatment group. The research hypothesis that the treatment reduces the incidence of the characteristic of interest can be expressed as

$$H_A : p < p_T.$$

The research hypothesis that the treatment increases the incidence of the characteristic of interest can be expressed as

$$H_A : p > p_T.$$

In either case, the corresponding null hypothesis that there is no treatment effect can be expressed as

$$H_0 : p = p_T.$$

The H_0 sampling box model has a ball for each of the N test subjects, a 0-ball for each test subject in the control group and a 1-ball for each test subject in the treatment group.

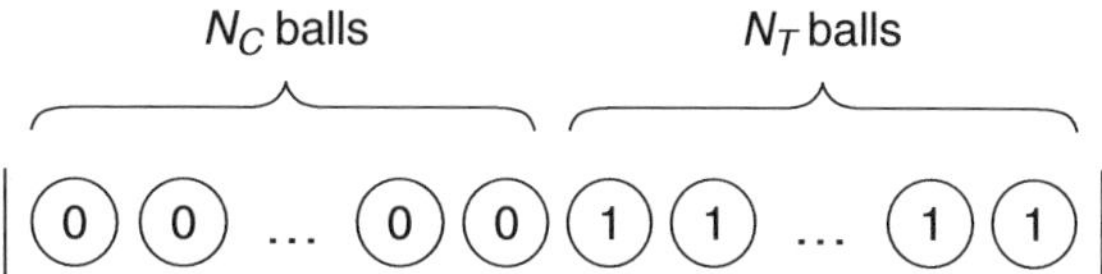

Given that there were a total of n cases observed (a total of n test subjects developed the characteristic of interest) we can simulate $\hat{p}$, the proportion of the cases that occur in the treatment group. We simulate under the null-hypothesis assumption of no treatment effect by drawing n times at random *without* replacement from the box and computing the proportion of 1-balls in the sample.

If the research hypothesis is

$$H_A : p < p_T,$$

then the estimated P-value is

$$\hat{P} = \hat{P}(\text{simulated } \hat{p} \leq \hat{p}_{\text{OBS}} \mid H_0),$$

where $\hat{p}_{\text{OBS}}$ denotes the real-data observed proportion of the cases that were in the treatment group. If the research hypothesis is

$$H_A : p > p_T,$$

then the estimated P-value is

$$\hat{P} = \hat{P}(\text{simulated } \hat{p} \geq \hat{p}_{\text{OBS}} \mid H_0).$$

In either case, we reject H_0 if $\hat{P} \leq 0.05$.

We can *always* accurately estimate P by simulating 10,000 $\hat{p}$-values and determining the experimental probability of the event of interest, either

$$(\text{simulated } \hat{p} \leq \hat{p}_{\text{OBS}}) \text{ or } (\text{simulated } \hat{p} \geq \hat{p}_{\text{OBS}}).$$

If the number of observed cases, n, is less than 5% of the number of test subjects, N (that is, $n \leq 0.05N$), and the large-sample conditions

$$np_T \geq 10 \quad \text{and} \quad n(1 - p_T) \geq 10$$

are both satisfied, then we can also estimate P fairly accurately by using the CLT on the Z-test statistic,

$$Z = \frac{\hat{p} - p_T}{\sqrt{p_T(1 - p_T)}/\sqrt{n}}.$$

If z_{OBS} denotes the observed value of Z, namely

$$z_{\text{OBS}} = \frac{\hat{p}_{\text{OBS}} - p_T}{\sqrt{p_T(1 - p_T)}/\sqrt{n}},$$

and if $H_A : p < p_T$, then

$$P = P(Z \leq z_{\text{OBS}}) \approx \text{area under standard normal curve to the left of } z_{\text{OBS}}.$$

If $H_A : p > p_T$, then

$$P = P(Z \geq z_{\text{OBS}}) \approx \text{area under the standard normal curve to the right of } z_{\text{OBS}}.$$

Note, however, that the CLT-based bell-curve approximation to P often *underestimates* P somewhat, especially if $n < 200$. (The intuitively mysterious two-sample Z-statistic, commonly used to analyze data from randomized controlled experiments, and presented by most other statistics textbooks, also underestimates P.) In conclusion, use of the simulation method is always recommended.

Section 9.3 Exercises

1. In conducting an experiment with a treatment group and a control group, why is it advantageous to randomly assign the experimental units to the two groups whenever possible?
2. In a randomized controlled experiment that uses proportions to measure the treatment effect, describe in words the proportion parameter of interest in the problem. What is the value of this parameter under the null hypothesis?
3. In words, describe the meaning of the null hypothesis in a randomized controlled experiment as described in this section. What does it mean if the null hypothesis is rejected?
4. Describe, in words, the difference in interpretation of the rejection of the null hypothesis in Section 9.3 as compared to Section 9.2.
5. Suppose a nutritional supplement has been invented that is claimed to improve the cognitive performance of rats. The new supplement is tried out in a laboratory with a group of 20 rats. The experimenters randomly assign half of the rats to the treatment group and half to the control group. The treatment group is fed the usual rat diet plus the new supplement. The control group is fed the same diet as the treatment group, but without the new supplement. The two groups of rats run a maze, and the experimenters find that none were successful in the first try but that 13 out of the 20 were successful on the second try. Of the 13 successes, 10 were by members of the treatment group.
   a. What proportion of all the rats were treatment group members? What proportion of the successful rats were treatment group members?
   b. Of the two proportions in part (a), which proportion is a sample statistic and which proportion is a constant whose value was set by the experimenter?

c. In words, what is the claim that is made about the treatment? How does that claim relate to the proportion parameter mentioned in part (b)?
d. In words, what is the null hypothesis that nullifies the claim? Write the null hypothesis using standard statistical notations for the proportion parameters involved.
e. To test the null hypothesis, we need to generate simulated data samples that conform to the null hypothesis. Describe the stage II sampling box model. How many 0-balls are there in the box? How many 1-balls? On each simulation, how many balls do we draw? Do we draw with or without replacement? What sample statistic will we measure on the simulated sample that we generate on each simulation?
f. The EOI will be based on comparing the simulated sample statistic to something. To what will we compare it? Why do we make this comparison?
g. The results from 10,000 simulations under H_0 are shown in the following table. What are the most frequently occurring values of $\hat{p}$ in the simulated samples? What are the least frequently occurring values of $\hat{p}$ in the simulated samples? What are the least frequently appearing values in the direction that, if observed in a real data sample, would be evidence for rejecting the null hypothesis?
h. Is the value of $\hat{p}$ observed in the real data sample among the most infrequently observed simulated $\hat{p}$-values?
i. Decide whether the null hypothesis should be rejected and explain why you made that decision.

$\hat{p}$	Frequency of occurrence	$\hat{p}$	Frequency of occurrence
0/13	0	7/13	3246
1/13	0	8/13	1511
2/13	0	9/13	255
3/13	16	10/13	21
4/13	270	11/13	0
5/13	1420	12/13	0
6/13	3261	13/13	0

6. In Exercise 5, suppose the proportion $\hat{p}$ observed in the real data sample had been 9/13. Use the table from Exercise 5 to estimate the probability of observing a simulated proportion as large as, or larger than, 9/13 under the null hypothesis. Would you reject the null hypothesis in this case?

7. The risk of acquiring Lyme disease is high in many areas of the United States; thus, a safe and effective vaccine is greatly needed. A 1998 article in the *New England Journal of Medicine* reported results from a clinical trial for a new vaccine. For this study, 10,936 subjects who lived in areas where there is a high risk of acquiring Lyme disease were randomly assigned to two groups of 5468 each. The treatment group received the vaccine, and the control group received a placebo. In the first year of the trial, 65 subjects acquired Lyme disease, with 22 of them being from the treatment group. Should we decide from these data that the vaccine is effective in preventing Lyme disease?
a. Determine p_T, the proportion of the test subjects assigned to the treatment group.
b. Determine $\hat{p}_{\text{OBS}}$, the proportion of the Lyme disease victims who happened to be in the treatment group.
c. State H_A and H_0 in terms of p_T and p, the proportion of the Lyme disease victims expected to be in the treatment group.
d. Describe the stage II sampling box model. How many 0-balls are there in the box? How many 1-balls? On each simulation, how many balls do we draw? Explain why it makes little difference whether we draw with or without replacement.
e. Use the CLT to estimate

$$P = P(\text{simulated } \hat{p} \leq \hat{p}_{\text{OBS}} \mid H_0).$$

f. Estimate P by simulating 10,000 values of $\hat{p}$, the proportion of Lyme disease victims who happen to be in the treatment group. Draw from the box

On each simulation, how many times do you draw?

g. Do we reject H_0 or not?

8. Children who play violent video games are claimed to display more violent behavior ("violent" behavior must be carefully defined) than children who merely watch violent television programs.

Suppose that a randomized controlled experiment obtains 50 volunteer children and randomly assigns 25 children to a treatment group who play violent video games for 1 hour and assigns the remaining 25 children to a control group who only watch a violent television program for 1 hour.

The two groups of children are then separately observed at play for 30 minutes immediately afterward. During the 30 minutes of play, a total of 34 children are observed to display violent behavior, with 21 of them being from the treatment group. Should we decide from these data that playing violent video games is more likely to lead to violent behavior in children than merely watching violent television programs?

a. Determine p_T, the proportion of the test subjects assigned to the treatment group.

b. Determine $\hat{p}_{OBS}$ the proportion of the violent children who happened to be in the treatment group.

c. State H_A and H_0, the assumption of no treatment effect, in terms of p_T and p, the proportion of the violent children expected to be in the treatment group.

d. Describe the stage II sampling box model. How many 0-balls are there in the box? How many 1-balls? On each simulation, how many balls do we draw? Do we draw with or without replacement?

e. We simulated 10,000 values of $\hat{p}$, the proportion of 1-balls in the sample. Estimate the P-value. Do we reject H_0 or not?

Note: These numbers are purely hypothetical for this exercise. A real study should take into account how much violent behavior the children engage in. An example of a real study done at the University of Missouri at Columbia is reported in the *Journal of Personality and Social Psychology*, vol. 78, no. 4, pp. 772–790.

$\hat{p}$	Frequency of occurrence	$\hat{p}$	Frequency of occurrence
11/34	4	18/34	1975
12/34	25	19/34	1149
13/34	128	20/34	479
14/34	488	21/34	131
15/34	1204	22/34	29
16/34	2042	23/34	2
17/34	2344	24/34	0

CHAPTER 9 SUMMARY

A statistical test attempts to answer the question: Is the pattern we observe in our data "real,"or is it just chance variation? Can we be sure that the observed pattern will be repeated if we repeat the chance experiment that produced the data? If so, the pattern is "real." If not, then the observed pattern was just chance variation. For example, in a recent clinical trial of a new experimental HIV vaccine, the treatment group had a slightly lower infection rate than the placebo group. But careful medical and statistical analysis showed that the observed difference was due to chance variation alone. (The vaccinated individuals did not develop antibodies—a sign that the vaccine was ineffective.)

The "chance"explanation is called the null hypothesis and is denoted H_0. The explanation that the observed pattern is due to more than chance is called the alternative hypothesis and is denoted H_A. To make a test of significance, we have to express H_0 using a box model for the data.

In this chapter we discuss tests for an unknown population proportion, p. Our test statistic is the sample proportion $\hat{p}$ computed from a random sample of size n. We shall assume one of the following two scenarios:

- The sample proportion $\hat{p}$ is the experimental success rate in n independent success/failure trials, each with success probability p.
- The sample is a simple random sample of size n drawn (without replacement) from a finite population of size N, out of which a proportion p has a characteristic of interest. The sample proportion $\hat{p}$ is the proportion of the sample with the characteristic of interest.

We view a statistical test for p as a four-stage decision procedure.

Stage I: Formulate the Null Hypothesis H_0 and the Alternative Hypothesis H_A

The "alternative hypothesis" is often the research hypothesis (claim) that someone wants to prove. It could be $H_A : p < p_0$ or $H_A : p > p_0$. The null hypothesis nullifies the research hypothesis and should have the form

$$H_0 : p = p_0,$$

a single specified value for p, this single value allowing us to create an H_0 box model for the sample data.

Stage II: Choose a Null-Hypothesis-Based Sampling Box Model

If $\hat{p}$ is the experimental success rate in n independent success/failure trials, then we can sample with replacement from any box of 0s and 1s for which the proportion of 1-balls is p_0. If we sample without replacement from a finite population of size N, then the obvious choice of box model has N balls, one for each member of the population, with a 1-ball for each population member with the characteristic of interest and a 0-ball for each member without the characteristic. Under H_0, the proportion of 1-balls equals p_0:

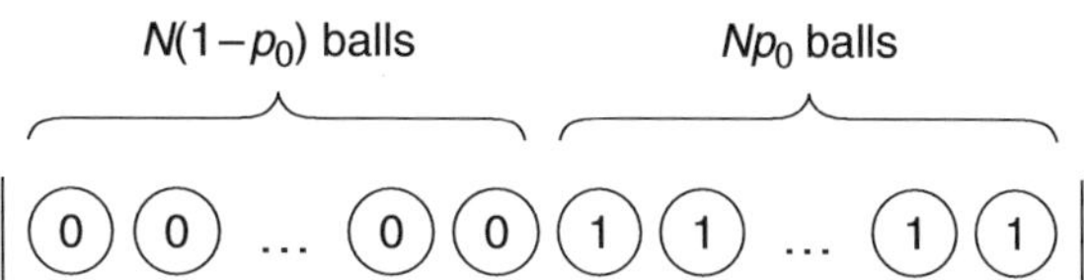

We can simulate the sample proportion by drawing n times (n denoting the sample size) at random without replacement from the box and computing the proportion of 1-balls in the sample.

If the sample size n is less than 5% of the population size N, then we will get almost the same results if we use sampling *with* replacement instead. If we sample with replacement, then we can use any box of 0s and 1s for which the proportion of 1-balls equals p_0, as specified by H_0.

Stage III: Estimate the P-Value of the Hypothesis Test

We measure the strength of the evidence against H_0 by computing the P-value (also called the observed significance level) of the test. The P-value is the chance, when we are doing simulations that presume H_0 to be true, of getting evidence against H_0 as strong as or even stronger than the real data evidence at hand. As stated, we run the simulations assuming that H_0 is true, that is, using the H_0 box model of stage II. If

$$H_A : p > p_0,$$

then

$$P\text{-value} = P(\text{simulated } \hat{p} \geq \text{observed } \hat{p} \mid p = p_0).$$

If the H_A inequality is in the other direction, namely

$$H_A : p < p_0,$$

then

$$P\text{-value} = P(\text{simulated } \hat{p} \leq \text{observed } \hat{p} \mid p = p_0).$$

That is, the direction of the H_A inequality is the same as the direction of the inequality for the event of interest. The P-value is the probability of getting evidence against H_0 and in favor of H_A as strong as or stronger than the evidence at hand—if H_0 were true. The smaller the P-value is, the stronger the evidence against H_0.

We can estimate the P-value by simulating 10,000 values of $\hat{p}$, using the stage II null-hypothesis box model. If the sample is so large that

$$np_0 \geq 10 \quad \text{and} \quad n(1 - p_0) \geq 10,$$

then we can also use the bell curve to estimate the P-value—provided we are either sampling with replacement, or sampling without replacement but the sample size n is less than 5% of the number of balls in the box. Under these assumptions, the Z-statistic

$$Z = \frac{\text{statistic} - \text{parameter}}{\text{SD of statistic}} = \frac{\hat{p} - p_0}{\sqrt{p_0(1 - p_0)}/\sqrt{n}}$$

is approximately standard normal.

Let z_{OBS} denote the observed real data value of the Z-statistic:

$$z_{\text{OBS}} = \frac{\text{observed } \hat{p} - p_0}{\sqrt{p_0(1 - p_0)}/\sqrt{n}}.$$

We can estimate the P-value of the hypothesis test as follows. If the alternative hypothesis is $H_A : p > p_0$, then

$$\begin{aligned} P\text{-value} &= P(\text{simulated } \hat{p} \geq \text{observed } \hat{p} \mid p = p_0) \\ &= P(\text{simulated } Z \geq z_{\text{OBS}}) \\ &\approx \text{area under the standard normal curve to the right of } z_{\text{OBS}}. \end{aligned}$$

Here the $\hat{p}$ random variable is viewed as being simulated. However, there is no simulation done because the CLT calculation of the P-value stands in place of doing simulations. That is, the CLT approach, when applicable, bypasses box model simulation.

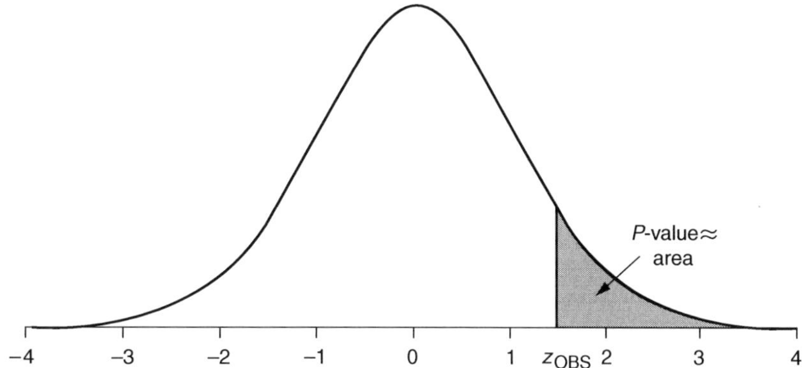

Estimating the P-value for $H_A : p > p_0$.

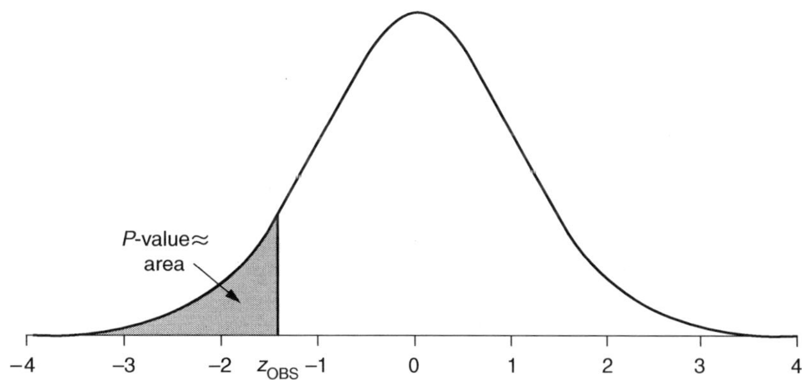

Estimating the P-value for $H_A : p < p_0$.

If the alternative hypothesis is $H_A : p < p_0$, then

$$\begin{aligned} P\text{-value} &= P(\text{simulated } \hat{p} \leq \text{observed } \hat{p} \mid p = p_0) \\ &= P(\text{simulated } Z \leq z_{\text{OBS}}) \\ &\approx \text{area under the standard normal curve to the left of } z_{\text{OBS}}. \end{aligned}$$

Stage IV: Decide Whether to Reject H_0

How small should P be before we reject H_0? The standard procedure is to compare the P-value with some fixed threshold called the "level of significance." In this book we shall use a 0.05 threshold:

- If $P \leq 0.05$, the result is called **statistically significant,** and we reject H_0.
- If $P > 0.05$, the result is *not statistically significant*, and we do *not reject* H_0.

The smaller P is, the stronger the evidence against H_0.

- If $P \leq 0.01$, the result is called **highly significant** or **statistically significant at level 0.01.**
- If $P \leq 0.001$, the result is **very highly significant**, or **statistically significant at level 0.001.**

Randomized Controlled Experiments

Consider a randomized controlled experiment with N test subjects (or experimental units) in which each test subject (or experimental unit) at the end of the experiment either will have or will not have a characteristic of interest (such as being a heart attack victim or a suffering a drug side effect). Assume that N_T of the test subjects are randomly selected to make up the treatment group, and thus the remaining $N_C = N - N_T$ make up the control group. The proportion of the test subjects (experimental units) assigned to the treatment group is

$$p_T = \frac{N_T}{N}$$

Assume that at the end of the experiment we have observed a total of n test subjects with the characteristic of interest, x_T in the treatment group and x_C in the control group. The proportion of the casses that occurred in the treatment group is

$$\hat{p} = \frac{x_T}{n}.$$

If $\hat{p} < p_T$, then we have evidence suggesting that the treatment reduces the incidence of the characteristic of interest. If $\hat{p} > p_T$, then we have evidence suggesting that the treatment increases the incidence of the characteristic of interest. Is the evidence statistically strong?

Let p denote the proportion of the cases expected to occur in the treatment group. The research hypothesis that the treatment reduces the incidence of the characteristic of interest can be expressed as

$$H_A : p < p_T.$$

The research hypotheses that the treatment increases the incidence of the characteristic of interest can be expressed as

$$H_A : p > p_T.$$

In either case, the corresponding null hypothesis that there is no treatment effect can be expressed as

$$H_0 : p = p_T.$$

The H_0 sample box model has a ball for each of the N test subjects, a 0-ball for each test subject in the control group and a 1-ball for each test subject in the treatment group.

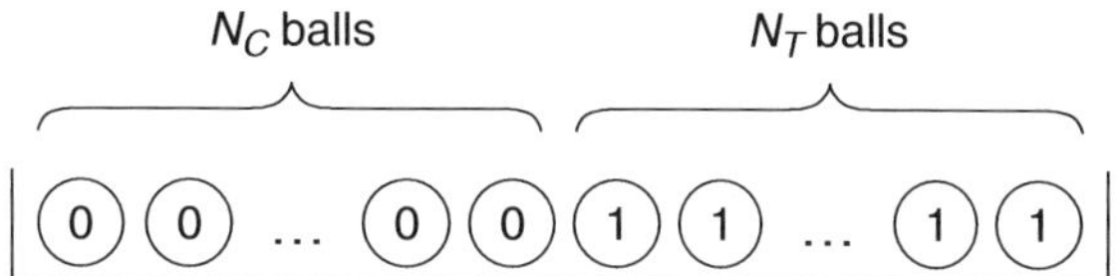

Given that there were a total of n cases observed (a total of n test subjects developed the characteristic of interest), we can simulate $\hat{p}$, the proportion of acases that occur in the treatment group. We simulate under null-hypothesis assumption of no treatment effect by drawing n times at random *without* replacement from the box and computing the proportion of 1-balls in the sample.

If the research hypothesis is

$$H_A : p < p_T,$$

then the estimated P-value is

$$\hat{P} = \hat{P}(\text{simulated } \hat{p} \leq \hat{p}_{\text{OBS}} \mid H_0),$$

where $\hat{p}_{\text{OBS}}$ denotes the real-data observed proportion of the cases that were in the treatment group. If the research hypothesis is

$$H_A : p > p_T,$$

then the estimated P-value is

$$\hat{P} = \hat{P}(\text{simulated } \hat{p} \geq \hat{p}_{\text{OBS}} \mid H_0).$$

In either case, we reject H_0 if $\hat{P} \leq 0.05$.

We can *always* accurately estimate P by simulating 10,000 $\hat{p}$-values and determining the experimental probability of the event of interest, either

$$(\text{simulated } \hat{p} \leq \hat{p}_{\text{OBS}}) \quad \text{or} \quad (\text{simulated } \hat{p} \geq \hat{p}_{\text{OBS}})$$

If the number of observed cases, n, is less than 5% of the number of test subjects, N (that is $n \leq 0.05N$), and the large-sample conditions

$$np_T \geq 10 \quad \text{and} \quad n(1 - p_T) \geq 10$$

are both satisfied, then we can also estimate P fairly accurately by using the CLT on the Z-test statistic,

$$Z = \frac{\hat{p} - p_T}{\sqrt{p_T(1 - p_0)}/\sqrt{n}},$$

If z_{OBS} denotes the observed value of Z, namely

$$z_{\text{OBS}} = \frac{\hat{p}_{\text{OBS}} - p_T}{\sqrt{p_T(1 - p_0)}/\sqrt{n}},$$

and if $H_A : p < p_T$, then

$$P = P(Z \leq z_{\text{OBS}}) \approx \text{area under standard normal curve to the left of } z_{\text{OBS}}.$$

If $H_A : p > p_T$, then

$$P = P(Z \geq z_{\text{OBS}}) \approx \text{area under standard normal curve to the right of } z_{\text{OBS}}.$$

Note, however that the CLT-based bell-curve approximation to P often *underestimates* P somewhat, especially if $n < 200$. (The intuitively mysterious two-sample Z-statistic, commonly used to analyze data from randomized controlled experiments, and presented by most other statistics textbooks, also underestimates P.) In conclusion, use of the simulation method is always recommended.

CHAPTER REVIEW EXERCISES

1. Suppose it is claimed that in a certain state the proportion p of adults who are divorced is less than 0.1.
   a. The many-valued null hypothesis that nullifies H_A is (choose one option)

   $p > 0.1$ $\quad$ $p < 0.1$ $\quad$ $p \leq 0.1$ $\quad$ $p \neq 0.1$ $\quad$ $p \geq 0.1$

   b. What is the single-valued null hypothesis we need so that we can construct a null-hypothesis box model?
2. Suppose it is claimed that in a certain state the proportion p of adults who are married exceeds 0.60. Suppose that the proportion of a simple random sample who are married is $\hat{p} = 0.64$.
   a. The single-valued null hypothesis we need so that we can construct a null-hypothesis box model is (choose one option)

   $p > 0.60$ $\quad$ $p \leq 0.60$ $\quad$ $p = 0.64$ $\quad$ $p = 0.60$ $\quad$ $p \geq 0.64$

   b. The event of interest in testing H_0 is (choose one option)

   simulated $\hat{p} \geq 0.60$ $\quad$ simulated $\hat{p} \geq 0.64$ $\quad$ simulated $\hat{p} \leq 0.64$ $\quad$ simulated $\hat{p} \leq 0.60$

3. Suppose $H_0 : p = 0.4$ with the claim being $H_A : p < 0.4$. Suppose that the sample proportion $\hat{p} = 0.37$ is observed. If 10,000 simulations yield

   $$\hat{P}(\text{simulated } \hat{p} \leq 0.37 \mid H_0) = 0.0951,$$

   what is the implication of the experimental P-value? (Choose one option.)
   i. We reject H_0.
   ii. We fail to reject H_0.
   iii. We have proved that H_0 is true.

(Choose one option.)

4. In Exercise 3, suppose that

$$\hat{P}(\text{simulated } \hat{p} \leq 0.37 \mid H_0) = 0.0032.$$

What is the implication of the experimental P-value? Do we reject H_0, do we fail to reject H_0, or have we proved that H_0 is true?

5. A manufacturer of semiconductor wafers claims that at most 1% of the wafers are defective. In a simple random sample of 100 wafers, 4 defective wafers are found. Does this prove that more than 1% of all wafers are defective? Or could the high rate of defective wafers in the sample, $\hat{p} = 0.04$, be due to sampling variation alone?
   a. State H_0 and H_A in terms of the proportion p of all wafers that are defective.
   b. State the EOI (event of interest) for testing H_0.
   c. Explain why we cannot use the CLT method to estimate the P-value in this example.
   d. We simulated 10,000 values of the proportion $\hat{p}$ of defective wafers in a sample of 100 wafers, assuming H_0 to be true. Determine the experimental P-value.

$\hat{p}$ observed in simulated sample	Frequency of occurrence	$\hat{p}$ observed in simulated sample	Frequency of occurrence
0.00	3688	0.04	143
0.01	3660	0.05	31
0.02	1888	0.06	3
0.03	586	0.07	1

   e. Do the data prove that more than 1% of the wafers are defective?

6. A medical researcher performed a clinical trial of a new treatment for poison ivy against the traditional ointment. The experiment was double-blind and used randomized controls. In the experiment the new treatment had a higher success rate than the traditional ointment. The P-value was 13%. Which of the following statements most accurately describes the implication of the P-value? (Choose one option.)
   i. The test proves that the new treatment is better than the traditional ointment.
   ii. We can reject H_0.
   iii. It is quite possible that the superior performance of the new treatment was due to chance.
   iv. The test proves that the new treatment is no better than the traditional ointment.

7. Some Scandinavian researchers studied whether surgery can prolong life among men suffering from prostate cancer, a type of cancer that usually develops and spreads very slowly. A total of 695 men diagnosed with prostate cancer were randomly assigned either to undergo surgery or not. Of the 347 men who had surgery, 16 eventually died of prostate cancer, compared with 31 of the 348 men who did not have surgery.
   a. Was this an observational study or a randomized experiment?
   b. Was the study double-blind?
   c. Determine p_T, the proportion of the test subjects assigned to the treatment group.
   d. Determine $\hat{p}_{OBS}$ the proportion of the prostate cancer deaths that occurred in the treatment group.
   e. State H_A and H_0 in terms of p_T and p, the proportion of the prostate cancer deaths expected to occur in the treatment group.
   f. Describe the stage II sampling box model. How many 0-balls are there in the box? How many 1-balls? On each simulation, how many balls do we draw? Do we draw with or without replacement?
   g. Simulate 10,000 values of $\hat{p}$, the proportion of the prostate cancer deaths that occur in the treatment group, assuming H_0 to be true and given that there were a total of 47 prostate cancer deaths among the test subjects. What is the EOI (event of interest)? What experimental P-value do you get?

h. Does prostate cancer surgery prolong lives? Is the result highly significant? Explain why we need a larger study.

8. The championship series of the National Basketball Association in June 2000 was played between the Los Angeles Lakers and the Indiana Pacers. The Lakers won the first two games, with Shaquille O'Neal, their star center, leading the way. Because O'Neal makes only 50% of his foul shots on average in regular season games, one of the strategies of Indiana was to foul O'Neal many times, hypothesizing that O'Neal would have an even lower proportion in playoff games.

   When Indiana won the third game, O'Neal had made only 3 out of 13 foul shots in the game. Considering this performance as a sample proportion and considering O'Neal's lifetime average of about 0.5 as a population proportion for regular season games, was this sample indicative of O'Neal having a lower population proportion in a championship series game?

   a. In terms of a population proportion, what is the null hypothesis?
   b. To test the null hypothesis, we need to generate simulated samples that conform to the null hypothesis. On each simulation, we will generate a sample having the same size as the original real-data sample. The simulated data samples will be generated from a box model. What is the choice of the sampling box? What is the definition of one simulation? What sample statistic will we measure on each simulated sample that we generate?
   c. What will be the event of interest for testing the null hypothesis?
   d. The results from 1000 simulations from the null-hypothesis box model are shown in the following table. What is the experimental probability of the event of interest?
   e. Decide whether to reject the null hypothesis and explain why you made that decision.

$\hat{p}$ observed in simulated sample	Frequency of occurrence	$\hat{p}$ observed in simulated sample	Frequency of occurrence
0/13	0	7/13	207
1/13	2	8/13	155
2/13	11	9/13	86
3/13	36	10/13	33
4/13	84	11/13	10
5/13	161	12/13	2
6/13	213	13/13	0

9. In a random sample of 250 residents of a large city, 114 people in the sample would favor raising the bar entrance age from 19 to 21. Test whether the true population proportion is equal to 0.5, versus the claim that it is less.

10. Between the 1999 and 2000 baseball seasons, Ken Griffey, Jr., was traded from an American League team, the Seattle Mariners, to a National League team, the Cincinnati Reds. Over Griffey's last 2 years in Seattle, his batting average (i.e., the proportion of his at-bats that produced hits) was .285 (353 hits in 1239 at-bats). Because Griffey switched from one league to another, the claim might be made that his batting average is going to be lower, at least for the first season.

    At a certain point in the 2000 season, Griffey had 50 hits in 219 tries, a batting average of .228. Use the four-stage hypothesis-testing approach to decide whether this sample is evidence of a drop in Griffey's batting average after the switch relative to a population proportion defined by his batting average during his last 2 years in Seattle.

    a. What is the null hypothesis?
    b. What is the choice of sampling box to be used to generate simulated samples corresponding to the null hypothesis?
    c. What is the definition of one simulation from the sampling box?
    d. Define the event of interest and explain why this is of interest in regard to testing the null hypothesis.
    e. In 1000 simulations, 33 of them had a value as small as, or smaller than, .228. Explain what this means and whether or not the null hypothesis should be rejected.

11. A magazine claims to a particular advertiser that the advertiser's ad will be remembered by more than 50% of the readers who see it.

    A statistician tests the claim by surveying a random sample of 25 subscribers. The proportion in the sample who remember the ad is 0.8.

Apply the four-stage approach of hypothesis testing, using the simulated data for 1000 simulations of a population with $p = 0.5$ in the following table to make a decision about this claim.

$\hat{p}$ observed in simulated sample	Frequency of occurrence	$\hat{p}$ observed in simulated sample	Frequency of occurrence
0.04	0	0.56	111
0.08	0	0.60	79
0.12	0	0.64	74
0.16	1	0.68	53
0.20	2	0.72	14
0.24	5	0.76	9
0.28	11	0.80	2
0.32	19	0.84	2
0.36	69	0.88	0
0.40	87	0.92	0
0.44	140	0.96	0
0.48	178	1.00	0
0.52	144		

12. A double-blind randomized controlled experiment from 1985 to 1986 studied the effectiveness of the drug enalapril in healing patients with congestive heart failure. From a convenience sample of 253 patients, 127 were randomly assigned to the treatment group and the remaining 126 were assigned to the control group. The treatment group received enalapril, whereas the control group received a placebo. Over 6 months, 33 of the patients in the treatment group died, compared to 55 of the patients in the control group. Could this difference be explained by chance variation alone, or can we conclude that enalapril reduces the 6-month mortality rate for patients with congestive heart failure?

    Simulate 10,000 values of $\hat{p}$, the proportion of the deaths that occur in the treatment group, assuming no treatment effect and given that there were a total of 88 deaths among the test subjects. What is the event of interest? What experimental P-value do you get? Does enalapril reduce the mortality rate?

13. In the 1954 polio vaccine field trials, 400,000 children were randomly assigned to equal-size treatment and control groups. The treatment group received the polio vaccine, and the control group received a placebo. After one year, 199 children in the study had contracted polio, with 57 of the polio cases being in the treatment group. Conduct a hypothesis test to determine if the vaccine was effective in reducing the risk of getting polio. Describe the null-hypothesis box model. How many 0-balls are there in the box? How many 1-balls? On each siumlation, how many balls do we draw? Do we draw with or without replacement? Describe the event of interest. Run 10,000 simulations or do a Z-test (if appropriate) to estimate the P-value. What do you conclude?

14. A coin suspected of being biased toward heads is tossed 100 times, yielding 65 heads. Test whether this is a fair coin, that is, whether the chance of heads, p, equals $1/2$.

15. In clinical studies of an allergy medication, 70 of the 781 subjects experienced drowsiness. Evaluate the manufacturer's claim that only 8% of the allergy medication users experience drowsiness, as opposed to a higher percentage.

16. In 1998 an article in the *Journal of Consulting and Clinical Psychology* (vol. 66, pp. 715–730) reported on an evaluation of a new parenting program. It was hoped that the parenting program would result in significant improvements in parenting skills for high-risk mothers. A total of 235 high-risk mothers at nine Head Start centers were randomly assigned to two groups, 158 to the treatment group and 77 to the control group. The results of the study revealed that 109 mothers in the treatment group showed the desired improvement in parenting skills as compared to 40 mothers in the control group.

Conduct a hypothesis test to determine whether the difference in percentages can be accounted for by chance variation alone or whether the difference is so large that we should reject the null-hypothesis of no treatment effect. Describe the null hypothesis box model. How many 0-balls are there in the box? How many 1-balls? On each siumlation, how many balls do we draw? Do we draw with or without replacement? Describe the event of interest. Run 10,000 simulations or do a Z-test (if appropriate) to estimate the P-value. What do you conclude?

17. The National Transportation Safety Board has reported that 30% of all accidents involve fatalities. Suppose a claim has been made that the proportion of accidents involving fatalities has recently dropped. If a random sample of 250 accidents reveals that 59 involved fatalities, use hypothesis testing to evaluate if this sample indicates a drop in the population proportion.
   a. State H_0 and H_A in terms of the proportion p of all accidents that involve fatalities.
   b. Compute the observed value of the test statistic.
   c. Estimate the P-value.
   d. Do you reject H_0 or fail to reject H_0?

PROFESSIONAL PROFILE

George W. Snedecor, 1881–1974, statistical pioneer.

Individuals working in 1930 for the Mathematics Statistical Service (the forerunner of the Statistical Service) at Iowa State College. Note the early punch card tabulator in the back of the room. Photo courtesy of the Iowa State Press.

George W. Snedecor. Portrait courtesy of the Iowa State Press.

George W. Snedecor was a statistical pioneer of the first rank. He founded the first academic department of statistics and the first statistics laboratory in the United States. He was an early leader in applying statistics to practical problems. In the process, he discovered the fundamental F distribution that forms the theoretical heart of the statistical comparison of population means, such as comparing the average yields of six varieties of wheat.

Snedecor joined the mathematics faculty at Iowa State College (as it was then known) in 1913. Nine years later, Snedecor teamed up with publisher Henry Wallace to focus on using new statistical tools in agricultural research. Out of this collaboration came a weekly seminar to study multiple regression and the pioneering use of punch card tabulators in data analysis. Together, Wallace and Snedecor published *Correlation and Machine Calculation* in 1925, a foundational paper concerning the statistical computation required to carry out certain kinds of statistical inference on (possibly large) data sets. While Snedecor continued his leadership role in statistics throughout his career, Wallace went on to become U.S. Secretary of Agriculture and U.S. Vice-President under President Franklin Roosevelt.

In 1927, Snedecor and a colleague, A. E. Brandt, established the Mathematical Statistical Service at Iowa State, the first such statistical consulting service in America. It provided statistical consultation for academic departments throughout Iowa State. The service grew in size and importance following the visit of R. A. Fisher in 1931 and became the Iowa State Statistics Laboratory in 1933; Snedecor was its first director.

Snedecor's famous general statistics text, Statistical Methods, was published in 1938. It rapidly became the most widely used and accepted statistical reference for researchers in many fields. In particular, it helped expose many researchers and practitioners to the new and powerful method of statistical inference called confidence interval estimation, the subject of this chapter. Ultimately co-authored with William G. Cochran, Statistical Methods went through seven editions, was translated into seven or more languages, and sold over 100,000 copies. It has been credited with being the most frequently cited reference book across many scientific fields.

10

Estimating with Confidence

Statistics may be defined as "a body of methods for making wise decisions in the face of uncertainty."

W. A. Wallis

Objectives

After studying this chapter, you will understand the following:

- Confidence interval estimation
- Confidence intervals for the population mean, median, and proportion
- Mean and SD formulas for $\overline{X} - \overline{Y}$ and $\hat{p}_{\hat{1}} - \hat{p}_{\hat{2}}$.
- The SE-based bootstrap method for confidence intervals
- Confidence interval for the difference of two population means
- Confidence interval for the difference of two population proportions
- Confidence interval for the difference of two means in the matched-pairs case

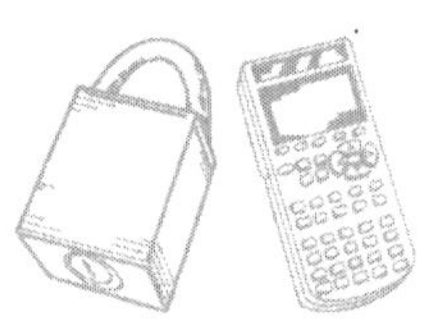

KEY PROBLEM

When a National Football League team needs a new football stadium with more seats and especially more luxury boxes, the team owner often goes to the local city council and sometimes the state legislature with the following message: I will move my team to another state unless the taxpayers subsidize the construction of a new stadium. The argument is that the taxpayers benefit from having a major professional sports franchise, which creates jobs, attracts visitors, and thereby generates tax revenue. But the taxpayers do not always buy the argument.

When a new indoor football stadium was proposed for the St. Louis Cardinals football team in 1986, a poll found that 45% of St. Louis County residents opposed the new stadium, 28% supported it, and 27% were undecided. The poll interviewed 301 registered voters in the county. The newspaper article ("Majority Opposes Stadium, Poll Shows," *St. Louis Post Dispatch,* October 12, 1986) states, "The county poll is accurate within plus or minus 5.7 percentage points at a confidence level of 95%. That means if the survey were taken 100 times, the results for the [random] group of respondents would each vary no more that 5.7% in either direction [from the true county population percentage opposing the stadium] about 95% of these times."

The *St. Louis Post Dispatch* is to be commended for attempting to communicate an important statistical concept: a confidence interval for a population parameter. The Key Problem for this chapter is to understand how to construct confidence intervals such as the 95% confidence interval (39.3%, 50.7%) for the proportion p of members of the population of registered voters who opposed the stadium construction and to clearly understand why we call such an interval a 95% confidence interval.

Postscript: In 1988 the owner of the St. Louis Cardinals football team got so fed up with the lack of civic cooperation for a new stadium that he moved the team to Arizona (where the team continued to lose until going to the Super Bowl in 2009). In 1994, with a new domed stadium under construction, the Los Angeles Rams were persuaded to move to St. Louis. Ten years later the St. Louis Cardinals baseball team received almost $100 million in public assistance to help finance a new $400 million baseball stadium.

In this chapter we shall continue our discussion of the accuracy of estimators that we began briefly in Chapter 4 and covered in greater detail in Chapter 8. We shall explain how statisticians use so-called confidence intervals, often of the form

$$\text{estimate} \pm \text{margin of error},$$

indicating our best guess (the estimate) and the expected accuracy for our guess.

A *confidence interval* (or *interval estimate*) for an unknown parameter is an interval, such as (38.3%, 50.7%) in the Key Problem), computed from sample data. The interval will vary from sample to sample and hence each resulting interval may or may not capture the true parameter value. The *confidence level* C of the interval is the probability that the interval will capture the true parameter value. If we were to repeat the estimation procedure again and again, independently, using the same method and the same number of observations each time, then C would equal the long-run proportion of times that the confidence interval captures the true parameter value.

10.1 CONFIDENCE INTERVALS FOR A POPULATION MEAN μ WHEN σ IS KNOWN

Suppose that we make n repeated, independent measurements of the same unknown physical quantity μ, using the same procedure each time. For example, when a loaded truck drives across the scales at a weigh station along a freeway, the truck's weight μ is automatically measured several times. Assume that the measurement procedure is unbiased and gives us readings that are neither systematically too high nor systematically too low. Also assume that the accuracy of the measurement procedure is known and does not vary with μ. That is, the typical random scale error is the same for repeated measurements of a light or heavy truck. Under these assumptions we can model the n measurements $X_1, X_2, \ldots, X_n$ for the truck in question as n independent random variables from a distribution with known standard deviation σ but unknown mean μ. The standard deviation is known because of past experience with repeated measurements of other trucks, all with the *same* σ.

Suppose one (or both) of the following two assumptions is realistic: Assume either that the individual measurements $X_1, X_2, \ldots, X_n$ are approximately normal, or that the number of observations n is large. Under either assumption, the sample mean

$$\overline{X} = \frac{1}{n}(X_1 + X_2 + \ldots + X_n)$$

is approximately normal with mean

$$\mu_{\overline{X}} = E(\overline{X}) = \mu$$

and standard deviation

$$\sigma_{\overline{X}} = \mathrm{SD}(\overline{X}) = \sigma/\sqrt{n}.$$

When n is large, we know this important and useful fact from the central limit theorem of Section 8.2. Since $\overline{X}$ is likely to differ from its mean by roughly one standard deviation, we estimate μ to be around $\overline{X}$, give or take $\sigma/\sqrt{n}$ or so. We know σ, so we can compute the "give-or-take number" $\sigma/\sqrt{n}$. Note that while the random variables X_i have standard deviation σ, by contrast $\overline{X}$ has standard deviation $\sigma/\sqrt{n}$, much smaller than σ when n is large.

The z_α Notation

The symbol z_α is used to denote the z-score for which the area under the standard normal curve to the right of z_α equals α, as illustrated in Figure 10.1. Read "z_α" as "z sub alpha."

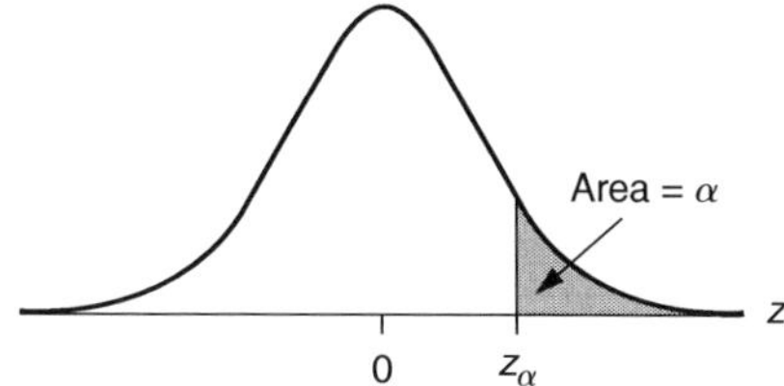

Figure 10.1 The z_α notation.

Table 10.1 gives some commonly used values of z_α. These values can be estimated from Table E. The t-table (Table F) in the back of the book has additional values of z_α in the bottom row (labeled ∞) just above the percentages.

Table 10.1 Some Commonly Used Values of z_α

$z_{0.10}$	$z_{0.05}$	$z_{0.025}$	$z_{0.01}$	$z_{0.005}$
1.282	1.645	1.960	2.326	2.576

For example, the area to the right of 1.645 is 0.05. It follows that if Z is a standard normal random variable, then $P(Z \geq 1.645) = 0.05$. In general, for any number α between 0 and 1,

$$P(Z \geq z_\alpha) = \alpha$$

as Figure 10.1 illustrates. By the symmetry of the normal density curve on either side of 0,

$$P(Z \leq -z_{\alpha/2}) = P(Z \geq z_{\alpha/2}) = \alpha/2$$

as Figure 10.2 illustrates. It follows that the remaining area in the middle must equal $1 - \alpha$. Hence,

$$P(-z_{\alpha/2} < Z < z_{\alpha/2}) = 1 - \alpha,$$

useful for constructing confidence intervals.

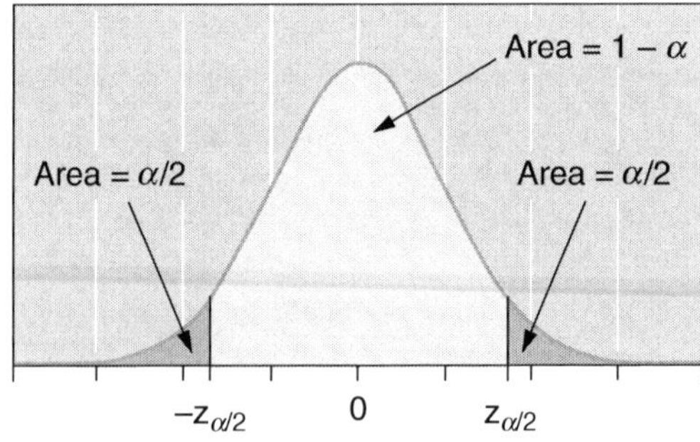

Figure 10.2 There is a $1 - \alpha$ probability that Z falls between $-z_{\alpha/2}$ and $z_{\alpha/2}$.

If we let $\alpha = 0.05$, then $1 - \alpha = 0.95$, or 95% when one uses percentages for probabilities. Since

$$z_{\alpha/2} = z_{0.025} = 1.96,$$

we see that there is a 95% probability that Z falls between -1.96 and 1.96.

If we let $\alpha = 0.10$, then $1 - \alpha = 0.90$, or 90%. Since

$$z_{\alpha/2} = z_{0.05} = 1.645,$$

we see that there is a 90% probability that Z falls between -1.645 and 1.645.

If we let $\alpha = 0.20$, then $1 - \alpha = 0.80$, or 80%. Since

$$z_{\alpha/2} = z_{0.10} = 1.282,$$

we see that there is an 80% probability that Z falls between -1.282 and 1.282.

Usually we express probabilities on the [0,1] scale. However, it is customary to use the [0%,100%] scale to talk about confidence levels or confidence probabilities as percentages.

We learned in Section 7.4 that if X is *exactly* normal with mean μ and standard deviation σ_x, then the standardized variable $Z = (X - \mu)/\sigma_X$ is *exactly* standard normal. Hence, it is not surprising that if a random variable X is approximately normal with mean μ and standard deviation σ_X, then the random variable

$$Z = (X - \mu)/\sigma_X$$

is approximately standard normal. Denoting multiplication by a dot $(\cdot)$, it follows that the chance that X will differ from μ by less than $z_{\alpha/2} \cdot \sigma_X$ is

$$\begin{aligned} &P(-z_{\alpha/2} \cdot \sigma_X < X - \mu < z_{\alpha/2} \cdot \sigma_X) \\ &= P(-z_{\alpha/2} < (X - \mu)/\sigma_X < z_{\alpha/2}) \approx 1 - \alpha. \end{aligned}$$

In particular, when X is approximately normal, there is approximately

- a 95% chance that X will differ from μ by less (in magnitude) than $1.96\sigma_X$
- a 90% chance that X will differ from μ by less than $1.645\sigma_X$
- an 80% chance that X will differ from μ by less than $1.282\sigma_X$

If we apply this to the sample mean $\overline{X}$ of n repeated, independent measurements with mean μ and standard deviation σ, then X has standard deviation of $\sigma/\sqrt{n}$, and we are approximately

- 95% "confident" that $\overline{X}$ will differ from μ by less (in magnitude) than $1.96\sigma/\sqrt{n}$
- 90% confident that $\overline{X}$ will differ from μ by less than $1.645\sigma/\sqrt{n}$
- 80% confident that $\overline{X}$ will differ from μ by less than $1.282\sigma/\sqrt{n}$

Then, we are approximately 95% confident that μ will be in the interval

$$(\overline{X} - 1.96\sigma/\sqrt{n},\ \overline{X} + 1.96\sigma/\sqrt{n})$$

95% confidence interval

and call this interval a **95% confidence interval** (CI) for μ.

We are 90% confident that μ will be in the interval

$$(\overline{X} - 1.645\sigma/\sqrt{n},\ \overline{X} + 1.645\sigma/\sqrt{n})$$

90% confidence interval

and call this interval a **90% confidence interval** for μ.

We are 80% confident that μ will be in the interval

$$(\overline{X} - 1.282\sigma/\sqrt{n},\ \overline{X} + 1.282\sigma/\sqrt{n})$$

80% confidence interval

and call this interval an **80% confidence interval** for μ. We note the above claims are approximate because $\overline{X}$ is only approximately normal.

Thus, if σ is known, confidence intervals for μ have the form

$$(\overline{X} - z^* \cdot \sigma/\sqrt{n},\ \overline{X} + z^* \cdot \sigma/\sqrt{n}),$$

for a suitable choice of the multiplier z^*. The *confidence level C* is

$$\begin{aligned} C &= P(\overline{X} - z^* \cdot \sigma/\sqrt{n} < \mu < \overline{X} + z^* \cdot \sigma/\sqrt{n}) \\ &= P(-z^* < \sqrt{n}(\overline{X} - \mu)/\sigma < z^*) \\ &\approx \text{area under standard normal curve from } -z^* \text{ to } z^*, \end{aligned}$$

because $Z = \sqrt{n}(\overline{X} - \mu)/\sigma$ is approximately standard normal. Figure 10.3 illustrates how z^* and C are related.

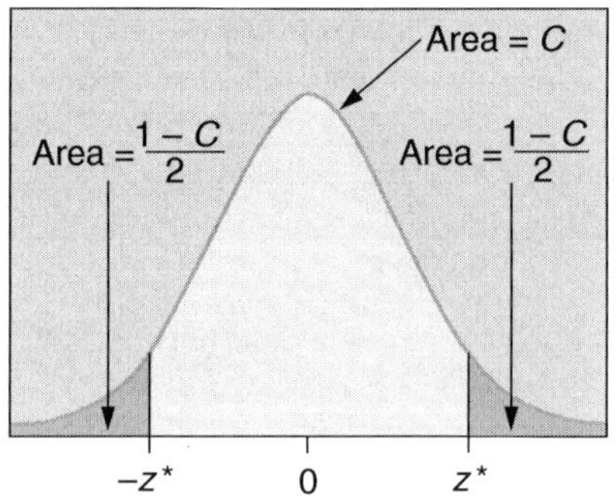

Figure 10.3 The area between $-z^*$ and z^* under the standard normal curve is C.

Table 10.2 gives some commonly used confidence levels C and the corresponding value of z^*. Additional values can be found in the last two rows of Table F. The most common choice of confidence level C is 0.95, or 95%. This choice offers a balance between precision (a narrow interval) and reliability (a high level of confidence).

Table 10.2 Some Commonly Used Confidence Levels C and Corresponding Value of z^*

z^*	1.282	1.645	1.960	2.326	2.576
C	80%	90%	95%	98%	99%

We note that the larger the multiplier z^* is, the more confident we are that the interval

$$\overline{X} \pm z^* \cdot \sigma/\sqrt{n}$$

will capture μ. But the larger z^* is, the less precise (that is, wider) the confidence interval will be.

Figure 10.4 illustrates the fact that a 95% confidence interval for μ may or may not capture μ. We simulated 25 random samples of the same size n from the same normal population with known mean μ and known standard deviation σ. Each sample consisted of n independent observations from the population. Since the sample mean $\overline{x}$ varied from sample to sample, so did the resulting 95% confidence interval for μ,

$$\overline{x} \pm 1.96\sigma/\sqrt{n}$$

In Figure 10.4 the 25 confidence intervals are represented by double arrows. A dot marks the sample mean $\overline{x}$ at the center of each interval. Out of the 25 confidence intervals, 24 intervals captured μ and one interval did not. In the long run, if we were to simulate more and more 95% confidence intervals, 95% of the intervals would capture μ and 5% would not. The practicing statistician builds only one confidence interval, but she draws encouragement from this imagining of a very large number of confidence intervals, where 95% succeed in placing μ in the interval.

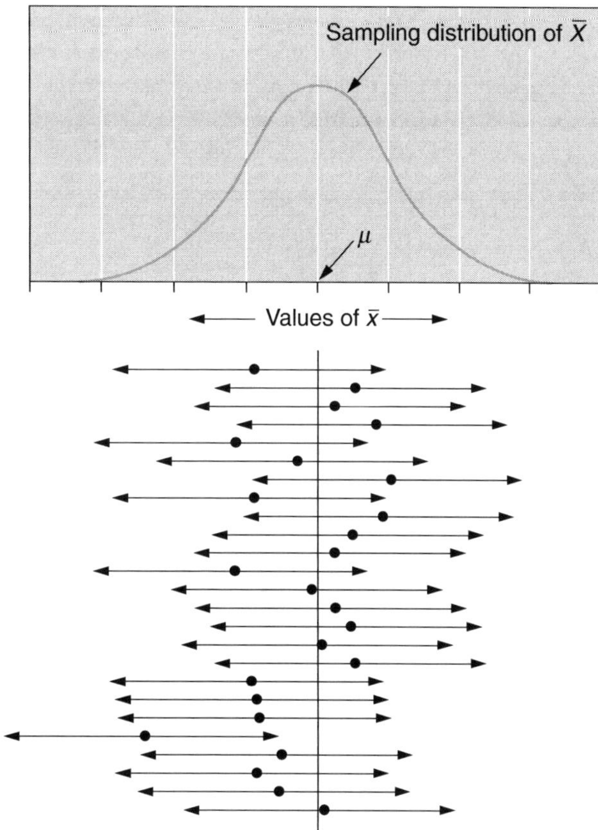

Figure 10.4 Twenty-four out of 25 simulated 95% confidence intervals for μ captured μ. One interval did not, giving a success rate of 96/100, or 96%.

Example 10.1

A jeweler's precision weighing scale is calibrated to produce unbiased readings that follow a normal curve centered at the true weight with standard deviation $\sigma = 0.006$ grams. The accuracy (typical error size) of the scale does not vary with the weight of the jewels being weighed. The jeweler weighs a pair of 22 karat gold bangles $n = 4$ times on the scale, getting the readings (in grams)

$$35.493 \quad 35.504 \quad 35.495 \quad 35.496$$

The jeweler estimates the weight of the bangles to be $\bar{x} = 35.497$ grams, give or take (using $\sigma = 0.006$ known)

$$\sigma/\sqrt{n} = 0.006/\sqrt{4} \text{ grams} = 0.003 \text{ grams}$$

or so. Since $1.96\sigma/\sqrt{n} = 0.006$ grams, the interval 35.497 grams ± 0.006 grams, or

$$(35.491 \text{ grams}, 35.503 \text{ grams})$$

is a 95% confidence interval for the true weight μ of the two bangles.

Can we be sure that μ is somewhere between 35.491 grams and 35.503 grams?

We are pretty confident that μ is between 35.491 grams and 35.503 grams. But we cannot be 100% sure. All we know is that in the long run nineteen out of twenty 95% confidence intervals will capture the true parameter value and only one out of twenty 95% confidence intervals will fail to do so. We cannot tell whether the interval (35.491, 35.503) is one of the nineteen good intervals that capture μ or the one bad interval that fails to capture μ.

So far we have constructed confidence intervals from independent observations $X_1, X_2, \ldots, X_n$ drawn from a conceptual population where observations come from measurement or experimental trials. Our formulas for confidence intervals are also valid if we construct them from a simple random sample $X_1, X_2, \ldots, X_n$ drawn without replacement from a real population whose size N is more than 20 times larger than the sample size n. To ensure that the sampling distribution of the sample mean $\overline{X}$ is approximately normal, we shall assume that either the population distribution is very close to normal or, if the sample size n is large, that the population distribution is not badly skewed and does not have outliers.

Example 10.2

Between 1960 and 1962 the Health Examination Survey examined a representative cross section of several thousand Americans, getting baseline data on the prevalence of diseases, demographic variables such as education and income, and physiological variables such as height, weight, and blood pressure. The average height of the almost 3600 adult women in the study was 63.1 inches with a standard deviation of 2.6 inches. The heights of the women followed a normal distribution without outliers.

Twenty years later (around 1980) the Health and Nutrition Examination Survey found that the height distribution of adult American women followed a normal curve with mean 63.7 inches and the same standard deviation of 2.6 inches. This survey examined a representative cross section of almost 6600 women.

Assume today that we take a simple random sample of 92 adult women and find that their average height is 64.6 inches. Does this provide strong evidence that, on average, adult women today are taller than 63.7 inches (the 1980 mean)?

Solution: Based on the information from the two surveys mentioned, we have sound reason to assume that today's height distribution for American women follows a normal curve with standard deviation $\sigma = 2.6$. We estimate μ, the average height of American women today, to be around 64.6 inches, give or take

$$\sigma/\sqrt{n} = 2.6/\sqrt{92} \text{ inches} = 0.27 \text{ inches}$$

or so. Using the last column of Table 10.2, we are 99% confident that μ is in the interval

$$\overline{x} \pm 2.576\,\sigma/\sqrt{n},$$

that is, between 63.9 inches and 65.3 inches. Thus we are more than 99% confident that, on average, adult women today are taller than 63.7 inches (the 1980 mean that falls to the left of the confidence interval).

Section 10.1 Summary

Assume that the data $x_1, x_2, \ldots, x_n$ are either the result of repeated, independent measurements (a sample from an experimental conceptual population) or a simple random sample (that is, a random sample drawn *without* replacement) from a real population that is at least 20 times larger than the sample size n. Assume that the population mean μ is unknown, but that the population standard deviation σ is known.

We can use the normal curve to construct confidence intervals for μ if the sampling distribution of the sample mean $\overline{X}$ is close to normal. To justify this assumption, check the data set for skewness and outliers. The more skewed the data set is, the larger the sample size n has to be before the approximation is good.

- If the sample size n is small (less than 20), then the population distribution has to be close to normal. Construct a normal probability plot to justify the assumption of normality. If the data are clearly nonnormal, do not use the normal curve to construct confidence intervals for μ.
- If the sample size n is at least 20, then we can use the normal curve to construct confidence intervals, as long as the data set is not too skewed (as seen by graphing its histogram) and does not have outliers. The more skewed the data set is, the larger the sample size n has to be.

In Chapter 8 we learned to estimate μ to be around x, give or take $\sigma/\sqrt{n}$ or so. Confidence intervals for μ convey the same information: our extimate $\overline{X}$ and the likely size of our estimation error, $\sigma/\sqrt{n}$.

Confidence intervals *for* μ have the form

$$\overline{x} \pm z^* \sigma/\sqrt{n}$$

for some positive constant z^*. The confidence level C, corresponding to a given choice of z^*, is

$$\begin{aligned} C &= P(\overline{X} - z^*\sigma/\sqrt{n} < \mu < \overline{X} + z^*\sigma/\sqrt{n}) \\ &\approx \text{area under standard normal curve from } -z^* \text{ to } z^*. \end{aligned}$$

In particular,

- $\overline{x} \pm 1.282\sigma/\sqrt{n}$ is an approximate 80% CI for μ,
- $\overline{x} \pm 1.645\sigma/\sqrt{n}$ is an approximate 90% CI for μ,
- $\overline{x} \pm 1.96\sigma/\sqrt{n}$ is an approximate 95% CI for μ,
- $\overline{x} \pm 2.576\sigma/\sqrt{n}$ is an approximate 99% CI for μ,

where the abbreviation CI stands for "confidence interval."

Section 10.1 Exercises

We shall use the abbreviation CI for "confidence interval."

1. Construct a 90% CI for the weight μ of the pair of gold bangles in Example 10.1. Which is wider, the 95% CI for μ or the 90% CI?
2. Suppose the jeweler in Example 10.1 weighed the pair of bangles nine times and not just four times. Would a 95% CI based on nine readings be wider or narrower than a 95% CI based on four readings? Explain your reasoning.
3. Use the data in Example 10.2 to construct a 95% CI for the average height of women today.
4. Suppose that a simple random sample of $n = 104$ American women was taken in 2003–04 and that their average height was $\overline{x} = 64.1$ inches. It is reasonable to assume that the height distribution of American women in 2003–04 followed a normal curve with standard deviation $\sigma = 2.6$ inches (as discussed in Example 10.2). Construct a 95% CI for the average height μ of American women in 2003–04.

5. Suppose that a simple random sample of $n = 138$ American men was taken in 2003–04 and that their average height was $\overline{x} = 69.6$ inches. There is reason to assume that the height distribution of American men in 2003–04 followed a normal curve with standard deviation $\sigma = 3.0$ inches. Construct a 95% CI for the average height μ of American men in 2003–04.

6. Let $\overline{x}$ denote the sample mean of a random sample of size n from an approximately normal population with unknown mean μ but known standard deviation σ. What is the confidence level C of the confidence interval

$$\overline{x} \pm 2.17\sigma/\sqrt{n}$$

for μ?

10.2 CONFIDENCE INTERVALS FOR A POPULATION MEAN μ WHEN σ IS UNKNOWN

Assume, as we did in Section 10.1, that the random observations $X_1, X_2, \ldots, X_n$ are either repeated, independent measurements (a sample from a conceptual population) or a simple random sample from a many times larger real population. Assume that the sampling distribution of the sample mean $\overline{X}$ is close to normal—either because the population is close to normal or because the sample size n is sufficiently large.

In Section 10.1 we assumed that the population mean μ was unknown, but somehow we knew the population standard deviation σ. More often we know neither μ nor σ.

If σ is known, then confidence intervals for μ have the form

$$\overline{x} \pm z^*\sigma/\sqrt{n},$$

for some positive constant z^*. If σ is unknown as well as μ, then we have to replace σ with the sample standard deviation s in this expression. Also, because of the variablity of S, we have to replace the multiplier z^*, derived from Table E, with a slightly larger multiplier t^*.

If σ is unknown, confidence intervals for μ take the form

$$\overline{x} \pm t^* s/\sqrt{n},$$

for some positive constant t^*. The confidence level C corresponding to a given choice of t^* is

$$\begin{aligned} C &= P(\overline{X} - t^*S/\sqrt{n} < \mu < \overline{X} + t^*S/\sqrt{n}) \\ &= P\left(-t^* < \sqrt{n}\left(\overline{X} - \mu\right)/S < t^*\right) \end{aligned}$$

because both events inside the two $P(\cdot)$s are exactly the same, as some algebra involving inequalities shows. Thus the confidence level C depends on the sampling distribution of the statistic

$$T = \sqrt{n}(\overline{X} - \mu)/S.$$

This sampling distribution is not normal. It is a new distribution, similar in shape to the standard normal distribution, a so-called t-distribution. To complicate matters, there is a different t-distribution for each sample size n. We call the parameter $df = n - 1$ the "degrees of freedom of the t-distribution." We identify a particular t-distribution by specifying its *degrees of freedom (df)*.

The One-Sample t-Statistic and the t-Distribution

Let $X_1, X_2, \ldots, X_n$ be independent observations from a normal population with mean μ and standard deviation σ. The *one-sample standardized t-statistic*

$$T = \frac{\overline{X} - \mu}{S / \sqrt{n}}$$

has a so-called t-distribution with $n - 1$ degrees of freedom.

Like normally distributed random variables, a t-distributed random variable has a density curve, called a *t-curve*. Each t-curve shares the following properties of the standard normal curve:

- The curve is bell-shaped and symmetric around 0.
- The curve extends indefinitely in both directions, always above the horizontal axis, approaching but never touching the axis.
- The total area under the curve is 1.

The t-distributions have more probability in the tails and less in the center (near 0) than the standard normal distribution does. As the number of degrees of freedom becomes larger and larger, the t-curves approach the standard normal curve. Figure 10.5 compares t-curves (also called t densities) with 1, 3, and 7 degrees of freedom, respectively, and the standard normal curve (or density).

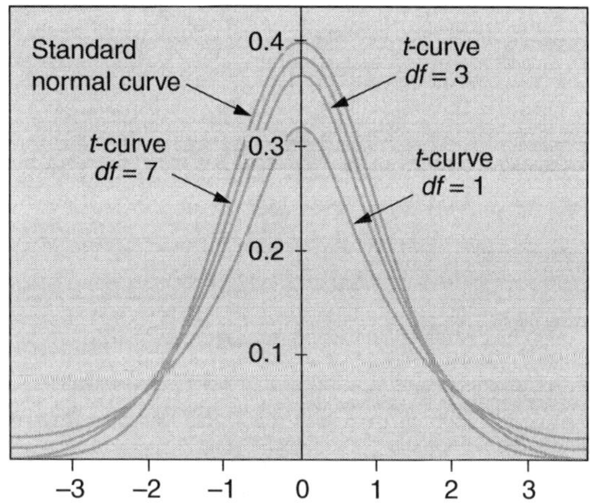

Figure 10.5 The standard normal curve and t-curves with 1, 3, and 7 degrees of freedom, respectively. All are bell-shaped and symmetric around 0. The t-curves have more area (probability) in the tails than the standard normal curve has. As the number of degrees of freedom increases, the t-curves approach the standard normal curve.

Now let $X_1, X_2, \ldots, X_n$ denote independent observations from a normal population with unknown mean μ and unknown standard deviation σ. To find the so-called critical value t^* such that the interval

$$\overline{X} \pm t^* S / \sqrt{n}$$

is a 95% confidence interval for μ, locate the column in the t-table (Table F) with "95%" at the bottom. Go up to the 95% column. The value t^* can be found in the 95% column in the horizontal line that begins with the number of degrees of freedom, namely $df = n - 1$. For example, if $n = 10$ and thus $df = 10 - 1 = 9$, then $t^* = 2.262$. If $n = 30$ and $df = 29$, then $t^* = 2.045$. Notice that as *df* increases as we go down the column, t^* decreases from 12.71 down toward 1.960—the normal curve value for a 95% confidence interval.

Example 10.3

Twenty-five independent measurements of the tensile strength of a composite material were made. The data are given in Table 8.3 in Section 8.4. The sample mean and sample standard deviation were $\bar{x} = 203.9336$ and $s = 18.320$ thousands of pounds per square inch, respectively. As a normal probability plot would confirm, the data look like a sample from a normal population. Since $n = 25$, $df = 25 - 1 = 24$ and we get $t^* = 2.064$ and the 95% confidence interval

$$203.9336 \pm 2.064 \cdot 18.320/\sqrt{25}$$

or

$$203.9336 \pm 7.5625$$

for the average tensile strength of the composite material.

❒ ❒

In Example 10.3 we assumed that the measurements were normally distributed. That raises the question: Since no real-world distribution is exactly normal, what happens to the confidence level C of our t confidence interval if the measurements are not normally distributed? The answer is: That depends! Outliers usually distort the confidence level. In the absence of outliers or strong skewness, however, nonnormality usually does not distort the confidence level very much. Confidence intervals based on the t-curve are fairly *robust*: Usually the confidence level does not change very much if the normality assumptions are violated, a fortunate fact learned from computer simulations.

Assume, as we did in Section 10.1, that the observations $X_1, X_2, \ldots, X_n$ are either repeated, independent measurements (a sample from a conceptual population) or a simple random sample from a many times larger real population. Assume that the sampling distribution of the sample mean $\bar{X}$ is close to normal—either because the population is close to normal, or because the sample size n is sufficiently large. Let the population have mean μ and standard deviation σ. Under these assumptions the one-sample t-statistic

$$T = \frac{\bar{X} - \mu}{S/\sqrt{n}}$$

will be approximately t-distributed with $n - 1$ degrees of freedom. If μ and σ are unknown, then the confidence level C of the interval

$$\bar{x} \pm t^* s/\sqrt{n}$$

is

$$C = P(\bar{X} - t^* S/\sqrt{n} < \mu < \bar{X} + t^* S/\sqrt{n})$$
$$\approx \text{area under } t \text{ curve from } -t^* \text{ to } t^*$$

In the following example we are considering a population distribution that is skewed to the right. But since the sample is large we can still use the t-curve to construct confidence intervals.

Example 10.4

Imagine that the average weight of a random sample of 90 men age 18 and over was $\overline{x} =$ 193.9 lb with a standard deviation of $s = 42.0$ lb. We wish to construct a 95% confidence interval for the average weight of all men age 18 and over. We know this distribution is skewed to the right, a consequence of obesity in the United States. Here, $df = 90 - 1 = 89$. Interpolating between 1.984 and 1.990 in Table F, we get $t^* = 1.987$ and the 95% confidence interval

$$193.9 \text{ lb} \pm 1.987 \cdot 42.0/\sqrt{90} \text{ lb}$$

or

$$193.9 \text{ lb} \pm 8.8 \text{ lb}$$

for the average weight of all men age 18 and over.

❒ ❒

We finally note that if the number of degrees of freedom is very large because n is very large, then we use the critical values derived from the normal curve (in the bottom row, labeled ∞ (imageine $n = \infty$), of the t-table).

Example 10.5

In Section 2.4 we examined a data set of the heights of 8585 nineteenth-century British men for which $\overline{x} = 67.02$ inches and $s = 2.564$ inches. If we pretend that these men are a simple random sample, then we get the 95% confidence interval

$$67.02 \text{ inches} \pm 1.96 \cdot 2.564/\sqrt{8585} \text{ inches}$$

or

$$67.02 \text{ inches} \pm 0.05 \text{ inches}$$

for the average height of all British men at the time. Note that the average estimated height of 5 feet 7 inches is a couple of inches less than today's average height.

❒ ❒

As long as the sample data set $x_1, x_2, \ldots, x_n$ is not too skewed and does not have outliers and the sample size n is reasonably large, we can use the t-curve to construct confidence intervals for μ. Outliers, however, often distort confidence intervals, as the following example illustrates.

Example 10.6

Since $\overline{x}$ and especially s are affected by outliers, so are confidence intervals. In Newcomb's data (Examples 2.8 and 2.11) there are 2 outliers. The inclusion of these outliers doubles the sample standard deviation s and therefore doubles the width of any confidence interval

$$\overline{x} \pm t^* s/\sqrt{n}.$$

Section 10.2 Summary

Assume that the data $x_1, x_2, \ldots, x_n$ are either the result of repeated, independent measurements (a sample from a conceptual population) or a simple random sample from a real population that is at least 20 times larger than the sample size n. Assume that both the population mean μ and the population standard deviation σ are unknown.

We can use the t-curve to construct confidence intervals for μ if the sampling distribution of the sample mean $\overline{X}$ is close to normal. To justify this assumption, check the data set for skewness and outliers. The more skewed the data set is, the larger the sample size n has to be.

- If the sample size n is small (less than 20), then the population distribution has to be close to normal. Construct a normal probability plot to justify the assumption of normality. If the data set is clearly nonnormal, do not use the t curve to construct confidence intervals for μ.
- If the sample size n is at least 20, then we can use the t-curve to construct confidence intervals for μ as long as the data set is not too skewed and does not have outliers. The more skewed the data set is, the larger the sample size n has to be.

Confidence intervals for μ with σ unknown have the form

$$\overline{x} \pm t^* s / \sqrt{n}$$

for some positive constant t^*. The confidence level C corresponding to a given choice of t^* is

$$C = P(\overline{X} - t^* S \,/\, \sqrt{n} < \mu < \overline{X} + t^* S \,/\, \sqrt{n})$$

$$\approx \text{area under } t \text{ curve with } df = n - 1 \text{ from } -t^* \text{ to } t^*.$$

For a specified confidence level C (such as 95% or 99%) the corresponding critical value t^* can be found in the t-table (Table F). Locate the number of degrees of freedom, $df = n - 1$, in the leftmost column and the confidence level C in the bottom row. Use interpolation (see Table D) if Table F does not have the value df equal to $n - 1$. The critical value t^* in the bottom row (labeled ∞) is the normal curve value.

As Figure 10.5 illustrates, t-curves have "fatter tails" and less area in the center than the standard normal curve. Consequently, to achieve a specified confidence level C (say 95%), the t-curve critical value t^* has to be greater than the corresponding normal curve value z^* (at the bottom of the column in Table F). Notice, however, that as we go down the column, increasing the number of degrees of freedom, t^* decrease toward the normal curve value z^*.

Section 10.2 Exercises

We shall use the abbreviation CI for "confidence interval."

1. Nine independent measurements were made on the tensile strength of a certain metal alloy. In increasing order, the measurements were 72, 88, 97, 105, 111, 118, 129, 139, 153. Here $\overline{x} = 112.4$ and $s = 25.5$.
   a. Construct a normal probability plot. Do the data look like independent observations from a normal population?
   b. Determine df, the number of degrees of freedom.
   c. Construct a 95% CI for the average tensile strength μ of the metal alloy.

2. Suppose that the average weight of a random sample of 90 women age 18–74 was $\overline{x} = 165.3$ lb with a standard deviation of $s = 44.2$ lb. Construct a 95% CI for the average weight of all women age 18–74.

3. Suppose that the average height of a random sample of 15 men age 18–24 was $\overline{x} = 69.6$ inches with a standard deviation of $s = 3.2$ inches. Construct a 95% CI for the average height of all 18–24 year old men. What assumptions are you making about the height distribution of all 18-24 year old men?

4. Suppose that the average weight of a random sample of 30 men age 18–24 was $\overline{x} = 177.3$ lb with a standard deviation of $s = 46.8$ lb. Construct a 95% CI for the average weight of all 18–24-year-old men. What assumptions are you making about the weight distribution of 18–24-year-old men?

5. A set of five independent measurements of the length of a certain specimen of microorganism is made. The sample mean value of the measurements is $\overline{x} = 27.5$ micrometers, with a sample SD of $s = 3.2$ micrometers. Find a 90% and a 99% CI for the true length of the microorganism. Which interval is wider? What assumptions are you making?

6. A random sample of 40 undergraduate students at a large university found that the mean study time was $\overline{x} = 19.5$ hours per week with a sample SD of $s = 4.05$ hours. Find a 90% CI for the mean study time per week for all undergraduates. Explain the meaning of your CI by imagining 100 students each independently doing such a survey.

7. Thirty cars were tested to determine the fuel economy of a certain model. The mean fuel economy was $\overline{x} = 19.45$ miles per gallon, with sample SD of $s = 1.14$ miles per gallon. Find a 95% CI for the mean fuel economy μ of the car model.

8. Explain the meaning of a 99% confidence level. Imagine constructing a very large number of CIs.

9. Which CI, when the two are based on the same data, is wider: an 80% or an 85% CI? Explain.

10. A British medical study found the average birth weight of 121 newborn babies born to non-smoking, non-drinking mothers to be $\overline{x} = 7.8$ pounds with a standard deviation of $s = 1.1$ pounds. The birth weights follow a normal curve. Give a 95% CI for the average birth weight of all newborn British babies born to non-smoking, non-drinking mothers.

11. The following data are from the article "Sensory and Mechanical Assessment of the Quality of Frankfurters" (*Journal of Texture Studies*, 1990, pp. 345–409). A random sample of $n = 9$ hot dogs yielded the following fat content values (as percentages):

25.2	21.3	17.0	29.8	21.0	25.5	16.0	20.9	19.5

Construct a normal probability plot to check whether the population is normally distributed. If so, construct a 95% CI for the population mean fat content μ.

12. Observing the seemingly large number of Dutch women on American college volleyball teams, it is remarked that Dutch women seem taller on average than American women in general. A random sample of five Dutch women yields heights in inches of 67, 70, 66, 69, 67. Suppose it is known from past studies of Dutch female heights that $\sigma = 2.7$ inches. Find a 95% CI for μ_{DUTCH}. (*Hint*: Recall that heights approximately follow a normal distribution.)

13. A random sample of 100 women was drawn from the population of all female students at a large university. The heights of the 100 women were plotted, and the density histogram followed a bell-shaped curve. The average height of the 100 women was $\overline{x} = 64.5$ inches, with a sample SD of $s = 2.5$ inches. Give an approximate 95% CI for the average height of all the female students at the university.

14. A random sample of 625 women aged 25 to 34 had average systolic (upper) blood pressure $\overline{x} = 121.5$ mmHg with a sample SD of $s = 12.5$ mmHg. The density histogram for the 625 blood pressure measurements followed a bell curve.
   a. Give a 68% CI for the average systolic blood pressure of all 25- to 34-year-old women. (Note that 0.68 is an unusual choice of confidence level.)
   b. Give a 95% CI for the average systolic blood pressure of all 25- to 34-year-old women.

10.3 LARGE-SAMPLE CONFIDENCE INTERVALS FOR A POPULATION PROPORTION p

In the Key Problem for this chapter the *St. Louis Post Dispatch* presented a 95% confidence interval for the proportion p of St. Louis County residents who opposed the construction of a new indoor football stadium. In this section we shall learn how to derive such confidence intervals.

Consider the following two scenarios:

(i) $\hat{p}$ is the experimental success rate in n independent success/failure trials with success probability p in each trial.

(ii) A proportion p of a large population have a certain attribute of interest. A simple random sample is drawn from the population, and $\hat{p}$ denotes the proportion of the sample that have the attribute. We assume that the population size N is at least 20 times greater than the sample size n, which is often the case in real-world applications.

Under either scenario, $\hat{p}$ is an unbiased estimator of p if p is unknown. Formally, this means that $E(\hat{p}) = p$. More intuitively, it means that if a large number of statisticians each carry out the same estimation, then the average of their individual estimates will be approximately p. That is,

$$\overline{\hat{p}} \approx p.$$

If the sample size n is so large that $np \geq 10$ and $n(1-p) \geq 10$, then, according the the Central Limit Theorem, the sampling distribution of $\hat{p}$ is approximately normal with mean p and standard deviation

$$\sqrt{p(1-p)}/\sqrt{n}.$$

(In the case of a simple random sample we are ignoring the finite population correction factor, which has little effect when the population size N is much larger than the size n of the random sample.) As we discussed in Section 9.2, the standardized test statistic

$$Z = \frac{\text{statistic} - \text{parameter}}{\text{SD of statistic}} = \frac{\hat{p} - p}{\text{SD}(\hat{p})} = \frac{\hat{p} - p}{\sqrt{p(1-p)}/\sqrt{n}}$$

is approximately standard normal. For large sample size n, the standard error of $\hat{p}$,

$$\text{SE}(\hat{p}) = \sqrt{\hat{p}(1-\hat{p})}/\sqrt{n},$$

will be very close but only approximately equal to the standard deviation of $\hat{p}$,

$$\text{SD}(\hat{p}) = \sqrt{p(1-p)}/\sqrt{n}.$$

It follows that for any positive constant z^*, the confidence interval for p,

$$\hat{p} \pm z^* \, \text{SE}(\hat{p})$$

has confidence level

$$\begin{aligned} C &= P(\hat{p} - z^*\text{SE}(\hat{p}) < p < \hat{p} + z^*\text{SE}(\hat{p})) \\ &\approx P(-z^* < (\hat{p} - p)/\text{SD}(\hat{p}) < z^*) \end{aligned}$$

$$\approx \text{area under standard normal curve from } -z^* \text{ to } z^*$$

where the above two equalities are only approximate because, as just stated, $SD(\hat{p}) \approx SE(\hat{p})$ and $(\hat{p} - p) / SD(\hat{p})$ is only approximately standard normal.

In particular, for large n, referring to Table 10.2 or the ∞ row of Table F,

- $\hat{p} \pm 1.282\ SE(\hat{p})$ is an approximate 80% CI for p,
- $\hat{p} \pm 1.645\ SE(\hat{p})$ is an approximate 90% CI for p,
- $\hat{p} \pm 1.96\ SE(\hat{p})$ is an approximate 95% CI for p,
- $\hat{p} \pm 2.576\ SE(\hat{p})$ is an approximate 99% CI for p,

where the abbreviation CI stands for "confidence interval." How large should n be? Since we cannot check whether $np \geq 10$ and $n(1-p) \geq 10$, if p is unknown, our rule of thumb is

$$n\hat{p} \geq 10 \text{ and } n(1-\hat{p}) \geq 10$$

Technical Comment

For values of np and/or $n(1-p)$ between 10 and 100, the confidence interval

$$\hat{p} \pm z^* SE(\hat{p})$$

often fails to achieve the claimed coverage probability C. In many instances the coverage probability of a 95% CI is only 94% or 93%. In some cases the coverage probability is as low as 92%. (Using the t-curve multiplier t^* instead of the normal-curve multiplier z^* would not solve this problem. We shall not present the solution proposed by Agresti and Coull in 1998.) The larger np and $n(1-p)$ are, however, the closer the coverage probability C will be to the nominal confidence level, namely

"area under standard normal curve from $-z^*$ to z^*."

So, in practice one can use the presented CI, but with the knowledge that its true coverage probability could be somewhat less, but not seriously less.

Example 10.7

Pepsi and Coca-Cola have recently been in competition to secure exclusive rights for their products on various college campuses and in large companies across the nation. Suppose a large college takes a random sample of 100 students on campus and finds that 60 students prefer Pepsi to Coke. Find a 95% CI for that college's population proportion p of students who prefer Pepsi.

Solution: Because $n\hat{p} = 100 \times 0.6 \geq 10$ and $n(1-\hat{p}) = 100 \times 0.4 \geq 10$, we can obtain a large-sample CLT-based 95% CI by using (as bulleted above)

$$\hat{p} \pm 1.96\ SE(\hat{p}).$$

Plugging $\hat{p} = 0.60$ into the formula for $SE(\hat{p})$, we get

$$0.60 \pm 1.96\sqrt{\frac{0.60(0.40)}{100}},$$

and the resulting 95% CI for p is (0.504, 0.696). We are therefore 95% confident that a majority, indeed at least 50.4% of the student population, prefer Pepsi. Perhaps the college's decision could be influenced by this information.

Example 10.8

A spectrophotometer can be used to measure the concentration of carbon monoxide in the air. In a calibration test, 50 measurements were taken of a laboratory gas sample with known CO concentration. Of the 50 independent measurements, 37 were within 5 parts per million (ppm) of the true concentration. We estimate the probability p that a measurement made with this instrument will be within 5 ppm of the true concentration to be around

$$\hat{p} = 37/50 = 0.740$$

give or take

$$\text{SE}(\hat{p}) = \sqrt{0.74(0.26)}/\sqrt{50} = 0.062$$

or so. Since $n\hat{p} = 37$ and $n(1 - \hat{p}) = 13$ are both greater than 10, we find the approximate 90% CI (recall $z^* = 1.645$) for p to be

$$\hat{p} \pm 1.645\ \text{SE}(\hat{p})$$

or, carrying out the computation,

$$0.740 \pm 0.102$$

We are roughly 90% confident that p is between 0.64 and 0.84.

Section 10.3 Exercises

1. In a random sample of 72 people in a city, 24 were in favor of lowering the drunk-driving blood alcohol level from 0.08 to 0.06. Find a 99% CI for the population proportion in favor of lowering the drunk-driving blood alcohol level from 0.08 to 0.06. Is it very unlikely that a majority of the population are in favor of the change?
2. In a random sample of 2000 U.S. citizens over the age of 18, 53.5% approve of the president. Find a 99% CI for the population proportion who approve of the president. If this is the president's first term and the next election were to occur soon after this poll was taken, would the president be very likely to be reelected?
3. In a random sample of 316 residents of the state of Illinois (population 12 million), 237 believe in the fairness of the jury system. Find a 90% CI for the proportion of people in the state of Illinois who believe the jury system is fair.
4. In a random sample of 1,872 U.S. households, 1,405 had home Internet access. Find a 95% CI for the proportion of all U.S. households with home Internet access.
5. In a random sample of 500 Illinois residents age 18 to 24 years old, 173 wsere enrolled in college. Find a 95% CI for the proportion of all 18–24-year-old residents of Illinois enrolled in college.
6. Answer the question in the Chapter 8 Key Problem.
7. Look at the formula for the standard deviation of the sample proportion $\hat{p}$, namely $\text{SD}(\hat{p}) = \sqrt{p(1-p)}/\sqrt{n})$. What value(s) of p will give the standard deviation its largest value? What value(s) of p will give the standard deviation its smallest value? (*Hint:* Try various $0 < p < 1$ in the formula.)
8. Suppose you believe that the value of the population proportion $0 < p < 1$ is very close to 0.5. How large should your random sample be to ensure that the length of a 95% CI for p is no larger than 0.08?
9. In a random sample of 16 college freshmen, 75% were in favor of lowering the drinking age from 21 to 18. If it is possible, based on what you have been taught, explain how you would find a 90% CI for the proportion p of college freshmen on campus who favor lowering the drinking age from 21 to 18. If it is not possible, explain your answer.
10. A study of electromechanical protection devices used in electrical power systems showed that of 193 devices that failed when tested, 75 failures were due to mechanical parts failures.

a. Find a point estimate for p, the proportion of all failures that are due to mechanical failures.
b. Find a 95% CI for p.

11. Power line noise, voltage variation, and power outages all can affect computer performance. When noise enters a television set, the result is static and snow; when noise enters a computer, errors can occur and circuits can be damaged. One study at a particular computer site found that out of 160 line disturbances that occurred during the study, 143 were due to line noise.
a. Estimate the proportion p of all line disturbances at that site that are due to line noise.
b. Find a 98% CI for p.

12. In a calibration test, a voltmeter is used to make 100 independent measurements of a known standard voltage. Of the 100 readings, 86 were within 0.01 V of the true voltage. Find a 95% CI for the probability p that a measurement made with this instrument is within 0.01 V of the true voltage.

10.4 USING BOOTSTRAPPING TO OBTAIN LARGE-SAMPLE CONFIDENCE INTERVALS

Three properties of the sample mean $\overline{X}$ and the sample proportion $\hat{p}$ enabled us to construct large-sample CIs for μ and p:

- If the sample size n is large, then the sampling distributions of $\overline{X}$ and $\hat{p}$ are almost normal.
- We know the bias of each estimator. (They are both unbiased.)
- We have simple formulas for their standard errors, namely

$$\text{SE}(\overline{X}) = \widehat{\text{SD}}(\overline{X}) = \frac{S}{\sqrt{n}} \quad \text{and} \quad \text{SE}(\hat{p}) = \widehat{\text{SD}}(\hat{p}) = \frac{\sqrt{\hat{p}(1-\hat{p})}}{\sqrt{n}}.$$

In Section 8.3 we looked at additional estimators with almost normal sampling distributions, such as the sample median and the sample standard deviation. We did not have a formula for the bias or the SD of these estimators, but in Section 8.4 we showed how we could use bootstrapping to estimate both the bias and SD of these estimators. Hence, reviewing Section 8.4 would be appropriate at this point.

Let $x_1, x_2, x_3, \ldots, x_n$ denote data obtained either from a repeated-trials random experiment (repeated independent observations from a conceptual population) or from a simple random sample (from a real population). Let θ denote a parameter of interest that describes an important characteristic of the population (conceptual or real). Assume that θ is unknown and let $\hat{\theta}$ denote an estimate of θ computed from the data $x_1, x_2, x_3, \ldots, x_n$. Assume that the sampling distribution of $\hat{\theta}$ is approximately normal with mean $\theta + \beta$, where β thus denotes the bias of $\hat{\theta}$. That is, $\beta = E(\hat{\theta}) - \theta$.

We can approximate the sampling distribution of $\hat{\theta}$ and estimate β and the SD of $\hat{\theta}$ by bootstrapping as learned in Section 8.4. We create a bootstrap box containing the original data set

$$\boxed{(x_1)\ (x_2)\ (x_3)\ \cdots\ (x_n)}$$

We then draw n times at random *with replacement* from the bootstrap box to get the simulated observations

$$x_1^*, x_2^*, x_3^*, \ldots, x_n^*$$

Keep in mind that the key to drawing such a bootstrap sample is to sample with replacement. We compute the simulated estimate $\hat{\theta}^*$ from the simulated observations using the same formula or computational procedure that gave us our original estimate $\hat{\theta}$ from the original data set. If we bootstrap at least 10,000 $\hat{\theta}^*$-values by drawing at least 10,000 bootstrap samples $x_1^*, x_2^*, x_3^*, \ldots, x_n^*$, then we can use

$$\mathrm{SE}(\hat{\theta}) = \text{SD of bootstrap-simulated } \hat{\theta}^*\text{-values}$$

to estimate $\mathrm{SD}(\hat{\theta})$.

Let θ_{box} denote the value of the parameter for the bootstrap box. For example, if $\theta = \sigma$, then $\theta_{\text{box}} = \sigma_{\text{box}}$. (See Section 6.3.) If $\overline{\theta}^*$ denotes the sample mean of the simulated $\hat{\theta}^*$-values, then

$$\hat{\beta} = \overline{\theta}^* - \theta_{\text{box}}$$

is our estimate of the bias β of $\hat{\theta}$.

If the bias β is positive, for example, this means that $\hat{\theta}$ is expected to be larger than θ and hence we might want to estimate θ not by the biased $\hat{\theta}$ but by the less biased $\hat{\theta} - \hat{\beta}$. Thus, our CI for θ should not be

$$\hat{\theta} \pm z^* \ \mathrm{SE}(\hat{\theta})$$

but rather

$$\hat{\theta} - \hat{\beta} \pm z^* \ \mathrm{SE}(\hat{\theta}).$$

In particular, we get the approximate 95% CI for θ ,

$$\hat{\theta} - \hat{\beta} \pm 1.96 \ \mathrm{SE}(\hat{\theta})$$

In general, for any positive constant z^* the CI

$$\hat{\theta} - \hat{\beta} \pm z^* \ \mathrm{SE}(\hat{\theta})$$

has approximate confidence level

"area under standard normal curve from $-z^*$ to z^*."

Example 10.9

The data in Table 8.3 in Example 8.23 represent 25 independent measurements of the tensile strength of a composite material. We are interested in the variability in tensile strength as measured by the standard deviation σ. We bootstrapped 50,000 values of the sample standard deviation S, getting the bell-shaped experimental histogram in Figure 8.13. Our bootstrap estimate of the bias of S was (noting that $\overline{s}^*$ denotes $\overline{\hat{s}}^*$, the sample mean if the simulated $\hat{s}^*$ values)

$$\hat{\beta} = \overline{s}^* - \sigma_{\text{bootstrap box}} = 17.75 - 17.95 = -0.20$$

and the standard error of S was

$$\mathrm{SE}(S) = \text{SD of simulated } s^*\text{-values} = 2.70.$$

We get the approximate 95% CI for σ:

$$s - \hat{\beta} \pm 1.96\ \mathrm{SE}(S)$$

or

$$18.3 + 0.2 \pm 5.3,$$

that is,

$$18.5 \pm 5.3.$$

Section 10.4 Exercises

1. Consider the 30 observations in Chapter 8 Review Exercise 10.
   a. Estimate the median force required to pull out one of the studs.
   b. Bootstrap 10,000 sample medians. Describe the shape of the experimental histogram for the 10,000 simulated values. List the mean and the SD of the simulated values.
   c. Estimate the bias of the sample median.
   d. Construct an approximate 95% CI for the median force required.
2. a. Estimate the standard deviation σ of the force required to pull out one of the studs in Chapter 8 Review Exercise 10.
   b. Bootstrap 10,000 sample standard deviations. Describe the shape of the experimental histogram for the 10,000 simulated values. List the mean and the SD of the simulated values.
   c. Estimate the bias of S. (*Hint:* $\sigma_{\text{box}} = 6.24$.)
   d. Construct an approximate 95% CI for σ.
3. Consider the 25 independent tensile strength measurements given in Table 8.3 in Example 8.23.
   a. Estimate the median tensile strength of the material.
   b. Bootstrap 10,000 sample medians. Describe the shape of the experimental histogram for the 10,000 simulated values. List the mean and the SD of the simulated values.
   c. Estimate the bias of the sample median.
   d. Construct an approximate 95% CI for the median tensile strength of the material.

10.5 CONFIDENCE INTERVAL FOR THE DIFFERENCE BETWEEN TWO POPULATION MEANS, $\mu_X - \mu_Y$

We often want to compare the means of two distinct populations. We might, for example, want to compare the average SAT score in critical reading for the freshman classes at Yale and the University of Illinois. How much harder is it to get into an Ivy League school? Or we might want to compare the average student loan debt of recent graduates from public and private universities. How much more expensive is an Ivy League education?

Suppose we want to estimate how much the average selling price of a two-bedroom home in Phoenix, Arizona, fell between 2005 and 2010. Let μ_X denote the average price in 2005, and let μ_Y denote the average price in 2010. To estimate the difference

$$\mu_X - \mu_Y,$$

we take a simple random sample of n_X homes sold in 2005 and an independent simple random sample of n_Y homes sold in 2010. Let $\overline{x}$ denote the average selling price of the sample of homes sold in 2005 and s_X the corresponding sample standard deviation. Let $\overline{y}$ denote the average selling price of the sample of homes sold in 2010 and s_Y the corresponding sample standard deviation. Our estimate of $\mu_X - \mu_Y$ is

$$\overline{x} - \overline{y}.$$

To determine the standard error of $\overline{x} - \overline{y}$ we shall need the following square root law:

> The standard deviation of the difference of two independent random variables is $\sqrt{a^2 + b^2}$, where
>
> - a is the SD of the first variable
> - b is the SD of the second variable

Since standard errors are estimates of standard deviations, the same square root law holds for standard errors.

Here, the independent sample means $\overline{X}$ and $\overline{Y}$ have standard errors

$$a = \text{SE}(\overline{X}) = S_X/\sqrt{n_X}$$

and

$$b = \text{SE}(\overline{Y}) = S_Y/\sqrt{n_Y}.$$

So $\text{SE}(\overline{X} - \overline{Y})$ equals

$$\sqrt{a^2 + b^2} = \sqrt{S_X^2/n_X + S_Y^2/n_Y}$$

As usual, we denote this standard error as SE. Let df denote the smaller of the numbers $n_X - 1$ and $n_Y - 1$. Consider the t curve with df degrees of freedom. The confidence interval for the average decline in house prices $\mu_X - \mu_Y$, given by

$$\overline{x} - \overline{y} \pm t^* \cdot \text{SE},$$

has confidence level C close to the area under the t curve from $-t^*$ to t^*.

Example 10.10

Between 2003 and 2004 a large health-related survey examined a random sample of $n_X = 2696$ women age 18 and over. Their average weight was $\overline{x} = 165.3$ lb with a standard deviation of $s_X = 44.2$ lb. Twenty-five years earlier an independent survey examined a random sample of $n_Y = 6588$ women age 18 and over. The average weight of these women, surveyed between 1976 and 1980, was $\overline{y} = 144.2$ lb with a standard deviation of $s_Y = 32.3$ lb. If we pretend that the two samples are simple random samples, then the standard error for $\overline{x} - \overline{y}$ is

$$\text{SE} = \sqrt{\frac{(44.2)^2}{2696} + \frac{(32.3)^2}{6588}}\ \text{lb} = 0.94\ \text{lb}.$$

The t curve with $df = 2696 - 1 = 2695$ degrees of freedom (using the smaller sample size as required) is essentially the standard normal curve. We estimate that the average weight of women age 18 and up increased about

$$\overline{x} - \overline{y} = 165.3\ \text{lb} - 144.2\ \text{lb} = 21.1\ \text{lb}.$$

We get the 95% confidence interval for the increase as

$$\overline{x} - \overline{y} \pm 1.96 \cdot \text{SE}$$

or

$$21.1 \text{ lb} \pm 1.8 \text{ lb}.$$

Section 10.5 Summary

Consider two *independent* random samples from two distinct populations with unknown population means μ_X and μ_Y and unknown population standard deviations of σ_X and σ_Y. Assume that the samples are either repeated, independent measurements (samples from conceptual populations) or simple random samples from much larger real populations (at least 20 times larger than the samples). Assume that the sampling distributions of the sample means $\overline{X}$ and $\overline{Y}$ are close to the normal—either because the populations are close to normal or because the sample sizes n_X and n_Y are sufficiently large that the Central Limit Theorem guarantees appropriate normality. Then the sampling distribution of $\overline{X} - \overline{Y}$ will be close to normal with mean

$$E(\overline{X} - \overline{Y}) = \mu_X - \mu_Y$$

(the "mean") and standard deviation

$$\text{SD}(\overline{X} - \overline{Y}) = \sqrt{\frac{\sigma_X^2}{n_X} + \frac{\sigma_Y^2}{n_Y}},$$

by the square root law.

The *two-sample z statistic*

$$Z = \frac{\overline{X} - \overline{Y} - \text{mean}}{\text{SD}} = \frac{\overline{X} - \overline{Y} - (\mu_X - \mu_Y)}{\sqrt{\frac{\sigma_X^2}{n_X} + \frac{\sigma_Y^2}{n_Y}}}$$

will have (approximately) a standard normal sampling distribution.

If the population standard deviations σ_X and σ_Y are unknown, we estimate them using the corresponding sample standard deviations, S_X and S_Y. When we plug these estimates into the formula for $\text{SD}(\overline{X} - \overline{Y})$, we get the *standard error,*

$$\text{SE}(\overline{X} - \overline{Y}) = \sqrt{\frac{S_X^2}{n_X} + \frac{S_Y^2}{n_Y}}$$

and the two-sample z statistic becomes the *two-sample t statistic*:

$$T = \frac{\overline{X} - \overline{Y} - \text{mean}}{\text{SE}} = \frac{\overline{X} - \overline{Y} - (\mu_X - \mu_Y)}{\sqrt{\frac{S_X^2}{n_X} + \frac{S_Y^2}{n_Y}}}$$

Let $\overline{x}$ and $\overline{y}$ denote the observed sample means, and s_X and s_Y the observed sample standard deviations. Let df denote the smaller of the numbers $n_X - 1$ and $n_Y - 1$. Consider the t curve with df degrees of freedom. The confidence interval for $\mu_X - \mu_Y$, given by

$$(\overline{x} - \overline{y}) \pm t^* \cdot \text{SE},$$

has confidence level C close to the area under the t curve (with degrees of freedom $= df$) from $-t^*$ to t^*.

Technical Comment

Even if the two population distributions are normal, the two-sample t statistic is (usually) not t distributed. But the approximation is quite good if the sample sizes n_X and n_Y are at least 10, especially if $n_X = n_Y$ and the two population distributions being compared have similar shapes.

Section 10.5 Exercises

1. Rima and Sridevi each toss a fair six-sided die five times. Let $\overline{X}$ denote Rima's average score and $\overline{Y}$ Sridevi's average score. Here, $\sigma_X = \sigma_Y = 1.708$. (See Section 6.3.) Compute $\text{SD}(\overline{X} - \overline{Y})$, the standard deviation of the difference between their two averages.

2. A mining company wants to compare the average heat-producing capacities, μ_X and μ_Y, of coal produced at two mines (in millions of calories per ton). At the first mine the average of 15 measurements is $\overline{x} = 8250.33$ with $s_X = 124.5$. At the second mine the average of 16 measurements is $\overline{y} = 7918.75$ with $s_Y = 109.8$.
   a. Estimate $\mu_X - \mu_Y$.
   b. Construct an approximate 99% confidence interval for $\mu_X - \mu_Y$.

3. A scientist investigates how much the resistance of electrical wiring can be reduced by alloying. A random sample of 25 standard wires had average resistance $\overline{x} = 0.138$ ohm with $s_X = 0.003$ ohm. An independent random sample of 25 alloyed wires had average resistance $\overline{y} = 0.081$ ohm with $s_Y = 0.004$ ohm. Let μ_X and μ_Y denote the average resistance of the two types of wire. Construct an approximate 95% confidence interval for $\mu_X - \mu_Y$.

4. The denier is a unit of measure for the density of fibers. A single strand of silk is 1 denier. To compare the nylon yarn produced by two spinning machines, 10 random samples were taken from each machine. The samples from the first machine had a mean of $\overline{x} = 9.72$ deniers with $s_X = 1.81$ deniers, while the samples from the second machine had a mean of $\overline{y} = 7.41$ deniers with $s_Y = 1.48$ deniers. Let μ_X and μ_Y denote the mean denier measure of all nylon yarn produced by the two spinning machines. Construct an approximate 95% confidence interval for $\mu_X - \mu_Y$.

5. In a random sample of size 64, $\overline{x} = 50.3$ with $\sigma_X = 2.5$. In another independent random sample, of size 64 from a second population, $\overline{y} = 49.25$ with $\sigma_Y = 3.1$. Find the standard deviation of the difference of the sample means, $\overline{X} - \overline{Y}$.

6. In a random sample of size 75 at one large private school, the mean SAT verbal score was $\overline{x} = 680.33$ with sample SD of $s_X = 65.12$. In a random sample of size 75 from a large state school, the mean SAT verbal score was $\overline{y} = 619.67$ with sample SD of $s_Y = 80.91$. Find a 90% CI for the difference between the two schools mean verbal SAT scores.

7. Twenty-five independent measurements of the weight of a sample of a chemical compound were made on each of two scales. On the first scale, the mean of the 25 measurements was $\overline{x} = 19.45$ grams with sample SD of $s_X = 0.49$ grams. On the second scale the mean of the 25 measurements was $\overline{y} = 18.42$ grams with sample SD of $s_Y = 0.27$ grams. Find a 95% CI for the difference between the mean measurements of the two scales, $\mu_X - \mu_Y$. Is it likely that the two scales weigh objects the same? Or is there a systematic difference?

8. In a batch chemical process, two catalysts are being compared for their effect on the output of the process reaction. A sample of 12 batches is prepared using catalyst 1, and a sample of 10 batches is obtained using catalyst 2. The 12 batches for which catalyst 1 was used give an average yield of $\overline{x} = 85$ with a sample SD of $s_X = 4$, whereas the second sample, prepared with catalyst 2, gives an average of $\overline{y} = 81$ and a sample SD of $s_Y = 5$. Find a 90% CI for the difference between the population means, $\mu_X - \mu_Y$.

9. The tire lives of 12 randomly selected Goodyear X5W tires driven in Seattle had a mean of $\overline{x} = 48{,}100$ miles with $s_X = 8{,}000$ miles. The tire lives of 12 randomly selected X5W tires driven in Chicago had a mean of $\overline{y} = 44,000$ miles with standard deviation of $s_Y = 7{,}200$ miles. Find a 95% CI for the difference between the population means, $\mu_X - \mu_Y$.

10.6 LARGE-SAMPLE CONFIDENCE INTERVAL FOR THE DIFFERENCE $p_1 - p_2$ BETWEEN TWO POPULATION PROPORTIONS

Frequently, we want to construct a CI for the difference between two population proportions, $p_1 - p_2$. For example, we might want to compare the divorce rates for old and young couples, the cure rates of two medications, the proportions of Native American students graduating at two universities, the proportions of defective units produced on two assembly lines, and so on.

As in Section 10.5, we compare two distinct populations by examining *independent* samples from the two populations. We shall consider the following two scenarios that lead to differences of proportions:

(i) We are sampling from conceptual populations: $\hat{p}_1$ is the experimental success rate in n_1 independent success/failure trials with (unknown) success probability p_1 in each trial, and $\hat{p}_2$ is the experimental success rate in an *independent* series of n_2 independent success/failure trials with (unknown) success probability p_2 in each trial.

(ii) We are sampling from two large real populations. An unknown proportion p_1 of population 1 have a certain attribute of interest, and an unknown proportion p_2 of population 2 have the attribute of interest. We draw a simple random sample of size n_1 from population 1 and an *independent* simple random sample of size n_2 from population 2. We assume that the samples represent at most 5% of the respective populations. We let $\hat{p}_1$ and $\hat{p}_2$ denote the proportions of the samples from populations 1 and 2 that have the attribute of interest.

Our estimator of

$$p_1 - p_2$$

is the unbiased estimator

$$\hat{p}_1 - \hat{p}_2.$$

If the sample sizes n_1 and n_2 are so large that $n_1\hat{p}_1$, $n_1(1-\hat{p}_1)$, $n_2\hat{p}_2$, and $n_2(1-\hat{p}_2)$ are at least 10, then the sampling distribution of $\hat{p}_1 - \hat{p}_2$ will be close to the normal with standard deviation

$$\mathrm{SD}(\hat{p}_1 - \hat{p}_2) = \sqrt{\frac{p_1(1-p_1)}{n_1} + \frac{p_2(1-p_2)}{n_2}}.$$

by the square root law.

Our estimate of the standard deviation is the standard error

$$\mathrm{SE}(\hat{p}_1 - \hat{p}_2) = \sqrt{\frac{\hat{p}_1(1-\hat{p}_1)}{n_1} + \frac{\hat{p}_2(1-\hat{p}_2)}{n_2}}.$$

The large-sample confidence interval for $p_1 - p_2$, given by

$$\hat{p}_1 - \hat{p}_2 \pm z^* \cdot \text{SE},$$

has confidence level C close to the area under the standard normal curve from $-z^*$ to z^*.

Example 10.11

Two methods of producing rain by "seeding" clouds were compared. Method 1, with unknown success probability p_1, was successful in 20 out of 99 attempts; while method 2, with unknown success probability p_2, was successful 10 out of 102 attempts. Here, $\hat{p}_1 = 20/99 = 0.202$ and $\hat{p}_2 = 10/102 = 0.098$. Our estimate of $p_1 - p_2$ is

$$\hat{p}_1 - \hat{p}_2 = 0.202 - 0.098 = 0.104.$$

The standard error of $\hat{p}_1 - \hat{p}_2$ is

$$\text{SE} = \sqrt{\frac{0.202 \cdot 0.798}{99} + \frac{0.098 \cdot 0.902}{102}} = 0.050.$$

Because $n_1\hat{p}_1 = 20$, $n_1(1 - \hat{p}_1) = 79$, $n_2\hat{p}_2 = 10$, and $n_2(1 - \hat{p}_2) = 92$, we can use the normal curve to construct an approximate 95% confidence interval for $p_1 - p_2$:

$$(\hat{p}_1 - \hat{p}_2) \pm 1.96 \cdot SE$$

or

$$0.104 \pm 0.098.$$

We are approximately 95% confident that

$$0.006 \leq p_1 - p_2 \leq 0.202.$$

In particular, we are about 95% confident that in the long run method 1 would be more effective than method 2.

Example 10.12

A simple random sample of 100 voters in a large city school district produces $\hat{p}_1 = 0.44$ in favor of a tax increase to build a new high school. Worried by this low level of support, the school district pays a public relations (PR) firm to make their case in the community. After the PR firm has carried out its campaign, a second independent simple random sample of size 100 yields $\hat{p}_2 = 0.48$.

The PR firm argues that this is a "real" improvement and proposes a further (and expensive) attempt to "bring it above 50%." That is, the firm claims that p_2, the proportion of the population supporting the tax increase after the PR campaign, is greater than p_1, the proportion that supported a tax increase before the PR campaign. Construct a 95% CI for $p_1 - p_2$ to help evaluate the PR company's claim.

Solution: Note first that we are treating this problem as a difference in proportions between two populations. At first glance, it appears that this is a one-population problem because we are dealing with a single community. However, the issue in question is whether the PR campaign has changed the population proportion. Note that the two random samples are independent of each other, as needed.

We use, noting that $n_1\hat{p}_1 \geq 10, n_1(1-\hat{p}_1) \geq 10$, $n_2\hat{p}_2 \geq 10$, and $n_2(1-\hat{p}_2) \geq 10$ as required, the following CI:

$$(\hat{p}_1 - \hat{p}_2) \pm z^* \sqrt{\frac{\hat{p}_1(1-\hat{p}_1)}{n_1} + \frac{\hat{p}_2(1-\hat{p}_2)}{n_2}},$$

where $z^* = 1.96$ is determined by $P(-z^* < Z < z^*) = 0.95$. Thus we obtain

$$(0.44 - 0.48) \pm 1.96\sqrt{\frac{0.44(0.56)}{100} + \frac{0.48(0.52)}{100}},$$

which gives

$$-0.04 \pm 0.138$$

or equivalently

$$-0.178 \leq p_1 - p_2 \leq 0.098.$$

Thus this CI for $p_1 - p_2$ allows the possibility that $p_1 - p_2 = 0$, indicating that there is no strong statistical evidence of an increase in community support. The school district may not wish to spend the money on further PR efforts from this firm!

Section 10.6 Exercises

1. Two detergents were tested for their ability to remove grape juice stains. An inspector determined detergent 1 to be successful on 63 out of 91 independent stain trials and detergent 2 to be successful on 42 out of 79 independent stain trials. Find a 95% CI for the difference $p_1 - p_2$ in success rates of detergent 1 and detergent 2. Is it likely that one detergent is better than the other one for removing grape juice stains?

2. Superplasticized concrete is formed by adding chemicals to conventional concrete to make it more fluid so that it can be placed more easily. Suppose that a random sample of 50 new construction projects in the Dallas–Fort Worth area yields 15 that are using this type of concrete. A random sample of 60 new projects in the Boston area also yields 15 using superplasticized concrete.
   a. Let p_1 and p_2 denote the proportion of new construction projects in Dallas–Fort Worth and Boston, respectively, that are using superplasticized concrete. Find point estimates for p_1, p_2 and $p_1 - p_2$.
   b. Find a 95% CI for $p_1 - p_2$.

3. A recent newspaper/TV network survey was conducted to determine whether the proportion of adult males who favor the death penalty is greater than the percentage of adult females who favor the death penalty. A random sample of 800 adult males produced 480 who favored the death penalty, whereas a random sample of 800 adult females produced 410 who favored the death penalty.

   Construct a 99% CI for the difference between the proportions of adult males and adult females who favor the death penalty. Based on your CI, do you believe that the percentage of adult males who favor the death penalty is greater than the percentage of adult females? How confident are you?

4. Two types of metal detectors are in use in airports around the world. One is called a continuous wave detector, and the other is called a pulse field wave detector. The two devices are equally efficient at detecting large metal objects such as guns or knives. However, it is thought that the continuous wave detector tends to be less efficient in that it can be triggered more easily by objects such as coins, lipstick holders, and other small harmless metal objects.

   Let p_1 and p_2 denote the proportions of passengers passing through the continuous wave and the pulse wave detectors, respectively, who trigger the device. The contention is that the continuous wave detector will be triggered by a higher proportion of passengers than will the pulse wave device.

Random samples of 175 passengers are observed passing through each of these types of devices. Of those passing through the continuous wave detector, 113 triggered a warning. However, only 14 of those passing through the pulse field detector activated an alarm.

a. Estimate p_1, p_2, and $p_1 - p_2$.
b. Construct a 98% CI for $p_1 - p_2$.
c. Do the data *prove* that $p_1 > p_2$?

5. A large automobile company wants to determine whether the proportion of male policyholders who do not submit claims of under \$500 is the same as the proportion of female policyholders who do not submit claims of under \$500. A random sample of 400 male policyholders produced 320 who had not submitted claims of under \$500, whereas a random sample of 300 female policyholders produced 210 who had not submitted claims of under \$500.

   Construct a 95% CI for the difference between the proportions of males and of females who had not submitted auto insurance claims of under \$500. Based on your CI, do you believe that the two proportions differ? How confident are you?

6. The use of optical fibers in telecommunications, the military, and industry is increasing rapidly. These fibers must be strong, durable, able to operate over a wide temperature range, and insensitive to radiation. Most fiber failures are due to a brittle fracture that grows into a complete crack. Two different fiber-drawing heat sources are being studied. These are carbon furnaces and CO_2 laser heating.

   Let p_1 and p_2 denote proportions of failures occurring using the carbon furnace and CO_2 laser heating, respectively. Suppose that of 500 test fibers produced using the carbon furnace, 23 failed, whereas only 10 of the 600 fibers produced using laser heating failed.

   a. Estimate p_1, p_2, and $p_1 - p_2$.
   b. Construct a 95% CI for $p_1 - p_2$.

10.7 CONFIDENCE INTERVAL FOR THE DIFFERENCE OF TWO POPULATION MEANS IN THE MATCHED-PAIRS DESIGN CASE: $\mu_D = \mu_X - \mu_Y$

During the second half of the nineteenth century, improvements in nutrition, sanitation, and overall standard of living in England resulted in increased life expectancy and a one-inch increase in the average population height. The data in Pearson and Lee's 1903 paper "On the Laws of Inheritance in Man" reflect this increase in heights. (See Example 3.12 in Section 3.3.) Pearson and Lee studied the correlation between the heights of 1078 fathers and their adult sons. The average height of the fathers was $\overline{x} = 67.68$ inches with $s_X = 2.7$ inches. The average height of the sons was $\overline{y} = 68.65$ inches with $s_Y = 2.7$ inches. The sample standard deviation for the difference,

$$(\text{height of son}) - (\text{height of father})$$

for the 1078 father-son pairs was also 2.7 inches: $s_{Y-X} = 2.7$ inches. Let us pretend that Pearson and Lee studied a simple random sample. Then the standard error for $\overline{y} - \overline{x}$ is

$$S_{Y-X} / \sqrt{1078} = 0.082 \text{ inches.}$$

We get the 95% confidence interval for the average increase in height from the fathers' generation to the sons' generation, $\mu_Y - \mu_X$:

$$(68.65 - 67.68) \pm 1.962 \cdot 2.7 / \sqrt{1078} \text{ inches}$$

or

$$0.97 \pm 0.16 \text{ inches.}$$

Clearly sons have gotten quite a bit taller.

Pearson and Lee's study was a "matched-pairs" study: each man of the older generation was matched with a man from the younger generation, namely his son. The benefit of a matched-pair comparison of two populations is that variation within the population is less likely to obscure the differences between the populations. If Pearson and Lee had taken independent samples of size 1078 from the older and the younger generations to estimate $\mu_Y - \mu_X$, then the standard error would have been 50% larger:

$$\text{SE} = \sqrt{\frac{(2.7)^2}{1078} + \frac{(2.7)^2}{1078}} \text{ inches} = 0.116 \text{ inches.}$$

Our 95% confidence interval for $\mu_Y - \mu_X$ would have been 50% wider and therefore less precise:

$$0.97 \pm 0.23 \text{ inches.}$$

In many applications, two measurements are taken on the same subject (or unit). Often these are "before" and "after" measurements, such as the weight before and after a particular weight-loss program. This is a superior design because the design prevents the random variability between units from obscuring the effect of the factor we are interested in.

Example 10.13

A sample of 24 ninth- and tenth-grade girls were put on an intensive rope-jumping program in an effort to improve their track performance. The time difference for each girl for the 40-yard dash was measured:

$$\begin{aligned} D_i &= (\text{"before program" time}) - (\text{"after program" time}) \\ &\equiv X_i - Y_i \end{aligned}$$

where i indexes the ith person. Note that a positive number indicates an improvement in racing ability. The following data (D_i's) were obtained:

0.38	0.21	0.13	0.33	−0.03	0.17	−0.18	−0.14
−0.33	0.01	0.22	0.29	−0.08	0.23	0.00	0.04
0.20	−0.08	0.09	0.70	0.33	−0.14	0.50	0.36

We will assume that the distribution of a difference $D_i = X_i - Y_i$ in race times for a randomly sampled girl has mean μ_D and standard deviation σ_D. If the conditioning program does not help, then $\mu_D = 0$ will hold. A point estimate for the mean difference (improvement) in race times is the sample mean, $\overline{d}_{\text{OBS}} = 0.13$, and the sample standard deviation for the difference in race times is $s_D = 0.24$.

We now construct a 95% CI for the difference in mean race times. We can use the one-sample CI formula given by

$$\overline{d} \pm t^* s_D / \sqrt{n},$$

from Section 10.2.

Here, from Appendix F with $df = n - 1 = 23$, $t^* = t(23) = 2.069$, only slightly greater than the z^* value of 1.96. Moreover, $\text{SE}(\overline{D}) = s_D/\sqrt{n} = 0.049$, which gives the interval (0.03, 0.23). Therefore, the statistical evidence supports the effectiveness of the rope-jumping program, at the 95% confidence level.

Also because we were able to work with the D values, what at first appeared to be a two-sample problem (X_{before} and X_{after}) can and should be treated as a one-sample problem.

❐ ❐

Let's revisit Example 10.13 to make a point about its matched-pairs design. Suppose instead of 24 girls measured twice, we have two independent random samples of 24 girls each, with the first sample being the 24 "before program" girls (X-values) and the second sample being the 24 "after program" girls (Y-values). Let, as before,

$$D_i = X_i - Y_i$$

But here X_i and Y_i are independent as contrasted with X_i and Y_i both being measured from the *same* girl in the matched-pairs case and, hence, dependent random variables. In the case of two independent samples, recall by the square root law that

$$\text{SD}(X - Y) = \sqrt{(\text{SD}(X))^2 + (\text{SD}(Y))^2}$$

Thus the random variability of $X_i - Y_i$ *adds* the random variability of X_i to that of Y_i.

By contrast, in the case of the matched-pairs design, there is an advantage over the two independent-samples design. The fact that X_i and Y_i are dependent actually makes $\text{SD}(X_i - Y_i)$ *smaller* than $\sqrt{(\text{SD}(X))^2 + (\text{SD}(Y))^2}$ (often by a lot!). This is a good thing because a smaller SD means a shorter CI for $\mu_D = \mu_{X-Y}$. Instead of X_i and Y_i being independent, in the matched-pairs case X_i and Y_i tend to both be small (an athletic girl) or both be large (a non-athletic girl) and hence $X_i - Y_i$ tends to be small for all the girls. By contrast, in the independent-samples case $X_i - Y_i$ tends to be quite variable—for example, especially large if X_i is from a non-athletic girl and Y_i is from an athletic girl. Because $\text{SD}(X_i - Y_i)$ is smaller in the matched-pairs case, shorter and hence better CIs are produced, making the matched-pairs design superior to the two-independent-samples design.

Section 10.7 Exercise

1. Twelve students participated in a semester-long weight-training program. Students bench-pressed their maximum weight at the beginning (X_i) and the end (Y_i) of the program. Let $D_i = Y_i - X_i$ be the difference of ("after" − "before") maximum bench press weights for the ith student. The sample mean difference $d = 10.2$ pounds, and the sample $\text{SD}(D) = 5$ pounds. Assume that the population of (after versus before) D_i's is normally distributed. Calculate a 90% CI for the mean difference in maximum bench press weight. Was the program effective? Explain.

CHAPTER 10 SUMMARY

A confidence interval (CI) is often of the form

$$\text{estimator} \pm (\text{critical value}) \cdot \text{SE (estimator)},$$

where the critical value is obtained from an appropriate distributional table.

C denotes the confidence probability, namely $C = P$ (lower limit $\le$ parameter $\le$ upper limit), where the upper limit and lower limit are determined from the data and from the critical value. C gives the long-run proportion of the statisticians whose CI captures the parameter being estimated.

When σ is known and n is large ($n \ge 20$), and the data set is not too skewed and does not have outliers, then a $C\%$ CI for μ is given by

$$\overline{X} - z^*\sigma/\sqrt{n} \le \mu \le \overline{X} + z^*\sigma/\sqrt{n}$$

where z^* satisfies $p(-z^* \le Z \le z^*) = C$.

When n is small ($n < 20$), the same CI is used, *provided* the population distribution is close to normal.

If σ is unknown and n is large, then, assuming the data set is not too skewed and possesses no outliers, the CI is of the form

$$\overline{X} \pm t^*S/\sqrt{n}$$

where $df = n - 1$ for obtaining t^* from the Table F t-table. The t^*-based CIs are wider than the z^*-based ones.

We note that if the X's are a random sample from a real population, then for all the CIs above to be correct, the population size N must satisfy $N \ge 20n$, that is, N much larger than n.

The CI for a population proportion is

$$\hat{p} \pm z^* \sqrt{\hat{p}(1-\hat{p})}/\sqrt{n}$$

provided n is large, namely $n\hat{p} \ge 10$ and $n(1-\hat{p}) \ge 10$. If the sampling is from a real population, then, as before, $N \ge 20n$ is required.

Large-sample CIs can be found via bootstrapping where the CI for θ is of the form

$$\hat{\theta} - \hat{\beta} \pm z^* \, \mathrm{SE}(\hat{\theta})$$

where the estimated bias $\hat{\beta}$ and the estimated SE, namely $\mathrm{SE}(\hat{\theta})$, are found via bootstrapping.

Other CI cases covered in the chapter are the two-population case of $\mu_X - \mu_Y$ (both large-sample and small-sample normal population case), difference of two proportions $p_1 - p_2$ (large sample case), and the matched-pair $\mu_D = \mu_X - \mu_Y$ case.

Consider two *independent* random samples of sizes n_X and n_Y, drawn from two distinct populations with unknown population means μ_X and μ_Y and unknown standard deviations. Let $\overline{x}$ and $\overline{y}$ denote the sample means and s_X and s_Y the sample standard deviations. Assume that the sample sizes n_X and n_Y are at least 10, but less than 5% of the population sizes. Our estimate of $\mu_X - \mu_Y$ is $\overline{x} - \overline{y}$. The standard error is

$$\mathrm{SE} = \sqrt{\frac{S_X^2}{n_X} + \frac{S_Y^2}{n_Y}}.$$

Let df denote the smaller of the numbers $n_X - 1$ and $n_Y - 1$. Consider the t curve with df degrees of freedom. The confidence interval for $\mu_X - \mu_Y$, given by

$$(\overline{x}-\overline{y}) \pm t^* \cdot \text{SE},$$

has confidence level C close to area under t curve from $-t^*$ to t^*.

Consider two *independent* random samples of sizes n_1 and n_2, drawn from two distinct populations of which unknown proportions p_1 and p_2 have a certain attribute of interest. Let $\hat{p}_1$ and $\hat{p}_2$ denote the proportions of the samples that have the attribute of interest. Assume that n_1 and n_2 are less than 5% of their respective population sizes, but large enough that $n_1\hat{p}_1$, $n_1(1-\hat{p}_1)$, $n_2\hat{p}_2$, and $n_2(1-\hat{p}_2)$ are each at least 10. Our estimate of $p_1 - p_2$ is $\hat{p}_1 - \hat{p}_2$. The standard error is

$$\text{SE} = \sqrt{\frac{\hat{p}_1(1-\hat{p}_1)}{n_1} + \frac{\hat{p}_2(1-\hat{p}_2)}{n_2}}.$$

The confidence interval for $p_1 - p_2$, given by

$$(\hat{p}_1 - \hat{p}_2) \pm z^* \cdot \text{SE}$$

has confidence level C close to area under standard normal curve from $-z^*$ to z^*.

CHAPTER REVIEW EXERCISES

1. Explain the meaning of a 90% confidence level.
2. Which density histogram will look more like a normal curve: (i) one based on a sample of 10 observations, (ii) one based on a sample of 100 observations, or (iii) in general, neither?
3. Which density histogram will look more like a normal curve: (i) one based on 50 sample means, each of which is based on a sample of 10 observations; (ii) one based on 50 sample means, each of which is based on a sample of 100 observations; or (iii) in general, neither?
4. What are the mean and SD of a sample mean $\overline{X}$ of size 100 taken from a population with mean value $\mu = 65.25$ and SD $\sigma = 10.52$? What approximate shape does the density histogram of this sample mean have?
5. Explain how to estimate the standard deviation of a sample variance for a sample of size 30 by using the bootstrap method.
6. In a random sample of 100 people, 70 were regular viewers of the television show *American Idol*. Find the estimated standard error of this sample population.
7. In a random sample of 100 teenage boys, the mean amount of time spent watching television was $\overline{x} = 3.6$ hours with sample SD of $s_X = 0.9$ hours. In a random sample of 100 teenage girls, the mean amount of time spent watching television was $\overline{y} = 4.2$ hours with sample SD of $s_Y = 1.2$ hours. Find the estimated standard error of the difference between the two samples $(\overline{X} - \overline{Y})$.
8. See Exercise 6. Find a 95% CI for the population proportion of people who watch *American Idol* on a regular basis.
9. See Exercise 7. Find a 99% CI for the difference in the population mean television hours between boys and girls: $\mu_X - \mu_Y$.
10. Fifty measurements of the length of a bone were made. The mean of the measurements was $\overline{x} = 34.4$ centimeters with sample SD of $S = 0.25$ centimeters. Find a 90% CI for the true bone length. Is it likely that the bone is 35 centimeters long?
11. In a random sample of 250 residents of a large city, 48% of the people in the sample would favor raising the bar entrance age from 19 to 21. Find a 95% CI for the population proportion of people who would favor raising the

age. If a referendum were held, is it likely that more than one-half of the people would vote to raise the bar entrance age?

12. In a random sample of 50 women, the mean height was $\overline{x} = 64.45$ inches with sample SD of $s = 3.5$ inches. Find a 90% CI for the true mean height of women.

13. In a random sample of 100 women, the mean number of alcoholic drinks consumed per week was $\overline{y} = 3.2$ with sample SD of $s_X = 1.85$. In a random sample of 100 men, the mean number of alcoholic drinks consumed per week was $\overline{y} = 5.4$ with sample SD of $s_Y = 1.25$. Find a 99% CI for the difference between the two population means. Is it likely that in the population men and women consume the same amount of alcoholic drinks per week?

14. Complete the following table, which summarizes the information in this chapter on CIs.

				Confidence interval	
Parameter	**Population variance**	**Standard deviation of estimator**	**Estimated standard error of estimator (if needed)**	**Lower bound**	**Upper bound**
Mean	Known				
Mean	Unknown				
Proportion	Unknown				
Difference between two means	Known				
Difference between two means	Unknown				

15. The following is a stem-and-leaf plot of bootstrapped sample medians. Find the bootstrap-estimated standard error of the sample median.

Stem	Leaf
232	0
232	
233	0,0,0,0
233	5,5,5,5,5
234	0,0,0,0,0,0,0,0,0,0,0,0,0,0,0,0
234	5,5,5,5,5,5,5,5,5
235	0,0,0,0,0,0,0,0,0,0,0,0,0,0,0,0,0
235	5,5,5,5,5,5,5,5,5,5,5,5,5,5,5,5,5,5,5,5,5,5
236	0,0,0,0,0,0,0,0,0,0,0,0,0,0,0,0,0,0,0
236	5,5,5,5,5,5,5
237	0,0

16. A sample of 25 sapling trees from a large tree farm had a mean height of 6 feet, a median height of $M = 6.3$ feet, and $\sum(X_i - \overline{X})^2 = 97$.
   a. Find the standard error of the sample mean, if possible.
   b. Find the standard error of the sample median, if possible.
   c. Find an estimate of the population SD.

17. A candidate for president wishes to take a sample to assess support for her. She wants to take a simple random sample and have the standard error of the sample proportion be less than 0.02. Assuming that her support in the

population is about $p = 0.4$, what size sample will provide the desired standard error: $n = 100, n = 400, n = 625, n = 900, n = 1600$, or $n = 2500$?

18. A random sample of size 6 is observed:

12	15	10	13	13	18

a. Estimate $SD(\overline{X})$ using six bootstrap samples.
b. Estimate $SD(\overline{X})$ using the theoretical formula for $SD(\overline{X})$ and s computed from the sample.
c. Form an approximate 95% CI for μ using the SD-based bootstrapping approach.

PROFESSIONAL PROFILE

William Sealy Gossett 1876–1937, known as "Student," developer of the Student t-test

Gate to the Guinness brewery in Dublin, Ireland, where William Gossett worked in quality control.

William Sealy Gossett ("Student") in 1908.

The Guinness brewery in Dublin, Ireland, pioneered several quality control programs to maintain the taste and other characteristics of its popular products. In 1899, as a part of its quality control campaign, Guinness hired statistician William Sealy Gossett to collect and analyze data on the barley grown on its farms and on the production of its brews. Gossett acquired his statistical knowledge from his study, by gathering data and making mistakes, and by spending two periods (during 1906–1907) in the biometric laboratory of Karl Pearson. Gossett and Pearson developed a strong personal relationship, and the two worked together on statistical papers.

In 1908, with the help of Pearson, Gossett published papers addressing the issue of small samples. The papers were published under the pseudonym Student. There was a practical reason for this. Another Guinness researcher had published a paper revealing trade secrets of the brewery. To prevent further such disclosures, Guinness banned employees from publishing papers of any kind, no matter what the content. The pseudonym *Student* was used to prevent Guinness from knowing one of its employees had written the articles.

Almost all of Gossett'papers, including the influential "The probable error of a mean," were published in Pearson's journal *Biometrika* under the pseudonym *Student*. "The probable error of a mean" introduces use of the t-statistic, for which Gossett is famous, and provides the distribution of the t-statistic, now called the t-distribution. Both the Student t-statistic and the Studentized residuals are named in Gossett's honor.

Gossett worked for over 30 years in the area of barley cultivation. This work led him to conclude that in the study of crops the goal of experiments should not be limited to improving average yield but should also be to develop varieties that produce good yields under wide variations in soil and climate. This experimental design principle anticipates the later work of R. A. Fisher and Genichi Taguchi.

In 1935 Gossett became the head brewer in charge of the scientific aspects of brewing at the Guinness brewery in Park Royal, in northwest London. He died in 1937 in Beaconsfield, England.

11

More on Hypothesis Testing

The only useful function of a statistician is to make predictions, and thus to provide a basis for action.

W. E. Deming

Objectives

After studying this chapter, you will understand the following:

- Formulating a null hypothesis and its alternative
- Using a test statistic to assess the strength of evidence supporting a real-world claim
- Understanding level of significance, P-value, and power
- Testing hypotheses about one and two population proportions
- Testing hypotheses about one and two population means
- Testing hypotheses via normal population theory, central limit theorem, bootstrapping and other five-step simulation methods
- Z-test versus t-test
- Testing the equality of matched-pairs means
- Significance testing versus acceptance-rejection testing
- One-sided versus two-sided hypothesis test

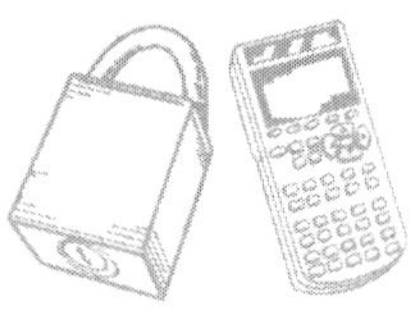

KEY PROBLEM

Data from the National Center for Health Statistics show that the average age of first-time mothers in the United States increased from 21.4 years in 1970 to 25.4 years in 2007. How long a woman waits before starting a family obviously depends on her individual circumstances. Research suggests that the trend toward later maternity is strongest among women with better educational qualifications, with some women postponing child rearing to pursue their careers. The average age of first-time mothers also varies from region to region. In the United States, first-time mothers living in New England are the oldest, with an average age at first birth of around 28 years. First-time mothers living in Arkansas, Louisiana, and Mississippi are the youngest, with an average age at first birth of around 23 years.

A researcher examined the records of a random sample of 41 British Columbian women who became first-time mothers in 2007. Their average age at time of delivery was 29.0 years with a standard deviation of 6.4 years. That is older than the U.S. national average. But since the sample size is small, the question is: Could the high sample mean be the result of nothing more than sample variation? Or is the British Columbian sample mean (29.0 years) so much higher than the U.S. population mean (25.4 years) that we can conclude that first-time British Columbian mothers (like first-time New England mothers) are older than the U.S. national average?

11.1 TESTS FOR A POPULATION MEAN

In Section 9.2 we performed tests of significance for a population proportion p. Such tests involve a null hypothesis

$$H_0 : p = p_0,$$

where p_0 is a specified value, and involve an alternative hypothesis H_A, usually of the form

$$H_A : p < p_0$$

or

$$H_A : p > p_0.$$

H_A is often a research hypothesis that we wish to prove.

The test is based on a *test statistic*. In Section 9.2 we used either the sample proportion $\hat{p}$ or took a large-sample approach using the Z-statistic,

$$Z = \frac{\text{statistic} - \text{parameter}}{\text{SD of statistic}} = \frac{\hat{p} - p_0}{\sqrt{p_0(1 - p_0)}/\sqrt{n}},$$

as our standardized test statistic.

The logic of hypothesis testing is straightfoward: If the sample data are inconsistent with H_0 (in the direction of H_A), then we reject H_0 in favor of H_A. If the sample data are consistent with H_0, then we cannot rule out (reject) H_0.

We decide whether or not to reject H_0 in favor of H_A based on the P-value. The P- value measures the strength of the evidence against H_0. The smaller P is, the stronger the evidence against H_0 is. The P-value is the probability of getting a test statistic value as extreme as, or even more extreme than, the test statistic value computed from the sample data. We estimate P under the assumption that H_0 is correct, using either box model simulations or calculations based on the test statistic's theoretical probability distribution, its so-called **sampling distribution.**

sampling distribution

In this section we shall perform tests of significance for a population mean μ. Using sample data we shall try to answer questions such as

- Do American adult females, on average, get their recommended daily allowance (RDA) of 18 mg iron? Do they, on average, get their RDA of 1 gram of calcium?
- In 1996, the average passenger car in the United States was driven 11,300 miles, according to government statistics. Do Americans, on average, drive more or less today?
- More and more newborn babies end up in intensive care because they are born prematurely. Is the average length of pregnancy for American women getting shorter, or are pediatricians just practicing "heroic medicine," trying to save extremely premature babies that standard medical practice would not have tried to save 20 years ago?

Assume, as we did in Sections 10.1 and 10.2, that the observations $X_1, X_2, \ldots, X_n$ are either repeated, independent measurements (a random sample from a conceptual population) or a simple random sample from a real population whose size is at least 20 times larger than the sample size n. Assume that the sampling distribution of the sample mean $\overline{X}$ is close to normal—either because the population is close to normal, or because the sample size n is sufficiently large. Let μ and σ denote the population mean and population standard deviation, respectively.

The null hypothesis for a test of significance concerning the population mean μ has the form

$$H_0 : \mu = \mu_0,$$

where μ_0 is a specified value. For example, the research hypothesis that adult American women, on average, get less than the recommended 18 mg of iron per day can be expressed as

$$H_A : \mu < 18,$$

where μ denotes the average daily iron intake (in milligrams) for adult American women. The corresponding null hypothesis is

$$H_0 : \mu = 18.$$

Z-statistic

In the rare cases where σ is known we shall use the so-called one-sample **Z-statistic,**

$$Z = \frac{\text{statistic} - \text{parameter}}{\text{SD of statistic}} = \frac{\overline{X} - \mu_0}{\sigma / \sqrt{n}}$$

using the fact from Chapter 8 that $\text{SD}(\overline{X}) = \sigma/\sqrt{n}$.

In the more common cases where σ is unknown, we estimate the standard deviation of the sample mean, $\text{SD}(\overline{X}) = \sigma/\sqrt{n}$, by the standard error of $\overline{X}$,

$$\text{SE}(\overline{X}) = S/\sqrt{n}$$

t-statistic

and use the so-called one-sample ***t*-statistic,**

$$T = \frac{\text{statistic} - \text{parameter}}{\text{SE of statistic}} = \frac{\overline{X} - \mu_0}{S/\sqrt{n}}$$

To be able to use either the standard normal density (Appendix E) or the t densities (Appendix F) to accurately estimate P-values, we shall have to make the following assumptions:

- If the sample size n is small (less than 20), then the population distribution has to be close to normal. To justify the assumption or normality, check the observed data set $x_1, x_2, \ldots, x_n$ for skewness and outliers. Construct a normal probability plot (often called a Q-Q plot). If the data are clearly nonnormal, do not use the z-test or the t-test.
- If the sample size n is at least 20, then we can use the z-test or the t-test as long as the data set is not too skewed and does not have outliers. The more skewed the data set is, the larger the sample size n has to be.

Distribution of the Z- and T-Statistics Computed from a Random Sample

Under H_0 and the stated assumption that either n is large or the data are (approximately) normal, Z is approximately standard normal and T is approximately t-distributed with $n - 1$ degrees of freedom. (See Section 10.2.) If $n \geq 30$, then T is almost standard normal because the t-distribution is almost standard normal. Thus one can use either a t-table (Appendix F) or a standard normal table (Appendix E) when $n \geq 30$.

First we consider a z-test for the mean of a normal population where σ is known.

Example 11.1

small sample z-test example of $H_0 : \mu = \mu_0$ when σ is known and the sampled population is normal

Between 1960 and 1962 the Health Examination Survey examined a representative cross section of several thousand Americans, getting baseline data on the prevalence of diseases, demographic variables such as education and income, and physiological variables such as height, weight, and blood pressure. The average height of the almost 3600 adult women in the study was 63.1 inches with a standard deviation of 2.6 inches. The heights of the women followed a normal distribution (approximately) without outliers.

Twenty years later (around 1980) the Health and Nutrition Examination Survey found that the height distribution of adult American women followed a normal curve with mean 63.7 inches and the same standard deviation of 2.6 inches. This survey examined a representative cross section of almost 6600 women.

Assume today that we take a simple random sample of 25 adult women and find that their average height is 64.6 inches. Does this prove that, on average, adult women today are taller than 63.7 inches (the 1980 mean)?

Solution: Based on the information from the two surveys mentioned, we have sound reason to assume that today's height distribution for American women follows a normal curve with standard deviation $\sigma = 2.6$. Thus, as we learned in the beginning of Section 8.2, $\overline{X}$ will be normally distributed, even for very small n and hence certainly for $n = 25$. Let μ denote the average height of American women today. Our research hypothesis (claim) is

$$H_A : \mu > 63.7.$$

The corresponding null hypothesis, which nullifies H_A, is

$$H_0 : \mu \leq 63.7,$$

which is replaced by

$$H_0 : \mu = 63.7$$

because 63.7 is the hardest-to-reject parameter value.

Because $\overline{X}$ is $N(63.7, 2.6/\sqrt{25})$ when H_0 is true, it follows that the Z-statistic, which standardizes $\overline{X}$,

$$Z = \frac{\overline{X} - \mu_0}{\sigma/\sqrt{n}} = \frac{\overline{X} - 63.7}{2.6/\sqrt{25}},$$

is standard normal under H_0, and hence the Appendix E standard normal tables can be used. The observed value of the Z-test statistic is

$$z = \frac{\overline{x} - \mu_0}{\sigma/\sqrt{n}} = \frac{64.6 - 63.7}{2.6/\sqrt{25}} = 1.73.$$

According to Appendix E,

$$P = P(Z \geq 1.73 \mid H_0) = 0.0418.$$

Here P is computed under the assumption that H_0 is true. Because P is less than our 0.05 threshold, we reject H_0 in favor of H_A and conclude that women today are taller than they were 30 years ago. The result is statistically significant, but not highly significant—because of the small sample size!

❒ ❒

Next we consider a t-test for the mean of a population.

Example 11.2

The Public Health Service examined a representative cross section of several thousand men aged 18 to 74. The systolic blood pressure of these men followed an almost normal distribution with mean $\mu = 129$ and standard deviation $\sigma = 18$.

Suppose that we take a simple random sample of 24 women aged 18 to 74 and find that their average systolic blood pressure is $\overline{x} = 120$ with a standard deviation of $s = 21$. Does this prove that, on average, women have lower systolic blood pressure than men?

Solution: The distribution of systolic blood pressure for women aged 18 to 74 is assumed to follow an almost normal distribution (slightly skewed to the right) with unknown mean μ and unknown standard deviation σ. (There is no reason that the standard deviation should be the same for men and women.) Our claim is

$$H_A : \mu < 129.$$

The corresponding null hypothesis is

$$H_0 : \mu \geq 129,$$

which is replaced by

$$H_0 : \mu = 129$$

because 129 is the closest null hypothesis value (and the hardest value to reject in favor of H_A) to the μ values of H_A. The observed value of the test statistic T is

$$t = \frac{\overline{x} - \mu_0}{s/\sqrt{n}} = \frac{120 - 129}{21/\sqrt{24}} = -2.10.$$

Under H_0, T is t-distributed with $n - 1 = 23$ degrees of freedom. To be able to look up the P-value in the t-table, we note that

$$P = P(T \leq -2.10 \mid H_0) = P(T \geq 2.10 \mid H_0).$$

Here we are using the fact that the t density, like the standard normal density, is symmetric around 0. As Figure 11.1 shows, the area under the t curve to the left of –2.10 is identical to the area under the t curve to the right of 2.10.

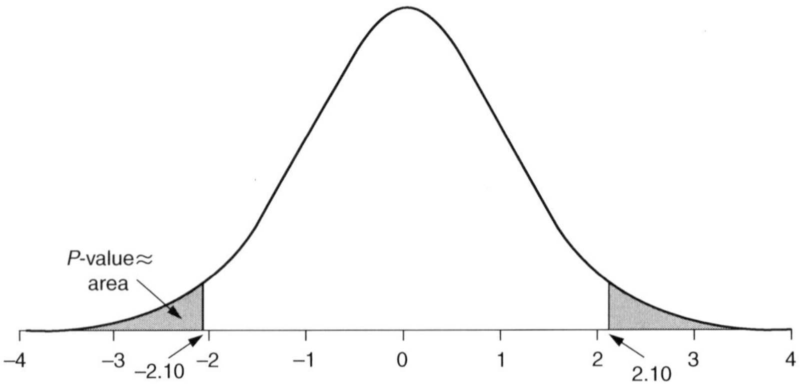

Figure 11.1 The t density with 23 degrees of freedom.

Appendix F allows us to estimate the area under the t curve to the right of 2.10. Locate the line in Table F with $df = 23$, that is, 23 degrees of freedom. You can't find the value 2.10 on that line. But you can find the values 2.069 and 2.177. To illustrate, we have reproduced a portion of Appendix F in Table 11.1.

Table 11.1 t-Distribution Probabilities from Appendix F

df	0.10	0.05	0.025	0.02	0.01
21	1.323	1.721	2.080	2.189	2.518
22	1.321	1.717	2.074	2.183	2.508
23	1.319	1.714	2.069	2.177	2.500
24	1.318	1.711	2.064	2.172	2.492
25	1.316	1.708	2.060	2.167	2.485

Because $2.069 < 2.10 < 2.177$, it follows that

$$0.025 > P > 0.02,$$

according to Appendix F. Because P is less than our 0.05 threshold, we reject H_0 in favor of H_A and conclude that women, on average, have lower blood pressure than men.

In summary, when we test $H_0 : \mu = \mu_0$ we perform a z-test if σ is known, using the test statistic

$$Z = (\overline{X} - \mu_0) / (\sigma / \sqrt{n})$$

and the standard normal table (Table E). If σ is unknown, then we perform a t-test, using the test statistic

$$T = (\overline{X} - \mu_0) / (S / \sqrt{n})$$

and the t-table (Table F).

We shall now answer the question in the Key Problem: Are first-time mothers in British Columbia older than the U.S. national average?

Example 11.3

The age distribution of British Columbian first-time mothers is assumed to follow a not-too-skewed distribution with unknown mean μ and unknown standard deviation σ. Our suspicion is

$$H_A : \mu > 25.4.$$

The corresponding null hypothesis, which nullifies H_A, is

$$H_0 : \mu \leq 25.4,$$

which is replaced by

$$H_0 : \mu = 25.4$$

because 25.4 is the hardest-to-reject H_0 parameter value. Let us pretend that the sample is a simple random sample. The observed value of the test statistic T is

$$t = \frac{\overline{x} - \mu_0}{s / \sqrt{n}} = \frac{29.0 - 25.4}{6.4 / \sqrt{41}} = 3.60.$$

Under H_0, T is apprximately t-distributed with $df = 41 - 1 = 40$. Using Table F, we get the P-value

$$P = P(T \geq 3.60 \mid H_0) < P(T \geq 3.551 \mid H_0) = 0.0005.$$

The result is very highly significant, and we reject H_0 in favor of H_A. We conclude that British Columbian first-time mothers (like first-time New England mothers) are older than the U.S. national average.

Let us consider the following example of a large-sample test for the mean μ using a T-statistic.

Example 11.4

You have been taught that 98.6 degrees Fahrenheit is the "normal" adult body temperature. Biologists, however, have come to suspect that the average temperature for healthy adults is actually less than 98.6 degrees Fahrenheit. Figure 11.2 is a histogram of temperature readings of 130 randomly sampled people from a large population of adults (data are from Allen L. Shoemaker, "What's Normal?—Temperature, Gender, and Heart Rate," *Journal of Statistics Education*, July 1996). Notice that much of the histogram is to the left of 98.6, the supposed normal temperature.

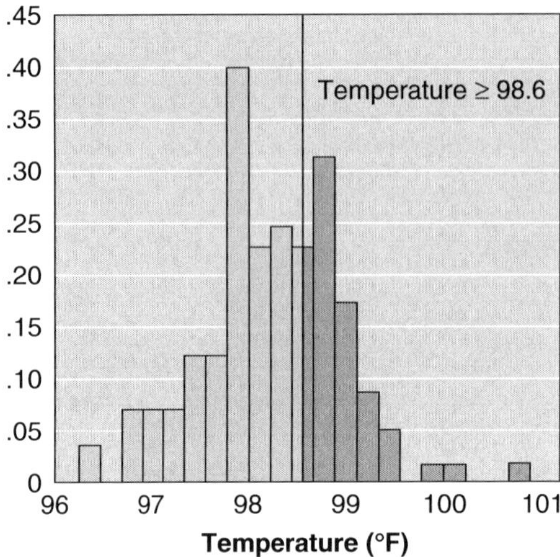

Figure 11.2 Density histogram of 130 body temperatures.

The average of the 130 readings is $\overline{x} = 98.25$, which is less than 98.6. The sample standard deviation was $s = 0.73$ °F. The question is whether this sample average of 98.25 is close enough to 98.6 that the difference could be due to mere chance variation. Or is the difference too big to be explained by chance variation alone? Can we conclude that the average body temperature μ for healthy adults is less than 98.6 degrees? Our suspicion is

$$H_A : \mu < 98.6\ ^\circ\text{F}.$$

The negation of H_A is

$$H_0 : \mu \geq 98.6\ ^\circ\text{F},$$

which is replaced by

$$H_0 : \mu = 98.6\ ^\circ\text{F}$$

because 98.6 is the hardest-to-reject H_0 parameter value. Let us pretend that the sample is a simple random sample.

The observed value of the test statistic T is

$$t = \frac{\overline{x} - \mu_0}{s/\sqrt{n}} = \frac{98.25 - 98.6}{0.73/\sqrt{130}} = -5.47.$$

Under H_0, T is approximately t-distributed with $n - 1 = 129$ degrees of freedom, a distribution that is almost indistinguishable from the standard normal distribution. Thus

$$P = P(T \leq -5.47 \mid H_0) \approx P(Z \leq -5.47) \approx 0.0000,$$

according to the footnote in Appendix E (or the ∞ line of the Appendix F table). Because $P < 0.001$, the result is *very highly significant*, and we conclude that the average adult body temperature is indeed less than 98.6 degrees Fahrenheit, as suspected.

Comment: The body temperature controversy is actually based on a misunderstanding. Outside the United States, where people use the metric system and measure temperatures in centigrade (°C), children are taught that the normal body temperature is around 37 °C. This is interpreted as meaning between 36 °C and 38 °C. Because the conversion formula is

$$°F = 32° + \frac{9}{5}\,°C,$$

the range from 36 °C to 38 °C translates into the range from 96.8 °F to 100.4 °F. As Figure 11.2 shows, the vast majority of body temperatures do indeed fall in this range. When we convert the "typical body temperature" of 37 °C into Fahrenheit, we get the "typical body temperature" of 98.6 °F (try it). Because of the decimal, we are misled into interpreting this as meaning: The majority of body temperatures are between 98.5 °F and 98.7 °F. And, as Figure 11.2 shows, that is a misinterpretation.

❐ ❐

We shall next see how we can use one-sample t-tests to study the mean difference between matched pairs of observations.

Tests for the Mean Difference of Matched Pairs

matched pairs

We are often interested in the mean difference of two random variables, often denoted X and Y, measured on each unit of a population. In fact, in Section 10.7, we developed a confidence interval for such a mean difference. The data are sometimes referred to as **matched pairs** because the two variables are paired within each unit in the population. A common example is the comparison of medical subjects before and after a treatment—for example, blood pressure before and after an exercise or a weight-loss program. Because this hypothesis test involves the difference between two variables, it is tempting to automatically think that the test is about the difference between the means of two populations. Therefore, it is important to clearly see that we are actually assessing the mean of a single random variable $D = X - Y$ (associated with a single population). This type of problem and its solution are illustrated in the following example. As we learned in Sections 4.4 and 10.7, there are advantages to a matched-pairs design because it allows us to examine the difference between X and Y for each population unit without having that difference obscured by the variation between units.

Example 11.5

Figure 11.3 shows the density histograms of the numbers of years of education of husbands and of wives for 177 Illinois couples in 1989 (data from the 1989 Current Population Survey). Suppose it is claimed that husbands tend to have more formal education. The distributions look reasonably similar, although it appears that more husbands than wives have at least some college education. Also, husbands go beyond a bachelor's degree more frequently than wives. The average number of years of schooling for the 177 husbands is $\overline{x} = 12.89$, and that for the 177 wives is $\overline{y} = 12.65$. The difference between the husbands' average educational attainment and the wives' is $\overline{d} = \overline{x-y} (= \overline{x} - \overline{y}) = 0.24$. Here $\overline{x-y}$ indicates that we compute each difference $d = x - y$ and then compute the mean of the differences. Can this observed difference be explained as nothing more than chance variation, or do husbands, on average, have more years of education than their wives in the Illinois population?

Solution: This problem is a matched-pairs comparison of two theoretical means, because the (X, Y) values are paired (a husband, X, is paired with his wife, Y). Thus, the married couples make up the units in the population, and it is the (husband – wife) difference that is measured on each couple. In other words, the variable we look at is

$$\text{difference} = (\text{husband's years of education}) - (\text{wife's years of education})$$

(i.e., $D_i = X_i - Y_i$ for the ith couple).

Figure 11.4 shows the density histogram of the 177 observed differences. One wife (with 14 years) has 11 more years than her husband (with 3)—hence the outlying rectangle centered at −11. Otherwise, the largest difference is +7 years. The average of these differences is $\overline{d} = 0.24$, and the sample SD of these differences is $s_D = 2.58$.

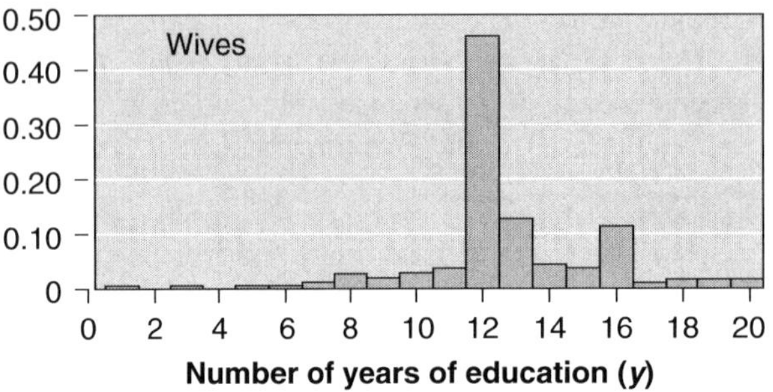

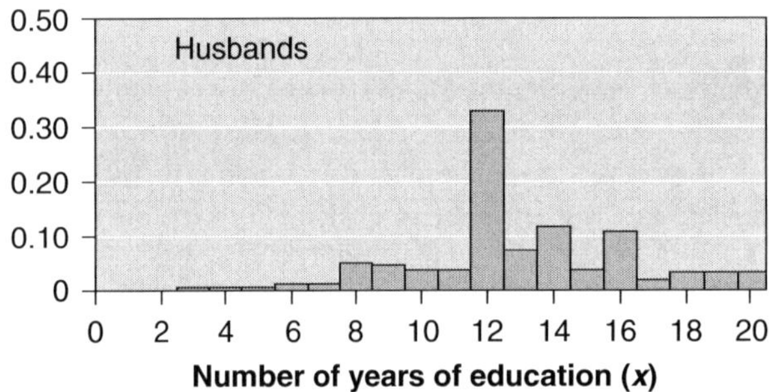

Figure 11.3 Density histograms of years of education for Illinois husbands and wives in 1989.

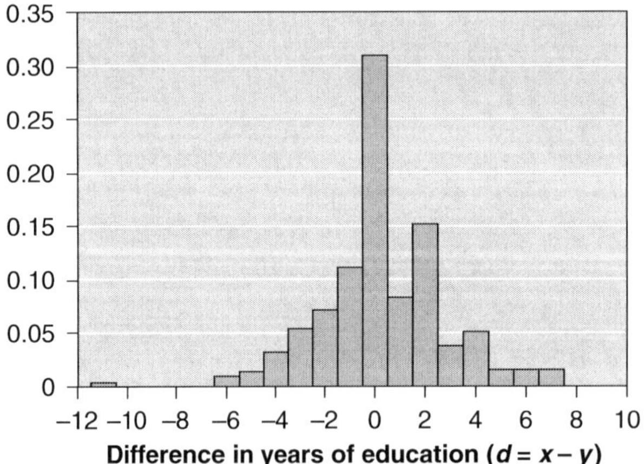

Figure 11.4 Density histogram of the observed differences in years of education for Illinois married couples in 1989.

Is it plausible that these differences could be a sample from a population of differences in which the average difference $\mu_D = 0$; that is, on average, is there no difference between the husband's and wife's numbers of years of education in the population of Illinois couples?

Note that we have made the problem into a single-population problem where the random variable $D = X - Y$ is being observed. We wish to test the following null hypothesis:

$$H_0 : \mu_D = 0 \quad \text{versus} \quad H_A : \mu_D > 0.$$

Let us pretend that the sample is a simple random sample. Under H_0, the test statistic

$$T = \frac{\overline{D} - 0}{S_D / \sqrt{n}}$$

is approximately t-distributed with $n - 1 = 176$ degrees of freedom, that is, approximately standard normal because n is very large. The observed value of the test statistic is

$$t = \frac{\overline{d} - 0}{S_D / \sqrt{n}} = \frac{0.24 - 0}{2.58 / \sqrt{177}} = 1.24.$$

The observed significance level or P-value is

$$P = P(T \geq 1.24 \mid H_0) \approx 0.1075,$$

according to Appendix E. An online P-value calculator would give $P = 0.1083$ for the t-distribution with $df = 176$. Because P is greater than our 0.05 threshold, we don't have enough evidence to reject H_0 in favor of H_A. We cannot prove statistically that the typical husband's number of years of schooling exceeds his wife's.

We shall next compare the responses to two treatments in a matched-pairs design by applying the one-sample t-test to the observed differences. The example was used by W. S. Gossett in 1908 to introduce the t-test and hence is of historical importance to the field of statistics.

Example 11.6

small-sample t-test example of $H_0 : \mu_0 = 0$ when the sampled matched pairs are normal

Two medications intended to increase sleeping time at night were administered to each of 10 patients, and for each patient the number of hours of sleep gained after taking each medication was measured. The treatments were dextro hyoscyamine hydrobromide and laevo hyoscyamine hydrobromide (two kinds of molecules that are mirror images of each other). The data for this study are presented in Table 11.2.

On average, the 10 patients gained 45 minutes of sleep after taking "dextro" and 2 hours and 20 minutes of sleep after taking "laevo." Can this difference be explained as nothing more than chance variation, or can we conclude that laevo is a more effective sleep aid than dextro?

Let μ_D denote the mean difference in the responses to the two treatments for the entire population; that is, μ_D would be the population mean if for each population member we measured the difference

$$D = (\text{average laevo sleep gain}) - (\text{average dextro sleep gain}).$$

Table 11.2 Sleep Gains for Two Medications

Patient	Dextro sleep gain	Laevo sleep gain	Difference (laevo gain − dextro gain)
A	0.7	1.9	1.2
B	−1.6	0.8	2.4
C	−0.2	1.1	1.3
D	−1.2	0.1	1.3
E	−0.1	−0.1	0.0
F	3.4	4.4	1.0
G	3.7	5.5	1.8
H	0.8	1.6	0.8
I	0.0	4.6	4.6
J	2.0	3.4	1.4

Student, "The Probable Error of a Mean," *Biometrika*, vol. 6 (1908), pp. 1-25.

Our research hypothesis is

$$H_A : \mu_D > 0 \text{ (laevo more effective).}$$

The corresponding null hypothesis is

$$H_0 : \mu_D = 0.$$

If we pretend that the sample is a simple random sample of $n = 10$ test subjects and assume that the D-values are close to normally distributed, the test statistic

$$T = \frac{\overline{D} - 0}{S_D/\sqrt{n}}$$

is approximately t-distributed with $n - 1 = 9$ degrees of freedom, under H_0. The observed value of the test statistic is

$$t = \frac{\overline{d} - 0}{S_D/\sqrt{n}} = \frac{1.58 - 0}{1.23/\sqrt{10}} = 4.06.$$

Using Appendix F, we get (noting the 3.690 entry for $df = 9$ with tail probability 0.0025)

$$P = P(T \geq 4.06 \mid H_0) < 0.0025.$$

Because P is less than our 0.05 threshold, we reject H_0 and conclude that laevo does give more hours of additional sleep than dextro. The result is highly statistically significant.

Bootstrap Hypothesis Testing of One Population Mean

Testing $H_0 : \mu = \mu_0$ Using a Bootstrap Approach

Let us return to the problem of Example 11.4, where we asked the following question about the population mean: Does the sample average of $\overline{x}_{\text{OBS}} = 98.25$ compel us to reject the null hypothesis $H_0 : \mu = 98.6$ in favor of the claimed $\mu < 98.6$? To answer that question, we asked what the chance is that, assuming the null hypothesis is true, we observe a sample mean temperature as low as or lower than the observed 98.25. To decide, we calculated a T-statistic and used the fact that n was large to justify doing a t-test.

But in applications where n is small to moderate and our sample data indicate that the population is far from normal, a t-test may not be a valid option because n is not large enough to justify using the t density. For example, the differences in Table 11.2 do not seem to be normally distributed. They seem to follow a fat-tailed, possibly skewed, distribution For this situation we introduce bootstrap-based hypothesis testing. For convenience we will illustrate the method for the problem of Example 11.4, even though bootstrap hypothesis testing is not needed for this large-sample problem. This method allows us to compare the t-test and the bootstrap approach in a situation where both are valid and, hence, both should produce similar results.

Bootstrap-based hypothesis testing is a special case of our five-step simulation approach to hypothesis testing. Recall from Section 9.2 that a key step in our simulation approach to statistical hypothesis testing is making a realistic choice of the null-hypothesis box model from which random samples can be drawn. Because we are testing a hypothesis, the specified box model must satisfy the null hypothesis as well as be a realistic model in terms of the shape and spread of the population distribution producing the observed sample if H_0 is true.

When we introduced bootstrapping in Section 8.4 as a method for estimating the standard error of an estimate, we learned that the central idea of bootstrapping is to use the sample data as the box model to substitute for the unknown population (probability model) that the data is viewed as being randomly sampled from. That is, the sample data becomes the box model. Because the random sample is representative of the population, its distribution will (approximately) have the shape, spread, and location of the unknown population distribution. Bootstrapping makes statistical inferences using the data alone.

Let us return to the setting of Example 11.4 to see how bootstrap hypothesis testing works. Recall the null hypothesis:

$$H_0 : \text{The population mean is 98.6; that is, } H_0 : \mu = 98.6.$$

We now consider the five steps required in Stage III (estimating P) for a **simulation-based bootstrap hypothesis test of $H_0 : \mu = \mu_0$.**

simulation-based bootstrap hypothesis test of $H_0 : \mu = \mu_0$

Step 1. Choose a Box Model (Bootstrap Definition of the Null-Hypothesis Box Model) We must choose a realistic box model, one that conforms with the null hypothesis and that approximates the shape and spread of the unknown population distribution under H_0.

In our study of hypothesis testing concerning a population proportion p, where the population consists only of 0s and 1s, the null hypothesis alone (specifying the proportion p_0 of 1s) was enough to determine the null-hypothesis box model. This will hold true again for chi-square testing of a fair die, as we will learn in Chapter 12. However, the null hypothesis alone is *not* enough to determine the box model for hypothesis testing of a population mean. Not only must the box model population have an average of 98.6, as assumed by the null hypothesis, but the box model population must also approximate the distribution of body temperatures of the real population of adults from which the actual sample was taken. Unlimited numbers of populations have a theoretical mean of 98.6. Indeed, the value at which a population is centered is independent of its spread about its center and independent of its general shape (symmetric, skewed, U-shaped, etc.). Thus, we do not know what model, and hence what numbered balls to put in the box, to assume for the null-hypothesis box model!

There are many cases in which the null hypothesis does not determine the null-hypothesis box model, and the CLT (central limit theorem) is not available to tell us the sampling distribution of the appropriate test statistic, as it does with T or Z when the sample size n is large. In such cases, bootstrap hypothesis testing, as described in this section, becomes a method of choice.

The basic bootstrap sampling presented in Sections 8.4 and 10.4 simply takes the observed sample as the box model. The justification for this is that the sample will have approximately the location, spread, and shape of the population, and the approximation will be quite good for large sample sizes and reasonably good for moderate sample sizes. Even for relatively small samples sizes it provides some valid statistical information. The box model that we want to create for testing a hypothesis concerning a population mean, however, must not only have the population's spread and general shape; it must also have as its theoretical mean the mean specified by the null hypothesis ($\mu = 98.6$ in Example 11.4). Taking the sample itself as the box model would approximate the population's spread and shape, as desired, but would result in a bootstrap sampling box with mean 98.25 (because $\overline{x}_{\text{OBS}} = 98.25$) rather than the desired 98.6 specified by $H_0 : \mu = 98.6$.

To construct the bootstrap-sampling box model for testing a hypothesis about the population mean, we simply add the difference between the null-hypothesis population mean and the observed sample mean ($0.35 = 98.6 - 98.25$) to each sample value. Then the box model has a new translated mean of $98.25 + 0.35 = 98.6$, as desired, but its spread and shape stay the same as the observed random sample. Thus the spread and shape are approximately those of the larger population being sampled from, as desired. This is precisely what we wanted as our null-hypothesis box model! That is, we construct a bootstrap box model for $H_0 : \mu = \mu_0$ as follows.

constructing a bootstrap box model for $H_0 : \mu = \mu_0$

Constructing a Bootstrap Box Model for $H_0 : \mu = \mu_0$

Suppose the observed sample $x_1, \ldots, x_n$ produces the sample mean $\overline{x}_{\text{OBS}}$.

Basic Principle The box model consists of $x_1^*, x_2^*, \ldots, x_n^*$ defined by

$$x_i^* = x_i + (\mu_0 - \overline{x}_{\text{OBS}}) .$$

Fact $\overline{x}^* = \frac{\sum x^*}{n} = \mu_0$, using the fact that $\sum x_i / n = \overline{x}_{\text{OBS}}$.

Thus the box model population mean is μ_0, as desired, and hence H_0 holds for the box model. Moreover, if n is moderate to large, the shape and spread of the box model will approximate the spread and shape of the unknown population model when H_0 is true.

The advantage of this nonparametric bootstrap approach is that we do not need to assume any particular theoretical distribution shape for the population (for example, a bell shape), but we let the data alone determine our estimate of the population's shape. The nonparametric approach to statistical inference is very powerful because the user takes no risk of being deceived by assuming an incorrect shape for the population distribution.

Step 2. Define One Simulation (One Bootstrap Sample) One simulation consists of randomly sampling, *with replacement*, 130 readings from the null-hypothesis box model (consisting of the null-hypothesis shifted sample). As we learned in Chapter 8, such random sampling with replacement from the bootstrap population model, defined by the actually observed data set, is the proper way to conduct bootstrap sampling.

Step 3. Define the Statistic of Interest The statistic of interest (i.e., the simulated test statistic to be used to decide whether to reject H_0) is the bootstrap sample mean, denoted $\overline{X}^*$. The asterisk, *, indicates that $\overline{X}^*$ is the simulated mean of a sample from a bootstrap population (box) model, not to be confused with the original $\overline{X}$ (without an asterisk), often denoted by $\overline{X}_{\text{OBS}}$, the observed sample mean.

Step 4. Repeat Simulations We perform the bootstrap sampling 100 times, each time obtaining a new mean. The stem-and-leaf plot in Table 11.3 contains these bootstrap sample means (i.e., $\overline{x}^*$-values). The average of the 100 $\overline{x}^*$-values is 98.6048, very close to the null-hypothesis value. Based on the law of large numbers, we expected these values to be close because the box model population mean is 98.6 and hence $E(\overline{X}^*) = 98.6$. The sample SD of the bootstrap sampled means is computed: $\mathrm{SE}(\overline{X}^*) = 0.0606$.

Table 11.3 Sample Mean Temperatures from 100 Bootstrap Samples of Size 130

Stem	**Leaf**
984	567
985	1222223334444444
985	55555667777778888889999
986	00000000001111112222222333344444
986	55555555666777788899
987	00014
987	55

Key: "984 567" stands for 98.45, 98.46, 98.47 degrees.

Step 5. Estimate the *P*-Value The P-value is approximated by $\hat{P}(\text{bootstrapped } \overline{X}^* \le 98.25 \mid H_0^*)$, using the 100 bootstrapped $\overline{x}^*$-values. Here we write H_0^* with the asterisk (indicating that it is the sample, shifted to make H_0 true, that stands in for the true unknown H_0 population), from which we obtain bootstrapped samples and their corresponding $\overline{x}^*$-values. It turns out that all the simulated sample means of $\overline{x}^*$-values in Step 4 were greater than 98.25: They ranged from 98.45 to 98.75. Thus the estimated P-value,

$$\hat{P}(\overline{X}^* \le 98.25 \mid H_0^*) = \frac{0}{100} = 0,$$

is much less than the conventional level-of-significance value of 0.05. We thus reject the null hypothesis, believing the evidence is very strong that, for the population from which the data were sampled, the population mean temperature μ is less than 98.6 degrees. We note that the CLT-based T-statistic approach of Example 11.4 and the bootstrapping approach yield almost identical results, namely P-values of approximately 0.

Note, however, that we ought to have bootstrapped 10,000 $\overline{x}^*$-values. Since we did only 100 bootstrap simulations, the experimental P-value

$$\hat{P} = 0$$

tells us only that the result is statistically significant, not how significant it is. With only 100 bootstrap samples the true P-value could easily be as high as 0.03, for all we know.

In the next example, we carry out Steps 1–5 in detail, focusing especially on Step 1, the crucial formation of the null-hypothesis-based box model in a simple case for instructional purposes.

Example 11.7

Let us simplify the problem of Example 11.4 by assuming that the sample is of 5 people rather than 130. Let us suppose because of Figure 11.2 that we do not believe the population is normal. Thus, in this small-sample case we certainly could not justify doing a t-test. That is, when we do not know if the population is approximately normal and when n is small, bootstrapping is a better choice for carrying out the hypothesis test.

Suppose the observed temperatures are

$$97.3, 97.5, 98.4, 98.6, \text{ and } 99.2.$$

The goal is to create a null-hypothesis bootstrap population (from which samples are to be drawn with replacement) with a spread and shape roughly like the observed data but with a mean of 98.6. The observed sample mean $\overline{X}$ of the five points is 98.2. To start building the bootstrap box model population, we have to adjust each sampled value so that the mean of the five values is 98.6. By the basic principle for forming a bootstrap box model, we have to add $\mu_0 - \overline{x}_{\text{OBS}} = 98.6 - 98.2 = 0.4$ to each sample value. For example, $97.3 + 0.4 = 97.7$, the first ball of the bootstrap box model. We then have the following bootstrap box model (check that these five values have a mean of 98.6, as required) as our null-hypothesis population (box model) from which simulated samples are to be drawn:

Now our plan is to randomly sample five numbers repeatedly with replacement from this box model null-hypothesis population of size 5. That is, we randomly draw one ball, record its value, and then put it back. We then randomly choose another (which could be the same as the first), record its value, and then put it back. We do this five times to obtain each Step 2 bootstrap simulation sample. For example, suppose we draw a size 5 random bootstrap sample with replacement and obtain

$$98.8, 99.6, 98.8, 97.9, \text{ and } 97.7.$$

This first bootstrap sample has a mean of $\overline{x}^* = 98.56$. For accuracy, we would then choose 100 or even 10,000 such bootstrap samples to obtain 100 or 10,000 $\overline{x}^*$-values, and from those $\overline{x}^*$-values we obtain our estimate of the P-value,

$$\hat{P}(\overline{X}^* \le 98.2 \mid H_0^*)$$

to decide whether to reject H_0 or not.

Let us illustrate bootstrap hypothesis testing with five bootstrap samples and their bootstrapped $\overline{x}^*$-values:

Sample number	Bootstrap samples of size 5					$\overline{x}^*$
1	98.8	97.6	98.8	99.9	97.7	98.56
2	97.9	99.0	97.9	99.6	99.6	98.80
3	97.9	97.7	99.6	99.0	98.8	98.60
4	97.7	99.0	97.7	99.6	97.7	98.38
5	98.8	97.9	97.7	98.8	99.0	98.44

Thus,

$$\hat{P}(\overline{X}^* \le 98.2 \mid H_0^*) = \frac{0}{5} = 0$$

because all bootstrapped $\overline{x}^* > 98.2$.

Our decision, based on these five bootstraps and hence *not* to be trusted (in fact, we need at least 10,000 bootstrap samples to trust our answer), is thus to reject H_0 because the estimated P-value is ≤ 0.05.

❐ ❐

In cases in which the population distribution shape is not known and the sample size is too small for a valid application of the CLT (i.e., <30), for instance in Example 11.7, the bootstrap approach is very appealing. But when the sample size used to compute the sample mean is large, as in Example 11.4, a t-test is valid.

If the hypothesis test involves a statistic of interest other than $\overline{X}$ or $\hat{P}$, bootstrapping may be necessary even if the sample size is large, because not all test statistics used in hypothesis testing are approximately normally distributed as a consequence of the CLT. Further, even when a test statistic, Y say, is approximately normal, we may have no formula for SE(Y) and, hence, no way to standardize Y to obtain the needed standardized test statistic from which to obtain a normal-distribution-based P-value. Thus, bootstrap hypothesis testing is a very valuable statistical tool.

Summary of When Use of the t-Test Is Legitimate from Robustness Perspective

Assume either that we have n independent measurements of a physical quantity μ with standard deviation σ, or that we have a simple random sample of size n from a much larger population with unknown mean μ and unknown standard deviation σ. To test the null hypothesis

$$H_0 : \mu = \mu_0,$$

we often use the test statistic

$$T = \frac{\overline{X} - \mu_0}{S/\sqrt{n}},$$

where $\overline{X}$ denotes the sample mean, and S denotes the sample standard deviation.

Under H_0, T is exactly t-distributed with $n - 1$ degrees of freedom if the distribution of the population is exactly normal. Because no population is exactly normal, it is an important fact that t-tests are **robust** in the sense that in many cases T is still approximately t-distributed even if the population is not exactly normal, and consequently our t-distribution-based (Appendix F) estimates of the P-values are still approximately correct. But the more *nonnormal* the population is, the larger n has to be before we can justify using the t-test. We offer the following guidelines for the use of the t-test:

robust

- The sample size n is less than 20 (small): Use the t-test only if the data are very close to normal, that is, the distribution is roughly bell-shaped with no outliers. Check for normality by creating a normal probability plot (Q-Q plot).
- The sample size n is between 20 and 30 (moderate): The t-test can be used unless the data distribution is strongly skewed or has outliers.
- The sample size is at least 30 (large): The t-test can be used unless there are outliers. (Skewing of the data, and hence of the population producing the data, is permitted.)

If the alternative hypothesis is $H_A : \mu < \mu_0$, then one can use the t-distribution (Appendix F) for T, obtaining

$$P = P(T \leq t \mid H_0),$$

where t denotes the observed value of the test statistic T. If $H_A : \mu > \mu_0$, then again one can use the t-distribution for T, obtaining

$$P = P(T \geq t \mid H_0).$$

As usual, we reject H_0 in favor of H_A if $P \leq 0.05$.

Section 11.1 Exercises

1. It is claimed that the average height of men in a particular population exceeds 69 inches. Suppose a sample of 100 heights of men has a mean of $\overline{x} = 70.53$ inches and standard deviation of $s = 3.22$ inches. Use the t-test to test the null hypothesis that the mean height of men in the population is 69 inches.
2. The design of an instrument's controls can determine how easy it is for people to use the instrument. An experiment investigated this by asking 35 right-handed people to turn an instrument knob with their right hand to advance a pointer by screw action. The test device had two knob-and-pointer sets that were identical except that one knob turned clockwise to advance the pointer and the other knob turned counterclockwise. Each of the 35 subjects turned both knobs, and the time in seconds that it took to move each pointer a fixed distance was measured. The observed mean difference of clockwise (X) minus counterclockwise (Y) times ($D = X - Y$) was $\overline{d} = 1.4$ seconds, and the SD of the differences was $s_D = 6.62$ seconds. Certain ergonomic engineers believe that right-handed people can turn the counterclockwise knob faster. Test the null hypothesis that negates this claim, namely that the population mean time difference $\mu_D = 0$.
3. For the data in Table 11.2, test the null hypothesis that the mean population increase in hours due to the "laevo" drug is 0. For convenience, the mean $\overline{d}$ of the 10 laevo differences is 2.33, and the standard deviation s_D is 1.90.
4. The average length of adult largemouth bass in a certain lake had been 14 inches. A 2-year program was conducted, introducing panfish and building up underwater weed beds in the lake, in the hope of increasing the average adult bass length in the population. It known from past experience that the lengths of bass are normally distributed and that the population standard deviation, σ, was 2 inches. Twelve adult bass were randomly sampled at the end of the program, and their mean length was $\overline{x} = 15.5$. State the appropriate null hypothesis and carry out the appropriate hypothesis test.
5. Suppose a sample of 10 heights of men has a mean of $\overline{x} = 70.53$ inches and standard deviation of $s = 3.22$ inches. Explain in detail how to use bootstrapping to test the hypothesis that the mean height of men in the population is 69 inches, where the goal is to assess whether there is strong statistical evidence that the men are taller on average than 69 inches. Be very clear about how to force the box model population to satisfy the null hypothesis.
6. It is claimed that the length of a particular species of microorganism exceeds 25.5 micrometers. A set of 64 independent measurements of the length of the microorganism are made. The mean value of the measurements is $\overline{x} = 27.5$ micrometers with standard deviation of $s = 3.2$ micrometers. Is there strong statistical evidence that the population average length of the microorganism really exceeds 25.5 micrometers as claimed?
7. A large company conducted a study to test the effectiveness of a proposed in-service training program for its sales staff. Forty randomly selected salespersons participated in the program. Their sales totals in thousands of dollars

for the 6-month period prior to attending the program and for the 6-month period following the program were used as the data. The mean difference in "after program" – "before program" ($D = X - Y$) sales was $\overline{d} = \overline{x} - \overline{y} = 5.22$, and the standard deviation of the difference was $s_D = 6.77$. Of course, company management hopes for evidence that the training program improves sales effectiveness. Assuming that the economic situation remained constant during this period, test the null hypothesis that the population mean difference $\mu_D = 0$.

8. Because it has been a very bad year, a farmer believes the corn yield from his land will have a mean of less than 50 bushels per acre, in which case his crop insurance will compensate him for his losses. In a random sample of eight plots, the farmer finds a mean yield of $\overline{x} = 49.25$ bushels per acre with standard deviation of $s = 3.25$ bushels per acre. Test the null hypothesis that $\mu = 50$, which negates the claim that $\mu < 50$. (The data have an approximately normal density histogram.)

9. Solve Exercise 8 assuming $\sigma = 3.25$ is known.

10. Suppose for Exercise 1 that we have 100 bootstrapped means ($\overline{x}^*$-values), each from a bootstrap sample of size 10 (with replacement) taken from the box model population with a mean height of 69 inches. The bootstrapped means are recorded in the following stem-and-leaf plot. Is there strong statistical evidence that the mean height of the men in the population is greater than 69 inches?

Stem	Leaf	Stem	Leaf
681	2	690	001233459
682	2	691	000113333799
683	08	692	2223339
684	2	693	1234444478
685	01117	694	1
686	233459	695	
687	01112334444559	696	26
688	03357778899	697	5
689	1122333445555678		

Key: "6975" stands for 69.75 inches.

11. An administrator of the University of Illinois believes that undergraduates study, on average, less than 20 hours per week. State the null hypothesis denying this claim. A random sample of 40 students showed the mean study time was $\overline{x} = 19.5$ hours per week with standard deviation of $s = 4.05$ hours. Do you believe the administrator is correct? (*Hint*: If needed, reduce the null hypothesis to one specifying a fixed value of μ.)

12. A feed company claims that cattle switched to its feed have a mean weight gain greater than 160 pounds. Suppose that weight gain is known to have a normal distribution. A rancher is highly skeptical of the claim. To test whether there is strong evidence against the claim, he gave the feed to 17 head of cattle. They gained the following amounts of weight:

62	83	90	101	104	106
109	109	109	127	143	187
204	205	209	266	277	

Use these data to test $H_0 : \mu = 160$, the negation of the feed company's claim.

13. Does aerobic exercise tend to lower the resting heart rate? To answer this question, the resting heart rates of 14 randomly selected female subjects in the 20–30 age range were measured before and after a 3-month intensive aerobic exercise program. The mean difference ("before" – "after" $= X - Y$) is $\overline{D} = \overline{X - Y} = 5.0$ beats/minute. The sample standard deviation of the difference is $s_D = 3.40$. Evaluate the claim that there was a significant

decrease in resting heart rate after the aerobic exercise program. Suppose it is known that such differences are normally distributed.

14. Suppose 20 small specimens were randomly taken from a certain batch of concrete. The mean compression strength of the specimens was $\overline{x} = 4129.58$ pounds with a standard deviation of $s = 164.12$ pounds. Explain in detail how to use bootstrapping to test the null hypothesis that the mean compression strength of the batch of concrete is 4200 pounds (the industry standard), as opposed to the possibility that the batch is defective and has a lower compression strength. Be very clear how you would force the null hypothesis to be true for the bootstrap sampling box model.

11.2 TESTS FOR EQUALITY OF TWO POPULATION PROPORTIONS AND EQUALITY OF TWO POPULATION MEANS

Until now we have tested hypotheses concerning one population mean. A more common statistical problem is to compare two population means. We have already discussed matched pairs, which we saw was really a one-population-mean problem in disguise. Alternatively, when we are sampling from two different populations and observing the same specific characteristic in each population, the samples are collected independently of each other; that is, each X (from the first population) is independent of each Y (from the second population). The first part of this section deals with the special case when the two population means being compared are proportions, and the second part deals with population means in general.

Testing the Equality of Two Population Proportions

Large-Samples Case We tested several hypotheses concerning a single population proportion p in Section 9.2. For example, we compared the breast cancer rate p for Puerto Rican–born women living in New York City with the known breast cancer rate for the entire female population of Puerto Rico and with the rate for the entire female population of the United States. A more widely occurring statistical problem is testing the equality of two unknown population proportions (i.e., testing whether $p_1 = p_2$). For example, we could ask whether a new manufacturing method produces a smaller proportion of defectives than the old method. The hypotheses are then written as

$$H_0 : p_1 = p_2 \quad \text{versus} \quad H_A : p_1 > p_2,$$

where p_1 represents the proportion of defectives if we use the old method and p_2 the proportion of defectives if we use the new method.

We illustrate the two-population-proportions testing problem with an example, where we use the large-sample CLT approach.

Example 11.8

A preventive-medicine study was designed to study the influence of being married on the likelihood of an individual coming down with the common cold. The research claim was that being married protects people against catching colds. In the statistical study, a simple random sample of 50 married individuals and an independent simple random sample of 50 single individuals were chosen from the community where the study was done. The number of subjects coming down with at least one cold in the 6 months of the study was observed for each group. The results appear in Table 11.4.

Table 11.4 Results of Test of Marital Status Effect on Catching Colds

Group	Number in group	Number catching a cold
Married	$n_M = 50$	$x_M = 15$
Single	$n_S = 50$	$x_S = 25$

We are interested in determining whether being married prevents colds—the research hypothesis. Thus, the null hypothesis is "being married does not prevent colds." More specifically, the null hypothesis is

$$H_0 : p_M = p_S \quad \text{versus} \quad H_A : p_M < p_S,$$

where p_M and p_S denote the rates of infection for the two populations (M = married, S = single). The null hypothesis can also be written $H_0 : p_M - p_S = 0$. How are we going to be able to test whether there is a significant difference between p_M and p_S? The natural statistic for assessing whether there is a difference between p_M and p_S is

$$\begin{aligned}\hat{p}_M - \hat{p}_S &= (\text{married sample proportion catching a cold}) \\ &\quad - (\text{single sample proportion catching a cold}) \\ &= \frac{15}{50} - \frac{25}{50} = -0.2,\end{aligned}$$

using the data in Table 11.4. Our goal, then, is to answer the question of whether the difference between $\hat{p}_M - \hat{p}_S = -0.2$ and the null hypothesis claim $p_M - p_S = 0$ can be explained by chance alone, and thus to determine whether we should reject or fail to reject the null hypothesis.

Note that $\hat{p}_M - \hat{p}_S$ is the difference of two sample proportions obtained from independent random samples. In the process of constructing a CI for the difference of two population proportions in Section 10.6, we learned that the difference of two independent sample proportions, here $\hat{p}_M - \hat{p}_S$, is approximately normal provided the sample sizes, here n_M and n_S, each satisfies our usual rule for being large, namely $n_M\hat{p}_M \geq 10, n_M(1 - \hat{p}_M) \geq 10, n_S\hat{p}_S \geq 10, n_S(1 - \hat{p}_S) \geq 10$. Thus, under H_0 (i.e., when $p_M = p_S = p$) when n_M and n_S are large, the standardized $\hat{p}_M - \hat{p}_S$, given by

$$\frac{(\hat{p}_M - \hat{p}_S) - (p_M - p_S)}{\text{SD}(\hat{p}_M - \hat{p}_S)} = \frac{(\hat{p}_M - \hat{p}_S) - (p_M - p_S)}{\sqrt{\dfrac{p_M(1 - p_M)}{n_M} + \dfrac{p_S(1 - p_S)}{n_S}}} = \frac{\hat{p}_M - \hat{p}_S}{\sqrt{p(1 - p)\left(\dfrac{1}{n_M} + \dfrac{1}{n_S}\right)}},$$

is approximately standard normal. However, under H_0, we do not know the common population proportion p. But since $\hat{p}_M$ and $\hat{p}_S$ both estimate p, we can combine or pool the information they provide to get

$$\hat{p} = \frac{x_M + x_S}{n_M + n_S} = \frac{15 + 25}{50 + 50} = \frac{40}{100} = 0.4.$$

Thus, the standard error of the statistic $\hat{p}_M - \hat{p}_S$ is

$$\text{SE}(\hat{p}_M - \hat{p}_S) = \widehat{\text{SD}}(\hat{p}_M - \hat{p}_S) = \sqrt{\hat{p}(1-\hat{p})\left(\frac{1}{n_M} + \frac{1}{n_S}\right)}$$

$$= \sqrt{(0.4)(0.6)\left(\frac{1}{50} + \frac{1}{50}\right)} = 0.098.$$

Plugging in the estimates, the observed value for the standardized test statistic Z is

$$z_{\text{OBS}} = \frac{\hat{p}_M - \hat{p}_S}{\text{SE}(\hat{p}_M - \hat{p}_S)} = \frac{-0.2}{0.098} = -2.04.$$

Thus, the P-value is $P(Z \leq -2.04)$, where Z is approximately standard normal. We can look the probability up in Appendix E; that is, $P(Z \leq -2.04) = 0.0207$. Note that we were looking for values of the test statistic Z as extreme or more so than -2.04, in the direction $(\leq)$ of supporting the research (i.e., alternative) hypothesis.

Thus, we reject the null hypothesis that $p_M - p_S = 0$. That is, this experiment supports the research claim that being married protects one against catching colds. This study is just an example of the general claim that a strong social network protects one against illness, perhaps in part through a strengthened immune system.

There are two physical settings leading to testing the equality of two theoretical proportions.

Scenario 1: We have n_1 independent success/failure trials (a repeated-trials random experiment) with success probability p_1 in each trial. We count the total number of successes, X_1, in the n_1 trials (see Section 7.1). Similarly, under a different set of experimental conditions, we have n_2 independent success/failure trials (a second repeated-trials experiment) with success probability p_2 in each trial. We count the total number of successes, X_2, in the second set of trials.

Scenario 2: Consider two large real populations, 1 and 2, for which unknown proportions, p_1 and p_2, respectively, have a characteristic of interest. We take independent simple random samples of sizes n_1 and n_2, respectively, from the two populations. Assume that the sample sizes are less than 5% of the respective population sizes. Let X_1 and X_2 denote the respective numbers of population members with the characteristic of interest in the two samples.

Under either scenario, the sample proportions

$$\hat{p}_1 = \frac{X_1}{n_1} \quad \text{and} \quad \hat{p}_2 = \frac{X_2}{n_2}$$

are unbiased estimates of p_1 and p_2, respectively. In the "large-samples case," that is, if the four conditions $n_1\hat{p}_1 \geq 10, n_1(1-\hat{p}_1) \geq 10$, $n_2\hat{p}_2 \geq 10$ and $n_2(1-\hat{p}_2) \geq 10$, are satisfied, then we can test the null hypothesis

$$H_0 : p_1 = p_2$$

using the test statistic

$$Z = \frac{\hat{p}_1 - \hat{p}_2}{\sqrt{\hat{p}(1-\hat{p})\left(\frac{1}{n_1} + \frac{1}{n_2}\right)}},$$

where

$$\hat{p} = (X_1 + X_2)/(n_1 + n_2).$$

Under H_0 and the four stated "large-sample" conditions for n_1 and n_2, Z is approximately standard normal via the CLT for Z and we can estimate the P-value using Appendix E.

If we are not in the "large-samples case," that is, if one or more of the four conditions fail to be satisfied, then we should not use the z-statistic with Appendix E to estimate the P-value. Instead we can run box model simulations to test H_0. For example, in Example 11.9 below we have

$$n_1\hat{p}_1 = 5 \quad \text{and} \quad n_2\hat{p}_2 = 4,$$

which are too small to allow us to apply the CLT to the Z-statistic to test H_0.

Small-Samples Case (of Testing) $H_0 : p_1 = p_2$ We present an example of testing $H_0 : p_1 = p_2$ using five-step simulation.

Example 11.9

A study of a random sample of 63 current stroke victims over the age of 60 found that out of the 53 patients with no history of previous strokes, there were 5 who suffered from dementia. Out of the 10 patients who had already had one stroke before the current stroke, 4 suffered from dementia. Does this prove that the risk of dementia is lower for first-time stroke victims than for victims with a history of several previous strokes?

Solution: Let population 1 consist of all first-time stroke victims and population 2 all victims with a history of strokes. The characteristic of interest is dementia. Let p_1 denote the probability of dementia (proportion with dementia in population 1) for first-time stroke victims, and let p_2 denote the probability of dementia (proportion with dementia in population 2) for those with a history of several previous strokes. We want to test

$$H_0 : p_1 = p_2 \quad \text{versus} \quad H_A : p_1 < p_2.$$

We shall use the following model: We selected $n_1 = 53$ patients at random with replacement from population 1 and independently selected $n_2 = 10$ patients at random with replacement from population 2. Let q_0 denote the size of sample 1 relative to the combined size of the two samples, namely

$$q_0 = \frac{n_1}{n_1 + n_2} = \frac{53}{53 + 10} = \frac{53}{63}.$$

Let q denote the proportion of all the sampled dementia patients ($5 + 4 = 9$ dementia patients in this example) expected to be in sample 1. Then when H_0 is true, namely $p_1 = p_2$, it is clear that, equivalently, $q = q_0$ must hold. That is,

$$p_1 = p_2 \text{ if and only if } q = q_0.$$

From this it is also clear that

$$p_1 < p_2 \text{ if and only if } q < q_0,$$

$$p_1 > p_2 \text{ if and only if } q > q_0.$$

We can therefore rephrase the null hypothesis and the alternative (research) hypothesis:

$$H_0 : q = q_0 \text{ versus } H_A : q < q_0.$$

Let $\hat{q}$ denote the proportion of the sampled dementia patients who happen to be in sample 1. Here the observed value of $\hat{q}$ is

$$\hat{q}_{\text{OBS}} = \frac{5}{5+4} = \frac{5}{9}.$$

Given that there were a total of 9 dementia cases in the study, we can simulate $\hat{q}$ under H_0 by drawing nine times at random *without* replacement from the null-hypothesis box (whose proportion of 1s is $q_0 = 53/63$):

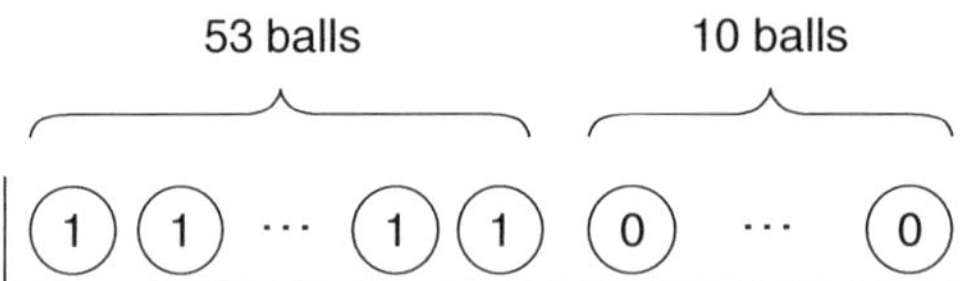

We then compute the proportion $\hat{q}$ of 1-balls in the sample. Note that the box has a 1-ball for each patient in sample 1 and a 0-ball for each patient in sample 2. Since the alternative hypothesis is

$$H_A : q < q_0,$$

the event of interest has the inequality in the same direction, namely

$$\text{EOI} = (\text{simulated } \hat{q} \le \hat{q}_{\text{OBS}}).$$

We ran 10,000 computer simulations and got

$$\begin{aligned} \hat{P} &= \hat{P}(\text{simulated } \hat{q} \le \hat{q}_{\text{OBS}} \mid H_0) \\ &= \hat{P}(\text{simulated } \hat{q} \le 5/9 \mid H_0) = 0.0279. \end{aligned}$$

Because the estimated P-value is less than our 0.05 threshold, we reject H_0 in favor of H_A and conclude that the risk of dementia is lower for first-time stroke victims than for victims with a history of previous strokes. The result is statistically significant, but with so few patients in the study the result is not highly significant.

The next example is not really a "small-samples" example; rather, it is a "rare event" example. But, in any case, it is analogous to Example 11.9 in that we cannot use the large-sample CLT and Appendix E to estimate the P-value because so few balls are sampled from the box.

Example 11.10

A company manufactures several types of integrated circuits. A quality control engineer suspects that the defect rate of the company's ALS (advanced lower-power Schottky) circuits is lower than the defect rate of the company's LPS (lower-power Schottky) circuits. Suppose that 2000 circuits of each type are randomly selected (two populations each randomly sampled from) and tested, and it is found that 2 of the ALS circuits and 9 of the LPS circuits are defective. Does this prove that the proportion of defective ALS circuits produced by the company is lower than the proportion of defective LPS circuits?

Solution: Let p_1 and p_2 denote the proportions of ALS circuits and LPS circuits produced, respectively, that are defective. We want to test

$$H_0 : p_1 = p_2 \quad \text{versus} \quad H_A : p_1 < p_2.$$

Then $n_1 = 2000$ ALS circuits and $n_2 = 2000$ LPS circuits were sampled. The size of the ALS sample relative to the combined size of the two samples is

$$q_0 = \frac{n_1}{n_1 + n_2} = \frac{2000}{2000 + 2000} = \frac{1}{2}$$

Let q denote the proportion of the defective sample circuits expected to be ALS. As we learned above, we can rephrase the null hypothesis and the alternative hypothesis:

$$H_0 : q = q_0 \quad \text{versus} \quad H_A : q < q_0.$$

Let $\hat{q}$ denote the observed proportion of the defective sample circuits that happen to be ALS. Here the observed value of $\hat{q}$ is

$$\hat{q}_{\text{OBS}} = \frac{2}{2+9} = \frac{2}{11}.$$

Given that there were a total of 11 defective circuits in the two samples combined, we can simulate $\hat{q}$ under H_0 by drawing 11 times at random *without* replacement from the box

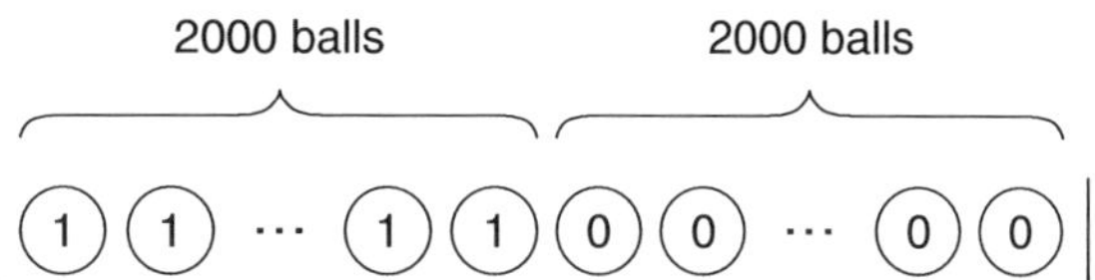

and computing the proportion $\hat{q}$ of 1-balls in the sample. Note that the box has a 1-ball for each ALS circuit tested and a 0-ball for each LPS circuit tested. Because we are sampling so few balls relative to the box size, it won't make any practical difference if we instead draw 11 times at random *with* replacement from the box

and for each simulation compute the simulated proportion of times $\hat{q}$ that we draw the 1-ball. Since the alternative hypothesis is

$$H_A : q < q_0,$$

the event of interest is

$$\text{EOI} = (\text{simulated } \hat{q} \le \hat{q}_{\text{OBS}}).$$

We ran 10,000 computer simulations and got

$$\begin{aligned}\hat{P} &= \hat{P}(\text{simulated } \hat{q} \le \hat{q}_{\text{OBS}} \mid H_0) \\ &= \hat{P}(\text{simulated } \hat{q} \le 2/11 \mid H_0) = 0.0325.\end{aligned}$$

Because the estimated P-value is less than our 0.05 threshold, we reject H_0 in favor of H_A and conclude that the proportion of ALS circuits produced that are defective, p_1, is smaller than the proportion of LPS circuits produced that are defective, p_2.

We finally note that we could also have tested $H_0 : p_M = p_S$ versus $H_A : p_M < p_S$ in Example 11.8 using box model simulations. In fact, we can test the equality of two population proportions using box model simulations in both the large-samples case and the small-samples case, as long as the assumptions of either Scenario 1 or Scenario 2 are satisfied. (Note that the large-samples Z-test also requires that the assumptions of either Scenario 1 or Scenario 2 be satisfied in addition to requiring large samples.)

Let us perform the test of significance in Example 11.8 using box model simulations. The proportion of married individuals in the study was

$$q_0 = \frac{50}{50+50} = \frac{1}{2}.$$

Let q denote the proportion of cases with colds expected to be in the married group. We can rephrase the null hypothesis and the alternative hypothesis:

$$H_0 : q = q_0 \quad \text{versus} \quad H_A : q < q_0.$$

Let $\hat{q}$ denote the observed proportion of the cases with colds that happen to be in the married group. The observed value of $\hat{q}$ was

$$\hat{q}_{\text{OBS}} = \frac{15}{15+25} = \frac{3}{8}.$$

Given that there were a total of 40 cold cases in the two groups combined, we can simulate $\hat{q}$ under H_0 by drawing 40 times at random *without* replacement from the box and computing the simulated proportion $\hat{q}$ of 1-balls in the sample.

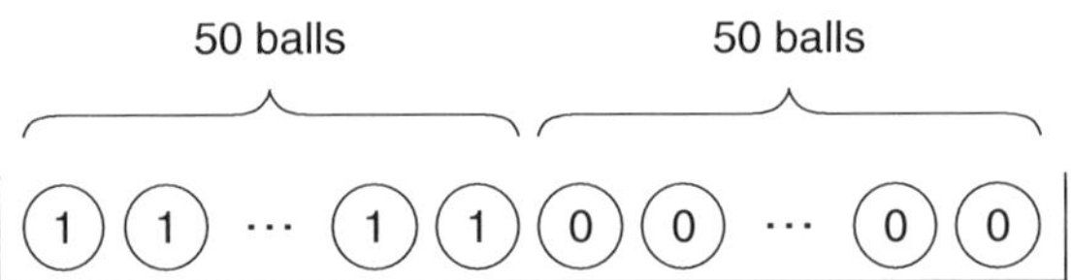

Note that the box has a 1-ball for each married test subject and a 0-ball for each single test subject. We run 10,000 computer simulations and get

$$\hat{P} = \hat{P}(\text{simulated } \hat{q} \le 3/8 \mid H_0) = 0.0336.$$

When we compare the highly accurate simulation-based estimate of the P-value (i.e., 0.0336) with the classical bell-curve estimate of P (i.e., 0.0207), we note that once again the bell-curve method underestimates P when we have samples that are only moderately large. The reason for this is that we (and everyone else) are ignoring the so-called correction for continuity, which we mentioned in Section 9.2.

simulation-based test of $p_1 = p_2$ is the approach of choice in the small-samples case

Summary of Box Model Five-Step Simulation–Based Test for Equality of Two Population Proportions Assume that the assumptions of either Scenario 1 or Scenario 2 are satisfied. Loosely speaking, under both scenarios we consider two large populations 1 and 2 (infinite or conceptual in Scenario 1), out of which unknown proportions p_1 and p_2, respectively, have a characteristic of interest. We take independent random samples of sizes n_1 and n_2, respectively, from the two populations. Let X_1 and X_2 denote the respective numbers of population members with the characteristic of interest in the two samples.

The number of population members with the characteristic of interest in the two samples combined is

$$n = X_1 + X_2.$$

Let q denote the proportion of the n cases expected to have come from population 1. Let q_0 denote the size of sample 1 relative to the combined size of the two samples, namely

$$q_0 = \frac{n_1}{n_1 + n_2}.$$

We then have

$$p_1 < p_2 \text{ if and only if } q < q_0;$$

$$p_1 = p_2 \text{ if and only if } q = q_0; \text{ and}$$

$$p_1 > p_2 \text{ if and only if } q > q_0.$$

We can therefore rephrase the null hypothesis $H_0 : p_1 = p_2$ as $H_0 : q = q_0$. Let

$$\hat{q} = \frac{X_1}{n}$$

denote the proportion of the n cases that came from population 1. Given that there were a total of n cases in the two samples combined, we can simulate $\hat{q}$ under H_0 by drawing n times at random *without* replacement from the null-hypothesis box [its proportion of 1s is $q_0 = n_1/(n_1 + n_2)$]

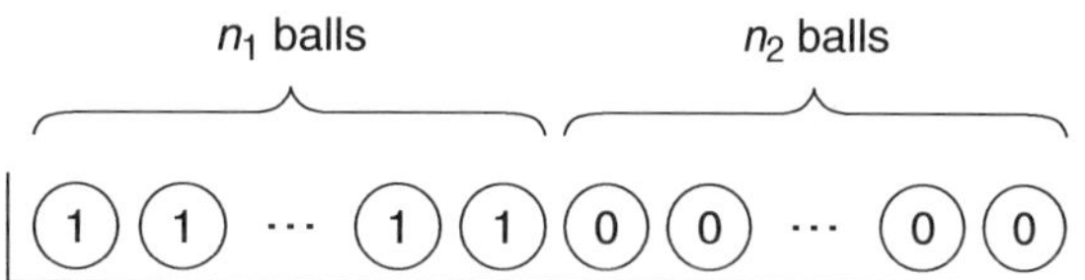

and computing the simulated proportion $\hat{q}$ of 1-balls among the n balls randomly drawn. If

$$H_A : p_1 < p_2, \text{ that is, } H_A : q < q_0,$$

then

$$\hat{P} = \hat{P}\left(\text{simulated } \hat{q} \le \frac{X_1}{n} \middle| H_0\right).$$

If

$$H_A : p_1 > p_2, \text{ that is, } H_A : q > q_0,$$

then

$$\hat{P} = \hat{P}\left(\text{simulated } \hat{q} \ge \frac{X_1}{n} \middle| H_0\right).$$

Testing for the Equality of Two Population Means

We shall now discuss the t-test for comparing two means. Our assumptions are the same as the assumptions in Section 10.5: Suppose we have two *independent* samples from two distinct populations with unknown means μ_X and μ_Y and unknown standard derivations σ_X and σ_Y. Assume that the samples are either repeated independent measurements (samples from conceptual populations) or simple random samples from real populations at least 20 times larger than the samples. Assume that the sampling distributions of the sample means $\overline{X}$ and $\overline{Y}$ are close to normal—either because the populations are close to normal or because the sample sizes n_X and n_Y are sufficiently large. Let S_X and S_Y denote the respective sample standard derivations.

Let df denote the smaller of the numbers $n_X - 1$ and $n_Y - 1$. Under the null hypothesis

$$H_0 : \mu_X = \mu_Y,$$

the two-sample t-statistic

$$T = \frac{\overline{X} - \overline{Y} - \text{mean}}{\text{SE}} = \frac{\overline{X} - \overline{Y} - 0}{\sqrt{S_X^2 / n_X + S_Y^2 / n_Y}}$$

is roughly t-distributed with df degrees of freedom if the sample sizes n_X and n_Y are at least 10, especially if $n_X = n_Y$ and the two populations being compared have similar shapes.

Example 11.11

Suppose we want to compare the average Brinell hardness values μ_X and μ_Y of two magnesium alloys, alloy X and alloy Y. Suppose that 15 independent hardness measurements of alloy X yielded an average hardness of $\overline{x} = 64.18$ with $s_X = 3.29$, and 15 independent hardness measurements for alloy Y yielded an average hardness $\overline{y} = 67.58$ with $s_Y = 2.85$. We shall test

$$H_0 : \mu_X = \mu_Y$$

against

$$H_A : \mu_X < \mu_Y.$$

The observed value of the two-sample t-statistic is

$$\begin{aligned} t &= \frac{\overline{x} - \overline{y} - (\mu_X - \mu_Y)}{\sqrt{s_X^2 / n_X + s_Y^2 / n_Y}} \\ &= \frac{64.18 - 67.58 - 0}{\sqrt{(3.29)^2 / 15 + (2.85)^2 / 15}} \\ &= -3.025. \end{aligned}$$

Using the t-curve with 14 degrees of freedom, we estimate the P-value to be between 0.0025 and 0.005. We therefore reject H_0 In favor of H_A.

❐ ❐ ❐ ❐ ❐ ❐ ❐ ❐ ❐ ❐ ❐ ❐ ❐ ❐ ❐ ❐ ❐ ❐ ❐ ❐

Surprisingly, the two-sample t-test is also valid for randomized comparative experiments—even though such experiments do not compare independent random samples from distinct populations.

Assume that we have a total of $n = n_X + n_Y$ test subjects or test units. Assume that n_X of these are randomly selected to receive treatment X and that the remaining n_Y test subjects or test units receive treatment Y. Let $\overline{X}$ denote the average response of those test subjects or test units receiving treatment X, and let S_X denote the corresponding standard deviation. Let $\overline{Y}$ denote the average response of those receiving treatment Y and S_Y the corresponding standard deviation. Let μ_X and μ_Y denote the long-run average responses to the two treatments. To test

$$H_0 : \mu_X - \mu_Y = 0$$

we can use the two-sample t-statistic.

Example 11.12

A clinical trial studied 40 children with severe cases of shigellosis. Of the 40 children, $n_X = 20$ were randomly selected to be treated with the antibiotic ampicillin. Their average recovery time was $\overline{x} = 10.20$ days, with $s_X = 9.6$ days. The remaining $n_Y = 20$ children received ceftriaxone. Their average recovery time was $\overline{y} = 1.85$ days, with $s_Y = 0.6$ days. The null hypothesis was that the two antibiotics were equally effective.

The observed value of the two-sample t-statistic was

$$t = \frac{10.20 - 1.85}{\sqrt{(9.6)^2/20 + (0.6)^2/20}} = 3.88.$$

With 19 degrees of freedom we estimate the P-value to be around 0.0005. We reject the null hypothesis. Ceftriaxone was obviously much more effective.

Section 11.2 Exercises

1. Scientists want to study the effect of exercise on weight loss. One hundred people are randomly divided into two equal groups. Both groups follow the same diet plan, but the first group also follows an exercise program. It is suspected that the exercise group may lose more weight on average. In the exercise group the mean weight loss over 3 months was $\overline{x} = 25.2$ pounds, with SD of $s_X = 10$ pounds. In the nonexercise group the mean weight loss was $\overline{y} = 20.4$ pounds, with SD of $s_Y = 6.3$ pounds. Test the null hypothesis that the difference between the population means is $\mu_X - \mu_Y = 0$.

2. A study was performed to compare the voting behaviors of women and men during an election. An exit poll was carried out, and the results are summarized in the following table.

	Total number polled	Number who voted for candidate A
Men	45	21
Women	51	36

Let p_M denote the proportion of men who voted for candidate A, and let p_W denote the proportion of women who voted for candidate A. It is suspected that $p_W > p_M$ because of candidate A's views on various social issues.

a. What are your estimates of p_M and p_W?
b. Using the formula given in this section, determine an estimate of the value of $\mathrm{SD}(\hat{p}_M - \hat{p}_W)$.
c. Using the values you found in parts (a) and (b), find the value of the statistic used to test whether there is a significant difference in p_M versus p_W.
d. Is there a significant difference in how the women and men voted in this race?

3. On a college campus, a professor wanted to test the effect on exam scores from her use of a computer program to teach calculus. In her first class, of 43 students, she used the computer as an instructional aid. In her second class of, 35 students, she taught without using the computer. Students were randomly assigned to the two sections. It was claimed that the computer-aided class would perform better. On the final exam the noncomputer class scored a mean of $\overline{x} = 78.32$ with SD of $s_X = 8.07$, whereas the computer-aided class scored a mean of $\overline{y} = 80.41$ with SD of $s_Y = 8.53$. Test the claim that the difference of the population means is 0.

4. In a study of people who stop to help drivers with disabled cars, some researchers thought that more people would stop if they first saw another driver getting help. In one experiment, 2000 drivers first saw a woman being helped to fix a flat tire, and 2.9% of them stopped to help a second woman fix a flat tire. Out of 2000 drivers who did not see the first helper, 1.75% stopped to help the second woman fix her flat tire. Test whether the two proportions are equal. (From McCarthy, "Help on the Highway," *Psychology Today*, July 1987.)

5. Fifty independent measurements of the weight of a chemical compound were made with each of two scales. There is concern that the second scale is measuring low. On the first scale the mean of the 50 measurements was $\overline{x} = 19.45$ grams with SD of $s_X = 0.49$ gram. On the second scale, the mean of the 50 measurements was $\overline{y} = 18.42$ grams with SD of $s_Y = 0.27$ gram. Test the claim that the difference between the population means is $\mu_X - \mu_Y = 0$.

6. In both 1978 and 2010, a survey was taken of first-year college students, asking whether or not they opposed the death penalty. The results were as follows:

Year	Number opposing death penalty
1978	165
2010	105

Suppose in both cases the survey polled 500 students. Assess whether there is a significant decrease over time in those opposing the death penalty.

7. One hundred and fifty high school students were randomly assigned to two groups of 75. Group A used a new math text, and group B used the old math text. The sample mean and SD of the SAT math scores for group A were 549.45 and 21.12, respectively, and the sample mean and SD for group B were 539.25 and 20.91, respectively. Test the null hypothesis that the difference between the population means is 0 (as opposed to the hope that the new math text would improve SAT math scores).

8. In a survey similar to that in Exercise 6, first-year college students were asked in both 1970 and 1990 what their political orientation was. The following table shows how the percentage of students who replied "liberal/far left" changed:

Year	Number saying they were "liberal" or "far left"
1970	190
1990	125

Once again, assume that 500 students were randomly surveyed in each case. Is there a significant decrease in the percentage?

9. Twelve independent batches from each of two competing cold medicines are tested for the amount of acetaminophen in milligrams. It is believed the second cold medicine (Brand Y) contains more acetaminophen, on average. For Brand X, $\overline{x} = 494.67$ and $s_X^2 = 233.15$, and for Brand Y, $\overline{y} = 523.67$, $s_Y^2 = 303.33$. Test the claim that the mean amount of acetaminophen is the same in each brand.

10. Conduct your own poll of students at your school as to whether men and women agree with the death penalty. What claim would you make about the proportions p_X (of women) and p_Y (of men who favor the death penalty)? Test the null hypothesis $H_0 : p_X = p_Y$ against the claim.

11. The World Health Organization air quality monitoring project measures suspended particles in micrograms per cubic meter. Let X and Y equal the concentrations of suspended particles in Melbourne and Houston, respectively. The $n_X = 13$ observations of X from Melbourne resulted in an average of $\overline{x} = 72.9$ with SD of $s_X = 25.6$. The $n_Y = 16$ observations of Y from Houston resulted in an average of $\overline{y} = 81.7$ and SD of $s_Y = 28.3$. Test the equality of the population means. Assume the goal is to detect whether Houston has more suspended particulate matter.

12. To test the effectiveness of the new weight-loss medication Redux, 40 women were separated into two groups: Group A, the control group, took a placebo, and group B, the experimental group, took Redux. The amount of weight lost by each member of the groups over a 6-month period is given in the following tables. Test the null hypothesis that the population mean weight loss for each of the two groups is the same (i.e., that Redux is ineffective).

Group A: Total weight lost (in pounds) over a 6-month period for each group member

3	4	5	6	7
10	11	12	15	18
19	20	23	24	25
30	33	38	40	42

Group B: Total weight lost (in pounds) over a 6-month period for each group member

5	5	7	7	10
10	10	10	15	18
20	24	28	29	38
42	44	49	50	55

13. A company is studying two alternative heat sources for manufacturing optical fibers, a carbon furnace and a CO_2 laser. Suppose that out of 200 test fibers produced using the carbon furnace, 10 were defective, whereas only 2 out of 200 test fibers produced using the laser were defective. Does this prove that the carbon furnace method results in a higher proportion of defective fibers than the laser method? Or could the observed difference be explained as nothing more than chance variation?
    a. State H_0 and H_A in terms of p_1, the proportion of fibers produced using the carbon furnace that are defective, and p_2, the proportion of fibers produced using the laser that are defective.
    b. Explain why we should not use a Z-test to test H_0.
    c. Determine q_0, the proportion of the test fibers that were produced using the carbon furnace.
    d. Determine $\hat{q}_{OBS}$, the proportion of the defective test fibers that were produced using the carbon furnace.

e. Let q denote the proportion of the defective test fibers expected to have been produced with the carbon furnace. Restate H_0 and H_A in terms of q and q_0.
f. Describe the box model for testing H_0. How many 0-balls should there be in the box? How many 1-balls should there be? On each simulation, how many balls do we draw? Do we draw with or without replacement? What is the variable $\hat{q}$ of interest?
g. Describe the event of interest in terms of $\hat{q}$ and $\hat{q}_{\text{OBS}}$.
h. Run 10,000 simulations to estimate P.
i. Do we reject H_0 or do we fail to reject H_0?

14. Shot peening is used to compress the surface area of metal parts to make them more resistant to fractures. It is done by bombarding the surface with small particles hurled at high velocity. Each time a particle hits, it puts a small dent in the surface and compresses the area directly beneath the surface. The bombardment continues until eventually the entire surface is compressed. Tests are conducted on a particular part to see if shot peening reduces the proportion of parts that fracture when put into use.

Suppose that we test a random sample of 85 shot-peened parts and a random sample of 15 parts that were not shot peened, and find that 4 parts in each category fracture when put into use.

a. State H_0 and H_A in terms of p_1, the proportion of shot-peened parts that fracture, and p_2, the proportion of parts not shot peened that fracture.
b. Explain why we should not use a Z-test to test H_0.
c. Determine q_0, the proportion of the tested parts that were shot peened.
d. Determine $\hat{q}_{\text{OBS}}$, the proportion of the fractured parts that were shot peened.
e. Let q denote the proportion of the fractured parts expected to have been peened. Restate H_0 and H_A in terms of q and q_0.
f. Describe the box model for testing H_0. How many 0-balls are there in the box? How many 1-balls? On each simulation, how many balls do we draw? Do we draw with or without replacement? Define the variable $\hat{q}$ of interest.
g. Describe the event of interest in terms of $\hat{q}$ and $\hat{q}_{\text{OBS}}$.
h. Run 10,000 simulations to estimate P.
i. Do we reject H_0 or do we fail to reject H_0?

11.3 ONE-SIDED AND TWO-SIDED HYPOTHESIS TESTING

In Example 9.4 we looked at the question: What determines a woman's breast cancer risk? If heredity were the only factor that mattered, then there would not be much that a woman could do to protect herself. But if environment and lifestyle also contribute to a woman's cancer risk, then there might be something that she could do.

To find out whether environment and lifestyle play a role, researchers 25 years ago decided to study the population of Puerto Rican–born women who had moved from Puerto Rico to New York City 30 years earlier. Genetically, these women were like their female friends and relatives still living in Puerto Rico. But their environment and lifestyle were more like those of the rest of the female population of the United States. The researchers wanted to compare the breast cancer rate p of these Puerto Rican–born women living in New York City with the corresponding (age-adjusted) breast cancer rate p_0 for the female population of Puerto Rico. The null hypothesis was the claim that there is no environmental or lifestyle effect on breast cancer rate. The alternative hypothesis, which the researchers wanted to prove, was that the circumstances and lifestyle of the affluent West (early onset of puberty, use of birth control pills, excess weight, too little exercise) increase a woman's breast cancer risk. In the context of the Puerto Rican–born women living in New York City, the hypotheses become

$$H_0 : p = p_0$$
$$H_A : p > p_0.$$

one-sided hypothesis

We say that H_A is a **one-sided hypothesis** because we are considering only the possibility that p might be higher than p_0, not lower.

Note that if the researchers had not already known that breast cancer rates were higher in the affluent West than in less developed countries, then the research claim that environment and lifestyle affect cancer rates would have become

$$H_A : p \neq p_0.$$

two-sided hypothesis

In that case H_A would have been a **two-sided hypothesis,** meaning that we would be considering both the possibility that p might be higher than p_0 and the possibility that p might be lower than p_0. That is, the lifestyle and environmental impact of living in New York City could possibly lower the rate of breast cancer.

In the next example the alternative hypothesis is two-sided, which alters how the P-value is computed.

Example 11.13

computing the P-value when H_A is two-sided

Suppose we toss a coin 10 times and get 8 heads instead of the expected 5 heads. Could the difference be explained as nothing more than chance variation, or do we have strong evidence that the coin is biased?

Solution: As usual, the null hypothesis is that the observed difference is due to chance alone. Under H_0 the coin is fair. If p denotes the probability that the coin lands heads up, then

$$H_0 : p = 0.5.$$

Because we have no preconceived notion (prior to data collection) that the coin favors a particular side, heads or tails, the alternative hypothesis is

$$H_A : p \neq 0.5.$$

Note that we are careful not to let the data influence our choice of H_A.

As usual, the P-value is the probability under H_0 of getting evidence against H_0 as strong as or even stronger than the evidence at hand. In the current setup, it would have been just as strong evidence against H_0 if we had gotten 2 heads instead of 8 heads. And getting 1 or no heads or getting 9 or 10 heads would have been even stronger evidence against H_0.

We can simulate the proportion of heads in 10 tosses of a fair coin (H_0 true) by drawing 10 times at random with replacement from the null-hypothesis box

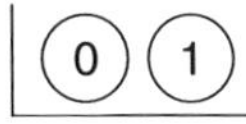

and computing the proportion $\hat{p}$ of 1-balls in the sample. We run 10,000 simulations and get

$$\begin{aligned}\hat{P} &= \hat{P}(\text{simulated } \hat{p} \leq 0.2 \text{ or simulated } \hat{p} \geq 0.8 \mid H_0)\\ &= 1 - \hat{P}(0.2 < \text{simulated } \hat{p} < 0.8 \mid H_0)\\ &= 1 - 0.8935 = 0.1065.\end{aligned}$$

Since $\hat{P}$ is greater than our 0.05 threshold, we fail to reject H_0. The result is not statistically significant. We don't have enough evidence to prove that the coin is biased.

In the next example, H_A is again two-sided because we have no preconceived conjecture about the alternative to H_0.

Example 11.14

Water treatment facilities monitor the quality of their drinking water constantly. One variable being monitored is pH, which measures the degree of acidity or alkalinity of the water. A pH of 7.0 is neutral, one below 7.0 is acidic, and a pH above 7.0 is alkaline. An engineer wants to measure the bias in a pH meter. He uses the meter to measure the pH of a neutral substance (pH = 7.0) 20 times, obtaining the following readings:

6.98	7.05	7.02	6.96	7.03	6.96	7.01	7.04	6.97	7.00
6.99	6.97	7.03	7.04	7.01	7.00	6.99	7.04	7.07	6.97

Let μ denote the long-run average reading of this meter when applied repeatedly to a neutral substance. We want to test whether the meter is correctly calibrated by testing

$$H_0 : \mu = 7.0 \quad \text{versus} \quad H_A : \mu \neq 7.0.$$

Here the alternative hypothesis is two-sided because we don't have any preconceived notion that the readings are systematically too high or systematically too low.

Let $\overline{X}$ and S denote the sample mean and the sample standard deviation, respectively, of 20 independent readings of the pH of a neutral substance with this meter. The measurements are known to be approximately normal. Under H_0 and because of population-sampling normality, the test statistic (a t-statistic)

$$T = \frac{\overline{X} - 7.0}{S \,/\, \sqrt{20}}$$

is approximately t-distributed with 19 degrees of freedom. The observed value of T is

$$t = \frac{7.0065 - 7.0000}{0.03265 \,/\, \sqrt{20}} = 0.89.$$

The P-value is the probability under H_0 of getting evidence against H_0 as strong as or even stronger than the evidence at hand. Here, a t-value of –0.89 would have been just as strong evidence against H_0 as a t-value of 0.89. And a t-value smaller than –0.89 or greater than 0.89 would have been even stronger evidence against H_0. As Figure 11.5 shows, the probability that T is at least as far from zero *in either direction* as the observed t can be estimated by the combined area under the t curve of the two tails. Because the t curve is symmetric around the y-axis,

$$\begin{aligned} P &= P(T \leq -0.89 \text{ or } T \geq 0.89 \mid H_0) \\ &\approx (\text{area under } t \text{ curve to the left of } -0.89) \\ &\quad + (\text{area under } t \text{ curve to the right of } 0.89) \\ &= 2 \cdot (\text{area under } t \text{ curve to the right of } 0.89). \end{aligned}$$

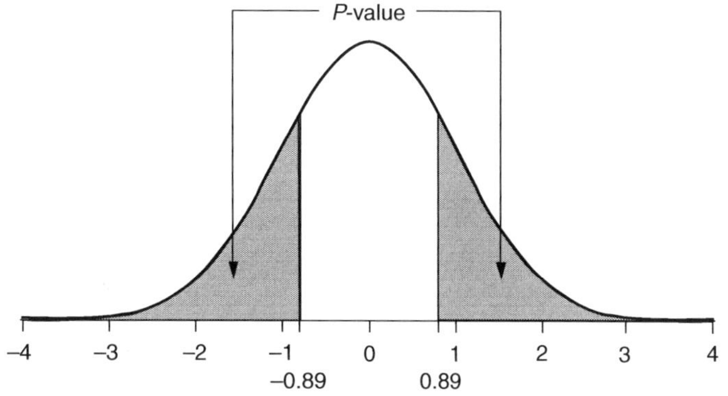

Figure 11.5 P-value for t-test with two-sided alternative.

To estimate the P-value, locate the line in Appendix F with $df = n - 1 = 20 - 1 = 19$, that is, 19 degrees of freedom. Because $P(T \geq 0.861) = 0.20$ and $P(T \geq 1.066) = 0.15$ from Appendix F and

$$0.861 < 0.89 < 1.066,$$

the identical tail areas are each between 0.15 and 0.20. It follows that P is between 0.30 and 0.40. With such a large P-value we do not reject H_0. We have no reason to doubt that the meter is correctly calibrated.

❐ ❐

In a test for one or two population proportions or one or two population means, it is not always easy to decide whether H_A should be one-sided or two-sided. It depends on whether or not we have a preconceived notion about the alternative to H_0. If we don't, then H_A *should be* two-sided. If we do have such a notion, then H_A should express the hopes and suspicions we bring to the data. And we should *not* let the data change H_A from being a two-sided to a one-sided alternative. We shall give one last illustration.

Example 11.15

Ten years ago an article in the *Journal of Marriage and the Family* compared the divorce rate p_1 for couples who lived together before they got married with the divorce rate p_2 for couples who did not live together before they got married. The question was: Does cohabitation prior to marriage affect the risk of divorce? Here the null hypothesis is that cohabitation has no effect,

$$H_0 : p_1 = p_2.$$

The alternative hypothesis H_A depends on your perspective. For example, if you are a dispassionate researcher, curious about whether cohabitation prior to marriage has any effect, good or bad, on divorce rates, then your alternative hypothesis is

$$H_A : p_1 \neq p_2.$$

In other words, if you have no preconception of the effect of premarital cohabitation on the likelihood of a subsequent divorce, then H_A is two-sided. If, on the other hand, you believe that cohabitation before marriage teaches a couple that marital commitment is unnecessary and thereby encourages the couple to split at the first sign of marital discord, then your research hypothesis H_A is one-sided, namely that premarital cohabitation increases the risk of divorce,

$$H_A : p_1 > p_2.$$

Finally, if you believe that the main effect of cohabitation before marriage is that it allows couples who are unsuited for each other to realize that they are incompatible before they make the mistake of getting married, then your research hypothesis H_A is that premarital cohabitation decreases the risk of divorce,

$$H_A : p_1 < p_2.$$

Of course, we need sample data to test H_0, regardless of our choice of H_A. And, clearly, we have to be careful not to jump from strong statistical evidence that $p_1 \neq p_2$ to being confident that the cause of the difference in divorce rates was the cohabitation. This is an observational study, and couples who cohabitate before marriage versus those who don't may be very different on average in other ways that influence p_1 and p_2.

decide on whether H_A should be one- or two-sided

In summary, we offer the following rule of thumb for deciding whether the alternative hypothesis H_A should be one-sided or two-sided in a test of significance for population means or proportions:

- If you have no preconceived notion that H_0 is wrong in a particular direction—for example, if you think that H_0 might be approximately correct—then H_A should be two-sided.
- If you suspect that the population parameter differs from the H_0 value in a particular direction, then H_A should express this conjecture and should therefore be one-sided.

In the one-sided case, sometimes the null hypothesis will contain an inequality, such as $H_0 : \mu \leq \mu_0$ or $H_0 : p \geq p_0$. Remember that we then replace the inequality by equality.

Section 11.3 Exercises

1. A demographer conjectures that in Illinois the average difference in age between husband and wife is less than 3 years. For a simple random sample of 177 married couples from Illinois, the average age of the husbands was $\overline{x} = 45.80$ years with a standard deviation of $s_X = 13.60$ years. The average age of the wives was $\overline{y} = 43.21$ years with a standard deviation of $s_Y = 13.08$ years. Finally, the average difference in age between husband and wife (husband's age minus wife's age) for the 177 couples was $\overline{d} = 2.59$ years with a standard deviation of $s_D = 4.75$ years. Does this prove that the average difference between husband's age and wife's age in Illinois is less than 3 years?
   a. State H_0 and H_A in terms of μ_X and μ_Y, the average age of husbands and wives in Illinois, respectively.
   b. Is H_A one-sided or two-sided?
   c. Compute the observed value of the test statistic.
   d. Estimate the P-value.

e. Do you reject H_0 or fail to reject H_0?

2. A certain pain killer is supposed to contain 220 mg naproxen sodium. A pharmacist randomly selects 20 tablets of the pain killer and measures the amount of naproxen sodium in each tablet, obtaining the readings shown here.

219.9	219.9	224.9	218.1	224.5
222.5	221.8	219.5	220.3	221.6
221.3	222.9	226.3	218.5	218.1
222.5	218.3	219.0	217.1	216.8

Test whether or not μ, the average amount of naproxen sodium in this brand of pain killer, equals 220 mg.

a. State H_0 and H_A in terms of μ.
b. Is H_A one-sided or two-sided?
c. Compute the observed value of the test statistic.
d. Estimate the P-value.
e. Do you reject H_0 or fail to reject H_0?

3. A quality control inspector takes a simple random sample of 25 one-pound pretzel bags packaged by a certain machine. The contents of the bags are weighed, resulting in the following readings, in grams.

465	456	438	454	447
449	442	449	446	447
468	433	454	463	450
446	447	456	452	444
447	456	456	435	450

One pound equals 453.6 grams. Let μ denote the average weight, in grams, of all one-pound bags of pretzels packaged by that machine. Test whether or not the packaging machine, on average, is putting one pound of pretzels into each bag.

a. State H_0 and H_A in terms of μ.
b. Is H_A one-sided or two-sided?
c. Compute the observed value of the test statistic.
d. Estimate the P-value.
e. Do you reject H_0 or fail to reject H_0?

4. A thermocouple is a temperature sensor used to measure the temperature inside a furnace. A researcher wanted to know whether two thermocouples gave systematically different readings. She wrapped the thermocouples together, placed them in an oven, and recorded the temperatures (in °C) reported by the two thermocouples. Twelve matched pairs of readings were obtained:

Thermocouple 1	102.18	159.89	211.05	247.93	290.01	314.62	325.97	361.24
Thermocouple 2	102.15	159.87	211.02	247.91	289.99	314.62	325.95	361.23
Thermocouple 1	385.36	436.68	470.40	511.32				
Thermocouple 2	385.35	436.64	470.39	511.29				

Let μ denote the long-run average of the difference,

$$(\text{thermocouple 1 reading}) - (\text{thermocouple 2 reading}).$$

a. State H_0 and H_A in terms of μ.
b. Is H_A one-sided or two-sided?
c. Compute the observed value of the test statistic.
d. Estimate the P-value.
e. Do you reject H_0 or fail to reject H_0?

11.4 SIGNIFICANCE TESTING VERSUS ACCEPTANCE/REJECTION TESTING: CONCEPTS AND METHODS

The version of hypothesis testing we have dealt with thus far is known as significance testing. In this type of testing there are one of two decisions made:

1. We reject H_0.
2. We fail to reject H_0.

If H_0 is true, it would obviously be an error to reject H_0. If H_0 is false, it would be an error of omission to fail to reject H_0.

Whenever we "fail to reject H_0," this means only that the data do not provide strong evidence that H_0 should be rejected! In particular, this is not an endorsement of the truth of H_0. It means that either H_0 or H_A may be true; strong evidence in favor of rejecting H_0 has simply not been found (even though the researcher probably had hoped to reject H_0). In other words, our test was inconclusive. From an academic point of view, we do not make an error when we fail to reject H_0, because we have not decided in favor of either H_0 or H_A. In the real world, however, a failure to reject H_0 when H_0 is false may lead to wrong decisions and wrong actions. We shall illustrate this by looking at a criminal trial as a hypothesis test.

Example 11.16

In a criminal trial the defendant is presumed to be innocent until proven guilty. In other words, the null hypothesis is

$$H_0 : \text{The defendant is innocent.}$$

The alternative hypothesis, which the prosecutor hopes to prove, is

$$H_A : \text{The defendant is guilty.}$$

In the case that the evidence is inconclusive, that is, we cannot rule out H_0, the defendant is given the benefit of the doubt and is found "not guilty." This does not mean that the defendant has been proven innocent; it only means that the prosecutor failed to prove guilt beyond a reasonable doubt. The defendant may still be guilty. Note, however, that the action taken by the court as a result of a "not guilty" verdict is to treat the defendant as if he or she had been found "innocent." In other words, the court does not distinguish between "not guilty" and "innocent."

Obviously, the court reaches the wrong decision if it convicts an innocent person or acquits a guilty person. In formal statistical jargon, the conviction of an innocent person (rejection

Type I error
Type II error

of H_0 when H_0 is true) would be called a **Type I error.** The acquittal of a guilty person (failure to reject H_0 when H_0 is false) would be called a **Type II error.** The only way that the court could be absolutely sure that it never made a Type I error (convicted an innocent person) would be to acquit every single defendant. The only way that the court could be absolutely sure that it never made a Type II error (acquitted a guilty person) would be to convict every single defendant. Obviously, neither of these two extremes is acceptable, and the court tries to balance the risk of Type I error against the risk of Type II error.

❒ ❒

In the next example we shall view the screening of donated blood as a hypothesis-testing problem.

Example 11.17

ELISA (enzyme-linked immunosorbent assay) tests are used to screen donated blood for the presence of HIV, the virus that causes AIDS. Because blood donors are prescreened for HIV risk factors, each unit of donated blood is presumed to be HIV-free unless the test shows otherwise. Our null and alternative hypotheses are

- H_0: The unit of blood is HIV-free.
- H_A: The unit of blood is HIV-contaminated.

In this context, we commit a Type I error if we reject (and subsequently destroy) a perfectly good unit of donated blood. We commit a Type II error if we fail to reject (and subsequently use) a contaminated unit of blood. If we commit a Type II error, then the recipient of the blood will be exposed to the AIDS virus. So we want a test with the highest **sensitivity** possible. (Recall that the sensitivity is the probability that a contaminated unit tests positive.) Statisticians call the sensitivity of a test the **power of a hypothesis test.**

sensitivity

power of a hypothesis test

On the other hand, if we reject and destroy too many perfectly good units of blood (commit too many Type I errors), then we might create a shortage of donated blood—not to mention the waste of resources involved in collecting the blood we end up throwing away. So we want a test with the highest **specificity** possible. (The specificity is the probability that an HIV-free unit of blood tests negative.)

specificity

Statisticians use the Greek letters α and β to denote the probability of a Type I error and the risk of a Type II error, respectively.

$$\begin{aligned}\alpha &= \mathbf{P(Type\ I\ error)}\\ &= P(\text{we reject } H_0 \mid H_0 \text{ is true})\\ &= 1 - P(\text{we accept } H_0 \mid H_0 \text{ is true})\\ &= 1 - \text{specificity of test}\end{aligned}$$

$$\begin{aligned}\beta &= \mathbf{P(Type\ II\ error)}\\ &= P(\text{we accept } H_0 \mid H_0 \text{ is false})\\ &= 1 - P(\text{we reject } H_0 \mid H_0 \text{ is false})\\ &= 1 - \text{sensitivity of test}\end{aligned}$$

Studies have found that if the blood screening for HIV is carried out by experienced laboratories, then

$$0.0060 < \alpha < 0.0150$$

$$0.0015 < \beta < 0.0030.$$

Note that the more serious error's (contaminated blood distributed) probability β is the smaller one, as it should be.

❐ ❐

acceptance/rejection testing

The hypothesis tests discussed in Examples 11.14, 11.16, and 11.17 were examples of **acceptance/rejection testing.** In each case, if we fail to reject H_0, then we cannot just say that the test was inconclusive; rather, we must accept H_0 to be true. For example, if the prosecutor in a criminal trial fails to prove the guilt of the defendant, then the court *accepts* the defendant's claim of innocence, that is, accepts the null hypothesis of innocence. Thus, it is important to find the guilty to be guilty (reject H_0) with high probability and to find the innocent to be innocent (accept H_0) with high probability. Similarly, if the ELISA test fails to detect any HIV antibodies in a unit of donated blood, then we will accept the unit and *actually use it*. Thus we want as high sensitivity (rejecting H_0 when H_0 is false) and as high specificity (accepting H_0 when H_0 is true) as possible.

Often, we are in this sort of hypothesis-testing situation where the setting requires us to decide between two alternatives. Consider the following examples:

- The Federal Reserve Board looks at economic data and then either lowers the prime lending (interest) rate (believing the economy needs further "stimulating") or does not lower it (believing that "stimulating" the economy could do more harm than good).
- You get the day's weather report ("data") and then either go on a picnic (believing it will not rain) or do not go on a picnic (believing it will rain).
- A manufactured product is judged either to meet product standards or not, based on performance-testing data.

From the hypothesis-testing viewpoint, the last example becomes

$$H_0\text{: product meets standards} \quad \text{versus} \quad H_A\text{: product does not meet standards.}$$

We thus must either accept H_0 (which is the same as rejecting H_A) or accept H_A (which is the same as rejecting H_0) in such cases. That is, these are examples of acceptance/rejection testing. The option of "failing to reject H_0 " is not open to us, because the setting requires us to approve the product or not approve it. Real-world risks and costs ride on both possible decisions. If we accept H_0 when H_A is true, the company ships a defective product, may lose customers, and may suffer a lawsuit. If we accept H_A when H_0 is true, the company loses money, time, and materials on needless extra work. We are not allowed the dodge that we do not know whether the product is acceptable or not!

In settings like these, where a decision in favor of H_0 or in favor of H_A must be made, clearly the null hypothesis does not have a special status. Rather, H_0 and H_A are in parallel roles. After analyzing the data statistically, we either embrace H_0 as true or embrace H_A as true. Indeed, H_0 and H_A are often equated to two contrasting real-world decisions, like acquitting or convicting a criminal defendant. Clearly, there are two possible correct decisions and two possible erroneous decisions. These are often expressed in terms of the standard diagram in Table 11.5. Because either H_0 is true or H_A is true, but not both, only one column in this table is active in an actual hypothesis-testing setting. That is, only one of the two errors is possible. For example, if it does rain today, the only possible error is going on the picnic. However, the statistician does not know the state of the real world (otherwise there would be no need for statistical inference), so which error is possible is not known. The standard terminology of "Type I" and "Type II" errors appearing in Table 11.5 is not very intuitive. The student of statistics simply has to memorize what each term means.

Significance testing, by contrast, focuses only on the probability of the Type I error of rejecting a null hypothesis that is in fact true. Hence, in significance testing, a failure to reject H_0 must never be interpreted as an acceptance of H_0. For example, if a coin suspected of favoring heads, with H_0 being the null hypothesis of a fair coin $(p = 1/2)$, is tossed 10 times and produces 7 heads, then the P-value is $P = P(\geq 7 \text{ heads} \mid H_0) = 1 - P(\leq 6 \text{ heads} \mid p = 1/2, n = 10) = 1 - 0.8281 = 0.1719$, according to Appendix F's binomial probability tables.

Table 11.5 Decisions and Errors in Acceptance/Rejection Testing

Statistical decision based on data (the inference)	Truth about the population (the reality)	
	H_0 true	**H_A true**
Reject H_0	Type I error	Correct decision
Accept H_0	Correct decision	Type II error

Because $0.1719 > 0.05$, we fail to reject H_0. But we *do not* accept H_0, which would be an incorrect assertion that there is strong statistical evidence the coin is fair. Indeed, 7 heads yields $\hat{p} = 0.7$, weak statistical evidence that the coin is unfair. In this hypothesis-testing situation, if it *is important* to the person doing the hypothesis testing to know that the coin is fair when it is fair, then one should not be using the significance-testing paradigm but rather should switch to the acceptance/rejection paradigm of this section.

acceptance/rejection hypothesis testing

Now, in **acceptance/rejection hypothesis testing,** where we are required either to accept or to reject H_0 (often because we must take one of two possible actions that each have costs and risks), we have to consider both $\alpha = P(\text{Type I error} \mid H_0) = P(\text{reject } H_0 \mid H_0 \text{ true})$ and $\beta = P(\text{Type II error} \mid H_A) = P(\text{accept } H_0 \mid H_0 \text{ false})$.

Thus we *must* also control the probability of making a Type II error, in contrast to controlling only the probability of Type I error, as is done in a significance-testing setting. The user who needs to have an acceptance/rejection hypothesis test carried out would like to turn his data over to the statistician, have her carry out a hypothesis test, and have *both* error probabilities be small. But this desire runs up against the famous (and almost always true) principle: There is no such thing as a free lunch.

quality control acceptance sampling
significance level α

As the following example of **quality control acceptance sampling** illustrates, in a test about a parameter, the risk α of a Type I error and the risk β of a Type II error depend on the parameter value. The example also illustrates that for a specified **significance level α** (probability of Type I error), the only way to reduce the risk of Type II error is to increase the sample size.

Example 11.18

A manufacturer receives a shipment of 100,000 parts from an outside subcontractor, who has guaranteed in writing that at most 0.5% (1 out of 200) of the parts are defective. When the shipment arrives, the manufacturer wants assurance that the parts meet agreed-to specifications. If p denotes the (unknown) proportion of defective parts among the 100,000 parts in the shipment, the manufacturer wants to test

$$H_0 : p \leq 0.005 \quad \text{versus} \quad H_A : p > 0.005$$

at the **significance level** $\alpha = 0.05$. Furthermore, the manufacturer wants the power of the test to be at least 0.99 if in fact $p = 0.02$ (2% of the parts are defective). In other words, if the shipment meets specifications (if H_0 is true), the risk of rejecting the shipment (a Type I error) should be, at most, $\alpha = 0.05$. But if the proportion of defective parts p is four times higher than promised, then there should be a 99% probability (power $= 0.99$) that the shipment is rejected; that is, if $p = 0.02$, the probability β of a Type II error should at most be 0.01. Thus, this is an acceptance/rejection hypothesis-testing setting where both $\alpha = P(\text{Type I error})$ and $\beta = P(\text{Type II error})$ are being made very small. Thus both possible decisions (rejecting H_0, accepting H_0) can be trusted.

The company statistician goes to work. She checks whether a simple random sample of $n = 400$ parts is large enough to give the desired high power at the desired low significance level. Her analysis goes as follows: Suppose that a simple random sample of $n = 400$ parts is inspected. Let $\hat{p}$ denote the proportion of defective parts among the 400 inspected. Suppose that the whole shipment is accepted if $\hat{p} \leq 0.01$ and rejected if $\hat{p} > 0.01$. (She got this threshold by playing around with the numbers.) The region of $\hat{p}$-values leading to rejection of H_0, namely ($\hat{p} > 0.01$), is called the **critical region** by statisticians. For different values of the proportion p of defective parts in the shipment ($p = 0.0025, 0.0050, 0.0075, \ldots, 0.0275$), the statistician finds the probability $\pi(p)$ that the shipment will be rejected and returned to the subcontractor. We define

critical region

$$\begin{aligned} \pi(p) &= P(\text{we reject } H_0 \mid p) \\ &= P(\hat{p} > 0.01 \mid \text{proportion of defective parts in shipment} = p) \end{aligned}$$

For each value of p, we can estimate the rejection probability $\pi(p)$ by using box model simulations (see Section 9.2). We used an online statistical calculator to derive the results in Table 11.6. Looking at Table 11.6, we see that if H_0 is true, that is, if $p \leq 0.005$, then the risk of a Type I error is at most 0.052, which is very close to the desired significance level $\alpha = 0.05$. We also note that if $p = 0.02$, then the rejection probability is $\pi(0.02) = 0.90$, which is not quite the power of 0.99 that we had hoped for. Unfortunately, with a sample size of $n = 400$ it is not possible to increase the power $\pi(0.02)$ without also increasing the significance level α of the test, that is, the risk of Type I error. For example, if we change the test of critical region to

$$\text{Reject } H_0 \text{ if } \hat{p} > 0.0075,$$

then the rejection probability $\pi(p)$ gets larger for all values of p. Specifically, the risk of Type I error becomes

$$\alpha = \pi(0.005) = 0.142 \text{ (too high)}$$

and the power $\pi(0.02)$ increases to 0.959 (better than before).

The only way to increase the power $\pi(p)$ for $p > 0.005$ without increasing the risk of Type I error for $p \leq 0.005$ is to increase the sample size n. For example, if $n = 800$ and we decide to accept the whole shipment if $\hat{p} \leq 0.00875$ and reject the shipment if $\hat{p} > 0.00875$, then our test has significance level

$$\alpha = \pi(0.005) = 0.050$$

and the power $\pi(0.020) = 0.991$, as desired.

Table 11.6 The Probability the Shipment Will Be Rejected as a Function of the Proportion p of Defective Parts

p	$\pi(p)$	
0.0025	0.004	H_0 true
0.0050	0.052	
0.0075	0.184	H_0 false
0.0100	0.371	
0.0125	0.561	
0.0150	0.718	
0.0175	0.830	
0.0200	0.903	
0.0225	0.947	
0.0250	0.972	
0.0275	0.986	

power function

In Figure 11.6 we have graphed the **power function** $\pi(p)$ for our first test, based on a sample size of $n = 400$ (and tabulated in Table 11.6) and using the critical region ($\hat{p} > 0.01$). We have also graphed the power function for the test of H_0 versus H_A, based on a sample size $n = 800$ and the critical region ($\hat{p} > 0.00875$) mentioned earlier.

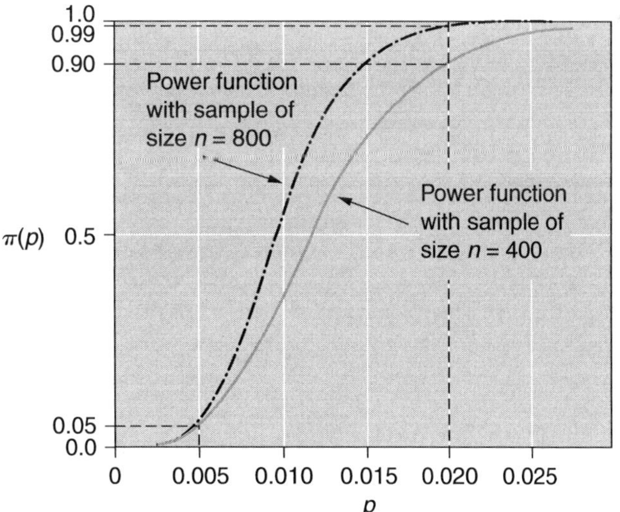

Figure 11.6 The probability that the shipment will be rejected as a function of p, the proportion of defective parts in the shipment.

As Figure 11.6 illustrates, for a specified significance level α (risk of Type I error), the way to reduce the risk of Type II error is to increase the sample size.

The Critical-Region Approach to Acceptance/Rejection Testing

critical-region approach to acceptance/rejection testing

Acceptance/rejection testing uses the so-called **critical-region approach to acceptance/rejection testing,** which we shall now explain in some detail.

The z_α Notation

z_α notation

The symbol z_α is used to denote the z-score for which the area under the standard normal curve to the right of z_α equals α, as illustrated in Figure 11.7. Read "z_α" as "z sub alpha."

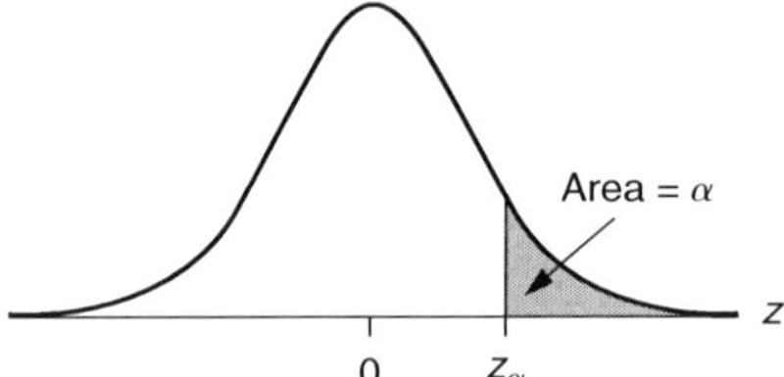

Figure 11.7 The z_α notation.

Table 11.7 gives some commonly used values of z_α. The t-table in the back of the book has additional values of z_α in the bottom line (labeled ∞) just above the percentages.

Table 11.7 Some Commonly Used Values of z_α

$z_{0.10}$	$z_{0.05}$	$z_{0.025}$	$z_{0.01}$	$z_{0.005}$
1.282	1.645	1.960	2.326	2.576

Consider a one-sample Z-test for a population mean μ. We make certain assumptions:

1. The population standard deviation σ is known.
2. Either the population is (very close to) normal or the sample size n is large ($n \geq 30$).

Suppose we want to test the null hypothesis

$$H_0 : \mu = \mu_0$$

at the user-specified significance level α. This, of course, means that we reject H_0 if and only if the P-value (also called **observed level of significance**) satifies $P \leq \alpha$, α being specified by the user. We compute the observed value of the Z-test statistic,

observed level of significance

$$Z = \frac{\overline{X} - \mu_0}{\sigma / \sqrt{n}}.$$

Under H_0, Z is (approximately) standard normal. In particular,

$$P(Z \geq z_\alpha) = \alpha.$$

If the alternative hypothesis is $H_A : \mu > \mu_0$, then the observed level of significance $P \leq \alpha$ if and only if we observe $Z \geq z_\alpha$. So we reject H_0 at the specified significance level α if and only if $Z \geq z_\alpha$. If $Z < z_\alpha$, we accept H_0. As illustrated in Figure 11.8, the set of values for the test statistic that leads us to reject H_0 is called the **critical region** (or **rejection region**). The set of values for the test statistic that leads us to accept H_0 is called the **acceptance region**.

critical region
rejection region
acceptance region

critical value

The value (or values) of the test statistic that separate the rejection and acceptance regions is (are) called the **critical value(s).**

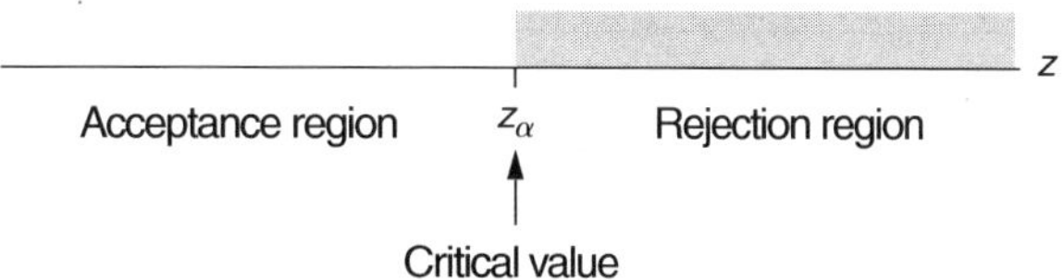

Figure 11.8 Rejection region, acceptance region, and critical value z_α for Z-test of $H_0 : \mu = \mu_0$ against the one-sided alternative $H_A : \mu > \mu_0$ at the significance level α.

If the alternative hypothesis is $H_A : \mu < \mu_0$, then the P-value $P \leq \alpha$ if and only if $Z \leq -z_\alpha$, noting that $P(Z \leq -z_\alpha) = P(Z \geq z_\alpha) = \alpha$ when H_0 is true and hence Z is standard normal. Figure 11.9 displays the rejection region, acceptance region, and critical value for this case.

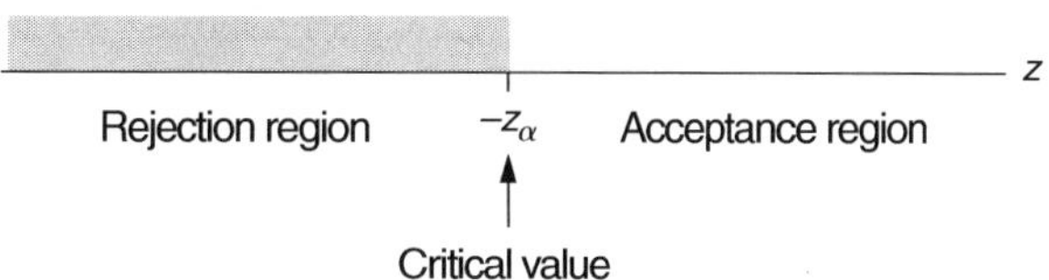

Figure 11.9 Rejection region, acceptance region, and critical value $-z_\alpha$ for Z-test of $H_0 : \mu = \mu_0$ against the one-sided alternative $H_A : \mu < \mu_0$ at the significance level α.

Finally, if the alternative hypothesis is the two-sided $H_A : \mu \neq \mu_0$, then the P-value $P \leq \alpha$ if and only if $Z \leq -z_{\alpha/2}$ or $Z \geq z_{\alpha/2}$. When H_0 is true, $P(Z \leq -z_{\alpha/2} \text{ or } Z \geq z_{\alpha/2}) = P(Z \leq -z_{\alpha/2}) + P(Z \geq z_{\alpha/2}) = P(Z \geq z_{\alpha/2}) + P(Z \geq z_{\alpha/2}) = \alpha/2 + \alpha/2 = \alpha$. Figure 11.10 displays the rejection region, acceptance region, and critical values $z_{\alpha/2}$ and $-z_{\alpha/2}$ for this case.

Figure 11.10 Rejection region, acceptance region, and critical values $-z_{\alpha/2}, z_{\alpha/2}$ for Z-test of $H_0 : \mu = \mu_0$ against the two-sided alternative $H_A : \mu \neq \mu_0$ at the significance level α.

In Figure 11.11, we summarize our discussion of the three critical regions, as determined by which H_A is chosen.

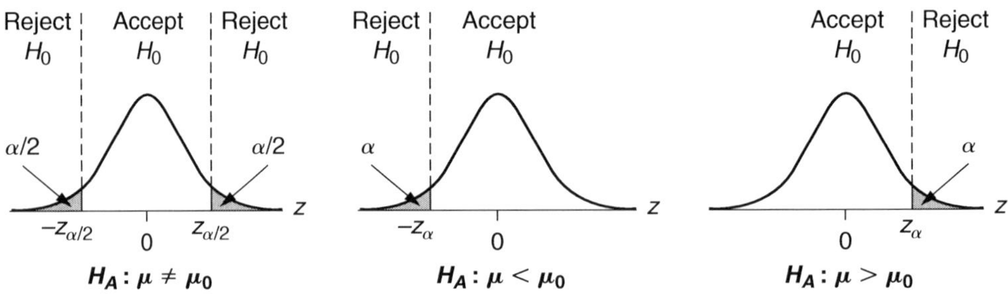

Figure 11.11 Critical-region approach to one-sample Z-test for $H_0 : \mu = \mu_0$ at the significance level α.

As an illustration of the critical-region approach to acceptance/rejection testing, we shall revisit Example 11.1 and determine the power function $\pi(\mu)$ of a one-sided Z-test.

Example 11.19 (Example 11.1 revisited)

power computations and critical-region determination illustrated for a Z-test of $H_0 : \mu = \mu_0$

Today's height distribution for American women follows a normal curve with unknown mean μ, but with the same standard deviation $\sigma = 2.6$ inches as 30 years and 50 years ago. Fifty years ago the average height of American women was estimated to be 63.1 inches. Thirty years ago the average height was estimated to be 63.7 inches. Consider a Z-test of

$$H_0 : \mu = 63.7 \text{ versus } H_A : \mu > 63.7$$

at the specified significance level $\alpha = 0.05$, based on the average height $\overline{X}$ of a simple random sample of 25 women. That is, we require that $\alpha = P(\text{Type I error}) = 0.05$. Under H_0, the test statistic

$$Z = \frac{\overline{X} - \mu_0}{\sigma / \sqrt{n}}$$

is standard normal. Here $\mu_0 = 63.7$, $\sigma = 2.6$, and $n = 25$. Let x and z_{OBS} denote the observed values of $\overline{x}$ and Z, respectively. We shall reject H_0 if

$$Z \geq z_{0.05} = 1.645$$

or, equivalently, if

$$\overline{X} \geq \frac{\sigma}{\sqrt{n}} \cdot 1.645 + \mu_0 = 64.5554.$$

Thus, the level $\alpha = 0.05$ critical region is $(\overline{X} \geq 64.5554)$. For each parameter value μ, the probability $\pi(\mu)$ that H_0 will be rejected is

$$\pi(\mu) = P(\overline{X} \geq 64.5554 \mid \mu).$$

To compute this, we use the fact that $\overline{X}$ is $N\left(\mu, \sigma/\sqrt{n}\right)$. As an example we shall compute

$$\pi(64.0) = P(\overline{X} \geq 64.5554 \mid \mu = 64.0).$$

As usual with normal probability computations, we shall proceed in three steps. We shall use the fact that $\text{SD}(\overline{X}) = \sigma/\sqrt{n} = 2.6/\sqrt{25} = 0.52$.

Step 1. Draw a number line and shade the interval (the critical region) of interest.

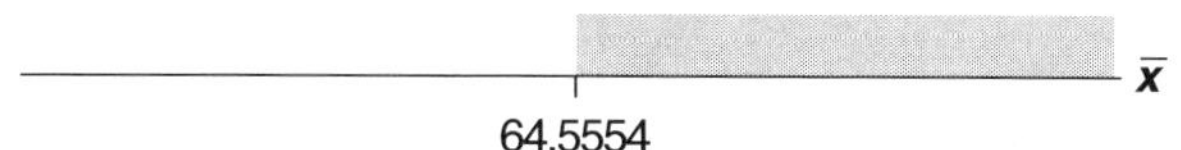

Step 2. Convert the interval endpoint into a z-score.

$$z_{\text{OBS}} = (64.5554 - \mu_0) / \text{SD}(\overline{X}) = (64.5554 - 64.0) / 0.52 = 1.068.$$

Step 3. Find the area under the standard normal curve over the appropriate z-score interval.

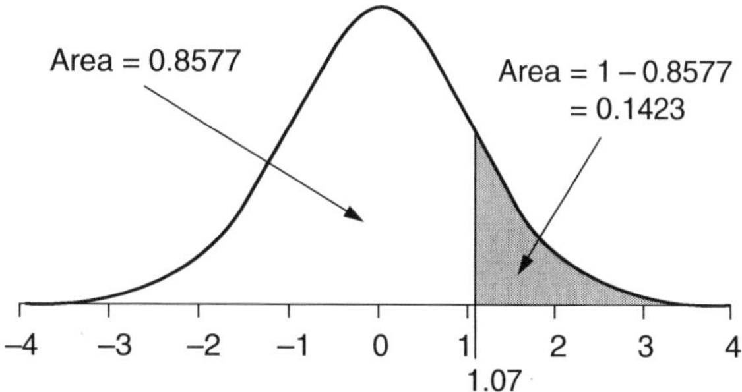

Using Appendix E, we find $\pi(64.0) = 0.1423$, for example.

The probability $\pi(\mu)$ that H_0 will be rejected was computed for $\mu = 63.7, 64.0, 64.3, \ldots,$ 65.8 and the results are listed in Table 11.8. The power function is graphed in Figure 11.12.

Table 11.8 The Probability $\pi(\mu)$ that H_0 Will Be Rejected as a Function of the Population Mean μ

μ	$(64.5554 - \mu)/0.52$	$\pi(\mu)$
63.7	1.645	0.05
64.0	1.07	0.14
64.3	0.49	0.31
64.6	−0.09	0.54
64.9	−0.66	0.75
65.2	−1.24	0.89
65.5	−1.82	0.97
65.8	−2.39	0.99

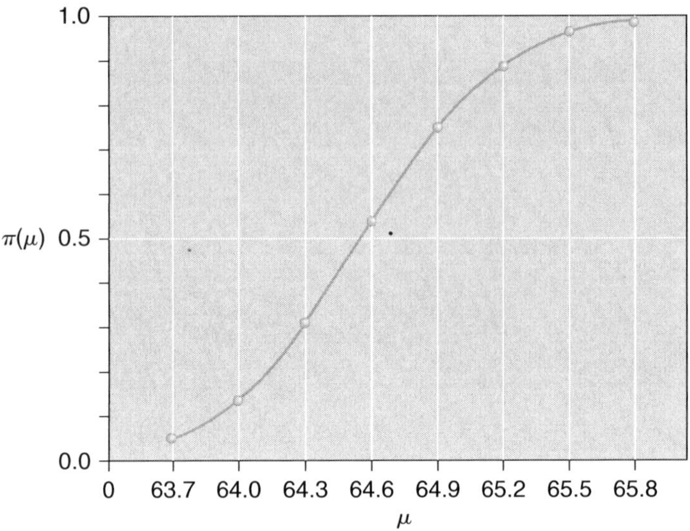

Figure 11.12 The probability that H_0 will be rejected as a function of the population mean μ (the power curve).

Finally we briefly outline the critical-region approach to doing a t-test. It is very similar to the critical-region approach to doing a Z-test.

t_α notation

The t_α Notation

The symbol t_α is used to denote the t-distribution-based t-value for which the area under the t density to the right of t_α equals α, as illustrated in Figure 11.13. When the value α is understood, we sometimes write "t_α" as "t^*."

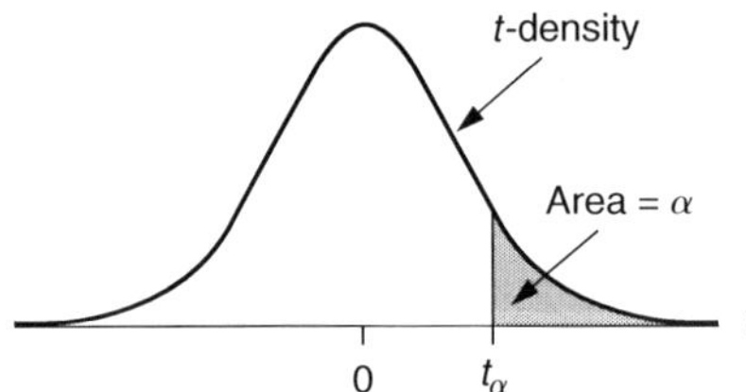

Figure 11.13 The t_α notation.

For a fixed α there is a different t_α for each t density as the degrees of freedom (df) change. For example, we find the critical values $t_{0.05}$ in the t-table in the column below the tail probability 0.05. In particular, if $df = 4$, that is, there are four degrees of freedom, then

$t_{0.05} = 2.132$, as shown in Table 11.9, excerpted from the t-table in Appendix F.

Table 11.9 The Top Portion of Appendix F Table

	t distribution critical values											
	Tail probability *p*											
df	.25	.20	.15	.10	.05	.025	.02	.01	.005	.0025	.001	.0005
1	1.000	1.376	1.963	3.078	6.314	12.71	15.89	31.82	63.66	127.3	318.3	636.6
2	.816	1.061	1.386	1.886	2.920	4.303	4.849	6.965	9.925	14.09	22.33	31.60
3	.765	.978	1.250	1.638	2.353	3.182	3.482	4.541	5.841	7.453	10.21	12.92
4	.741	.941	1.190	1.533	2.132	2.776	2.999	3.747	4.604	5.598	7.173	8.610
5	.727	.920	1.156	1.476	2.015	2.571	2.757	3.365	4.032	4.773	5.893	6.869

finding the critical region for a level α t-test of $H_0 : \mu = \mu_0$

Now consider a one-sample t-test for a population mean μ. We make the following assumptions:

1. The population standard deviation σ is unknown.
2. Either the population is (very close to) normal or the sample size n is large ($n \geq 30$).

Suppose we want to test the null hypothesis

$$H_0 : \mu = \mu_0$$

at the significance level α. This, of course, means that we reject H_0 if and only if the P-value or observed level of significance $P \leq \alpha$ (specified by the user). The critical-region approach is summarized in Figure 11.14. In each case the level α critical region is shaded.

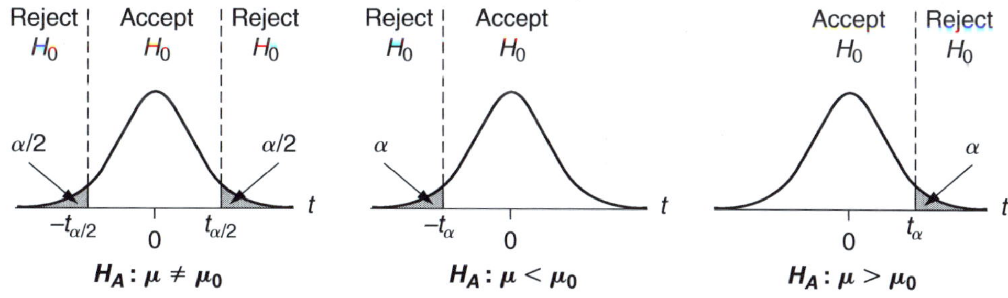

Figure 11.14 Critical-region approach to a one-sample t-test, for $H_0 : \mu = \mu_0$ at the significance level α.

Section 11.4 Exercises

1. What are some common values for the probability of a Type I error?
2. Suppose you decide to lower the probability of a Type I error from 0.05 to 0.01 without obtaining any additional data to base your test on. What happens to the probability of a Type II error?

3. What must you do to decrease the probability of Type I and Type II errors simultaneously?

4. In a court of law in the United States, a conviction for a crime requires "guilt beyond a reasonable doubt." What level of certainty should be required for reasonable doubt? Should you be 99% certain of a person's guilt to convict? Explain your (subjective) answer.

5. Find the critical region for the hypothesis test in Exercise 1 of Section 11.3 for a level of significance of 0.05.

6. Find the critical region for the hypothesis test in Exercise 2 of Section 11.3 for a level of significance of 0.05.

7. Find the power for the hypothesis test in Exercise 5 above for $\mu = \mu_X - \mu_Y = 2.00, 2.25, 2.50, 2.75$, and 3.00. Sketch the power function. (*Hint:* With the very large sample size $n = 177$, you may assume that $\sigma_D = s_D = 4.75$.)

8. A candidate commissions a poll, wishing to reject the null hypothesis that the proportion of the population who support him is $p = 1/2$, in favor of showing that a majority support him. The poll will have $n = 400$ people in it.
   a. What is the alternative hypothesis?
   b. Find c so that $P(\hat{p} \geq c) = 0.05$ when the null hypothesis $H_0 : p = 1/2$ is true.
   c. If $\hat{p} = 0.47$, do you accept or reject the null hypothesis?
   d. If $\hat{p} = 0.53$, do you accept or reject the null hypothesis?
   e. If $\hat{p} = 0.61$, do you accept or reject the null hypothesis? (Use level of significance $\alpha = 0.01$.)

9. A random sample of $n = 100$ boxes of cereal was obtained to test whether the population mean weight of the contents is $\mu = 16$ ounces. The alternative is that the population mean weight is greater than 16 ounces. The manufacturer does not want to put more cereal in the boxes than necessary.

   The critical region is to reject the null hypothesis if the sample mean $\overline{X} \geq 16.2$. Assume the population SD is $\sigma = 0.5$.
   a. What is $P(\overline{X} \geq 16.2)$ when the null hypothesis is true?
   b. Suppose the population mean is $\mu = 16.1$. What is $P(\overline{X} \geq 16.2)$?
   c. Suppose the population mean is $\mu = 16.2$. What is $P(\overline{X} \geq 16.2)$?
   d. Suppose the population mean is $\mu = 16.3$. What is $P(\overline{X} \geq 16.2)$?
   e. If the population mean is $\mu = 16.1$, is there a high probability that one would reject the null hypothesis? What if the population mean is $\mu = 16.3$?

CHAPTER REVIEW EXERCISES

1. It is claimed that the mean height of U.S. women is greater than 65 inches. In a random sample of 50 women, the mean height was 64.45 inches with SD of 3.5 inches. Test the null hypothesis that the population mean height of U.S. women is 65 inches.

2. In a random sample of 100 teenage boys, the mean amount of time spent watching television was $\overline{x} = 3.6$ hours with a standard deviation s_X of 0.9 hour. In a random sample of 100 teenage girls, the mean amount of time spent watching television was $\overline{y} = 4.2$ hours with SD of $s_Y = 1.2$ hours. Test whether the difference between the population means is zero versus the alternative that on average girls spend more time watching television.

3. In a random sample of eight people, the mean amount of sleep per night is 7.45 hours with SD of 1.15 hours. Test whether the population mean amount of sleep per night is 8 hours. Assume sampling from a normal population and that we are assessing a claim that sleep time actually averages less than 8 hours.

4. A sample of five body temperatures yields

98.7	98.5	98.6	98.3	99.0

a. Find the sample mean.
b. Create the population for bootstrap-testing the null hypothesis that the population mean is 98.6.

5. A study was conducted to investigate the effectiveness of hypnotism in reducing pain. Results for eight randomly selected subjects are given below. (The data are based on "An Analysis of Factors that Contribute to the Efficacy of Hypnotic Analgesia," by Price and Barber, *Journal of Abnormal Psychology*, Vol. 96, No. 1.) Note that this is a (before-after) matched-pairs study.

Before	−5.5	−5.0	−6.6	−9.7	−4.0	−7.0	−7.0	−8.4
After	−1.4	−0.5	0.7	1.0	2.0	0.0	−0.6	−1.8

Test the claim that the affective responses to pain are the same before and after hypnosis.

6. Three junior men and three senior men were sampled from a college. Their heights are

Juniors:	68	72	75	Seniors:	70	71	73

a. Find the sample means of the two groups.
b. Create the populations for bootstrap-testing that the population means are equal.

7. In a random sample of four people, the mean body temperature was 98.25 with SD of 0.73 degrees. Test whether the population mean temperature is 98.6 degrees versus less. Assume that the population distribution is bell-shaped.

8. Fifty measurements of the length of a bone were made. The mean of the measurements was $\overline{x} = 34.4$ centimeters with SD of $s_X = 0.25$ centimeter. Test whether the true length of the bone is $\mu = 35$ centimeters. Suppose there are no claims or conjectures about which side of 35 μ will be on if the null hypothesis is false.

9. Two ambulance services are tested for response time. A sample of 50 response times from the first service produces a mean $\overline{x} = 12.2$ minutes and SD of $s_X = 1.5$ minutes. A sample of 50 responses the second service produces a mean of $\overline{y} = 14.0$ minutes with SD of $s_Y = 2.1$ minutes. Test the claim that the two services have the same mean response time. It is not known or claimed that either service might be better.

10. In a random sample of 100 men, the mean number of alcoholic drinks consumed per week was $\overline{x} = 5.4$ with SD of $s_X = 1.25$. In a random sample of 100 women, the mean number of alcoholic drinks consumed per week was $\overline{y} = 3.2$ with SD of $s_Y = 1.85$. Test whether the difference between the population means is zero versus the claim that women drink less on average.

11. In a random sample of 300 people, 58% reported owning at least one cat. Test whether the true population proportion of people who own cats is 0.6. (Test this using $H_A : p < 0.6$.)

12. In a survey of members of the Biopharmaceutical Section of the American Statistical Association, individuals were asked to indentify their type of employer and what their current salary was. Out of 188 respondents who said they were working in academia, 41 said that they made more than $91,000 a year. Of the 529 people who worked for pharmaceutical companies, 180 said that they made more than $91,000 a year. Test the null hypothesis that the pro-

portion of people making over \$91,000 is the same for those working in academia and those working in pharmaceutical companies, versus the alternative hypothesis that the proportion is higher for those working in pharmaceutical companies.

13. A leading candy bar is claimed to have, on average, 170 calories. However, a competitor believes that the candy bar contains more calories on average than 170. The average number of calories in a sample of 40 of the candy bars in question is $\overline{x} = 175.5$, with SD of $s_X = 7.5$ calories. Perform a hypothesis test using a significance level of 0.01. What is the P-value?

14. The average height of women in the United States is often listed as 5 feet, 5 inches (65 inches). Suppose you believe this number is incorrect and that women are on average taller than 65 inches. In a sample of 200 women, the average height is $\overline{x} = 65.35$ inches with SD of $s_X = 3.5$ inches.
    a. Perform a hypothesis test with the alternative hypothesis that the average height of U.S. women is not 65 inches.
    b. Perform a hypothesis test with the alternative hypothesis that the average height of U.S. women is greater than 65 inches.

15. Find the critical region for the hypothesis test in Exercise 13 using a level of significance of 0.05. (Assume $s = 7.5$.)

16. Find the critical region for the test of hypothesis in part (b) of Exercise 14 using a level of significance of 0.05. (Assume $s = 3.5$.)

17. Find the power of the test in Exercise 16 for $\mu = 66$.

18. According to the manufacturer, a certain gasoline additive will "significantly" increase gas mileage. Eight cars of the same model were randomly selected and driven by the same professional driver, with and without the additive. The results are given in the table below. Assuming that the population is normally distributed, test whether there was a significant improvement in gas mileage.

	Mileage (miles per gallon)	
Car	**Without additive**	**With additive**
1	16	17
2	15	15
3	13	16
4	15	14
5	17	16
6	13	16
7	16	17
8	10	12

19. Proponents of alternative medicine might claim that the proportion of people using alternative medicine increased during the 1990s. In a 1990 survey of a sample of 1539 people, 520 had used alternative medicine. In a recent survey of 2055 people, 865 had used alternative medicine. Conduct a hypothesis test to determine whether the claim of increased use is supported.

20. The enrollment of children in preschool programs is claimed to result in long-term societal benefits. To investigate this claim, the Hi/Scope Educational Research Association studied two samples of "at risk" Michigan students: a sample of 62 from the population of those who attended preschool as children and a sample of 61 from the population of those who did not. For each sample they recorded the proportion who needed social services as adults. The sample statistic was 0.803 for those who did not attend preschool and 0.613 for those who did. Conduct a hypothesis test of the claim that attending preschool reduces the probability of needing social services as an adult.

PROFESSIONAL PROFILE

Karl Pearson, 1857–1936, originator of the discipline of mathematical statistics and developer of the chi-square test.

Bicyclist on the streets on the campus of the University of Illinois at Urbana-Champaign. Photo by William F. Stout.

Karl Pearson

Karl Pearson, credited with being the founder of mathematical statistics (the effort to make statistics a rigorous science undergirded by mathematics), founded the first university department of statistics (University College London, 1911). Along with Francis Galton and Walter Weldon, he founded (in 1901) Biometrika, one of today's most prestigious statistics journals.

Pearson's contributions to the field of statistics include many commonly used methods for analyzing data, including the use of the correlation coefficient to measure association in large data sets, the method of moments and the related system of continuous probability density curves, the foundations of statistical hypothesis testing and statistical decision theory, and principal component analysis. Of particular importance in today's statistical environment is his chi-square (χ^2) test.

Pearson's chi-square test is applicable when it is important to determine whether the data fits an assumed probability model. Chi-square goodness of fit testing postulates a null hypothesis that a given box model (using the textbook's terminology) is consistent with the observed data. The classic example is testing whether a die is fair based on a large number of rolls. However, Pearsons chi-square test has applications in a wide range of fields; it is fundamental to the practice of statistics.

Suppose university police are planning a campaign to have bicyclists obey traffic rules, such as stopping for red lights and not riding on sidewalks. Some claim that there are fewer bicyclists on Saturday and Sunday, when there are no classes, and that the campaign should target Monday–Friday. This claim needs to be statistically tested. The null hypothesis is that bicycle usage is uniform for all seven days. You can use Pearson's chi-square (χ^2) test to examine the frequency of bicycles observed on each day of the week to see whether weekend usage seems statistically less.

Areas where Pearson's chi-square (χ^2) test can provide valuable insights include marketing, the biological sciences, engineering, and physics, among many others. The application of Pearsonχ^2 chi-square (χ^2) test is very broad and has helped statisticians make major contributions to individual lives, to science, and to society at large.

12

Chi-Square Testing

It's Monday again, and I feel sad.
A commonly expressed sentiment that can be statistically tested.

Objectives

After studying this chapter, you will understand the following:

- ❒ Is the die fair?—the fundamental goodness-of-fit question
- ❒ Goodness-of-fit testing using five-step simulation and the intuitive D statistic
- ❒ The chi-square (χ^2) goodness-of-fit statistic
- ❒ How to use the five-step simulation approach to test a goodness-of-fit hypothesis using the chi-square statistic
- ❒ How and when to use the theoretical chi-square distribution in place of five-step simulation
- ❒ Chi-square goodness-of-fit testing in the unequal expected frequencies (unfair die) case
- ❒ Chi-square tests of independence of two multicategory population characteristics and homogeneity of multicategory characteristics for several pupulations
- ❒ Why chi-square testing is such a widely used hypothesis testing procedure

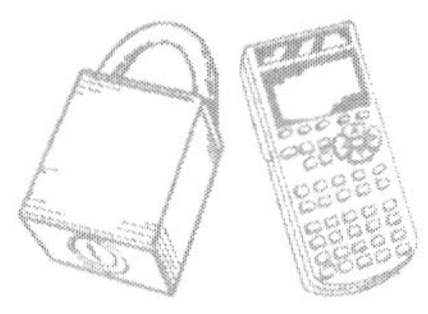

KEY PROBLEM

The quote at the beginning of this chapter suggests that one's emotional well-being may tend to be worse on Monday, the back-to-work day, etc., than on any other day. Social scientists have studied whether certain behaviors related to positive or negative emotional well-being occur more often on particular days of the week. Our Key Problem is a striking example of such research. Zung and Green (1974) conducted a study to see if suicides in North Carolina tend to occur more frequently on particular days of the week than on others rather than evenly over all days. Out of the population of all suicides in North Carolina, Zong and Green selected as their sample those that occurred from 1965 to 1971. The following table shows the 3672 suicides in the sample sorted by the day of the week on which they occurred.

Day	Sun	Mon	Tues	Wed	Thurs	Fri	Sat
Number	550	569	527	553	491	496	486

More suicides occurred on Mondays. Is Monday's larger number just by chance? Or are there subtle factors favoring slightly more suicides on Monday?

When a hypothesis or claim is made about the proportion of a population that falls into one of two categories, the testing methods described in Chapter 9 can be applied to test the claim. For example, we could use the methods described in Chapter 9 to test whether or not a majority of Americans favor the United States continuing military action in Afghanistan. In this example, each (adult) American is categorized as either favoring or not favoring the war in Afghanistan.

By contrast, the problems in this chapter (e.g., the one posed in the Key Problem) involve *more than* two categories (in the Key Problem case, the 7 days of the week). To test hypotheses regarding more than two categories, we will use the four-stage approach to hypothesis testing we introduced in Section 9.2. Stage II will consist of choosing a null-hypothesis-based sampling box model with balls labeled 1, 2, 3, 4, 5, 6, 7 to identify the 7 days of the week in the Key Problem. Our box models will need to contain as many types of balls as there are categories into which the units in the population are grouped. Analogous to Section 9.2's five-step-simulation-based test for a population proportion, we will take the five-step simulation approach using an intuitive "D statistic" or using a statistic called the chi-square statistic (χ^2). Next, analogous to the large-sample CLT (central limit theorem) approach of Section 9.2, we will learn a theoretical approach, using the chi-square probability distribution (with its probabilities tabulated in Table C in the Appendix) to approximate the unknown discrete distribution of the χ^2 statistic when H_0 is true. When the sample size is sufficiently large, this approach can be used instead of five-step simulation. This is analogous to the central limit theorem's allowing use of the normal distribution and Table E when testing $H_0 : p = p_0$ in Chapter 9.

Finally, we shall learn how to use the chi-square statistic to test whether two different ways of categorizing population members, such as categorizing people over age 24 by religious affiliation and by highest degree earned, are independent or associated. (Are certain religious groups more educated than other religious groups?)

12.1 IS THE DIE FAIR?

We start with the simple, instructive, but somewhat largely impractical problem of assessing a die's fairness. People use dice in games of chance, from a family playing a board game around the kitchen table to a Friday night casino visit. We expect the dice we use to be fair—that is, to turn up each face in equal proportions in the long run. The more money that rides on the throw, and the more times the dice are thrown, as in casino craps, the more important it is that the dice be absolutely fair. Casinos use carefully manufactured dice to ensure their fairness. So it is natural to question whether a particular die that we believe to be fair actually is fair. In fact, solving this problem will lead to our understanding of powerful statistical methods that can solve many important real-world problems.

Suppose someone wishes to confirm (or perhaps challenge) the fairness of a particular six-sided die and asks us to conduct a statistical study. We roll the die 60 times and obtain the outcomes in Table 12.1. We see from the table that 1 was obtained 7 times, 2 was obtained 9 times, and so on. We want to test the claim that the die may not be fair, noting that 5 being obtained 18 times suggests unfairness.

Table 12.1 Sixty Rolls of a Six-Sided Die

Outcome of die	f	$\hat{P}$(outcome)
1	7	$7/60 = 0.117$
2	9	$9/60 = 0.150$
3	14	$14/60 = 0.233$
4	7	$7/60 = 0.117$
5	18	$18/60 = 0.300$
6	5	$5/60 = 0.083$
Total	60	1.000

As discussed above, this statistical decision-making situation is similar to the ones we studied in Chapter 9. We will use the same four-stage approach to hypothesis testing that was used there. Also, we will use the same five-step simulation in stages II and III as we did there.

In stage I, we formulate the null hypothesis, which in the present case is that the data were produced by a fair die. Depending on the setting, the practitioner's hope may be to produce statistical evidence supporting the claim that the die is fair or to confirm a claim that the die is unfair. Both viewpoints produce the same null hypothesis that the die is fair.

In stages II and III, we choose a null-hypothesis-based box model and carry out the rest of the five-step simulation. We estimate the strength of evidence against the null hypothesis by estimating the probability, under the assumption the null hypothesis is true, of getting evidence against the null hypothesis as strong as, or even stronger than, the evidence at hand. In stage IV, we decide whether to reject the null hypothesis based on the size of the P-value.

goodness-of-fit test

The test for whether the hypothesized model (for example, a fair die) fits the data is called a **goodness-of-fit test.** Although much of the focus of this chapter is on variations of the fair-die goodness-of-fit question, we will discover that many fascinating and important real-life problems can be solved with chi-square goodness-of-fit testing methods. For example, we can test whether a given data set is well modeled by a null-hypothesis-specified, many-sided, loaded (i.e., unfair) die. Thus, we require neither six categories nor equal probabilities for the categories of the null-hypothesis model being tested. Many real-world settings can be modeled as a null hypothesis that in effect postulates a particular many-sided die. Hence, goodness-of-fit testing of such models, usually using a chi-square statistic, is a method of enormously wide applicability.

Consider two illustrations. The question in the Key Problem, whether suicides are equally likely on each day of the week, is statistically equivalent to the question of whether a seven-sided die is fair. The fact that a seven-sided fair die cannot physically be built is irrelevant—imagining such a die is all that is needed! We will address this problem in Example 12.3. As a second example, in a famous genetics experiment the proportions of four kinds of peas were observed in an attempt to determine whether they occurred with the probabilities of 1/16, 3/16, 3/16, and 9/16 predicted by Mendelian genetic theory. This hypothesized set of proportions corresponds statistically to a four-sided, loaded die.

Interestingly, in the Key Problem the researchers were hoping for evidence to reject the null hypothesis, whereas in the case of Mendel's data the researchers hoped to find *no evidence* to reject the null hypothesis, thus strengthening the scientific evidence that Mendel's genetic theory is valid. Statisticians refer to a probability law for the number of times each face occurs in n tosses of a many-sided, loaded die as the multinomial probability law or **multinomial distribution.** We will analyze Mendel's famous data set in Exercise 2 of Section 12.6. Once again, the fundamental issue here is one of goodness of fit: how well the null-hypothesis model produced by the Mendelian genetic theory fits the observed data.

multinomial distribution

Let us return now to the problem of the die whose fairness is the null hypothesis. Think about what sort of outcome (category) frequencies we would expect from a fair die. A natural first question to ask is: "What does 'fair' mean?" Clearly, a fair die is one for *which each side has an equal probability, 1/6, of occurring on a roll.* This definition is based on theoretical probabilities. From an experimental probability point of view, a fair die is one for which, when it is rolled more and more times, the experimental probabilities of the sides appearing become closer and closer to 1/6. This we can test.

In terms of Table 12.1, the null-hypothesis model of fairness means that the proportion of times we obtained each side (i.e., the experimental probabilities) should each be fairly close to 1/6 (0.167) in the 60 rolls. That is,

$$H_0 : p(1) = 1/6,\ p(2) = 1/6, \ldots, p(6) = 1/6$$

is the null hypothesis of a fair die, where $p(i)$ denotes the probability of outcome (face or side) i. This is our first example of a null hypothesis specifying a particular multinominal distribution.

In a frequency table such as Table 12.1, and often in this chapter, f denotes *frequency*, or the number of times the outcome occurs. The f value in the "1" row in Table 12.1 is 7; that is, we got a 1 seven times in 60 rolls of the die. So

$$\hat{p}(1) = 7/60 = 0.117 \text{ (rounded)}$$

compared to $p(1) = 1/6 = 0.167$ under the null-hypothesis, fair-die model. Similarly, $\hat{p}(2) = 9/60 = 0.15$ can be compared with $p(2) = 1/6 = 0.167$, and so on. The density histogram (discussed in Section 1.5) in Figure 12.1 presents the observed proportions (experimental probabilities) obtained in the 60 rolls of the die in Table 12.1. Note that the areas of the rectangles do add to 1, as required for a density histogram. The horizontal line gives the theoretical probabilities of the outcomes for a fair die. An off-the-cuff response to the observed frequencies of Table 12.1 might be something like "We got too few sixes and too many fives. Each outcome should occur about $60 \times (1/6) = 10$ times. Since 18 fives seems rather large, it looks as if there is statistical evidence that the die is not fair." But of course, even if the die is fair, these observed frequencies are random and highly unlikely to be equal to their expected frequencies of 10. Just as we do not necessarily get 30 heads and 30 tails when we toss a fair coin 60 times, we can also expect some deviation from 10 ones, 10 twos, and so on, when we roll a six-sided die 60 times, *even when the die is absolutely fair*. So the answer to the question of whether the die of Table 12.1 is fair is unclear, even though the observed and theoretical frequencies are far apart. We can start to get a statistical answer to this question by computing a statistical index that measures how much the observed frequencies depart from the expected frequencies for a fair die. In search of such a statistic, let us expand Table 12.1, as shown in Table 12.2, and discuss the meaning of each column.

Outcome of die: Which side fell face up?

Observed frequency (O): This is the observed frequency of each of the given outcomes of the die in 60 rolls—that is, the number of times that each side of the die appeared. The symbol O replaces the symbol f used in Table 12.1 to emphasize that this is an *observed* frequency.

Expected frequency (E): This is the theoretical or expected frequency (i.e., the theoretical expected value when the die is fair) of the given face of the die in 60 rolls. We find E using the rule that we learned in Chapter 6 for the expected frequency in n independent trials of an event that has probability p on each trial:

$$E = n \times p = 60 \times (1/6) = 10.$$

We will make frequent use of this $E = n \times p$ formula in this chapter.

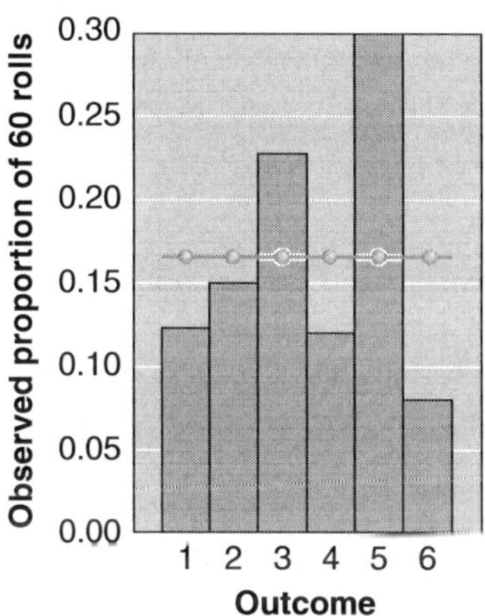

Figure 12.1 Density histogram comparison of experimental (O) and theoretical (E) probabilities for 60 rolls of a fair die.

Table 12.2 Observed Frequencies for 60 Rolls of a Six-Sided Die and Expected Frequencies, Assuming the Null Hypothesis That It Is Fair

Outcome of die	Observed frequency (O)	Expected frequency (E)	$O - E$	$\|O - E\|$
1	7	10	−3	3
2	9	10	−1	1
3	14	10	4	4
4	7	10	−3	3
5	18	10	8	8
6	5	10	−5	5
Total	60	60	0	24

***O* – *E*:** This is the difference, or deviation, between the observed and the expected frequencies of the outcome. For example, in row 1 we see that $O - E = 7 - 10 = -3$. This tells us that when the die was tossed 60 times, the 1 appeared three fewer times than expected for a fair die. On the other hand, in row 5 we see that $O - E = 18 - 10 = 8$. That is, 5 was observed eight more times than expected.

|*O* – *E*|: That is, we ignore whether $O - E$ is positive or negative and measure only the size of the deviation. This is the absolute value of $O - E$. For example, for row 6, $O - E = -5$ and $|O - E| = 5$.

Why include a column for absolute values? Look at the sum of the $O - E$ column. The positive and negative values in column $O - E$ cancel each other out, giving a sum of zero (and this is always true whatever the observed and expected frequencies). This sum is thus not informative about excessive deviations of the data from the fair-die model! However, the sum of $|O - E|$ column entries, which we see from the table is equal to 24, adds up the size of the deviations and is thus informative. It tells us that in the 60 rolls of the die, which may not be fair, the outcomes differed from what was expected for a fair die by a total difference of 24.

D statistic

We have found a simple and intuitive statistical index, or statistic, which we will call the ***D* statistic** and define as follows:

$$D = \sum |O - E|.$$

Mathematically, the D statistic is the sum of the absolute deviations of the observed from the estimated frequencies, which is the sum of the $|O - E|$ entries in the last column of Table 12.2. We note that for each of the six outcomes of the die where the observed frequency is far from the expected frequency, the corresponding term $|O - E|$ is large; when the observed and expected frequencies are close, the corresponding $|O - E|$ term is small. Large values of D indicate that, on the whole, the observed frequencies are far from the expected ones. Small values of D mean the opposite: the observed frequencies are close to the expected ones. The statistic D thus gives a measure of the distance between the observed and expected frequencies. The larger the value of D, the stronger the evidence that the die is unfair. We now need a way to decide whether $d_{\text{OBSERVED}} = 24$, read "d observed $= 24$," provides strong *enough* evidence to contradict the null-hypothesis assertion that the die is fair, leading us to reject the null hypothesis. We leave it to the next section to learn, via five-step simulation, how to determine whether 24 is "small" or "large" and whether we have weak or strong statistical evidence that the null hypothesis should be rejected. Note that D, which adds up the size of the deviations, is perhaps the simplest statistic one could develop to test the hypothesis of a fair die.

12.2 GOODNESS-OF-FIT TESTING USING THE *D* STATISTIC

At the end of Section 12.1, we were left with the question of how to decide whether an observed value of the statistic D is large enough to cause us to reject the null hypothesis of a fair die ($H_0 : p(1) = p(2) = \cdots = p(6)$). In other words, we need a way of deciding whether our observed value of the D statistic ($d_{\text{OBS}} = 24$), is larger than the value we would have gotten with a fair die. Just as in Chapter 9, we need a P-value—in this case $P(D \geq 24 \mid H_0)$. Our notation "$|H_0$" is to remind us that the probability is to be computed using the null-hypothesis model, in our case that of a fair die.

Suppose we have a six-sided die that we know is fair (for example, the manufacturer has carefully constructed it to be a fair die). We roll it 60 times and obtain the results shown in Table 12.3. Let us look at the observed value of D, which happens to be 20; that is, the observed frequencies differed from what was expected by a total of 20. Clearly, this simulation needs to be done many times—at least 10,000, in fact—in order to accurately represent the typical values, or more generally, the distribution of D for 60 tosses of a fair die.

Table 12.3 Results of 60 Rolls of a Six-Sided Die Known to Be Fair

Outcome of die	Observed frequency (O)	Expected frequency (E)	$O-E$	$\|O-E\|$
1	4	10	−6	6
2	6	10	−4	4
3	11	10	1	1
4	10	10	0	0
5	15	10	5	5
6	14	10	4	4
Total	60	60	0	20

Rather than tediously rolling a real fair die 60 times over and over, we instead do five-step simulation using the textbook's software. We repeat these 60 rolls many times for a simulated die that is fair, and each time we calculate a value of D. We will have, as a result, a table that shows the frequency distribution for the possible values of D obtainable from many simulations of 60 rolls of a fair die. From this frequency table we will be able to determine the value of $\hat{P}(\text{simulated } D \geq 24 \mid H_0)$. If an observed D value of 24 or more is unusually large [i.e., $\hat{P}(\text{simulated } D \geq 24 \mid H_0)$ is very small], we would conclude that the real data of Table 12.1, which yielded $d_{\text{OBS}} = 24$, indicate a die that is not fair.

What we have just described is essentially the four-stage hypothesis-testing approach of Chapter 9, applied to a null-hypothesis box model of three or more categories, using the D statistic. Let us review in detail how this four-stage approach would be applied in this problem, using the five-step simulation approach:

Stage I. Formulate the Null Hypothesis H_0

The null hypothesis is that the theoretical probabilities of the six sides of the die are equal. That is, the null hypothesis is the following (multinominal) probability distribution:

x	1	2	3	4	5	6
$p(x)$	1/6	1/6	1/6	1/6	1/6	1/6

We note that this null hypothesis distribution can be written

$$H_0 : p(1) = \cdots = p(6) = 1/6.$$

In goodness-of-fit testing it is not necessary to formulate the alternative hypothesis.

Stages II and III. Choose a Null-Hypothesis-Based Sampling Box Model and Estimate the P-Value Use the five-step method to calculate the experimental probability of getting a simulated value of D greater or equal to 24, using a box model determined by the null hypothesis. We will next lay out the five steps of the five-step method carefully.

Step 1. Choose the Null-Hypothesis-Based Sampling Box Model The box will have six balls in it: one labeled 1, one labeled 2, one labeled 3, one labeled 4, one labeled 5, and one labeled 6. Note that sampling with replacement from this box is equivalent to throwing a six-sided fair die.

Step 2. Define One Simulation One simulation is defined as the drawing (with replacement) of 60 balls, the same sample size as occurred in the real data sample. Note that this simulation is equivalent to 60 rolls of a six-sided fair die.

Step 3. Define the Event of Interest (EOI) We want to estimate the probability that a fair die produces a value of D equal to or more extreme than the value of D that was computed from the real data sample. More extreme values are values that would provide even stronger evidence against the null hypothesis than our observed value $d_{\text{OBS}} = 24$ does. In the case of D, "more extreme" always means "larger." If we let EOI stand for the "event of interest," then

$$\text{EOI} = (\text{simulated } D \geq 24).$$

Step 4. Repeat Simulations We conduct a large number N of simulations and keep track of the number of times (N_{EOI}) the event of interest occurred. Table 12.4 is a table of the frequencies of each value of D for $N = 10{,}000$. Choosing $N = 10{,}000$ provides great accuracy in estimating the P-value in Step 5.

From Table 12.4 we can obtain N_{EOI} by adding up the frequencies (f) of all the columns in which D is 24 or greater. Note that this table contains only even values of D; the D statistic cannot be odd when the outcomes are consecutive integers.

Table 12.4 Frequency Table for D Statistic for 10,000 Simulations of 60 Fair Die Tosses

D	2	4	6	8	10	12	14	16	18	20	22	24	26	28	30	32	34
f	26	157	477	958	1369	1759	1632	1334	1003	589	382	169	92	32	13	6	2

Step 5. Calculate the Experimental Probability of the Event of Interest, Obtaining the *P*-value

According to Table 12.4, there were $169 + 92 + 32 + 13 + 6 + 2 = 314$ occurrences of a value of D as large as, or larger than 24 in our 10,000 simulations. Therefore,

$$\hat{P}(\text{EOI} \mid H_0) = \hat{P}(\text{simulated } D \geq 24 \mid H_0) = \frac{314}{10{,}000} = 0.0314.$$

We note that the five-step simulation software finds the P-value directly once one sets up the null-hypothesis box model, sets up one simulation, defines the event of interest, and requests $N = 10{,}000$ simulations to find P. If more simulations are needed for even greater accuracy, one keeps requesting 10,000 simulations *without* resetting the software in between. You will quickly learn to solve goodness-of-fit hypothesis-testing problems in just this manner.

Stage IV. Decide Whether or Not to Reject the Null Hypothesis We found that, on the basis of 10,000 simulations, $\hat{P}(\text{simulated } D \geq 24 \mid H_0) = 0.0314$. Since a P-value of 0.0314 is less than our usual threshold of 0.05, we conclude that it is unlikely that we would obtain $D \geq 24$ if the die were fair. So we conclude that the die of Table 12.2 is not fair. That is, we reject the null-hypothesis model of a fair die.

highly significant

Note that if the user has selected the much less commonly chosen and more strict 0.01 or **"highly significant"** rejection criterion, the null hypothesis would not have been rejected. Thus, the evidence for rejection is "significant" but not "highly significant."

Section 12.2 Exercises

1. It is suggested that a particular six-sided die may be loaded, that is, an unfair die. Here are the observed results of rolling the die 60 times. Use the four-stage hypothesis-testing approach to decide whether the die is fair by answering the following questions.

Outcome of roll	Frequency
1	4
2	17
3	14
4	6
5	18
6	1
Total	60

a. What is the null hypothesis?
b. Calculate the D statistic for the real data results.
c. To determine whether the D statistic for the real data is unusual under the null hypothesis, we must have some idea of the kind of values of D to expect when the null hypothesis is true. We obtain these values by generating *simulated data samples* from a box model based on the null hypothesis. Describe the box model and define one simulation from it.
d. What are the test statistic and the event of interest used to obtain an estimated P-value? Use the five-step method with 10,000 simulations to estimate the P-value. (*Hint:* Select the five-step method from the software. Don't forget to sample with replacement. Select the D statistic.)
e. Based on the experimental probability of the event of interest (that is, the estimated P-value), decide whether the die is fair. Use level of significance 0.05.

2. Suppose we roll a six-sided die that we assume is fair. How many times would we expect each side to occur if we roll the die
a. 150 times?
b. 300 times?
c. 600 times?

3. Needing random digits for a class project, Nancy and Pete go through one page of a telephone book and write down the last digit of each of 50 telephone numbers. Here are their data:

Digit	0	1	2	3	4	5	6	7	8	9
f	12	4	3	1	4	11	2	5	3	5

Their instructor is skeptical that their method produces valid random numbers.
a. Use the simulation-based instructional software to apply the four-stage hypothesis-testing approach to decide whether this concern is supported by the data. Think of the problem as asking whether a 10-sided die is fair.

b. Obtain a phone book, randomly choose 50 phone numbers, and repeat the exercise.

4. A breakfast cereal company features a special offer by including one of four differently colored ballpoint pens in each cereal box. In a random sampling of 50 boxes of cereal, the following numbers of pens were obtained. Consider the null hypothesis that the company distributes the four colors in equal numbers. What is the value of D? Do you think that the company is distributing the four colors in equal numbers, or are some colors more likely to be obtained than others?

Color	f
Blue	17
Yellow	1
Red	16
White	16
Total	50

5. Explain why large values of D suggest that a die may not be fair.

6. Write down the null-hypothesis expected frequencies (the E's) for 30 rolls of a fair die, as given in Table 12.2. Suppose that in reality the six-sided die being tossed is unfair with theoretical probabilities for faces $1, 2, \ldots, 6$ given by $p(1) = 1/20, p(2) = 17/60, p(3) = p(4) = p(5) = p(6) = 1/6$. Find E for each face if we roll this unfair die 60 times. Now suppose that the observed O's for this unfair die come out exactly equal to the E's for faces 1 and 2 and that $|O - E| = 2$ for faces 3, 4, 5, and 6. That is, the observed frequencies for the unfair die are quite close to their expected values.
   a. What is D for the fair-die null hypothesis?
   b. Is the value of D large enough to yield strong evidence (i.e., use 0.05) that the die is unfair?
   c. Suppose instead that $p(1) = 1/30, p(2) = 18/60, p(3) = p(4) = p(5) = p(6) = 1/6$. Repeat part (b), except do 30,000 simulations for high accuracy.

7. Consider the experiment of 60 rolls of a fair six-sided die. For each trial the D statistic is computed. Use the simulation software to estimate the indicated experimental probabilities, which could be P-values.
   a. $\hat{P}(\text{simulated } D \geq 16 \mid H_0)$
   b. $\hat{P}(\text{simulated } D \geq 17 \mid H_0)$
   c. $\hat{P}(\text{simulated } D \geq 20 \mid H_0)$
   d. $\hat{P}(\text{simulated } D \geq 8 \mid H_0)$

12.3 GOODNESS-OF-FIT TESTING USING THE CHI-SQUARE STATISTIC

In Section 12.2, we introduced a goodness-of-fit test statistic, which we called the D statistic, for testing whether a die is fair or not. D measures the distance between the observed frequencies and what is expected of a fair die. We compared the observed value of D from the real data set, d_{OBS}, to the values of D a fair die would have given us. The D statistic provided us with a convenient yardstick for telling how strong the evidence in the real data is against the fair-die null hypothesis.

chi-square statistic

However, the usual way statisticians measure the strength of evidence against a fair-die null hypothesis is to compute a different test statistic than D, called the **chi-square statistic,** written as χ^2. The symbol is the Greek letter chi (pronounced like "sky," but without the "s"). Like the D statistic, the chi-square statistic measures the distance between the observed frequencies and what is expected under the null hypothesis.

In this section we focus on describing the calculation of this new χ^2 statistic and on its usefulness for goodness-of-fit testing via the simulation approach. First, let's compare how the D statistic and the χ^2 statistic are computed when they are both applied to the real data from the fair-die problem. Table 12.5, called a χ^2 frequency table, shows both the D statistic (the sum of the $|O-E|$ column entries) and the χ^2 statistic (the sum of the $(O-E)^2/E$ column entries) for the same data, 60 rolls of a six-sided die. (In practice, a χ^2 table does not show the value of the D statistic.) The value of χ^2 is given by

$$\chi^2 = \sum \frac{(O-E)^2}{E}.$$

Therefore,

$$\begin{aligned}\chi^2_{\text{OBS}} &= \frac{(-3)^2}{10} + \frac{(-1)^2}{10} + \frac{4^2}{10} + \frac{(-3)^2}{10} + \frac{8^2}{10} + \frac{(-5)^2}{10} \\ &= \frac{9}{10} + \frac{1}{10} + \frac{16}{10} + \frac{9}{10} + \frac{64}{10} + \frac{25}{10} = 12.4\end{aligned}$$

There are similarities, but also important differences, between how we calculate D and how we calculate χ^2. First, instead of taking the absolute value $|O-E|$ [i.e., the magnitude of $(O-E)$], we square $(O-E)$, obtaining the square of the magnitude of $(O-E)$. Like the absolute value, the square of $(O-E)$ is always nonnegative, regardless of the sign of $(O-E)$. We then divide each value of $(O-E)^2$ by its corresponding expected value E before adding across categories.

Table 12.5 χ^2 Frequency Table for Die Fairness Data

Outcome of die	Obtained frequency (O)	Expected frequency (E)	$O-E$	$\lvert O-E\rvert$	$(O-E)^2$	$(O-E)^2/E$
1	7	10	−3	3	9	0.9
2	9	10	−1	1	1	0.1
3	14	10	4	4	16	1.6
4	7	10	−3	3	9	0.9
5	18	10	8	8	64	6.4
6	5	10	−5	5	25	2.5
Total	60	60	0	24	124	12.4

Now let's focus on using the χ^2 statistic in the same four-stage simulation approach to hypothesis testing that was used earlier with D. The process is identical except D is replaced by χ^2. The goal here is the same as with the D statistic: We are trying to determine whether the observed chi-square value, 12.4, is too large to have been produced by a fair die. To answer that question we have to determine the likelihood that a fair die would produce a χ^2 value as large as or larger than the observed χ^2 value, $\chi^2_{\text{OBS}} = 12.4$. In other words, we want to estimate $P(\chi^2 \geq 12.4 \mid H_0)$, namely the P-value. We again use the five-step method.

Step 1. Choose the Null-Hypothesis-Based Sampling Box Model We use the same box model as we did when estimating $P(D \geq 24 \mid H_0)$: six balls marked as 1, 2, 3, 4, 5, 6.

Step 2. Define One Simulation We draw one ball with replacement 60 times and note the frequency of occurrence of each number, denoted O. We will use this simulation to compute χ^2 in Step 3.

Step 3. Define the Event of Interest (EOI) The event of interest here is (simulated $\chi^2 \geq$ observed χ^2), that is, (simulated $\chi^2 \geq 12.4$).

Step 4. Repeat Simulations. We simulated 50,000 values of the discrete χ^2 statistic. Figure 12.2 displays the experimental distribution. (This is how the theoretical probability histogram for the discrete χ^2 statistic actually looks!)

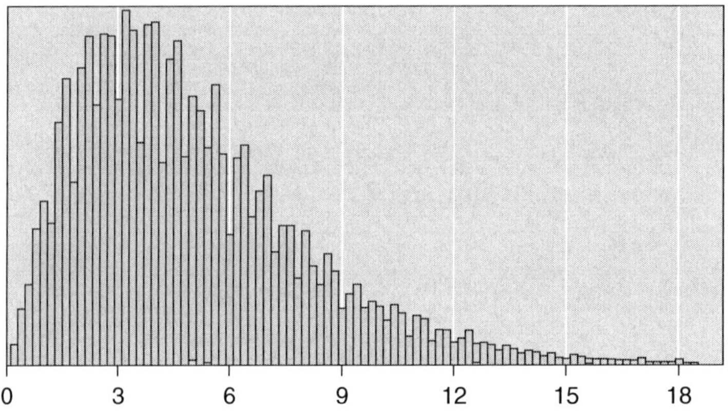

Figure 12.2 Experimental histogram for 50,000 simulated values of the discrete χ^2 statistic for 60 tosses of a fair die.

As Figure 12.2 illustrates, with a fair die it is possible to get a χ^2 value as high as 12.4, but not very likely.

Step 5. Estimate the *P*-value (Find the Experimental Probability of the Event of Interest) out of 50,000 simulations, 1501 yielded a χ^2-value equal to or greater than 12.4. so our experimental P-value was

$$\hat{P}(\text{simulated } \chi^2 \geq 12.4 \mid H_0) = \frac{1501}{50{,}000} = 0.03.$$

So the χ^2 test leads to the same conclusion as the D test: we reject the fair-die null hypothesis, but the result is not highly significant.

Section 12.3 Exercises

1. Consider the simulation of 60 rolls of a fair six-sided die. For each simulation the χ^2 statistic is computed. Run 10,000 such simulations using the simulation software to estimate each of the indicated probabilities:
   a. $P(\chi^2 \geq 6)$
   b. $P(\chi^2 \geq 8)$
   c. $P(\chi^2 \geq 10)$
   d. $P(\chi^2 \geq 14)$

2. A six-sided die was rolled 60 times with the following results:

Outcome of die	Observed frequency (O)	Expected frequency (E)
1	8	10
2	7	10
3	13	10
4	11	10
5	15	10
6	6	10
Total	60	60

a. Calculate D for these data. Note that the expected values (E) are based on a null hypothesis that assumes a fair die.
b. Using Table 12.4 or the siumation software, estimate the theoretical probability of the event of interest [i.e., getting a value of D under the null hypothesis that is as large as, or larger than, the value you obtained with the real data in part (a)]. Do you have strong statistical evidence that the die used in part (a) is unfair? Why or why not?
c. Calculate χ^2 for these data. Note that the expected values are based on a null hypothesis that assumes a fair die.
d. Using simulation software, estimate the theoretical probability of the event of interest [i.e., the probability of getting a value of χ^2 under the null hypothesis that is as large as, or larger than, the value you obtained with the real data in part (a)]. Do you have strong statistical evidence that the die used in part (a) is unfair? Why or why not?

3. In a colored-pen offer similar to those in Example 6.10 and Exercise 4 of Section 12.2, this time involving six pen colors, the following numbers of pens were obtained from 90 cereal boxes:

Color of pen	Number obtained
Sky blue	17
Passionate pink	31
Deep purple	7
Burnt orange	10
Boring brown	9
Anemic ash	16
Total	90

a. Use the textbook software to test the null hypothesis that the company distributes the six colors in equal numbers using the χ^2 statistic.
b. Repeat part (a) using the D statistic instead. Are the P-values similar?

4. During a busy day in a large city, 90 traffic tickets were issued at six locations that each have the same amount of traffic. Test the null hypothesis, using both the χ^2 and the D statistic, that the rate of ticket issuing is the same at all six locations.

Location	Number of tickets given
A	12
B	7
C	21
D	15
E	11
F	24
Total	90

5. The following table gives the outcomes of rolling an actual four-sided die 40 times.

Outcome	Expected number	Obtained number
1	10	12
2	10	7
3	10	14
4	10	7
Total	40	40

a. Calculate χ^2 for these data under the null hypothesis that the die is fair.
b. Test the null hypothesis.

6. The following table tells how many in a sample of 40 persons prefer each of four kinds of orange juice. Find D and χ^2 for these data, assuming the null hypothesis is that each juice is equally preferred. Explain why this null hypothesis is equivalent to a four-sided fair die. Then use the five-step method to help you decide if the sample χ^2 statistic provides convincing evidence that the population of all people prefer some kind of orange juice over others.

Kind of orange juice	Number of persons preferring
Fresh	13
Freeze-dried	11
Frozen	8
Canned	8
Total	40

7. Suppose that a certain unfair six-sided die has probabilities given by $p(1) = 1/20$, $p(2) = 17/60$, and the remaining probabilities are each 1/6. Find E for each face if we roll this unfair die 60 times. Now suppose that the observed O's for this unfair die come out exactly equal to their E's for faces 1 and 2 and that $(O - E)^2 = 4$ for faces 3, 4, 5, and 6.
a. What is χ^2 for the fair-die null hypothesis? (*Hint:* Each $E = 10$ for a fair die.)
b. Is the value of χ^2 large enough to yield strong evidence the die is unfair? Compare your answer with Exercise 6

of Section 12.2.

c. Which produces a value of χ^2 large enough to yield strong evidence that the die is unfair? Compare your answer with Exercise 6 of Section 12.2.

8. Which produces a larger value of χ^2 for a fair six-sided die tossed 60 times: every O differing from E by 2 (in some direction), or two O's differing by 4 from E and the rest each equaling E?

9. Explain the only case where an observed $\chi^2 = 0$ is possible. Is this likely?

10. Could you use a χ^2 test statistic to test whether boys and girls are born with equal likelihood in a certain country? (*Hint:* Using the die analogy, what fair "die" might work here?)

11. Consider the following two sets of data for colors of randomly selected objects in two different situations:

	Set A	
Color	**Expected**	**Observed**
Red	100	105
Blue	100	97
Green	100	98
	Set B	
Color	**Expected**	**Observed**
Red	5	10
Blue	5	2
Green	5	3

a. By just looking at the data, do you think the evidence is strong that the observed values follow the same distribution as the expected values in the case of data set A? What about data set B?

b. Calculate

$$(\text{observed red} - \text{expected red})^2 + (\text{observed blue} - \text{expected blue})^2 + (\text{observed green} - \text{expected green})^2$$

for both data sets A and B.

c. Using your answers from parts (a) and (b), explain why dividing by the expected number is important in the calculation of the χ^2 statistic.

d. Do a chi-square test in both cases.

12. Redo Exercise 1 in Section 12.2 using the χ^2 statistic. Use 10,000 simulations.

13. Redo Exercise 3 in Section 12.2 using the χ^2 statistic. Do 20,000 simulations.

14. Redo Exercise 4 in Section 12.2 using the χ^2 statistic.

12.4 CHI-SQUARE GOODNESS-OF-FIT EXAMPLES

Now we provide several applications of χ^2 goodness-of-fit testing to real-life examples, including the Key Problem from the beginning of this chapter. In each case, the real-world setting is analogous to a many-sided fair-die null hypothesis, but not necessarily a six-sided fair die. Note that one could continue to use the D statistic via the textbook's simulation software.

Example 12.1 Bicycle Accidents

In the course of research on possible ways to reduce bicycle accidents in a large city, it is hypothesized that accidents may occur more frequently on particular days of the week than on others. If this hypothesis were to prove true, better-focused expenditures of time and money based on the day of the week could perhaps result in reduced accident rates without an increased financial burden on the city. The data for 910 randomly selected bicycle accidents are given in the "Observed (O)" column of Table 12.6. (Adapted from Insurance Institute for Highway Safety.)

Table 12.6 χ^2 for Bike Accidents by Day of Week

Day of the week	Expected (E)	Observed (O)	$(O-E)^2$	$(O-E)^2/E$
Sunday	130	118	$(-12)^2 = 144$	1.11
Monday	130	119	$(-11)^2 = 121$	0.93
Tuesday	130	127	$(-3)^2 = 9$	0.07
Wednesday	130	137	$(7)^2 = 49$	0.38
Thursday	130	128	$(-2)^2 = 4$	0.03
Friday	130	146	$(16)^2 = 256$	1.97
Saturday	130	135	$(5)^2 = 25$	0.19
Total	910	910		4.68

Solution: Table 12.6 shows that $\chi^2 = 4.68$.

Stage I. Formulate the Null Hypothesis We want to determine whether the sample provides statistical evidence that bicycle accidents are more likely to happen on particular days of the week than on others. We thus begin by assuming the null-hypothesis model that accidents are equally likely to happen on each day. Under the H_0, one-seventh of the accidents are expected to occur on each of the seven days of the week.

Stages II and III. Choose the Null-Hypothesis-Based Sampling Box Model and Estimate the P-Value We use the five-step method:

Step 1. Choose a Null-Hypothesis-Based Sampling Box Model To model the null hypothesis of bicycle accidents being equally likely on each day of the week, use a box containing seven balls, each labeled to indicate a different day of the week. The 1-ball represents the first day of the week (Sunday), the 2-ball represents Monday, the 3-ball represents Tuesday, and so on.

The day of an accident is determined by the draw of a ball.

Now we must determine whether values of χ^2 as large as or larger than the one obtained from the data in Table 12.6 $(\chi^2_{OBSERVED} = 4.68)$ occur rarely or often under the null-hypothesis model.

Step 2. Define One Simulation A simulation consists of drawing 910 times with replacement from the box, one draw for each of the 910 accidents reported in the real data of Table 12.6, for a sample size of 910.

Step 3. Define the Event of Interest For each simulated data sample we calculate the χ^2 test statistic. The event of interest needed to produce the P-value is (simulated $\chi^2 \geq 4.68$). The following table gives the results for the first simulation, yielding $\chi^2 = 5.71$. This simulated χ^2 exceeds the observed 4.68, which is an occurrence of the event of interest (check the calculation of this simulated χ^2 by completing the $(O-E)^2/E$ column and totaling its values). That is, the EOI is $(\chi^2 \geq 4.68)$.

	Number of accidents			
Day of the week	**Expected** (E)	**Observed** (O)	$(O-E)^2$	$(O-E)^2/E$
Sunday	130	141	121	
Monday	130	139	81	
Tuesday	130	123	49	
Wednesday	130	119	121	
Thursday	130	145	225	
Friday	130	122	64	
Saturday	130	121	81	
Total	910	910		

Step 4. Repeat Simulations We performed a total of 10,000 simulations using the textbook's instructional software.

Step 5. Estimate the Probability of the Event of Interest (Estimating the *P* Value) Out of the 10,000 simulations, 5907 yielded a χ^2-value equal to or greater than the observed value, 4.68. So the estimated P-value is

$$\hat{P}(\text{simulated } \chi^2 \geq 4.68 \mid H_0) = 5907/10{,}000 = 0.5907.$$

Stage IV. Decide Whether to Reject the Null Hypothesis We found that (simulated $\chi^2 \geq 4.68$) occurred rather often (over one-half of the time) by chance for the null-hypothesis model of equal probabilities for each day of the week. We thus have no evidence to rule out the null-hypothesis model. So we fail to reject the null hypothesis: We conclude that there is no statistical evidence that some days of the week have higher rates of bicycle accidents than others in the long run. Thus, the observed differences in occurrence of accidents on the different days of the week are typical of what we might expect from ordinary chance variation.

Example 12.2 Twins

According to a genetic model, one-third of all pairs of twins consist of a pair of girls, one-third consist of a pair of boys, and one-third consist of a girl and a boy. (The model assumes that boys and girls are equally likely and that one-third of all twin pairs are identical.)

Hospital records for a simple random sample of 600 pairs of twins yielded the numbers shown in Table 12.7. (Hospital records do not reveal whether the twins were identical or fraternal. But the records do specify gender.) Do the data support the model?

Table 12.7 Gender of Twins

Gender	Observed frequency
GG	182
GB	206
BB	212
Total	600

Here, GG stands for a pair of girls, GB stands for a girl and a boy, and BB stands for a pair of boys.

Solution:

Stage I. Formulate the Null Hypothesis We assume a null hypothesis that the three types of twins (a pair of girls, a girl and a boy, a pair of boys) are equally likely. Under this assumption we would expect about one-third of all twin pairs to be of each type.

The expected frequency is

$$\frac{1}{3} \times 600 = 200$$

for each of the three categories. The observed value of the χ^2 statistic is

$$\begin{aligned} \chi^2_{\text{OBS}} &= \frac{(182-200)^2}{200} + \frac{(206-200)^2}{200} + \frac{(212-200)^2}{200} \\ &= 1.62 + 0.18 + 0.72 = 2.52. \end{aligned}$$

Stages II and III. Choose the Null-Hypothesis-Based Sampling Box Model and Estimate the *P*-Value We use the five-step method:

Step 1. Choose the Null-Hypothesis-Based Sampling Box Model We can use a box model with balls 1, 2, and 3 in it, corresponding to the three types of twins, drawing with replacement.

Step 2. Define One Simulation A simulation would consist of doing 600 draws with replacement from the box, one draw for each pair of twins in the real data sample. Each time we draw the 1-ball we get a pair of girls, each time we draw the 2-ball we get a girl and a boy, and each time we draw the 3-ball we get a pair of boys. We would record how many times each of the three balls was drawn.

Step 3. Define the Event of Interest For each simulated data sample, we calculate the χ^2 statistic. The event of interest is (simulated $\chi^2 \geq 2.52$).

Step 4. Repeat Simulations We did 10,000 simulations.

Step 5. Estimate the Probability of Occurrence of the Event of Interest Out of 10,000 simulations, 2808 yielded a χ^2-value equal to or greater than the observed value, 2.52. So our experimental P-value was 0.28.

Stage IV. Decide Whether to Reject the Null Hypothesis Since $\hat{P} = 0.28$ is greater than 0.05, the observed data do not contradict the genetic model. We do not reject the null hypothesis that the three types of twins are equally likely.

Example 12.3 Key Problem: Suicide Rates

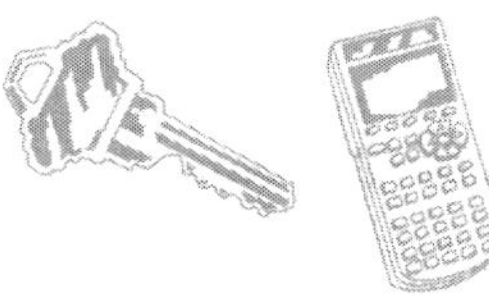

As mentioned at the beginning of this chapter, Zung and Green (1974) conducted a study to see whether suicides in North Carolina occur more frequently on particular days of the week. They used recorded suicide data from 1965 to 1971 as their data sample. There is a common belief that people are particularly vulnerable to depressive emotions on Mondays, and Zung and Green wanted to see whether their suicide data would confirm this common belief. If suicide data reveal significantly more suicides on particular days, then this information might lead to proactive measures to prevent suicides, and it might also shed light on the validity of the theory that Mondays are generally more depressing than other days. The data for 3672 sampled suicides classified by day of the week are as follows:

Day	Sunday	Monday	Tuesday	Wednesday	Thursday	Friday	Saturday
Number	550	569	527	533	491	496	486

We do indeed observe the most suicides on Monday. But, as we are beginning to understand well, this fact *by itself* does not prove that the daily suicide rates will differ in the long run. The question is: Can the observed differences be explained by chance alone?

Solution:

The suicide data are entered in the "Observed number (O)" column of the χ^2 table, Table 12.8. If the number of suicides per day does not depend on the day of the week, we expect about one-seventh of the suicides to happen on each day. That is, for each day of the week the expected number of suicides (in the "Expected number (E)" column of Table 12.8) is $1/7 \times 3672 = 524.6$, rounded to the nearest tenth. These expected values can now be used to calculate the χ^2 value for the real data:

$$\chi^2 = \frac{645.16}{524.6} + \frac{1971.36}{524.6} + \frac{5.76}{524.6} + \frac{806.56}{524.6} + \frac{1128.96}{524.6} + \frac{1489.96}{524.6} = 13.09$$

Note: The fact that the E column and the O column do not have exactly the same sum is due only to roundoff error.

Stage I. Formulate the Null Hypothesis We want to determine whether the sample data provide statistical evidence that there is a greater likelihood for a suicide to occur on a particular day of the week than on other days. We begin by assuming the null-hypothesis model that suicides are equally likely to happen on each day of the week. So our box model represents a fair seven-sided die.

Table 12.8 Number of Suicides

Day of the week	Expected number $(E)^*$	Observed number (O)	$(O-E)^2$
Sunday	524.6	550	$(25.4)^2 = 645.16$
Monday	524.6	569	$(44.4)^2 = 1971.36$
Tuesday	524.6	527	$(2.4)^2 = 5.76$
Wednesday	524.6	553	$(28.4)^2 = 806.56$
Thursday	524.6	491	$(-33.6)^2 = 1128.96$
Friday	524.6	496	$(-28.6)^2 = 817.96$
Saturday	524.6	486	$(-38.6)^2 = 1489.96$
Total	3672.2	3672	

*The total does not sum to 3672 because of roundoff errors.

Next we must determine whether $\chi^2 \geq 13.09$ occurs rarely or often under the null-hypothesis model.

Stages II and III. Choose the Null-Hypothesis-Based Box Model and Estimate the *P*-Value We use the five-step method:

Step 1. Choose a Null-Hypothesis-Based Sampling Box Model To model the null hypothesis of suicides being equally likely on each day of the week, use a box containing seven balls, each ball labeled to indicate a different day of the week as in Example 12.1.

Step 2. Define One Simulation A simulation consists of drawing a ball from the box with replacement 3672 times, once for each of the reported suicides in the real data.

Step 3. Define the Event of Interest For each simulated data sample we calculate χ^2. The event of interest is $(\text{simulated } \chi^2 \geq \chi^2_{\text{OBS}}) = (\text{simulated } \chi^2 \geq 13.09)$.

Step 4. Repeat Simulations We perform a total of 5000 simulations.

Step 5. Estimate the Probability of the Event of Interest From our original data we obtained $\chi^2 = 13.09$. The EOI occurred 203 times. Thus

$$\hat{P}(\text{simulated } \chi^2 \geq 13.09 \mid H_0) = 203/5000 = 0.0406.$$

Stage IV. Decide Whether to Reject the Null Hypothesis We found in stages II and III that, under the null-hypothesis model of equal probabilities for the seven days of the week, a simulated χ^2 greater or equal to 13.09 occurred rather infrequently by chance ($0.0406 < 0.05$). We thus have some statistical evidence to reject the null-hypothesis model of equal theoretical probabilities for the seven days of the week in the population of all suicides in North Carolina. Moreover, the one day that is most different from the null-hypothesis model rate of 1/7 is Monday! We conclude that the differences between the different days of the week are statistically significant. Because the number of Monday suicides was greater than expected in the sample, the results lend support to the suggestion that the likelihood of suicides on Monday is greater than on any other day of the week.

However, we should also note that the differences between Monday and other days having higher-than-expected numbers of suicides are not very big. In the real data sample, 15.5% of the suicides occurred on Mondays, 15.1% on Wednesdays, and 15.0% on Sundays. Furthermore, even though we have strong statistical evidence of unequal suicide rates, even the largest rate (15.5% for Mondays) differed by only 1.2% from the 14.3% rate expected under the null hypothesis. Such a small difference or effect size does not seem worthy of further study or to warrant the targeting of mental health interventions toward Mondays. Thus the example illustrates that with enough data we can detect differences too small to be of practical importance.

Section 12.4 Exercises

1. The following table is a frequency table of χ^2 values based on 1000 simulations of a 10-sided die tossed 200 times, under the null hypothesis of a 10-sided fair die. Use this table to estimate the following probabilities.

χ^2 values	Frequency	χ^2 values	Frequency
0–0.99	3	10–10.99	82
1–1.99	6	11–11.99	60
2–2.99	28	12–12.99	46
3–3.99	54	13–13.99	44
4–4.99	82	14–14.99	40
5–5.99	88	15–15.99	23
6–6.99	101	16–16.99	22
7–7.99	104	17–17.99	14
8–8.99	89	18–18.99	10
9–9.99	81	19 or more	23
		Total	1000

a. $P(\chi^2 \geq 6.0 \mid H_0)$
b. $P(\chi^2 \geq 4.8 \mid H_0)$
[*Hint:* The partial interval (4, 4.8) is what proportion of the entire (4, 4.99) interval?]
c. $P(\chi^2 \geq 11.2 \mid H_0)$
d. $P(\chi^2 \geq 14 \mid H_0)$
e. $P(\chi^2 \geq 15.4 \mid H_0)$
f. $P(\chi^2 \geq 16 \mid H_0)$

Use the four-stage hypothesis-testing approach to solve Exercises 2 through 4 using the instructional software.

2. A suspicious-looking octahedral die is tossed 168 times, yielding $\chi^2_{\text{OBSERVED}} = 17.4$. Each of 50 students in a statistics class at Johnson Community College rolled a fair octahedral die 168 times and got the χ^2 statistic results shown in the following table.
a. Use the table to decide whether the suspect die was loaded.

b. Redo the hypothesis test using the five-step simulation instructional software.

Stem	Leaf	f
0		0
1	5,7,8	3
2	7	1
3	0,4,6,8	4
4	1,2,3,3,3,4,6,6,8	9
5	0,0,1,2,3,8	6
6	1,4,5,5,6,6,9	7
7	0,2,6,9	4
8		0
9	0,0,4,6,9,9	6
10	0,9	2
11	2,6	2
12	7	1
13	5,5,7	3
14		0
15		0
16	2	1
17		0
18	0	1
Total		50

Key: "4 2" stands for 4.2.

3. The statistics class at Buffalo Grove College interviewed 100 students at random and asked each of them to give his or her favorite number between 0 and 9. The results are shown in the table that follows. Use the simulations of χ^2 values for a fair 10-sided die from the table and the survey data to decide whether, among the population of all students, certain numbers are preferred over others.

Outcome	f
0	5
1	3
2	11
3	10
4	19
5	9
6	11
7	15
8	13
9	4
Total	100

4. A clothing store stocks a large number of men's ties that are identical except that they are in four different colors. The records of 40 sales are shown in the following table. The 40 sales can be considered a sample of the population

of all such sales. Using these sample data, decide whether, in the population of all sales, some colors are preferred over others.

Color of tie	Number sold
Amber	7
Blue	9
Orange	14
Maroon	10
Total	40

5. The last digit of each of 100 residential telephone numbers was taken (in order) from one page of a telephone book. The frequencies were as follows:

Digit	f
0	3
1	8
2	15
3	14
4	10
5	7
6	8
7	9
8	11
9	15
Total	100

This sample can be considered to be a sample of the population of all the last digits of telephone numbers in the entire residential telephone book. Calculate χ^2 for these sample data. Is the telephone book ruled out as a good source of random data? Why or why not?

6. Using the stem-and-leaf table based on 50 simulations of 168 rolls each of a fair eight-sided die from Exercise 2, find the following experimental probabilities:
   a. $\hat{P}(\text{simulated } \chi^2 \leq 1.7)$
   b. $\hat{P}(\text{simulated } \chi^2 \leq 3.5)$
   c. $\hat{P}(\text{simulated } \chi^2 \leq 13.5)$
   d. $\hat{P}(4.6 \leq \text{ simulated } \chi^2 \leq 12.7)$
   e. $\hat{P}(\text{simulated } \chi^2 \geq 14)$
   f. Repeat parts (a) through (e) using the simulation software and 10,000 trials. Which method is likely more accurate and why?

7. Sometimes dishonest researchers or lazy students fake data. When fair-die data are faked, there is often a tendency to make the O's too close to the E's to be likely for an actual fair die, which is *very* random in its behavior. John is supposed to toss a six-sided fair die 90 times for homework, but decides to fake the data instead. The following are his "observed" data:

Digit	1	2	3	4	5	6
Frequency	14	16	15	13	16	16

a. Calculate a χ^2 statistic for the null hypothesis of a fair die.
b. Denote the answer to part (a) by C. Very small χ^2 values will provide strong evidence of this kind of data faking. However, the event of interest is (simulated $\chi^2 \leq$ observed χ^2). Now we'll look at (simulated $\chi^2 \leq C$) as our P-value and use the usual 0.05 criterion for rejection of the null hypothesis. What do you conclude?

8. Refer to Exercise 7. John thinks his data look a little too good to be true and changes the 14 ones to 12 ones and the 16 twos to 18 twos. Now redo Exercise 7.

9. Explain why χ^2 testing is important in real-world problems of interest with respect to a cynic's claim that all we have learned to test is whether a many-sided die is fair.

12.5 THE CHI-SQUARE CURVE

In this chapter we look at categorical data. We shall assume that our data are either the result of repeated, *independent observations* of a categorical variable (like the score on a die) or a *simple random sample* from a many-times-larger real population whose members can be classified by a categorical variable (such as marital status). We perform goodness-of-fit tests to see how well a null-hypothesis probability model H_0 fits our data. The chi-square goodness-of-fit test statistic was introduced by Karl Pearson around 1900. How did he estimate the P-value

$$P = P(\text{simulated}\, \chi^2 \geq \text{observed}\, \chi^2 \mid H_0),$$

chi-square curve

without today's fast computers and simulation software? Pearson's method involved a new curve, called the **chi-square curve.** Just as we learned in Chapter 8 to superimpose a bell curve on histograms for sample means and sample proportions, Pearson figured out that it was possible to superimpose a different smooth curve on the histogram of Figure 12.2. This curve is the chi-square curve. In fact, there is more than one chi-square cure. There is a different curve of each number of *degrees of freedom*. The curves for 3, 5, and 9 degrees of freedom are shown in Figure 12.3.

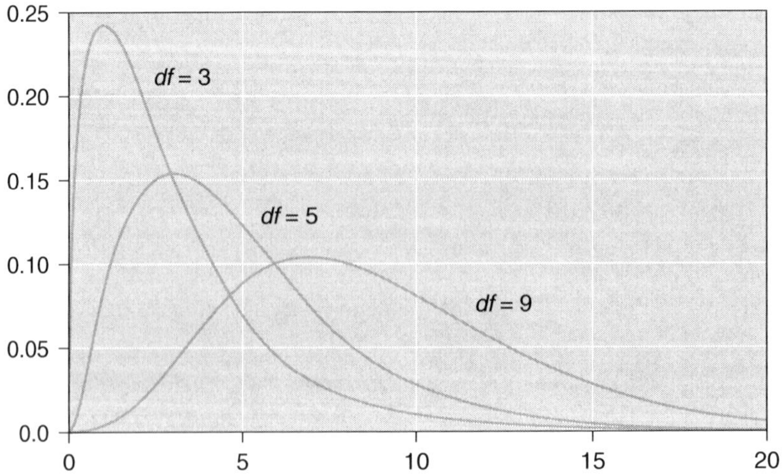

Figure 12.3 Chi-square curves for 3, 5, and 9 degrees of freedom.

Which chi-square curve should we superimpose on the histogram in Figure 12.2? In all the examples we have considered so far in this chapter, the null hypothesis model was *fully specified:* there were no parameters to estimate from the data; the null hypothesis told us all the expected frequencies. When the model is fully specified, we have a simple formula for computing the number of degrees of freedom:

$$df = \text{number of terms in } \chi^2 - 1.$$

For the six-sided die, we get $6 - 1 = 5$ degrees of freedom. Figure 12.4 shows Pearson's approximation. The chi-square curve with 5 degrees of freedom is superimposed on the probability histogram for the χ^2 statistic for 60 tosses of a fair die.

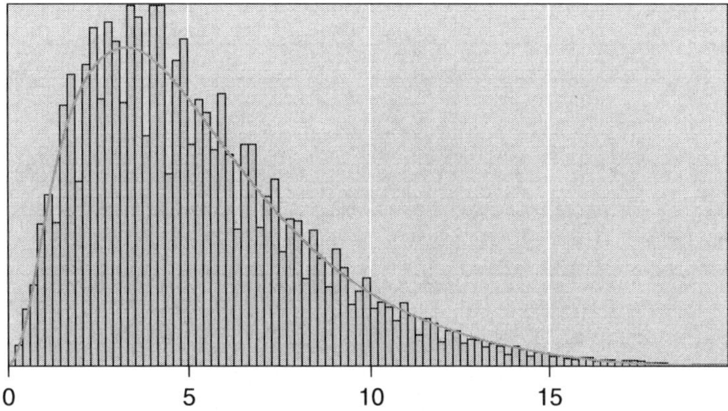

Figure 12.4 Pearson's approximation. The χ^2 curve with 5 degrees of freedom is superimposed on the probability histogram for the discrete χ^2 statistic.

As Figure 12.2 and Figure 12.4 show, the probability histogram for the discrete χ^2 statistic is much more irregular than the smooth superimposed χ^2 curve in Figure 12.4. But the right-tail areas are almost the same. For the chi-square test, the P-value is approximately equal to the area to the right of the observed value, χ^2_{OBS}, under the χ^2 curve with the appropriate number of degrees of freedom, as indicated in Figure 12.5.

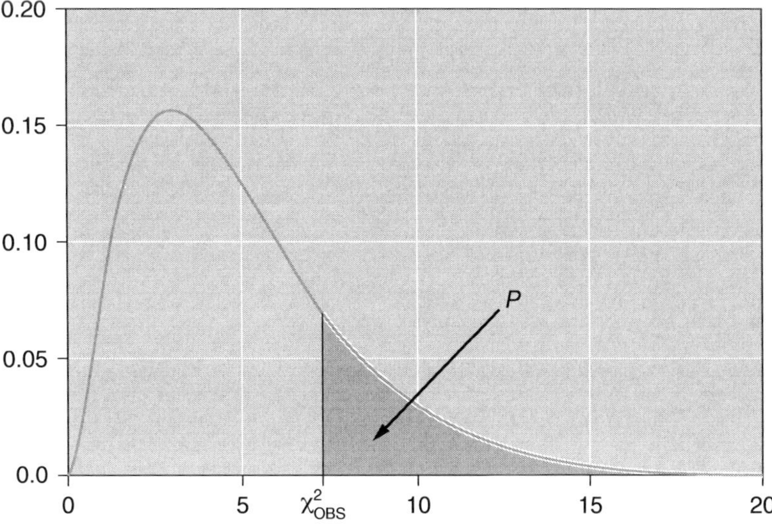

Figure 12.5 P is approximately equal to the shaded area to the right of χ^2_{OBS}, under the χ^2 curve.

This area can be estimated from statistical tables. We can use a statistical calculator to compute the exact value.

As Figure 12.6 illustrates, statistical tables display the right-tail area under the χ^2 curve in the top line and give the corresponding threshold in the table below.

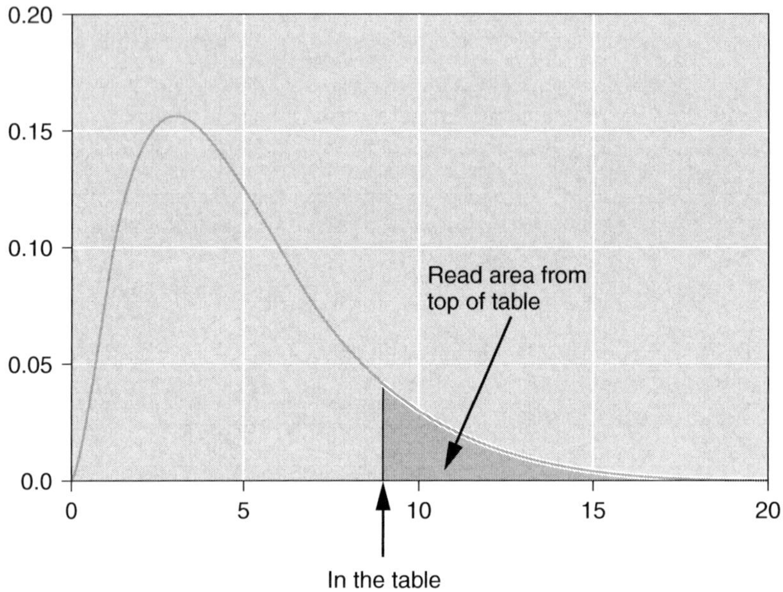

Figure 12.6 Finding areas under the χ^2 curve using statistical tables.

Look at Table 12.9. In the fifth row of the .05 column you find 11.07. This means that the area to the right of 11.07 under the χ^2 curve with 5 degrees of freedom is .05. Similarly, we see that the area to the right of 12.83 under the χ^2 curve with 5 degrees of freedom is .025.

Let us go back to the six-sided die we tested in section 12.3 The observed value of the chi-square statistic was 12.4. We can use Table 12.9 (or Table C in the back of the book) to estimate

$$P = P(\text{simulated } \chi^2 \geq 12.4 \mid H_0).$$

Table 12.9 gives areas (probabilities) under the χ^2 curve in the right-hand tail, which is almost always what we need for chi-square testing. Because different df values produce different distributions of χ^2 values, each row in the table is associated with a χ^2 curve for a particular number of degrees of freedom. For example, row 5 would be our choice for a χ^2 problem involving a fair six-sided die. An expanded version of this table, with many more choices for the number of degrees of freedom, is given by Table C of Appendix C.

Table 12.9 Areas under the χ^2 Curve

	Probability (area) to the right of the interior table entry						
Degrees of freedom (df)	0.25	0.20	0.15	0.10	0.05	0.025	0.01
1	1.32	1.64	2.07	2.71	3.84	5.02	6.63
2	2.77	3.22	3.79	4.61	5.99	7.38	9.21
3	4.11	4.64	5.32	6.25	7.81	9.35	11.34
4	5.39	5.99	6.74	7.78	9.49	11.14	13.28
5	6.63	7.29	8.12	9.24	11.07	12.83	15.09

Look at row 5. Since

$$11.07 < \chi^2_{\text{OBS}} < 12.83,$$

we estimate that

$$0.05 > P > 0.025.$$

An online chi-square calculator gave the area to the right of 12.4 under the χ^2 curve with 5 degrees of freedom as 0.0297, which is very close to our experimental P-value, $\hat{P} = 0.03$.

Sample Size Restriction

Pearson's χ^2 curve approximation is related to the central limit theorem, covered in Chapter 8. The bell curve approximation in the central limit theorem is accurate only if the sample size is large enough. Similarly, Pearson's χ^2 curve is accurate only if the sample size is large enough. Specifically, we shall require that all of the expected frequencies—all of the E's that go into the computation of the χ^2 statistic—be at least 5. If this condition is satisfied, then we have two ways of estimating the P-value,

$$P = P(\text{simulated } \chi^2 \geq \chi^2_{\text{OBS}} \mid H_0).$$

We can either use computer simulation or estimate areas under the χ^2 curve. If the condition is not satisfied, then we can always use computer simulation.

Now let us work through a new example. We use the notation χ^2_r to denote a random variable that is chi-square distributed with r degrees of freedom.

Example 12.4 The Random Number Generator on a Calculator

Many years ago (1988), a colleague of one of the authors purchased a calculator with a random number generator function. To test the accuracy of this random number generator, he generated digits in groups of three, placing a decimal point to the left of each such grouping. For example, 527 generated becomes 0.527. This process is supposed to produce uniformly distributed random numbers between 0.000 and 0.999, each with probability 1/1000 (a thousand-sided die). Based on experience over the years, the owner of the calculator has suspected that this random number generator is a little off: on average, the numbers have seemed slightly smaller than they should be. To test this claim, he generated a sample of 1500 "random" values (like 0.527) and counted how many of them fell into each of 10 evenly spaced intervals: 0.000–0.099, 0.100–0.199, ..., 0.900–0.999. The results are shown in Table 12.10.

Table 12.10 Distribution of 1500 "Random" Numbers from a Calculator

Outcome	Frequency
0.000–0.099	166
0.100–0.199	156
0.200–0.299	178
0.300–0.399	153
0.400–0.499	127
0.500–0.599	127
0.600–0.699	164
0.700–0.799	166
0.800–0.899	132
0.900–0.999	131
Total	1500

The claim can now be more precisely worded: In the infinite population of numbers that could be generated with the calculator, it is believed that not all 10 of the categories created by the 10 intervals will contain one tenth of the numbers generated. Does the sample of 1500 data values provide strong statistical evidence that there is a problem with the calculator's random number generator?

Stage I. Formulate the Null Hypothesis The null hypothesis is that there is nothing wrong with the random number generator—more precisely, that the numbers it produces are uniformly (that is, evenly) distributed among the intervals 0.000–0.099, 0.100–0.199, ..., 0.900–0.999. That is, the random number generator functions like a 10-sided die. Under the null hypothesis, the expected frequency for each of the 10 intervals is equal: $E = 1/10 \times 1500 = 150$ for all intervals.

Stages II and III. Choose the Null-Hypothesis-Based Sampling Box Model and Estimate the *P*-Value Using 150 as the expected frequency E for each interval, $\chi^2 = 21.867$. Is $\chi^2_{\text{OBS}} = 21.867$ an unusually large value for a random number generator that is working properly, the null-hypothesis model? Thus, we seek $P(\chi^2 \geq 21.867 \mid H_0)$. We use a χ^2 density table. Since we have 10 categories, the χ^2 statistic has nine degrees of freedom. Since $E = 150$ is much larger than 5, we can use the χ^2 density in Table C in the back of this book to get $P(\chi^2_9 \geq 21.67) = 0.01$. Thus, the P-value, namely $P(\chi^2_9 \geq 21.867 \mid H_0)$, is less than 0.01. We note that with 50,000 simulations using the null-hypothesis model we got the very accurate estimate of $\hat{P}(\chi^2 \geq \chi^2_{\text{OBS}}) = 0.0097$.

Stage IV. Decide Whether to Reject the Null Hypothesis With $P < 0.01$ we reject H_0. Because the observed value of the χ^2 statistic for the calculator's random number generator is highly unlikely to occur with a properly functioning random number generator, we conclude that the calculator's random number generator is not working the way that it is supposed to.

Section 12.5 Summary

Assume that our data are either the result of repeated, independent observations of a categorical variable (like the outcome of the spin of a roulette wheel) or a simple random sample from a many-times-larger real population whose members can be classified by a categorical variable (such as marital status: married, never married, separated, divorced, widowed). The

χ^2 statistic can be used to test the hypothesis H_0 that our data were generated according to a particular chance model. The χ^2 statistic measures the difference between the observed frequencies and what is expected under H_0.

Assume that all the expected frequencies are at least 5. Then the P-value,

$$P = P(\text{simulated } \chi^2 \geq \chi^2_{\text{OBS}} \mid H_0),$$

is approximately equal to the area to the right of χ^2_{OBS} under the χ^2 curve with the approximate number of degrees of freedom. When the null-hypothesis is fully specified (there are no parameters to be estimated from the data), then

$$\begin{aligned} df &= \text{number of terms in } \chi^2 - 1 \\ &= \text{number of categories} - 1. \end{aligned}$$

Section 12.5 Exercises

For Exercises 1 through 4, estimate for an ordinary fair six-sided die the probability indicated using Table 12.9 (or Table C) and then find the corresponding experimental probability using computer simulation. Note that $df = 5$.

1. $\hat{P}(\text{simulated } \chi^2 \geq 4.4 \mid H_0)$
2. $\hat{P}(\text{simulated } \chi^2 \geq 7.4 \mid H_0)$
3. $\hat{P}(\text{simulated } \chi^2 \geq 9.2 \mid H_0)$
4. $\hat{P}(\text{simulated } \chi^2 \geq 11.2 \mid H_0)$

For Exercises 5 and 6, find for a four-sided fair die the probability indicated using Table 12.9. Then find the corresponding experimental probability using computer simulation. Note that $df = 3$.

5. $\hat{P}(\text{simulated } \chi^2 \geq 4.6 \mid H_0)$
6. $\hat{P}(\text{simulated } \chi^2 \geq 6.5 \mid H_0)$

In the following exercises we use the notation χ^2_r to denote a random variable that is chi-square distributed with r degrees of freedom.

7. Find these probabilities. Use the expanded table of χ^2 probabilities in Appendix C. Note these have nothing to do with χ^2 goodness-of-fit testing.
   a. $P(\chi^2_4 \geq 9.5)$
   b. $P(\chi^2_4 \geq 13.3)$

8. Find these probabilities. Use the expanded table of χ^2 probabilities in Appendix C.
   a. $P(\chi^2_7 \geq 12.0)$
   b. $P(\chi^2_7 \geq 14.1)$

9. Find these probabilities. Use the expanded table of χ^2 probabilities in Appendix C.
   a. $P(\chi^2_{10} \geq 12.0)$
   b. $P(\chi^2_{10} \geq 12.0)$

10. Find these probabilities. Use the expanded table of χ^2 probabilities in Appendix C.
   a. $P(\chi^2_{20} \geq 18.0)$
   b. $P(\chi^2_{20} \geq 40.0)$

For Exercises 11 through 14, use the χ^2 table of probabilities in Appendix C to do the χ^2 test. In each case justify using the χ^2 density according to the $E \geq 5$ rule.

11. The class instructor, with an apparent gleam in her eye, brought a special die into her statistics class at Iowa State University. The class rolled the six-sided die 96 times and obtained the following results:

Outcome	f
1	26
2	7
3	17
4	19
5	14
6	13
Total	96

Calculate the χ^2 statistic for this experiment. How likely are you to get a χ^2 value as large as or larger than the one obtained here, by chance, if the die was fair? Do you believe the die used in this exercise was fair? Explain.

12. A statistics class at the University of Illinois interviewed 380 students (at random, we assume) and asked each of them for his or her favorite number between 0 and 9. Here are the results:

Outcome	f
0	0
1	12
2	27
3	56
4	37
5	40
6	58
7	103
8	31
9	16
Total	380

Calculate χ^2 for these sampled data. Do you think that the population of all UI students prefer some numbers over others? Why?

13. The following table shows the number of eighths in the fractional part on a closing stock price for a sample of 319 "low-priced" stocks from the American Stock Exchange (done before the conversion to decimal reporting). For example, a stock priced at 9 3/8 yield 3/8. Assess whether these sampled data indicate that some fractions occur more commonly than others in the population of all stock prices.

Fraction	Frequency
0	60
1/8	30
2/8(1/4)	29
3/8	27
4/8(1/2)	47
5/8	49
6/8(3/4)	39
7/8	38
	319

14. In a study of the pricing of merchandise a random sample of 174 low-priced items were examined, and for each item the last digit of the price was recorded (for example, $2.79 produces a 9). Here are the results. Would you conclude that, in the infinite population of all such low-priced articles, some last digits are more likely than others?

Last digit	Frequency
0	11
1	10
2	10
3	34
4	7
5	25
6	4
7	18
8	6
9	49
	174

12.6 UNEQUAL EXPECTED FREQUENCIES

The χ^2 statistic can be used in a remarkable variety of ways to solve vital problems for science, medicine, and industry. Indeed, it played a key role in helping confirm Mendel's theory of genetics, a triumph of modern science. (See Exercise 2 of this section.)

Recall from Section 12.3 that a basic pattern for a χ^2 table is the following:

Outcome	**Expected frequency (E)**	**Obtained frequency (O)**	$(O - E)^2/E$

So far, all of the chance models we have tested have had equally likely outcomes and all the expected frequencies have been equal. We shall now consider examples where the outcomes are not equally likely.

Example 12.5 Effectiveness of Highway Warning Signs

(Adapted from Insurance Institute for Highway Safety.) To reduce highway deaths due to automobiles hitting stationary objects on the sides of highways, highway warning signs were extensively installed to warn drivers of the presence of such objects. After the signs were installed, the question was raised as to whether the signs were more effective for some types of roadside objects than for others. Using data gathered over several years before the signs were installed, eight different types of roadside objects were identified as the ones most frequently hit in fatal accidents, and the proportion of fatal accidents involving each type of object was computed. The proportions for the nine categories (the eight types of objects plus the "other" category) are given in Table 12.11.

The relative frequencies in Table 12.11 are based on so many observations that we can consider Table 12.11 the "true" distribution. Next, after the highway warning sign campaign had been in effect for a while, a sample of 691 fatal accidents was gathered, and the observed frequencies of accidents involving each of the types of objects hit were recorded. Those data are presented in Table 12.12.

Table 12.11 Distribution of Struck Objects in Fatal Collisions (Prior to Sign Installation)

Type of object struck	*P*(object struck)
Tree	0.28
Embankment	0.10
Utility pole	0.10
Sign/post/fence	0.10
Guardrail	0.09
Ditch	0.07
Curb	0.06
Culvert	0.05
Other	0.15

Table 12.12 Distribution of Struck Objects in Sample of Fatal Collisions (After Sign Installation)

Type of object struck	Frequency of fatal accidents (O)
Tree	179
Embankment	100
Utility pole	107
Sign/post/fence	68
Guardrail	57
Ditch	36
Curb	43
Culvert	28
Other	73
Total	691

The statistical analysis uses the sample data to decide whether the installation of the signs has caused the proportions of fatal accidents per type of object hit to be different from the proportions before the signs were installed. The suspicion is that the signs were more effective in some locations than in other locations.

Solution:

Stage I. Formulate the Null Hypothesis The null hypothesis is that the proportions for the different types of objects hit in fatal collisions after the signs were installed are the same as the proportions of fatal collisions before the signs were installed. That is, H_0 asserts that the sample is drawn from the multinomial distribution (or unfair-die distribution) with $(p(1), p(2), \ldots, p(9))$ given by $(0.28, 0.1, 0.1, 0.1, 0.09, 0.07, 0.06, 0.05, 0.15)$.

Stages II and III. Choose the Null-Hypothesis-Based Sampling Box Model and Estimate the *P*-Value Given the null hypothesis, the expected values for the frequencies in the sample of 691 fatal accidents are $0.28 \times 691 = 193.5$ for trees, $0.10 \times 691 = 69.1$ for embankments, and so forth. Using this null hypothesis model, we then set up the χ^2 frequency table of outcomes in the usual way (Table 12.13). We compute χ^2 using the now familiar formula:

$$\chi^2_{\text{OBS}} = \sum \frac{(O-E)^2}{E} = 49.723$$

Now, we want to compute the likelihood of observing a χ^2 value ≥ 49.723 (the event of interest) when the null hypothesis is true. So the probability we want to calculate is $P(\chi^2 \geq 49.723 \mid H_0)$. We will do this by using the χ^2 density Table C. We need to know the number of degrees of freedom so that we know which row to use in the table. Since we have nine categories, the χ^2 statistic has eight degrees of freedom. Thus, the probability we want to calculate, $P(\chi^2 \geq 49.723 \mid H_0)$, can be rewritten as $P(\chi^2_8 \geq 49.723)$. By using Appendix C, we see that $P(\chi^2_8 \geq 27.87) = 0.0005$. Thus, $P(\chi^2_8 \geq 49.723) < 0.0005$.

Stage IV. Decide Whether to Reject the Null Hypothesis So the difference between the expected number of accidents per type of object hit (based on the null hypothesis) and the observed number of accidents per type of object hit, from the real data sample, provides very strong evidence against H_0.

Thus, we conclude that the proportions of objects hit in fatal collisions have changed since the warning signs were installed. It appears that the signs were more effective for certain roadside objects than for others. Further investigation could lead to improved highway safety methods that could be used in addition to signs to reduce the proportion of accidents involving, for example, embankments and utility poles.

Table 12.13 χ^2 Frequency Table for Roadside Collision Data

Outcome (type of object hit)	Expected number of fatal accidents (E)	Observed number of fatal accidents (O)	$O - E$
Tree	193.5	179	−14.5
Embankment	69.1	100	30.9
Utility pole	69.1	107	37.9
Sign/post/fence	69.1	68	−1.1
Guardrail	62.2	57	−5.2
Ditch	48.4	36	−12.4
Curb	41.5	43	1.5
Culvert	34.6	28	−6.6
Other	103.7	73	−30.7
Total	691.2	691	

Section 12.6 Exercises

Note: One can always use the five-step χ^2 approach. But when all $E \geq 5$ we can also use the χ^2 table lookup distribution approach.

1. Is it true, as is often assumed, that people prefer the flakiest (lightest) pie crust? An experiment was done at Cornell University to test this assumption. Thirty people were involved. Each person was blindfolded and asked to state the order of preference from most to least favorite for three pieces of pastry—one light (L), one medium (M), and one heavy (H)—which were presented in random order. There are six different ways in which the order of preference could be given, as shown in the following table. Also given is the number of persons choosing each of the six.

Preference ordering	Number of persons choosing (obtained frequencies)
LMH	10
LHM	3
MLH	3
MHL	8
HLM	1
HML	5
	30

   a. Under the assumption that each of the six orderings is preferred equally (that is, the crust type does not matter), what is the expected number of persons choosing each?
   b. Test the assumption of equal preference using χ^2.
   c. A model was developed under the assumption of preference for lighter pastry. According to this model, the probability of each of the six outcomes being chosen is as follows:

Preference ordering	Probability of outcome under assumption of preference for light pastry
LMH	0.23
LHM	0.18
MLH	0.18
MHL	0.15
HLM	0.15
HML	0.11

   d. Now what is the expected number for each of the outcomes (i.e., expected number of persons preferring each ordering of pie crusts)?
   e. Test the assumption of preference for light pastry, using the model given in part (c).

2. In the study of genetics, scientists are interested in determining how characteristics of living things are passed down from one generation to the next. The fundamental principles of heredity were discovered by Gregor Johann Mendel (1822–1884) when he proved by crossbreeding garden peas that there are definite patterns in the way characteristics

such as seed size, shape, and surface texture are passed on from generation to generation. One of Mendel's classic experiments involved crossbreeding two kinds of pea plants, one pure strain with round, yellow seeds and the other with wrinkled, green seeds. All of the seeds resulting from this crossbreeding were round, yellow seeds. But when he produced a second offspring generation by crossing plants from the first offspring generation, he got the results shown in the table below. The theoretical probabilities that are predicted according to Mendel's law of heredity are given, and the observed results obtained from 556 trials (556 next-generation pea plants) are also given. *Note:* This is one of science's truly famous data sets!

a. Find the expected frequencies.
b. Did the seed-growing experiment support or cast doubt on his genetic theory?

Type of seed	Theory-predicted probability	Expected frequency (E)	Observed number (O)
Round and yellow	9/16		315
Wrinkled and yellow	3/16		101
Round and green	3/16		108
Wrinkled and green	1/16		32
			556

3. A biological experiment studying two characteristics of flowers yielded the following results. The expected values are based on Mendel's genetic theory of inheritance. Do the results support the theory?

Characteristic of flower	Expected (E)	Observed (O)
AB	180	164
Ab	60	78
aB	60	65
ab	20	13

4. The results of the election were that 60% of the voters favored Mr. Alpha, 30% favored Mr. Beta, and 10% favored Ms. Gamma. A preelection poll of 653 likely voters produced these "votes":

Candidate	Number of votes
Alpha	503
Beta	115
Gamma	35

Could a simple random sample of 653 voters have produced these numbers?

5. A census in a certain city shows the racial composition to be 30% white, 55% black, and 15% other. The city council has the following racial composition (number of persons of each race):

White	5
Black	4
Other	1

Could a simple random sample of 10 residents have produced these numbers?

6. A typist prepared a manuscript of 167 pages, of which 103 contained no errors. Forty-five pages had one typing error each. Sixteen pages had two errors each. Three pages had three errors each. None had more than three errors. The table summarizes this information.

Number of typing errors	Number of such pages (observed outcomes)
0	103
1	45
2	16
3 or more	3
Total	167

The probabilistic model called the *Poisson model* (see Chapter 7) has sometimes been used to predict the number of typing errors per page. Using this model, the following probabilities have been calculated:

Number of typing errors	Probability
0	0.59
1	0.31
2	0.08
3 or more	0.02

a. Calculate the expected number of pages having errors totaling 0, 1, 2, and so on, according to this model.
b. Use χ^2 to test whether there is evidence that this model is an inappropriate one.

7. The distribution of ABO blood types among African Americans is reported to be as follows: type A, 27%; type B, 20%; type AB, 4%; and type O, 49%. Suppose that a simple random sample of 200 African Americans yielded 64 of type A, 34 of type B, 11 of type AB, and 91 of type O.
   a. Assuming that the reported ABO blood type distribution is correct, what are the expected frequencies of the four blood types for a random sample of 200 African Americans?
   b. Compute the observed value of the χ^2 statistic for testing the reported blood type distribution.
   c. What is the number of degrees of freedom?
   d. Estimate the P-value.
   e. Are the observed frequencies consistent with the reported ABO blood type distribution?

8. Four-o'clocks are ornamental plants. According to genetic theory, the flower color is determined by a single gene pair. If we cross two pink-flowered plants, the chance that an offspring will have red flowers is 1/4, pink flowers 1/2, and white flowers 1/4. In an experiment, 408 progeny were produced by crossing pink-flowered plants; of these, 111 had red flowers, 188 had pink flowers, and 109 had white flowers.
   a. If we assume the genetic model is correct, what are the expected frequencies of the three colors?
   b. Compute the observed value of the χ^2 statistic for testing the genetic model.

c. What is the number of degrees of freedom?
d. Estimate the P-value.
e. Are the observed frequencies consistent with the reported genetic model?

12.7 TWO-WAY CONTINGENCY TABLE ANALYSIS

The statistician in Example 12.4 compared the random number generators on two statistical calculators: the one purchased in 1988 and the latest version of that calculator, purchased in 2010. His suspicion was that the 2010 calculator used the same flawed random number generator as the 1988 calculator. He generated 13,500 "random" digits with the 1988 calculator and 1,500 "random" digits with the 2010 calculator. Table 12.14 shows the observed frequencies of the digits 0, 1, 2, . . . , 9 for each of the calculators. We have also computed the column totals and the row totals. This kind of table is usually referred to as a **contingency table** or a **two-way table.** The observed frequencies in Table 12.14 are arranged in 10 rows and 2 columns. (We do not count the column of totals or the row of totals.) This is a "10 × 2 table."

contingency table
two-way table

Table 12.14 Distribution of "Random" Digits Generated by Two Calculators

Digit	1988 frequency	2010 frequency	Total
0	1,451	164	1,615
1	1,500	178	1,678
2	1,540	165	1,705
3	1,245	132	1,377
4	1,188	142	1,330
5	1,361	167	1,528
6	1,392	129	1,521
7	1,507	164	1,671
8	1,127	122	1,249
9	1,189	137	1,326
Total	13,500	1,500	15,000

The null hypothesis is that the distribution of "random" digits produced by the 2010 calculator is the same as the distribution of "random" digits produced by the 1988 calculator. We shall perform a χ^2 test to see whether the observed frequencies in Table 12.14 support or contradict the null hypothesis.

To compute the χ^2 statistic we shall need the expected frequencies. These expected frequencies are computed under H_0, the assumption that the two calculators use the same random number generator. Notice that the 1988 calculator generated 90% of all the 15,000 random digits and the 2010 calculator generated 10% of the random digits. So if the two calculators use the same random number generator, then we would expect the 1988 calculator to have generated 90% of all the 0s, 90% of all the 1s, 90% of all the 2s, and so forth. And we would expect the 2010 calculator to have generated 10% of all the 0s, 10% of all the 1s, 10% of all the 2s, and so forth. Given that the two calculators combined generated 1615 zeros, we would expect the 1988 calculator to have generated

$$90\% \text{ of } 1615 \text{ zeros} = 1453.5 \text{ zeros},$$

and we would expect the 2010 calculator to have generated

$$10\% \text{ of } 1615 \text{ zeros} = 161.5 \text{ zeros}.$$

Table 12.15 displays the expected frequencies under H_0.

Table 12.15 Expected Frequencies for the Two Calculators

Digit	1988 frequency	2010 frequency	Total
0	1,453.5	161.5	1,615
1	1,510.2	167.8	1,678
2	1,534.5	170.5	1,705
3	1,239.3	137.7	1,377
4	1,197.0	133.0	1,330
5	1,375.2	152.8	1,528
6	1,368.9	152.1	1,521
7	1,503.9	167.1	1,671
8	1,124.1	124.9	1,249
9	1,193.4	132.6	1,326
Total	13,500.0	1,500.0	15,000

We can now compute

$$\chi^2_{\text{OBS}} = \sum \frac{(O-E)^2}{E} = \frac{(1451-1453.5)^2}{1453.5} + \cdots + \frac{(137-132.6)^2}{132.6} = 7.533.$$

How many degrees of freedom are there? In testing homogeneity or independence in an $R \times C$ table (with R rows and C columns) there are $(R-1) \times (C-1)$ degrees of freedom. Here we are testing homogeneity in a 10×2 table: we are testing whether the distribution of "random" digits is independent of which calculator we use. So there are

$$(10-1) \times (2-1) = 9 \text{ degrees of freedom.}$$

Since all of the expected frequencies are greater than 5, we can use the χ^2 curve with 9 degrees of freedom to estimate the P-value,

$$P = P(\text{simulated } \chi^2 \geq 7.533 \mid H_0).$$

According to Table C,

$$P(\text{simulated } \chi^2 \geq 11.39 \mid H_0) \approx 0.25).$$

It follows that P is greater than 0.25. An online chi-square calculator gave the area to the right of 7.533 under the χ^2 curve with 9 degrees of freedom to be 0.58. With such a large P-value we do not reject H_0. We have no reason to doubt that the two calculators use the same random number generator.

chi-square test of homogeneity

The test we just performed was a so-called **chi-square test of homogeneity.** The test is used to determine whether a categorical variable of interest is identically distributed across different populations, that is, whether the distribution of the categorical variable is independent of which population we are looking at. In our example, the two populations were the two calculators and the categorical variable of interest was a "random" number generated by one of the calculators.

chi-square test for independence

Very closely related to the chi-square test of homogeneity is the **chi-square test for independence.** The test is used to determine whether two (or more) categorical variables from a single population are independent.

Both tests assume that we either have repeated, independent categorical observations or (independent) simple random sample(s).

Let us take another look at how we computed the expected frequencies in Table 12.15. Let N_r denote the rth row total, N_c the cth column total, and N the overall total. Then the expected frequency for row r in column c is

$$E_{rc} = N_r \times N_c / N,$$

under the null hypothesis of either homogeneity or independence. For example, in Table 12.15 the expected number of zeros generated by the 1988 calculator is

$$E_{11} = 1615 \times 13,500 / 15,000 = 1453.5,$$

under the null hypothesis that the two calculators used the same random number generator.

In example 12.6 we test whether young people who are engaged in religious activities are happier than young people who are not. The null hypothesis is that personal happiness is independent of religious involvement.

Example 12.6

In a random sample of high school and college students, the following contingency table (Table 12.16) resulted for participation in religious activities, regarded as an explanatory variable, and level of happiness, viewed as a response variable, as assessed by student responses to "How happy has your home life been?"

Table 12.16 Student Happiness Responses

Religious activities participation	Very happy	Fairly happy	Unhappy	Marginal row totals
None	105	78	25	208
Very little	257	149	30	436
Somewhat	368	153	24	545
Very much	151	52	13	216
Marginal column totals	881	432	92	1405

Using our rule of

$$E_{rc} = \frac{N_r \times N_c}{N},$$

we obtain, in $O(E)$ tabular format, Table 12.17. Thus,

$$\chi^2_{\text{OBS}} = \frac{(105 - 130.43)^2}{1130.43} + \frac{(273.39 - 257)^2}{273.39} + \cdots + \frac{(14.14 - 13)^2}{14.14} = 32.537.$$

Table 12.17 Student Happiness Responses and Expected Frequencies

Religious activities participation	Very happy	Fairly happy	Unhappy	Marginal row totals
None	105 (130.43)	78 (23.95)	25 (13.62)	208
Very little	257 (273.39)	149 (134.06)	30 (28.55)	436
Somewhat	368 (342.37)	153 (167.88)	24 (35.69)	545
Very much	151 (134.81)	52 (66.41)	13 (14.14)	216
Marginal column totals	**881**	**432**	**92**	**1405**

The number of degrees of freedom satisfies

$$df = (R - 1) \times (C - 1) = 3 \times 2 = 6.$$

Hence the P-value is approximated by

$$P = \hat{P}(\chi_6^2 \geq 32.537) < 0.0005$$

because for Table C,

$$P(\chi_6^2 \geq 24.10) = 0.0005.$$

Hence the evidence is extremely strong that happiness (as assessed in the study) and religious activity participation (as assessed in the study) are associated rather than independent. But, one must be careful *not* to conclude that relgious participation causes more happiness. For example, there might be a lurking "hidden variable" that facilitates both more participation and more happiness.

Example 12.7

Recently the "Mediterranean diet" has gotten much attention. For some time the American Heart Association (AHA) has recommended a low-fat diet, which is certainly lower in fat than the Mediterranean diet. In an important study, 605 heart attack survivors were randomly assigned to "prudent diet 1" (low-fat diet of the AHA) or a slightly modified Mediterranean diet over four years. The tracked medical results, below, were reported in 1998.

Diet	Cancer	Fatal hart disease	Non-fatal heart disease	Healthy	Total
AHA	15	24	25	239	303
Mediterranean	7	14	8	273	302
Total	22	38	33	512	605

Clearly the data suggest that the Mediterranean diet is superior with respect to cancer and heart disease among heart attack survivors. But is there statistical evidence of this? First we note that this is clearly a study of stratified random sampling with two strata based on diet. So we, technically, will do a test for homogeneity of proportions:

$$H_0 : \text{four-sided die No. 1 (AHA diet) is same as}$$
$$\text{four-sided die No. 2 (Mediterranean diet).}$$

The observed and expected frequencies are shown in Table 12.18, with the expected frequencies in parentheses $[11.02 = (22 \times 303)/605]$.

Table 12.18 Observed and Expected Frequencies for Diet and Health Study

Diet	Outcome				
	Cancer	Fatal hart disease	Non-fatal heart disease	Healthy	Total
AHA	15	24	25	239	303
	(11.02)	(19.03)	(16.53)	(256.42)	
Mediterranean	7	14	8	273	302
	(10.98)	(18.97)	(16.47)	(255.58)	
Total	22	38	33	512	605

$$df = (2-1)(4-1) = 3$$

$$\chi^2 = \frac{\Sigma(O-E)^2}{E} = 16.55$$

All $E \geq 5$, so we may use the chi-square distribution for computing probabilities for the χ^2 statistic. Via Table C, $P(\chi_3^2 \geq 16.27) = 0.001$, and and hence $P(\chi_3^2 \geq 16.55) < 0.001$. Hence the two diets differ with very high significance. And, clearly, the difference is the supiority of the Mediterranean diet with respect to the occurrence of cancer and of fatal and non-fatal heart disease. Note that the five-step approach could have been used.

It should be noted that this was a randomized comparitive experiment, as covered in Section 9.3. The only difference is that there are four outcome categories, whereas in Section 9.3 there were only two outcome categories, such as heart attack versus no heart attack. This difference forced us to use a different approach than in Section 9.3.

Use of the Chi-Square Statistic to Test for Independence of Two Characteristics in a Two-Way Table

The null and alternative hypotheses are

H_0: The two variables under consideration are not associated.

H_A: The two variables are associated.

Let

$$N_r = r\text{th row total}$$
$$N_c = c\text{th column total}$$
$$N = \text{overall total}$$

Under H_0, the expected frequency for row r in column c is

$$E_{rc} = \frac{N_r \times N_c}{N}$$

for all row and column (r, c) combinations,

$$df = (R - 1) \times (C - 1)$$

where R = number of rows and C = number of columns,

$$\chi^2_{\text{OBS}} = \sum \frac{(O_{rc} - E_{rc})^2}{E_{rc}}$$

where O_{rc} denotes the observed frequency in row r and column c, and the sum is over all (r, c) combinations.

To find the P-value, either use the five-step simulation or if all $E_{rc} \geq 5$ use the area under the chi-square curve with $df = (R - 1) \times (C - 1)$ degrees of freedom:

$$P = P(\chi^2_{df} \geq \chi^2_{\text{OBS}}).$$

Section 12.7 Exercises

1. Tell how many degrees of freedom there would be in a χ^2 statistic based on a table having
   a. Four rows and three columns
   b. Three rows and three columns
   c. Five rows and two columns

2. A survey asked a simple random sample of 320 high school boys whether they were right- or left-handed and whether they participated in sports. The results were as follows:

	Participate in sports?		
	Yes	**No**	**Total**
Right-handed	86	194	280
Left-handed	22	18	40
Total	108	212	320

   a. Using the row and column totals given in the table, what are the values for
      i. $\hat{P}$(student is left-handed)
      ii. $\hat{P}$(student is right-handed)
      iii. $\hat{P}$(student participates in sports)
      iv. $\hat{P}$(student does not participate in sports)
      v. $\hat{P}$(left-handed | participates in sports)
      vi. $\hat{P}$(left-handed | does not participate in sports)
      vii. Discuss the results of parts (i), (v), and (vi).
   b. To determine the expected frequencies in each of the four positions in the table of Exercise 2, begin by calculating the following probabilities, assuming independence between the two factors (use the law of theoretical independence):
      i. $\hat{P}$(student is right-handed and participates in sports)
      ii. $\hat{P}$(student is left-handed and participates in sports)
      iii. $\hat{P}$(student is right-handed and does not participate in sports)
      iv. $\hat{P}$(student is left-handed and does not participate in sports)
      v. Discuss the result of part (i).

c. Calculate the expected frequencies for the four cells in the table, using the experimental probabilities you found in part (b) and multiply by the sample size $N = 320$.
d. Repeat part (c) using the standard rule for finding expected frequencies in a contingency table [using $N_r \times N_c) / N$]. Verify that the two are the same.
e. Calculate the χ^2 statistic and test the claim of independence.

3. On March 25, 1997, the weather forecast for 98 North American cities was compiled and two factors were looked at: the prediction for high temperature and the prediction for the sky conditions (cloudy, clear, or precipitation). The results are summarized in the following table.
   a. Write out the table of the expected values, assuming that temperature is independent of sky conditions.
   b. Perform the χ^2 test of independence. How many degrees of freedom are there in this case? What is your conclusion?

	Clear	**Cloudy**	**Precipitation**	**Total**
Over 60 degrees	24	28	11	63
50–59 degrees	4	7	7	18
Less than 50 degrees	2	12	3	17
Total	30	47	21	98

4. In simple random sample of 66 former students, a survey was conducted to determine students' feelings about the professor. The table that follows gives a summary of the results of that survey and the numbers of students passing or failing the course.
   a. Write out the table of expected values, assuming that whether a student likes the professor is independent of whether the student passes.
   b. Perform the χ^2 test for independence. What is your conclusion?

	Passed course	**Failed course**	**Total**
Liked professor	38	3	41
Disliked professor	4	7	11
No opinion	2	12	14
Total	44	22	66

5. Suppose a simple random sample of 500 college students collected the data on field of study and political party preference found in the following table.

	Democratic	Republican	Other	Total
Engineering	72	31	5	108
Science	90	22	2	114
Business	62	61	5	128
Fine Arts	30	3	1	34
Liberal Arts	64	48	4	116
Total	318	165	17	500

a. Write out the table of expected values, assuming that the two factors are independent.
b. What is the value of the χ^2 statistic? What is your conclusion?
c. Suppose the table consisted of 250 Democrats and 250 Republicans. Is this a test of independence? Discuss.

6. Suppose a car dealership interested in whether men or women prefer different colors obtained the following data from a simple random sample of 400 single men and women who purchased a car:

	Gender		
Color	Men	Women	Total
Black	56	29	85
White	29	27	56
Red	72	32	104
Blue	52	40	92
All others	44	19	63
Total	251	149	400

a. Write out the table of expected values, assuming the two factors are independent.
b. What is the value of the χ^2 statistic? What is your conclusion?

7. a. Find the conditional relative frequency table for Example 12.6.
b. Does it appear that student happiness and religious activity participation are independent, based on the table in part (a)?

8. a. An intelligence test for dogs, pigs, and chimpanzees is carried out where intelligence is categorized as low, medium, or high (carefully defined in the context of the experiment). A total of 10 dogs, 10 pigs, and 10 chimpanzees were each selected for the study. Is this study of association between species and intelligence a test of independence or of homogeneity?
b. Will the χ^2 analysis be different depending on whether it is viewed as a test of independence versus homogeneity?

9. In the data set "Popular Kids," students in grades 4–6 were asked whether good grades, athletic ability, or popularity was most important to them. A two-way table separating the students by grade and by choice of most important factor is shown below:

	Grade			
Goals	**4**	**5**	**6**	**Total**
Grades	49	50	69	168
Popular	24	36	38	98
Sports	19	22	28	69
Total	92	108	135	335

a. Construct the conditional relative frequency table for goals | grade level. Does there appear to be evidence of an association? How would you characterize it if you found one?
b. Is there statistical evidence of an association between goals and grade level?

10. The "Popular Kids" data set also divided the students' responses into "Urban," "Suburban," and "Rural" school areas. Is there an association between the type of school area and the students' choice of good grades, athletic ability, or popularity as most important?
a. Construct the conditional relative frequency table for grades | school area. Does there appear to be evidence of an association? How would you characterize it if you found evidence?
b. Is there statistical evidence of an association?

CHAPTER 12 SUMMARY

In this chapter we learned how to test goodness-of-fit hypotheses about whether a particular loaded-die model fits a data set. Such data are sometimes called a **one-way table.**

We assume that our data are either the result of repeated independent observations of a categorical variable or a simple random sample from a many-times-larger real population whose members can be classified by a categorical variable.

The goodness-of-fit hypothesis testing followed the **four-stage hypothesis-testing approach** that was introduced in Chapter 9. One goodness-of-fit test statistic used is the ***D* statistic.** The preferred test statistic is the **chi-square statistic,** which is written as χ^2. In doing five-step simulation, it is just as easy to use D as it is to use χ^2.

$$\chi^2 = \sum \frac{(O - E)^2}{E},$$

where we sum over all categories, and O and E denote the observed and expected frequencies of a given category.

Hypothesis testing of a many-sided, loaded die (or some real-life analog) with the χ^2 statistic is referred to as **χ^2 goodness-of-fit testing** because we are testing how well a hypothesized set of category proportions fits the observed data. The estimation of the P-value is accomplished in two different ways. The first way is to again use the box-model simulation approach for either a D or χ^2 test statistic, which works for any number of trials (die throws). The second way is to use a **χ^2 table** (Appendix C), which gives probability information for the χ^2 distribution. The χ^2 density approximation for evaluating the probabilities of events involving a χ^2 statistic is effective *only* when all the expected frequencies satisfy $E \geq 5$. The simulation approach is always effective!

The use of the χ^2 table requires knowing the **number of degrees of freedom (df)** for the χ^2 statistic for each specific situation. For the goodness-of-fit of loaded-die problems in this chapter, the number of degrees of freedom always equals **(number of categories) – 1.**

A **test of independence** for a two-way contingency table with R rows and C columns uses $\boldsymbol{df = (R-1) \times (C-1), E_{rc} = (N_r \times N_c)/N}$, and $\boldsymbol{\chi^2 = \Sigma(O_{rc} - E_{rc})^2/E_{rc}}$.

A **test of homogeneity of population proportions** occurs when the random sampling producing a contingency table is stratified using the levels of one of the two variables. The analysis is carried out in a manner identically to that used for a test of independence.

CHAPTER REVIEW EXERCISES

Note: If there are multiple ways of performing a goodness-of-fit test (simulating the D statistic, simulating the χ^2 statistic, finding areas under the χ^2 curve) it is useful to do the same problem in more than one way.

1. At a restaurant there are six choices on the menu. On a certain day, the items are ordered the following numbers of times:

Grilled chicken sandwich	12	Garden salad	5
Patty melt	14	Fish sandwich	11
Steakburger	6	Grilled steak	12

Is there statistical evidence that some items are preferred over others? Use a χ^2 test.

2. Calculate the following probabilities from Table C:
   a. $P(\chi_5^2 \geq 14.3)$
   b. $P(\chi_5^2 \geq 20.1)$
   c. $P(\chi_5^2 \geq 4.3)$
   d. $P(\chi_5^2 \geq 22.9)$
3. You are told that $P(\chi^2 \geq 9.24) = 0.1$ for a fair die tossed 50 times. Without looking at a χ^2 table, what will this probability be (at least approximately) if the die is tossed 200 times, 1000 times, or 10,000 times? Defend your answers.
4. Explain intuitively why the number of degrees of freedom is 1 less than the number of possible outcomes.
5. A weather model predicts the following for a certain month in your area:
   Fifteen days will have no rain.
   Ten days will have rain, but less than an inch.
   Six days will have more than an inch of rain.
   The rainfall during the month is looked back on after the month is over, and the actual results were as follows:
   Thirteen days had no rain.
   Eleven days had rain, but less than an inch.
   Seven days had more than an inch of rain.
   a. How many degrees of freedom are there in this case?
   b. What is the χ^2 statistic?
   c. Did the model predict well what the actual rainfall would be?

6. Find the following probabilities using an online statistical calculator.
   a. $P(\chi_3^2 \geq 4.2)$
   b. $P(\chi_5^2 \geq 5.3)$
   c. $P(\chi_9^2 \geq 15.2)$

7. Suppose we classify one-half of men and one-half of women as extroverts according to some psychological scale. Assume the rest are classified as introverts. We ask whether being extroverted affects one's choice of marriage partner, and similarly for being introverted. Suppose this personality trait has no influence on the choice of mate. There are four (male, female) combinations: (E, E), (E, I), (I, E), (I, I).
   a. If there is no influence, what is the probability of each combination?
   b. Suppose that the following data have been found for 100 marriages:

Classification	Frequency
(E,E)	35
(E,I)	15
(I,E)	18
(I,I)	32

   c. Test whether one's trait is associated with the trait of one's spouse.

8. In 1000 poker hands, you would expect (based on theoretical probabilities) to see the following hands these numbers of times:

Poker hand	Expected number in 1000 hands
Nothing	501.6
One pair	422.6
Two pairs	47.5
Three of a kind	21.1
Other (the better hands)	7.2

You and your friends play games of poker over an extended period of time, and in 1000 hands you see the following distribution:

Poker hand	Observed number in 1000 hands (O)
Nothing	488
One pair	438
Two pairs	51
Three of a kind	16
Other (the better hands)	7

   a. How many degrees of freedom are there in this case?
   b. Did your results follow the expected distribution? Use the χ^2 test.

9. Explain why we would be more likely to conclude that a die is loaded the larger the χ^2 statistic is, under the fair-die null hypothesis.

10. In 1987 the racial makeup of college campuses was as follows:

White	79%
Hispanic	5%
Native American / Native Alaskan	1%
African American	9%
Asian Pacific Islander	4%
Nonresident	2%

The total number of bachelor's degrees awarded, broken down by racial categories, was as follows:

Group	Thousands of bachelor's degrees
White	841
Hispanic	27
Native American / Native Alaskan	4
African American	57
Asian Pacific Islander	33
Nonresident	29

Source: U.S. Department of Education. Cited in *QEM, Education that Works: An Action Plan for the Education of Minorities*, MIT, Cambridge, Mass., 1990.

a. Out of the total of 991,000 degrees awarded, how many could be expected in each of the racial groups, if you assume they follow the same distribution as the student population?
b. Do a χ^2 test to determine whether the actual number of bachelor's degrees awarded has the same distribution as the student population.

11. You quickly shuffle a deck of cards 40 times and draw one card.
a. How many times do you expect each of the four suits (spades, clubs, diamonds, and hearts) to show up?
b. This test was performed, and the actual results were as follows:

Hearts	12 times
Diamonds	7 times
Clubs	11 times
Spades	10 times

Do a χ^2 test to determine whether the cards were being drawn randomly from a well-shuffled deck so that each suit is equally likely.
c. Suppose somebody draws 10 cards and gets 5 hearts, 1 diamond, no clubs, and 4 spades. You suspect that the (1/4, 1/4, 1/4, 1/4) model is wrong but realize that you cannot use the χ^2 density (Table C) because $10 \times 1/4 = 2.5 < 5$. There is a way to test this. Do so.

12. A car dealer believes that each of the five available colors of a certain model of car is equally likely to be chosen by the customer. At the end of the month, these are the results the dealer sees:

Color	Frequency
Red	21
Blue	10
White	15
Black	17
Light taupe	7
Total	70

Use a χ^2 statistic to test whether the dealer is correct that each color is equally likely to be chosen.

13. A clothing store kept track of how often certain sizes of clothing were bought during a day. The results were as follows:

Size	Number of articles sold
Small	32
Medium	40
Large	21
Extra large	15

a. If you believed that each size is equally likely to be sold, how many of each size would you have expected to be sold?
b. Using the χ^2 statistic, test the belief that each size is equally likely to be sold.

14. A researcher is interested in determining whether infants are more likely to be born on certain days of the year than on others (ignoring leap years). If the researcher gathered the data for the number of births on every day of the year, what would be the number of degrees of freedom for the χ^2 statistic?

15. a. Find the following probabilities using an online statistical calculator:

$$P(\chi_2^2 \geq 10.2), P(\chi_5^2 \geq 10.2), \text{ and } P(\chi_6^2 \geq 10.2).$$

b. What trend do you see in your results to part (a)? As the number of degrees of freedom goes up, what happens to the probability of having a χ^2 statistic greater than 10.2?

16. If you were merely using five-step simulation to make your decision in these "dice" problems, would it be as easy or easier to use the D statistic instead of the χ^2 statistic? Explain why people prefer to use the χ^2 statistic when not doing simulations.

17. In a class of 197 students, the professor looks at the first digit of every student's social security number. The results are as follows:

Digit	0	1	2	3	4	5	6	7	8	9
Frequency	0	2	2	179	7	7	0	0	0	0

Is there evidence the first digit of the social security number is not a good source of random numbers? Perform a χ^2 test (noting that you *know* what your conclusion is going to be!).

18. Recall Exercise 17. This time the professor looks at the last digit of every student's social security number. Now the results are as follows:

Digit	0	1	2	3	4	5	6	7	8	9
Frequency	25	23	13	16	16	20	26	22	16	20

Is there evidence the last digit of the social security number is not a good source of random numbers? Perform a χ^2 test.

19. In the 1968 presidential election the popular vote was divided as follows:

Richard Nixon	31,770,237 votes
Hubert Humphrey	21,270,533 votes
George Wallace	9,906,141 votes

a. What percentage of the vote did each candidate win?
b. The actual winner of the election is determined by the votes of the Electoral College. There were 538 members. If the members' votes reflected how the people voted, how many of the electoral votes should each candidate have won?
c. The real Electoral College results were as follows:

Richard Nixon	301 electoral votes
Hubert Humphrey	191 electoral votes
George Wallace	46 electoral votes

Pretending that the electoral college votes are a random sample (they are not), test the expected versus the actual results using the χ^2 test.

20. Researchers conducted a study of 58 children admitted to the child psychiatry ward of the University of Iowa for aggressive behavior. The purpose of the study was to determine the influence for social class, sex, and age on the clinical characteristics of the children. A small portion of their data is shown in the table below, which lists the numbers of children exhibiting antisocial behavior for two categories of social class.

	Displays "antisocial behavior"		
Social class	**Yes**	**No**	**Row totals**
Middle-class families	24	4	28
Poor families	17	13	30
Column totals	41	17	58

a. Construct the conditional relative frequency table of social class | antisocial behavior. Interpret it.

21. One of the questions in the 1991 General Social Survey attempted to determine whether exposure to television weakens or strengthens confidence in the television news reporting. The results are presented in the table below. A total of 993 people were sampled.

	Average hours of daily TV watching			
As far as the people running the press, would you have...	**0–1**	**2–4**	**5 or more**	**Row totals**
A good deal of confidence	276	41	17	334
Only some confidence	196	174	47	417
Hardly any confidence	130	97	15	242
Column totals	602	312	79	993

a. Construct the relative frequency table for confidence | amount of TV watching.
b. Carry out the chi-square test. Is it a test of independence or one of equality of two populations?

PROFESSIONAL PROFILE

Carl Friedrich Gauss, 1777–1855, often called the "Prince of Mathematicians," developer of the normal or Gaussian distribution (bell curve).

This German 10-mark banknote depicting Carl Friedrich Gauss was in use from 1993 to December 31, 2001. A banknote of similar design was issued in 1989. The note depicts the Gaussian distribution and historic buildings in Göttingen where Gauss was director of the astronomical observatory from 1807 until his death.

Portrait of Carl Friedrich Gauss painted by Christian Albert Jensen.

Carl Friedrich Gauss was a child prodigy who became known as the "Prince of Mathematicians" (*Princeps mathematicorum*). His magnum opus, *Disquisitiones Arithmeticae,* was completed when he was only 21 (1798), though it was not published until 1801. This book became the basis for the consolidation of number theory as a discipline and continues to be important in this field today.

For statisticians, Gauss's major contributions center around his insights into normally distributed errors and the statistical theory underlying the method of least squares, the latter an outgrowth of his appointment as director of the observatory at Göttingen. On January 1, 1801, the dwarf planet Ceres (1 Ceres) was discovered by Giuseppe Piazzi. Located in the asteroid belt between Mars and Jupiter, its position made observation very difficult. Gauss developed the method of least squares (published in 1809) to determine when Ceres would return to view. The method of least squares is the usual approach to linear regression, the topic of Chapter 13.

Whenever you study the data resulting from many randomly sampled observations, you find that, though a few observations are far from the mean, most fall in a cluster fairly close to the mean. Gauss was not the first to study how data clusters in a bell shaped curve around a mean (Abraham de Moivre is credited with the first statement of the central limit theorem), but he relied on this characteristic in his analysis of astronomical data and determined the formula for the probability density function appearing in the statement of the central limit theorem. Thus, another name for a normal distribution or bell curve is the Gaussian distribution.

Gauss was willing to tackle many practical problems in "real life," including the theory behind insurance, the optimum number of jurors and witnesses, and infant mortality. An extraordinarily hard worker and dedicated perfectionist, Gauss delayed publication of his ideas or theories until he felt they were complete. Study of his personal diary reveals that he made a number of important mathematical discoveries long before his contemporaries claimed these results as their own, but this did not bother Gauss; he truly believed in his personal motto, *pauca sed matura* ("few but ripe").

13

Statistical Inference about Regression

Like father, like son (almost).

Objectives

After studying this chapter, you will understand:

- A probability model of linear regression for bivariate (x, Y) data.
  - Assessing the truth of the model's assumptions
- Regression SD and Regression SE
- The standard error of the linear regression slope estimator
  - Assesses when the regression analysis is accurate
- Hypothesis testing whether x and Y are linearly related
  - Large-sample approach
  - Normal-distribution-based small-sample approach
  - Five-step-method-based permutation test approach
- Learning about regression inference by studying simulated regression-model-based bivariate data
- Obtaining a confidence interval for the slope parameter
- Obtaining a linear regression prediction interval for a new observation Y at x, or a linear regression confidence interval for the expected value of Y given x
- Transforming bivariate data variables to extend the scope of linear regression to certain nonlinear relationships

KEY PROBLEM

Does a Child's Age at First Word Suggest Her Later Mental Aptitude Score?

The Gesell adaptive score is designed to scientifically measure the mental aptitude of children. Some researchers have been interested in finding out how well, if at all, the Gesell adaptive score can be predicted from the age at which a child utters his or her first word. Table 13.1 gives the Gesell adaptive scores obtained at a much later age and the ages at first word (in months) for 20 randomly selected children. One data point had been removed because the child's score was unusually high for the age at which he uttered his first word. Note that the points (10, 83) and (10, 100) each occur twice. The scatterplot Figure 13.1 clearly shows a negative correlation between age at first word (x) and the Gesell adaptive score (x). Its least squares linear regression equation, also shown in Figure 13.1, is

$$y = 109.3 - 1.2x.$$

Because the slope is -1.2, this fitted line predicts that the Gesell adaptive score will be lower by 1.2 units for each month later that a child utters his or her first word. Suppose that we want to apply this regression line to other children or that we want to use this relationship in a child development study. Thus, we want to know whether such a relationship can be confirmed statistically and hence is not just due to chance. In particular, is there strong statistical evidence in the data that the age at which the first word is spoken really helps predict the Gesell adaptive score? If so, how accurate are predictions given by the line? In this regard, we certainly would not expect the age at first word to be more than moderately predictive of something as psychologically complex as the mental aptitude the Gesell score is designed to measure. However, there does seem to be a negative linear trend unlikely to exist just by chance.

Finally, for this (at best) moderate association between age and the Gesell score, how accurately has the true regression line been estimated? For example, if a new random sample of 20 children were obtained, would the new regression line likely be about the same or not?

Table 13.1 Age at First Word and Gesell Score for 20 Children

Age at first word (in months)	Gesell score	Age at first word (in months)	Gesell score
15	95	7	113
26	71	9	96
10	83	10	83
9	91	11	84
15	102	11	102
20	87	10	100
18	93	12	105
11	100	42	57
8	104	11	86
20	94	10	100

For those of you who studied the $*$ subsection of Section 3.3 of Chapter 3 entitled **Outliers, Leverage Points, and Influence Points,** note that Figure 13.1 shows that the child whose first word occurred at 42 months is a leverage point because $x = 42$ is so extreme. Thus, this point could be very influential in that its removal could greatly change the location of the regression line. Recall that influential points are often removed from a regression analysis. Removal of an influential point is done because the new line obtained as a result is without the inappropriately large influence of a single data point upon the line's location. This new line is thus seen as more representative of the data set.

13.1 INFERENCE ABOUT THE REGRESSION LINE SLOPE

bivariate data

In chapter 3, we learned how to summarize (x, y) data, called **bivariate data,** when its scatterplot suggests y is linearly related to x—for instance, as indicated by the points lying in a roughly elliptically shaped cloud that is tilted up or down from the horizontal direction. We summarized such bivariate data using specialized statistics and formulas for the estimator $\hat{m}$ of the slope and the estimator $\hat{b}$ of the y-intercept, these parameter estimates together yielding the regression line, $y = \hat{m}x + \hat{b}$. The y-intercept is where the line crosses the vertical y-axis. Note that the $\hat{m}$ of this chapter was written as m in Chapter 3. This is done so that now m denotes a probability model parameter and $\hat{m}$ the estimator for estimating m from the collected data. We do the same for other parameters and their estimators such as b and $\hat{b}$.

In the examples from Chapter 3 we analyzed various bivariate data sets such as gas mileage (y) as a function of car weight (x), breast cancer mortality (y) as a function of animal fat consumption (x), age of wife (y) as a function of age of husband (x), and height of son (y) as a function of height of father (x), among other linear relationships for which a linear regression analysis was done. Indeed, a brief review of Chapter 3 will make the study of statistical inference concerning regression in Chapter 13 not only easier to understand but also more informative.

Suppose bivariate data forms a roughly elliptically shaped cloud, which suggests that fitting a regression line to it is appropriate. Then usually the most important inferential question is whether there is statistical evidence showing that as x changes y will also change. Recalling that the regression line is given by $\hat{y} = \hat{m}x + \hat{b}$, this question reduces to whether there is statistical evidence that the slope parameter $m \neq 0$.

When we conclude that $m \neq 0$, this means there is statistical evidence that as x increases either y will increase (if $m > 0$) or y will decrease (if $m < 0$). When the inference is that $m \neq 0$, then we describe the linear relationship between x and y using the estimated regression line $\hat{y} = \hat{m}x + \hat{b}$. If $m = 0$ is the conclusion, then it is of course inferred that the variables x and y have no linear relationship at all.

Because information about the slope m is so crucial, this section discusses making statistical decisions about m in some detail.

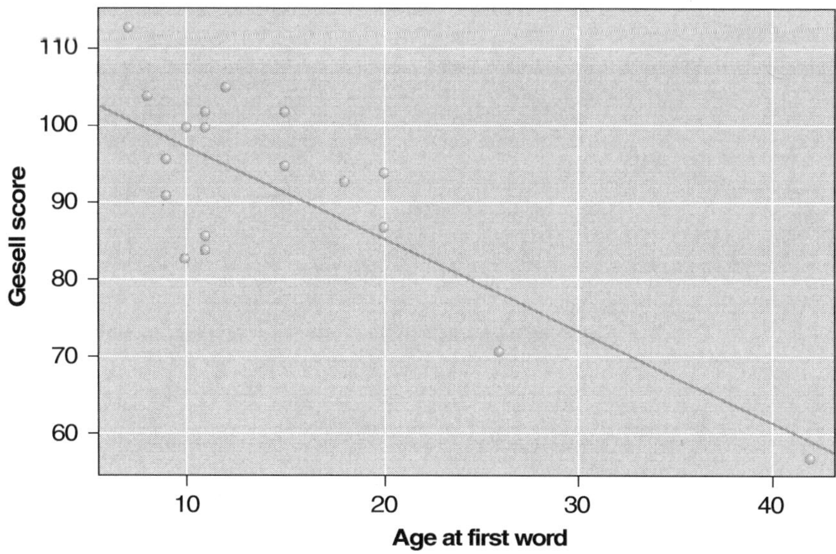

Figure 13.1 Gesell score versus age (in months) at first word spoken.

Interpreting Rejecting and Failing to Reject H_0: $m = 0$

When a well-designed statistical hypothesis test fails to reject H_0: $m = 0$, then one infers that $m = 0$ or m is close to 0. Either conclusion means that the influence of x on y is of no practical importance. We say the influence of x on y is of no practical importance if the true slope m of the regression line is close enough to zero that as x varies over its natural range the change in y is so small that it is insignificant from a practical perspective.

We illustrate this with an example where the slope $m \neq 0$ but m is so small in magnitude that it is not of practical importance. Suppose that in a two-month dietary supplement study, where the supplement is varied from 0 (placebo) to 100 milligrams/day, the true value of m is such that the change in predicted weight loss goes from $y = 0$ to $y = 1/3$ pound as the supplement dosage is varied from 0 to 100. If this is the true relationship between x and y, then a well-designed hypothesis test will usually fail to reject H_0: $m = 0$, as it should.

When a well-designed hypothesis test concludes that $m \neq 0$ (by rejecting H_0: $m = 0$), one usually infers that m is not close to zero in the sense that as x varies over its natural range, the change in y is of practical importance. Moreover, the regression line tells the practitioner what this practically important relationship is.

Returning to the weight loss illustration, suppose that the predicted true value of m is such that the predicted weight loss goes from 0 to 10 pounds as x goes from 0 to 100 milligrams, its specified range. Then a well-designed study will usually conclude that $m \neq 0$ by by rejecting H_0: $m = 0$. Moreover, the estimated regression line will usually inform the statistician that the weight loss goes from 0 to approximately 10 pounds as x goes from 0 to 100. A loss of 10 pounds is certainly a practically important amount.

In summary, when $m = 0$ or m is small in magnitude in the sense that the predicted change in y as x moves over its natural range is not of practical importance, then, as desired, H_0: $m = 0$ will usually not be rejected. However, when m is large enough that x has a practically important influence on y, then for a well designed study H_0: $m = 0$ will usually be rejected as it should be and, moreover, $\hat{y} = \hat{m}x + \hat{b}$ will usually give an accurate description of the relationship between x and y.

Understanding and Confirming the Standard Regression Model Assumptions

In order to understand and properly use statistical inference concerning regression, we must start by understanding the usual probability model for producing bivariate data that display an underlying (but noisy) linear relationship. This model is important to us because doing a regression analysis is justified only when it is determined that the model's three basic assumptions all are true.

Suppose a scatterplot is roughly elliptical and slopes either up or down, which suggests a linear relationship between x and y. From Chapter 3 we know how to compute the slope and y-intercept of a least squares regression line. Thus, we know how to compute the regression line equation. The estimated slope $\hat{m}$ is of particular interest. When nonzero, its value tells us the predicted change in y when x is increased one unit, as was illustrated in the Key Problem.

The estimated slope helps us infer whether the true slope is zero or not zero. We refer to the true m and b as the regression parameters and $\hat{m}$ and $\hat{b}$ as their estimates. Note that, unlike the case in Chapter 3, we now distinguish between estimated regression parameters and their true population values. As the new notation suggests, in the probability model we develop for linear regression, the treatment variable x is considered nonrandom, whereas the response variable Y is considered random. To emphasize its being a random variable, Y is capitalized from here on.

Before we can discuss inference about the true slope in a regression analysis, we need an appropriate probability model for how the bivariate data were generated. This is consistent with our approach to statistical inference throughout the book, as shown in Figure 13.2.

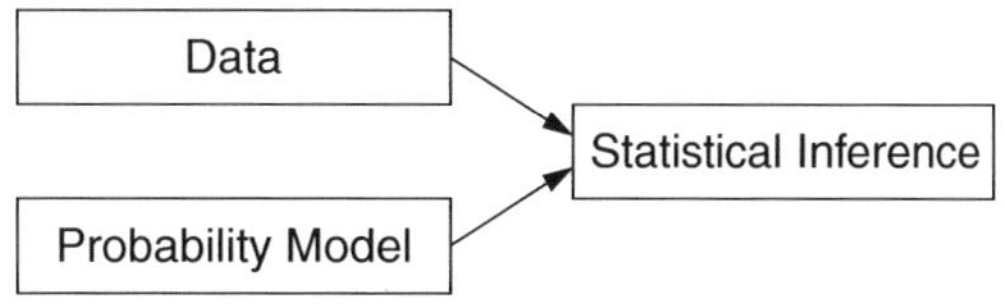

Figure 13.2 Model-based statistical inference.

The standard linear regression model provides a probability model for a random sample of points (x, Y) where, as already discussed, x is fixed (nonrandom) and Y is random. To describe the probability distribution of the random variable Y for each fixed value x, and hence model the (x, Y) values of the scatterplot, we make three **fundamental linear regression model assumptions,** which are stated in the box below. These regression model assumptions, when true, allow us to make statistical inferences using the estimated least squares regression line of Chapter 3, $\hat{Y} = \hat{m}x + \hat{b}$ where $\hat{b}$ is the estimator of the true y-intercept parameter b and $\hat{m}$ is the estimator of the true slope parameter m. Formulas for the $\hat{m}$ and $\hat{b}$ estimators, as computed from the observed (x, Y) values are given in Chapter 3.

fundamental linear regression model assumptions

Regression Model Assumptions

1. The Y random variables are all independent of each other. That is, we assume that the data are collected in such a way that the value of an observed Y does not depend in any way on the other observed Y-values.

theoretical regression line

2. The expected value of Y at x, denoted $E(Y \mid x)$ and called the **theoretical regression line,** is linear in x and can be written

$$E(Y \mid x) = mx + b,$$

where b and m are the unknown intercept and slope (theoretical) parameters. That is, we assume that the data can be described by an underlying linear relationship.

regression SD
residual SD

3. The standard deviation of Y for fixed x, denoted $\mathrm{SD}(Y \mid x)$, *is the same for every x.* It is denoted by σ and is called the **regression SD** or the **residual SD.**

The first assumption states that the Y's in the (x, Y) points of the scatterplot are independent random variables. It is useful to examine two common physical settings for regression that produce such independent random Y's.

One common setting is the random sampling of bivariate (X, Y) data from a large real-world population such as father and adult son heights in America. Thus, both X and Y are random. Often the X's and Y's are assumed to be normally distributed in the random sampling setting. In order to do a regression analysis in this setting, in our regression model we treat the random X's as if they are fixed. This seemingly unusual approach is theoretically legitimate. A second setting is an experimental one where the Y's are independent trials for various experimentally fixed x's, such as observing crop yield Y for various planned levels of fertilizer x. Note that x is not random.

The second regression model assumption—informally, that Y tends to be linear in x except for random error—has a useful empirical interpretation: If we were able to observe Y a large number of independent times for a fixed x_1, we would obtain an observed mean $\overline{y}_1$. Suppose we graph the point $(x_1, \overline{y}_1)$, and then repeat the process at a different x_2 to get $\overline{y}_2$ and point $(x_2, \overline{y}_2)$, and so on for the other $(x, \overline{y})$ values. If the second assumption holds, these points will lie close to the regression line, as was shown in Figure 3.22 for paired father and son heights and as discussed in detail below.

However, in a linear regression experiment, usually there is only one observed y for each value of x. Further, the x-values are often nonsystematically distributed on the x-axis, producing the typical elliptical cloud that, when present, suggests assumption 2 is true. In this case, we can use a nice trick to produce the above experiment of looking at $(x, \overline{y})$s. First, choose uniformly separated $x_1 < x_2 < x_3$, and so on, over the range of x, where the x's are reasonably close to one another. Then for x_1 we average all the y's whose x's are closer to x_1 than to x_2. Then for x_2, we average all the y's whose x's are closer to x_2 than to x_1 or x_3, and so on. This produces $(x_1, \overline{y}_1)$, $(x_2, \overline{y}_2)$, and so on. In this case, if these points are in an approximately straight line, then assumption 2 of linearity holds.

We consider a real data setting where there is an elliptical cloud of points and in particular assumption 2 holds. This famous data set shown in Figure 13.3 and collected by Karl Pearson consists of 1078 randomly chosen $(\text{father}, \text{son}) = (x, y)$ heights obtained around the year 1900 when the heights of men and women were much less than today. Note that the cloud of points looks quite elliptical in shape. To confirm assumption 2, first choose sons' heights $x_1 = 59$, $x_2 = 60, \ldots, x_{17} = 75$. In each case we will take the average of all the fathers' heights within 1/2 inch of the targeted x, obtaining $(x_1, \overline{y}_1)$, $(x_2, \overline{y}_2)$, and so on. We thus obtain 17 $(x, \overline{y})$ pairs. The result is shown in Figure 3.22 in Section 3.3 of Chapter 3. Clearly the 17 points lie approximately on the drawn line, confirming that assumption 2 holds.

In contrast to the father-son setting just discussed, in some cases the x-values are fixed according to the design of an experiment. In fact, for a few cases, the design of the regression experiment calls for observing multiple y-values for each fixed x of the experiment, as shown in Figure 13.6 of the simulated regression of Example 13.1, where, by design, there are 15 randomly generated y-values for each fixed $x = 0, 1, 2, 3, 4$. Note that the $(x, \overline{y})$ points are marked bold, and they lie almost in a straight line, confirming that assumption 2 seems to hold, as in fact it did in the simulation model.

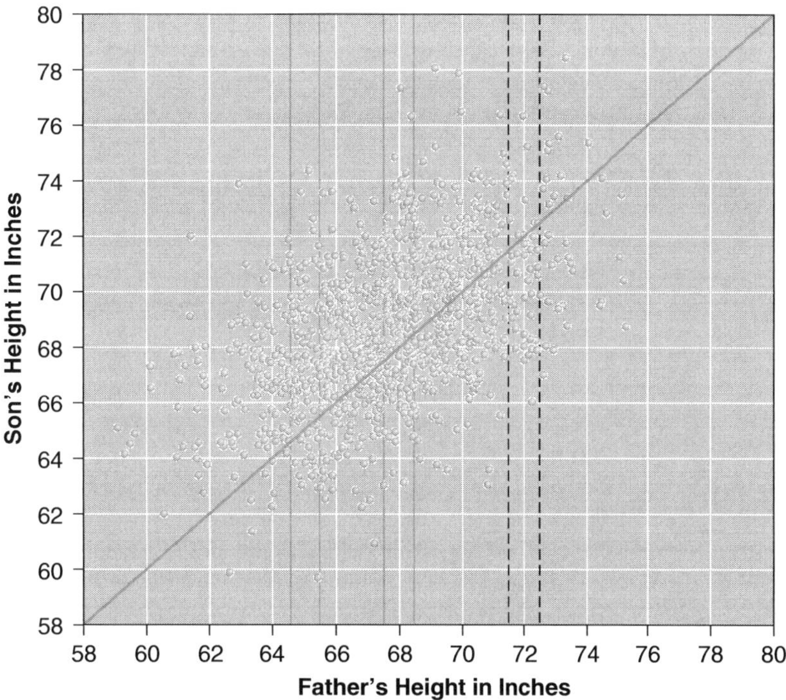

Figure 13.3 Heights of 1078 randomly chosen father-son pairs.

The second fundamental assumption rules out the possibility that the true underlying relationship between x and Y is nonlinear. However, we could observe or simulate data sets where there is an underlying (x, y) relationship but one that is curved (non-linear). Thus, we cannot make this assumption unless we have good reason to believe that the underlying relationship is linear. That is, we cannot do a linear regression unless we have evidence that the underlying (x, y) relationship is at least approximately linear.

We now consider assumption 3, which tells us that $SD(Y \mid X = x)$, the standard deviation of Y for each given x, is the same. That is, the typical Y variation about its regression line at x is the same regardless of the value of x. This common value is called the regression SD or sometimes the residual SD. The regression SD σ of assumption 3 tells us how close we can typically expect random variables Y to be to the true regression line. From the viewpoint of the practitioner, the regression SD σ measures how useful x is in predicting its paired Y. For example, if the relationship between x and Y were not used in predicting Y, then one's only way of predicting Y for a particular x value would be to predict Y at its x using the observed average $\overline{Y}$ of *all* the observed Y's. Recall Example 3.11, where we learned that the age of a husband is very useful in predicting the age of his wife, as Figure 3.20 shows. If we did not know the husband's age, all we could do to predict his wife's age would be to use $\overline{y} = 43$ years, the average age of *all* the husbands. However, knowing that $x = 30$, say, we use the regression line to predict $\hat{y} = 0.95(30) - 0.7 = 27.8$, expecting rather good accuracy. Here the estimate of the regression SD σ, using an estimator we will study later in the chapter, is $\hat{\sigma} \approx 0.31$. So, the typical variation of women's ages is $\sigma_Y = 13$ years, while the regression SD that gives the typical error in using the regression line to estimate the wife's age is about 1/3 of 1 year. Example 3.1, with r^2 correlation close to 1 and the correspondingly small σ, is an excellent illustration of the predictive power of regression when r^2 is close to 1. The regression SD σ of assumption 3 is usually unknown, so we will have to estimate it from the scatterplot data.

If one is careful, one can roughly assess the truth of assumption 3 by examining the scatterplot. For example, consider the father-son heights of Figure 13.3. Let's look at the observed $Y \mid x$ distribution for $x \approx 65$, 68, and 72 by looking at the three corresponding bars of width one inch. One must adjust for there being more points in the $x \approx 68$ bar than in the $x \approx 72$ bar. The key is that the average variation among the points in a bar appears to be about the same even though the $x \approx 68$ bar has more points and hence is more spread out. Assumption 3 appears to be correct. Indeed, if one computed all 17 sample standard deviations, one for each bar, these 17 values would be fairly similar, thus confirming assumption 3. When this assumption fails to be true for bivariate data, the standard regression analysis being discussed here must be replaced by a more sophisticated procedure carried out by a statistical expert. Such a case is illustrated next.

Figure 13.4 is a scatterplot that gives the diastolic blood pressure for each of 57 adult women as a function of age. (Diastolic blood pressure is the pressure when the heart is at rest.) Also displayed is the fitted least squares regression line $(\overline{Y} = 56.1 + 0.59x)$. Clearly the observed y-values vary much less about the regression line for x small (e.g., near age 20) than for x large (e.g., near age 50). Imagine a bar corresponding to y's near $x = 25$ compared with a bar for y's near $x = 50$. Clearly there is much more variability shown by the y's in the $x = 50$ bar. This data set thus clearly violates the assumption that $\text{SD}(Y \mid x)$ is the same regardless of the value of x, and the cloud of bivariate data is not even crudely elliptical. Hence, the third fundamental assumption fails, and we are not allowed to use the standard formulas we learned in Chapter 3 for estimating the regression line. In fact, a statistical expert would find a regression line somewhat different from the usual least squares regression line.

In Chapter 3, we learned that plotting residuals (see ***Residual Plots** in Section 3.3) for a least squares regression analysis sometimes reveals that the regression line cannot be trusted for one of several possible reasons, such as the scatterplot's having an influential outlier or displaying a nonlinear pattern. In addition, a residual plot can help one discover when the residual SD is not constant as x varies. In the case of female blood pressures, the pattern of increasing residual SD as x increases is clear, as the residual plot in Figure 13.5 shows. Although we can see this by examining the scatterplot, the residual plot often makes results more obvious. If a pattern of systematic increase or systematic decrease in the residuals as x increases is absent from the residual plot, then it is safe to assume $\text{SD}(Y \mid x)$ does not depend on x. In such a case, it is correct to do a least squares regression provided we believe assumptions 1 and 2 are also true.

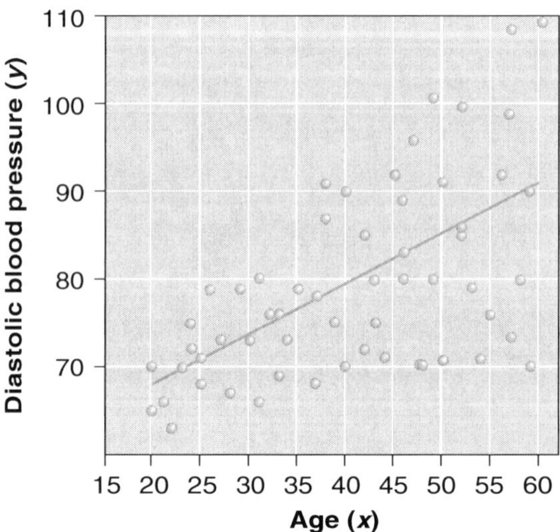

Figure 13.4 Scatterplot plot of female blood pressure versus age.

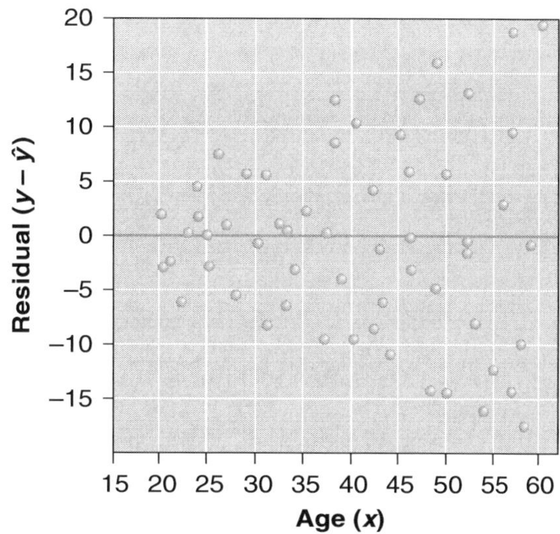

Figure 13.5 Residual plot of female blood pressure versus age.

In summary, before we are permitted to do a standard regression analysis as presented in Chapter 3, and before we are permitted to use the least squares regression line and carry out various statistical inferences as shown in this chapter, we need evidence that the three basic regression assumptions hold. As already discussed, whether the Y's are independent is confirmed by studying how the Y's were chosen. In particular, the value of a Y for an (x, Y) data point cannot be influenced by the other, already chosen Y's. Assumption 2 is determined by graphing the scatterplot. If it is roughly elliptical, then the assumption is judged to hold. Alternatively, one can break the x-axis up into fairly narrow intervals (allowing several y's for each interval) and graph $(x, \bar{y})$ for each interval. If these $(x, \bar{y})$ points roughly form a line, then this provides evidence that assumption 2 holds. Assumption 3 is also confirmed by examining the scatterplot. In particular, the variation of the Y's cannot go from large to small or from small to large as one goes from one end to the other end of the range of the x's.

Note that deciding whether we can assume that these assumptions hold or not is somewhat subjective. One should be neither too strict nor too lenient!

Regression Standard Deviation σ as Estimated by the Regression Standard Error S_E

standard error of the residuals
residual SE
regression SE

When we assume assumption 3 holds, the residual standard deviation σ can be well estimated by what is called the **standard error of the residuals, residual SE,** or **regression SE,** defined by

$$S_E = \sqrt{\frac{\Sigma(Y - \hat{Y})^2}{n - 2}},$$

Be careful to note that $\hat{Y}$ denotes the value of the fitted regression line that corresponds to Y in the $(Y - \hat{Y})^2$ term, noting in particular that both Y and $\hat{Y}$ in $(Y - \hat{Y})^2$ have the same x-value.

Another very illuminating formula for S_E (the estimator of σ) is given by

$$\hat{\sigma} = S_E = S_Y \sqrt{1 - r^2},$$

where S_Y denotes the sample standard deviation of all the Y's. When r^2 is close to 1, S_E will be quite small and hence the $\hat{y}$ at x computed from the regression equation will be an accurate predictor of Y at x.

When the sample size n is large, a sophisticated version of the law of large numbers tells us that

$$S_E \approx \sigma,$$

namely that the standard error of the residuals estimates the residual SD σ well. As we often have done in this textbook, in the next example we use simulation to demonstrate a statistical concept. In Example 13.1 we study the third assumption: that the residual SD σ is constant as x varies. In particular, we study the role of σ in the regression model in defining the typical amount of variation (spread) of the Y-values about the theoretical regression for fixed x.

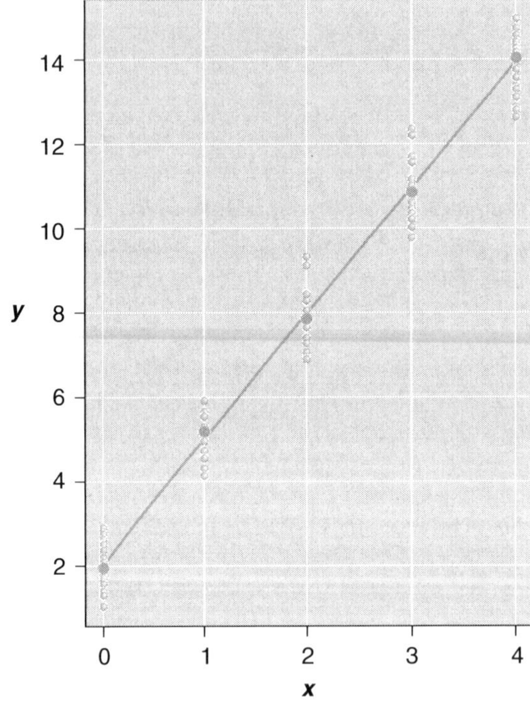

Figure 13.6 Simulation of a scatterplot under regression model assumptions, with $\bar{y}$ for each x included.

Example 13.1

Consider a regression model satisfying

$$E(Y \mid x) = 3x + 2 \qquad \text{and} \qquad \text{SD}(Y \mid x) = \sigma = 0.6$$

for every x. Suppose the experimental design calls for repeated observation of y at $x = 0, 1, 2, 3, 4$. To satisfy assumption 1, 75 independent Y-values were simulated, with 15 Y-values each at $x = 0, 1, 2, 3, 4$. Note that Assumptions 2 and 3 also hold with $m = 3$, $b = 2$, and $\sigma = 0.6$. In Figure 13.6, the resulting scatterplot together with the true regression line are shown (one cannot see all the points because of overlap). From a box model perspective, we needed five boxes to generate these data, one for each x of the simulated experiment. We assumed a normal population of Y-values for each x (hence the computer software was used to do the normal-distribution-based box model sampling rather than specifying balls in a box, an inappropriate approach when the random variables are continuous as in the normal and chi-square distributions). Thus the five boxes and the normal distributions producing the 75 data points are:

Box 1 ($x = 0$):	$Y_{0,1}, \ldots, Y_{0,15}$	$N(2, 0.6)$
Box 2 ($x = 1$):	$Y_{1,1}, \ldots, Y_{1,15}$	$N(5, 0.6)$
Box 3 ($x = 2$):	$Y_{2,1}, \ldots, Y_{2,15}$	$N(8, 0.6)$
Box 4 ($x = 3$):	$Y_{3,1}, \ldots, Y_{3,15}$	$N(11, 0.6)$
Box 5 ($x = 4$):	$Y_{4,1}, \ldots, Y_{4,15}$	$N(14, 0.6)$

Here, in each case, $E(Y \mid x) = 3x + 2$ determines the mean of the box model's normal distribution; for example $E(Y \mid 2) = 8$ for box 3. Further for each box $\text{SD}(Y \mid x) = 0.6$; for example, $\text{SD}(Y \mid 2) = 0.6$ for box 3, the common σ of assumption 3. Thus, assumptions 2 and 3 hold.

It is instructive to compare Figure 13.4, where larger x-values have larger $\text{SD}(Y \mid x)$ values, with Figure 13.6, where we notice that the spread, or typical variation, of the y-values about the line is nearly the same for all x's. Note in this figure that when the x's are systematically separated, even though assumptions 2 and 3 hold, we do not see an elliptical pattern. The three regression model assumptions hold, in particular the second one guaranteeing the linearity of the (x, Y) relationship. Hence, it is acceptable to use our regression least squares formulas to solve for the estimated slope $\hat{m}$ and the estimated intercept $\hat{b}$ and thus solve for the estimated regression line.

This simulation is instructive in other ways. From Chapters 5 and 10 we have learned that if the sample size is large, a good estimator, like $\overline{X}$ for μ_X, will be close to the parameter it estimates, by the law of large numbers. Here the estimator S_E of the regression error σ was computed from the 75 data points, yielding

$$S_E - 0.65$$

Note that $\sigma = 0.6$ and hence that $S_E \approx \sigma$, just as the law of large numbers suggested. This confirms that S_E is a good estimator of the residual SD σ when the sample size is large, as claimed. In fact, $\text{SD}(S_E) = 0.059$ can be shown theoretically. Recall that the SD of an estimator gives us the typical estimation error size when using that estimator. Hence the typical error in estimating σ by S_E is about 0.059 in magnitude, as was the case with $S_E - \sigma = 0.65 - 0.6 = 0.05$, the estimation error. Finally, $S_E = 0.65$ suggests that the typical distance of an observed y from the regression line is about 0.65; Figure 13.6 crudely supports this.

As suggested above, regression models arise in two ways. In the preceding simulation, the x-values were experimentally fixed and hence are not random. This is often the case. Replication at each x is more uncommon. For example, one might study crop yield as a function of fertilizer level x fixed at 0, 500, 1000, and 1500 pounds per acre, where the experimental design is to have two fields at each predetermined fertilizer level, producing eight independent observed y-values. Alternatively, one might study yield as a function of annual rainfall for 25 widely separated fields, where the rainfall level (X) is obviously random. Thus the observed x-values will be nonsystematically (haphazardly) smeared in an interval, say from 27 inches to 49 inches. As stated above, such data often produce the typical elliptical cloud of data points, seen with father and son heights, for example. One important consequence of the theory and application of regression is that the regression model assumptions can apply regardless of whether the x-values are fixed by the experimental design or are random as a result of random sampling (X, Y) values from an actual real-world population. This important principle must be strongly emphasized.

The SD and SE of $\hat{m}$ and Their Determination of Regression Accuracy

Recall from Chapter 3 that the slope estimator $\hat{m}$ is defined by

$$\hat{m} = r\frac{S_Y}{s_x} = \frac{\Sigma(x-\overline{x})(Y-\overline{Y})}{\Sigma(x-\overline{x})^2}$$

Here S_Y and s_x denote the usual sample standard deviations for Y and x, respectively. For example,

$$S_Y = \sqrt{\Sigma(Y-\overline{Y})^2/(n-1)}, \qquad s_x = \sqrt{\Sigma(x-\overline{x})^2/(n-1)}.$$

Also, recall from Chapter 3 that the computationally convenient formula (usefule for small data sets and hand calculations) for the slope estimate $\hat{m}$ is based on five simple sums computed from the x- and y-values:

$$\hat{m} = \frac{\sum xy - (\sum x)(\sum y)/n}{\sum x^2 - (\sum x)^2/n}.$$

residual
estimation error

Recall from Chapter 3 that a **residual** $Y - \hat{Y}$ is also called an **estimation error** because Y is the observed value and $\hat{Y}$ is the predicted value at x using the regression line equation. The standard deviation of the estimator $\hat{m}$ is a function of the regression SD σ and can be shown to be

$$\text{SD}(\hat{m}) = \frac{\sigma}{\sqrt{\Sigma(x-\overline{x})^2}} = \frac{\sigma}{s_x\sqrt{n-1}}.$$

We saw in the simulation of Example 13.1 that $S_E \approx \sigma$ when n is large. Also, $s_x = 142$ was computed from the 75 x-values. Thus, substituting S_E for σ in the formula for SD$(\hat{m})$ above, a good estimator of SD$(\hat{m})$ is a function of the regression standard error S_E and is given by the standard error of $\hat{m}$:

definition of SE($\hat{m}$)

$$\mathbf{SE}(\hat{\boldsymbol{m}}) = \frac{S_E}{\sqrt{\Sigma(x-\overline{x})^2}} = \frac{S_E}{s_x\sqrt{n-1}}.$$

Note that in Example 13.1,

$$\text{SE}(\hat{m}) = \frac{S_E}{s_x\sqrt{n-1}} = \frac{0.65}{142\sqrt{74}} = 0.00053,$$

which is very small relative to the value of the true slope, $m = 3$, indicating that the estimated slope $\hat{m}$ should come quite close to the true slope m, relative to the magnitude of m. Because the number of data points, 75, is large, such a small typical error for the estimate $\hat{m}$ of the slope parameter m is expected.

A close examination of the right-hand expression for the standard error $\hat{m}$, namely $\text{SE}(\hat{m})$. shows the three components that influence its size. These are (1) the sample standard deviation s_x of the x-values; (2) the sample size, n; and (3) the standard error S_E of the residuals. In particular, $\text{SE}(\hat{m})$ can be made smaller as desired by

- increasing the sample size n,
- decreasing the size of the residual errors $Y - \hat{Y}$ and hence decreasing S_E; that is, improving the capability of the regression line to predict the Y's, a difficult challenge not usually possible,
- increasing the spread of the x-values, thus making s_x larger

In practice, can we do these things? We can design a study in which n is large if cost allows, and we sometimes have control over the x-values and, hence, are able to spread them out widely, thus increasing s_x. But how close the observed data points typically are to the regression line, which is measured by σ, is almost never under our experimental control. Rather, it is a consequence of how capable x is of predicting Y accurately via the linear regression real-world equation. That is, the closeness of the Y-values to the regression line varies depending on the strength of the linear relationship between x and Y. For example, the age of husband x predicts the age of wife Y well, thereby producing a small σ, whereas height x predicts weight Y less well, thereby producing a large σ. To summarize, we almost never have control over the size of the regression SD parameter σ, which tells us the typical variation of observed Y-values about the regression line at any x.

The standard error of $\hat{m}$, namely $\text{SE}(\hat{m})$, estimates for us the typical error size $\hat{m} - m$. The mean (expected value) of $\hat{m}$ is actually equal to the true slope parameter m, that is,

$$E(\hat{m}) = m.$$

In summary, $\hat{m}$ is an unbiased estimator of m with the standard deviation of $\hat{m}$ being $\text{SD}(\hat{m}) = \sigma / (s_x \sqrt{n-1})$, and standard error of $\hat{m}$ being $\text{SE}\ (\hat{m}) = S_E / (s_x \sqrt{n-1})$.

In the following example, we use a simulation study to allow us to directly experience how sample size affects the estimation behavior of $\hat{m}$. Recall from Chapter 3 that the intercept b is estimated by $\hat{b} = \overline{y} - \hat{m}\overline{x}$.

Example 13.2

In Example 3.12, the regression line for predicting a son's height Y (in inches) from his father's height x, around the year 1900, is given by (using 1078 pairs)

$$\hat{Y} = 33.73 + 0.516x.$$

Computation shows that

$$\overline{x} = 67.68, \overline{y} = 68.65, S_E = 2.325, s_x = 2.70, s_Y = 2.71, r = 0.514.$$

Hence, the estimated slope $\hat{m}$ and intercept $\hat{b}$ are given by

$$\hat{m} = r\frac{s_Y}{s_x} = 0.516 \quad \text{and} \quad \hat{b} = \overline{y} - \hat{m}\overline{x} = 33.73.$$

We will ignore estimation error and treat these sample values as population parameters, and we will use the estimated regression line as the true regression line to allow a two-stage box-model-based simulation experiment to be used to simulate the three regression settings created by choosing three different sample sizes. The required parameters and resulting regression line, using $m = 0.516$ and $b = 33.73$, are

$$\mu_x = 67.68,\ \sigma_x = 2.70,\ \mu_Y = 68.65,\ E(Y \mid x) = b + mx = 33.73 + 0.516x,$$
$$\rho = 0.514,\ SD(Y \mid x) = \sigma = 2.325.$$

In this manner, we will observe the influence of the sample size on the accuracy of the estimated slope $\hat{m}$.

In detail, in stage 1 of the simulation experiment each x of a simulated (x, Y) pair is obtained by a computer-based drawing from a $N(67.68, 2.70)$ box model for the random father's height. Then, in stage 2, using the true regression line, the corresponding random son's height Y is gotten by drawing from a $N(33.73 + 0.516x, 2.325)$ box, with the observed x value plugged in. Note that, as required for carrying out a linear regression, these Y's are independent because each Y is the result of an independent X being chosen.

For each of the three prescribed sample sizes n, we will simulate estimating the regression line 1000 times (think of 1000 statisticians each doing the regression analysis on a set of simulated (X, Y) data gotten as just described) using sample sizes

$$n = 4, 25, \text{ and } 100, \text{ respectively.}$$

In Figure 13.7, one of the 1000 simulated scatter plots of size $n = 100$ is shown. A positive association is exhibited but not a strong one (as expected, noting that $r \approx 1/2$). In each of the three cases, 1000 regression line estimations are simulated, and in each case the variability and the bias of $\hat{m}$ are studied. Note the roughly elliptical shape of the scatterplot. Further note that it is only roughly elliptical even though the three regression assumptions hold exactly.

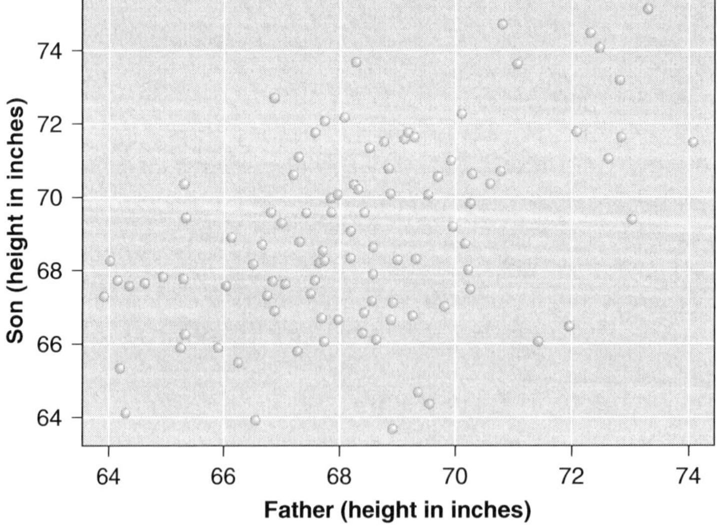

Figure 13.7 Scatterplot of 100 simulated son versus father heights.

Keep firmly in mind that our purpose here is to study the effectiveness of the estimator $\hat{m}$ in specifying the regression line's true slope m.

What can we learn from our simulated experiment of watching 1000 statisticians estimate the regression slope $m = 0.516$ when there are only 4 (father's height, son's height) data points, when there are 25 data points, and when there are the large number of 100 such data points? The (x, y) correlation is around 0.5, representing a moderate association, as Figure 13.7 itself suggests.

We learned in Chapter 5 that for sampling a large number of independent replications, such as 1000 $\hat{m}$'s, the law of large numbers holds for the average of the $\hat{m}$'s. In our case (with the bar over $\hat{m}$ meaning the average of 1000 $\hat{m}$'s),

$$\overline{\hat{m}} \approx E(\hat{m}).$$

It was remarked above that $\hat{m}$ is unbiased, and thus $E(\hat{m}) = m$. Hence, the simulation should confirm that

$$\overline{\hat{m}} \approx m = 0.516.$$

As the following table confirms, this is indeed true in each of these simulation studies, with the largest observed bias being for $n = 4$, namely

$$\overline{\hat{m}} - m = 0.511 - 0.516 = -0.005,$$

a very small error relative to the true $m = 0.516$.

n	4	25	100	True m
$\overline{\hat{m}}$	0.511	0.512	0.516	0.516

The theory says that

$$\text{SD}(\hat{m}) = \frac{\sigma}{s_x\sqrt{n-1}} = \frac{2.325}{2.70\sqrt{n-1}}.$$

The following table results:

	$n = 4$	$n = 25$	$n = 100$
$\text{SD}(\hat{m})$	0.4972	0.1758	0.0865 .

Keeping in mind that the true $m = 0.516$, this table tells us the typical error that a statistician will experience in estimating $\hat{m}$, depending on the sample size. When $n = 4$, according to the table, the typical size of the estimation error $(\hat{m} - m)$ is given by $\text{SD}(\hat{m}) \approx 0.5$. Thus the typical variation in using $\hat{m}$ as an estimator of m is 0.5, almost as large as the true value of the m it is estimating. Hence, $\hat{m}$ is a very poor estimator of m when $n = 4$. Even though it is unbiased, it is far too variable. To illustrate, one statistician could easily obtain $\hat{m} = 0.1$

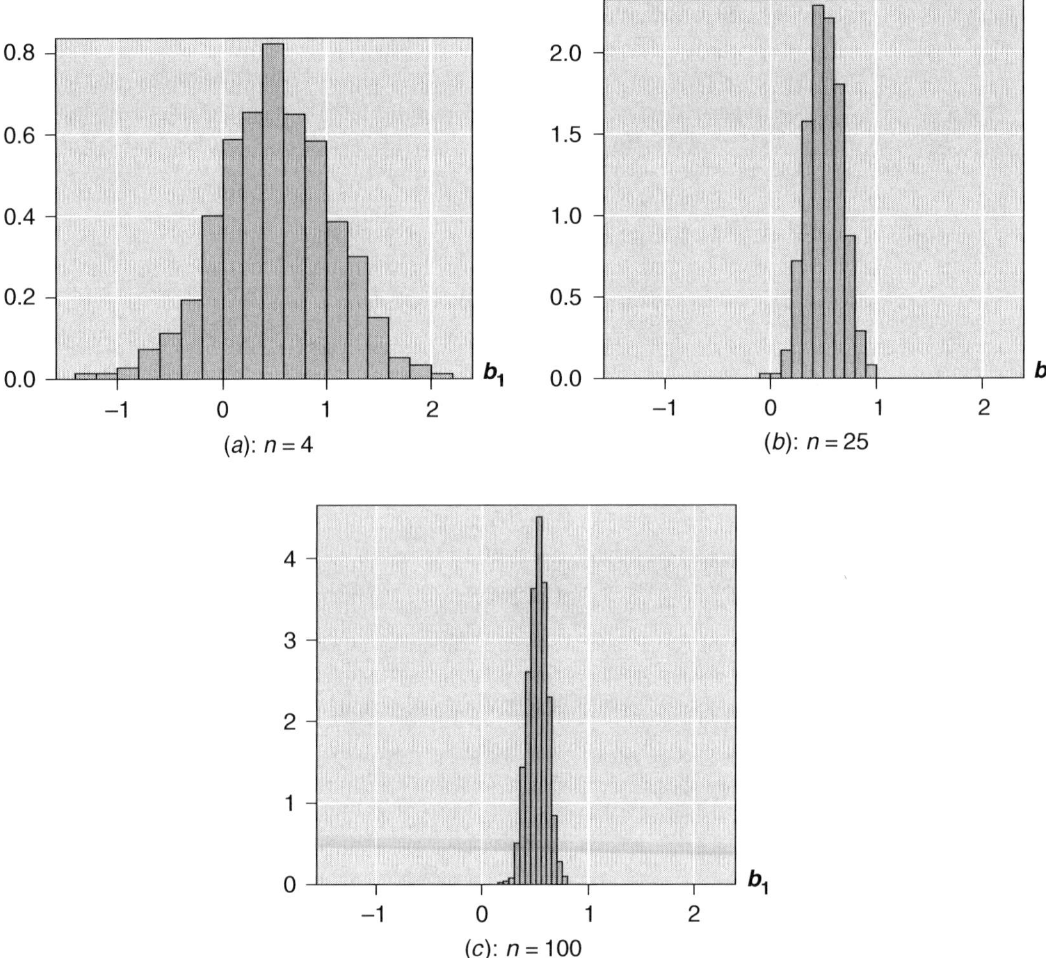

Figure 13.8 Density histogram of 1000 simulated $\hat{m}$-values for heights of fathers and sons when $n = 4$, 25, and 100, respectively.

and another could easily obtain $\hat{m} \approx 0.9$. Indeed, the density histogram in Figure 13.8(a) shows the variability of the 1000 simulated estimated $\hat{m}$ slopes—that is, the estimates of m by 1000 simulated statisticians each having a sample size of only four (x, y) values and with the resulting 1000 $\hat{m}$ estimates varying from less than -1 to greater than 2. Thus, doing a regression analysis when $n = 4$ is worthless in this setting.

By contrast, when $n = 25$, $\mathrm{SD}(\hat{m}) \approx 0.18$, which produces $\hat{m}$ estimates varying as in Figure 13.8(b) This is better, although not good, in that typical values of $\hat{m}$ range from $0.516 - 0.18 = 0.336$ to $0.516 + 0.18 = 0.696$. Thus, using $\hat{m}$ as an estimator of m when $n = 25$ is of minimal use because $\hat{m}$ is quite variable (from one statistician to the next) and can easily be fairly far away from the true value of m.

Finally, when $n = 100$, $\mathrm{SD}(\hat{m}) \approx 0.09$, producing typical variations in $\hat{m}$ from one statistician doing a regression to the next of $0.516 - 0.09 = 0.426$ to $0.516 + 0.09 = 0.606$, which is fairly good, as displayed in Figure 13.8(c). Suppose it is judged essential to obtain a typical variation given by $\mathrm{SD}(\hat{m}) = 0.04$, say. Then one can show that the statistician would need $n = 464$ observations (a very large sample size). We have thus learned that when the correlation between a variable X and the Y it is predicting only moderate (such as $\rho = 0.5$), then to be confident of accurately estimating m, the sample size must be quite large. Even $n = 100$ could fail to produce the necessary accuracy in some settings.

We now contrast our ability to make accurate inferences about the slope m in Example 13.2(b), where the correlation was moderate ($\rho = 0.514$), with a situation where the correlation is close to 1 as in the next example.

Example 13.3

The regression wife's age Y as a function of her husband's age x was studied in Examples 3.11 and 3.13. A high sample correlation ($r = 0.95$) was found between the ages of the roughly 300 husband and wife pairs. The fitted line was

$$\hat{Y} = 0.95x - 0.7.$$

Computation showed that

$$\overline{x} = 4.06,\ \overline{y} = 43,\ s_X = s_Y = 13,\ S_E = 4.06.$$

We will use these statistics to produce our simulation model:

$$\mu_X = 4.06,\ \sigma_X = 13,\ \mu_Y = 43,\ E(Y \mid x) = b + mx = -0.7 + 0.95x,\ \mathrm{SE}(Y \mid x) = \sigma = 4 \text{ years.}$$

As in Example 13.2, we simulate the regression line estimation process repeatedly (1000 times) for sample sizes $n = 4, 25, 100$. Figure 13.9 shows one (x, y) scatterplot for $n = 100$. Note that the scatterplot is roughly elliptical, as desired, but also note that the elliptical shape is very narrow. Clearly, with 100 observations we should estimate the line rather accurately.

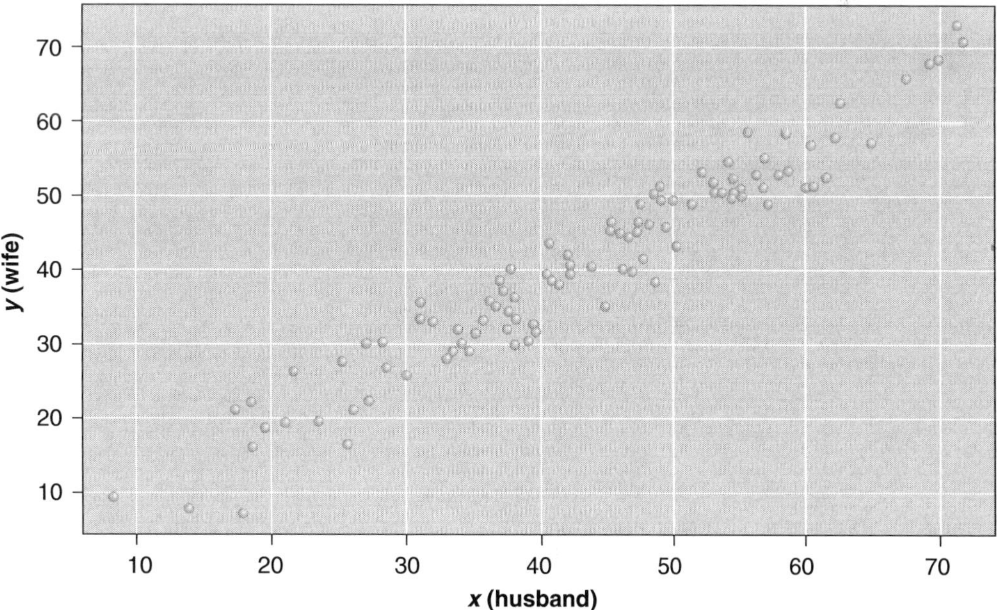

Figure 13.9 Scatterplot of 100 simulated wife versus husband ages.

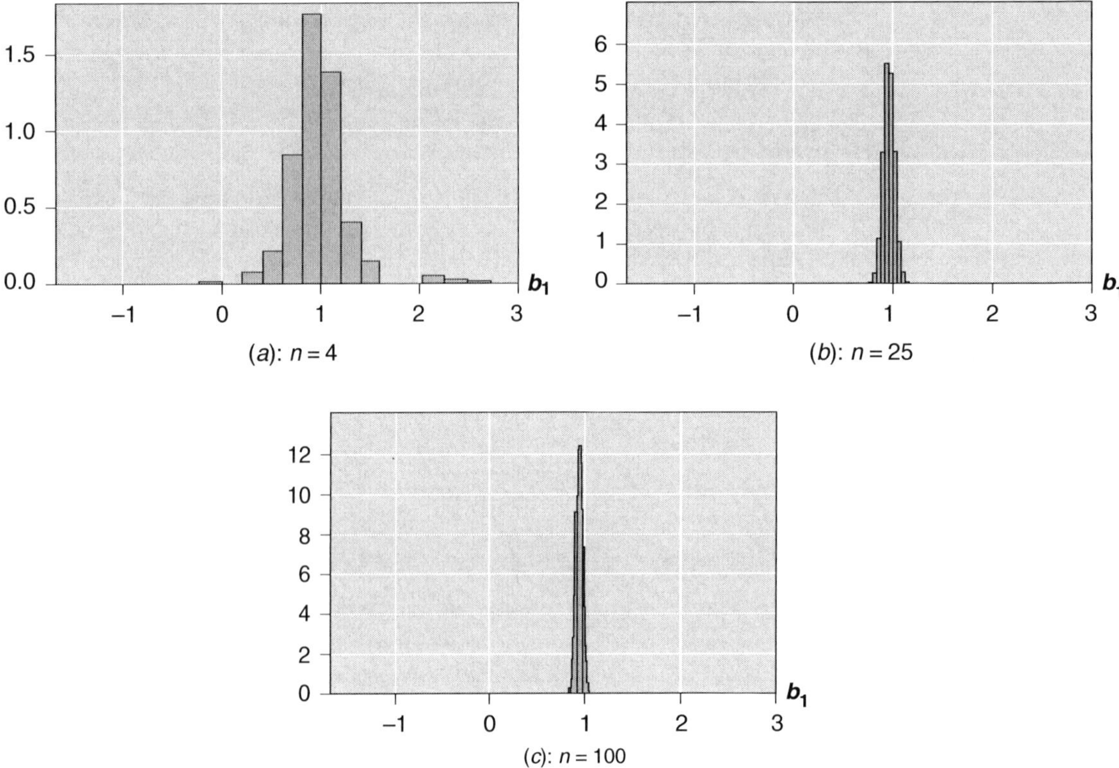

Figure 13.10 Distribution of 1000 simulated $\hat{m}$-values for ages of husbands and wives when $n = 4$, 25, and 100, respectively.

As before, the average of the 1000 $\hat{m}$'s is very close to the true m, regardless of the sample size n:

n	4	25	100	True m
$\overline{\hat{m}}$	0.950	0.951	0.949	0.950

Proceeding as before, we obtain for $\text{SD}(\hat{m})$:

	$n = 4$	$n = 25$	$n = 100$
$\text{SD}(\hat{m})$	0.1803	0.0637	0.0314

This table is more encouraging than the corresponding table in the moderate correlation case. The typical $\hat{m} - m$ error ($\text{SD}(\hat{m})$) a statistician expects, noting that the true $m = 0.95$, runs from approximately 0.18 to 0.03 as n runs from $n = 4$ to $n = 25$ to $n = 100$. Clearly, 0.03 is very good, 0.06 is okay, and even 0.18 is of some minimal use. Figure 13.10(a)–(c) again shows the improved capability of $\hat{m}$ to estimate m accurately as n ranges from 4 to 25 to 100, which was sown in Example 13.2 as well.

❐ ❐

Another interesting comparison across the two examples with their widely different correlations ($r = 0.514$ versus $r = 0.95$) is to compare

$$\frac{\text{SD}(\hat{m})}{m} \cdot 100\%$$

values for different n values and r values. This formula gives (on the percent scale) the typical variation in $\hat{m}$, *relative to its true value* m, a ratio we would like to be small. It is tabulated as

		n		
		4	**25**	**100**
r	0.514	96.4%	34.1%	16.8%
	0.95	19%	6.7%	3.3%

When r is at least somewhat close to 1, the estimated slope tends to display much less error for each n ($n = 4$, 16, 100) than when r is not close to 1. For example, in the $n = 4$ case the 19% error for $r = 0.95$ is on average fairly large, but perhaps marginally tolerable. By contrast the 96% error for $r = 0.514$ is surely unacceptable! The performance of the index with $n = 4$ in the $r = 0.95$ case is almost as good as for $n = 100$ when $r = 0.514 \approx 0.5$. Clearly, if one suspects that ρ is only moderate and yet one wants an accurate estimate of regression slope, then using a large sample size is required.

Clearly a smaller standard deviation of $\hat{m}$ means both a more accurate estimate of the true regression line $E(Y \mid x)$ for any given x and a more accurate prediction of a newly observed Y for any given x. Indeed, as discussed in Section 3.3, two basic statistical problems arise in regression: (i) estimate the line $E(Y \mid x)$ at x, thus estimating the underlying straight-line relationship between Y and x (recall Example 3.12), and (ii) at a particular x predict a new Y that will be observed at a future time (recall Examples 3.10 and 3.11).

We now take a look at our Key Problem.

Example 13.4

Consider the Gesell score data of Table 13.1 of the Key Problem. Recall from Chapter 3 that for ease in computing b_1, five sums should be computed:

$$\sum x = 285, \sum y = 1846, \sum x^2 = 5317, \sum y^2 = 173{,}514, \sum xy = 24{,}807$$

Recall that the sample size $n = 20$; therefore, by our computational formula for $\hat{m}$,

$$\hat{m} = \frac{\sum xy - (\sum x)(\sum y)/n}{\sum x^2 - (\sum x)^2/n} = -1.193.$$

Also,

$$S_E = \sqrt{\frac{\sum(Y - \hat{Y})^2}{n-2}} = 8.6 \quad \text{and} \quad s_x = \sqrt{\frac{\sum(x - \overline{x})^2}{n-1}} = 8.13$$

Thus the estimated standard error of $\hat{m}$ is given by

$$\text{SE}(\hat{m}) = \frac{S_E}{s_x\sqrt{n-1}} = \frac{8.61}{8.13\sqrt{19}} = 0.243.$$

Now, for the moderately large $n = 20$, $\hat{m}$ is approximately normally distributed. (Using advanced probability theory, this normality can be shown to follow from our three fundamental regression model assumptions.) Thus, according to the 95% rule for the standard error of an estimator that is approximately normal, about 95% of the time we would expect $\hat{m}$ to be within $2 \cdot \mathrm{SE}(\hat{m}) = 2 \cdot 0.243 = 0.486$ of the true slope m. In particular, noting that $\hat{m} = -1.2$, $\mathrm{SE}(\hat{m}) = 0.243$ seems to provide strong evidence that the true $m \neq 0$. That is, our standard error computation provides strong evidence that useful prediction of Y given x is possible and, in particular, that the Gesell score is somewhat predictable from the age at which a child's first word was spoken. This suggests that we should be able to confirm that x and Y are associated via a formal hypothesis test about m, which we learn how to do next.

Hypothesis Testing of Whether x Predicts Y

Testing whether x is useful in predicting Y amounts, when the three regression assumptions hold, to testing whether the true slope $m = 0$ and hoping to reject the posssibility that $m = 0$. Thus, the null hypothesis is

$$\mathrm{H}_0\colon m = 0.$$

This null hypothesis is of central importance because it asserts that there is no information about predicting Y using the value of x. In most applications the statistician hopes to find sufficient statistical evidence to reject this hypothesis, thus establishing evidence that there is a predictive linear relationship.

In Chapter 11 we considered the one-population test of the mean ($\mathrm{H}_0 : \mu = \mu_0$) when the sample size is large (say $n \geq 20$), the population standard deviation is unknown, and the observations are not necessarily normally distributed. There we learned to use the standardized test statistic,

$$T = \frac{\overline{X} - \mu_0}{\mathrm{SE}(\overline{X})},$$

which we learned was approximately t-distributed with $n - 1$ degrees of freedom. Here T is approximately t-distributed rather than approximately standardized normal ($N(0, 1)$) because the population standard deviation is unknown. If n is very large (say $n \geq 1000$), then T is approximately $N(0, 1)$ because the t-distribution approximates the $N(0, 1)$ distribution very well, when its df ($n - 1$ here) is very large. Thus, we are usually guided to Table F (t-distribution) when $20 \leq n \leq 1000$, but occasionally to Table E ($N(0, 1)$ distribution), when $n > 1000$.

We have just reviewed this one-population case in some detail because we will proceed similarly here for linear regression inference about the slope m in order to assess whether we have statistical evidence that there is a linear relationship between the random Y and the nonrandom x. To do so, we will use a standardized test built up from the estimated slope. Similar to the standardized T-statistic used in the case of $\mathrm{H}_0 : \mu = \mu_0$, we will be led to a standardized estimator of the regression slope using the estimator's standard error, and then to the t-distribution and occasionally to the standardized normal distribution, depending on how large the sample size is.

To carry out the hypothesis test for the slope, as for any other statistic of interest we find the P-value, also called the observed level of significance. This is the probability that, if the null hypothesis were true, the statistic would be at least as extreme as its observed value. The meaning of "extreme" depends on which alternative hypothesis we have chosen. As was learned in Chapter 11, there are three possible alternative hypotheses:

$$H_{1a}: m > 0, H_{1b}: m < 0, H_{1c}: m \neq 0.$$

Which one is appropriate depends on the investigation. Sometimes the investigator can assert the one-sided alternative (H_{1a}: $m > 0$) because it is known that if x has any influence on Y at all, it causes Y to increase. For example, since increasing the fertilizer level x should not decrease the crop yield Y, only H_{1a} needs to be considered. Similarly, the investigator may be interested only in evidence that $m < 0$. For example, in the Key Problem for Chapter 3, since it cannot be that heavier cars get better gas mileage, we would choose H_{1b}.

Recall that a standardized test statistic is given by

$$\frac{\text{statistic} - \text{parameter (under } H_0\text{)},}{\text{SE of statistic}}$$

Thus, the standardized test statistic for the slope is

$$T = \frac{\hat{m} - 0}{\text{SE}(\hat{m})}$$

where $m = 0$, the hypothesized value under H_0.

There are various settings that can occur. First, it is possible that the regression SD σ could be known to the practitioner. However, this occurs so rarely in typical regression analysis settings that we will always assume σ is unknown.

Sometimes a fourth assumption is justified, namely that $Y \mid x$ is normally distributed for every x. Thus, using assumptions 2 and 3 to get the mean and SD of the normal distribution.

$$Y \mid x \text{ is distributed } N(mx + b, \sigma)$$

can be assumed when the fourth assumption is justified. We will consider three settings below. The first two are the large-sample setting where the Y's are not necessarily normal and the small-sample setting where the Y's are judged to be normally distributed. The third setting is where there is no assumption of normality and no assumption about sample size, just the basic three regression assumptions. In particular, n could be small and the Y's not necessarily normal.

The Large-Sample t-Test Approach

large-sample t-test of $m = 0$

Suppose n is large, say $n \geq 20$. The standardized test statistic T can be shown to be approximately t-distributed with $n - 2$ degrees of freedom (**d.f.**) under H_0. Therefore, we can do a large-sample t-test by using the t-table in Appendix F to find the probability that a t-distributed random variable T is at least as extreme as the observed value t_{OBS} of the statistic T. The P-value is, then, approximately

$$\begin{array}{ll} P(T > t_{\text{OBS}}) & \text{for } H_{1a}; \\ P(T < t_{\text{OBS}}) & \text{for } H_{1b}; \text{ or} \\ \left.\begin{array}{l} 2P(T > t_{\text{OBS}}) \text{ if } t_{\text{OBS}} > 0 \\ 2P(T < t_{\text{OBS}}) \text{ if } t_{\text{OBS}} < 0 \end{array}\right\} & \text{for } H_{1c}. \end{array}$$

using $n-2$ d.f. Note that the two-sided H_{1c}: $m \neq 0$ requires multiplication by 2 in order to correctly compute the P-value. A decision of whether to reject the null hypothesis, H_0, is then made based on the P-value, usually using $P \leq 0.05$ as the basis for rejecting H_0. The test statistic $\hat{m}$ was standardized by using the standard error of $\hat{m}$, namely $\text{SE}(\hat{m})$.

Example 13.4 (continued)

Continuing our analysis of the Key Problem, we obtain $t_{\text{OBS}} = \hat{m}/\text{SE}(\hat{m}) = -1.193/0.243 = -4.909$. To test the hypothesis that H_0: $m = 0$ versus H_A: $m < 0$, we approximate the P-value for a t-distribution with $n-2 = 10-2 = 8$ d.f. by (using Table F)

$$p(T < -4.909) \approx P(T < -5.041) = 0.005,$$

which is very highly significant. Thus we have strong statistical evidence that there is a predictive relationship between a child's age at first word spoken and Gesell score.

small-sample-based test of $m_1 = 0$ when Y-values are normal

The Small-Sample, Normal-Distribution-Based t-Test Approach

In many regression problems, it is reasonable to make a fourth regression model assumption, namely that Y, given x, is normally distributed for every given x. So the statistician sometimes needs to decide when it is reasonable to assume the Y's are normal. Suppose the Y's are normal. For fixed x, standardizing and using regression model assumption 3,

$$\frac{Y - E(Y \mid x)}{\text{SD}(Y \mid x)} = \frac{Y - E(Y \mid x)}{\sigma}$$

will be a standard normal random variable. Since the estimated regression line $\hat{Y}$ value at x estimates $E(Y \mid x)$,

$$\frac{Y - E(Y \mid x)}{\sigma} \approx \frac{Y - \hat{Y}}{\sigma}$$

Thus, roughly, $(Y - \hat{Y})/\sigma$ will be $N(0,1)$ or, equivalently, Y-$\hat{Y}$ will be roughly $N(0,\sigma)$ where the assumption of normally distributed Y-values is valid. Thus the assumption of normality of the Y-values amounts to asserting, approximately, that the residuals $Y - \hat{Y}$ are sampled from the same normal population with mean 0 and standard deviation σ regardless of the value of x. As a rough check of the validity of this assumption, we can draw a normal probability plot, as explained in Section 7.4.

Assuming random sampling of random variables from a normal population, we know from Section 10.2 that their sample mean, when standardized using its theoretical mean and standard error, has a t-distribution. A similar analysis based on the three regression model assumptions, the assumption of Y's being normal, and the assumption that the null hypothesis $m = 0$ is true, shows that the statistic

$$T = \frac{\hat{m} - m}{\text{SE}(\hat{m})} = \frac{\hat{m}}{\text{SE}(\hat{m})}$$

has a t distribution with $n-2$ degrees of freedom.

Therefore, based on these assumptions, using the t-table in Appendix F we can compute under H_0 the P-value according to

$$\begin{array}{ll} P(T > t_{\text{OBS}}) & \text{for } H_{1a}; \\ P(T < t_{\text{OBS}}) & \text{for } H_{1b}; \text{ or} \\ \left.\begin{array}{l} 2P(T > t_{\text{OBS}}) \text{ if } t_{\text{OBS}} > 0 \\ 2P(T < t_{\text{OBS}}) \text{ if } t_{\text{OBS}} < 0 \end{array}\right\} & \text{for } H_{1c}. \end{array}$$

where t_{OBS} is the observed value of T, and T is t-distributed with $n-2$ degrees of freedom. By contrast, for a given level of significance, say 0.05, the critical value of the test (recall one rejects H_0 if the statistic is more extreme than the critical value) can also be found from Appendix F. For example, we have $n = 20$ for our Key Problem. Assuming normal data, the critical value for testing $m = 0$ versus $m < 0$ at the standard 0.05 level of significance can be found to be $t(18) = 1.734$, based on the t distribution with $n - 2 = 18$ degrees of freedom, shown in Figure 13.11. That is, we reject the null hypothesis provided that $T < -1.734$ is observed. Since $T = -5.0 < -1.734$, we reject H_0 at the 0.05 level. Indeed, note from Table F for a t-distributed random variable that $P(T < -5.0) < P(T < -3.922) = 0.0005$, which provides us much stronger evidence—namely very high significance—than just noting that we reject at $\alpha = 0.05$! In fact, the P-value is 0.000 rounded to three significant figures, according to more complete t-distribution tabulations.

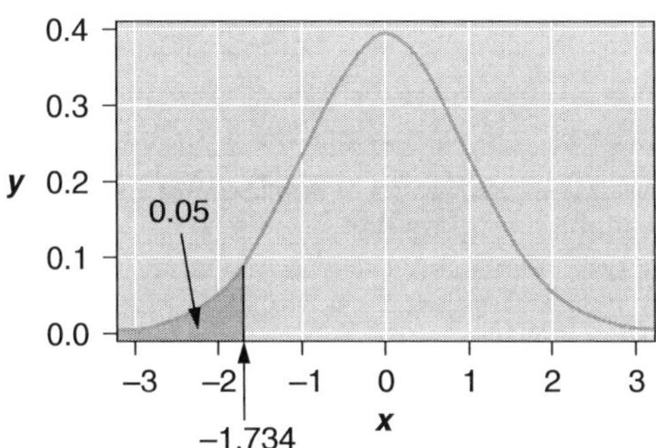

Figure 13.11 t distribution with 18 degrees of freedom and showing left tail with area 0.05.

regression effect

When the null hypothesis (H_0) is rejected, we say that there is a **regression effect.** That is, we have strong statistical evidence that the regression line slope $m \neq 0$. Thus, the hypothesis test tells us that knowledge about x can help predict Y, using the regression line. But it does not tell us how accurately we can do such prediction, which we will take up shortly. Measures like r^2 and the residual SD help provide some quantitative answers in this regard.

If the null hypothesis (H_0) is not rejected, we say that there is no evidence of a regression effect. That is, we are not sure whether the observed positive or negative slope $\hat{m}$ is merely due to random noise in the data or due to a theoretical $m \neq 0$ that we cannot detect.

simulation-based test of no relationship between x and Y

The Five-Step Simulation–Based Permutation Test Approach

The textbook emphasizes doing statistical inference (such as hypothesis testing, confidence intervals) using box model simulation methods. Indeed, simulation-based inference is an important part of the modern statistician's toolbox. In particular, there is an effective simulation-based hypothesis test of whether there is a predictive linear relationship for bivariate (x, y) data. This approach does not require either that the sample size be large or that the Y's be normal. Of course, it can be applied when either one of or both of these assumptions are true. However, if the sample size is small ($n < 20$) and if there is no evidence that the Y-values are normally distributed, then this five-step permutation test is the only valid choice for us.

Consider bivariate data, namely an (x, y) scatterplot. Suppose there is no indication of a nonlinear pattern. Then the question of whether x can successfully predict Y is the question of whether the least squares regression line has a nonzero slope m. Recall that the estimator of m is $\hat{m} = rS_Y / s_X$. We know from Chapter 5 that a sample mean will be close to its theoretical mean and a sample SD will be close to its theoretical SD when the number of independent trials (the sample size) is large. In the same manner, when the number of sampled (x, y) pairs (the sample size), such as all the father-son pairs in England in Example 13.2, is large, then the correlation parameter, which we denote by ρ—or sometimes $\rho(X, Y)$—and called the theoretical correlation of X and Y, satisfies a version of the law of large numbers in that the sample correlation r satisfies

$$r \approx \rho$$

if the sample size is large.

We note that the parameter ρ is defined only when both X and Y are random. However, when x is fixed and only Y is random, r is still defined and measures the linear association between the two variables regardless of whether x is random or x is fixed. That is, r measures the strength of a possible underlying linear relationship between X and Y (or x and Y). The five-step procedure presented below uses the statistic r. The procedure works in exactly the same manner in both the cases of fixed x and random X. For simplicity we present the procedure for the case of random X's, where the parameter ρ makes sense as the correlation between X and Y.

The issue of whether x can be effectively used to predict Y has been presented above as the issue of whether slope $m = 0$, namely that there is no underlying linear relationship. But this issue can be equivalently expressed when sampling from a bivariate (X, Y) population as whether $\rho = 0$, which also means there is no underlying linear relationship. Indeed, theoretical reasoning tells us that $\rho = 0$ if and only if $m = 0$, as is somewhat intuitively clear. Thus we seek a test of $H_0 : \rho = 0$ versus one of the alternative hypotheses

$$H_{1a} : \rho > 0, H_{1b} : \rho < 0, \text{ or } H_{1c} : \rho \neq 0.$$

Since ρ is estimated by r, when r is large in magnitude we have strong statistical evidence that $\rho \neq 0$. By contrast, when r is small in magnitude, such evidence is weak or totally lacking. Thus, we can base a test of whether X can successfully predict Y (using least squares linear regression) on the size of r. The idea is to reject H_0 if r is sufficiently large that it could not easily occur when $\rho = 0$. The five-step method presented in the example below does assume that the three regression assumptions hold.

Unlike the two hypothesis-testing procedures presented above, where we knew the probability law for the test statistic $\hat{m}$ when the null hypothesis H_0 was true, here we have not studied the distribution of the test statistic r when H_0 is true. However, five-step simulation leads us out of this dilemma, as the following example demonstrates.

Example 13.5 Crickets, Anyone?

Crickets make their chirping sounds by rapidly sliding one wing over the other. They do this very quickly about four times to produce a single chirp. Scientists have shown that crickets move their wings faster in warm temperatures than in cold temperatures. Data from one experiment designed to show the relation between wing speed and ambient temperature are shown in Table 13.2. The pulse rate provided in the table is a measure (obtained electronically) of the cricket wing speed. The table gives the pulse rate and the temperature recorded on 15 different days. A scatterplot of the data is given in Figure 13.12.

Table 13.2 The Cricket Data

Pulses per second	14	20	16	20	18	17	16	15	17	15	16	15	17	16	17
Temperature (°F)	76	89	72	93	84	81	75	70	82	69	83	80	83	81	84

Source: George Pierce, *The Songs of Insects* (Cambridge, Mass: Harvard University Press, 1949), pp. 12–21.

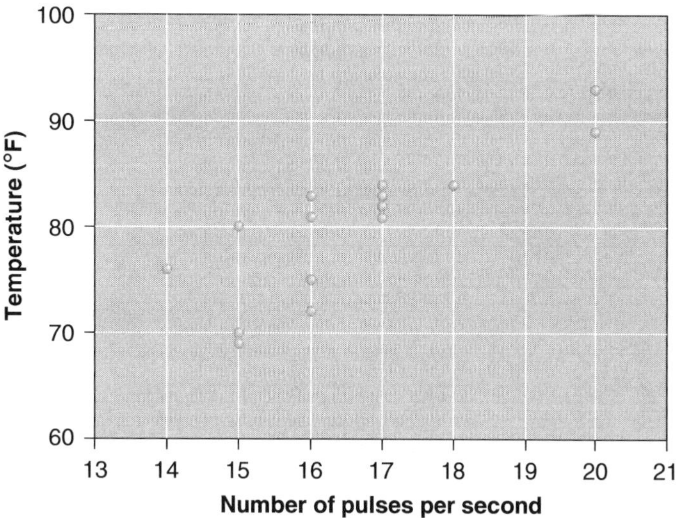

Figure 13.12 Cricket data.

Although nobody would ever use cricket chirps to predict the temperature, it is nonetheless quite interesting that there appears to be a predictive linear relationship. Perhaps the relationship is caused by crickets being cold blooded and hence their muscles move more slowly when the temperature is low.

The usual least squares regression analysis produces $\hat{Y} = 26.74 + 3.216x$. Also the observed $r = 0.823$. A skeptic might say that there is no relationship between the cricket pulse rate and the temperature. The least squares line suggests that there may be a relationship, but maybe the fact that y appears to increase with x is a random accident in the data, as when a fair coin produces 8 heads in 10 tosses. The positive slope may be just due to chance. After all, there are only 15 data points! If the skeptic is correct, then knowing the number of pulses per second x does not help at all in predicting the temperature y. That means if we were given a very large number of (x, y) data points, we would find no evidence that the true regression line slope is greater than 0. In fact, the positive slope linear pattern in Figure 13.11 would seem to strongly argue that $\rho > 0$. But we will carry out the test anyhow, to statistically evaluate our strong impression.

We wish to test

$$H_0 : \rho = 0 \quad \text{versus} \quad H_A : \rho > 0.$$

Because $n < 20$ and we have no reason to assume that the Y-values are normally distributed, we have the five-step permutation test (explained below) as our only defensible option. The basic idea of this test is that when there is a positive relationship between x and y (the alternative hypothesis) then larger x-values tend to be paired with larger y-values. By contrast, when there is no relationship between the x's and y's, then, given the 15 (x, y) pairs actually observed for each observed x, any value among the 15 observed y-values is just as likely as any other to have been paired with that x.

permutation test

The idea of the **permutation test** is based on this just-presented idea. We randomly assign the 15 observed y-values to the 15 observed x-values, providing a new data set of 15 (x, y^*) pairs generated under the null hypothesis assumption of no relationship between the x- and y-values. (Think of drawing the 15 observed y-values randomly without replacement. Then assign the first y-value drawn, y_1^* to the first x value, x_1, and so on.) Then we can compute the sample correlation r^* from this new (generated) data set, thus producing an estimator of ρ for a shuffling of the y's producing (x, y^*) values that are just as likely as the observed (x, y) points under the assumption that $H_0 : \rho = 0$ is true. If we were to repeat this simulation process many times, we could then find

$$\hat{P}(\text{simulated } r^* \geq \text{observed } r \mid H_0 \text{ true}).$$

If this P-value is sufficiently small, this will constitute strong evidence that $\rho > 0$, that is, that we should reject $H_0 : \rho = 0$. We now carry out this permutation test via our five-step simulation approach.

Step 1. Choose a Model The goal is to generate 15 pairs: $(14, y_1^*), (20, y_2^*), \ldots, (17, y_{15}^*)$ where the y^*-values are randomly chosen from the 15 observed y-values. Thus, the box model is the 15 sampled y-values.

Step 2. Define One Simulation We randomly sample 15 times *without replacement*, obtaining $y_1^*, y_2^*, \ldots, y_{15}^*$, which are used to form the 15 pairs shown in step 1. We obtained

y^*	84	81	69	80	72	76	84	83	81	89	75	83	93	82	70

Thus the first such bivariate sample of 15 simulated (x, y^*) values is

x	14	20	16	20	18	17	16	15	17	15	16	15	17	16	17
y^*	84	81	69	80	72	76	84	83	81	89	75	83	93	82	70

Note that the order of the x's is unchanged.

Step 3. Define the Event of Interest (EOI) Using the (x, y^*) simulated pairs, one computes the statistic of interest, the sample correlation coefficient r^*. In the given simulation, $r^* = -0.22$ as compared with the observed $r = 0.823$. The event of interest is $(\text{simulated } r^* \geq \text{observed } r) = (\text{simulated } r^* \geq 0.823)$, which does not happen for this simulation.

Step 4. Repeat Simulations One repeats the simulation a large number of times, say 1000, as we did, or even 10,000 or 50,000.

Step 5. Compute the *P*-Value (the Experimental Probability of the EOI) $N = 1000$ simulations were done. The result was

$$\hat{P}(\text{EOI}) = \hat{P}(\text{simulated } r^* \geq 0.823) = \frac{N_{\text{EOI}}}{N} = \frac{2}{1000} = 0.002 \leq 0.01,$$

where N_{EOI} denotes the number of times EOI occurs. Thus the null hypothesis is soundly rejected with high significance, and it is concluded that one can predict temperature from cricket chirps.

Confidence Interval for m

Confidence interval for m when Y-values normal

Often, instead of merely testing whether the slope m is zero, we are interested in inferences about its size. For example, how much increase in the wheat yield is likely as we try various concentrations of fertilizer? This suggests computing a confidence interval for m. Recall from Chapter 10 that the general form of a confidence interval (CI) for a parameter is

$$\text{statistic} \pm (\text{critical value}) \cdot (\text{SE of the statistic}),$$

where the statistic is a good estimator (that is, approximately unbiased and with a small standard error) of the parameter. We can consider both the normally distributed case and the large-sample case.

Normally Distributed Case

Suppose it is known that the Y-values are normally distributed at each x. If the true slope is m, then under our usual regression model assumptions (as we saw for hypothesis testing of $m = 0$) the standardized statistic

$$T = \frac{\hat{m} - m}{\text{SE}(\hat{m})}$$

is t-distributed with $n - 2$ degrees of freedom. Based on this, analogous to what we learned in Section 10.2 for the t distribution–based CI for μ using $\overline{X}$ and $\text{SE}(\overline{X})$, we can obtain a $C\%$ CI for m:

$$\hat{m} \pm t(n-2) \cdot \text{SE}(\hat{m})$$

or

$$\hat{m} - t^*(C, n-2) \cdot \text{SE}(\hat{m}) \leq m \leq \hat{m} + t^*(C, n-2) \cdot \text{SE}(\hat{m}),$$

where $t^*(C, k)$ is the needed confidence interval value obtained using the t distribution with k degrees of freedom as given in Table F as shown in Figure 13.13 for $C = 95\%$ and $n = 20$, *with* T being $t(n-2)$-distributed.

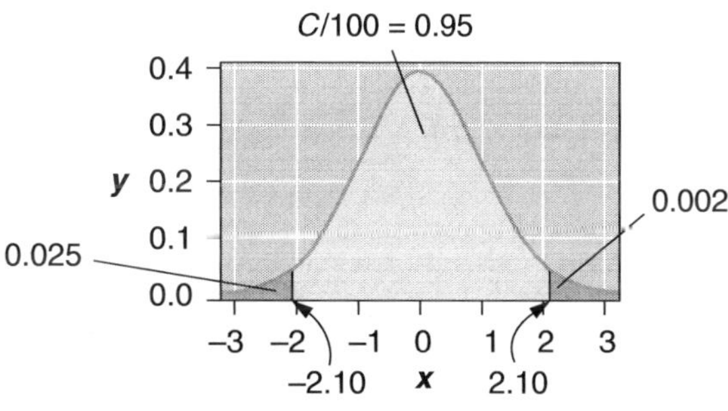

Figure 13.13 t-distribution with 18 degrees of freedom and showing 95% of the area between −2.10 and 2.10.

Example 13.6

Returning to our Key Problem, suppose an expert on the Gesell test informs us that for children of any fixed age, it is a sound modeling assumption that Gesell test scores are normally distributed. Then we can obtain a 95% CI for the slope parameter using *the* t-distribution-based CI given above. Recall that $\hat{m} = -1.2$ and $\text{SE}(\hat{m}) = 0.24$. We obtain:

$$-1.2 - (2.10 \cdot 0.24) < m < -1.2 + (2.10 - \cdot 0.24)$$

or

$$-1.704 < m < -0.696.$$

Here, $t^*(95\%, n-2) = t^*(95\%, 18) = 2.10$ from Figure 13.13 or from the t-table in Appendix F. Since the 95% CI lies entirely on the negative side of 0, this is another statistical indication that the slope of the linear regression between the age at first word spoken (x) and Gesell adaptive score (Y) is negative. However, we now also know it is very likely (approximately 95%) that m lies between −1.7 and −0.7.

Large-Sample Case

Confidence interval for m when sample size large

Of course, if $n \geq 20$, then as long as the distribution of Y's does not have outliers and is not too skewed, then

$$\frac{\hat{m} - m}{\mathrm{SE}(\hat{m})}$$

is approximately $t(n-2)$-distributed. Thus, the above confidence interval formulation applies here too; note, however, that the resulting confidence interval level C supplied in part by the computed t-table probabilities is, as usual, only approximately correct.

Section 13.1 Exercises

Assume normal population sampling as needed in the following exercises.

1. The following table records the times required to cook turkeys of various weights. We are interested in the relationship between the weight of a turkey (x) and required cooking time (Y).

Weight (pounds)	Cooking time (hours)
5	3.5
7	4
10	4.4
14	5.6
18	6.7
22	8.4

   a. Give the least squares regression equation for $\hat{Y}$ at x.
   b. Give the standard error of the slope estimate $\hat{m}$.
   c. Suppose the assumption of normality of Y given x is valid. Give a 90% CI for the slope m.
   d. Test at the 10% level of significance whether the cooking time increases with the weight of a turkey. Do this in two distinct ways: one with normality of the Y assumed, and one without.
   e. Do part (c) assuming the normal population is not justified (using the textbook's software).

2. Good runners take more steps per second as they speed up. Below are the average numbers of steps per second for a group of women runners at different speeds. The speeds are in feet per second.

Speed (ft/s)	15.8	16.9	17.5	18.7	19.2	20.2	22.5	24.0
Steps per second	3.1	3.2	3.2	3.3	3.4	3.4	3.6	3.8

   a. We want to predict steps per second from running speed for a particular athlete who is not part of this study. Give a prediction equation. If her speed is 23.0 feet per second, predict her number of steps per second.

b. When her speed increases by 1 foot per second, how many more or fewer steps per second do you expect the runner to take?
c. Give the standard error of your estimate in part (b).
d. Assuming normal population sampling, test at the 5% level of significance whether the model is worthwhile for prediction. (*Hint*: It seems safe to assume that $m < 0$ cannot occur.)
e. Without assuming normality, do the 5% level hypothesis test of part (d).

3. The following table gives the math exam scores for pairs of twins from eight families. Test at the 5% level of significance to determine whether there is any correlation between the scores of set of twins. State your null hypothesis and alternative hypothesis and your assumptions. (*Hint*: Recall that the issue of whether there is a nonzero correlation is equivalent to whether $m \neq 0$.)

Score of twin 1	75	86	67	88	76	56	60	92
Score of twin 2	66	75	70	78	82	66	71	88

4. The t-statistic for testing whether the slope parameter is zero has another expression based only on the correlation coefficient r between x and Y. It is

$$T = \frac{\sqrt{n-2}}{\sqrt{1-r^2}} r.$$

Use this formula to recalculate T for the problem in Exercise 3.

5. The data in the following table are the number of push-ups y that could be done by a sample of 12 male instructors at Howard Community College. (The 38-year-old teacher was a track coach.) Assume normal population sampling for the Y-values. Find a 95% CI for the slope parameter m when regressing the number of push-ups on the age of the instructors.

Age	Number of push-ups	Age	Number of push-ups
21	10	22	9
25	8	27	6
22	11	44	4
28	6	48	3
30	7	35	8
38	15	48	5

6. In some cases we may wish to fit a model line to the data (x, Y) that is known to pass through the origin. The equation of the line then takes the form of $\hat{Y} = mx$. The estimated least squares–based slope can be computed by

$$\hat{m} = \frac{\sum xY}{\sum x^2},$$

and its standard error is

$$\mathrm{SE}(\hat{m}) = \frac{S_E}{\sqrt{\sum x^2}},$$

where the residual standard error S_E is now defined by $S_E = \sqrt{\Sigma(Y - \hat{Y})^2/(n-1)}$, where the division is by $n - 1$ (this is theoretically correct) and not $n - 2$ as was done in Section 13.1. If the distribution of Y given x is normal with mean $E(Y \mid x) = mx$ and the constant residual standard deviation σ holds, it can be shown that

$$\frac{\hat{m} - m}{\text{SE}(\hat{m})}$$

has a t distribution with $n - 1$ degrees of freedom (note that $df = n - 1$, not $n - 2$). Give a formula for a $100(1 - \alpha)\%$ CI for m.

7. Consider the *Consumer Reports* numerical ratings of 77 cereals and the number of grams of sugar in each serving. (*Data source:* Free publication available in many grocery stores. Data set available through the *Statlib Data and Story Library* (DASL).) A high rating suggests the cereal is healthy. The correlation between the two variables is $r = -0.760$, suggesting a strong negative association. A scatterplot of the two variables indicates a linear relationship. Suppose $\hat{m} = -2.40$, $\hat{b} = 59.3$, and $\text{SE}(\hat{m}) = 0.2373$.
   a. What is the regression line?
   b. Find a CI for m.
   c. Test the hypothesis H_0: $m = 0$ versus H_1: $m < 0$.

8. A study was done to see if there is a positive correlation between the number of times per month that college students call home and the amount of money that their parents contribute toward their education. A total of 175 students were surveyed and the sample correlation was found to be $r = 0.18$.
   a. Test $H_0 : \rho = 0$ versus $H_1 : \rho > 0$ using a large-sample approach.

9. Refer to the cricket chirping Example 13.5. Suppose you do the usual regression analysis for y = temperature and x = pulses per second and you predict the temperature based on 19 chirps per second. State two reasons why your prediction is likely to be a poor one. (*Hint*: Both reasons concern problems in effectively predicting y from x when certain things are true. What things are true here?)

13.2 PREDICTION INTERVAL (PI) FOR REGRESSION-BASED PREDICTION OF Y GIVEN x AND CONFIDENCE INTERVAL (CI) FOR ESTIMATION OF THE LINE $E(Y \mid x)$

In this section we will learn how to construct confidence intervals (CIs) for the value of the estimated regression line at x, that is, a CI for $E(Y \mid x) = mx + b$, where x is a specified number. We will learn to construct a CI for a predicted new value (not yet observed) of Y at x where the prediction also uses the already estimated regression line. That is, we will construct a CI for Y at x where Y has not yet been observed. For example, we might be predicting the age of the wife of Bob, who is 42.

It is important to understand that $E(Y \mid x)$ is a population mean. For example, in the Key Problem, $E(Y \mid 15)$ is shorthand for $E(Y \mid x = 15)$ and is the mean of all the Y's (Gesell adaptive scores) in the population of children who utter their first word at approximately 15 months. This is very different from predicting Zoe's Gesell score knowing that her first word was uttered at $x = 15$ months.

Both the CI for $E(Y \mid x)$ and the CI for the newly observed Y at x will be built from the estimated regression line

$$\hat{Y} = \hat{m}x + \hat{b}.$$

We will consider the more basic problem of constructing a CI for the parameter $E(Y \mid x)$ first. We will see that predicting Zoe's score should have a much larger typical error than estimating

$E(Y \mid x = 15)$, which is the true regression line's point at $x = 15$ represents the expected or average score of all children for whom $x = 15$.

CI Estimate of the Mean Value of Y Given x (denoted either $E(Y \mid x)$ or $\mu_{Y|x}$)

We introduce the notation $\mu_{Y|x}$ because use of the symbol μ for a theoretical mean is almost universal.

Suppose that in our Key Problem we would like to estimate the average Gesell adaptive score for the *population* of all children who utter their first word at $x = 30$ months. That is, we want to estimate $E(Y \mid x = 30)$, which we denote by $\mu_{Y|x}$ for convenience and to stress that it is a population mean, as just discussed. Based on the least squares line $Y = 109.3 - 1.2x$, we estimate $\mu_{Y|30}$ by

$$\hat{\mu}_{Y|30} = 109.3 - (1.2 \cdot 30) = 73.3.$$

How accurate is this estimate? We can answer this question by calculating a 90% CI for $\mu_{Y|30}$. We solve for this CI under the assumption that the Y-values (at each x) are normally distributed. As one might anticipate from Section 13.1, a probability argument can establish that the desired CI is based on the t-distribution. As usual, the CI uses our general form of a CI,

$$(\text{statistic}) \pm (\text{critical value}) \times (\text{SE of the statistic}).$$

Specialized to our regression setting of estimating $\mu_{Y|x}$ this yields

$$\hat{\mu}_{Y|x} - t^*\text{SE}(\hat{\mu}_{Y|x}) \leq \mu_{Y|x} \leq \hat{\mu}_{Y|x} + t^*\text{SE}(\hat{\mu}_{Y|x}),$$

with the critical value t^* found in a t-table with $n - 2$ degrees of freedom. Here it can be proven that the standard error of $\hat{\mu}_{Y|x}$ is given by the formula

$$\text{SE}(\hat{\mu}_{Y|x}) = S_E\sqrt{\frac{1}{n} + \frac{(x-\overline{x})^2}{\sum(x-\overline{x})^2}} = S_E\sqrt{\frac{1}{n} + \frac{(x-\overline{x})^2}{s_x^2(n-1)}},$$

As usual, S_E denotes the residual standard error computed as part of the regression analysis and n the sample size. Thus, the $C\%$ CI for $\hat{\mu}_{Y|x}$ is given by

$$\hat{\mu}_{Y|x} - t^*(C, n-2)S_E\sqrt{\frac{1}{n} + \frac{(x-\overline{x})^2}{s_x^2(n-1)}} \leq \mu_{Y|x} \leq \hat{\mu}_{Y|x} + t^*(C, n-2)S_E\sqrt{\frac{1}{n} + \frac{(x-\overline{x})^2}{s_x^2(n-1)}},$$

CI for $\mu_{Y|x}$ when Y-values normal

where $t^*(C, k)$ is the $C\%$ CT value with k degrees of freedom from Table F of the t-distribution.

For the Key Problem, we have $\hat{\mu}_{Y|30} = 73.3$ at $x = 30$, $\overline{x} = 14.25$, $s_x^2 = 66.09$, $n = 20$, $t(18) = 1.734$, and $S_E = 8.6$. Here

$$\sqrt{\frac{1}{n} + \frac{(x-\overline{x})^2}{s_x^2(n-1)}} = \sqrt{\frac{1}{20} + \frac{1}{5.06}} = 0.4976,$$

Thus, a 90% CI for the expected score Y at $x = 30$ (for $\mu_{Y|30}$) is

$$73.3 \pm 1.734 \cdot 8.6 \cdot 0.4976;$$

that is, the interval 73.3 ± 7.41, or $(65.9, 80.7)$. Thus, based on the amount of data we have available (only 20 children), there is a fair amount of uncertainty in estimating $\mu_{Y|30}$, noting we are really trying to estimate one point on the true regression line. If we had 200 children instead of 20, then, if for simplicity SE and s_x^2 take the same values, we would have obtained the much more accurate interval 73.3 ± 2.30, or $(71, 75.6)$. Our estimate of the regression line $E(Y \mid x)$ will be very accurate if n is very large (e.g., for the totally impractical $n = 10,000$, we get 73.3 ± 0.33), since $1/n$ and $1/(n-1)$ both are almost 0 for very large n. Thus, we can estimate the $E(Y \mid x)$ of the regression line as accurately as we wish by having enough data points n—a valuable insight!

Predicting an Individual New Observation, Y_{new}

Given an estimated regression line $\hat{Y} = \hat{m}X + \hat{b}$, we may attempt to predict a "new" observation Y at x, which we denote as Y_{new} to emphasize that it is an (as yet) unobserved random variable. This variable is one whose unknown value played no part in the computation of $\hat{b}$ and $\hat{m}$ used to produce the estimated line. For example, we could try to predict the temperature given a cricket pulse rate of 18 per second. A natural predictor of $\hat{Y}_{\text{new}}$ given x is $\hat{Y}_{\text{new}} = \hat{m}x + \hat{b}$. This is also the estimate of the regression line at x, usually denoted by $E(Y \mid x)$. Thus, throughout this section, our prediction of a new Y at x is the *same* as our estimate of the regression line at x. An observed value of the new random variable Y_{new} will have $\mu_{Y|x}$ as its $\mu_{Y|x}$ as its theoretical mean, but clearly it will randomly deviate from $\mu_{Y|x}$. Thus, even if we know the true mean $E(Y \mid x)$ of the regression line with great accuracy, we still cannot know exactly what Y_{new} at x will be. (Recall how much Y deviated about the estimated regression line in Figure 13.1.) Clearly, predicting Y_{new} at x accurately may not be possible even if estimating $\mu_{Y|x}$ accurately is possible.

In the Key Problem we use $\hat{Y} = \hat{m}X + \hat{b}$ as the prediction of an individual childs random response Y_{new} at x. The standard error $\text{SE}(Y_{\text{new}})$ we will observe is larger than the standard error $\text{SE}(\hat{E}(Y \mid x))$ for the regression line's mean estimator. In addition to the variability in the estimation of the regression line mean, the natural variability of Y about the mean $E(Y \mid x)$ adds random noise in the Y_{new} being predicted. This added noise is quantified by the residual standard error, SE, its true unknown value being the residual or regression SD σ.

prediction interval PI

An interval used to predict a new individual response Y_{new} is technically called a **prediction interval PI** rather than a confidence interval (CI). Note that we use "confidence interval" for the *fixed* target population mean $\mu_{Y|x}$ (the value of the true regression line at x), and we use "prediction interval" for the random target Y_{new} at x. So for the rest of this section we will not refer to a CI for predicting a new Y, but rather use the correct term of prediction interval. A C% prediction interval for Y at x is given by

$$\hat{\mu}_{Y|x} - t^* \text{SE}(\hat{Y}_{\text{new}}) \leq \hat{Y}_{\text{new}} \leq \hat{\mu}_{Y|x} + t^* \text{SE}(\hat{Y}_{\text{new}})$$

Prediction interval for Y_{new} when Y-values normal

where

$$\text{SE}(\hat{Y}_{\text{new}}) = S_E \sqrt{1 + \frac{1}{n} + \frac{(x - \bar{x})^2}{s_x^2 (n-1)}}$$

and $t^* = t(n-2)$ is the same as the t-distribution-based critical value in the CI for $\mu_{Y|x}$.

Comparing this prediction interval with the CI for $\mu_{Y|x}$, we see that the only difference is that its standard error term has the extra constant 1 added to the expression under the square root sign to account for the random variation of Y_{new} around its mean $\mu_{Y|x}$. But note that when n is large the contribution from the 1 does *not* get smaller.

For our Key Problem we have

$$\text{SE}(\hat{Y}_{\text{new}}) = 8.6\sqrt{1 + \frac{1}{20} + \frac{(30 - 14.25)^2}{66.09 \times 19}} = 9.606,$$

so a 90% prediction interval for Y_{new} at $x = 30$ is $73.3 \pm (1.734 \times 9.606)$, that is, 73.3 ± 16.66 or (56.64, 89.96). This is quite a bit wider than the 90% CI of $\mu_{Y|30}$, namely (65.9, 80.7). Increasing the number of children studied is of limited value in improving prediction accuracy. For example, using 10,000 children with S_E and s_x^2 the same yields

$$73.3 \pm 14.91,$$

a small gain over the prediction interval 73.3 ± 16.66 we found based on 20 children. That extra 1 in the square root term makes all the difference. Recall for $n = 10{,}000$ that the CI for $E(Y \mid 30)$ is 73.3 ± 0.33. Compare 0.33 with 14.91, the typical error sizes for the CI versus the PI.

There is a key lesson here: One can estimate $\mu_{Y|x}$ as accurately as one wants by collecting enough data, but the accuracy of predicting a new observation Y_{new} at x cannot be increased beyond the inherent variability of the Y's around the regression line, as controlled by the residual standard deviation σ, estimated by SE, regardless of how many data points are collected!

Note that both the CI for $\mu_{Y|x}$ and the PI for Y_{new} depend not only on SE, which is a measure of the inherent variability of Y about the true line, but also on how far the mean $\overline{x}$ is from the x-value for which the estimation or prediction is done. The farther x is from the center $\overline{x}$ of the observed x-values, the less accurate the estimate or prediction will be [note $(x - x)^2$ in the numerator under the square root sign, present in both the CI and the PI].

extrapolation

If we need to make a prediction for x beyond the range of the data (say, for children whose first word comes at less than 7 months or more than 42 months in our Key Problem), we have an **extrapolation** problem. For an extrapolation to be useful, we need sound reasons to believe that the linear relationship we are using between the two variables will continue to hold in the region of x that is now of interest, (e.g., $x \approx 7$ and $x \approx 42$). Unless we can justify such an assertion on the basis of deep knowledge of the subject matter, it is extremely risky to assume linearity outside the range of x-values actually observed.

A humorous example of faulty extrapolation is found in Mark Twain's *Life on the Mississippi*. The lower Mississippi River (from southern Illinois to New Orleans) weaves back and forth many times across its floodplain, and occasionally the water breaks through the gap between one bend and the next, jumping the old "oxbow" bend and making the river shorter overall. Twain noted that the river was 242 miles shorter than it was 176 years ago and, by assuming linearity of shortening over time, predicted that in 742 years the lower Mississippi would be only 1.75 miles long. Of course, as the river shortens, it also straightens, so oxbow jumping becomes less and less frequent! Mark Twain clearly understood (in his humorous way) the dangers of extrapolation beyond the data.

Section 13.2 Exercises

1. Refer to Exercise 1 of Section 13.1. We wish to estimate the cooking time required for turkeys weighing 12 pounds. Assume a normal distribution of Y given x.
   a. Find a 90% interval for the mean cooking time required, which the meat supplier might state on the label supplied with its turkeys. Here the meat supplier has very large number of turkeys in mind for each x-value.
   b. Find a 90% interval for the cooking time required for your turkey, which weighs 12 pounds. Is this a CI or a PI?
   c. Which of the intervals in parts (a) and (b) is wider? Why?
2. The following table gives the time (in seconds) it took for seven randomly chosen decathlon participants in the 2004 Olympic Gamess to finish the 100-meter race (x) and the 110-meter hurdles (y). We wish to use the data for the

100-meter race to predict the time it takes to complete the 110-meter hurdles among top men athletes. Assume normal population sampling.

x	11.14	10.55	10.89	10.89	11.06	11.10	11.29
y	14.37	14.39	14.56	14.77	14.78	15.38	15.54

a. For decathlon athletes (imagine thousands of them) who take 11 seconds to finish the 100-meter race, how long do you expect it takes them on average to finish the 110-meter hurdles? Give a 90% interval for this mean time. Is your interval a CI or a PI?
b. For one athlete who just finished the 100-meter race in 11 seconds, give a 90% interval for the time it will take to finish the 110-meter hurdles. Is your interval a CI or a PI?
c. Examine the scatterplot and the residual plot and comment on the validity of your interval.

3. In part (c) of Exercise 1, if we increase the sample size to 25 but assume the same standard error of the residuals, how wide will the 90% prediction interval be at $x = \overline{x}$?

4. Assume normal population sampling of (X, Y) values. In a linear regression analysis with 16 observed of (x, y) pairs, we find the five sums $\sum x = -1.6, \sum y = 0.6, \sum x^2 = 14.0, \sum y^2 = 38.0$, and $\sum xy = 13.0$.
a. Find the 95% CI for the mean value of Y when $x = -0.1$ (the value of $\overline{x}$).
b. Do part (a) for $x = 2.0$ and compare the two intervals, commenting on the comparison.
c. If the range of the x's is from -2.5 to 2.2, can we trust a prediction of y at $x = 2.0$?

5. Assume a large sample. A study to find the mail survey response rate among the elderly (ages 60 years to 90 years) estimated the following relationship between age (x) and percentage of people responding (y):

$$\hat{Y} = 90.2 - 0.6x.$$

a. What is the estimated percentage of those responding to a mail survey among 75-year-olds?
b. Find a 95% confidence interval for the percentage responding among 80-year-olds. Assume that the standard error of the slope is 0.14 and that $n = 200, \overline{x} = 68, s_x = 5$.

6. The following table provides the approximate percentage of the state's vote for Candidate Senator Kerry in 2004 and Candidate Senator Obama in 2008.

State	WY	KS	AZ	NY	ME	CT	IL	RI	MA
Kerry (%)	30	38	45	49	55	56	56	61	63
Obama (%)	33	42	45	55	59	61	63	65	63

a. Regress the Obama percentages on the Kerry percentages. Graph the resulting line and the points from the table.
b. Compute the regression SE, that is, S_E. Interpret the accuracy of your regression line.
c. The Ohio percentage for Kerry was 49%. Find a 95% prediction interval (PI) for the Ohio Obama vote. (FYI: Obama got 52% of the Ohio vote.)
d. Find the CI for an Obama percentage given a typical state percentage of (a) 35% and (b) 50%. Which interval is wider and why?

*13.3 APPLYING REGRESSION TO NONLINEAR RELATIONSHIPS BY TRANSFORMING THE VARIABLES

So far we have learned how to model a linear relationship, make inferences about the slope parameter, and use the resulting linear equation to predict a new Y or to estimate $E(Y \mid x)$. Not all bivariate relationships are linear, as we learned in Chapter 3. However, some nonlinear regression models can be converted into linear regression models by a simple transformation of one or both variables, as we also learned in Chapter 3. Reviewing the Section 3.3 material on nonlinear relationships is a prerequisite to reading this section. Now we continue the Section 3.3 investigation of nonlinear relationships with a series of worked-out examples. This section requires a good working knowledge of algebra and in particular of logarithmic and exponential functions.

Example 13.7 (a good data collection example)

Nonlinear power relationship example

Suppose we have measurements of the volume (v) and circumference (c) of several roundish pebbles–some flattened (pancake-shaped, roughly cylindrical) and some fairly spherical—collected in one area. Note that rounded pebbles or gravel are usually somewhat flattened and hence not spherical. To measure the circumference of a random pebble, one can take a string or thread and wrap it around the circular portion of the pebble. (For example, the pancake-shaped stone would have the string going around its circular edge.) One can argue that the underlying relationship between these two variables is of the form $v = Ac^B$, where, as we know from studying geometry, the exponent B is equal to 3 if the stones are roughly spherical. For a set of flattened pebbles, $B < 3$ seems likely. Our goal is to estimate A and especially B for an average pebble from the area being sampled from.

Let's do some algebraic analysis to better understand the problem. First, consider stones that are roughly spherical. For simplicity, we will assume they are exactly spherical. Then the length of the string, regardless of where it will wrap around a stone, will be c (circumference) as the string's length. Recall from geometry, with r being radius, that

$$v = \pi r^3 \quad \text{and} \quad c = 2\pi r; \quad \text{thus} \quad r = c/2\pi.$$

Therefore,

$$v = \pi \left(\frac{c}{2\pi}\right)^3 = \frac{1}{8\pi^2} c^3$$

Thus $A = 1/(8\pi^2)$ and, importantly, $B = 3$. Thus, if stones are roughly spherical $B \approx 3$.

Consider the other extreme of pancake-shaped (approximated as a cylinder) stones, all 1/4 inch thick for simplicity. For simplicity, assume they are all perfectly circular. Then, because these are roughly cylinders (if the edges were not rounded, they would be perfect cylinders),

$$v = (\text{height}) \times (\text{area}) = \frac{1}{4}\pi r^2 \quad \text{and} \quad c \approx 2\pi r$$

Thus

$$v = \frac{1}{4}\pi \left(\frac{c}{2\pi}\right)^2 = \frac{1}{16\pi} c^2$$

Thus, using $v = Ac^B$, $A = 1/(16\pi)$ and, importantly, $B = 2$. Thus, for stones that are flattened (and all about the same thickness) $B \approx 2$. Hence, $2 \leq B \leq 3$ "seems" to be the case. But one cannot simply do a linear regression to solve for $\hat{A}$ and $\hat{B}$, because the relationship $v = Ac^B$ is not a linear one.

Recall from Section 3.3 that we can transform one or both of the variables to obtain a linear relationship. So x is transformed to x' and y is transformed to y' (or maybe just one is transformed). Then even though y is not linearly related to x, $y' = b + mx'$. So next one does the linear regression on the (x', y') terms, estimating b and m. Then, one solves for the nonlinear relationship y in terms of x using the estimated b and m.

To try to find a linear model resulting from the transformation of one or both of the variables, and noting that the relationship between v and c is a power relationship (recall the Section 3.3 discussion), we take the logarithm (i.e., the natural logarithm) of v and see what happens to the equation. Then, recalling the general rule for logarithms that $\ln(uw) = \ln(u) + \ln(w)$ and that $\ln(u^w) = w\ln(u)$, this yields, recalling $v = Ac^B$,

$$\ln(v) = \ln(A) + B\ln(c).$$

Thus we get a linear equation by defining $y' = b + mx'$, where $x' = \ln(c)$, $y' = \ln(v)$, $b = \ln(A)$, and $m = B$. Here $\ln A$ is the natural logarithm *of* A. We have succeeded in converting the original nonlinear equation into a linear one, as desired. Given the transformed (x', y') data, we can fit the least squares line to the (x', y') points and get an estimate of the intercept b and the slope m using the standard least squares equations of Chapter 3. We can then solve for $\hat{A}$ and $\hat{B}$ in the original model ($\hat{A} = \exp(b)$, $\hat{B} = \hat{m}$). The pebble problem is an interesting data collection exercise for one to try.

❒ ❒

The natural logarithm (ln) transformation is often helpful in producing a linear equation for a positive variable v that increases or decreases exponentially fast with the other variable u. Many situations display such so-called exponential growth or decay. The model for such exponential change is

$$v = AB^u$$

with $B > 0$ resulting in exponential growth and $B < 0$ exponential decay. Try the special case of $v = 2^u$ for $u = 1, 2, 3, 4, \ldots$, to see how fast exponential growth can be! [Do not confuse this with $v = u^2$ or $v = u^3$ (power growth, as in the pebble example), where v grows much more slowly.] Variables such as the early growth (v) in size or weight over time (t) of a living organism having ample nutrients, or the number (v) of organisms produced over time (t) if conditions are ideal for the breeding of the organism, often display exponential growth. (Charles Darwin once talked about the world being rapidly covered with elephants if such exponential growth were possible for elephants.) Another example is money that increases in value by a fixed percentage every time period, with the profits reinvested (so-called compound interest). For a given data set, examination of its scatterplot is helpful in determining whether the growth in y as x varies displays exponential growth, as was discussed in Section 3.3.

Consider the exponential growth equation

$$v = AB^u.$$

Again, do not confuse this with the visually similar $v = Au^B$ we considered in the preceding pebble example. Let us take the logarithm of y in the above exponential growth equation and see what happens to the equation:

$$\ln v = \ln A + u\ln B.$$

This result shows that we can linearize the new equation by transforming the data and parameters as follows: $y' = \ln(v), b = \ln(A), x' = u, m = \ln(B)$. This yields

$$y' = b + mx',$$

which is linear, as desired.

We will use this approach in the next example.

Example 13.8

Table 13.3 gives the annual sales in billions of dollars (s) and return on assets (a) for seven telecommunications companies in the United States for the year 1999. (Of course the world of telecommunications companies had become very different in a couple years!)

Table 13.3 Sales and Returns for U.S. Companies in 1999

Company	Sales (s)	Return on assets (a)
AT&T	64.1	2.7%
MCI WorldCom	38.1	5.5%
Bell South	25.7	9.3%
Sprint	17.2	8.7%
GTE	25.6	8.5%
Net2Phone	0.5	35.5%
IDT Corp.	1.0	19.8%

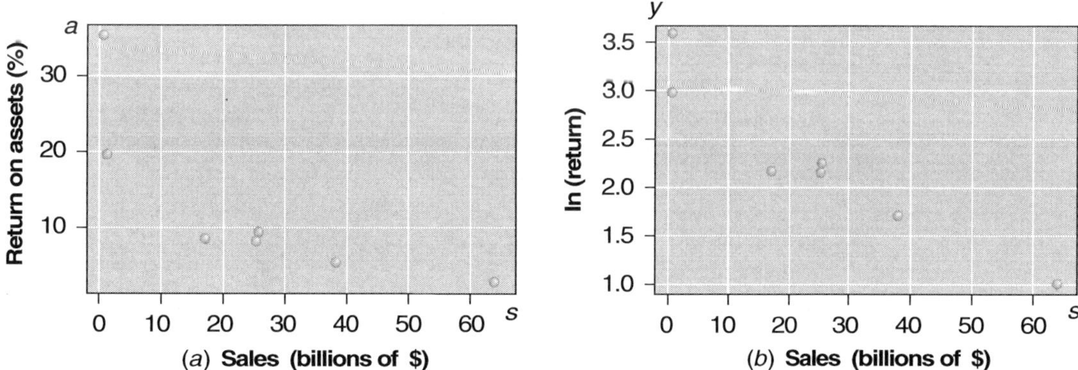

Figure 13.14 Return on assets versus sales for telecommunication companies.

Figure 13.14(a) gives the scatterplot of the data, indicating higher returns on assets for companies with lower annual sales. The increase in returns for companies as sales decline [moving from right to left in Figure 13.14(a)] is faster than a linear trend would suggest. In fact, it is more like a power or exponential rate of increase as *sales decrease*. The graph in Figure 13.14(a) has increasing sales as the horizontal axis, so we have a negative rate of return on assets (denoted by a) rather than growth as sales (denoted by s) increase. Somewhat arbitrarily, we'll explore the possibility of exponential decay, rather than a negative power decay, assuming $a = AB^s$ as the model. As argued above, the possible exponential decay in a as s increases suggests that we take the natural logarithm transformation of the a variable and not transform the s variable. We do so and examine the scatterplot to see whether it appears to be linear. The scatterplot of $y = \ln(\text{return}) = \ln a$ (vertical axis) versus sales $= s$ (horizontal axis) given in Figure 13.14(b) does have points lying roughly in a line. This suggests a linear trend and leads us to the linearized model

$$\ln a = \ln A + s \ln B, \quad \text{producing } y' = b + mx'$$

where $y' = \ln a$, $b = \ln A$, $x' = s$, and $m = \ln B$.

The standard least squares regression calculations give $\hat{b} = 3.14$ and $\hat{m} = -0.036$ with $r^2 = -0.91$; that is, x' does a good job of linearly predicting y', as the (x', y') scatterplot shows. Moreover, the nonlinear exponential decay equation proposed, $a = AB^s$, gives $\hat{A} = \exp(\hat{b}) = exp(3.14) = 23.10$ and $\hat{B} = \exp(\hat{m}) = \exp(-0.036) = 0.965$. Thus our least squares estimated equation expressed in terms of the original variables a and s becomes

$$a = 23.10 \times (0.965)^s.$$

We note that taking a power approach ($a = As^B$) should work well too—sometimes there is no unique transformation that is clearly best. Because the exponential growth approach worked well, we won't try a power approach.

❐ ❐

The following example shows how in a power relationship ($v = Au^B$) approach we can actually find the power that best fits the data. It is also a nice example of how the removal of an outlier can greatly increase the correlation r, even when the outlier is not a leverage point (see Chapter 3).

Example 13.9

Figure 13.15 plots the average planets temperature $t = {}^\circ\text{F} + 360^\circ$, where °F is the planets average Fahrenheit temperature, against d, the average distance from the Sun in tens of millions of miles, for the nine planets. Earth is about 92,900,000 miles from the Sun, so its $d = 9.29$. (See Table 13.4 for the actual data.) The temperature has been converted to $t = {}^\circ\text{F} + 360^\circ$ so that the apparent horizontal asymptote becomes the horizontal axis. This amounts to approximating the lower asymptote of planet temperatures by −360°F. (It is very cold on the outer planets that are far from the Sun.)

Clearly, the farther a planet is from the Sun, the colder it is. The line shown is the least squares regression line. It loosely tracks the points, but one can see that the points are not close to the line; rather, they clearly have a curved pattern. The correlation $r = 0.62$, which measures the degree of fit of the data to a straight line, is well below 1, even though the curved relationship

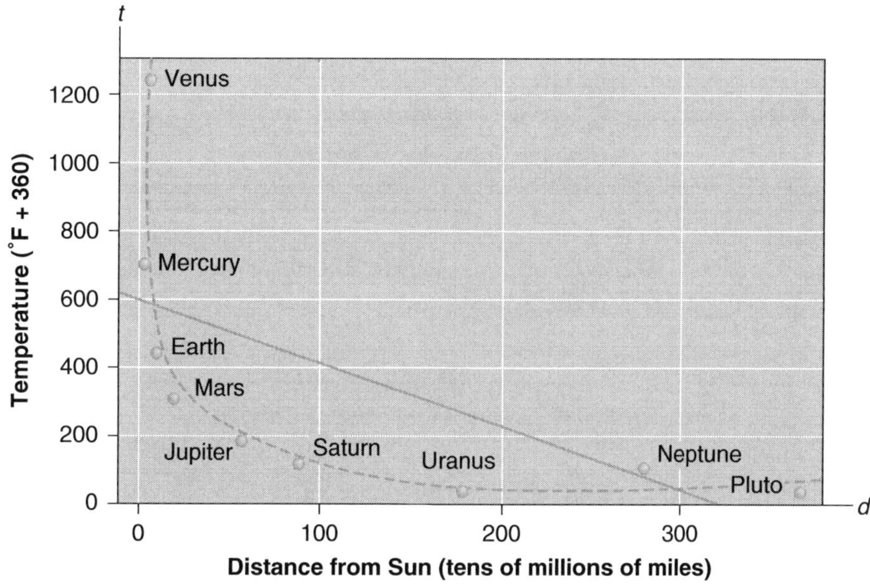

Figure 13.15 Temperature as a function of distance from the Sun.

shown by the dashed curve produces an excellent fit. It is ingrained in casual statistical thinking to view r as a measure of how well the response variable (here t) can be predicted by the explanatory variable (here d). But r is really a measure of how well the response variable can be predicted by a *straight-line* equation involving the explanatory variable. Hence, the value of r, if moderately large, can be very deceiving if it is erroneously interpreted as measuring how well the underlying (but possibly curved) best-fitting relationship can fit the data.

Table 13.4 Temperature as a function of Distance from the Sun

	d = Distance (tens of millions of miles)	Temperature (°F)	$t = (°\text{F} + 360°)$
Mercury	3.596	332	692
Venus	6.720	854	1214
Earth	9.290	59	419
Mars	14.15	−67	293
Jupiter	48.33	−162	198
Saturn	88.67	−208	152
Uranus	178.20	−344	16
Neptune	279.30	−261	99
Pluto	366.40	−355	5

The (d, t) data in Figure 13.15 appear to be asymptotic to both the vertical and horizontal axes. As noted in Section 3.3, this suggests a power relationship for (d, t) of the form

$$t = Ad^B,$$

although one could just as reasonably try an exponential decay relationship, as we did in Example 13.8. As stated in Section 3.3, we can convert the problem to a linear regression by taking the logarithm of t:

$$\ln t = \ln(Ad^B) = \ln A + B \ln d$$

The result of such a linear regression analysis, which expresses $y' = \ln t$ as a straight-line relationship with the explanatory variable $\ln d$, is

$$\widehat{\ln t} = 8.249 - 0.88(\ln d), \quad r = -0.87,$$

where r is the sample correlation between $\ln t$ and $\ln d$.

Figure 13.16 shows this linear regression. The fit does not look extremely good, although the magnitude of $r = 0.87$ is much greater than the previous magnitude of $r = 0.62$ for the poorly fitting linear regression of Figure 13.15.

Let's solve the equation for t using algebra (it is good algebraic exercise to work out the details). The above equation becomes

$$\hat{t} = 3823.8 \times d^{-0.88} = 3823.8x'$$

where we let $x' = d^{-0.88}$, with the sample correlation between x' and t being $r_{x',t} = 0.78$.

This linear relationship between t and x' is plotted in Figure 13.17. Using this rescaled set of axes, Venus is clearly an outlier, with the remaining points lying roughly along a line. Note here that the distance increases from right to left because of how the axis is scaled as a function of $d^{-0.88}$. This suggests that we redo the linear regression analysis of

$$\widehat{\ln t} = \ln A + B \ln d$$

analysis with $x' = \ln d$ and $y' = \ln t$ but with Venus removed. This yields, proceeding just as above, t with Venus removed. Proceeding just as above, this yields,

$$\ln t = 7.966 - 0.84$$

which, using algebra as before, becomes

$$\hat{t} = 2881.3x'$$

where $x' = d^{-0.84}$ with the sample correlation between x' and t being $r_{x',t} = 0.97$. This linear relationship between t and $d^{-0.84}$ with Venus removed as an outlier is graphed in Figure 13.18.

The regression line fits the data extremely well, as $r_{x',t} = 0.97$ indicates. Of course, our real goal was to estimate the nonlinear relationship of t with d, not t with $x' = d^{-0.84}$. The nonlinear equation as a function of the explanatory variable d is

$\hat{t} = 2.881.3d^{-0.84}$

and is plotted in Figure 13.19. This is the desired solution of our regression analysis (recall $t = T + 360°$, where T is each plant's true Fahrenheit temperature).

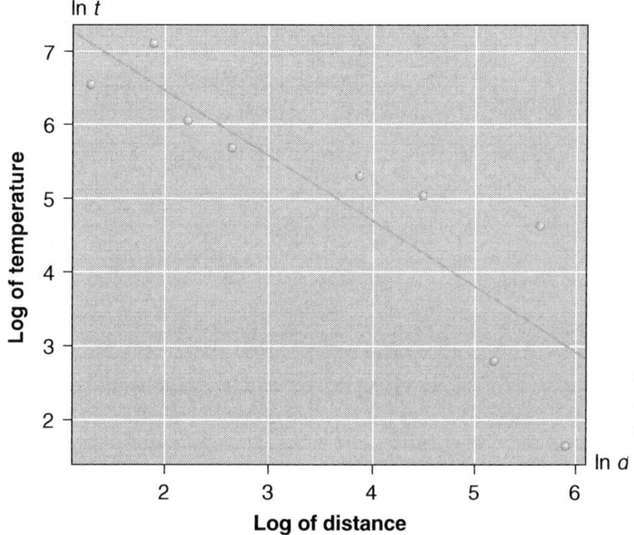

Figure 13.16 Linear regression of $\ln r$ versus $\ln d$.

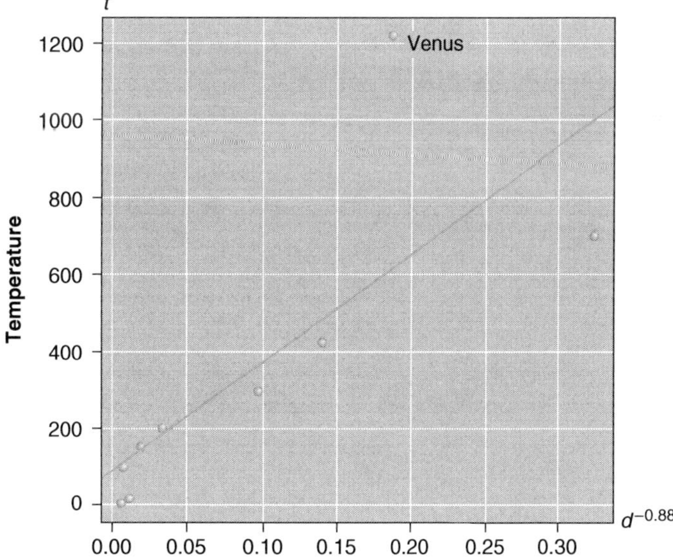

Figure 13.17 Linear relationship between t and $d^{-0.88}$.

Since the most distant planet, Pluto, satisfies $t \approx 0°$ as shown on the graph, its actual temperature is solved to be $0 = T + 360°$ and hence $T = -360°$. Venus was not used in deriving the final equation. Of course, acting as a scientist, one needs to explore why all of the planets except Venus fit this relationship so very well. Of course, Pluto was recently demoted from being one of the planets of our solar system to being only a "dwarf planet."

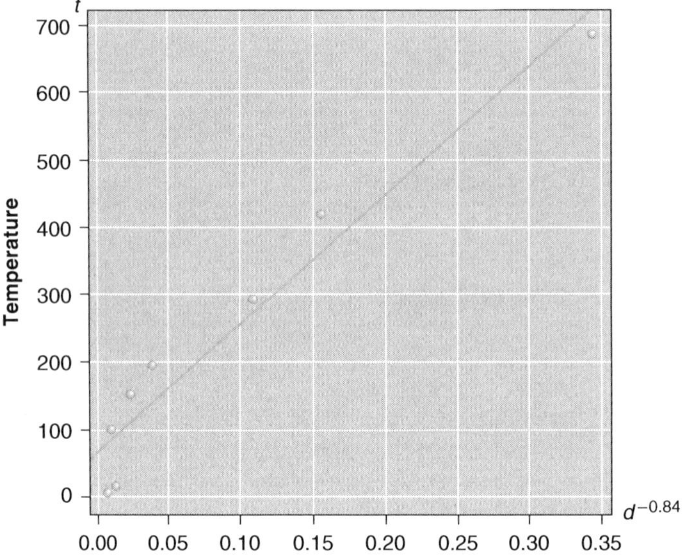

Figure 13.18 Linear relationship between t and $d^{-0.84}$ with Venus removed.

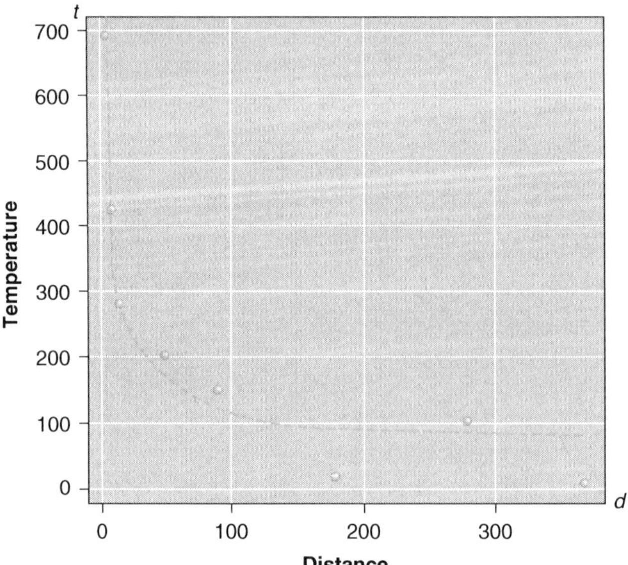

Figure 13.19 Estimated nonlinear relationship of t with d.

❒ ❒ ❒ ❒ ❒ ❒ ❒ ❒ ❒ ❒ ❒ ❒ ❒ ❒ ❒ ❒ ❒ ❒ ❒ ❒

It is important to realize that not all nonlinear models can be linearized by transformations that allow us to fit a nonlinear model using a linear regression analysis on the transformed variables. For example, $v = (A + u)^B$ cannot be linearized by transforming u and v. Indeed, we approximated the lower asymptote in Example 13.9 by −360°F because trying to estimate that lower asymptote would have made the problem impossible to linearize. Even the simple and often occurring quadratic relationship

$$v = A + Bu + Cu^2$$

cannot be linearized. Fortunately there are easily applied and effective methods to fit such nonlinear models learned in the area of statistics called "nonlinear regression."

Section 13.3 Exercises

1. A researcher asked a number of subjects to compare notes of various decibel (sound) levels against a standard (set at 80 decibels) and to assign each of them a loudness rating. Subjects were told to assign the standard note of 80 decibels a loudness rating (L) of 10.

Stimulus (d)	30	50	60	70	75	80	85	90	95	100
Response (L)	0.2	1.0	3.0	5.0	8.5	10.0	14.0	20.0	29.0	43.0

a. Is a linear regression between d and L appropriate?
b. Consider the natural logarithm transformation of L. Is a linear regression appropriate between d and $\ln(L)$?
c. Give a 95% prediction interval for L_{new} when $d = 55$. [*Hint*: You can get a 95% prediction interval for $\ln(L_{\text{new}})$ first.]

2. Refer to Exercise 2 of Section 13.2. If we are interested in the speed of the athletes in the 100-meter race (call it s) and in the 110-meter hurdles (call it h), do you think a linear relationship exists between s and h? If not, can you find a transformation that allows you to find a prediction equation for h?

3. The following table gives the length of the planetary year in days (y) and the distance from the Sun in hundreds of millions of miles (x) for the nine planets of the Solar System.

x	0.360	0.672	0.929	1.415	4.833	8.867	17.820	27.930	36.640
y	88.00	224.70	365.26	687.00	4332.60	10759.20	30685.40	60189.00	90465.00

a. Find the least squares estimate for regressing y on x.
b. Examine the residual plot. Do you see a nonlinear pattern?
c. Perform a linear regression of y on $x^{1.5}$ (i.e., $y = a + bx^{1.5}$) and examine the residual plot.
d. Compare the standard error of the residuals S_E of the two regressions.
e. Compare $100r^2$% for the two regressions.

4. A product is manufactured in batches at a certain factory. The batch size can be varied (within a certain range). A machine manufactures the product at a constant rate. Before the production of a new batch begins, the machine has

to be set up. The setup time of the machine can be assumed to be constant. A certain number of units of the product were produced each day for the last 14 days. The batch sizes, however, were different on each of these 14 days. The following table gives the details regarding the batch size and the total time (setup time plus production time) required for production.

a. Regress total time on batch size.
b. Regress total time on the reciprocal of batch size [1/(batch size)].
c. You will find that the second regression is better than the first. Give reasons why the second regression is superior.

Day	Batch size (number of units)	Total time (minutes)	Day	Batch size (number of units)	Total time (minutes)
1	10,000	795	8	24,000	675
2	12,000	760	9	26,000	690
3	14,000	710	10	28,000	675
4	16,000	730	11	30,000	680
5	18,000	680	12	32,000	660
6	20,000	705	13	34,000	660
7	22,000	680	14	36,000	655

5. *Elasticity of demand* is a useful microeconomics concept. Elasticity of demand of a product measures the percentage change in the quantity bought in response to a 1% change in the price of the product. The following table shows the quantity of a certain product demanded at various prices. Estimate the elasticity of demand using regression, after appropriate transformation of the variables.

Price	Quantity demanded
25	38
28	31
31	31
33	28
36	27
40	25
42	24
45	23

CHAPTER 13 SUMMARY

The regression model assumptions are that the observed Y-values are independent, $E(Y \mid x) = b + mx$, and the standard deviation of Y given x, $\mathrm{SD}(Y \mid x)$, does not change with x and hence is a constant, denoted by σ. σ is called the **residual SD** or **regression SD.** The residual SD is well estimated by

$$s_E = \sqrt{\frac{\Sigma(Y - \hat{Y})^2}{n - 2}},$$

called the **standard error of the residuals** or the **standard error of regression.**

The standard deviation of $\hat{m}$ is given by

$$SD(\hat{m}) = \frac{\sigma}{s_x \sqrt{n-1}},$$

with s_x denoting the sample SD of the x's and σ being the regression SD, which is estimated by SE. Thus, the SE of $\hat{m}$ is given by

$$SE(\hat{m}) = \frac{S_E}{s_x \sqrt{n-1}}.$$

The estimated slope $\hat{m}$ is an unbiased estimator of m; that is, $E(\hat{m}) = m$. The desirable outcome of $SE(\hat{m})$ being small can be accomplished by increasing n, decreasing S_E, or increasing s_x, although S_E cannot usually be decreased, even by modifying the experiment.

The hypothesis H_0: $m = 0$ (no linear relationship between x and Y) can be tested by a large-sample t-test, a small-sample t-test (provided the Y-values are normally distributed), or a permutation test carried out via five-step simulation (always allowed).

A $C\%$ **confidence interval for m** when the Y-values are normally distributed or when n is large is given by

$$\hat{m} \pm t^*(C, n-2)SE(\hat{m})$$

A **confidence interval for $E(Y \mid x)$** and a **prediction interval for a new Y-value (denoted Y_{new}) at x** are both possible when the Y-values are normally distributed or when n is large.

CI (letting $E(Y \mid x)$ be denoted by $\mu_{Y|x}$):

$$\hat{\mu}_{Y|x} - t^*(C, n-2)SE(\hat{\mu}_{Y|x}) \leq \hat{\mu}_{Y|x} \leq \hat{\mu}_{Y|x} + t^*(C, n-2)SE(\hat{\mu}_{Y|x})$$

where

$$SE(\hat{\mu}_{Y|x}) = S_E \sqrt{\frac{1}{n} + \frac{(x-\overline{x})^2}{s_x^2(n-1)}}$$

Prediction interval for a "new" Y:

$$\hat{\mu}_{Y|x} - t^*(C, n-2)SE(\hat{Y}_{new}) \leq \hat{Y}_{new} \leq \hat{\mu}_{Y|x} + t^*(C, n-2)SE(\hat{Y}_{new})$$

where

$$SE(Y_{new}) = S_E \sqrt{1 + \frac{1}{n} + \frac{(x-\overline{x})^2}{s_x^2(n-1)}}$$

Various nonlinear relationships, but not all, can be analyzed using linear regression, often by first taking the logarithm of x or of y or of both. Often a careful multi-step analysis, where at each step the results are evaluated for possible improvement, can produce a much better fit and increasingly large r^2.

CHAPTER REVIEW EXERCISES

1. Nine households are surveyed for their monthly expenditures on food (y). The following table gives the data together with the number of adults in each household (x).

x	2	3	3	5	6	6	6	6	6
y	220	234	297	304	419	481	569	651	652

   a. Find the least squares line to estimate y from x (either estimating $\mu_{Y|x}$ or predicting Y_{new}). Is the model significant at the 5% level of significance, assuming normal population sampling?
   b. Find a 95% CI for the expected (average over many people) monthly food expenditure for all households with four adults, assuming normal population sampling.
   c. Find a 95% prediction interval for the monthly food expenditure for a particular household with four adults, assuming normal population sampling.

2. Refer to Exercise 1.
   a. If we randomly change the order of the y-values before doing the regression, would we expect the model to be significant?
   b. Use the set of nine integers randomly reordered to change the order of the y-values randomly. For example, if the first chosen integer is 3, then $x_1 = 2$ is paired with $y_3 = 297$. Then perform the hypothesis test that the slope is 0 against the two-sided alternative (using either a permutation test or a normal population–based test). Does the result surprise you?

3. The following table gives population (x) and the number of telephones (y) for each of six zones. We wish to see how the population size relates to the number of telephones.

x	4041	2200	30,148	60,324	65,468	30,988
y	1332	690	11,476	18,368	22,044	10,686

   a. Does the scatterplot suggest a linear relationship between x and y?
   b. Even if the scatterplot did not suggest the need for it, consider plotting $\sqrt{y}$ versus $\sqrt{x}$. Fit a linear model ($\sqrt{y} = a + b\sqrt{x}$) using least squares ($y' = \sqrt{y}, x' = \sqrt{x}$).
   c. Give a 95% CI for the slope parameter of the model in part (b).
   d. Predict the number of telephones for a zone with population size 10,000, using the model in part (b). Give a 95% prediction interval, assuming normal population sampling.
   e. What would you get in part (d) if you used a linear model of y versus x? What would you get in part (d) for your prediction if you used a linear model of y versus x?

4. The rate of soil water evaporation (Y) depends on the wind speed of the air (x). Suppose the estimated relationship is given by the equation

$$\hat{Y} = 4.5 + 0.0084x$$

and the standard error of the slope $SE(\hat{m}) = 0.022$. Find a 95% CI for the slope. Assume that the study was based on 11 pairs of data points.

5. The following table lists the winning times for the 100-meter run event at the Summer Olympics from 1896 to 2008.

Year	Winning time (in seconds)	Year	Winning time (in seconds)
1896	12.0	1960	10.2
1900	11.0	1964	10.0
1904	11.0	1968	9.95
1908	10.8	1972	10.14
1912	10.8	1976	10.06
1920	10.8	1980	10.25
1924	10.6	1984	9.99
1928	10.8	1988	9.92
1932	10.3	1992	9.96
1936	10.3	1996	9.84
1948	10.3	2000	9.87
1952	10.4	2004	9.85
1956	10.5	2008	9.69

Regress winning time on year. Find the 95% CI for the slope, assuming normal Y's at each x. Predict the winning time for 2012. Find the prediction interval.

6. The following data show the number of promises made by political candidates and the number of these promises that were fulfilled once they were elected. Find a linear relationship between promises made and promises fulfilled. (Please do not take the result seriously! These are not "real" data.) With this in mind, test H_0: $m = 0$ versus H_1: $m < 0$.

Promises made	Promises fulfilled	Promises made	Promises fulfilled
15	11	39	3
28	8	30	9
22	9	43	4
44	1	21	3
20	8	19	9
35	4	34	5

7. The following table shows the midterm grade and the final grade obtained by each of 15 students.

Midterm grade	Final grade	Midterm grade	Final grade
90	83	82	84
80	91	99	93
98	97	66	52
72	83	89	75
92	86	37	56
97	80	53	27
56	68	95	81
90	97		

a. Regress final grade on midterm grade.
b. Calculate the standard error of the slope estimate.
c. Find the 95% CI for the slope. What are your underlying assumptions?
d. What is the P-value for H_0: $m = 0$ versus H_1: $m = 0$?

8. Regression analysis can be used to find out whether the stock market has *momentum* (i.e., whether the market is likely to go up/down today because it went up/down yesterday). Do the following exercise: Note the Standard & Poor's 500 stock index for the last month. For each day, calculate the "daily return" as (Today's index – Yesterday's index) / (Yesterday's index). For each day's return (call it Today's return) find the corresponding previous day's return (call it Yesterday's return). That is, pair off two consecutive days' returns for every pair of days [(Day 1 return, Day 2 return), (Day 2 return, Day 3 return), etc.]. Test the hypothesis that today's return is not related to yesterday's return. If this null hypothesis is not rejected, then the stock market displays no statistical evidence of momentum. If the null hypothesis is rejected, there are two possibilities: (1) A positive relationship between the variables indicates market momentum. (2) A negative relationship indicates that the market has a tendency to "overreact" one day and correct itself the next day. (Be sure you understand this interesting concept of overreacting.)

9. Suppose the amount of oxygen consumed (Y) during the physical exercise period (x, in minutes) based on $n = 12$ is estimated as

$$\hat{Y} = 585 + 96.5x.$$

a. For a 10-minute exercise, what is the prediction of the amount of oxygen consumed by a particular athlete?
b. If the standard error $\text{SE}(\hat{m})$ of the slope is 1.5, find a 95% CI for the average amount of oxygen consumed during a 10-minute workout, assuming normal population sampling, and assuming $\overline{x} = 12$, with $s_x = 4$.
c. Find a 95% prediction interval for the amount of oxygen consumed by a person during a 15-minute workout, assuming normal population sampling, and assuming $\overline{x} = 12$, with $s_x^2 = 4$.

10. Regression is a very useful method for finding growth rates. The following table shows the population of the United States for different years.

Year	U.S. population
1909	90,490,000
1919	104,514,000
1929	121,767,000
1939	130,879,718
1949	149,188,130
1959	177,829,628
1969	202,676,946
1979	225,055,487
1989	246,819,230
1999	272,690,813
2009	305,529,237

Source: U.S. Census Bureau.

a. Plot the population of the United States on the y-axis and the year on the x-axis. What is the natural shape of the graph you observe?
b. If you plot the natural log of population on the y-axis and the year on the x-axis, what is the shape of the graph you observe?
c. What is the equation you would use in estimating the population growth rate? [Take a hint from part (b).]

Glossary

absolute value The magnitude or size of a number, disregarding whether the number is negative or positive. For positive numbers, the absolute value is simply the value of the number. For negative numbers, the absolute value is the value of the negative number without the minus sign.

acceptance/rejection hypothesis testing Hypothesis testing in which the two possible decisions are to reject the null hypothesis or to accept the null hypothesis. Contrasted with **significance testing** where one never accepts the null hypothesis.

actual (real) population See **population, actual (or real).**

addition rule The fact that $P(A \text{ or } B) = P(A) + P(B)$ when the events are mutually exclusive and $P(A \text{ or } B) = P(A) + P(B) - P(A \text{ and } B)$ in general.

additive two-way ANOVA model A two-way ANOVA model that assumes that no interaction is present.

alternative hypothesis The hypothesis that specifies the values of the *population parameter* that are possible when the *null hypothesis* fails; it can be one-sided or two-sided. Usually corresponds to a real-world research hypothesis. Denoted H_A. (See H_A.)

analysis of variance (ANOVA) A body of statistical methods for determining the influences of various postulated explanatory categorical variables (often called *factors*) on a continuous *response variable*, such as the influences of wheat variety and soil type on crop yield.

ANOVA See **analysis of variance.**

ANOVA table The standard tabular format for presenting the results of an analysis of variance. Also called *analysis of variance table*.

area fallacy A graph consisting of objects representing frequencies in which the relative areas of the objects are not proportional to the corresponding relative frequencies. Usually caused by making heights of objects proportional to relative frequencies and drawing the objects to scale, with the result that taller objects are also drawn as wider. The resulting distortion of the objects magnifies frequency differences via areas that are visually misleading.

arithmetic average Also called *average*. The sum of a set of numbers divided by the quantity of numbers in the set; same as the **mean** or **sample mean** of the set of numbers.

average Technically, the mean. Informally used to indicate the middle of a set of numbers. Use of "mean" over "average" preferred.

back-to-back stem-and-leaf plot Compares the stem-and-leaf plots of two data sets by using a common set of stems but two opposite facing sets of leaves.

bar chart A graph consisting of bars (rectangles) that count the number of data points in each of several nonnumerical categories. Also see **Pareto chart.**

believability The fact that a valid statistical inference can be trusted and relied on, as contrasted with the fact that an invalid statistical inference cannot be trusted.

bell-shaped distribution A histogram of data (or probability density function) that is shaped like a Gaussian (or normal) density.

Bernoulli trials The trials of a binomial experiment.

between-samples sum of squares In a one-way ANOVA, the sum of squared deviations, each such deviation being a sample mean minus the overall sample mean of all the observations. Each such squared deviation is multiplied by its sample size, and then these are added. Also called a *treatment sum of squares* and denoted SSTR. Measures variation in means between populations.

between-samples variability In analysis of variance, variability of observations between different samples caused by differences among the population means of the samples. Also called *treatment variability.*

bias Systematic error, as contrasted with random error.

biased estimator An estimator of a population parameter that systematically underestimates or systematically overestimates the parameter. For example, $\sum(X_i - \bar{X})^2/n$ systematically underestimates σ^2. By formula, $\hat{\theta}$ is a biased estimator of θ if $E(\hat{\theta}) \neq \theta$.

bimodal data set A data set with two **clusters** of high concentration, such as durations of Old Faithful geyser eruptions.

binomial distribution (or binomial pdf) The probability model for the number of successes x in a fixed number n of two-outcome ("success", "failure") trials that are independent and are of equal probability p; the distribution of a binomial random variable X is computed as

$$P(X = x) = \binom{n}{x} p^x (1-p)^{n-x} \qquad x = 0, 1, \ldots, n$$

binomial experiment Successive independent trials with a fixed number of trials and with two possible outcomes (designated 0 and 1 for convenience) per trial with $p(1) = p$ the same for all trials. If the statistic of interest is the number of $1s$, then this leads to the **binomial distribution.**

bivariate data Data involving two variables, such as height and weight, or amount of smoking and a measure of overall health; often graphed in a **scatterplot.**

blocking Converting a one-variable statistical design into a two-variable statistical design by introducing a second factor (variable) into the experimental design judged likely to contribute substantially to the observed variability. Levels of the .rst factor are randomly assigned within homogeneous "blocks" of the second factor.The blocking factor is usually a varying property of the units of the experiment, such as soil type of the various soil plots used in a study of the yields of different varieties of wheat. Commonly used design for ANOVA studies.

Bonferroni method In hypothesis testing of more than two populations, a method to simultaneously compare pairs of population parameters (often population means) when the hypothesis of the joint equality of all the parameters has been rejected. The confidence-interval-based paired comparisons produced by the method satisfy a specified overall statistical confidence percentage of being *simultaneously* correct.

bootstrap The technique of random resampling from a given sample, done in practice with replacement, but done in this textbook .rst without replacement using an extensive cloning of the given sample to produce a population-sized box to resample from. This was done to help students understand why bootstrapping works. From each bootstrapped sample drawn, the statistic of interest is computed. Because the given sample should be shaped approximately like the unknown population, the bootstrap-produced experimental distribution of the statistic of interest (as given by its density histogram) should be a good estimate of the unknown theoretical distribution of the statistic of interest. Many estimation and hypothesis-testing procedures can be constructed using bootstrapping.

bootstrap CI, standard error-based Usual CLT-based CI for θ using $\text{SE}(\hat{\theta}) = \widehat{\text{SD}}(\hat{\theta}^*)$, the bootstrap-based estimate of SD $(\hat{\theta})$.

bootstrap hypothesis testing Uses bootstrap resampling of the sample, but with the sample modified so that the null hypothesis holds for it.

bootstrapping to obtain the standard error of a statistic of interest Uses usual bootstrap resampling to generate a large number of bootstrapped copies of the statistic of interest, then computes their SD to estimate SE(statistic of interest).

box modeling (or box model) A method of simulating data by randomly drawing a number from a box repeatedly either with or without replacement between draws, depending on the application. The box may contain any number of real numbers, some or all of which may appear more than once. For example, the box containing 1, 2, 3, 4, 5, and 6 simulates the throwing of a fair die. The box also may be defined by supplying a list of numbers, each therefore having equal probability of being sampled from the box. Finally, one can specify a distribution for the box model, such as normal, binomial, and so forth. In order to carry out bootstrapping, the sample often becomes the box model.

boxplot A graphical way to display data that shows a box with the first and third quartiles as ends, with lines (*whiskers*) extending from the ends to the maximum and minimum values in the data set, and a vertical line through the middle of the box representing the median. Used to show the spread of a data set. Also called a *box-and-whisker plot* or a *five-number boxplot*.

boxplot, outlier A boxplot that also displays outliers.

categorical variable or data A variable or data that falls into a finite number of (often nonnumerical) categories; the emphasis is on the frequency or proportions in each category (an example is the number of men and women in a college class).

causality Causality between bivariate variables x and y holds when it is known that varying x causes a change in the value of y. A correlation, even when large, between x and y does *not* imply causation (although causation is often the reason for the correlation).

cause-and-effect relationship An (x,y) relationship where a change in the x-value *causes* a change in the y-value.

census An attempt to sample every member of a population.

central limit theorem (CLT) The result that the theoretical probability distribution of $\overline{X}$ is approximately normal with mean μ and standard deviation $\sigma / \sqrt{n}$ when the sample size n is large, where μ and σ are the population mean and SD, respectively, of the random sample that $\overline{X}$ is computed from. Applies to a simple random sample if sample size $n \geq 20$ and $n \leq 0.05N$, where N is the population size. Applies also to $\hat{p}$.

central tendency Measures of central tendency indicate the middle or center of a set of data. Important measures of central tendency include the sample mean and the sample median.

chance variation The random variation displayed by a random variable or statistic in a setting where the outcomes are, or seem, random. See **random**.

chi-square density The theoretical curve used to calculate chi-square probabilities via areas under the curve (pdf) for chi-square hypothesis testing; it is a continuous distribution.

chi-square goodness-of-fit test for one-way table Use of chi-square statistic to test a specific many-sided unfair die (or real-world analog) null hypothesis.

chi-square statistic $\chi^2 = \sum[(O - E)^2 / E]$, where the sum is over all categories, O is the observed frequency, and E is the expected frequency under the null hypothesis. χ^2 has a chi-square density, approximately, under the null hypothesis when the sample size is large relative to the number of categories. Applies to both one-way and two-way tables.

chi-square table A table that provides numerical values for probabilities involving right-hand tail areas under the chi-square density and that is needed for chi-square testing; provided in Table C of Appendix C.

chi-square test A statistical hypothesis test for situations in which an observed frequency and a theoretical frequency calculated from a null hypothesis are available for the possible categories and the data are from a large number of trials. A chi-square statistic (χ^2) is calculated and the null hypothesis is rejected if the statistic is large. One uses a simulation approach or uses the chi-square pdf depending on the setting.

chi-square test of homogeneity for a two-way table Use of a chi-square statistic to test whether several unfair many-sided die (or other real-world analog) have the same distribution.

chi-square test of independence for a two-way table Use of a chi-square statistic to test whether two categorical variables, such as gender and political preference, are independent or not.

cluster A portion of high concentration in a data set. Can be of inferential importance, as in the example of an unusually high number of cases of bacterial meningitis in a college town in a data set from many towns, indicating an epidemic in the college town.

coefficient of determination, r^2 The square of the correlation coefficient r of the best-fitting least squares regression line. The quantity $100r^2$ is interpreted as the percentage of the total variation in the y's around $\bar{y}$ that is explained by the best-fitting regression line.

complementary event Given an event A, the complement of A ("not A") consists of all the outcomes not in A. For example, if A is the event that two of three children are boys, then not A is the event that there are either zero, one, or three boys. The equation $P(\text{not } A) = 1 - P(A)$ is called the complement rule.

completely randomized design Assignment of treatments to units in a completely random manner. Contrasted with a randomized block design.

conceptual population See **population, conceptual**.

conditional probability The probability that one event will occur given that another has occurred. The conditional probability that event A will occur given that event B has occurred is given by $P(A \mid B) = P(A \text{ and } B) \ / \ P(B)$.

conditional relative frequency table Are-expression of a contingency table giving experimental conditional probability of each of the levels of one categorized variable given each of the levels of the other categorized variable.

confidence interval (CI) An interval computed from a random sample that is used to accurately estimate the location of some population parameter by the interval. The "confidence probability" that the random confidence interval contains the parameter is stated along with specifying the confidence interval, such as a 95% confidence interval (95% means a confidence probability of 0.95). Usually of the form: statistic±(critical value) · (SD of statistic).

confidence interval, general form of Statistic±(critical value) · SD(statistic), where SD(statistic) is often estimated either by bootstrapping or by use of a theoretical equation for SD(statistic).

confidence level The probability (typically 0.9, 0.95, 0.99) that the statistician's confidence interval contains the true, unknown population parameter. Usually stated as a percentage.

confounding Said of two or more possible variables potentially causing an observed effect on the response variable when the statistical analysis cannot determine which variable is the cause. Often a confounding variable (e.g., age) obscures the effect of a treatment variable (e.g., smoking).

contingency table A two-variable frequency table (like gender and religious preference). Variables are categorical.

continuity correction Used to produce greater accuracy when approximating a binomial distribution by the normal distribution.

continuous probability model or distribution (pdf, density) A probability law for a continuous random variable. The probability that the random variable falls within any given interval is given by the area of the interval under a specified curve called a *density* or continuous pdf. Examples include the normal and chi-square distributions.

continuous random variable A random variable whose values are continuous and flow together. This means there are no "gaps" between values. Two examples where modeling the situation with a continuous random variable is appropriate are heights of randomly selected people and measurements involving time. The probability of a continuous random variable falling on an interval is given by the area under the random variable's continuous pdf over the interval.

control group The group of individuals or "units" not receiving the treatment (perhaps receiving a placebo), allowing for a comparison between the effect of a treatment and that of no treatment.

controlled experiment Experiment in units where the treatment is compared with no treatment by having the units split into a treatment group and a control group. Does not necessarily use randomization, although very desirable to do so.

controls Members of the control group.

correlation coefficient, *r* A measure of how close two variables of a scatterplot are to being perfectly linearly related; computed by dividing the sample covariance by the product of the two variables' sample standard deviations. Also called the *Pearson correlation coefficient.*

counting random variable A random variable that counts the number of times over multiple trials an outcome of interest occurs.

covariance A measure of how closely two variables of a scatterplot are to being perfectly linearly related. A closely related but more easily interpreted measure is the **correlation coefficient.**

critical region Also called the **rejection region.** When doing acceptance rejection hypothesis testing or a test of significance with a specified level of significance, the set of values for a hypothesis-testing statistic for which we decide to reject the null hypothesis. For example, we might reject $H_0 : \mu = 0$ in favor of $H_1 : \mu > 0$ if $\overline{X} > 1.3$. This is the critical region.

critical value In a confidence interval, the distribution-based value (often Z- or t-table based) that provides the correct confidence level.

data Numerical information obtained by counting or measuring.

decision making See **hypothesis testing.**

degrees of freedom (df) (1) *For the chi-square distribution:* An integer that determines which chi-square density, and, hence, which row of the chi-square table (in Appendix C) to use. Different rules for a one-way and a two-way table apply. (2) *For the t distribution:* An integer that determines which t density, and, hence, which row of the t table (in Appendix F) to use. (3) *For the F distribution:* The F distribution has a numerator degrees of freedom and a denominator degrees of freedom that together determine which F density, and hence which entries of the F tables (in Appendix H) to use to test ANOVA or multiple-regression hypotheses.

density See **continuous distribution** and **continuous pdf.**

density histogram A frequency histogram vertically scaled so that the sum of the areas of its rectangles is 1. Used to estimate the population distribution producing the data. See also **experimental probability distribution.**

density polygon A piecewise linear graph obtained by joining the midpoints of the tops of the rectangles of a density histogram. Its shape approximates the probability law producing the data.

descriptive statistics Graphical and numerical techniques for describing or summarizing data that capture the essence of the data. As opposed to **inferential statistics.**

deviation from the (sample) mean The difference between the data value and the sample mean: $(x-\overline{x})$.

discrete probability model or distribution (pdf, density) Assigns a positive probability to each of a finite number or an infinite sequence of outcomes making up the sample space of a model.

discrete random variable A random variable whose values are discrete. This means that values have "gaps" between them. Random variables that involve counting objects (e.g., the number of heads in 10 coin tosses) are common examples of discrete random variables.

disjoint events Events with no outcomes in common. Also called *mutually exclusive events.*

distribution (1) Also called **probability distribution.** The probability law for a random variable or statistic of interest; for example, the height of a person chosen at random may follow the normal distribution. In particular, in the discrete case, a list of all possible values of a random variable together with their corresponding probabilities: usually denoted as pairs of $(x, p(x))$'s. In the continuous case, a pdf $f(x)$ for which probabilities correspond to areas under $f(x)$. (2) zAlso applies to a data set: The distribution of a data set refers to the general slope and location of the data points.

dotplot A plot of a data set on a number line where each dot locates a member of the data set. Multiples values that are the same are stacked vertically.

double-blind study A study in which neither the subjects nor the experts conducting the study and evaluating the data know who has received the treatment and who has received the placebo.

***D* statistic** A sum over all categories of absolute deviations, the $|O-E|$ values. The D statistic can be used in place of the chi-square statistic for measuring the size of the difference between the observed and expected frequencies under a null hypothesis for categorical data. $D=\sum|O-E|$ See **chi-square statistic** and **chi-square test**.

Empirical 68-95-99.7% Rule for the spread of approximately bell-shaped data States that when the distribution of the data is roughly normal (bell-shaped), approximately 68% of the data values lie within one standard deviation of the sample mean, approximately 95% of the data values lie within two standard deviations of the sample mean, and approximately 99.7% (almost all) of the data lie within three standard deviations of the sample mean (68% within $\pm s$ of $\overline{x}$, 95% within 2s of $\overline{x}$, 99.7% within $3s$ of $\overline{x}$).

equally likely outcomes A theoretical probability model, such as that for a fair die, in which every outcome has the same probability. Then $P(A)$ (number of outcomes in A)/(number of outcomes in S), where S = sample space.

error sum of squares (SSE) In linear regression, SSE is the sum of the squared vertical distances between the observed y-values and their corresponding fitted regression line $\hat{y}$ -values. It is the unexplained variation. Same as the residual sum of squares in linear regression and related to the within-samples sum of squares in ANOVA.

estimation Using data to make a data-informed inference about the magnitude of a population or probability model parameter.

estimation error $\hat{\theta} - \theta$

estimator A sample statistic that is intended to estimate a population parameter.

event A set of outcomes of interest in a probability model. For example, the event A that the number showing on a fair die is even is given as $A = \{2, 4, 6\}$. Usually the goal is to compute or estimate $P(A)$.

event (A and B) This event occurs if both A and B occur.

event (A or B) This event occurs if either A or B occurs.

event occurring An event occurs when the random experiment produces an outcome in the set of outcomes that define the event.

event of interest (when hypothesis testing) The event of interest is defined to occur if the nullhypothesis-generated value of the test statistic is as extreme as, or more extreme (in the direction providing evidence that H_A holds) than, the value of the statistic that occurred with the real data sample. Its probability is the ***P*-value.** Often denoted **EOI.**

effect size Size of the treatment effect in a randomized experiment.

exclusive events Events with no outcomes in common. Also called *mutually exclusive events.* Leads to $P(A \text{ or } B) = P(A) + P(B)$ when A, B exclusive.

expected frequency, E The number of times an outcome is *expected* to occur in a real data sample *based on a chi-square null-hypothesis model.* Also see **observed frequency, chi-square statistic,** and **chi-square test.**

expected value (experimental) The average value of a random quantity that has been repeatedly observed in replications of an experiment; possibly obtained through the five-step simulation method.

expected value (theoretical) Given a discrete random variable X with probability law $p(x)$, the expected value $E(X)$ is $E(X) = \sum xp(x)$. Also called the **theoretical mean of the random variable** X and often denoted by μ_X. For a box model of X, $\mu_x =$ box average. For a continuous random variable, the summation is replaced by a calculus integration.

experimental conditional probability $\hat{P}(B \mid A) = \hat{P}(A \text{ and } B) / \hat{P}(A)$.

experimental probability The estimated probability of an event based on many trials of the experiment producing the event; obtained by dividing the number of experiments (likely simulated) for which the event of interest occurs by the total number of experiments. It is often obtained by applying the five-step method to obtain simulated data.

experimental probability distribution (or histogram) The density histogram for all possible outcomes or all values of a random variable (the "statistic of interest" in five-step simulation) based on many realizations of the random variable, often on many simulations of the random variable via five-step simulation. This distribution estimates the unknown theoretical probability distribution.

explanatory variable In regression in particular, but more generally the independent variable that controls or explains the dependent or **response variable.**

extreme value An unusually large or small data point.

fair die A many-sided die for which each side has the same chance of occurring. See also **loaded die.**

***F* distribution** The continuous probability law of the ratio of two mean squares (such as the between-samples mean square divided by the within-samples mean square for a one-way ANOVA) used to carry out ANOVAs when sampling from normal populations.

finite population factor $\sqrt{(N-n)(N-1)}$, where N = population size, n = sample size. Proportional reduction in $\mathrm{SD}(\overline{X})$ when random sampling without replacement (single random sampling).

five-step method of simulation Basic method used in this book for carrying out a simulation study. The five steps are the choice of a model, the definition of a simulation, the definition of the event or statistic of interest associated with the simulation, repetitions of the simulation, and the calculation of the experimental probability of the event of interest, experimental distribution of the statistic of interest, or the average of the statistic of interest.

four-stage significance-testing approach Stage I: Form the null hypothesis and the alternative hypothesis. Stage II: Choose a null-hypothesis-based sampling box model. Stage III: Estimate the P-value of the hypothesis test. Stage IV: For significance testing, reject or fail to reject the null hypothesis based on the P-value from Stage III.

frequency Number of data points, usually in an interval.

frequency histogram See **histogram.**

frequency table A table giving the number of data points in a data set falling in each of a set of given intervals.

***F*-test** Standard ANOVA approach that divides treatment (between-samples) mean squares by mean square for error (within-samples) and then judges significance by using the F distribution.

gapping Occurrence of one or more intervals containing little to no data, caused by the fact that the probability model generating the data assigns a probability of 0 or close to 0 to the interval. For example, a study comparing tall and short men might randomly sample only men shorter than 5'6" and men taller than 6'0", creating a gap from 5'6" to 6'0".

gaps See **gapping.**

geometric distribution A discrete probability distribution useful for calculating the probability of the number of trials before a "success" occurs in Bernoulli trials. If X represents the number of independent trials before the first success, and the outcome of each trial is either success' or failure then

$$P(X=x)=(1-p)^{x-1}p$$

for $x = 1, 2, \ldots$, where p is the probability of success in a single trial.

goodness-of-fit test A statistical test for testing how well a hypothesized model fits a set of data. For one approach, see **chi-square testing** in Chapter 10. A *goodness-of-fit statistic* is a statistic on which one bases such an assessment, such as D or χ^2.

H_A Symbol used to denote the alternative hypothesis. Also see H_0.

H_0 Symbol used to denote the null hypothesis. It is usually followed by a ":" and then the statement of the null hypothesis in either symbols or words. For example, "$H_0 : p = 0.50$" indicates that the null hypothesis assumes a population proportion value of 0.50. Also see H_A.

hidden variable See **lurking variable.**

histogram A graph whose rectangles count the frequencies of occurrence of data points in specified intervals. Sometimes called a *frequency histogram.*

hypothesis A claim that is made about the value of a population parameter or about the values of several population parameters. The real-world claim of interest or research hypothesis (such as that the experimental drug is better than the traditional treatment) is turned into the alternative hypothesis H_A, whereas its negation becomes the null hypothesis, H_0.

hypothesis testing Deciding whether or not to reject the null hypothesis H_0 based on the strength of the statistical evidence in a real data sample. The statistical evidence is based on estimating the probability that the null hypothesis model would generate a value of an appropriately chosen sample statistic as extreme as, or more extreme than, its value that occurred in the real data sample. This is called the P-value. The *lower* this probability, the *greater* the evidence for rejecting the null hypothesis. Also see **four-stage significance-testing approach.** See **acceptance/rejection hypothesis testing** and **significance testing.**

independence Property of two events that holds if the occurrence of one does not affect the probability of the occurrence of the other $(P(A \mid B) = P(A))$; see also **theoretical independence, law of.**

independent events See **independence.**

independent identically distributed random variables Basic model for a random experiment or a large-population, small-sample $(n \leq 0.05N)$ survey sample. Called a random sample.

independent trials Holds when outcomes or events occurring in previous trials have no influence on the occurrance of outcomes or events in future trials.

inferential statistics Techniques used to draw conclusions from data using probability modeling and statistical techniques.

infinite populations See **conceptual population.**

influence The ability of a single data point to have a major effect on a statistical inference or on the computed value of a statistic like $\overline{X}$. A desirable property of a **resistant statistic** is that every single data point has, by itself, little influence on the value of the statistic. For example, the sample median is not overly influenced by very large data points. In regression, a point (x, y) whose removal greatly changes the resulting regression line.

interaction In a two-way ANOVA, the influence of the value of one explanatory variable on how varying the values of the other explanatory variable affects the response variable. For example, letting explanatory variables be rainfall and variety of wheat, varieties of wheat do poorly in extreme drought conditions, whereas with adequate rainfall and variety may be much better than the rest.

interquartile range (IQR) The difference between the third quartile and the first quartile of a set of data. Aresistant measure of spread.

judgment sampling Choosing a sample according to expert knowledge rather than doing probability sampling.

large-sample approach Use of a statistic to do hypothesis testing, build a confidence interval, or do some other type of inference, where the approximate probability law of that statistic is deduced from the fact that the sample size is large. For example, the CLT may have been applied.

law of large numbers The fact that if the size of a random sample (or a simple random sample if without replacement) is large, $\overline{X} \approx \mu$. This is the backbone of the five-step simulation in Chapter 5 using $\overline{X}$ to approximately estimate $\mu = E(X)$.

least squares regression The method of locating the best-fitting regression line for a scatterplot by minimizing the sum of the squared vertical distances between the points of the scatterplot and the line.

left-skewed distribution See **skewed distribution.**

level of significance User-specified probability of rejecting the null hypothesis when it is true; usually set at 0.05, sometimes at 0.01 or 0.1. Must be used with acceptance / rejection hypothesis testing. May be used with significance testing to set the standard for when evidence is strong for rejecting H_0. In both cases, supplying the P-value is recommended.

leverage point An (x, y) point with an extreme x-value, thus with the potential to be influential.

linear interpolation A technique for estimating the unknown value of a variable (e.g., y) based on the corresponding known value of a related second variable (e.g., x). It assumes a linear relationship between two known pairs of (x, y) values, one of which has an x-value less than the x for the unknown y and one of which has an x-value greater than the x for the unknown y. See Appendix D.

linear relationship A relationship between two variables whose scatterplot is well fit by a straight line; the equation $y = b_0 + b_1 x$ is often used, where b_1 is the slope and b_0 is the vertical-axis intercept value.

linear transformation (of data or of a random variable) A linear transformation of data or a random variable is one in which each data point or the random variable is multiplied by a constant value or has a constant value added to it.

loaded die A many-sided die for which some of the sides are more likely than others to land faceup when the die is rolled. See **fair die.** Often chi-square null hypotheses for a one-way table can be thought of as specifying the probabilities of a fair or perhaps a particular loaded die.

lurking variable A third variable that influences the response variable and is associated (correlated) with the explanatory variable, thus misleading the user to think the response variable is influenced by the explanatory variable. Also called a **hidden variable.**

matched-pairs design An apparent two-sample problem that reduces to a one-sample problem because the members of the two samples are naturally paired, such as measurements of the same person before and after treatment. Amounts to a randomized block design where each block is of size 2.

mean (sample) The sum of a set of numbers divided by the quantity of numbers n in the set; same as the– arithmetic average, and often defined by $X = \sum X / n$.

mean square An ANOVA sum of squares divided by its degrees of freedom.

mean (theoretical) The population or probability distribution mean. In the case of a discrete distribution given by $p(x)$, it equals $\sum xp(x)$. In the case of a continuous distribution, a calculus integration is needed. Same as the theoretical expected value of a random variable X sampled from the population. Sometimes called the *population mean.* Denoted by μ, μ_X, or $E(X)$.

measurement A special kind of estimation where a random experiment is carried out by repeatedly observing the "amount" of a property such as height, weight, speed, volume, and so forth, modeled by a random sample (independent identically distributed random variables).

median (population) In an actual population, the number such that half of the population values are smaller than this median number and half are greater. For a continuous probability distribution or continuous pdf, the number such that half of the area under the pdf lies on either side of the number. For a discrete pdf, the median is similarly defined.

median (sample) The middle value when a data set is arranged in order from smallest to largest value. When the number of observations in the sample is even, the median is the average of the two middle values.

missing baseline distortion Occurs when the vertical axis (y) of a graph starts not at 0 but close to where the changes in y are occurring. The effect is to greatly exaggerate the size of the changes in y occurring as x varies.

mode The value (or values) in a set of data that occurs the most frequently; a seldom-used measure of the center of a data set.

model A mathematical set of probability rules, a random physical mechanism, a box model, or a randomnumber-based simulation for producing data that are as similar as possible to actual real-world data. Also see **probability model.**

Monte Carlo simulation Refers to solving probability problems by repeatedly simulating the real-world experiment in question; for example, representing real-world random phenomenon by tossing a coin repeatedly, rolling a die repeatedly, repeatedly choosing random digits, or repeatedly drawing from a box model, likely by computer. The five-step simulation approach in this textbook is, thus, Monte Carlo simulation.

multiple comparisons Simultaneous confidence interval comparisons of differences of pairs of population means in an ANOVA to judge which are distinct, with some measure of overall statistical confidence attached to the results.

multiple correlation coefficient, R^2 A measure of the degree of fit of the best-fitting curve when the regression is nonlinear or linear with multiple explanatory variables; reduces to the ordinary squared correlation or coefficient of determination r^2 when the regression line is $Y = b_0 + b_1 x$.

multiple regression Linear regression where there are at least two explanatory variables influencing the response variable, such as amount of rainfall and amount of fertilizer affecting predicted crop yield.

multiplication rule for probabilities If events A and B are independent, $P(A \text{ and } B) = P(A) \times P(B)$. In general $P(A \text{ and } B) = P(A) \times P(B \mid A)$. For a probability tree, $P(\text{outcome})$ = product of probabilities along the branches producing the outcome.

mutually exclusive events Events with no outcomes in common.

(n choose x) Same as $\binom{n}{x}$; defined by $\binom{n}{x} = \frac{n!}{x!(n-x)!}$

negative relationship A relationship between two variables in which one decreases as the other increases. In the special case of a straight line, a negative relationship means that the slope of the regression line is negative and that the correlation coefficient is negative.

negative slope The slope of a line for which the y-value decreases as the x-value increases.

nominal data Data falling into named categories (such as hair colors) that are not ordered (the days of the week are ordered). In particular, the data do not represent measurements.

nonlinear regression Regression that seeks the relationship between bivariate (x, Y) pairs when the relationship is not linear, as in the case of $Y = B_0 + B_1 x + B_2 x^2 +$ random error. Some nonlinear models can be solved using linear regression applied to transformed x- and y-values.

nonparametric statistical inference Said of a statistical procedure that does not require detailed assumptions about the shape of the population distribution. Such detailed assumptions are usually specified by various population parameters.

nonresponse bias Polling bias resulting from the fact that, even if the intended sample is selected by random sampling, the subset of those actually responding to the survey may be quite different from those not responding.

normal distribution (or pdf or density) Also called the *bell-shaped curve* or the *Gaussian distribution*. The most widely used continuous distribution; often used to model biological measurements and errors of measurement. In general, for every choice of theoretical mean and SD there is a different normal density. Probabilities for all normal problems are computed using Table E (in Appendix E), which gives standard normal (mean of 0, SD of 1) probabilities.

normally distributed data set A data set whose density histogram is roughly bell-shaped. From a modeling viewpoint, the data is often assumed to have resulted from random sampling from a normal population. Obeys the 68-95-99.7% Rule.

null hypothesis The model (such as that of a fair coin) in hypothesis testing that negates the claim or research hypothesis (e.g., the coin favors heads). The data provide a measure of how weak or strong the evidence against this null hypothesis is. The formulation of the null hypothesis is the critical first step in using real data to do hypothesis testing. It usually is the negation of a real-world research hypothesis H_A. It is the first stage of the four-stage hypothesis-testing approach, and it is the first step of the five-step simulation method for estimating the probability of occurrence of the event of interest, namely the P-value. Denoted H_0. See $\boldsymbol{H_0}$.

observational study A study in which the units of the study were placed in the control and treatment groups according to the natural state of affairs (those who smoke and don't smoke have done so by choice rather than forced to by a statistically designed experiment) rather than being assigned (hopefully randomly) by the person doing the study. Such data are usually not very useful for statistical inference because there is no control over confounding variables, like age in some examples, differing greatly between the two groups.

observed frequency (denoted O) Given a set of outcome categories, the number of times an outcome is observed to have occurred in a real data sample in a statistical study (often a chi-square test). See **expected frequency.**

observed level of significance Same as ***P*-value.**

one-sided hypothesis test A hypothesis test in which the alternative hypothesis is that the population parameter lies to one specified side of its null hypothesis value.

one-way ANOVA Analysis of variance in which the population means vary because of the influence of a single explanatory variable or factor, as when wheat yields vary because of differing wheat varieties.

ordered categorical variable or data Numbers assigned to categories where the categories are ordered, such as, low, middle, high social economic status.

ordinal data Same as **ordered categorical data.**

outcomes The list of the possible ways a random experiment can occur.

outlier A data value that is extreme (very large or very small) relative to the rest of the data. Further than ±3 standard deviations s from the sample mean $\overline{x}$ is the most common outlier criterion for single-variable data (but boxplots use a different criterion and often distinguish between a potential outlier and outlier).

p Symbol used to denote "population proportion."

$\hat{p}$ Symbol used to denote the sample proportion, an estimator of the population proportion p.

pdf In the continuous case appropriate areas under the the representing pdf yield probabilities. In the discrete case $p(x) = P(X = x)$.

parameter A number describing a characteristic of a population or a probability model, such as the theoretical mean, μ, or the true slope of a regression line.

Pareto chart A bar chart where the bars are arranged from tallest to shortest as one moves to the right.

pie chart A graph for categorical data. The proportion of elements belonging to each category is proportionally represented as a pie-shaped section of a circle.

placebo A nonactive treatment that is administered to the control group so that the human subject cannot tell whether he or she has received the experimental treatment; for example, a shot of saline solution instead of a new vaccine.

point estimate An estimate of a population parameter that is one specific number (as opposed to an interval estimate).

Poisson distribution The discrete probability law of the number of occurrences X of a randomly occurring event in a fixed time or fixed area interval when the rate of occurrence is fixed across the interval, separate occurrences are independent, and occurrences are isolated from each other. An example is the number of phone calls arriving in a given period of time. The distribution is computed as

$$P(X = x) = \lambda^x e^{-\lambda}/x! < x = 0, 1, \ldots; \quad \lambda > 0$$

population The entire collection of objects or people under consideration for statistical study. Often the statistical goal is to use random sampling to make inferences about the population, such as about its center or spread. The population is modeled by a probability distribution. Can be real finite or infinite conceptual. Can be continuous or discrete.

population, actual (or real) A population consisting of actual objects that can, in principle, be measured or classified. An actual population always has finitely many members.

population, conceptual A population consisting of imagined members. For example, the population consisting of all possible tosses of a die in the future. In practice, this population is not just those currently suffering from asthma, but all future sufferers or potential sufferers as well. Conceptual populations have infinitely many members. Often large finite populations of continuous measurements, such as heights, are modeled as a conceptual population described by a continuous random variable, like the normal. Also called an **infinite population.**

population distribution The distribution of one randomly sampled member from a population. See **sampling distribution.**

population parameter See **parameter.**

population proportion A ratio whose numerator is equal to the number of objects in the population that display a particular property or quality of interest and whose denominator is the total number of objects in the population. It is a type of population parameter. Denoted usually by p.

population size The number of objects or people in a population, often denoted by N.

positive association See **positive relationship.**

positive relationship A relationship between two variables in which one tends to increase as the other increases. In the special case of a straight line, a positive relationship means that the slope of the regression line is positive and that the correlation coefficient is positive.

positive slope The slope of a line on which the y-value increases as the x-value increases.

power function Graph of the power of a test (P(reject $H_0 \mid H_0$ false)) as a function of the value of the parameter. See **power of a test.**

power of a test The probability of rejecting the null hypothesis when it is false. It usually increases as the distance increases between the true value of the parameter of interest and its null hypothesis values.

prediction Inference about what the value of a "new" random variable is likely to be. Different than estimating the value of a (nonrandom) regression parameter like $E(Y \mid x)$.

prediction error See **residual.**

prediction interval in regression analysis An interval computed from the data for which a new (yet unobserved) random observation of Y at x has a high probability of falling within. Contrasted with a confidence interval for $E(Y \mid X = x)$ at x.

probability See **conditional probability, experimental probability, frequency interpretation of probability, or theoretical probability.**

probability density function (pdf) A function that gives the theoretical probabilities associated with the values of a random variable. Discrete pdfs give the probabilities explicitly. Continuous pdfs require that one find the area in a specified interval underneath the curve graphed by the pdf in order to find the probability of the random variable being within the specified interval.

probability distribution See **distribution.**

probability histogram A histogram made up of rectangles of width 1 where a rectangle centered at integer x has height (and area) equal to $P(X = x)$. This histogram displays the shape of the probability distribution of any integer-valued random variable, like the sum of two dice.

probability model A list of all possible outcomes of a random phenomenon, called a chance experiment, and an assignment of probabilities to these outcomes. Often a physical or computerized mechanism like box-model simulation for simulating a random phenomenon, such as the simulation of the Chapter 5 cereal box problem. Can be a continuous pdf too.

probability sampling Any sampling scheme where the sample members are selected using a chance mechanism.

probability tree Graphical technique for computing the probability of an event that is the result of a multistage experiment where the probabilities are known for each outcome at each stage given an outcome from the previous stage.

***P*-value** The probability of observing a value of the test statistic that is as extreme or more extreme than the observed value of the test statistic in a direction suggesting the alternative hypothesis is true. This probability is computed under the assumption that the null hypothesis H_0 is true. Often denoted by P.

***Q-Q* plot** A graphical technique used to assess whether data is roughly normal. Not used explicitly in the textbook.

qualitative variable or data Same as categorical variable or data.

quantitative variable or data Having a numerical value (often the result of a measurement process) like weight or height as opposed to eye color.

quartiles Quartiles divide a data set into four parts. The first quartile is roughly the value that separates the data into the lower 25% and the upper 75% of the data. The third quartile is roughly the value that separates the data into the lower 75% and the upper 25% of the data. The second quartile divides the data into the lower 50% and the upper 50%. The second quartile is the same as the median. Precise definitions of quartiles vary slightly from textbook to textbook and computer program to program.

quota sampling Choosing a sample by guaranteeing that certain groups (like men and women) are represented in the same proportion in the sample as in the population; otherwise any individuals who fit the quotas can be selected.

r Symbol for correlation coefficient.

random Not predictable; occurring by chance. For example, the outcome of tossing a fair coin is random.

random experiment Any experiment (physical or phenomenon) that involves a random mechanism to produce its data. See repeated-trials random experiment for a special kind of random experiment.

randomization Using a random mechanism to choose which units receive which treatments in an experiment or which individuals to sample from a population, instead of having experts decide, for example. See also **randomized controlled experiment.**

randomized block design Experimental units are first divided into homgeneous blocks (usually based on differing levels of a confounding variable) and then *within* each block units are randomly assigned to treatment groups.

randomized controlled experiment A controlled experiment where units, often human subjects, are randomly assigned to the treatment and control groups. Hence this is a special kind of randomized experiment.

randomized experiment Amethod of data collection where a collection of units, often human subjects, are randomly assigned to one or more treatment groups, sometimes including a control group. A randomized experiment is a special kind of random experiment where the random assignment of units to treatments is the "random mechanism."

random numbers Also called *random digits.* Digits (most commonly $0, 1, \ldots 9$) that occur in equally likely and random fashion, as when produced by using a spinner having 10 equal sectors; often produced by a computer program, in which case the numbers merely look random.

random sample Results from random sampling with replacement or a repeated-trials random experiment. Mathematically, a sequence of independent and identically distributed random variables. Approximates random sampling without replacement (**simple random sampling**) when sample size n satisfies $n \leq 0.05N$, where N is population size.

random variable A numerical description of the outcome of a random phenomenon, like an automobile accident, a thunderstorm, or a game of roulette. Random variables are denoted by capital letters, usually towards the end of the alphabet, like U, X, and Y. $(X = x)$ denotes the event that the random variable X has an observed numerical value of x. Random variables are represented by a probability distribution function (pdf) that in the discrete case gives both the values (or ranges of values) and their associated probabilities and in the continuous case supplies probabilities as areas under the pdf.

range The measure of spread (variation) that is the difference between the largest value and the smallest value in a set of data. This is not a resistant measure of spread.

ranked data See **ordinal data.**

ranked variable or data See **ordered categorical variable or data.**

real data Data that are obtained from an actual experiment or survey designed to answer a specific research question about the underlying probability model. The data are used to obtain statistical evidence about the probability model underlying the data. Contrasted with **simulated data.**

real population See **population actual (or real).**

regression equation The equation of the (least squares) regression line.

regression line Related to **regression equation.** A straight line used to estimate the relationship between two variables, based on the points of a scatterplot; determined by a least squares analysis.

rejection of null hypothesis The decision that is made in hypothesis testing when the appropriate event of interest (EOI) occurs with very low probability (e.g., ≤ 0.05) under the null-hypothesis model. See **hypothesis testing** and ***P*-value.**

rejection region Same as critical region.

relative frequency interpretation of probability The interpretation of a theoretical probability as approximately predicting the proportion of occurrences of the event of interest (such as the proportion of heads being approximately 1 / 2) occurring in a large number of independent (replicated) trials of a random

experiment (such as repeated tossing a coin), often obtained by five-step simulation.

relative frequency (of an event) Same as **experimental probability (of an event).**

repeated measurements See **measurement.**

repeated-trials random experiment Repeated independent trials under identically experimental conditions. Produces a random sample (independent identically distributed random variables). Helps reduce variability and thus increases accuracy when the observations are measurements.

replication Name for the process of carrying out repeated trials of a random experiment. Increases accuracy and decreases the influence of confounding variables.

resampling methods Any one of several methods, bootstrap statistics being the best known and most emphasized in this textbook, that rely on repeated sampling using a box model that has been somehow determined by the sample or samples of actual collected data.

research hypothesis Areal-world claim that becomes the alternative hypothesis H_A and stands against the negation of the claim (the negation being the chance explanation) that became the null hypothesis

residual In the case of regression, the difference between the actual Y-value and the estimated regression line $\hat{Y}$-value: $Y - \hat{Y}$.

residual plot Plot of $y - \hat{y}$ versus x-values in a regression analysis.

residual (sample) SD The sample SD of the residuals $y - \hat{y}$, equals $s_Y \sqrt{1 - r^2}$.

residual sum of squares See **error sum of squares.**

resistant A statistic is resistant if it is not sensitive to the influence of individual extreme observations. The median is a resistant measure of the center of a set of data and the interquartile range (IQR) is a resistant measure of the variation (or spread) of a set of data.

response variable In regression but also more generally, the dependent variable that is being influenced by the independent or **explanatory variable.**

right-skewed distribution See **skewed distribution.**

robust Said of a statistical procedure that works well even if the assumptions supporting it (like a normal population being sampled from) are only crudely true.

S_E See **standard error of the residuals in a regression analysis.**

S_Y The sample standard deviation of a set of data whose values are denoted by Y. Denoted S when no ambiguity is caused by dropping the subscript.

sample proportion A ratio whose numerator is equal to the number of objects in the sample that display a particular property or quality of interest, and whose denominator is the total number of objects in the sample. It is a sample statistic used as an estimator of the population proportion p. Often denoted $\hat{p}$.

sample size The number of objects or people in a sample.

sample space The set of all possible outcomes of a probability experiment. For example, the probability experiment of throwing a die once has the sample space $S(1, 2, 3, 4, 5, 6)$.

sample statistic A statistic that has been computed from a (hopefully random) sample. Like $\overline{X}$.

sample survey A survey of a population, usually human, made by taking a sample judged to be representative of the population. Use of a random mechanism for choosing the sample is essential for validity and believability.

sample variance See **variance (sample).**

sampling distribution The probability distribution of a statistic, such as an estimator or test statistic that is computed from a sample.

scatterplot A graph of two-variable (bivariate) data in which each point is located by its coordinates (x, y).

SD See **standard deviation.**

selection bias Polling bias resulting from a sampling method that makes some population members more likely to be sampled than others. For example, the *Literary Digest* poll mentioned in Chapter 7 favored the more affluent.

significance level The specified (allowed) probability of Type I error when doing acceptance/rejection testing. The standard for having strong evidence that H_0 is false when doing significance testing. Usually 0.05. 0.01, or 0.1.

significance, statistical See **statistically significant.**

significance testing Hypothesis testing in which the two decisions are *rejecting* the null hypothesis and *failing* to reject the null hypothesis. That is, the null hypothesis is never accepted when doing significance testing. Contrasted with **acceptance/rejection hypothesis testing.**

simple random sample Random sampling without replacement. This expression terminology is used to distinguish random sampling **without replacement** from random sampling **with replacement.**

Simpson's paradox Nontechnically, this paradox allows a confounding variable to apparently lead us to the opposite of the correct conclusion about the comparison of a treatment versus a control. In the book's Chapter 7 example, the confounding variable age seemed to suggest that smoking increased life expectancy!

simulated data Contrasted with **real data** obtained from an actual statistical study.

simulation Any method for generating data from a given probability model. Methods used include physical models such as coins or dice, and simulation based on a specified box model, often done on the computer. See **five-step method of simulation.**

single-blind study A study in which the subjects do not know who has received the treatment and who has received the placebo.

skewed or skewed distribution Said of an asymmetric pdf or histogram of data that is stretched out in one direction. For example, the pdf of a chi-square distribution is *skewed to the right,* or *right-skewed,* and a histogram of incomes is typically skewed to the right. Causes the mean to be pulled away from the median in the direction of the skew.

slope The rate of change of the y-values with respect to the change of the x-values in a straight-line relationship.

spread (of data) See **variation (of data).**

standard deviation, sample The square root of the sample variance. The most widely used measure of the amount of spread (variation) in a set of data. Denoted by s.

standard deviation, theoretical The standard deviation of the population or distribution, given by the square root of the theoretical variance. Denoted by σ. In the case of a discrete distribution given by $p(x)$,

$$\sigma_x = \sqrt{\sum(-\mu_x)^2 p()}.$$

standard deviation of a sample proportion $\sqrt{p(1-p)} / \sqrt{n}$ where p is the population proportion of ones in a population consisting entirely of ones and zeroes; $n =$ the sample size. Estimated by substituting $\hat{p}$ for p.

standard deviation of a sample statistic For any sample statistic, Y, computed from a random sample, the standard deviation of Y, written $\mathrm{SD}(Y)$. This is not to be confused with the population SD, usually denoted by σ. $\mathrm{SD}(Y)$ often estimated by the standard error of Y, $\mathrm{SE}(Y)$. $\mathrm{SE}(Y)$ is often obtained by estimating parameters in the formula for $\mathrm{SD}(Y)$. Also obtained by bootstrapping.

standard error of a sample mean $s / \sqrt{n}$, where n is the number of observations used to compute the sample mean and σ is the population SD.

standard error of a sample statistic An estimate of the (theoretical) SD of the statistic. For example, $\mathrm{SE}(\overline{X}) = S / \sqrt{n}$ estimates $\mathrm{SD}(\overline{X}) = \sigma / \sqrt{n}$ and $\mathrm{SE}(\hat{p}) = \sqrt{\hat{p}(1-\hat{p})} / \sqrt{n}$ estimates $\mathrm{SD}(\hat{p}) = \sqrt{p(1-p)} / \sqrt{n}$.

standard error of $\hat{p}$ $\sqrt{\hat{p}(1-\hat{p})} / \sqrt{n}$

standard error of the residuals in a regression analysis $S_E = \sqrt{\Sigma(Y - \hat{Y})^2 / (n-2)}$. Used to estimate σ, the SD of Y given fixed x in a regression analysis.

standard error of $\overline{X}$ $S / \sqrt{n}$.

standardized score Also called *standard score*. Each score x is centered at $\overline{x}$ and this difference is divided by the sample SD, s, producing a standardized score $z = (x - \overline{x}) / s$. Also see ***z*-score.**

standardized test statistic Of the form: (statistic parameter – under H_0) / (SD of statistic). Often has a standard normal or t distribution.

standard normal density Same as **standard normal distribution.**

standard normal distribution A normal distribution with a mean of 0 and a variance of 1 (and, hence, an SD of 1).

statistic A piece of numerical information computed from a sample, such as $\overline{X}$. Because the sample is random, the statistic is a random variable and has a sample distribution. Also see **statistics (2).**

statistical decision making See **hypothesis testing.**

statistical evidence Numerical evidence, usually in the form of a probability or P-value, gathered from a statistical study for use in hypothesis testing. See **hypothesis testing.**

statistical inference Using statistical reasoning to draw a conclusion from data about the unknown character of the real world.

statistical literacy The state of being reasonably informed about the basics of statistical reasoning, both in "thinking statistically" and in understanding basic statistical methodology. Makes one resistant to being deceived by improper data representations or analyses.

statistically significant Said of data behavior that is too unusual to be attributable to chance under the probability model that is presumed to be producing the data under the null hypothesis. For example, if the sample mean is so large that its value cannot be reasonably attributed to chance under the null hypothesis of a population mean of 0, the difference of the sample mean from 0 is said to be statistically significant. See **hypothesis testing.** Usually if the P-value satisfies $P \leq 0.05$, the result is declared to be "statistically significant."

statistical regularity The empirical (real-world) fact that the experimental probability becomes closer and closer to a number, called the **theoretical probability**, as the number of trials (often five-step simulation repetitions in this textbook) becomes large; this law is the foundation of statistical reasoning. Also see **relative frequency interpretation of probability.**

statistics (1) The science of gathering, describing, and drawing conclusions from data; (2) reported numerical information, such as in a newspaper.

stem-and-leaf plot A graphical display of data using certain digits (such as those in the tens place) as stems and the remaining digits (such as those in the ones place) as leaves. Also called **stemplot.**

stemplot Same as **stem-and-leaf plot**.

straight-line equation The slope-intercept form is $y = b_0 + b_1 x$, where b_1 is the slope and b_0 is the y-intercept, namely the point at which the line intersects the y-axis. Same as **linear relationship.**

strata See **stratified random sampling.**

stratified random sampling Contrasted with simple random sampling, the units are first split into strata using one variable; then simple random sampling is carried out within each strata.

strength of evidence The degree to which statistical evidence from real data supports rejection of the null hypothesis (H_0). The evidence is usually the probability of occurrence of the event of interest (EOI) under H_0, called the P-value. The lower this probability is, the stronger the evidence for rejecting H_0. See **hypothesis testing, event of interest,** and **statistical evidence.**

sum of squares The sum of the squares of observed quantities computed in an ANOVA to estimate the contributions from various sources, such as the effect of one variable in a two-way ANOVA. Also occurs in the computation of a sample variance or a sample standard deviation and in other statistical settings.

survey sampling A study in which a sample is randomly chosen form the population so as to be representative of the population. Then one infers that characteristics of the sample are characteristics of the population. The only truly acceptable way to do this is to use probability sampling to obtain the sample, the simplest being random sampling without replacement (simple random sampling), modeled by independent identically distributed random variables when sample size is at most 5% of population size. Also called **sample survey.**

symmetric distribution of data When the shape of the data to the left of its center is similar to the shape to the right of the center. In this case the sample mean and sample median will be quite close.

***t* distribution (or pdf or density)** A bell-shaped density close to a standard density but with fatter tails. Sometimes called Student's t distribution.

test of significance See **significance testing.**

theoretical independence, law of The fact that if the events A and B are independent, then $P(A \text{ and } B) = P(A)\,P(B)$, where P denotes theoretical probability.

theoretical probability The true probability of an event; it is what the experimental probability will be close to in a very large number of trials. In the special case of an equally likely outcomes probability model, the theoretical probability is obtained by dividing the number of outcomes that produce the event of interest by the total number of possible outcomes.

theoretical probability histogram A graph of the pdf of a an integer-valued random variable.

theoretical 68-95-99.7% rule for the probability of intervals centered at μ States that when the distribution of a random variable X is roughly normal (bell-shaped) $P(\mu - \sigma \le X \le \mu + \sigma) \approx 0.68$, $P(\mu - 2\sigma \le X \le \mu + 2\sigma) \approx 0.95$, $P(\mu - 3\sigma \le X \le \mu + 3\sigma) \approx 1$. Here $\mu = E(X)$, $\sigma =$ SD(X).

transformation of data Use of x^a, log x, or some other transformation to produce a better-fitting and simpler model for two-variable (x, y) data. Such transformations can also be applied to y. Sometimes transformation converts a nonlinear regression problem into a linear regression problem solvable by the methods of Chapter 3 and 11.

treatment group The group of individuals or units in an experiment that receive the treatment being studied.

treatment sum of squares (SSTR) The ANOVA sum of squares due to differences in the population means. Also described as a between-samples sum of squares.

tree diagram Shows graphically all possible outcomes of a multistage experiment.

***t*-statistic** A statistic of the form (statistic−parameter)/(SE of the statistic). Often the statistic is approximately t-distributed. $(\overline{X} - \mu) \,/\, (S \,/\, \sqrt{n})$ is one example.

***t*-test** A special hypothesis test that is usually about the population mean (or about the equality of two population means) used when the population is known to be approximately normally distributed, even if the sample size is small, and the population standard deviation is unknown. Based on a t-statistic, obtained by centering at the null hypothesis mean and then dividing by the appropriate standard error. Uses the t distribution rather than the standard normal distribution. A test about the regression line slope is often a t-test.

two-sample problem Any inference problem about two populations in which the data consist of two independent random samples, one from each population.

two-sided hypothesis test A hypothesis test in which the alternative hypothesis is that the population parameter may lie on either side of its null hypothesis value.

two-way ANOVA Analysis of variance in which the population means vary because of the influence of two explanatory variables or factors, as when wheat yields vary because of differing wheat varieties and differing soil types.

two-way table Another name for a two-variable contingency table.

Type I (hypothesis test) error The error of incorrectly rejecting a null hypothesis when it is true.

Type II (hypothesis test) error The error of incorrectly accepting a null hypothesis when it is false.

typical variation The standard deviation is interpreted as a measure of the typical amount of variation from one observation to the next.

unbiased estimator If the expected value of a statistic $\hat{\theta}$ used to estimate a parameter is equal to the value of the parameter $(E\hat{\theta}) = \theta$, then that statistic is an unbiased estimator. Put empirically, as the () number of statisticians who apply a procedure gets larger and larger, if the average of the estimator over all the statisticians approximates the value of the parameter that the procedure intends to estimate, then the procedure is unbiased. The sample average is an example of an unbiased estimator of the population mean (assuming the observations are based on random samples).

uniform distribution A histogram of data (or probability function) that is relatively flat over its range.

U-shaped distribution A histogram of data (or probability density function) that is dense (high) in the extremes and sparse (low) in the middle.

validity The correctness of a statistical inference.

variable A quantity that varies, often randomly. For example, the weight of a randomly chosen member of a football team is a variable. Variables are usually represented by letters.

variance (sample) A measure of the variation (spread) in a set of data. The sample variance is almost the arithmetic average of the squared differences of all data values from their sample mean (one divides by $n - 1$ instead of n; otherwise it would be the arithmetic average). Often an intermediate step in the 2 calculation of the sample standard deviation. Denoted by s^2 or S^2.

variance (theoretical) The variance of the population or distribution. In the case of a discrete distribution given by $p(x)$, it equals $\sum(x-\mu)^2 p(x)$, where μ is the population mean. Denoted by σ^2.

variation (of data) The degree to which data are spread out around their center. Useful measures of spread include the sample SD and the interquartile range.

volunteer effect Bias that arises when response to a survey is voluntary; it happens because even if the survey is sent to a representative random sample, the sample of volunteers is not representative of the population, due to nonresponse bias.

within-samples sum of squares In ANOVA, the sum over all the samples of each sample's sum of squares. Also see **error sum of squares (SSE).**

within-sample variability In ANOVA, the variability of observations within a sample that is caused by the population variance.

without replacement Describes random sampling from a real population, or from a box, in which each drawn object is not replaced before the next random drawing.

with replacement Describes random sampling from a real population, or from a box, in which each drawn object is replaced before the next random drawing.

***y*-intercept** The value on the *y*-axis where the line crosses it.

***z*-score** Also called a **standardized score.** Tells how many SDs a data value is above or below the sample mean data value sample mean

$$z \text{ score} = \frac{\text{data value} - \text{sample mean}}{\text{sample standard deviation}} = \frac{x - \overline{x}}{s}.$$

***Z*-statistic** A statistic of the form (statistic−parameter)/(SD of statistic), where the SD is known. Often the *Z*-statistic is approximately standard normal. Also see ***Z*-test.** An example is $(\overline{X} - \mu) \,/\, (\sigma / \sqrt{n}$.

***Z*-test** A hypothesis test about the population mean μ (or about the equality of two population means) using either the fact that the population is normal or using the central limit theorem result that $(\overline{X} - \mu)/(\sigma - n)$, where σ is the population SD and n is the sample size, has an approximately standard normal distribution when n is large.

IV

Appendices

COMPUTATIONALLY GENERATED RANDOM DIGITS

Computers (and some calculators) can produce random digits rapidly and in very large quantities. Often it is not practical to require such devices to store long lists of random data. Therefore it is usually better to generate such random digits only as they are needed.

Formulas have been invented to compute random digits. At first, you might think this is impossible, because if you *compute* a number by a formula then you will *know* what the next digit will be, and thus such digits cannot be random. This is true, but although these integers are generated deterministically, they nonetheless appear to be true random digits for all practical purposes. Digits produced by such formulas are therefore sometimes called *pseudo-random.* They are *pseudo*-random because they are not random. For most purposes, such pseudo-random digits work extremely well in that they appear in every observable way to be random and are very convenient to use.

Some such formulas are very complicated; others are surprisingly simple. You can usually find out from a user manual or a consultant the formula that a particular computer program uses for generating its random numbers.

There is one particularly easy method for producing pseudo-random numbers that you can use on a calculator or computer with little or no programming. It has some flaws and hence is not often used, but it shows us how a deterministic approach can produce digits that appear to be random. It is called *mid-square method,* and it was suggested by the mathematician John von Neumann.

Take an arbitrary number of five (or more) digits (you could use the last five digits of your telephone number or social security number, for example.) Suppose we start with 63537. This is not a part of your list of random digits—it is only the starting value for the procedure. We square this number:

$$(63537)^2 = 403\mathbf{69503}69$$

We now take the middle five digits, 69503, shown in boldface in the above equation. These are our first five "random" digits. Next we square these:

$$(69503)2 = 48\mathbf{30667}009$$

Again, we take the middle five digits and square:

$$(30667)2 = 9\mathbf{40464}889$$

We repeat this process as long as desired. We then write down the string of random digits. For our example, we get

69503 30667 04648

This string of digits may be regarded as a set of random digits. The advantage is that it has been mechanically produced. In this case, we used an ordinary calculator to do the job. A caution is necessary about this method: after a while it can sometimes start generating all 0s. Nobody uses this method for actual production of random number tables. We have presented it because it illustrates the idea that a simple computation done repeatedly can produce digits that appear to be random. In addition, you can certainly use it if you wish.

B

RANDOM NUMBER TABLES

Table B.1 Random Number Table for Tossing a Fair Coin: {0, 1}

01110	10000	00010	10111	00010	11001	10011	10001	10110
11101	00111	10111	10011	11001	11011	00000	00000	01111
11111	11000	01101	01101	00000	00101	01001	11100	11001
10001	00101	00001	01100	11101	10001	11101	10011	11110
00101	11011	10100	00110	11110	00110	10111	10011	10101
00111	10000	11111	00010	10000	10011	10111	01100	11001
01110	00100	00001	11100	11010	01011	01100	01100	11010
10011	11100	00111	10010	01011	11010	10011	10111	11010
10100	10001	01101	01100	11100	10011	00001	11110	00011
00111	00100	00110	00101	01011	01110	10110	11101	01111
01001	10001	00101	00000	10111	00001	10000	00000	00110
11011	01010	11100	01100	01011	11111	10000	00110	10100
11010	10101	00001	10011	01110	11101	01110	01111	11011
01010	00001	11110	11101	00111	01100	01001	01000	10110
01111	11111	10000	10000	01000	00101	00100	00110	00110
11010	11000	00111	01010	01111	11101	01001	10100	01001
01100	11011	00111	01011	10010	10000	00110	01011	01010
00111	01011	10011	10100	11111	11111	01011	10011	11000
01001	11011	10011	00100	10110	01010	00110	10111	11000
10010	10111	10001	10100	00010	01000	10011	01110	11000
01001	01001	00100	00100	10110	10000	11011	00001	00101
10000	00011	00111	00010	10111	01111	00100	00000	01111
00011	11110	01101	10100	00010	10010	00001	01100	11100
10001	10000	01011	00010	01010	10000	11000	11100	01010
00111	00111	00000	11100	00110	11001	10101	01001	00010
01010	01010	10011	00000	11001	10101	10111	10011	00010
10100	11111	10001	10011	01010	11011	00101	11010	10011
01000	01100	01011	10110	00101	10100	11000	10000	01000
11010	11001	00110	00100	11101	01000	11011	10011	11110
10101	00010	01001	00111	10111	10100	00110	01010	11111
10000	10110	10110	00110	10001	01110	01010	00110	11001
01011	01001	10111	01011	10000	11111	00000	10000	00001
00000	00000	01111	00101	11111	01100	01111	10011	10110
10110	01111	01111	01000	01010	01000	00110	01001	11000
11100	11100	10010	00101	00111	01111	01010	11000	00010
11100	10001	10001	01010	00110	10111	11001	11001	11111
00110	00100	00001	11100	10001	10101	01110	01010	11100
01001	10011	10011	10010	01011	00111	11110	01010	00110
10001	01100	10001	00011	00111	00001	11110	10111	10010
00010	10100	01101	00110	00111	00000	01010	01000	10101

Table B.2 Random Number Table for Rolling a Fair Six-Sided Die: $\{1, 2, 3, 4, 5, 6\}$

66533	45332	24614	22231	26431	35541	12165	62116	16111
61261	22613	26252	14622	32262	33244	34614	13316	41136
61144	46631	56646	24544	36461	14612	21234	23335	16212
21341	66222	53246	24444	13311	44244	41643	54163	21243
15365	46135	23345	53331	46112	54655	65626	24216	11144
51612	21315	16156	56511	31516	26121	23151	66611	64242
31244	26623	33555	53333	24465	14566	54345	25532	43426
41452	65222	25316	44431	33141	64245	62514	45553	24234
25515	41332	25311	65143	61134	65443	45366	61566	63112
35226	33554	55613	54321	21434	22314	11416	64624	26565
41554	23265	43245	46443	26431	23326	51232	45243	54452
24356	54551	52551	64423	13134	25513	43312	43234	12543
56341	24435	44336	62665	26123	25311	55156	46545	45462
26452	16243	13225	66222	31311	41621	51634	63265	22422
41265	26326	23356	62315	32132	55212	26512	56646	42132
13355	53313	42141	56223	61515	22554	44363	42162	52564
22434	61332	13551	43314	45151	43646	15241	55526	61141
62616	22135	55522	62532	43455	14443	61544	55551	34455
53565	42466	22156	21551	13322	54653	62516	66132	36151
55634	65353	51535	66264	61332	11323	15613	44653	36232
64424	24211	63132	25346	41315	66415	33444	16552	56425
34324	62343	35623	22645	41232	35323	25646	52446	24233
44252	16335	35563	33652	12265	25133	56646	14452	45423
55613	35253	13233	26555	56412	23255	61144	36162	51462
34446	33554	16144	16156	23641	24226	52346	26545	55324
16263	13232	52551	32542	41514	52151	43454	32133	34634
34544	15413	14125	62245	13145	53112	52116	63164	42245
23235	36666	61125	44255	62636	53632	56521	26641	56613
32161	66226	66631	33144	12225	42633	61631	14334	45532
13423	54133	12612	15644	41562	36533	25533	22642	34621
66565	24634	23326	42543	55652	25144	12623	46551	42423
52544	45645	23641	64431	41455	21554	23525	26554	66453
21311	46163	14532	63152	12231	34425	41356	64213	62225
14163	41364	32442	55443	54236	45413	43531	12112	45356
63431	26466	35321	56544	25625	66235	56345	15116	65553
66142	14561	25443	53424	45123	32155	41112	45254	54661
56651	14313	51634	65631	45624	34654	32533	63314	11516
61641	44461	14462	35555	26443	61415	62343	22661	31154
25144	62266	63315	51561	26416	26626	46624	55541	64515
26423	66151	45645	44611	16624	56615	55414	36255	25265

Table B.3 Random Number Table for Rolling a Fair 10-Sided Die: {1, 2, 3, 4, 5, 6, 7, 8, 9, 0}*

32236	12683	41949	91807	57883	65394	35595	39198	75268
40336	50658	32089	78007	58644	73823	62854	31151	64726
88795	93736	22189	47004	48304	77410	78871	98387	44647
12807	65194	58586	78232	57097	01430	00304	32036	23671
65929	96713	94452	56211	85446	13656	32155	84455	38125
50339	82178	19650	41283	03944	13736	02627	41929	60613
73840	53838	90804	94332	63639	73187	87067	37557	29635
87062	66298	10731	40629	64955	08081	31443	72112	58006
48038	94580	55223	97799	10105	27952	62493	42176	69615
89830	54426	02692	21233	39553	33483	03141	90919	99219
72234	40065	24052	95658	98335	21125	45364	67989	32451
02833	78254	05112	95160	62546	85982	85567	27427	21436
20565	20846	63664	72162	75338	04022	77166	83339	99021
18090	91089	09799	75883	36480	37067	40933	65634	79883
11519	97203	70899	00697	84864	24470	07933	48202	15392
12732	61573	22852	32281	84871	13331	08947	09023	38248
26823	44530	84507	10396	12240	62603	13396	69378	37173
74622	88768	90819	54769	95306	07685	50369	13763	02205
58535	99062	55182	89858	67701	94838	37317	10432	75653
78551	56329	09024	81507	90137	19241	55198	74006	52851
41477	58940	04016	38081	45519	27559	92403	30967	86797
17004	22782	09508	37331	94994	67305	34040	91360	83009
36925	31844	12940	51503	24822	53594	72930	23342	88646
97569	75612	07237	92264	77989	09054	03863	83891	09041
35122	31549	19982	66024	68615	15959	40347	50052	35312
34358	17573	32838	68335	93497	17412	19850	98965	27357
09285	03384	61410	01932	26797	92577	42580	23354	38677
98363	76867	91821	26538	47181	50938	95676	45306	96725
91769	65764	52386	18551	22196	50282	19985	90730	95175
09393	34982	05654	51208	37731	33916	49063	76700	63094
33202	19891	95374	23650	64877	70661	53330	07223	03469
34111	18376	08231	95163	62837	56995	03022	93618	67560
04410	39026	99536	85083	34607	38979	47259	30921	90092
88209	36990	83284	50578	83549	19006	29501	04565	48865
89257	25109	26253	57523	99297	29901	29472	52817	66611
25581	36013	52215	47684	55094	93140	32969	05603	66922
31742	82956	36361	52786	79761	49819	41375	67628	81707
46012	07959	79667	95325	49142	99596	25691	85964	53568
00174	92988	49499	74089	99209	33816	51757	25031	35862
04861	63886	09763	03265	12748	77513	91010	15062	20270

*This 10-digit case is what is usually referred to when one refers to a random number table.

Table B.4 Random Number Table for a Standard Normal Variable

−0.3713	−0.2478	−0.6191	0.0207	−0.3093	−0.6102	−1.5955	−0.4378	0.6066
2.1352	−0.8940	−0.0062	0.6737	−0.6091	−0.3083	−0.0850	1.7532	−1.3563
0.9692	−0.6652	−1.9856	−0.6354	0.5742	−0.7707	0.9028	−0.4957	2.0585
−0.1241	−0.1783	−1.0843	−0.7939	−0.1232	−1.4296	2.7796	0.4968	0.1914
0.0297	0.3037	−1.3847	0.0486	0.2966	0.5970	0.2056	0.1413	−0.1678
−0.3438	1.2274	−1.1718	0.8335	0.7018	0.0750	0.3924	−0.6794	−0.0852
1.1178	1.8089	1.3865	−0.4378	0.6506	1.2432	0.1845	−0.6945	−0.2513
0.7023	−0.1139	0.3233	1.6031	−0.5919	−1.2176	0.7271	−0.0229	−0.1501
0.4092	0.3912	−1.4844	−0.1497	−1.3398	−0.3652	−1.7329	−1.0306	0.9890
0.2007	−2.3895	−0.5677	−1.3207	−1.3434	0.4915	1.5763	0.6146	1.3502
1.0997	−1.4403	−0.6747	−0.3954	−1.5214	1.9471	−0.4529	0.1710	0.1409
−0.8716	−0.9046	0.1550	2.4269	−2.0841	1.2182	−1.5685	−0.0411	1.9245
0.0647	0.3757	0.0207	0.8082	1.3206	0.4904	−0.8063	−1.4134	1.7442
−0.5876	1.0762	0.1382	1.1575	−0.1087	−1.3951	0.0209	−0.3103	−0.1289
0.9306	0.5654	−0.7672	1.4689	0.2111	0.6171	−2.2243	−0.7565	0.8861
0.2412	−0.8951	−0.7605	0.1465	−0.0505	1.0706	−0.0549	−0.3443	0.4225
2.4332	−0.2278	−1.2614	−1.4321	0.1858	−0.2124	0.0483	1.0501	0.3017
−1.3169	−0.9054	−1.7658	0.5712	−0.5247	−0.1075	−1.0862	0.4623	0.0951
1.0249	−1.4323	−0.5404	−1.0173	−0.4266	−1.0423	−0.9953	1.1717	−0.7675
−0.9134	0.3134	−0.9818	0.6456	−1.5494	−1.5576	−1.9423	−1.0402	0.9997
2.3278	0.5612	−1.4350	0.4842	0.2576	0.3703	0.4033	0.1137	0.7600
1.0979	1.4661	1.0422	1.6997	0.3377	1.3285	−0.8753	1.2995	0.6081
−0.0024	−0.1728	0.4132	−0.0027	0.2905	−2.3368	−0.5928	−0.7066	−0.3341
−1.5959	−0.5691	0.6664	1.1476	0.0528	−0.2079	0.4760	−0.3523	−0.6874
−1.2178	0.0393	0.9418	0.9675	−0.0114	−0.3220	−0.9746	−0.0663	0.1718
0.5636	0.8843	1.2293	0.8789	−0.4601	−0.9147	0.6869	−0.7901	−0.9107
−1.8032	1.1091	0.3521	−0.0354	−1.3890	−0.0706	−1.7638	−0.1438	−0.3393
−0.9270	0.2753	−0.3444	1.2687	−0.0046	−1.3528	1.3998	−0.4565	−0.4570
0.8453	0.6641	−0.0013	0.0300	0.0875	0.1401	0.3621	0.9879	−0.5743
−0.1557	−0.1358	−0.1314	0.6539	−0.9242	0.2399	0.0259	0.0477	−1.0546
−0.4832	−0.2027	1.2441	1.1753	−0.9627	0.6382	1.2822	−0.5381	1.3506
−1.0901	0.2983	−1.3574	−0.7218	−0.2277	0.0415	−0.4500	0.7018	0.4680
0.4859	2.1673	1.3597	−0.3802	0.5927	−2.0448	1.2126	0.9862	−0.0708
0.0313	−0.1177	−0.3527	−0.5585	0.5493	−0.7759	−1.0542	2.7646	1.4251
0.0114	−0.8899	1.1575	1.0503	0.5112	0.9398	0.1870	−1.0656	0.0948
−0.4968	−1.1686	−0.1628	−0.6773	0.8950	1.4502	0.3037	−1.9043	0.8748
2.1473	−2.6982	−0.3968	−0.6915	−1.3053	0.1680	0.9301	−0.1786	−0.2899
−1.6251	1.6827	0.5806	−1.5906	−0.9129	−1.1662	−0.9490	−1.3759	−1.6889
1.1831	−1.1058	−0.7076	0.7401	1.2219	1.1260	−0.6956	−0.2945	−0.6131
1.2256	−0.2552	1.4166	−3.4767	0.3159	−0.1370	0.0577	−1.1569	−0.2177

CHI-SQUARE RIGHT-HAND TAIL PROBABILITIES

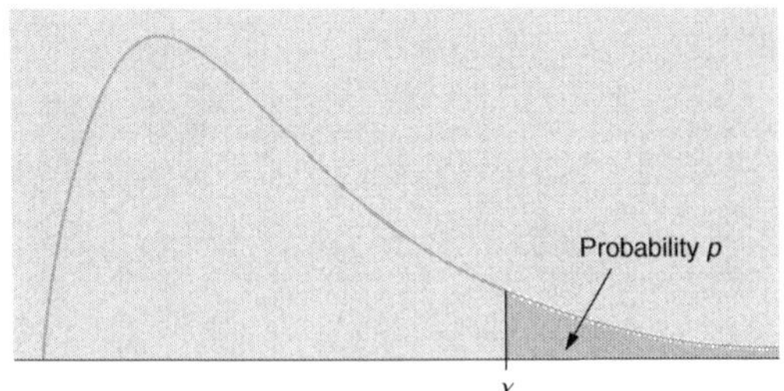

Table C.1 Right-Hand Tail Chi-Square Density Critical Values x

	Right-Hand Tail Probability p											
df	**.25**	**.20**	**.15**	**.10**	**.05**	**.025**	**.02**	**.01**	**.005**	**.0025**	**.001**	**.0005**
1	1.32	1.64	2.07	2.71	3.84	5.02	5.41	6.63	7.88	9.14	10.83	12.12
2	2.77	3.22	3.79	4.61	5.99	7.38	7.82	9.21	10.60	11.98	13.82	15.20
3	4.11	4.64	5.32	6.25	7.81	9.35	9.84	11.34	12.84	14.32	16.27	17.73
4	5.39	5.99	6.74	7.78	9.49	11.14	11.67	13.28	14.86	16.42	18.47	20.00
5	6.63	7.29	8.12	9.24	11.07	12.83	13.39	15.09	16.75	18.39	20.51	22.11
6	7.84	8.56	9.45	10.64	12.59	14.45	15.03	16.81	18.55	20.25	22.46	24.10
7	9.04	9.80	10.75	12.02	14.07	16.01	16.62	18.48	20.28	22.04	24.32	26.02
8	10.22	11.03	12.03	13.36	15.51	17.53	18.17	20.09	21.95	23.77	26.12	27.87
9	11.39	12.24	13.29	14.68	16.92	19.02	19.68	21.67	23.59	25.46	27.88	29.67
10	12.55	13.44	14.53	15.99	18.31	20.48	21.16	23.21	25.19	27.11	29.59	31.42
11	13.70	14.63	15.77	17.28	19.68	21.92	22.62	24.72	26.76	28.73	31.26	33.14
12	14.85	15.81	16.99	18.55	21.03	23.34	24.05	26.22	28.30	30.32	32.91	34.82
13	15.98	16.98	18.20	19.81	22.36	24.74	25.47	27.69	29.82	31.88	34.53	36.48
14	17.12	18.15	19.41	21.06	23.68	26.12	26.87	29.14	31.32	33.43	36.12	38.11
15	18.25	19.31	20.60	22.31	25.00	27.49	28.26	30.58	32.80	34.95	37.70	39.72
16	19.37	20.47	21.79	23.54	26.30	28.85	29.63	32.00	34.27	36.46	39.25	41.31
17	20.49	21.61	22.98	24.77	27.59	30.19	31.00	33.41	35.72	37.95	40.79	42.88
18	21.60	22.76	24.16	25.99	28.87	31.53	32.35	34.81	37.16	39.42	42.31	44.43
19	22.72	23.90	25.33	27.20	30.14	32.85	33.69	36.19	38.58	40.88	43.82	45.97
20	23.83	25.04	26.50	28.41	31.41	34.17	35.02	37.57	40.00	42.34	45.31	47.50
21	24.93	26.17	27.66	29.62	32.67	35.48	36.34	38.93	41.40	43.78	46.80	49.01
22	26.04	27.30	28.82	30.81	33.92	36.78	37.66	40.29	42.80	45.20	48.27	50.51
23	27.14	28.43	29.98	32.01	35.17	38.08	38.97	41.64	44.18	46.62	49.73	52.00
24	28.24	29.55	31.13	33.20	36.42	39.36	40.27	42.98	45.56	48.03	51.18	53.48
25	29.34	30.68	32.28	34.38	37.65	40.65	41.57	44.31	46.93	49.44	52.62	54.95
26	30.43	31.79	33.43	35.56	38.89	41.92	42.86	45.64	48.29	50.83	54.05	56.41
27	31.53	32.91	34.57	36.74	40.11	43.19	44.14	46.96	49.64	52.22	55.48	57.86
28	32.62	34.03	35.71	37.92	41.34	44.46	45.42	48.28	50.99	53.59	56.89	59.30
29	33.71	35.14	36.85	39.09	42.56	45.72	46.69	49.59	52.34	54.97	58.30	60.73
30	34.80	36.25	37.99	40.26	43.77	46.98	47.96	50.89	53.67	56.33	59.70	62.16
40	45.62	47.27	49.24	51.81	55.76	59.34	60.44	63.69	66.77	69.70	73.40	76.09
50	56.33	58.16	60.35	63.17	67.50	71.42	72.61	76.15	79.49	82.66	86.66	89.56
60	66.98	68.97	71.34	74.40	79.08	83.30	84.58	88.38	91.95	95.34	99.61	102.7
80	88.13	90.41	93.11	96.58	101.9	106.6	108.1	112.3	116.3	120.1	124.8	128.3
100	109.1	111.7	114.7	118.5	124.3	129.6	131.1	135.8	140.2	144.3	149.4	153.2

Table entry is the value of x (critical value) corresponding to $P(x_{\text{df}}^2 \geq x) = p$ for the chosen probability p (column) and number of df (row).

D

LINEAR INTERPOLATION

Interpolation can be used to help find a value that is not provided in a table. The procedure is called linear interpolation, since it assumes that the relations between the values involved are (or are nearly) linear—that is, they are straight-line relationships. This assumption gives very good estimates of desired values for the exercises in this book.

Example 1 **Find $P(\chi^2_{10} \geq 20.0)$.**

The chi-square table in Appendix C does not have an entry of 20.0 for a chi-square with 10 degrees of freedom. So interpolation must be used.

The chi-square values of 18.31 and 20.48 are entries in the table, with corresponding probabilities (areas) of 0.05 and 0.025, respectively (that is, $P(\chi^2_{10} \geq 18.31) = 0.05$ and $P(\chi^2_{10} \geq 20.48) = 0.025$). So $P(\chi^2_{10} \geq 20.0)$ is somewhere between 0.05 and 0.025. Note that $20.48 - 18.31 = 2.17$ and $20.0 - 18.31 = 1.69$. Now consider the following proportionality argument. The following table shows the proportions we are dealing with.

		Chi-square	Probability (area)		
2.17	1.69	18.31	0.05	d	0.025
		20.0	$P(\chi^2_{10} \geq 20.0)$		
		20.48	0.025		

The unknown probability $P(\chi^2_{10} \geq 20.0)$ is proportionally decreased from 0.05 by an amount d according to the proportionality equation

$$\frac{d}{0.025} = \frac{1.69}{2.17}$$

Thus,

$$d = \frac{1.69}{2.17}(0.025) \approx 0.019.$$

Thus,

$$P(\chi^2_{10} \geq 20.0) \approx 0.05 - 0.019 \approx 0.031.$$

Example 2 Find the value of a in the following:

The chi-square table in Appendix C does not have a column for a probability of 0.03. However, there are columns for 0.05 and 0.025, so we can use interpolation to find the required chi-square value a with 12 degrees of freedom.

		Chi-square	Probability (area)		
2.31	d	21.03	0.05	0.02	0.025
		a	0.03		
		23.34	0.025		

Thus the unknown a is proportionally increased from 21.03 by an amount d according to the proportionality equation

$$\frac{d}{2.31} = \frac{0.02}{0.025} = 0.8.$$

So we have

$$d = 0.8(2.31) \approx 1.85.$$

The unknown value a is thus

$$a \approx 21.03 + d \approx 21.03 + 1.85 = 22.88.$$

Therefore,

$$P(\chi^2_{12} \geq 22.88) \approx 0.03.$$

E

STANDARD NORMAL DENSITY LEFT-HAND TAIL PROBABILITIES

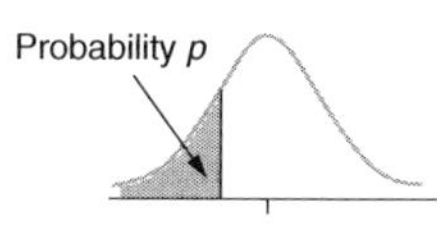

Table E.1 Standard Normal Probabilities to Left of Given z

z	0	1	2	3	4	5	6	7	8	9
−3.4*	.0003	.0003	.0003	.0003	.0003	.0003	.0003	.0003	.0003	.0002
−3.3	.0005	.0005	.0005	.0004	.0004	.0004	.0004	.0004	.0004	.0003
−3.2	.0007	.0007	.0006	.0006	.0006	.0006	.0006	.0005	.0005	.0005
−3.1	.0010	.0009	.0009	.0009	.0008	.0008	.0008	.0008	.0007	.0007
−3.0	.0013	.0013	.0013	.0012	.0012	.0011	.0011	.0011	.0010	.0010
−2.9	.0019	.0018	.0017	.0017	.0016	.0016	.0015	.0015	.0014	.0014
−2.8	.0026	.0025	.0024	.0023	.0023	.0022	.0021	.0021	.0020	.0019
−2.7	.0035	.0034	.0033	.0032	.0031	.0030	.0029	.0028	.0027	.0026
−2.6	.0047	.0045	.0044	.0043	.0041	.0040	.0039	.0038	.0037	.0036
−2.5	.0062	.0060	.0059	.0057	.0055	.0054	.0052	.0051	.0049	.0048
−2.4	.0082	.0080	.0078	.0075	.0073	.0071	.0069	.0068	.0066	.0064
−2.3	.0107	.0104	.0102	.0099	.0096	.0094	.0091	.0089	.0087	.0084
−2.2	.0139	.0136	.0132	.0129	.0125	.0122	.0119	.0116	.0113	.0110
−2.1	.0179	.0174	.0170	.0166	.0162	.0158	.0154	.0150	.0146	.0143
−2.0	.0228	.0222	.0217	.0212	.0207	.0202	.0197	.0192	.0188	.0183
−1.9	.0287	.0281	.0274	.0268	.0262	.0256	.0250	.0244	.0239	.0233
−1.8	.0359	.0351	.0344	.0336	.0329	.0322	.0314	.0307	.0301	.0294
−1.7	.0446	.0436	.0427	.0418	.0409	.0401	.0392	.0384	.0375	.0367
−1.6	.0548	.0537	.0526	.0516	.0505	.0495	.0485	.0475	.0465	.0455
−1.5	.0668	.0655	.0643	.0630	.0618	.0606	.0594	.0582	.0571	.0559
−1.4	.0808	.0793	.0778	.0764	.0749	.0735	.0721	.0708	.0694	.0681
−1.3	.0968	.0951	.0934	.0918	.0901	.0885	.0869	.0853	.0838	.0823
−1.2	.1151	.1131	.1112	.1093	.1075	.1056	.1038	.1020	.1003	.0985
−1.1	.1357	.1335	.1314	.1292	.1271	.1251	.1230	.1210	.1190	.1170
−1.0	.1587	.1562	.1539	.1515	.1492	.1469	.1446	.1423	.1401	.1379
−0.9	.1841	.1814	.1788	.1762	.1736	.1711	.1685	.1660	.1635	.1611
−0.8	.2119	.2090	.2061	.2033	.2005	.1977	.1949	.1922	.1894	.1867
−0.7	.2420	.2389	.2358	.2327	.2296	.2266	.2236	.2206	.2177	.2148
−0.6	.2743	.2709	.2676	.2643	.2611	.2578	.2546	.2514	.2483	.2451
−0.5	.3085	.3050	.3015	.2981	.2946	.2912	.2877	.2843	.2810	.2776
−0.4	.3446	.3409	.3372	.3336	.3300	.3264	.3228	.3192	.3156	.3121
−0.3	.3821	.3783	.3745	.3707	.3669	.3632	.3594	.3557	.3520	.3483
−0.2	.4207	.4168	.4129	.4090	.4052	.4013	.3974	.3936	.3897	.3859
−0.1	.4602	.4562	.4522	.4483	.4443	.4404	.4364	.4325	.4286	.4247
−0.0	.5000	.4960	.4920	.4880	.4840	.4801	.4761	.4721	.4681	.4641

Table entry is the left-hand tail probability corresponding to $P(Z \le z) = p$ for the chosen value of z given by the chosen row (tenths) and column (hundreths).

*For any $z \le -3.5$, the area is approximately 0; indeed, for any $z \le -4$, the area is 0 to four decimal places.

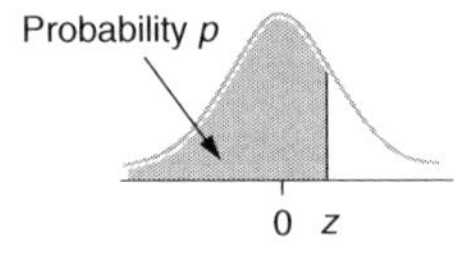

Table E.1 Standard Normal Probabilities to Left of Given *z* (*Continued*)

z	0	1	2	3	4	5	6	7	8	9
0.0	.5000	.5040	.5080	.5120	.5160	.5199	.5239	.5279	.5319	.5359
0.1	.5398	.5438	.5478	.5517	.5557	.5596	.5636	.5675	.5714	.5753
0.2	.5793	.5832	.5871	.5910	.5948	.5987	.6026	.6064	.6103	.6141
0.3	.6179	.6217	.6255	.6293	.6331	.6368	.6406	.6443	.6480	.6517
0.4	.6554	.6591	.6628	.6664	.6700	.6736	.6772	.6808	.6844	.6879
0.5	.6915	.6950	.6985	.7019	.7054	.7088	.7123	.7157	.7190	.7224
0.6	.7257	.7291	.7324	.7357	.7389	.7422	.7454	.7486	.7517	.7549
0.7	.7580	.7611	.7642	.7673	.7704	.7734	.7764	.7794	.7823	.7852
0.8	.7881	.7910	.7939	.7967	.7995	.8023	.8051	.8078	.8106	.8133
0.9	.8159	.8186	.8212	.8238	.8264	.8289	.8315	.8340	.8365	.8389
1.0	.8413	.8438	.8461	.8485	.8508	.8531	.8554	.8577	.8599	.8621
1.1	.8643	.8665	.8686	.8708	.8729	.8749	.8770	.8790	.8810	.8831
1.2	.8849	.8869	.8888	.8907	.8925	.8944	.8962	.8980	.8997	.9015
1.3	.9032	.9049	.9066	.9082	.9099	.9115	.9131	.9147	.9162	.9177
1.4	.9192	.9207	.9222	.9236	.9251	.9265	.9279	.9292	.9306	.9319
1.5	.9332	.9345	.9357	.9370	.9382	.9394	.9406	.9418	.9429	.9441
1.6	.9452	.9463	.9474	.9484	.9495	.9505	.9515	.9525	.9535	.9545
1.7	.9554	.9564	.9573	.9582	.9591	.9599	.9608	.9616	.9625	.9633
1.8	.9641	.9649	.9656	.9664	.9671	.9678	.9686	.9693	.9699	.9706
1.9	.9713	.9717	.9726	.9732	.9738	.9744	.9750	.9756	.9761	.9767
2.0	.9772	.9778	.9783	.9788	.9793	.9798	.9803	.9808	.9812	.9817
2.1	.9821	.9826	.9830	.9834	.9838	.9842	.9846	.9850	.9854	.9857
2.2	.9861	.9864	.9868	.9871	.9875	.9878	.9881	.9884	.9887	.9890
2.3	.9893	.9896	.9898	.9901	.9904	.9906	.9909	.9911	.9913	.9916
2.4	.9918	.9920	.9922	.9925	.9927	.9929	.9931	.9932	.9934	.9936
2.5	.9938	.9940	.9941	.9943	.9945	.9946	.9948	.9949	.9951	.9952
2.6	.9953	.9955	.9956	.9957	.9959	.9960	.9961	.9962	.9963	.9964
2.7	.9965	.9966	.9967	.9968	.9969	.9970	.9971	.9972	.9973	.9974
2.8	.9974	.9975	.9976	.9977	.9977	.9978	.9979	.9979	.9980	.9981
2.9	.9981	.9982	.9982	.9983	.9984	.9984	.9985	.9985	.9986	.9986
3.0	.9987	.9987	.9987	.9988	.9988	.9989	.9989	.9989	.9990	.9990
3.1	.9990	.9991	.9991	.9991	.9992	.9992	.9992	.9992	.9993	.9993
3.2	.9993	.9993	.9994	.9994	.9994	.9994	.9994	.9995	.9995	.9995
3.3	.9995	.9995	.9995	.9996	.9996	.9996	.9996	.9996	.9996	.9997
3.4*	.9997	.9997	.9997	.9997	.9997	.9997	.9997	.9997	.9997	.9998

Table entry is the left-hand tail probability corresponding to $P(Z \leq z) = p$ for the chosen value of z given by the chosen row (tenths) and column (hundredths).

*For any $z \geq -3.5$, the area is approximately 1; indeed, for any $z \geq -4$, the area is 1 to four decimal places.

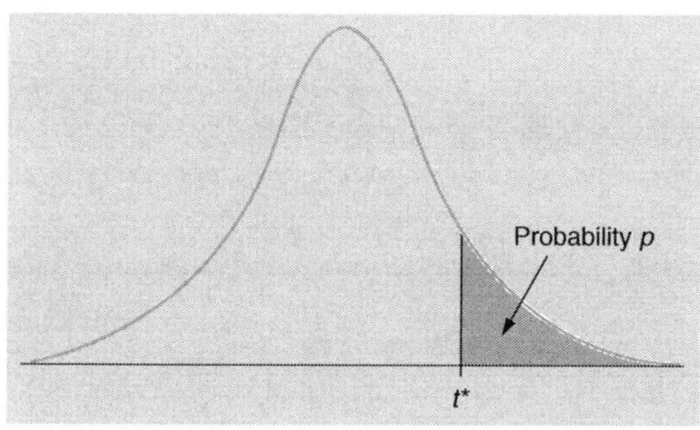

T-DISTRIBUTION RIGHT-HAND TAIL PROBABILITIES

Table F Table entry for p and C is the critical value t^* with probability p lying to its right and probability C lying between $-t^*$ and t^*

	Upper tail probability p											
df	**.25**	**.20**	**.15**	**.10**	**.05**	**.025**	**.02**	**.01**	**.005**	**.0025**	**.001**	**.0005**
1	1.000	1.376	1.963	3.078	6.314	12.71	15.89	31.82	63.66	127.3	318.3	636.6
2	0.816	1.061	1.386	1.886	2.920	4.303	4.849	6.965	9.925	14.09	22.33	31.60
3	0.765	0.978	1.250	1.638	2.353	3.182	3.482	4.541	5.841	7.453	10.21	12.92
4	0.741	0.941	1.190	1.533	2.132	2.776	2.999	3.747	4.604	5.598	7.173	8.610
5	0.727	0.920	1.156	1.476	2.015	2.571	2.757	3.365	4.032	4.773	5.893	6.869
6	0.718	0.906	1.134	1.440	1.943	2.447	2.612	3.143	3.707	4.317	5.208	5.959
7	0.711	0.896	1.119	1.415	1.895	2.365	2.517	2.998	3.499	4.029	4.785	5.408
8	0.706	0.889	1.108	1.397	1.860	2.306	2.449	2.896	3.355	3.833	4.501	5.041
9	0.703	0.883	1.100	1.383	1.833	2.262	2.398	2.821	3.250	3.690	4.297	4.781
10	0.700	0.879	1.093	1.372	1.812	2.228	2.359	2.764	3.169	3.581	4.144	4.587
11	0.697	0.876	1.088	1.363	1.796	2.201	2.328	2.718	3.106	3.497	4.025	4.437
12	0.695	0.873	1.083	1.356	1.782	2.179	2.303	2.681	3.055	3.428	3.930	4.318
13	0.694	0.870	1.079	1.350	1.771	2.160	2.282	2.650	3.012	3.372	3.852	4.221
14	0.692	0.868	1.076	1.345	1.761	2.145	2.264	2.624	2.977	3.326	3.787	4.140
15	0.691	0.866	1.074	1.341	1.753	2.131	2.249	2.602	2.947	3.286	3.733	4.073
16	0.690	0.865	1.071	1.337	1.746	2.120	2.235	2.583	2.921	3.252	3.686	4.015
17	0.689	0.863	1.069	1.333	1.740	2.110	2.224	2.567	2.898	3.222	3.646	3.965
18	0.688	0.862	1.067	1.330	1.734	2.101	2.214	2.552	2.878	3.197	3.611	3.922
19	0.688	0.861	1.066	1.328	1.729	2.093	2.205	2.539	2.861	3.174	3.579	3.883
20	0.687	0.860	1.064	1.325	1.725	2.086	2.197	2.528	2.845	3.153	3.552	3.850
21	0.686	0.859	1.063	1.323	1.721	2.080	2.189	2.518	2.831	3.135	3.527	3.819
22	0.686	0.858	1.061	1.321	1.717	2.074	2.183	2.508	2.819	3.119	3.505	3.792
23	0.685	0.858	1.060	1.319	1.714	2.069	2.177	2.500	2.807	3.104	3.485	3.768
24	0.685	0.857	1.059	1.318	1.711	2.064	2.172	2.492	2.797	3.091	3.467	3.745
25	0.684	0.856	1.058	1.316	1.708	2.060	2.167	2.485	2.787	3.078	3.450	3.725
26	0.684	0.856	1.058	1.315	1.706	2.056	2.162	2.479	2.779	3.067	3.435	3.707
27	0.684	0.855	1.057	1.314	1.703	2.052	2.158	2.473	2.771	3.057	3.421	3.690
28	0.683	0.855	1.056	1.313	1.701	2.048	2.154	2.467	2.763	3.047	3.408	3.674
29	0.683	0.854	1.055	1.311	1.699	2.045	2.150	2.462	2.756	3.038	3.396	3.659
30	0.683	0.854	1.055	1.310	1.697	2.042	2.147	2.457	2.750	3.030	3.385	3.646
40	0.681	0.851	1.050	1.303	1.684	2.021	2.123	2.423	2.704	2.971	3.307	3.551
50	0.679	0.849	1.047	1.299	1.676	2.009	2.109	2.403	2.678	2.937	3.261	3.496
60	0.679	0.848	1.045	1.296	1.671	2.000	2.099	2.390	2.660	2.915	3.232	3.460
80	0.678	0.846	1.043	1.292	1.664	1.990	2.088	2.374	2.639	2.887	3.195	3.416
100	0.677	0.845	1.042	1.290	1.660	1.984	2.081	2.364	2.626	2.871	3.174	3.390
120	0.677	0.845	1.041	1.289	1.658	1.980	2.076	2.358	2.617	2.860	3.160	3.373
150	0.676	0.844	1.040	1.287	1.655	1.976	2.072	2.351	2.609	2.849	3.145	3.357
200	0.676	0.843	1.039	1.286	1.653	1.972	2.067	2.345	2.601	2.838	3.131	3.340
300	0.675	0.843	1.038	1.284	1.650	1.968	2.063	2.339	2.592	2.828	3.118	3.323
500	0.675	0.842	1.038	1.283	1.648	1.965	2.059	2.334	2.586	2.820	3.107	3.310
1000	0.675	0.842	1.037	1.282	1.646	1.962	2.056	2.330	2.581	2.813	3.098	3.300
∞*	0.674	0.842	1.036	1.282	1.645	1.960	2.054	2.326	2.576	2.807	3.091	3.291
	50%	**60%**	**70%**	**80%**	**90%**	**95%**	**96%**	**98%**	**99%**	**99.5%**	**99.8%**	**99.9%**
	Confidence level C											

*This row lists the corresponding standard normal critical values.

CUMULATIVE BINOMIAL PROBABILITIES

The table gives the cumulative probability of obtaining x or fewer successes in n independent Bernoulli trials, namely $P(X \leq x)$, where $p =$ probability of success in a single trial.

Table G Cumulative Binomial Probabilities

		p									
n	*x*	**0.05**	**0.10**	**0.15**	**0.20**	**0.25**	**0.30**	**0.35**	**0.40**	**0.45**	**0.50**
2	0	0.9025	0.8100	0.7225	0.6400	0.5625	0.4900	0.4225	0.3600	0.3025	0.2500
	1	0.9975	0.9900	0.9775	0.9600	0.9375	0.9100	0.8775	0.8400	0.7975	0.7500
	2	1.0000	1.0000	1.0000	1.0000	1.0000	1.0000	1.0000	1.0000	1.0000	1.0000
3	0	0.8574	0.7290	0.6141	0.5120	0.4219	0.3430	0.2746	0.2160	0.1664	0.1250
	1	0.9928	0.9720	0.9393	0.8960	0.8438	0.7840	0.7183	0.6480	0.5748	0.5000
	2	0.9999	0.9990	0.9966	0.9920	0.9844	0.9730	0.9571	0.9360	0.9089	0.8750
	3	1.0000	1.0000	1.0000	1.0000	1.0000	1.0000	1.0000	1.0000	1.0000	1.0000
4	0	0.8145	0.6561	0.5220	0.4096	0.3164	0.2401	0.1785	0.1296	0.0915	0.0625
	1	0.9860	0.9477	0.8905	0.8192	0.7383	0.6517	0.5630	0.4752	0.3910	0.3125
	2	0.9995	0.9963	0.9880	0.9728	0.9492	0.9163	0.8735	0.8208	0.7585	0.6875
	3	1.0000	0.9999	0.9995	0.9984	0.9961	0.9919	0.9850	0.9744	0.9590	0.9375
	4	1.0000	1.0000	1.0000	1.0000	1.0000	1.0000	1.0000	1.0000	1.0000	1.0000
5	0	0.7738	0.5905	0.4437	0.3277	0.2373	0.1681	0.1160	0.0778	0.0503	0.0313
	1	0.9774	0.9185	0.8352	0.7373	0.6328	0.5282	0.4284	0.3370	0.2562	0.1875
	2	0.9988	0.9914	0.9734	0.9421	0.8965	0.8369	0.7648	0.6826	0.5931	0.5000
	3	1.0000	0.9995	0.9978	0.9933	0.9844	0.9692	0.9460	0.9130	0.8688	0.8125
	4	1.0000	1.0000	0.9999	0.9997	0.9990	0.9976	0.9947	0.9898	0.9815	0.9688
	5	1.0000	1.0000	1.0000	1.0000	1.0000	1.0000	1.0000	1.0000	1.0000	1.0000
6	0	0.7351	0.5314	0.3771	0.2621	0.1780	0.1176	0.0754	0.0467	0.0277	0.0156
	1	0.9672	0.8857	0.7765	0.6554	0.5339	0.4202	0.3191	0.2333	0.1636	0.1094
	2	0.9978	0.9842	0.9527	0.9011	0.8306	0.7443	0.6471	0.5443	0.4415	0.3438
	3	0.9999	0.9987	0.9941	0.9830	0.9624	0.9295	0.8826	0.8208	0.7447	0.6563
	4	1.0000	0.9999	0.9996	0.9984	0.9954	0.9891	0.9777	0.9590	0.9308	0.8906
	5	1.0000	1.0000	1.0000	0.9999	0.9998	0.9993	0.9982	0.9959	0.9917	0.9844
	6	1.0000	1.0000	1.0000	1.0000	1.0000	1.0000	1.0000	1.0000	1.0000	1.0000
7	0	0.6983	0.4783	0.3206	0.2097	0.1335	0.0824	0.0490	0.0280	0.0152	0.0078
	1	0.9556	0.8503	0.7166	0.5767	0.4449	0.3294	0.2338	0.1586	0.1024	0.0625
	2	0.9962	0.9743	0.9262	0.8520	0.7564	0.6471	0.5323	0.4199	0.3164	0.2266
	3	0.9998	0.9973	0.9879	0.9667	0.9294	0.8740	0.8002	0.7102	0.6083	0.5000
	4	1.0000	0.9998	0.9988	0.9953	0.9871	0.9712	0.9444	0.9037	0.8471	0.7734
	5	1.0000	1.0000	0.9999	0.9996	0.9987	0.9962	0.9910	0.9812	0.9643	0.9375
	6	1.0000	1.0000	1.0000	1.0000	0.9999	0.9998	0.9994	0.9984	0.9963	0.9922
	7	1.0000	1.0000	1.0000	1.0000	1.0000	1.0000	1.0000	1.0000	1.0000	1.0000

Table G Cumulative Binomial Probabilities *(Continued)*

		p									
n	*x*	**0.05**	**0.10**	**0.15**	**0.20**	**0.25**	**0.30**	**0.35**	**0.40**	**0.45**	**0.50**
8	0	0.6634	0.4305	0.2725	0.1678	0.1001	0.0576	0.0319	0.0168	0.0084	0.0039
	1	0.9428	0.8131	0.6572	0.5033	0.3671	0.2553	0.1691	0.1064	0.0632	0.0352
	2	0.9942	0.9619	0.8948	0.7969	0.6785	0.5518	0.4278	0.3154	0.2201	0.1445
	3	0.9996	0.9950	0.9786	0.9437	0.8862	0.8059	0.7064	0.5941	0.4770	0.3633
	4	1.0000	0.9996	0.9971	0.9896	0.9727	0.9420	0.8939	0.8263	0.7396	0.6367
	5	1.0000	1.0000	0.9998	0.9988	0.9958	0.9887	0.9747	0.9502	0.9115	0.8555
	6	1.0000	1.0000	1.0000	0.9999	0.9996	0.9987	0.9964	0.9915	0.9819	0.9648
	7	1.0000	1.0000	1.0000	1.0000	1.0000	0.9999	0.9998	0.9993	0.9983	0.9961
	8	1.0000	1.0000	1.0000	1.0000	1.0000	1.0000	1.0000	1.0000	1.0000	1.0000
9	0	0.6302	0.3874	0.2316	0.1342	0.0751	0.0404	0.0207	0.0101	0.0046	0.0020
	1	0.9288	0.7748	0.5995	0.4362	0.3003	0.1960	0.1211	0.0705	0.0385	0.0195
	2	0.9916	0.9470	0.8591	0.7382	0.6007	0.4628	0.3373	0.2318	0.1495	0.0898
	3	0.9994	0.9917	0.9661	0.9144	0.8343	0.7297	0.6089	0.4826	0.3614	0.2539
	4	1.0000	0.9991	0.9944	0.9804	0.9511	0.9012	0.8283	0.7334	0.6214	0.5000
	5	1.0000	0.9999	0.9994	0.9969	0.9900	0.9747	0.9464	0.9006	0.8342	0.7461
	6	1.0000	1.0000	1.0000	0.9997	0.9987	0.9957	0.9888	0.9750	0.9502	0.9102
	7	1.0000	1.0000	1.0000	1.0000	0.9999	0.9996	0.9986	0.9962	0.9909	0.9805
	8	1.0000	1.0000	1.0000	1.0000	1.0000	1.0000	0.9999	0.9997	0.9992	0.9980
	9	1.0000	1.0000	1.0000	1.0000	1.0000	1.0000	1.0000	1.0000	1.0000	1.0000
10	0	0.5987	0.3487	0.1969	0.1074	0.0563	0.0282	0.0135	0.0060	0.0025	0.0010
	1	0.9139	0.7361	0.5443	0.3758	0.2440	0.1493	0.0860	0.0464	0.0233	0.0107
	2	0.9885	0.9298	0.8202	0.6778	0.5256	0.3828	0.2616	0.1673	0.0996	0.0547
	3	0.9990	0.9872	0.9500	0.8791	0.7759	0.6496	0.5138	0.3823	0.2660	0.1719
	4	0.9999	0.9984	0.9901	0.9672	0.9219	0.8497	0.7515	0.6331	0.5044	0.3770
	5	1.0000	0.9999	0.9986	0.9936	0.9803	0.9527	0.9051	0.8338	0.7384	0.6230
	6	1.0000	1.0000	0.9999	0.9991	0.9965	0.9894	0.9740	0.9452	0.8980	0.8281
	7	1.0000	1.0000	1.0000	0.9999	0.9996	0.9984	0.9952	0.9877	0.9726	0.9453
	8	1.0000	1.0000	1.0000	1.0000	1.0000	0.9999	0.9995	0.9983	0.9955	0.9893
	9	1.0000	1.0000	1.0000	1.0000	1.0000	1.0000	1.0000	0.9999	0.9997	0.9990
	10	1.0000	1.0000	1.0000	1.0000	1.0000	1.0000	1.0000	1.0000	1.0000	1.0000
11	0	0.5688	0.3138	0.1673	0.0859	0.0422	0.0198	0.0088	0.0036	0.0014	0.0005
	1	0.8981	0.6974	0.4922	0.3221	0.1971	0.1130	0.0606	0.0302	0.0139	0.0059
	2	0.9848	0.9104	0.7788	0.6174	0.4552	0.3127	0.2001	0.1189	0.0652	0.0327
	3	0.9984	0.9815	0.9306	0.8389	0.7133	0.5696	0.4256	0.2963	0.1911	0.1133
	4	0.9999	0.9972	0.9841	0.9496	0.8854	0.7897	0.6683	0.5328	0.3971	0.2744
	5	1.0000	0.9997	0.9973	0.9883	0.9657	0.9218	0.8513	0.7535	0.6331	0.5000
	6	1.0000	1.0000	0.9997	0.9980	0.9924	0.9784	0.9499	0.9006	0.8262	0.7256
	7	1.0000	1.0000	1.0000	0.9998	0.9988	0.9957	0.9878	0.9707	0.9390	0.8867
	8	1.0000	1.0000	1.0000	1.0000	0.9999	0.9994	0.9980	0.9941	0.9852	0.9673
	9	1.0000	1.0000	1.0000	1.0000	1.0000	1.0000	0.9998	0.9993	0.9978	0.9941
	10	1.0000	1.0000	1.0000	1.0000	1.0000	1.0000	1.0000	1.0000	0.9998	0.9995
	11	1.0000	1.0000	1.0000	1.0000	1.0000	1.0000	1.0000	1.0000	1.0000	1.0000

Table G Cumulative Binomial Probabilities *(Continued)*

		p									
n	*x*	**0.05**	**0.10**	**0.15**	**0.20**	**0.25**	**0.30**	**0.35**	**0.40**	**0.45**	**0.50**
12	0	0.5404	0.2824	0.1422	0.0687	0.0317	0.0138	0.0057	0.0022	0.0008	0.0002
	1	0.8816	0.6590	0.4435	0.2749	0.1584	0.0850	0.0424	0.0196	0.0083	0.0032
	2	0.9804	0.8891	0.7358	0.5583	0.3907	0.2528	0.1513	0.0834	0.0421	0.0193
	3	0.9978	0.9744	0.9078	0.7946	0.6488	0.4925	0.3467	0.2253	0.1345	0.0730
	4	0.9998	0.9957	0.9761	0.9274	0.8424	0.7237	0.5833	0.4382	0.3044	0.1938
	5	1.0000	0.9995	0.9954	0.9806	0.9456	0.8822	0.7873	0.6652	0.5269	0.3872
	6	1.0000	0.9999	0.9993	0.9961	0.9857	0.9614	0.9154	0.8418	0.7393	0.6128
	7	1.0000	1.0000	0.9999	0.9994	0.9972	0.9905	0.9745	0.9427	0.8883	0.8062
	8	1.0000	1.0000	1.0000	0.9999	0.9996	0.9983	0.9944	0.9847	0.9644	0.9270
	9	1.0000	1.0000	1.0000	1.0000	1.0000	0.9998	0.9992	0.9972	0.9921	0.9807
	10	1.0000	1.0000	1.0000	1.0000	1.0000	1.0000	0.9999	0.9997	0.9989	0.9968
	11	1.0000	1.0000	1.0000	1.0000	1.0000	1.0000	1.0000	1.0000	0.9999	0.9998
	12	1.0000	1.0000	1.0000	1.0000	1.0000	1.0000	1.0000	1.0000	1.0000	1.0000
13	0	0.5133	0.2542	0.1209	0.0550	0.0238	0.0097	0.0037	0.0013	0.0004	0.0001
	1	0.8646	0.6213	0.3983	0.2336	0.1267	0.0637	0.0296	0.0126	0.0049	0.0017
	2	0.9755	0.8661	0.6920	0.5017	0.3326	0.2025	0.1132	0.0579	0.0269	0.0112
	3	0.9969	0.9658	0.8820	0.7473	0.5843	0.4206	0.2783	0.1686	0.0929	0.0461
	4	0.9997	0.9935	0.9658	0.9009	0.7940	0.6543	0.5005	0.3530	0.2279	0.1334
	5	1.0000	0.9991	0.9925	0.9700	0.9198	0.8346	0.7159	0.5744	0.4268	0.2905
	6	1.0000	0.9999	0.9987	0.9930	0.9757	0.9376	0.8705	0.7712	0.6437	0.5000
	7	1.0000	1.0000	0.9998	0.9988	0.9944	0.9818	0.9538	0.9023	0.8212	0.7095
	8	1.0000	1.0000	1.0000	0.9998	0.9990	0.9960	0.9874	0.9679	0.9302	0.8666
	9	1.0000	1.0000	1.0000	1.0000	0.9999	0.9993	0.9975	0.9922	0.9797	0.9539
	10	1.0000	1.0000	1.0000	1.0000	1.0000	0.9999	0.9997	0.9987	0.9959	0.9888
	11	1.0000	1.0000	1.0000	1.0000	1.0000	1.0000	1.0000	0.9999	0.9995	0.9983
	12	1.0000	1.0000	1.0000	1.0000	1.0000	1.0000	1.0000	1.0000	1.0000	0.9999
	13	1.0000	1.0000	1.0000	1.0000	1.0000	1.0000	1.0000	1.0000	1.0000	1.0000
14	0	0.4877	0.2288	0.1028	0.0440	0.0178	0.0068	0.0024	0.0008	0.0002	0.0001
	1	0.8470	0.5846	0.3567	0.1979	0.1010	0.0475	0.0205	0.0081	0.0029	0.0009
	2	0.9699	0.8416	0.6479	0.4481	0.2811	0.1608	0.0839	0.0398	0.0170	0.0065
	3	0.9958	0.9559	0.8535	0.6982	0.5213	0.3552	0.2205	0.1243	0.0632	0.0287
	4	0.9996	0.9908	0.9533	0.8702	0.7415	0.5842	0.4227	0.2793	0.1672	0.0898
	5	1.0000	0.9985	0.9885	0.9561	0.8883	0.7805	0.6405	0.4859	0.3373	0.2120
	6	1.0000	0.9998	0.9978	0.9884	0.9617	0.9067	0.8164	0.6925	0.5461	0.3953
	7	1.0000	1.0000	0.9997	0.9976	0.9897	0.9685	0.9247	0.8499	0.7414	0.6047
	8	1.0000	1.0000	1.0000	0.9996	0.9978	0.9917	0.9757	0.9417	0.8811	0.7880
	9	1.0000	1.0000	1.0000	1.0000	0.9997	0.9983	0.9940	0.9825	0.9574	0.9102
	10	1.0000	1.0000	1.0000	1.0000	1.0000	0.9998	0.9989	0.9961	0.9886	0.9713
	11	1.0000	1.0000	1.0000	1.0000	1.0000	1.0000	0.9999	0.9994	0.9978	0.9935
	12	1.0000	1.0000	1.0000	1.0000	1.0000	1.0000	1.0000	0.9999	0.9997	0.9991
	13	1.0000	1.0000	1.0000	1.0000	1.0000	1.0000	1.0000	1.0000	1.0000	0.9999
	14	1.0000	1.0000	1.0000	1.0000	1.0000	1.0000	1.0000	1.0000	1.0000	1.0000
15	0	0.4633	0.2059	0.0874	0.0352	0.0134	0.0047	0.0016	0.0005	0.0001	0.0000
	1	0.8290	0.5490	0.3186	0.1671	0.0802	0.0353	0.0142	0.0052	0.0017	0.0005
	2	0.9638	0.8159	0.6042	0.3980	0.2361	0.1268	0.0617	0.0271	0.0107	0.0037
	3	0.9945	0.9444	0.8227	0.6482	0.4613	0.2969	0.1727	0.0905	0.0424	0.0176

Table G Cumulative Binomial Probabilities *(Continued)*

		p									
n	*x*	0.05	0.10	0.15	0.20	0.25	0.30	0.35	0.40	0.45	0.50
15	4	0.9994	0.9873	0.9383	0.8358	0.6865	0.5155	0.3519	0.2173	0.1204	0.0592
	5	0.9999	0.9978	0.9832	0.9389	0.8516	0.7216	0.5643	0.4032	0.2608	0.1509
	6	1.0000	0.9997	0.9964	0.9819	0.9434	0.8689	0.7548	0.6098	0.4522	0.3036
	7	1.0000	1.0000	0.9994	0.9958	0.9827	0.9500	0.8868	0.7869	0.6535	0.5000
	8	1.0000	1.0000	0.9999	0.9992	0.9958	0.9848	0.9578	0.9050	0.8182	0.6964
	9	1.0000	1.0000	1.0000	0.9999	0.9992	0.9963	0.9876	0.9662	0.9231	0.8491
	10	1.0000	1.0000	1.0000	1.0000	0.9999	0.9993	0.9972	0.9907	0.9745	0.9408
	11	1.0000	1.0000	1.0000	1.0000	1.0000	0.9999	0.9995	0.9981	0.9937	0.9824
	12	1.0000	1.0000	1.0000	1.0000	1.0000	1.0000	0.9999	0.9997	0.9989	0.9963
	13	1.0000	1.0000	1.0000	1.0000	1.0000	1.0000	1.0000	1.0000	0.9999	0.9995
	14	1.0000	1.0000	1.0000	1.0000	1.0000	1.0000	1.0000	1.0000	1.0000	1.0000
	15	1.0000	1.0000	1.0000	1.0000	1.0000	1.0000	1.0000	1.0000	1.0000	1.0000
16	0	0.4401	0.1853	0.0743	0.0281	0.0100	0.0033	0.0010	0.0003	0.0001	0.0000
	1	0.8108	0.5147	0.2839	0.1407	0.0635	0.0261	0.0098	0.0033	0.0010	0.0003
	2	0.9571	0.7892	0.5614	0.3518	0.1971	0.0994	0.0451	0.0183	0.0066	0.0021
	3	0.9930	0.9316	0.7899	0.5981	0.4050	0.2459	0.1339	0.0651	0.0281	0.0106
	4	0.9991	0.9830	0.9209	0.7982	0.6302	0.4499	0.2892	0.1666	0.0853	0.0384
	5	0.9999	0.9967	0.9765	0.9183	0.8103	0.6598	0.4900	0.3288	0.1976	0.1051
	6	1.0000	0.9995	0.9944	0.9733	0.9204	0.8247	0.6881	0.5272	0.3660	0.2272
	7	1.0000	0.9999	0.9989	0.9930	0.9729	0.9256	0.8406	0.7161	0.5629	0.4018
	8	1.0000	1.0000	0.9998	0.9985	0.9925	0.9743	0.9329	0.8577	0.7441	0.5982
	9	1.0000	1.0000	1.0000	0.9998	0.9984	0.9929	0.9771	0.9417	0.8759	0.7728
	10	1.0000	1.0000	1.0000	1.0000	0.9997	0.9984	0.9938	0.9809	0.9514	0.8949
	11	1.0000	1.0000	1.0000	1.0000	1.0000	0.9997	0.9987	0.9951	0.9851	0.9616
	12	1.0000	1.0000	1.0000	1.0000	1.0000	1.0000	0.9998	0.9991	0.9965	0.9894
	13	1.0000	1.0000	1.0000	1.0000	1.0000	1.0000	1.0000	0.9999	0.9994	0.9979
	14	1.0000	1.0000	1.0000	1.0000	1.0000	1.0000	1.0000	1.0000	0.9999	0.9997
	15	1.0000	1.0000	1.0000	1.0000	1.0000	1.0000	1.0000	1.0000	1.0000	1.0000
	16	1.0000	1.0000	1.0000	1.0000	1.0000	1.0000	1.0000	1.0000	1.0000	1.0000
17	0	0.4181	0.1668	0.0631	0.0225	0.0075	0.0023	0.0007	0.0002	0.0000	0.0000
	1	0.7922	0.4818	0.2525	0.1182	0.0501	0.0193	0.0067	0.0021	0.0006	0.0001
	2	0.9497	0.7618	0.5198	0.3096	0.1637	0.0774	0.0327	0.0123	0.0041	0.0012
	3	0.9912	0.9174	0.7556	0.5489	0.3530	0.2019	0.1028	0.0464	0.0184	0.0064
	4	0.9988	0.9779	0.9013	0.7582	0.5739	0.3887	0.2348	0.1260	0.0596	0.0245
	5	0.9999	0.9953	0.9681	0.8943	0.7653	0.5968	0.4197	0.2639	0.1471	0.0717
	6	1.0000	0.9992	0.9917	0.9623	0.8929	0.7752	0.6188	0.4478	0.2902	0.1662
	7	1.0000	0.9999	0.9983	0.9891	0.9598	0.8954	0.7872	0.6405	0.4743	0.3145
	8	1.0000	1.0000	0.9997	0.9974	0.9876	0.9597	0.9006	0.8011	0.6626	0.5000
	9	1.0000	1.0000	1.0000	0.9995	0.9969	0.9873	0.9617	0.9081	0.8166	0.6855
	10	1.0000	1.0000	1.0000	0.9999	0.9994	0.9968	0.9880	0.9652	0.9174	0.8338
	11	1.0000	1.0000	1.0000	1.0000	0.9999	0.9993	0.9970	0.9894	0.9699	0.9283
	12	1.0000	1.0000	1.0000	1.0000	1.0000	0.9999	0.9994	0.9975	0.9914	0.9755
	13	1.0000	1.0000	1.0000	1.0000	1.0000	1.0000	0.9999	0.9995	0.9981	0.9936
	14	1.0000	1.0000	1.0000	1.0000	1.0000	1.0000	1.0000	0.9999	0.9997	0.9988
	15	1.0000	1.0000	1.0000	1.0000	1.0000	1.0000	1.0000	1.0000	1.0000	0.9999
	16	1.0000	1.0000	1.0000	1.0000	1.0000	1.0000	1.0000	1.0000	1.0000	1.0000
	17	1.0000	1.0000	1.0000	1.0000	1.0000	1.0000	1.0000	1.0000	1.0000	1.0000

Table G Cumulative Binomial Probabilities *(Continued)*

		p									
n	*x*	0.05	0.10	0.15	0.20	0.25	0.30	0.35	0.40	0.45	0.50
18	0	0.3972	0.1501	0.0536	0.0180	0.0056	0.0016	0.0004	0.0001	0.0000	0.0000
	1	0.7735	0.4503	0.2241	0.0991	0.0395	0.0142	0.0046	0.0013	0.0003	0.0001
	2	0.9419	0.7338	0.4797	0.2713	0.1353	0.0600	0.0236	0.0082	0.0025	0.0007
	3	0.9891	0.9018	0.7202	0.5010	0.3057	0.1646	0.0783	0.0328	0.0120	0.0038
	4	0.9985	0.9718	0.8794	0.7164	0.5187	0.3327	0.1886	0.0942	0.0411	0.0154
	5	0.9998	0.9936	0.9581	0.8671	0.7175	0.5344	0.3550	0.2088	0.1077	0.0481
	6	1.0000	0.9988	0.9882	0.9487	0.8610	0.7217	0.5491	0.3743	0.2258	0.1189
	7	1.0000	0.9998	0.9973	0.9837	0.9431	0.8593	0.7283	0.5634	0.3915	0.2403
	8	1.0000	1.0000	0.9995	0.9957	0.9807	0.9404	0.8609	0.7368	0.5778	0.4073
	9	1.0000	1.0000	0.9999	0.9991	0.9946	0.9790	0.9403	0.8653	0.7473	0.5927
	10	1.0000	1.0000	1.0000	0.9998	0.9988	0.9939	0.9788	0.9424	0.8720	0.7597
	11	1.0000	1.0000	1.0000	1.0000	0.9998	0.9986	0.9938	0.9797	0.9463	0.8811
	12	1.0000	1.0000	1.0000	1.0000	1.0000	0.9997	0.9986	0.9942	0.9817	0.9519
	13	1.0000	1.0000	1.0000	1.0000	1.0000	1.0000	0.9997	0.9987	0.9951	0.9846
	14	1.0000	1.0000	1.0000	1.0000	1.0000	1.0000	1.0000	0.9998	0.9990	0.9962
	15	1.0000	1.0000	1.0000	1.0000	1.0000	1.0000	1.0000	1.0000	0.9999	0.9993
	16	1.0000	1.0000	1.0000	1.0000	1.0000	1.0000	1.0000	1.0000	1.0000	0.9999
	17	1.0000	1.0000	1.0000	1.0000	1.0000	1.0000	1.0000	1.0000	1.0000	1.0000
	18	1.0000	1.0000	1.0000	1.0000	1.0000	1.0000	1.0000	1.0000	1.0000	1.0000
19	0	0.3774	0.1351	0.0456	0.0144	0.0042	0.0011	0.0003	0.0001	0.0000	0.0000
	1	0.7547	0.4203	0.1985	0.0829	0.0310	0.0104	0.0031	0.0008	0.0002	0.0000
	2	0.9335	0.7054	0.4413	0.2369	0.1113	0.0462	0.0170	0.0055	0.0015	0.0004
	3	0.9868	0.8850	0.6841	0.4551	0.2631	0.1332	0.0591	0.0230	0.0077	0.0022
	4	0.9980	0.9648	0.8556	0.6733	0.4654	0.2822	0.1500	0.0696	0.0280	0.0096
	5	0.9998	0.9914	0.9463	0.8369	0.6678	0.4739	0.2968	0.1629	0.0777	0.0318
	6	1.0000	0.9983	0.9837	0.9324	0.8251	0.6655	0.4812	0.3081	0.1727	0.0835
	7	1.0000	0.9997	0.9959	0.9767	0.9225	0.8180	0.6656	0.4878	0.3169	0.1796
	8	1.0000	1.0000	0.9992	0.9933	0.9713	0.9161	0.8145	0.6675	0.4940	0.3238
	9	1.0000	1.0000	0.9999	0.9984	0.9911	0.9674	0.9125	0.8139	0.6710	0.5000
	10	1.0000	1.0000	1.0000	0.9997	0.9977	0.9895	0.9653	0.9115	0.8159	0.6762
	11	1.0000	1.0000	1.0000	1.0000	0.9995	0.9972	0.9886	0.9648	0.9129	0.8204
	12	1.0000	1.0000	1.0000	1.0000	0.9999	0.9994	0.9969	0.9884	0.9658	0.9165
	13	1.0000	1.0000	1.0000	1.0000	1.0000	0.9999	0.9993	0.9969	0.9891	0.9682
	14	1.0000	1.0000	1.0000	1.0000	1.0000	1.0000	0.9999	0.9994	0.9972	0.9904
	15	1.0000	1.0000	1.0000	1.0000	1.0000	1.0000	1.0000	0.9999	0.9995	0.9978
	16	1.0000	1.0000	1.0000	1.0000	1.0000	1.0000	1.0000	1.0000	0.9999	0.9996
	17	1.0000	1.0000	1.0000	1.0000	1.0000	1.0000	1.0000	1.0000	1.0000	1.0000
	18	1.0000	1.0000	1.0000	1.0000	1.0000	1.0000	1.0000	1.0000	1.0000	1.0000
	19	1.0000	1.0000	1.0000	1.0000	1.0000	1.0000	1.0000	1.0000	1.0000	1.0000
20	0	0.3585	0.1216	0.0388	0.0115	0.0032	0.0008	0.0002	0.0000	0.0000	0.0000
	1	0.7358	0.3917	0.1756	0.0692	0.0243	0.0076	0.0021	0.0005	0.0001	0.0000
	2	0.9245	0.6769	0.4049	0.2061	0.0913	0.0355	0.0121	0.0036	0.0009	0.0002
	3	0.9841	0.8670	0.6477	0.4114	0.2252	0.1071	0.0444	0.0160	0.0049	0.0013
	4	0.9974	0.9568	0.8298	0.6296	0.4148	0.2375	0.1182	0.0510	0.0189	0.0059
	5	0.9997	0.9887	0.9327	0.8042	0.6172	0.4164	0.2454	0.1256	0.0553	0.0207
	6	1.0000	0.9976	0.9781	0.9133	0.7858	0.6080	0.4166	0.2500	0.1299	0.0577
	7	1.0000	0.9996	0.9941	0.9679	0.8982	0.7723	0.6010	0.4159	0.2520	0.1316

Table G Cumulative Binomial Probabilities *(Continued)*

		p									
n	*x*	**0.05**	**0.10**	**0.15**	**0.20**	**0.25**	**0.30**	**0.35**	**0.40**	**0.45**	**0.50**
20	8	1.0000	0.9999	0.9987	0.9900	0.9591	0.8867	0.7624	0.5956	0.4143	0.2517
	9	1.0000	1.0000	0.9998	0.9974	0.9861	0.9520	0.8782	0.7553	0.5914	0.4119
	10	1.0000	1.0000	1.0000	0.9994	0.9961	0.9829	0.9468	0.8725	0.7507	0.5881
	11	1.0000	1.0000	1.0000	0.9999	0.9991	0.9949	0.9804	0.9435	0.8692	0.7483
	12	1.0000	1.0000	1.0000	1.0000	0.9998	0.9987	0.9940	0.9790	0.9420	0.8684
	13	1.0000	1.0000	1.0000	1.0000	1.0000	0.9997	0.9985	0.9935	0.9786	0.9423
	14	1.0000	1.0000	1.0000	1.0000	1.0000	1.0000	0.9997	0.9984	0.9936	0.9793
	15	1.0000	1.0000	1.0000	1.0000	1.0000	1.0000	1.0000	0.9997	0.9985	0.9941
	16	1.0000	1.0000	1.0000	1.0000	1.0000	1.0000	1.0000	1.0000	0.9997	0.9987
	17	1.0000	1.0000	1.0000	1.0000	1.0000	1.0000	1.0000	1.0000	1.0000	0.9998
	18	1.0000	1.0000	1.0000	1.0000	1.0000	1.0000	1.0000	1.0000	1.0000	1.0000
	19	1.0000	1.0000	1.0000	1.0000	1.0000	1.0000	1.0000	1.0000	1.0000	1.0000
	20	1.0000	1.0000	1.0000	1.0000	1.0000	1.0000	1.0000	1.0000	1.0000	1.0000
21	0	0.3406	0.1094	0.0329	0.0092	0.0024	0.0006	0.0001	0.0000	0.0000	0.0000
	1	0.7170	0.3647	0.1550	0.0576	0.0190	0.0056	0.0014	0.0003	0.0001	0.0000
	2	0.9151	0.6484	0.3705	0.1787	0.0745	0.0271	0.0086	0.0024	0.0006	0.0001
	3	0.9811	0.8480	0.6113	0.3704	0.1917	0.0856	0.0331	0.0110	0.0031	0.0007
	4	0.9968	0.9478	0.8025	0.5860	0.3674	0.1984	0.0924	0.0370	0.0126	0.0036
	5	0.9996	0.9856	0.9173	0.7693	0.5666	0.3627	0.2009	0.0957	0.0389	0.0133
	6	1.0000	0.9967	0.9713	0.8915	0.7436	0.5505	0.3567	0.2002	0.0964	0.0392
	7	1.0000	0.9994	0.9917	0.9569	0.8701	0.7230	0.5365	0.3495	0.1971	0.0946
	8	1.0000	0.9999	0.9980	0.9856	0.9439	0.8523	0.7059	0.5237	0.3413	0.1917
	9	1.0000	1.0000	0.9996	0.9959	0.9794	0.9324	0.8377	0.6914	0.5117	0.3318
	10	1.0000	1.0000	0.9999	0.9990	0.9936	0.9736	0.9228	0.8256	0.6790	0.5000
	11	1.0000	1.0000	1.0000	0.9998	0.9983	0.9913	0.9687	0.9151	0.8159	0.6682
	12	1.0000	1.0000	1.0000	1.0000	0.9996	0.9976	0.9892	0.9648	0.9092	0.8083
	13	1.0000	1.0000	1.0000	1.0000	0.9999	0.9994	0.9969	0.9877	0.9621	0.9054
	14	1.0000	1.0000	1.0000	1.0000	1.0000	0.9999	0.9993	0.9964	0.9868	0.9608
	15	1.0000	1.0000	1.0000	1.0000	1.0000	1.0000	0.9999	0.9992	0.9963	0.9867
	16	1.0000	1.0000	1.0000	1.0000	1.0000	1.0000	1.0000	0.9998	0.9992	0.9964
	17	1.0000	1.0000	1.0000	1.0000	1.0000	1.0000	1.0000	1.0000	0.9999	0.9993
	18	1.0000	1.0000	1.0000	1.0000	1.0000	1.0000	1.0000	1.0000	1.0000	0.9999
	19	1.0000	1.0000	1.0000	1.0000	1.0000	1.0000	1.0000	1.0000	1.0000	1.0000
	20	1.0000	1.0000	1.0000	1.0000	1.0000	1.0000	1.0000	1.0000	1.0000	1.0000
	21	1.0000	1.0000	1.0000	1.0000	1.0000	1.0000	1.0000	1.0000	1.0000	1.0000
22	0	0.3235	0.0985	0.0280	0.0074	0.0018	0.0004	0.0001	0.0000	0.0000	0.0000
	1	0.6982	0.3392	0.1367	0.0480	0.0149	0.0041	0.0010	0.0002	0.0000	0.0000
	2	0.9052	0.6200	0.3382	0.1545	0.0606	0.0207	0.0061	0.0016	0.0003	0.0001
	3	0.9778	0.8281	0.5752	0.3320	0.1624	0.0681	0.0245	0.0076	0.0020	0.0004
	4	0.9960	0.9379	0.7738	0.5429	0.3235	0.1645	0.0716	0.0266	0.0083	0.0022
	5	0.9994	0.9818	0.9001	0.7326	0.5168	0.3134	0.1629	0.0722	0.0271	0.0085
	6	0.9999	0.9956	0.9632	0.8670	0.6994	0.4942	0.3022	0.1584	0.0705	0.0262
	7	1.0000	0.9991	0.9886	0.9439	0.8385	0.6713	0.4736	0.2898	0.1518	0.0669
	8	1.0000	0.9999	0.9970	0.9799	0.9254	0.8135	0.6466	0.4540	0.2764	0.1431
	9	1.0000	1.0000	0.9993	0.9939	0.9705	0.9084	0.7916	0.6244	0.4350	0.2617
	10	1.0000	1.0000	0.9999	0.9984	0.9900	0.9613	0.8930	0.7720	0.6037	0.4159
	11	1.0000	1.0000	1.0000	0.9997	0.9971	0.9860	0.9526	0.8793	0.7543	0.5841

Table G Cumulative Binomial Probabilities *(Continued)*

		p									
n	x	0.05	0.10	0.15	0.20	0.25	0.30	0.35	0.40	0.45	0.50
22	12	1.0000	1.0000	1.0000	0.9999	0.9993	0.9957	0.9820	0.9449	0.8672	0.7383
	13	1.0000	1.0000	1.0000	1.0000	0.9999	0.9989	0.9942	0.9785	0.9383	0.8569
	14	1.0000	1.0000	1.0000	1.0000	1.0000	0.9998	0.9984	0.9930	0.9757	0.9331
	15	1.0000	1.0000	1.0000	1.0000	1.0000	1.0000	0.9997	0.9981	0.9920	0.9738
	16	1.0000	1.0000	1.0000	1.0000	1.0000	1.0000	0.9999	0.9996	0.9979	0.9915
	17	1.0000	1.0000	1.0000	1.0000	1.0000	1.0000	1.0000	0.9999	0.9995	0.9978
	18	1.0000	1.0000	1.0000	1.0000	1.0000	1.0000	1.0000	1.0000	0.9999	0.9996
	19	1.0000	1.0000	1.0000	1.0000	1.0000	1.0000	1.0000	1.0000	1.0000	0.9999
	20	1.0000	1.0000	1.0000	1.0000	1.0000	1.0000	1.0000	1.0000	1.0000	1.0000
	21	1.0000	1.0000	1.0000	1.0000	1.0000	1.0000	1.0000	1.0000	1.0000	1.0000
	22	1.0000	1.0000	1.0000	1.0000	1.0000	1.0000	1.0000	1.0000	1.0000	1.0000
23	0	0.3074	0.0886	0.0238	0.0059	0.0013	0.0003	0.0000	0.0000	0.0000	0.0000
	1	0.6794	0.3151	0.1204	0.0398	0.0116	0.0030	0.0007	0.0001	0.0000	0.0000
	2	0.8948	0.5920	0.3080	0.1332	0.0492	0.0157	0.0043	0.0010	0.0002	0.0000
	3	0.9742	0.8073	0.5396	0.2965	0.1370	0.0538	0.0181	0.0052	0.0012	0.0002
	4	0.9951	0.9269	0.7440	0.5007	0.2832	0.1356	0.0551	0.0190	0.0055	0.0013
	5	0.9992	0.9774	0.8811	0.6947	0.4685	0.2688	0.1309	0.0540	0.0186	0.0053
	6	0.9999	0.9942	0.9537	0.8402	0.6537	0.4399	0.2534	0.1240	0.0510	0.0173
	7	1.0000	0.9988	0.9848	0.9285	0.8037	0.6181	0.4136	0.2373	0.1152	0.0466
	8	1.0000	0.9998	0.9958	0.9727	0.9037	0.7709	0.5860	0.3884	0.2203	0.1050
	9	1.0000	1.0000	0.9990	0.9911	0.9592	0.8799	0.7408	0.5562	0.3636	0.2024
	10	1.0000	1.0000	0.9998	0.9975	0.9851	0.9454	0.8575	0.7129	0.5278	0.3388
	11	1.0000	1.0000	1.0000	0.9994	0.9954	0.9786	0.9318	0.8364	0.6865	0.5000
	12	1.0000	1.0000	1.0000	0.9999	0.9988	0.9928	0.9717	0.9187	0.8164	0.6612
	13	1.0000	1.0000	1.0000	1.0000	0.9997	0.9979	0.9900	0.9651	0.9063	0.7976
	14	1.0000	1.0000	1.0000	1.0000	0.9999	0.9995	0.9970	0.9872	0.9589	0.8950
	15	1.0000	1.0000	1.0000	1.0000	1.0000	0.9999	0.9992	0.9960	0.9847	0.9534
	16	1.0000	1.0000	1.0000	1.0000	1.0000	1.0000	0.9998	0.9990	0.9952	0.9827
	17	1.0000	1.0000	1.0000	1.0000	1.0000	1.0000	1.0000	0.9998	0.9988	0.9947
	18	1.0000	1.0000	1.0000	1.0000	1.0000	1.0000	1.0000	1.0000	0.9998	0.9987
	19	1.0000	1.0000	1.0000	1.0000	1.0000	1.0000	1.0000	1.0000	1.0000	0.9998
	20	1.0000	1.0000	1.0000	1.0000	1.0000	1.0000	1.0000	1.0000	1.0000	1.0000
	21	1.0000	1.0000	1.0000	1.0000	1.0000	1.0000	1.0000	1.0000	1.0000	1.0000
	22	1.0000	1.0000	1.0000	1.0000	1.0000	1.0000	1.0000	1.0000	1.0000	1.0000
	23	1.0000	1.0000	1.0000	1.0000	1.0000	1.0000	1.0000	1.0000	1.0000	1.0000
24	0	0.2920	0.0798	0.0202	0.0047	0.0010	0.0002	0.0000	0.0000	0.0000	0.0000
	1	0.6608	0.2925	0.1059	0.0331	0.0090	0.0022	0.0005	0.0001	0.0000	0.0000
	2	0.8841	0.5643	0.2798	0.1145	0.0398	0.0119	0.0030	0.0007	0.0001	0.0000
	3	0.9702	0.7857	0.5049	0.2639	0.1150	0.0424	0.0133	0.0035	0.0008	0.0001
	4	0.9940	0.9149	0.7134	0.4599	0.2466	0.1111	0.0422	0.0134	0.0036	0.0008
	5	0.9990	0.9723	0.8606	0.6559	0.4222	0.2288	0.1044	0.0400	0.0127	0.0033
	6	0.9999	0.9925	0.9428	0.8111	0.6074	0.3886	0.2106	0.0960	0.0364	0.0113
	7	1.0000	0.9983	0.9801	0.9108	0.7662	0.5647	0.3575	0.1919	0.0863	0.0320
	8	1.0000	0.9997	0.9941	0.9638	0.8787	0.7250	0.5257	0.3279	0.1730	0.0758
	9	1.0000	0.9999	0.9985	0.9874	0.9453	0.8472	0.6866	0.4891	0.2991	0.1537
	10	1.0000	1.0000	0.9997	0.9962	0.9787	0.9258	0.8167	0.6502	0.4539	0.2706
	11	1.0000	1.0000	0.9999	0.9990	0.9928	0.9686	0.9058	0.7870	0.6151	0.4194

Table G Cumulative Binomial Probabilities *(Continued)*

		p									
n	*x*	**0.05**	**0.10**	**0.15**	**0.20**	**0.25**	**0.30**	**0.35**	**0.40**	**0.45**	**0.50**
24	12	1.0000	1.0000	1.0000	0.9998	0.9979	0.9885	0.9577	0.8857	0.7580	0.5806
	13	1.0000	1.0000	1.0000	1.0000	0.9995	0.9964	0.9836	0.9465	0.8659	0.7294
	14	1.0000	1.0000	1.0000	1.0000	0.9999	0.9990	0.9945	0.9783	0.9352	0.8463
	15	1.0000	1.0000	1.0000	1.0000	1.0000	0.9998	0.9984	0.9925	0.9731	0.9242
	16	1.0000	1.0000	1.0000	1.0000	1.0000	1.0000	0.9996	0.9978	0.9905	0.9680
	17	1.0000	1.0000	1.0000	1.0000	1.0000	1.0000	0.9999	0.9995	0.9972	0.9887
	18	1.0000	1.0000	1.0000	1.0000	1.0000	1.0000	1.0000	0.9999	0.9993	0.9967
	19	1.0000	1.0000	1.0000	1.0000	1.0000	1.0000	1.0000	1.0000	0.9999	0.9992
	20	1.0000	1.0000	1.0000	1.0000	1.0000	1.0000	1.0000	1.0000	1.0000	0.9999
	21	1.0000	1.0000	1.0000	1.0000	1.0000	1.0000	1.0000	1.0000	1.0000	1.0000
	22	1.0000	1.0000	1.0000	1.0000	1.0000	1.0000	1.0000	1.0000	1.0000	1.0000
	23	1.0000	1.0000	1.0000	1.0000	1.0000	1.0000	1.0000	1.0000	1.0000	1.0000
	24	1.0000	1.0000	1.0000	1.0000	1.0000	1.0000	1.0000	1.0000	1.0000	1.0000
25	0	0.2774	0.0718	0.0172	0.0038	0.0008	0.0001	0.0000	0.0000	0.0000	0.0000
	1	0.6424	0.2712	0.0931	0.0274	0.0070	0.0016	0.0003	0.0001	0.0000	0.0000
	2	0.8729	0.5371	0.2537	0.0982	0.0321	0.0090	0.0021	0.0004	0.0001	0.0000
	3	0.9659	0.7636	0.4711	0.2340	0.0962	0.0332	0.0097	0.0024	0.0005	0.0001
	4	0.9928	0.9020	0.6821	0.4207	0.2137	0.0905	0.0320	0.0095	0.0023	0.0005
	5	0.9988	0.9666	0.8385	0.6167	0.3783	0.1935	0.0826	0.0294	0.0086	0.0020
	6	0.9998	0.9905	0.9305	0.7800	0.5611	0.3407	0.1734	0.0736	0.0258	0.0073
	7	1.0000	0.9977	0.9745	0.8909	0.7265	0.5118	0.3061	0.1536	0.0639	0.0216
	8	1.0000	0.9995	0.9920	0.9532	0.8506	0.6769	0.4668	0.2735	0.1340	0.0539
	9	1.0000	0.9999	0.9979	0.9827	0.9287	0.8106	0.6303	0.4246	0.2424	0.1148
	10	1.0000	1.0000	0.9995	0.9944	0.9703	0.9022	0.7712	0.5858	0.3843	0.2122
	11	1.0000	1.0000	0.9999	0.9985	0.9893	0.9558	0.8746	0.7323	0.5426	0.3450
	12	1.0000	1.0000	1.0000	0.9996	0.9966	0.9825	0.9396	0.8462	0.6937	0.5000
	13	1.0000	1.0000	1.0000	0.9999	0.9991	0.9940	0.9745	0.9222	0.8173	0.6550
	14	1.0000	1.0000	1.0000	1.0000	0.9998	0.9982	0.9907	0.9656	0.9040	0.7878
	15	1.0000	1.0000	1.0000	1.0000	1.0000	0.9995	0.9971	0.9868	0.9560	0.8852
	16	1.0000	1.0000	1.0000	1.0000	1.0000	0.9999	0.9992	0.9957	0.9826	0.9461
	17	1.0000	1.0000	1.0000	1.0000	1.0000	1.0000	0.9998	0.9988	0.9942	0.9784
	18	1.0000	1.0000	1.0000	1.0000	1.0000	1.0000	1.0000	0.9997	0.9984	0.9927
	19	1.0000	1.0000	1.0000	1.0000	1.0000	1.0000	1.0000	0.9999	0.9996	0.9980
	20	1.0000	1.0000	1.0000	1.0000	1.0000	1.0000	1.0000	1.0000	0.9999	0.9995
	21	1.0000	1.0000	1.0000	1.0000	1.0000	1.0000	1.0000	1.0000	1.0000	0.9999
	22	1.0000	1.0000	1.0000	1.0000	1.0000	1.0000	1.0000	1.0000	1.0000	1.0000
	23	1.0000	1.0000	1.0000	1.0000	1.0000	1.0000	1.0000	1.0000	1.0000	1.0000
	24	1.0000	1.0000	1.0000	1.0000	1.0000	1.0000	1.0000	1.0000	1.0000	1.0000
	25	1.0000	1.0000	1.0000	1.0000	1.0000	1.0000	1.0000	1.0000	1.0000	1.0000

H

F-DISTRIBUTION RIGHT-HAND TAIL PROBABILITIES

Upper 5% Table

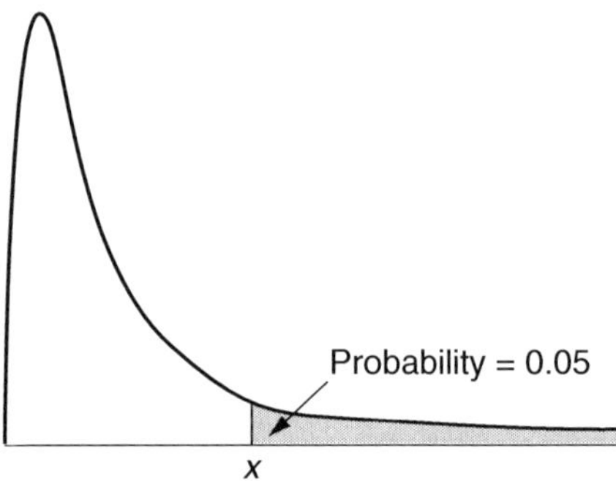

Upper 1% Table

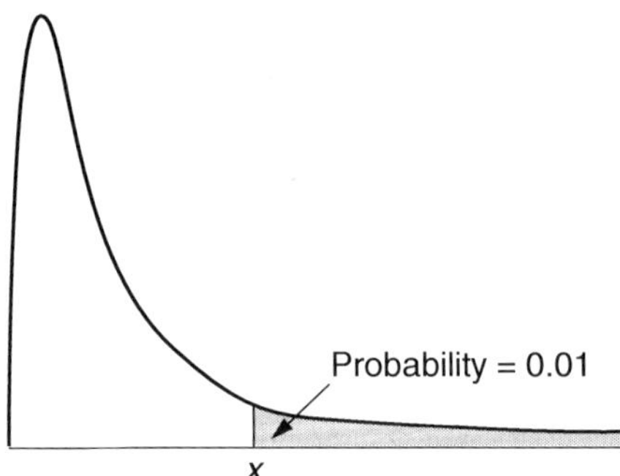

Table entry is the value of x (critical value) corresponding to $P(F \geq x) = 0.01$ or 0.05 for the given table (Table H.1:0.05; Table H.2:0.01) and number of numerator df (column) and denominator df (row).

Table H.1 Upper 5% of F Table: $P(F \geq x) = 0.05$

Denominator degrees of freedom (df)	Numerator degrees of freedom																				
	1	2	3	4	5	6	7	8	9	10	11	12	13	14	15	16	17	18	19	20	21
1	161.45	199.50	215.71	224.58	230.16	233.99	236.77	238.88	240.54	241.88	242.98	243.90	244.69	245.36	245.95	246.47	246.92	247.32	247.69	248.02	248.31
2	18.51	19.00	19.16	19.25	19.30	19.33	19.35	19.37	19.38	19.40	19.40	19.41	19.42	19.42	19.43	19.43	19.44	19.44	19.44	19.45	19.45
3	10.13	9.55	9.28	9.12	9.01	8.94	8.89	8.85	8.81	8.79	8.76	8.74	8.73	8.71	8.70	8.69	8.68	8.67	8.67	8.66	8.65
4	7.71	6.94	6.59	6.39	6.26	6.16	6.09	6.04	6.00	5.96	5.94	5.91	5.89	5.87	5.86	5.84	5.83	5.82	5.81	5.80	5.79
5	6.61	5.79	5.41	5.19	5.05	4.95	4.88	4.82	4.77	4.74	4.70	4.68	4.66	4.64	4.62	4.60	4.59	4.58	4.57	4.56	4.55
6	5.99	5.14	4.76	4.53	4.39	4.28	4.21	4.15	4.10	4.06	4.03	4.00	3.98	3.96	3.94	3.92	3.91	3.90	3.88	3.87	3.86
7	5.59	4.74	4.35	4.12	3.97	3.87	3.79	3.73	3.68	3.64	3.60	3.57	3.55	3.53	3.51	3.49	3.48	3.47	3.46	3.44	3.43
8	5.32	4.46	4.07	3.84	3.69	3.58	3.50	3.44	3.39	3.35	3.31	3.28	3.26	3.24	3.22	3.20	3.19	3.17	3.16	3.15	3.14
9	5.12	4.26	3.86	3.63	3.48	3.37	3.29	3.23	3.18	3.14	3.10	3.07	3.05	3.03	3.01	2.99	2.97	2.96	2.95	2.94	2.93
10	4.96	4.10	3.71	3.48	3.33	3.22	3.14	3.07	3.02	2.98	2.94	2.91	2.89	2.86	2.85	2.83	2.81	2.80	2.79	2.77	2.76
11	4.84	3.98	3.59	3.36	3.20	3.09	3.01	2.95	2.90	2.85	2.82	2.79	2.76	2.74	2.72	2.70	2.69	2.67	2.66	2.65	2.64
12	4.75	3.89	3.49	3.26	3.11	3.00	2.91	2.85	2.80	2.75	2.72	2.69	2.66	2.64	2.62	2.60	2.58	2.57	2.56	2.54	2.53
13	4.67	3.81	3.41	3.18	3.03	2.92	2.83	2.77	2.71	2.67	2.63	2.60	2.58	2.55	2.53	2.51	2.50	2.48	2.47	2.46	2.45
14	4.60	3.74	3.34	3.11	2.96	2.85	2.76	2.70	2.65	2.60	2.57	2.53	2.51	2.48	2.46	2.44	2.43	2.41	2.40	2.39	2.38
15	4.54	3.68	3.29	3.06	2.90	2.79	2.71	2.64	2.59	2.54	2.51	2.48	2.45	2.42	2.40	2.38	2.37	2.35	2.34	2.33	2.32
16	4.49	3.63	3.24	3.01	2.85	2.74	2.66	2.59	2.54	2.49	2.46	2.42	2.40	2.37	2.35	2.33	2.32	2.30	2.29	2.28	2.26
17	4.45	3.59	3.20	2.96	2.81	2.70	2.61	2.55	2.49	2.45	2.41	2.38	2.35	2.33	2.31	2.29	2.27	2.26	2.24	2.23	2.22
18	4.41	3.55	3.16	2.93	2.77	2.66	2.58	2.51	2.46	2.41	2.37	2.34	2.31	2.29	2.27	2.25	2.23	2.22	2.20	2.19	2.18
19	4.38	3.52	3.13	2.90	2.74	2.63	2.54	2.48	2.42	2.38	2.34	2.31	2.28	2.26	2.23	2.21	2.20	2.18	2.17	2.16	2.14
20	4.35	3.49	3.10	2.87	2.71	2.60	2.51	2.45	2.39	2.35	2.31	2.28	2.25	2.22	2.20	2.18	2.17	2.15	2.14	2.12	2.11
21	4.32	3.47	3.07	2.84	2.68	2.57	2.49	2.42	2.37	2.32	2.28	2.25	2.22	2.20	2.18	2.16	2.14	2.12	2.11	2.10	2.08
22	4.30	3.44	3.05	2.82	2.66	2.55	2.46	2.40	2.34	2.30	2.26	2.23	2.20	2.17	2.15	2.13	2.11	2.10	2.08	2.07	2.06
23	4.28	3.42	3.03	2.80	2.64	2.53	2.44	2.37	2.32	2.27	2.24	2.20	2.18	2.15	2.13	2.11	2.09	2.08	2.06	2.05	2.04
24	4.26	3.40	3.01	2.78	2.62	2.51	2.42	2.36	2.30	2.25	2.22	2.18	2.15	2.13	2.11	2.09	2.07	2.05	2.04	2.03	2.01
25	4.24	3.39	2.99	2.76	2.60	2.49	2.40	2.34	2.28	2.24	2.20	2.16	2.14	2.11	2.09	2.07	2.05	2.04	2.02	2.01	2.00
26	4.23	3.37	2.98	2.74	2.59	2.47	2.39	2.32	2.27	2.22	2.18	2.15	2.12	2.09	2.07	2.05	2.03	2.02	2.00	1.99	1.98
27	4.21	3.35	2.96	2.73	2.57	2.46	2.37	2.31	2.25	2.20	2.17	2.13	2.10	2.08	2.06	2.04	2.02	2.00	1.99	1.97	1.96
28	4.20	3.34	2.95	2.71	2.56	2.45	2.36	2.29	2.24	2.19	2.15	2.12	2.09	2.06	2.04	2.02	2.00	1.99	1.97	1.96	1.95
29	4.18	3.33	2.93	2.70	2.55	2.43	2.35	2.28	2.22	2.18	2.14	2.10	2.08	2.05	2.03	2.01	1.99	1.97	1.96	1.94	1.93
30	4.17	3.32	2.92	2.69	2.53	2.42	2.33	2.27	2.21	2.16	2.13	2.09	2.06	2.04	2.01	1.99	1.98	1.96	1.95	1.93	1.92
31	4.16	3.30	2.91	2.68	2.52	2.41	2.32	2.25	2.20	2.15	2.11	2.08	2.05	2.03	2.00	1.98	1.96	1.95	1.93	1.92	1.91
32	4.15	3.29	2.90	2.67	2.51	2.40	2.31	2.24	2.19	2.14	2.10	2.07	2.04	2.01	1.99	1.97	1.95	1.94	1.92	1.91	1.90
33	4.14	3.28	2.89	2.66	2.50	2.39	2.30	2.23	2.18	2.13	2.09	2.06	2.03	2.00	1.98	1.96	1.94	1.93	1.91	1.90	1.89
34	4.13	3.28	2.88	2.65	2.49	2.38	2.29	2.23	2.17	2.12	2.08	2.05	2.02	1.99	1.97	1.95	1.93	1.92	1.90	1.89	1.88
35	4.12	3.27	2.87	2.64	2.49	2.37	2.29	2.22	2.16	2.11	2.07	2.04	2.01	1.99	1.96	1.94	1.92	1.91	1.89	1.88	1.87
36	4.11	3.26	2.87	2.63	2.48	2.36	2.28	2.21	2.15	2.11	2.07	2.03	2.00	1.98	1.95	1.93	1.92	1.90	1.88	1.87	1.86
37	4.11	3.25	2.86	2.63	2.47	2.36	2.27	2.20	2.14	2.10	2.06	2.02	2.00	1.97	1.95	1.93	1.91	1.89	1.88	1.86	1.85
38	4.10	3.24	2.85	2.62	2.46	2.35	2.26	2.19	2.14	2.09	2.05	2.02	1.99	1.96	1.94	1.92	1.90	1.88	1.87	1.85	1.84
39	4.09	3.24	2.85	2.61	2.46	2.34	2.26	2.19	2.13	2.08	2.04	2.01	1.98	1.95	1.93	1.91	1.89	1.88	1.86	1.85	1.83
40	4.08	3.23	2.84	2.61	2.45	2.34	2.25	2.18	2.12	2.08	2.04	2.00	1.97	1.95	1.92	1.90	1.89	1.87	1.85	1.84	1.83

Table H.2 Upper 1% of F Table: $P(F \geq x) = 0.01$

Denominator degrees of freedom (df)	Numerator degrees of freedom																				
	1	2	3	4	5	6	7	8	9	10	11	12	13	14	15	16	17	18	19	20	21
1	4052.2	4999.3	5403.5	5624.3	5764.0	5859.0	5928.3	5981.0	6022.4	6055.9	6083.4	6106.7	6125.8	6143.0	6157.0	6170.0	6181.2	6191.4	6200.7	6208.7	6216.1
2	98.50	99.00	99.16	99.25	99.30	99.33	99.36	99.38	99.39	99.40	99.41	99.42	99.42	99.43	99.43	99.44	99.44	99.44	99.45	99.45	99.45
3	34.12	30.82	29.46	28.71	28.24	27.91	27.67	27.49	27.34	27.23	27.13	27.05	26.98	26.92	26.87	26.83	26.79	26.76	26.72	26.69	26.66
4	21.20	18.00	16.69	15.98	15.52	15.21	14.98	14.80	14.66	14.55	14.45	14.37	14.31	14.25	14.20	14.15	14.11	14.08	14.05	14.02	13.99
5	16.26	13.27	12.06	11.39	10.97	10.67	10.46	10.29	10.16	10.05	9.96	9.89	9.82	9.77	9.72	9.68	9.64	9.61	9.58	9.55	9.53
6	13.75	10.92	9.78	9.15	8.75	8.47	8.26	8.10	7.98	7.87	7.79	7.72	7.66	7.60	7.56	7.52	7.48	7.45	7.42	7.40	7.37
7	12.25	9.55	8.45	7.85	7.46	7.19	6.99	6.84	6.72	6.62	6.54	6.47	6.41	6.36	6.31	6.28	6.24	6.21	6.18	6.16	6.13
8	11.26	8.65	7.59	7.01	6.63	6.37	6.18	6.03	5.91	5.81	5.73	5.67	5.61	5.56	5.52	5.48	5.44	5.41	5.38	5.36	5.34
9	10.56	8.02	6.99	6.42	6.06	5.80	5.61	5.47	5.35	5.26	5.18	5.11	5.05	5.01	4.96	4.92	4.89	4.86	4.83	4.81	4.79
10	10.04	7.56	6.55	5.99	5.64	5.39	5.20	5.06	4.94	4.85	4.77	4.71	4.65	4.60	4.56	4.52	4.49	4.46	4.43	4.41	4.38
11	9.65	7.21	6.22	5.67	5.32	5.07	4.89	4.74	4.63	4.54	4.46	4.40	4.34	4.29	4.25	4.21	4.18	4.15	4.12	4.10	4.08
12	9.33	6.93	5.95	5.41	5.06	4.82	4.64	4.50	4.39	4.30	4.22	4.16	4.10	4.05	4.01	3.97	3.94	3.91	3.88	3.86	3.84
13	9.07	6.70	5.74	5.21	4.86	4.62	4.44	4.30	4.19	4.10	4.02	3.96	3.91	3.86	3.82	3.78	3.75	3.72	3.69	3.66	3.64
14	8.86	6.51	5.56	5.04	4.69	4.46	4.28	4.14	4.03	3.94	3.86	3.80	3.75	3.70	3.66	3.62	3.59	3.56	3.53	3.51	3.48
15	8.68	6.36	5.42	4.89	4.56	4.32	4.14	4.00	3.89	3.80	3.73	3.67	3.61	3.56	3.52	3.49	3.45	3.42	3.40	3.37	3.35
16	8.53	6.23	5.29	4.77	4.44	4.20	4.03	3.89	3.78	3.69	3.62	3.55	3.50	3.45	3.41	3.37	3.34	3.31	3.28	3.26	3.24
17	8.40	6.11	5.19	4.67	4.34	4.10	3.93	3.79	3.68	3.59	3.52	3.46	3.40	3.35	3.31	3.27	3.24	3.21	3.19	3.16	3.14
18	8.29	6.01	5.09	4.58	4.25	4.01	3.84	3.71	3.60	3.51	3.43	3.37	3.32	3.27	3.23	3.19	3.16	3.13	3.10	3.08	3.05
19	8.18	5.93	5.01	4.50	4.17	3.94	3.77	3.63	3.52	3.43	3.36	3.30	3.24	3.19	3.15	3.12	3.08	3.05	3.03	3.00	2.98
20	8.10	5.85	4.94	4.43	4.10	3.87	3.70	3.56	3.46	3.37	3.29	3.23	3.18	3.13	3.09	3.05	3.02	2.99	2.96	2.94	2.92
21	8.02	5.78	4.87	4.37	4.04	3.81	3.64	3.51	3.40	3.31	3.24	3.17	3.12	3.07	3.03	2.99	2.96	2.93	2.90	2.88	2.86
22	7.95	5.72	4.82	4.31	3.99	3.76	3.59	3.45	3.35	3.26	3.18	3.12	3.07	3.02	2.98	2.94	2.91	2.88	2.85	2.83	2.81
23	7.88	5.66	4.76	4.26	3.94	3.71	3.54	3.41	3.30	3.21	3.14	3.07	3.02	2.97	2.93	2.89	2.86	2.83	2.80	2.78	2.76
24	7.82	5.61	4.72	4.22	3.90	3.67	3.50	3.36	3.26	3.17	3.09	3.03	2.98	2.93	2.89	2.85	2.82	2.79	2.76	2.74	2.72
25	7.77	5.57	4.68	4.18	3.85	3.63	3.46	3.32	3.22	3.13	3.06	2.99	2.94	2.89	2.85	2.81	2.78	2.75	2.72	2.70	2.68
26	7.72	5.53	4.64	4.14	3.82	3.59	3.42	3.29	3.18	3.09	3.02	2.96	2.90	2.86	2.81	2.78	2.75	2.72	2.69	2.66	2.64
27	7.68	5.49	4.60	4.11	3.78	3.56	3.39	3.26	3.15	3.06	2.99	2.93	2.87	2.82	2.78	2.75	2.71	2.68	2.66	2.63	2.61
28	7.64	5.45	4.57	4.07	3.75	3.53	3.36	3.23	3.12	3.03	2.96	2.90	2.84	2.79	2.75	2.72	2.68	2.65	2.63	2.60	2.58
29	7.60	5.42	4.54	4.04	3.73	3.50	3.33	3.20	3.09	3.00	2.93	2.87	2.81	2.77	2.73	2.69	2.66	2.63	2.60	2.57	2.55
30	7.56	5.39	4.51	4.02	3.70	3.47	3.30	3.17	3.07	2.98	2.91	2.84	2.79	2.74	2.70	2.66	2.63	2.60	2.57	2.55	2.53
31	7.53	5.36	4.48	3.99	3.67	3.45	3.28	3.15	3.04	2.96	2.88	2.82	2.77	2.72	2.68	2.64	2.61	2.58	2.55	2.52	2.50
32	7.50	5.34	4.46	3.97	3.65	3.43	3.26	3.13	3.02	2.93	2.86	2.80	2.74	2.70	2.65	2.62	2.58	2.55	2.53	2.50	2.48
33	7.47	5.31	4.44	3.95	3.63	3.41	3.24	3.11	3.00	2.91	2.84	2.78	2.72	2.68	2.63	2.60	2.56	2.53	2.51	2.48	2.46
34	7.44	5.29	4.42	3.93	3.61	3.39	3.22	3.09	2.98	2.89	2.82	2.76	2.70	2.66	2.61	2.58	2.54	2.51	2.49	2.46	2.44
35	7.42	5.27	4.40	3.91	3.59	3.37	3.20	3.07	2.96	2.88	2.80	2.74	2.69	2.64	2.60	2.56	2.53	2.50	2.47	2.44	2.42
36	7.40	5.25	4.38	3.89	3.57	3.35	3.18	3.05	2.95	2.86	2.79	2.72	2.67	2.62	2.58	2.54	2.51	2.48	2.45	2.43	2.41
37	7.37	5.23	4.36	3.87	3.56	3.33	3.17	3.04	2.93	2.84	2.77	2.71	2.65	2.61	2.56	2.53	2.49	2.46	2.44	2.41	2.39
38	7.35	5.21	4.34	3.86	3.54	3.32	3.15	3.02	2.92	2.83	2.75	2.69	2.64	2.59	2.55	2.51	2.48	2.45	2.42	2.40	2.37
39	7.33	5.19	4.33	3.84	3.53	3.30	3.14	3.01	2.90	2.81	2.74	2.68	2.62	2.58	2.54	2.50	2.46	2.43	2.41	2.38	2.36
40	7.31	5.18	4.31	3.83	3.51	3.29	3.12	2.99	2.89	2.80	2.73	2.66	2.61	2.56	2.52	2.48	2.45	2.42	2.39	2.37	2.35

I

BONFERRONI SIMULTANEOUS CONFIDENCE INTERVALS

Table I Bonferroni Table for Overall 95% Confidence for Multiple CIs.

Degrees of freedom (df)	Number of comparisons														
	1	**2**	**3**	**4**	**5**	**6**	**7**	**8**	**9**	**10**	**11**	**12**	**13**	**14**	**15**
1	12.71	25.45	38.19	50.92	63.66	76.39	89.12	101.86	114.59	127.32	140.05	152.79	165.52	178.25	190.98
2	4.30	6.21	7.65	8.86	9.92	10.89	11.77	12.59	13.36	14.09	14.78	15.44	16.08	16.69	17.28
3	3.18	4.18	4.86	5.39	5.84	6.23	6.58	6.90	7.18	7.45	7.70	7.94	8.16	8.37	8.58
4	2.78	3.50	3.96	4.31	4.60	4.85	5.07	5.26	5.44	5.60	5.75	5.89	6.02	6.14	6.25
5	2.57	3.16	3.53	3.81	4.03	4.22	4.38	4.53	4.66	4.77	4.88	4.98	5.08	5.16	5.25
6	2.45	2.97	3.29	3.52	3.71	3.86	4.00	4.12	4.22	4.32	4.40	4.49	4.56	4.63	4.70
7	2.36	2.84	3.13	3.34	3.50	3.64	3.75	3.86	3.95	4.03	4.10	4.17	4.24	4.30	4.36
8	2.31	2.75	3.02	3.21	3.36	3.48	3.58	3.68	3.76	3.83	3.90	3.96	4.02	4.07	4.12
9	2.26	2.69	2.93	3.11	3.25	3.36	3.46	3.55	3.62	3.69	3.75	3.81	3.86	3.91	3.95
10	2.23	2.63	2.87	3.04	3.17	3.28	3.37	3.45	3.52	3.58	3.64	3.69	3.74	3.79	3.83
11	2.20	2.59	2.82	2.98	3.11	3.21	3.29	3.37	3.44	3.50	3.55	3.60	3.65	3.69	3.73
12	2.18	2.56	2.78	2.93	3.05	3.15	3.24	3.31	3.37	3.43	3.48	3.53	3.57	3.61	3.65
13	2.16	2.53	2.75	2.90	3.01	3.11	3.19	3.26	3.32	3.37	3.42	3.47	3.51	3.55	3.58
14	2.14	2.51	2.72	2.86	2.98	3.07	3.15	3.21	3.27	3.33	3.37	3.42	3.46	3.49	3.53
15	2.13	2.49	2.69	2.84	2.95	3.04	3.11	3.18	3.23	3.29	3.33	3.37	3.41	3.45	3.48
16	2.12	2.47	2.67	2.81	2.92	3.01	3.08	3.15	3.20	3.25	3.30	3.34	3.38	3.41	3.44
17	2.11	2.46	2.65	2.79	2.90	2.98	3.06	3.12	3.17	3.22	3.27	3.31	3.34	3.38	3.41
18	2.10	2.45	4.64	2.77	2.88	2.96	3.03	3.09	3.15	3.20	3.24	3.28	3.32	3.35	3.38
19	2.09	2.43	2.63	2.76	2.86	2.94	3.01	3.07	3.13	3.17	3.22	3.25	3.29	3.32	3.35
20	2.09	2.42	2.61	2.74	2.85	2.93	3.00	3.06	3.11	3.15	3.20	3.23	3.27	3.30	3.33
21	2.08	2.41	2.60	2.73	2.83	2.91	2.98	3.04	3.09	3.14	3.18	3.21	3.25	3.28	3.31
22	2.07	2.41	2.59	2.72	2.82	2.90	2.97	3.02	3.07	3.12	3.16	3.20	3.23	3.26	3.29
23	2.07	2.40	2.58	2.71	2.81	2.89	2.95	3.01	3.06	3.10	3.14	3.18	3.21	3.25	3.27
24	2.06	2.39	2.57	2.70	2.80	2.88	2.94	3.00	3.05	3.09	3.13	3.17	3.20	3.23	3.26
25	2.06	2.38	2.57	2.69	2.79	2.86	2.93	2.99	3.03	3.08	3.12	3.15	3.19	3.22	3.24
26	2.06	2.38	2.56	2.68	2.78	2.86	2.92	2.98	3.02	3.07	3.11	3.14	3.17	3.20	3.23
27	2.05	2.37	2.55	2.68	2.77	2.85	2.91	2.97	3.01	3.06	3.10	3.13	3.16	3.19	3.22
28	2.05	2.37	2.55	2.67	2.76	2.84	2.90	2.96	3.00	3.05	3.09	3.12	3.15	3.18	3.21
29	2.05	2.36	2.54	2.66	2.76	2.83	2.89	2.95	3.00	3.04	3.08	3.11	3.14	3.17	3.20
30	2.04	2.36	2.54	2.66	2.75	2.82	2.89	2.94	2.99	3.03	3.07	3.10	3.13	3.16	3.19
>30	1.96	2.24	2.40	2.50	2.58	2.64	2.70	2.74	2.78	2.81	2.84	2.87	2.90	2.92	2.94

*Table entry is the critical value tB to achieve 95% overall confidence for a chosen number of mean comparisons (column) and degrees of freedom (row).

CUMULATIVE POISSON PROBABILITIES

The table gives the probability of x or fewer events when the expected number of such events is λ.

Table J Cumulative Poisson Probabilities

	$\lambda = E(X)$									
x	**0.1**	**0.2**	**0.3**	**0.4**	**0.5**	**0.6**	**0.7**	**0.8**	**0.9**	**1.0**
0	0.905	0.819	0.741	0.670	0.607	0.549	0.497	0.449	0.407	0.368
1	0.995	0.982	0.963	0.938	0.910	0.878	0.844	0.809	0.772	0.736
2	1.000	0.999	0.996	0.992	0.986	0.977	0.966	0.953	0.937	0.920
3	1.000	1.000	1.000	0.999	0.998	0.997	0.994	0.991	0.987	0.981
4	1.000	1.000	1.000	1.000	1.000	1.000	0.999	0.999	0.998	0.996
5	1.000	1.000	1.000	1.000	1.000	1.000	1.000	1.000	1.000	0.999
6	1.000	1.000	1.000	1.000	1.000	1.000	1.000	1.000	1.000	1.000
x	**1.1**	**1.2**	**1.3**	**1.4**	**1.5**	**1.6**	**1.7**	**1.8**	**1.9**	**2.0**
0	0.333	0.301	0.273	0.247	0.223	0.202	0.183	0.165	0.150	0.135
1	0.699	0.663	0.627	0.592	0.558	0.525	0.493	0.463	0.434	0.406
2	0.900	0.879	0.857	0.833	0.809	0.783	0.757	0.731	0.704	0.677
3	0.974	0.966	0.957	0.946	0.934	0.921	0.907	0.891	0.875	0.857
4	0.995	0.992	0.989	0.986	0.981	0.976	0.970	0.964	0.956	0.947
5	0.999	0.998	0.998	0.997	0.996	0.994	0.992	0.990	0.987	0.983
6	1.000	1.000	1.000	0.999	0.999	0.999	0.998	0.997	0.997	0.995
7	1.000	1.000	1.000	1.000	1.000	1.000	1.000	0.999	0.999	0.999
8	1.000	1.000	1.000	1.000	1.000	1.000	1.000	1.000	1.000	1.000
x	**2.2**	**2.4**	**2.6**	**2.8**	**3.0**	**3.2**	**3.4**	**3.6**	**3.8**	**4.0**
0	0.111	0.091	0.074	0.061	0.050	0.041	0.033	0.027	0.022	0.018
1	0.355	0.308	0.267	0.231	0.199	0.171	0.147	0.126	0.107	0.092
2	0.623	0.570	0.518	0.469	0.423	0.380	0.340	0.303	0.269	0.238
3	0.819	0.779	0.736	0.692	0.647	0.603	0.558	0.515	0.473	0.433
4	0.928	0.904	0.877	0.848	0.815	0.781	0.744	0.706	0.668	0.629
5	0.975	0.964	0.951	0.935	0.916	0.895	0.871	0.844	0.816	0.785
6	0.993	0.988	0.983	0.976	0.966	0.955	0.942	0.927	0.909	0.889
7	0.998	0.997	0.995	0.992	0.988	0.983	0.977	0.969	0.960	0.949
8	1.000	0.999	0.999	0.998	0.996	0.994	0.992	0.988	0.984	0.979
9	1.000	1.000	1.000	0.999	0.999	0.998	0.997	0.996	0.994	0.992
10	1.000	1.000	1.000	1.000	1.000	1.000	0.999	0.999	0.998	0.997
11	1.000	1.000	1.000	1.000	1.000	1.000	1.000	1.000	0.999	0.999
12	1.000	1.000	1.000	1.000	1.000	1.000	1.000	1.000	1.000	1.000

Table J Cumulative Poisson Probabilities *(Continued)*

	$\lambda = E(X)$									
x	**4.2**	**4.4**	**4.6**	**4.8**	**5.0**	**5.2**	**5.4**	**5.6**	**5.8**	**6.0**
0	0.015	0.012	0.010	0.008	0.007	0.006	0.005	0.004	0.003	0.002
1	0.078	0.066	0.056	0.048	0.040	0.034	0.029	0.024	0.021	0.017
2	0.210	0.185	0.163	0.143	0.125	0.109	0.095	0.082	0.072	0.062
3	0.395	0.359	0.326	0.294	0.265	0.238	0.213	0.191	0.170	0.151
4	0.590	0.551	0.513	0.476	0.440	0.406	0.373	0.342	0.313	0.285
5	0.753	0.720	0.686	0.651	0.616	0.581	0.546	0.512	0.478	0.446
6	0.867	0.844	0.818	0.791	0.762	0.732	0.702	0.670	0.638	0.606
7	0.936	0.921	0.905	0.887	0.867	0.845	0.822	0.797	0.771	0.744
8	0.972	0.964	0.955	0.944	0.932	0.918	0.903	0.886	0.867	0.847
9	0.989	0.985	0.980	0.975	0.968	0.960	0.951	0.941	0.929	0.916
10	0.996	0.994	0.992	0.990	0.986	0.982	0.977	0.972	0.965	0.957
11	0.999	0.998	0.997	0.996	0.995	0.993	0.990	0.988	0.984	0.980
12	1.000	0.999	0.999	0.999	0.998	0.997	0.996	0.995	0.993	0.991
13	1.000	1.000	1.000	1.000	0.999	0.999	0.999	0.998	0.997	0.996
14	1.000	1.000	1.000	1.000	1.000	1.000	1.000	0.999	0.999	0.999
15	1.000	1.000	1.000	1.000	1.000	1.000	1.000	1.000	1.000	0.999
16	1.000	1.000	1.000	1.000	1.000	1.000	1.000	1.000	1.000	1.000
x	**6.5**	**7.0**	**7.5**	**8.0**	**8.5**	**9.0**	**9.5**	**10.0**	**10.5**	**11.0**
0	0.002	0.001	0.001	0.000	0.000	0.000	0.000	0.000	0.000	0.000
1	0.011	0.007	0.005	0.003	0.002	0.001	0.001	0.000	0.000	0.000
2	0.043	0.030	0.020	0.014	0.009	0.006	0.004	0.003	0.002	0.001
3	0.112	0.082	0.059	0.042	0.030	0.021	0.015	0.010	0.007	0.005
4	0.224	0.173	0.132	0.100	0.074	0.055	0.040	0.029	0.021	0.015
5	0.369	0.301	0.241	0.191	0.150	0.116	0.089	0.067	0.050	0.038
6	0.527	0.450	0.378	0.313	0.256	0.207	0.165	0.130	0.102	0.079
7	0.673	0.599	0.525	0.453	0.386	0.324	0.269	0.220	0.179	0.143
8	0.792	0.729	0.662	0.593	0.523	0.456	0.392	0.333	0.279	0.232
9	0.877	0.830	0.776	0.717	0.653	0.587	0.522	0.458	0.397	0.341
10	0.933	0.901	0.862	0.816	0.763	0.706	0.645	0.583	0.521	0.460
11	0.966	0.947	0.921	0.888	0.849	0.803	0.752	0.697	0.639	0.579
12	0.984	0.973	0.957	0.936	0.909	0.876	0.836	0.792	0.742	0.689
13	0.993	0.987	0.978	0.966	0.949	0.926	0.898	0.864	0.825	0.781
14	0.997	0.994	0.990	0.983	0.973	0.959	0.940	0.917	0.888	0.854
15	0.999	0.998	0.995	0.992	0.986	0.978	0.967	0.951	0.932	0.907
16	1.000	0.999	0.998	0.996	0.993	0.989	0.982	0.973	0.960	0.944
17	1.000	1.000	0.999	0.998	0.997	0.995	0.991	0.986	0.978	0.968
18	1.000	1.000	1.000	0.999	0.999	0.998	0.996	0.993	0.988	0.982
19	1.000	1.000	1.000	1.000	0.999	0.999	0.998	0.997	0.994	0.991
20	1.000	1.000	1.000	1.000	1.000	1.000	0.999	0.998	0.997	0.995
21	1.000	1.000	1.000	1.000	1.000	1.000	1.000	0.999	0.999	0.998
22	1.000	1.000	1.000	1.000	1.000	1.000	1.000	1.000	0.999	0.999
23	1.000	1.000	1.000	1.000	1.000	1.000	1.000	1.000	1.000	1.000

INDEX

Note: All page numbers followed by the letter "*f*" refer to figures, and all page numbers followed by the letter "*t*" refer to tables.